IMPORTANT FORMULAS

1 The **distance** $d(P_1, P_2)$ between $P_1(x_1, y_1)$ and $P_2(x_2, y_2)$:

$$d(P_1, P_2) = \sqrt{(x_2 - x_1)^2 + (y_2 - y_1)^2}$$

2 The **equation of a circle** with center $C(h,k)$ and radius r:

$$(x - h)^2 + (y - k)^2 = r^2$$

3 The **slope** m of the line through $P_1(x_1,y_1)$ and $P_2(x_2,y_2)$:

$$m = \frac{y_2 - y_1}{x_2 - x_1}$$

4 The **point-slope form** for the equation of a line through $P_1(x_1,y_1)$ with slope m:

$$y - y_1 = m(x - x_1)$$

5 The **slope-intercept form** for the equation of a line with slope m and y-intercept b:

$$y = mx + b$$

6 **Linear function** f: $f(x) = ax + b$

7 **Quadratic function** f: $f(x) = ax^2 + bx + c$

8 **Polynomial function** f: $f(x) = a_n x^n + a_{n-1}x^{n-1} + \cdots + a_1 x + a_0$

9 **Rational function** f: $f(x) = \dfrac{h(x)}{g(x)}$, where h and g are polynomial functions

10 **Exponential function** f with base $a > 0$: $f(x) = a^x$

11 **Logarithmic function** f with base $a > 0$: $f(x) = \log_a x$, where $y = \log_a x$ if and only if $a^y = x$

12 **Geometric formulas** for areas, circumference of a circle, volumes, and curved surface area are as follows, where r = radius, h = altitude, b (or a) = length of base, A = area, C = circumference, V = volume, and S = curved surface area:

Triangle	$A = \frac{1}{2}bh$
Circle	$A = \pi r^2, \quad C = 2\pi r$
Parallelogram	$A = bh$
Trapezoid	$A = \frac{1}{2}(a + b)h$
Right circular cylinder	$V = \pi r^2 h, \quad S = 2\pi rh$
Right circular cone	$V = \frac{1}{3}\pi r^2 h, \quad S = \pi r\sqrt{r^2 + h^2}$
Sphere	$V = \frac{4}{3}\pi r^3, \quad S = 4\pi r^2$

FUNCTIONS AND GRAPHS

FUNCTIONS AND GRAPHS

FOURTH EDITION

EARL W. SWOKOWSKI
Marquette University

PRINDLE, WEBER & SCHMIDT · Boston

PWS PUBLISHERS

Prindle, Weber & Schmidt • Duxbury Press • PWS Engineering •
Statler Office Building • 20 Park Plaza • Boston, Massachusetts 02116

PWS Publishers is a division of Wadsworth, Inc.

88 87 86 85 — 10 9 8 7 6 5 4

Printed in the United States of America.

ISBN 0-87150-460-X

Library of Congress Cataloging in Publication Data

Swokowski, Earl W.
Functions and graphs.

Includes index.
1. Functions. 2. Algebra—Graphic methods.
3. Trigonometry. I. Title.
QA331.3.S95 1984 512′.1 83-23048
ISBN 0-87150-460-X

Production and design: Kathi Townes
Text composition: Syntax International
Technical artwork: Vantage Art, Inc.
Cover photograph: Erich Hartmann/Magnum Photos, Inc.
Cover printing: Federated Lithographers-Printers, Inc.
Text printing and binding: Halliday Lithograph

PREFACE

The aim of *Functions and Graphs* is to meet the wide-ranging mathematical needs of students who enroll in courses at the precalculus level. As I worked on this revision, one of my goals was to maintain the mathematical soundness of earlier editions, but to make some of the discussions less formal by rewriting, placing more emphasis on graphing, and adding new examples, figures, and exercises. Another objective was to make the subject matter more relevant through the use of applied problems and references to calculus, a course that many students will take after completing this text. Finally, suggestions from previous users on how to improve the book led me to change the order in which certain concepts are presented. The comments that follow highlight some of the features of this edition.

Chapter 1 contains topics that are fundamental for the study of algebra and trigonometry. These include properties of real numbers, exponents and radicals, techniques for simplifying algebraic expressions, and solutions of equations and inequalities. An innovation in this edition is the use of *test values* as an aid to solving inequalities.

Chapter 2 begins with a discussion of coordinate systems and graphs of equations in two variables. Tests for symmetry are introduced in the first section, and hence are available for use throughout the remainder of the text. Section 2.2 consists of a detailed study of lines, including derivations of equations and results about the relationships between the slopes of parallel and perpendicular lines. Functions are considered in Sections 2.3 and 2.4, with special attention being given to horizontal and vertical shifts of graphs. In Section 2.5, this material is applied to the problem of sketching the graph of a quadratic function. Composite and inverse functions are studied in the last section of the chapter.

Chapter 3 contains an in-depth discussion of polynomial and rational functions. Considerable emphasis is given to the use of the *Intermediate Value Theorem* as an aid to sketching graphs. Complex numbers are defined in Section 3.4 and are then used in subsequent sections to establish results on the zeros of a polynomial function. In Section 3.7, guidelines are listed to help students proceed in a systematic manner when sketching the graph of a rational function. The concept of oblique asymptote has been added to this section. Partial fraction decompositions are discussed at the end of the chapter.

Chapter 4 has been extensively rewritten, and now contains much more material on properties and applications of the natural exponential and logarithmic functions. The use of calculators is emphasized in many examples and exercises. A discussion of interpolation is no longer included in this chapter, but now appears in Appendix I.

The approach to trigonometry has been changed in this edition. Angles are discussed at the beginning of Chapter 5, and the trigonometric functions are introduced by means of right triangles. This is followed with a unit circle formulation. Thus, all the standard ways of viewing trigonometric functions are available early in the chapter. The method of interpolation is, again, relegated to an appendix, and calculators are recommended for numerical work. A new section on harmonic motion has been added to provide some non-triangular applications of trigonometry.

The emphasis in Chapter 6 is on trigonometric identities and equations, and the presentation is similar to that in previous editions, with two exceptions. The trigonometric form for complex numbers and De Moivre's Theorem now appear in Sections 6.9 and 6.10, respectively. This insertion makes it unnecessary to include a separate chapter on complex numbers. The last section of the chapter contains an introduction to geometric and algebraic properties of vectors in two dimensions.

Matrices are introduced in Chapter 7 as a tool for solving systems of linear equations. Symbols for specifying row operations are used to make it easier to check manipulations that lead to the echelon form of a matrix. The last part of the chapter deals with determinants and linear programming.

Chapter 8 has sections on mathematical induction, the Binomial Theorem, summation notation, and sequences. The standard topics in analytic geometry are included in Chapter 9.

There is a review section at the end of each chapter consisting of a list of important topics and pertinent exercises. The review exercises are similar in scope to those that appear throughout the chapter and may be used by students to prepare for examinations. Answers to odd-numbered exercises are given at the end of the text. Instructors may obtain an answer booklet for the even-numbered exercises from the publisher.

This revision has benefited from comments and suggestions of users of previous editions. I wish to thank the following individuals, who reviewed all or parts of the manuscript and offered many helpful suggestions: E. Ray Bobo, Georgetown University; Dennis W. Brewer, University of Arkansas, Fayetteville; Louis Hoelzle, Bucks County Community College; John Hornsby, University of New Orleans; William H. Keils, San Antonio College; Mary L. Lindbloom, Central Michigan University; Richard Randell, University of Iowa; Carlos Rodriguez, San Antonio College; Ronald Rose, American River College; Loyd Wilcox, Golden West College. In addition, many instructors were kind enough to respond to a questionnaire from my publisher: Charles Applebaum, Bowling Green State University; Elson C. Baldwin, Prince George's Community College; William R. Ballard, University of Montana; Janet Buck, Seattle Pacific University; Dennis W. Brewer, University of Arkansas, Fayetteville; Graham Chalmers, California State University, Long Beach; Torgeir Haugland, Highline Community College; David Horowitz, Golden West College; Bettye Knox, San Antonio College; Norman Locksley, Prince George's Community College; Roger B. Nelson, Lewis and Clark College; James E. Nyman, University of Texas, El Paso; Carol W. Penney, University of Georgia; Daniel E. Scanlon, Orange Coast College; Douglas D. Smith, Central Michigan University; Lewis Steppe, Central Virginia Community College; Sally Thomas, Orange Coast College; Gerald Thompson, Augusta College; and James F. Thorpe, Saddleback College, were among those who contributed information and ideas.

I am also grateful for the excellent cooperation of the staff of PWS Publishers. Two people in the company deserve special mention. They are my copy editor, Kathi Townes, who did an outstanding job at every stage of production, and David Pallai, who contacted many reviewers and offered valuable advice.

In addition to all of the persons named here, I express my sincere appreciation to the many unnamed students and teachers who have helped shape my views about precalculus mathematics.

Earl W. Swokowski

TABLE OF CONTENTS

1

FUNDAMENTAL CONCEPTS *1*

1.1 Real Numbers *1*
1.2 Exponents and Radicals *10*
1.3 Algebraic Expressions *20*
1.4 Equations and Inequalities *34*
1.5 More on Inequalities *45*
1.6 Review *52*

2

FUNCTIONS *55*

2.1 Coordinate Systems in Two Dimensions *55*
2.2 Lines *65*
2.3 Definition of Function *73*
2.4 Graphs of Functions *82*
2.5 Quadratic Functions *90*
2.6 Composite and Inverse Functions *97*
2.7 Review *103*

3

POLYNOMIAL AND RATIONAL FUNCTIONS *106*

3.1 Graphs of Polynomial Functions *106*
3.2 Properties of Division *112*
3.3 Synthetic Division *116*
3.4 Complex Numbers *120*
3.5 The Zeros of a Polynomial *128*
3.6 Complex and Rational Zeros of Polynomials *135*
3.7 Rational Functions *141*
3.8 Partial Fractions *152*
3.9 Review *157*

4

EXPONENTIAL AND LOGARITHMIC FUNCTIONS *159*

4.1 Exponential Functions *159*
4.2 Logarithms *170*
4.3 Logarithmic Functions *176*
4.4 Common Logarithms *182*
4.5 Review *189*

5

THE TRIGONOMETRIC FUNCTIONS *190*

5.1 Angles *190*
5.2 Trigonometric Functions of Angles *197*
5.3 The Circular Functions *207*
5.4 Values of the Trigonometric Functions *213*
5.5 Graphs of the Trigonometric Functions *222*
5.6 Trigonometric Graphs *228*
5.7 Additional Graphical Techniques *232*
5.8 Applications Involving Right Triangles *235*
5.9 Harmonic Motion *241*
5.10 Review *245*

6

ANALYTIC TRIGONOMETRY *247*

6.1 Trigonometric Identities *247*
6.2 Trigonometric Equations *253*
6.3 The Addition and Subtraction Formulas *258*
6.4 Multiple-Angle Formulas *266*
6.5 Product and Factoring Formulas *272*
6.6 The Inverse Trigonometric Functions *276*
6.7 The Law of Sines *284*
6.8 The Law of Cosines *290*
6.9 Trigonometric Form for Complex Numbers *293*
6.10 De Moivre's Theorem and *n*th Roots of Complex Numbers *298*
6.11 Vectors *302*
6.12 Review *312*

7

SYSTEMS OF EQUATIONS AND INEQUALITIES *315*

7.1 Systems of Equations *315*
7.2 Systems of Linear Equations in Two Variables *321*
7.3 Systems of Linear Equations in More than Two Variables *327*
7.4 The Algebra of Matrices *339*
7.5 Determinants *348*
7.6 Properties of Determinants *354*
7.7 Cramer's Rule *359*
7.8 Systems of Inequalities *362*
7.9 Linear Programming *367*
7.10 Review *372*

8

SEQUENCES AND SERIES *375*

8.1 Mathematical Induction *375*
8.2 The Binomial Theorem *381*
8.3 Infinite Sequences and Summation Notation *387*

8.4 Arithmetic Sequences *394*
8.5 Geometric Sequences *399*
8.6 Review *405*

9

TOPICS IN ANALYTIC GEOMETRY *407*

9.1 Conic Sections *407*
9.2 Parabolas *409*
9.3 Ellipses *415*
9.4 Hyperbolas *421*
9.5 Rotation of Axes *427*
9.6 Polar Coordinate Systems *431*
9.7 Polar Equations of Conics *437*
9.8 Plane Curves and Parametric Equations *442*
9.9 Review *449*

APPENDICES *A1*

I Using Logarithmic and Trigonometric Tables *A1*
II Tables *A12*
Table 1 Common Logarithms *A12*
Table 2 Natural Exponential Function *A13*
Table 3 Natural Logarithms *A13*
Table 4 Values of the Trigonometric Functions *A14*
Table 5 Trigonometric Functions of Radians and Real Numbers *A18*

ANSWERS TO ODD-NUMBERED EXERCISES *A21*

INDEX *A45*

FUNDAMENTAL CONCEPTS

The material in this chapter is basic to the study of algebra. We begin by discussing properties of real numbers. Next we turn our attention to exponents and radicals, and how they may be used to simplify complicated algebraic expressions. We consider solutions of equations and inequalities in the last two sections.

SECTION 1.1
REAL NUMBERS

Real numbers are used in all phases of mathematics, and you are undoubtedly well acquainted with symbols that are used to represent them, such as

$$1, \quad 73, \quad -5, \quad \frac{49}{12}, \quad \sqrt{2}, \quad 0, \quad \sqrt[3]{-85}, \quad 0.33333\ldots, \quad \text{and} \quad 596.25.$$

The collection of all real numbers is said to be **closed** relative to operations of addition (denoted by $+$) and multiplication (denoted by $\cdot$). This means that to every pair a, b of real numbers there corresponds a unique real number $a + b$, called the **sum** of a and b, and a unique real number $a \cdot b$ (also written ab), called the **product** of a and b. If a and b denote the same real number, we write $a = b$ (read "a equals b"). An expression of this type is called an **equality.**

The special numbers 0 and 1, referred to as **zero** and **one,** respectively, have the properties $a + 0 = a$ and $a \cdot 1 = a$ for every real number a. Each real number a has a **negative,** denoted by $-a$, such that $a + (-a) = 0$, and each nonzero real number a has a **reciprocal,** $\frac{1}{a}$, such that $a\left(\frac{1}{a}\right) = 1$. These and other important properties are included in the following list, in which a, b, and c denote arbitrary real numbers.

COMMUTATIVE PROPERTIES

$$a + b = b + a, \qquad ab = ba$$

ASSOCIATIVE PROPERTIES

$$a + (b + c) = (a + b) + c, \qquad a(bc) = (ab)c$$

IDENTITIES

$$a + 0 = a = 0 + a, \qquad a \cdot 1 = a = 1 \cdot a$$

INVERSES

$$a + (-a) = 0 = (-a) + a, \qquad a\left(\frac{1}{a}\right) = 1 = \left(\frac{1}{a}\right)a \quad \text{if } a \neq 0$$

DISTRIBUTIVE PROPERTIES

$$a(b + c) = ab + ac, \qquad (a + b)c = ac + bc$$

The real numbers 0 and 1 are sometimes referred to as the **additive identity** and **multiplicative identity,** respectively. The negative, $-a$, is also called the **additive inverse** of a and, if $a \neq 0$, $\frac{1}{a}$ is called the **multiplicative inverse** of a. The symbol a^{-1} may be used in place of $\frac{1}{a}$, as indicated by the following definition.

DEFINITION OF a^{-1}

$$a^{-1} = \frac{1}{a}$$

If $a = b$ and $c = d$, then, since a and b are merely different names for the same real number, and c and d are different names for another real number, it follows that $a + c = b + d$ and $ac = bd$. As a special case, using the fact that $c = c$, we see that if $a = b$, then $a + c = b + c$ and $ac = bc$. We sometimes refer to these rules by the statements "any real number c may be added to both sides of an equality" and "both sides of an equality may be multiplied by the same real number c."

The following theorem can be proved. (See Exercises 54 and 55.)

THEOREM

$a \cdot 0 = 0$ for every real number a.

If $ab = 0$, then either $a = 0$ or $b = 0$.

The preceding theorem implies that $ab = 0$ *if and only if* either $a = 0$ or $b = 0$. The phrase "if and only if," which is used throughout mathematics, always has a twofold character. Here it means that if $ab = 0$, then $a = 0$ or $b = 0$ and, *conversely*, if $a = 0$ or $b = 0$, then $ab = 0$. Consequently, if both $a \neq 0$ and $b \neq 0$, then $ab \neq 0$; that is, *the product of two nonzero real numbers is always nonzero.*

The following properties can also be proved.

PROPERTIES OF NEGATIVES

$$-(-a) = a$$

$$(-a)b = -(ab) = a(-b)$$

$$(-a)(-b) = ab$$

$$(-1)a = -a$$

The operation of **subtraction** (denoted by $-$) is defined as follows:

DEFINITION OF SUBTRACTION

$$a - b = a + (-b)$$

If $b \neq 0$, then **division** (denoted by $\div$) is defined as follows:

DEFINITION OF DIVISION

$$a \div b = a\left(\frac{1}{b}\right) = ab^{-1}$$

The symbol a/b is often used in place of $a \div b$, and we refer to it as the **quotient of *a* by *b*** or the **fraction *a* over *b*.** The numbers a and b are called the **numerator** and **denominator,** respectively, of the fraction. It is important to note that since 0 has no multiplicative inverse, a/b is not defined if $b = 0$; that is, *division by zero is not permissible.* Also note that if $b \neq 0$, then

$$1 \div b = \frac{1}{b} = b^{-1}.$$

The following properties of quotients may be established, where all denominators are nonzero real numbers.

PROPERTIES OF QUOTIENTS

$$\frac{a}{b} = \frac{c}{d} \quad \text{if and only if} \quad ad = bc$$

$$\frac{a}{b} = \frac{ad}{bd}, \qquad \frac{a}{-b} = \frac{-a}{b} = -\frac{a}{b}$$

$$\frac{a}{b} + \frac{c}{b} = \frac{a+c}{b}, \qquad \frac{a}{b} + \frac{c}{d} = \frac{ad+bc}{bd}$$

$$\frac{a}{b} \cdot \frac{c}{d} = \frac{ac}{bd}, \qquad \frac{a}{b} \div \frac{c}{d} = \frac{a}{b} \cdot \frac{d}{c} = \frac{ad}{bc}$$

The **positive integers,** 1, 2, 3, 4, . . . , may be obtained by adding the real number 1 successively to itself. The negatives of the positive integers, -1, -2, -3, -4, . . . , are referred to as **negative integers.** The **integers** consist of the totality of positive and negative integers together with the real number 0.

If a, b, and c are integers and $c = ab$, then a and b are called **factors,** or **divisors,** of c. For example, the integer 6 may be written as

$$6 = 2 \cdot 3 = (-2)(-3) = 1 \cdot 6 = (-1)(-6).$$

Hence, 1, -1, 2, -2, 3, -3, 6, and -6 are factors of 6.

A positive integer p different from 1 is **prime** if its only positive factors are 1 and p. The first few primes are 2, 3, 5, 7, 11, 13, 17, and 19. One reason that primes are important is because every positive integer different from 1 can be expressed as a product of primes in one and only one way (except for order of factors). The proof of this result, which is called **The Fundamental Theorem of Arithmetic,** will not be given in this book. As examples, we have

$$12 = 2 \cdot 2 \cdot 3, \qquad 126 = 2 \cdot 3 \cdot 3 \cdot 7, \qquad 540 = 2 \cdot 2 \cdot 3 \cdot 3 \cdot 3 \cdot 5.$$

A real number is called a **rational number** if it can be written in the form a/b, where a and b are integers and $b \neq 0$. Real numbers that are not rational are called **irrational.** The ratio of the circumference of a circle to its diameter is irrational and is denoted by π. It is often approximated by the decimal 3.1416 or by the rational number 22/7. We use the notation $\pi \approx 3.1416$ to indicate that π *is approximately equal to* 3.1416. To cite another example, there is no rational number b such that $b^2 = 2$, where b^2 denotes $b \cdot b$. However, there is an *irrational* number, denoted by $\sqrt{2}$, such that $(\sqrt{2})^2 = 2$. In general, if a, is *any* nonnegative real number, then by definition, the **principal square root of *a*,** denoted by $\sqrt{a}$, is the *nonnegative* real number b such that $b^2 = a$. For simplicity $\sqrt{a}$ is referred to as the *square root of a*. Some square roots are rational. For example,

$$\sqrt{25} = 5, \qquad \sqrt{\frac{9}{4}} = \frac{3}{2}, \qquad \sqrt{16} = 4.$$

Other square roots, such as $\sqrt{3}$, $\sqrt{5}$, and $\sqrt{7/2}$, are irrational.

Real numbers may be represented by decimal expressions. Decimal representations for rational numbers are either terminating or nonterminating and repeating. For example, we can show by division that a decimal representation for 7434/2310 is 3.2181818 . . . , where the digits 1 and 8 repeat indefinitely. The rational number $\frac{5}{4}$ has the terminating decimal representation 1.25. We may also obtain decimal representations for irrational numbers; however, such decimals are always nonterminating and nonrepeating. The process of finding decimal representations for irrational numbers is usually difficult. Often some method of successive approximation is employed. For example, an arithmetic process can be used to find square roots. Using this technique we could approximate $\sqrt{2}$ by obtaining, successively, 1, 1.4, 1.41, 1.414, 1.4142, ...

It is possible to associate real numbers with points on a line l in such a way that for each real number a there corresponds one and only one point on l, and conversely, to each point P on l there corresponds precisely one real number. Such an association is called a **one-to-one correspondence.** We first choose an arbitrary point O, called the **origin,** and associate with it the real number 0. Points associated with the integers are then determined by laying off successive line segments of equal length on either side of O as illustrated in Figure 1.1. The points corresponding to rational numbers such as $\frac{23}{5}$ and $-\frac{1}{2}$ are obtained by subdividing the equal line segments. Points associated with certain irrational numbers, such as $\sqrt{2}$, can be found by geometric construction. For other irrational numbers such as π, no construction is possible. However, the point corresponding to π can be approximated to any degree of accuracy by locating successively the points corresponding to decimal representations 3, 3.1, 3.14, 3.141, 3.1415, 3.14159, ... It can be shown that to every irrational number there corresponds a unique point on l and conversely, every point that is not associated with a rational number corresponds to an irrational number.

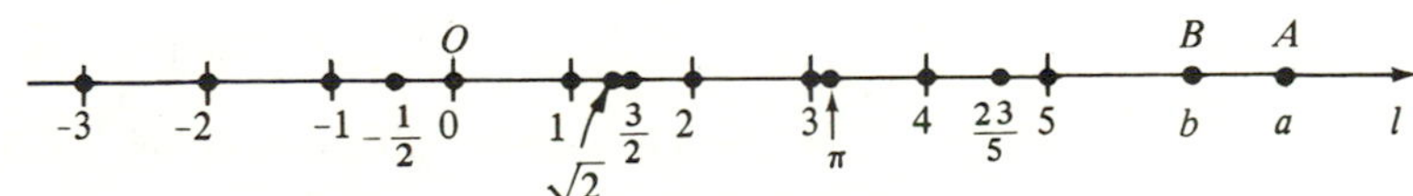

FIGURE 1.1

The number a that is associated with a point A on l is called the **coordinate** of A. An assignment of coordinates to points on l is called a **coordinate system** for l, and l is called a **coordinate line,** or a **real line.** A direction can be assigned to l by taking the **positive direction** along l to the right and the

negative direction to the left. The positive direction is noted by placing an arrowhead on l as shown in Figure 1.1.

The numbers that correspond to points to the right of O in Figure 1.1 are called **positive real numbers,** whereas numbers that correspond to points to the left of O are **negative real numbers.** *The real number* 0 *is neither positive nor negative.* The totality of positive real numbers is closed relative to addition and multiplication; that is, if a and b are positive, then so are the sum $a + b$ and the product ab.

Note that if a is positive, then $-a$ is negative. Similarly, if $-a$ is positive, then $-(-a) = a$ is negative. A common error is to think that $-a$ is always a negative number; however, this is not necessarily the case. For example, if $a = -3$, then $-a = -(-3) = 3$, which is positive. Indeed, whenever a is negative, the number $-a$ is positive.

If a and b are real numbers and $a - b$ is positive, we say that ***a* is greater than *b*** and we write $a > b$. An equivalent statement is that ***b* is less than *a*,** written $b < a$. The symbols $>$ or $<$ are called **inequality signs** and expressions such as $a > b$ or $b < a$ are called **inequalities.** From the manner in which we constructed the coordinate line l in Figure 1.1, we see that if A and B are points with coordinates a and b, respectively, then $a > b$ (or $b < a$) *if and only if A lies to the right of B.* The following definition is stated for reference, where a and b denote real numbers.

DEFINITIONS OF > AND <

$a > b$ means $a - b$ is positive.

$b < a$ means $a - b$ is positive.

Observe that $a > b$ and $b < a$ have exactly the same meaning. As illustrations we may write

$$
\begin{array}{rll}
5 > 3 & \text{since} & 5 - 3 = 2 \text{ is positive;} \\
-6 < -2 & \text{since} & -2 - (-6) = -2 + 6 = 4 \text{ is positive;} \\
-\sqrt{2} < 1 & \text{since} & 1 - (-\sqrt{2}) = 1 + \sqrt{2} \text{ is positive;} \\
2 > 0 & \text{since} & 2 - 0 = 2 \text{ is positive;} \\
-5 < 0 & \text{since} & 0 - (-5) = 5 \text{ is positive.}
\end{array}
$$

The last two illustrations are special cases of the following general properties.

$a > 0$ if and only if a is positive.

$a < 0$ if and only if a is negative.

It should be clear from our discussion that if a and b are real numbers, then *one and only one of the following statements is true:*

$$a = b, \quad a > b, \quad \text{or} \quad a < b.$$

We refer to the **sign** of a real number as being positive or negative if the number is positive or negative, respectively. We also say that two real numbers *have the same sign* if both are positive or both are negative. The numbers have *opposite signs* if one is positive and the other is negative. The next result about the signs of products and quotients of two real numbers a and b can be proved using properties of negatives and quotients.

LAWS OF SIGNS

(i) If a and b have the same sign, then ab and a/b are positive.

(ii) If a and b have opposite signs, then ab and a/b are negative.

The converses of the Laws of Signs are also true. For example, if a quotient is negative, then the numerator and denominator have opposite signs.

Several other useful symbols involve inequality signs. In particular, $a \geq b$, which is read ***a* is greater than or equal to *b*,** means that either $a > b$ or $a = b$ (but not both). As an illustration, we have $a^2 \geq 0$ for every real number a. The symbol $a \leq b$ is read ***a* is less than or equal to *b*** and means that either $a < b$ or $a = b$. The expression $a < b < c$ means that both $a < b$ and $b < c$, in which case we say that ***b* is between *a* and *c*.** Similarly, the expression $c > b > a$ means that both $c > b$ and $b > a$. Thus.

$$1 < 5 < \frac{11}{2}, \qquad -4 < \frac{2}{3} < \sqrt{2}, \qquad 3 > -6 > -10.$$

Other variations of the inequality notation are used. For example, $a < b \leq c$ means both $a < b$ and $b \leq c$. Similarly, $a \leq b < c$ means both $a \leq b$ and $b < c$. Finally, $a \leq b \leq c$ means both $a \leq b$ and $b \leq c$.

If a is a real number, then it is the coordinate of some point A on a coordinate line l, and the symbol $|a|$ is used to denote the number of units (or distance) between A and the origin, without regard to direction. The nonnegative number $|a|$ is called the *absolute value* of a. Referring to Figure 1.2, we see that for the point with coordinate -4, we have $|-4| = 4$. Similarly, $|4| = 4$. In general, if a is negative we change its sign to find $|a|$, whereas if a is nonnegative, then $|a| = a$. The next definition summarizes this discussion.

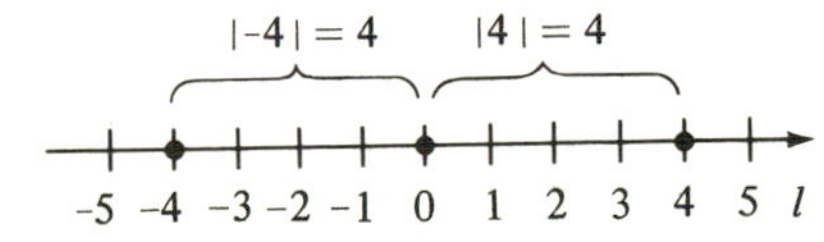

FIGURE 1.2

DEFINITION

If a is a real number, then the **absolute value** of a, denoted by $|a|$, is

$$|a| = \begin{cases} a & \text{if } a \geq 0. \\ -a & \text{if } a < 0. \end{cases}$$

The use of this definition is illustrated in the following example.

EXAMPLE 1 Find $|3|$, $|-3|$, $|0|$, $|\sqrt{2}-2|$, and $|2-\sqrt{2}|$.

Solution Since 3, $2-\sqrt{2}$, and 0 are nonnegative,

$$|3| = 3, \quad |2-\sqrt{2}| = 2-\sqrt{2}, \quad \text{and} \quad |0| = 0.$$

Since -3 and $\sqrt{2}-2$ are negative, we use the formula $|a| = -a$ to obtain

$$|-3| = -(-3) = 3 \quad \text{and} \quad |\sqrt{2}-2| = -(\sqrt{2}-2) = 2-\sqrt{2}.$$ ■

Note that in Example 1, $|-3| = |3|$ and $|2-\sqrt{2}| = |\sqrt{2}-2|$. It can be shown in general that

$$|a| = |-a| \text{ for every real number } a.$$

We shall use the concept of absolute value to define the distance between any two points on a coordinate line. Let us begin by noting that the distance between the points with coordinates 2 and 7, shown in Figure 1.3, equals 5 units on l. This distance is the difference, $7-2$, obtained by subtracting the smaller coordinate from the larger. If we employ absolute values, then since $|7-2| = |2-7|$, it is unnecessary to be concerned about the order of subtraction. We shall use this as our motivation for the next definition.

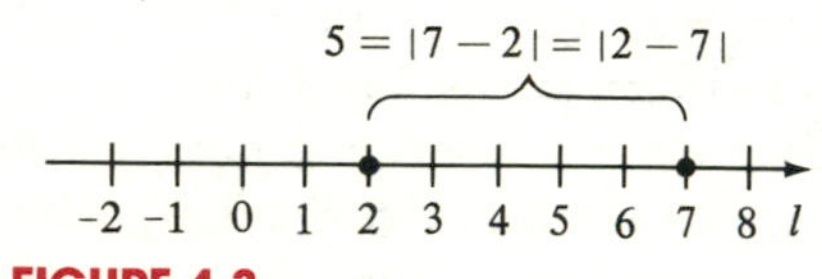

FIGURE 1.3

DEFINITION

Let a and b be the coordinates of two points A and B, respectively, on a coordinate line l. The **distance between A and B,** denoted by $d(A, B)$, is

$$d(A, B) = |b-a|.$$

The number $d(A, B)$ is also called the **length of the line segment AB.** Observe that since $d(B, A) = |a-b|$ and $|b-a| = |a-b|$,

$$d(A, B) = d(B, A).$$

Also note that the distance between the origin O and the point A is

$$d(O, A) = |a-0| = |a|,$$

which agrees with the geometric interpretation of absolute value illustrated in Figure 1.2. The formula $d(A, B) = |b-a|$ is true regardless of the signs of a and b, as illustrated in the next example.

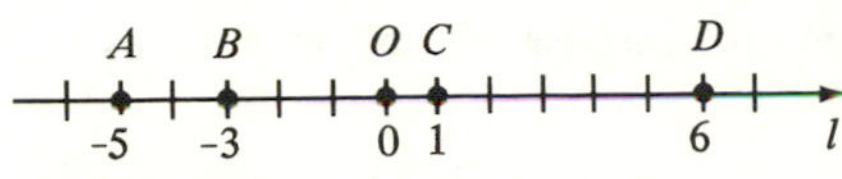

FIGURE 1.4

EXAMPLE 2 Let A, B, C, and D have coordinates -5, -3, 1, and 6, respectively, on a coordinate line l (see Figure 1.4). Find $d(A, B)$, $d(C, B)$, $d(O, A)$, and $d(C, D)$.

Solution Using the definition of the distance between points, we obtain

$$d(A, B) = |-3 - (-5)| = |-3 + 5| = |2| = 2$$
$$d(C, B) = |-3 - 1| = |-4| = 4$$
$$d(O, A) = |-5 - 0| = |-5| = 5$$
$$d(C, D) = |6 - 1| = |5| = 5.$$

These answers can be checked by referring to Figure 1.4. ■

The concept of absolute value has uses other than that of finding distances between points. Generally, it is employed whenever we are interested in the "magnitude" or "numerical value" of a real number without regard to its sign.

EXERCISES 1.1

In Exercises 1–10 justify the given equality by stating only one of the following properties: (a) commutative, (b) associative, (c) identity, (d) inverse, (e) distributive.

1 $(4 \cdot 5) \cdot 4 = 4 \cdot (4 \cdot 5)$

2 $3 \cdot (4 + 5) = (4 + 5) \cdot 3$

3 $(4 \cdot 5) \cdot 4 = 4 \cdot (5 \cdot 4)$

4 $(4 + 5) \cdot 3 = 4 \cdot 3 + 5 \cdot 3$

5 $3 \cdot (5 + 0) = 3 \cdot 5$

6 $3 + (-3) = 0$

7 $1 \cdot (2 + 3) = 2 + 3$

8 $(1 + 2) + 1 = 1 + (1 + 2)$

9 $(\frac{1}{4})4 = 1$

10 $0 \cdot 1 = 0$

In Exercises 11–20 write the expression as a rational number with least positive numerator.

11 $\dfrac{3}{4} + \dfrac{5}{3}$

12 $\dfrac{1}{5} + \dfrac{3}{2}$

13 $\dfrac{5}{6} \cdot \dfrac{2}{3}$

14 $\dfrac{4}{9}\left(\dfrac{1}{2} + \dfrac{3}{4}\right)$

15 $\dfrac{1}{2} + \dfrac{1}{4} + \dfrac{1}{6}$

16 $\dfrac{3}{-2} \cdot \dfrac{-5}{6}$

17 $\dfrac{13}{4} \div \left(\dfrac{3}{2} + 1\right)$

18 $\dfrac{2}{5} + \dfrac{7}{4} + \dfrac{3}{2}$

19 $\dfrac{5}{7}\left(\dfrac{3}{2} - \dfrac{5}{6}\right)$

20 $\dfrac{10}{11} \cdot \dfrac{11}{10}$

21 Show, by means of examples, that the operation of subtraction is neither commutative nor associative.

22 Use examples to show that the operation of division, as applied to nonzero real numbers, is neither commutative nor associative.

In Exercises 23–26 replace the symbol □ with either <, >, or =.

23 (a) $-7 \square -4$
(b) $3 \square -1$
(c) $1 + 3 \square 6 - 2$

24 (a) $-3 \square -5$
(b) $-6 \square 2$
(c) $\frac{1}{4} \square 0.25$

25 (a) $\frac{1}{3} \square 0.33$
(b) $\frac{125}{57} \square 2.193$
(c) $\frac{22}{7} \square \pi$

26 (a) $\frac{1}{7} \square 0.143$
(b) $\frac{3}{4} + \frac{2}{3} \square \frac{19}{12}$
(c) $\sqrt{2} \square 1.4$

Express the statements in Exercises 27–38 in terms of inequalities.

27 -8 is less than -5.

28 2 is greater than 1.9.

29 0 is greater than -1.

30 $\sqrt{2}$ is less than π.

31 x is negative.

32 y is positive.

33 a is between 5 and 3.

34 b is between $\frac{1}{10}$ and $\frac{1}{3}$.

35 b is greater than or equal to 2.

36 x is less than or equal to -5.

37 c is not greater than 1.

38 d is nonnegative.

Rewrite the numbers in Exercises 39–42 without using symbols for absolute value.

39 (a) $|4-9|$
(b) $|-4|-|-9|$
(c) $|4|+|-9|$

40 (a) $|3-6|$
(b) $|0.2-\frac{1}{5}|$
(c) $|-3|-|-4|$

41 (a) $3-|-3|$
(b) $|\pi-4|$
(c) $(-3)/|-3|$

42 (a) $|8-5|$
(b) $-5+|-7|$
(c) $(-2)|-2|$

In Exercises 43–46 the given numbers are coordinates of three points A, B, and C (in that order) on a coordinate line l. For each, find (a) $d(A, B)$; (b) $d(B, C)$; (c) $d(C, B)$; and (d) $d(A, C)$.

43 $-6, \quad -2, \quad 4$

44 $3, \quad 7, \quad -5$

45 $8, \quad -4, \quad -1$

46 $-9, \quad 1, \quad 10$

Rewrite the expressions in Exercises 47–50 without using symbols for absolute value.

47 $|5-x|$ if $x > 5$

48 $|x^2+1|$

49 $|-4-x^2|$

50 $|a-b|$ if $a < b$

In Exercises 51–60 all letters denote real numbers and no denominators are zero.

51 Prove or disprove: $\dfrac{1}{a}+\dfrac{1}{b}=\dfrac{1}{a+b}$.

52 Prove or disprove: $\dfrac{a}{b+c}=\dfrac{a}{b}+\dfrac{a}{c}$.

Prove the rules in Exercises 53–56.

53 $a=-a$ if and only if $a=0$.

54 $a\cdot 0=0$ (*Hint:* Write $a\cdot 0=a\cdot(0+0)=a\cdot 0+a\cdot 0$ and then add $-(a\cdot 0)$ to both sides.)

55 If $ab=0$, then either $a=0$ or $b=0$.

56 If $a+b=0$, then $b=-a$.

57 $a(b-c)=ab-ac$

58 $\left(\dfrac{a}{b}\right)^{-1}=\dfrac{b}{a}$

59 $\dfrac{-a}{-b}=\dfrac{a}{b}$

60 $\dfrac{a}{b}+\dfrac{c}{b}=\dfrac{a+c}{b}$ (*Hint:* Write $\dfrac{a}{b}+\dfrac{c}{b}=ab^{-1}+cb^{-1}$ and use the Distributive Properties.)

SECTION 1.2

EXPONENTS AND RADICALS

The symbol a^2 denotes the real number $a\cdot a$. Similarly, a^3 is used in place of $a\cdot a\cdot a$. In general, if n is any positive integer,

$$a^n = \underbrace{a\cdot a\cdot\cdots\cdot a}_{n\text{ factors}}$$

where n factors, all equal to a, appear on the right-hand side of the equal sign. The positive integer n is called the **exponent** of a in the expression a^n, and a^n is read "a to the nth **power**" or simply "a to the n." Note that $a^1 = a$. Some numerical examples are

$$\left(\frac{1}{2}\right)^5 = \frac{1}{2}\cdot\frac{1}{2}\cdot\frac{1}{2}\cdot\frac{1}{2}\cdot\frac{1}{2} = \frac{1}{32}$$
$$(-3)^3 = (-3)(-3)(-3) = -27$$
$$(\sqrt{2})^4 = \sqrt{2}\sqrt{2}\sqrt{2}\sqrt{2} = (\sqrt{2})^2(\sqrt{2})^2 = (2)(2) = 4.$$

If $a \neq 0$, the extension to nonpositive exponents is made by defining

$$a^0 = 1 \quad \text{and} \quad a^{-n} = \frac{1}{a^n}.$$

The following rules are indispensable when working with exponents.

LAWS OF EXPONENTS

If a and b are real numbers and m and n are integers, then

(i) $a^m a^n = a^{m+n}$; (ii) $(a^m)^n = a^{mn}$; (iii) $(ab)^n = a^n b^n$;

(iv) $\left(\frac{a}{b}\right)^n = \frac{a^n}{b^n}$, if $b \neq 0$; (v) If $a \neq 0$, then $\frac{a^m}{a^n} = a^{m-n} = \frac{1}{a^{n-m}}$.

Complete proofs of (i)–(v) require the use of the method of mathematical induction, which we will discuss in Section 8.1. However, if we are allowed to count an arbitrary number of factors, then it is easy to supply arguments that establish the laws for positive exponents. Thus, to prove (i) if m and n are positive integers, we may write

$$a^m a^n = \underbrace{a \cdot a \cdot \cdots \cdot a}_{m \text{ factors}}\ \underbrace{a \cdot a \cdot \cdots \cdot a}_{n \text{ factors}}.$$

Since the total number of factors a on the right is $m + n$, this expression is equal to a^{m+n}.

Similarly, to prove (ii) we write

$$(a^m)^n = \underbrace{a^m \cdot a^m \cdot \cdots \cdot a^m}_{n \text{ factors } a^m}$$

and count the number of times a appears as a factor on the right-hand side. Since $a^m = a \cdot a \cdot \cdots \cdot a$, where a occurs as a factor m times, and since the number of such groups of m factors is n, the total number of factors is $m \cdot n$. Proofs for the cases $m \leq 0$ and $n \leq 0$ are left to the reader.

Laws (iii) and (iv) can be obtained in similar fashion. Law (v) is clear if $m = n$. For the case $m > n$, the integer $m - n$ is positive, and we may write

$$\frac{a^m}{a^n} = \frac{a^n a^{m-n}}{a^n} = \frac{a^n}{a^n} \cdot a^{m-n} = 1 \cdot a^{m-n} = a^{m-n}.$$

A similar argument can be used if $n > m$.

We can, of course, use symbols for real numbers other than a and b when applying laws of exponents, as in the following illustrations.

$$x^5x^6 = x^{5+6} = x^{11} \qquad (y^5)^7 = y^{5\cdot 7} = y^{35}$$

$$(rs)^7 = r^7s^7 \qquad \left(\frac{p}{q}\right)^{10} = \frac{p^{10}}{q^{10}}$$

$$\frac{c^8}{c^3} = c^{8-3} = c^5 \qquad \frac{u^3}{u^8} = u^{3-8} = u^{-5} = \frac{1}{u^5}$$

The Laws of Exponents can be extended to rules such as $a^m a^n a^p = a^{m+n+p}$ and $(abc)^n = a^n b^n c^n$. To simplify statements, we shall assume that in all problems involving exponents, symbols that appear in denominators represent *nonzero* real numbers.

EXAMPLE 1 Simplify each expression:

(a) $(3x^3y^4)(4xy^5)$ (b) $(2a^2b^3c)^4$ (c) $\left(\frac{2r^3}{s}\right)^2\left(\frac{s}{r^3}\right)^3$

Solution

(a)

$$\begin{aligned}(3x^3y^4)(4xy^5) &= (3)(4)x^3xy^4y^5\\ &= 12x^4y^9\end{aligned}$$

(b)

$$\begin{aligned}(2a^2b^3c)^4 &= 2^4(a^2)^4(b^3)^4c^4\\ &= 16a^8b^{12}c^4\end{aligned}$$

(c)

$$\begin{aligned}\left(\frac{2r^3}{s}\right)^2\left(\frac{s}{r^3}\right)^3 &= \left(\frac{2^2r^6}{s^2}\right)\left(\frac{s^3}{r^9}\right)\\ &= 2^2\left(\frac{r^6}{r^9}\right)\left(\frac{s^3}{s^2}\right)\\ &= 4\left(\frac{1}{r^3}\right)(s) = \frac{4s}{r^3}\end{aligned}$$

■

EXAMPLE 2 Eliminate negative exponents and simplify:

(a) $(a^{-2}b^3)^{-3}$ (b) $\dfrac{8x^3y^{-5}}{4x^{-1}y^2}$

Solution

(a)
$$\begin{aligned}(a^{-2}b^3)^{-3} &= (a^{-2})^{-3}(b^3)^{-3}\\ &= a^6b^{-9}\\ &= a^6\cdot\frac{1}{b^9} = \frac{a^6}{b^9}\end{aligned}$$

(b)
$$\begin{aligned}\frac{8x^3y^{-5}}{4x^{-1}y^2} &= \frac{8}{4}\frac{x^3}{x^{-1}}\frac{y^{-5}}{y^2}\\ &= 2x^{3-(-1)}y^{-5-2}\\ &= 2x^4y^{-7} = \frac{2x^4}{y^7}\end{aligned}$$

Another method for simplifying (b) is to multiply numerator and denominator of the fraction by y^5x, thereby eliminating negative exponents. Thus,

$$\begin{aligned}\frac{8x^3y^{-5}}{4x^{-1}y^2} &= \frac{8x^3y^{-5}}{4x^{-1}y^2}\cdot\frac{y^5x}{y^5x}\\ &= \frac{8}{4}\frac{x^4y^0}{x^0y^7}\\ &= 2\frac{x^4(1)}{(1)y^7} = \frac{2x^4}{y^7}.\end{aligned}$$ ■

Roots of real numbers are defined by the statement

$$\sqrt[n]{a} = b \quad \text{if and only if} \quad b^n = a$$

provided that both a and b are nonnegative real numbers and n is a positive integer, or that both a and b are negative and n is an odd positive integer. The number $\sqrt[n]{a}$ is called the **principal *n*th root** of a. If $n = 2$, it is customary to write $\sqrt{a}$ instead of $\sqrt[2]{a}$ and to call $\sqrt{a}$ the (principal) **square root** of a. The number $\sqrt[3]{a}$ is referred to as the **cube root** of a. Examples of roots include

$$\begin{aligned}\sqrt[5]{\tfrac{1}{32}} &= \tfrac{1}{2} &&\text{since} & (\tfrac{1}{2})^5 &= \tfrac{1}{32}\\ \sqrt[4]{81} &= 3 &&\text{since} & 3^4 &= 81\\ \sqrt[3]{-8} &= -2 &&\text{since} & (-2)^3 &= -8\\ \sqrt{16} &= 4 &&\text{since} & 4^2 &= 16.\end{aligned}$$

Complex numbers (see Section 3.4) are needed to define $\sqrt[n]{a}$ if $a < 0$ and n is an *even* positive integer since for all real numbers b, $b^n \geq 0$ whenever n is even.

It is important to observe that if $\sqrt[n]{a}$ exists, it is a *unique* real number. More generally, if $b^n = a$ for a positive integer n, then n is called *an nth root* of a. For example, both 4 and -4 are square roots of 16 since $4^2 = 16$ and also $(-4)^2 = 16$. However, the *principal* square root of 16 is 4, and we write $\sqrt{16} = 4$.

To complete our terminology, the expression $\sqrt[n]{a}$ is called a **radical,** the number a is called the **radicand,** and n is the **index** of the radical. The symbol $\sqrt{}$ is called a **radical sign.**

If $\sqrt{a} = b$, then $b^2 = a$; that is, $(\sqrt{a})^2 = a$. Similarly, if $\sqrt[3]{a} = b$, then $b^3 = a$, or $(\sqrt[3]{a})^3 = a$. In general, if n is any positive integer, then

$$(\sqrt[n]{a})^n = a.$$

It also follows that if $a > 0$, or if $a < 0$ and n is an odd positive integer, then

$$\sqrt[n]{a^n} = a.$$

For example, $\sqrt{5^2} = 5, \quad \sqrt[3]{(-2)^3} = -2, \quad \sqrt[4]{3^4} = 3.$

A common algebraic error is to *always* replace $\sqrt{x^2}$ by x; however, this is not true if x is negative. For example,

$$\sqrt{(-3)^2} = \sqrt{9} = 3 = |-3|.$$

In general we may write

$$\sqrt{x^2} = |x| \quad \text{for every real number } x.$$

Thus, $\sqrt{x^2} = x$ if and only if $x \geq 0$.

The following laws can be proved, where m and n denote positive integers and where it is assumed that the indicated roots exist.

LAWS OF RADICALS

$$\text{(i)}\ \sqrt[n]{ab} = \sqrt[n]{a}\,\sqrt[n]{b}; \quad \text{(ii)}\ \sqrt[n]{\frac{a}{b}} = \frac{\sqrt[n]{a}}{\sqrt[n]{b}}; \quad \text{(iii)}\ \sqrt[m]{\sqrt[n]{a}} = \sqrt[mn]{a}.$$

EXAMPLE 3 Show that each statement is true.

(a) $\sqrt{50} = 5\sqrt{2}$ (b) $\sqrt[3]{-108} = -3\sqrt[3]{4}$ (c) $\sqrt[3]{\sqrt{64}} = 2$

Solution

(a) $$\sqrt{50} = \sqrt{25 \cdot 2} = \sqrt{25}\sqrt{2} = 5\sqrt{2}$$

(b) $$\sqrt[3]{-108} = \sqrt[3]{(-27)(4)} = \sqrt[3]{-27}\,\sqrt[3]{4} = -3\sqrt[3]{4}$$

(c) Applying (iii) of the Laws of Radicals with $m = 3$ and $n = 2$,

$$\sqrt[3]{\sqrt{64}} = \sqrt[6]{64} = 2.$$

As a check, we have $\sqrt[3]{\sqrt{64}} = \sqrt[3]{8} = 2.$ ■

EXAMPLE 4 Simplify the following, where all letters denote positive real numbers.

(a) $\sqrt[3]{16x^3y^8z^4}$ (b) $\sqrt{3a^2b^3}\sqrt{6a^5b}$

Solution

(a)
$$\begin{aligned}\sqrt[3]{16x^3y^8z^4} &= \sqrt[3]{(2^3x^3y^6z^3)(2y^2z)}\\ &= \sqrt[3]{(2xy^2z)^3(2y^2z)}\\ &= \sqrt[3]{(2xy^2z)^3}\sqrt[3]{2y^2z}\\ &= 2xy^2z\sqrt[3]{2y^2z}\end{aligned}$$

(b)
$$\begin{aligned}\sqrt{3a^2b^3}\sqrt{6a^5b} &= \sqrt{18a^7b^4}\\ &= \sqrt{(9a^6b^4)(2a)}\\ &= \sqrt{(3a^3b^2)^2(2a)}\\ &= \sqrt{(3a^3b^2)^2}\sqrt{2a}\\ &= 3a^3b^2\sqrt{2a}\end{aligned}$$
■

An important type of simplification involving radicals is **rationalizing a denominator.** The process involves beginning with a quotient that contains a radical in the denominator and then multiplying numerator and denominator by some expression so that the resulting denominator contains no radicals. The next example illustrates this technique.

EXAMPLE 5 Rationalize the denominator in each of the following.

(a) $\dfrac{1}{\sqrt{5}}$ (b) $\sqrt{\dfrac{2}{3}}$ (c) $\sqrt[3]{\dfrac{x}{y}}$

Solution We may proceed as follows. (Supply reasons.)

(a)
$$\frac{1}{\sqrt{5}} = \frac{1}{\sqrt{5}}\cdot\frac{\sqrt{5}}{\sqrt{5}} = \frac{\sqrt{5}}{(\sqrt{5})^2} = \frac{\sqrt{5}}{5}$$

(b)
$$\sqrt{\frac{2}{3}} = \sqrt{\frac{2}{3}\cdot\frac{3}{3}} = \sqrt{\frac{6}{3^2}} = \frac{\sqrt{6}}{\sqrt{3^2}} = \frac{\sqrt{6}}{3}$$

(c)
$$\sqrt[3]{\frac{x}{y}} = \sqrt[3]{\frac{x}{y}\cdot\frac{y^2}{y^2}} = \frac{\sqrt[3]{xy^2}}{\sqrt[3]{y^3}} = \frac{\sqrt[3]{xy^2}}{y}$$
■

If a calculator is used to find decimal approximations of radicals, there is no advantage in writing $1/\sqrt{5} = \sqrt{5}/5$ or $\sqrt{2/3} = \sqrt{6}/3$ as we did in the preceding example. However, for *algebraic* simplifications changing expressions to such forms is often desirable.

The denominators of certain fractional expressions contain sums or differences that involve radicals. In some cases denominators can be rationalized, as illustrated in the next example.

EXAMPLE 6 Rationalize the denominator of $\dfrac{1}{\sqrt{x} + \sqrt{y}}$.

Solution Multiplying numerator and denominator by $\sqrt{x} - \sqrt{y}$, we obtain

$$\begin{aligned}\frac{1}{\sqrt{x}+\sqrt{y}} &= \frac{1}{\sqrt{x}+\sqrt{y}} \cdot \frac{\sqrt{x}-\sqrt{y}}{\sqrt{x}-\sqrt{y}} \\ &= \frac{\sqrt{x}-\sqrt{y}}{(\sqrt{x})^2 + \sqrt{y}\sqrt{x} - \sqrt{x}\sqrt{y} - (\sqrt{y})^2} \\ &= \frac{\sqrt{x}-\sqrt{y}}{x-y}.\end{aligned}$$

■

For certain problems that occur in calculus, rationalizing a *numerator* is necessary, as illustrated in the following example.

EXAMPLE 7 Rationalize the numerator of the fraction

$$\frac{\sqrt{x+h}-\sqrt{x}}{h}, \qquad \text{where } x > 0 \text{ and } h > 0.$$

Solution We multiply numerator and denominator by $\sqrt{x+h} + \sqrt{x}$ and proceed as follows:

$$\begin{aligned}\frac{\sqrt{x+h}-\sqrt{x}}{h} &= \frac{\sqrt{x+h}-\sqrt{x}}{h} \cdot \frac{\sqrt{x+h}+\sqrt{x}}{\sqrt{x+h}+\sqrt{x}} \\ &= \frac{(\sqrt{x+h})^2 - \sqrt{x}\sqrt{x+h} + \sqrt{x+h}\sqrt{x} - (\sqrt{x})^2}{h(\sqrt{x+h}+\sqrt{x})} \\ &= \frac{(x+h)-x}{h(\sqrt{x+h}+\sqrt{x})} \\ &= \frac{h}{h(\sqrt{x+h}+\sqrt{x})} \\ &= \frac{1}{\sqrt{x+h}+\sqrt{x}}\end{aligned}$$

■

Radicals may be used to define *rational exponents*. First, if n is a positive integer and a is a real number, we define

$$a^{1/n} = \sqrt[n]{a}$$

provided that $\sqrt[n]{a}$ is a real number. Next we have the following definition.

DEFINITION OF RATIONAL EXPONENTS

If m/n is a rational number and n is a positive integer, and if a is a real number such that $\sqrt[n]{a}$ exists, then

$$a^{m/n} = (\sqrt[n]{a})^m = \sqrt[n]{a^m}.$$

We may also write

$$a^{m/n} = (a^{1/n})^m = (a^m)^{1/n}.$$

It can be shown that the Laws of Exponents are true for rational exponents.

EXAMPLE 8 Simplify each expression.

(a) $(-27)^{2/3}(4)^{-5/2}$ (b) $(4a^{1/3})(2a^{1/2})$

(c) $(r^2s^6)^{1/3}$ (d) $\left(\dfrac{2x^{2/3}}{y^{1/2}}\right)^2\left(\dfrac{3x^{-5/6}}{y^{1/3}}\right)$

Solution

(a) $(-27)^{2/3}(4)^{-5/2} = (\sqrt[3]{-27})^2(\sqrt{4})^{-5} = (-3)^2(2)^{-5} = \frac{9}{32}$

(b) $(4a^{1/3})(2a^{1/2}) = 8a^{1/3+1/2} = 8a^{5/6}$

(c) $(r^2s^6)^{1/3} = (r^2)^{1/3}(s^6)^{1/3} = r^{2/3}s^2$

(d) $\left(\dfrac{2x^{2/3}}{y^{1/2}}\right)^2\left(\dfrac{3x^{-5/6}}{y^{1/3}}\right) = \left(\dfrac{4x^{4/3}}{y}\right)\left(\dfrac{3x^{-5/6}}{y^{1/3}}\right) = \dfrac{12x^{1/2}}{y^{4/3}}$ ■

In Chapter 4 we shall extend our work to irrational exponents such as $3^{\sqrt{2}}$ or 5^{π}.

Many scientific disciplines involve numbers that are very large or very small. To simplify working with such numbers, it is customary to write them in the form $a \cdot 10^n$, where a is expressed as a decimal such that $1 \leq a < 10$ and n is an integer. This is referred to as the **scientific form** for real numbers. In scientific applications the symbol $\times$ is usually used for the multiplication symbol and we write $a \times 10^n$ instead of $a \cdot 10^n$. As an illustration, the distance a ray of light travels in one year is approximately 5,900,000,000,000 miles. This number may be written in scientific form as 5.9×10^{12}. The positive exponent 12 indicates that the decimal point should be moved 12 places to the *right*. The notation works equally well for small numbers. To illustrate, the weight of an oxygen molecule is estimated to be 0.0000000000000000000000053 grams or, in scientific form, 5.3×10^{-23} grams. The negative exponent indicates that the decimal point should be moved 23 places to the *left*. Some other illustrations of scientific form are

$$513 = 5.13 \times 10^2, \qquad 20{,}700 = 2.07 \times 10^4, \qquad 92{,}000{,}000 = 9.2 \times 10^7$$
$$0.000648 = 6.48 \times 10^{-4}, \qquad 0.00000000043 = 4.3 \times 10^{-10}.$$

This notation has a number of advantages. It is compact, and it enables the reader to see quickly (without counting zeros) the relative magnitudes of large or small quantities. It may also be used to simplify calculations that involve quantities such as those illustrated in the next example.

EXAMPLE 9 Calculate $\dfrac{(1{,}100{,}000)^2\sqrt{0.00000004}}{(8{,}000{,}000{,}000)^{2/3}}$.

Solution Writing each number in scientific form and using the Laws of Exponents, we have

$$\begin{aligned}\frac{(1.1 \times 10^6)^2(4 \times 10^{-8})^{1/2}}{(8 \times 10^9)^{2/3}} &= \frac{[(1.1)^2 \times 10^{12}][4^{1/2} \times 10^{-4}]}{8^{2/3} \times 10^6}\\ &= \frac{[1.21 \times 10^{12}][2 \times 10^{-4}]}{4 \times 10^6}\\ &= \frac{(1.21)(2)}{4} \times 10^{12-4-6}\\ &= (0.605) \times 10^2\\ &= 6.05 \times 10.\end{aligned}$$

■

Calculators often employ scientific notation in their display panels. In this case the number 10 used in the scientific notation $a \times 10^n$ is suppressed and only the exponent is shown. For example, to find $(4{,}500{,}000)^2$ on a typical calculator, we would enter the integer 4,500,000 and press the x^2 (or squaring) key. The display panel would show

2.025 13

which is translated as 2.025×10^{13}. Thus

$$(4{,}500{,}000)^2 = 20{,}250{,}000{,}000{,}000.$$

Many calculators also allow the entry of a number in scientific form. The owner of a calculator should consult the user's manual for details.

As a final remark, in applied problems, numbers are often obtained by various types of measurements and, hence, are approximations to exact values. In such cases answers should be rounded off since the final result of a calculation cannot be more accurate than the data that has been used. For example, if the length and width of a rectangle are measured to two-decimal-place accuracy, we cannot expect more that two-decimal-place accuracy in the calculated value of the area. If a number x is written in scientific form as $x = a \times 10^k$, where $1 \le a < 10$, and if a is rounded off to n decimal places, then we say that x is accurate (or has been rounded off) to $n + 1$ **significant figures.** For example, given $x = 37.2638$, we have the following:

Number of significant figures	5	4	3	2	1
Approximation to x	37.264	37.26	37.3	37	40

In Exercises 1.2, whenever an index of a radical is even (or a rational exponent m/n with n even is employed), we will assume that the letters that appear underneath the corresponding radical sign denote positive real numbers.

EXERCISES 1.2

Express the numbers in Exercises 1–10 in the form a/b, where a and b are integers.

1 $(-\frac{2}{3})^4$

2 $(-3)^3$

3 $\dfrac{2^{-3}}{3^{-2}}$

4 $\dfrac{2^0 + 0^2}{2 + 0}$

5 $(-2)^3 + 3^{-2}$

6 $(-\frac{3}{2})^4 - 2^{-4}$

7 $16^{-3/4}$

8 $9^{5/2}$

9 $(-0.008)^{2/3}$

10 $(0.008)^{-2/3}$

For Exercises 11–46 eliminate negative exponents and simplify.

11 $(\frac{1}{2}x^4)(16x^5)$

12 $(-3x^{-2})(4x^4)$

13 $\dfrac{(2x^3)(3x^2)}{(x^2)^3}$

14 $\dfrac{(2x^2)^3}{4x^4}$

15 $(\frac{1}{6}a^5)(-3a^2)(4a^7)$

16 $(-4b^3)(\frac{1}{6}b^2)(-9b^4)$

17 $\dfrac{(6x^3)^2}{(2x^2)^3}$

18 $\dfrac{(3y^3)(2y^2)^2}{(y^4)^3}$

19 $(3u^7v^3)(4u^4v^{-5})$

20 $(x^2yz^3)(-2xz^2)(x^3y^{-2})$

21 $(8x^4y^{-3})(\frac{1}{2}x^{-5}y^2)$

22 $\left(\dfrac{4a^2b}{a^3b^2}\right)\left(\dfrac{5a^2b}{2b^4}\right)$

23 $(\frac{1}{3}x^4y^{-3})^{-2}$

24 $(-2xy^2)^5\left(\dfrac{x^7}{8y^3}\right)$

25 $(3y^3)^4(4y^2)^{-3}$

26 $(-3a^2b^{-5})^3$

27 $(-2r^4s^{-3})^{-2}$

28 $(2x^2y^{-5})(6x^{-3}y)(\frac{1}{3}x^{-1}y^3)$

29 $(5x^2y^{-3})(4x^{-5}y^4)$

30 $(-2r^2s)^5(3r^{-1}s^3)^2$

31 $\left(\dfrac{3x^5y^4}{x^0y^{-3}}\right)^2$

32 $(4a^2b)^4\left(\dfrac{-a^3}{2b}\right)^2$

33 $(4a^{3/2})(2a^{1/2})$

34 $(-6x^{7/5})(2x^{8/5})$

35 $(3x^{5/6})(8x^{2/3})$

36 $(8r)^{1/3}(2r^{1/2})$

37 $(27a^6)^{-2/3}$

38 $(25z^4)^{-3/2}$

39 $(8x^{-2/3})x^{1/6}$

40 $(3x^{1/2})(-2x^{5/2})$

41 $\left(\dfrac{-8x^3}{y^{-6}}\right)^{2/3}$

42 $\left(\dfrac{-y^{3/2}}{y^{-1/3}}\right)^3$

43 $\left(\dfrac{x^6}{9y^{-4}}\right)^{-1/2}$

44 $\left(\dfrac{c^{-4}}{16d^8}\right)^{3/4}$

45 $\dfrac{(x^6y^3)^{-1/3}}{(x^4y^2)^{-1/2}}$

46 $a^{4/3}a^{-3/2}a^{1/6}$

Simplify the expressions in Exercises 47–66.

47 $\sqrt{81}$

48 $\sqrt[3]{-125}$

49 $\sqrt[5]{-64}$

50 $\sqrt[4]{256}$

51 $\dfrac{1}{\sqrt[3]{2}}$

52 $\sqrt{\dfrac{1}{7}}$

53 $\sqrt{9x^{-4}y^6}$

54 $\sqrt{16a^8b^{-2}}$

55 $\sqrt[3]{8a^6b^{-3}}$

56 $\sqrt[4]{81r^5s^8}$

57 $\sqrt[3]{\dfrac{54a^7}{b^2}}$

58 $\sqrt[5]{\dfrac{-96x^7}{y^3}}$

59 $\sqrt{\dfrac{1}{3u^3v}}$

60 $\sqrt[3]{\dfrac{1}{4x^5y^2}}$

61 $\sqrt[4]{(3x^5y^{-2})^4}$

62 $\sqrt[6]{(2u^{-3}v^4)^6}$

63 $\sqrt[5]{\dfrac{8x^3}{y^4}}\sqrt[5]{\dfrac{4x^4}{y^2}}$

64 $\sqrt{5xy^7}\sqrt{10x^3y^3}$

65 $\sqrt[3]{3t^4v^2}\sqrt[3]{-9t^{-1}v^4}$

66 $\sqrt[3]{(2r-s)^3}$

Rationalize the denominators in Exercises 67–70.

67 $\dfrac{\sqrt{a}}{\sqrt{a}-\sqrt{b}}$

68 $\dfrac{1}{\sqrt{c}+d}$

69 $\dfrac{\sqrt{t}-4}{\sqrt{t}+4}$

70 $\dfrac{4}{3-\sqrt{w}}$

Rationalize the numerators in Exercises 71 and 72.

71 $\dfrac{\sqrt{2(x+h)+1}-\sqrt{2x+1}}{h}$

72 $\dfrac{\sqrt{x}-\sqrt{x+h}}{h\sqrt{x}\sqrt{x+h}}$

73 The mass of a hydrogen atom is approximately 0.0000000000000000000000017 grams. Express this number in scientific form.

74 The mass of an electron is approximately 9.1×10^{-31} kilograms. Express this number in decimal form.

Express the numbers in Exercises 75 and 76 in scientific form.

75 (a) 427,000
(b) 0.000000098
(c) 810,000,000

76 (a) 85,200
(b) 0.0000055
(c) 24,900,000

Express the numbers in Exercises 77 and 78 in decimal form.

77 (a) 8.3×10^5
(b) 2.9×10^{-12}
(c) 5.63×10^8

78 (a) 2.3×10^7
(b) 7.01×10^{-9}
(c) 1.23×10^{10}

Use scientific form to find the numbers in Exercises 79–84.

79 (2,100,000,000)(0.0000000033)

80 (840,000,000)(0.00000021)

81 $\dfrac{\sqrt{81{,}000{,}000}}{(0.0000002)^3}$

82 $\dfrac{(300{,}000)^4}{(8{,}000{,}000)^{2/3}}$

83 $\dfrac{\sqrt[5]{(0.00243)^2(20{,}000)^3}}{\sqrt[3]{8{,}000{,}000}}$

84 $\dfrac{(160{,}000)^{3/2}(12{,}100{,}000)}{(0.0000011)^2}$

85 In astronomy, distances to stars are measured in light years, where 1 light year is the distance a ray of light travels in one year. If the speed of light is 186,000 miles per second, approximate 1 light year.

86 (a) It is estimated that the Milky Way galaxy contains 100 billion stars. Express this number in scientific form.
(b) The diameter d of the Milky Way galaxy is estimated as 100,000 light years. Express d in miles (refer to Exercise 85).

SECTION 1.3
ALGEBRAIC EXPRESSIONS

Certain explanations can be shortened by using the notation and terminology of sets. A **set** may be thought of as a collection of objects of some type. The objects are called **elements** of the set. Capital letters A, B, C, R, S, . . . are often used to denote sets. Lowercase letters a, b, x, y, . . . usually represent elements of sets. Throughout our work $\mathbb{R}$ will denote the set of real numbers, and $\mathbb{Z}$ the set of integers. If S is a set, then $a \in S$ means that a is an element of S, whereas $a \notin S$ signifies that a is not an element of S. If every element of a set S is also an element of a set T, then S is called a **subset** of T. For example, $\mathbb{Z}$ is a subset of $\mathbb{R}$. Two sets S and T are said to be **equal,** written $S = T$, if S and T contain precisely the same elements. The notation $S \neq T$ means that S and T are not equal.

If the elements of a set S have a certain property, then we sometimes write $S = \{x: \quad\}$, where the property describing the arbitrary element x is stated in the space after the colon. For example, $\{x : x > 3\}$ represents the set of all real numbers greater than 3.

Finite sets are sometimes denoted by listing all the elements within braces. For example, if S consists of the first five positive integers, we write $S = \{1, 2, 3, 4, 5\}$. When sets are described in this way, the order used in listing the elements is considered irrelevant, and we could also write $S = \{1, 3, 2, 4, 5\}$, $S = \{4, 3, 2, 5, 1\}$, etc.

We frequently use symbols to denote arbitrary elements of a set. For example, we may use x to denote a real number, although no *particular* real number is specified. A letter that is used to represent any element of a given set is sometimes called a **variable.** Throughout this text, unless otherwise specified, variables will represent real numbers. The **domain of a variable** is the set of real numbers represented by the variable. To illustrate, given the expression $\sqrt{x}$, we note that to obtain a real number we must have $x \geq 0$, and hence in this case the domain of x is assumed to be the set of nonnegative real numbers. Similarly, given $1/(x - 2)$, we must exclude $x = 2$ (Why?), and consequently, we take the domain of x as the set of all real numbers different from 2.

If x is a variable we shall often substitute real numbers for x; that is, we shall replace x by numbers from its domain. A symbol that represents a *specific* real number is called a **constant.** In most of our work, letters near the end of the alphabet, such as x, y, and z, will be used for variables, whereas letters such as a, b, and c will denote constants.

If we begin with any collection of variables and real numbers, then an **algebraic expression** is the result obtained by applying additions, subtractions, multiplications, divisions, or the taking of roots. The following are examples of algebraic expressions:

$$x^3 - 2x + \frac{3^{1/9}}{\sqrt{2x}}, \qquad \frac{2xy + 3x}{y - 1}, \qquad \frac{4yz^{-2} + \left(\dfrac{-7}{x + w}\right)^5}{\sqrt[3]{y^2 + 5z}},$$

where x, y, z, and w are variables. If specific numbers are substituted for the variables in an algebraic expression, the resulting real number is called the **value** of the expression for these numbers. To illustrate, the value of the second expression above for $x = -2$ and $y = 3$ is

$$\frac{2(-2)(3) + 3(-2)}{3 - 1} = \frac{-12 - 6}{2} = -9.$$

When we work with algebraic expressions, we will assume that the domains are chosen so that variables do not represent numbers that make the expressions meaningless. Thus, we assume that denominators are not zero, roots always exist, etc. For simplicity, we shall not always state domains of variables. If meaningless expressions occur when specific numbers are substituted for the variables, then these numbers are *not* in the domains of the variables.

Certain algebraic expressions are given special names. If x is a variable, then a **monomial** in x is an expression of the form ax^n, where a is a real number and n is a nonnegative integer. The number a is called the **coefficient** of x^n. A *polynomial in* x is a sum of monomials in x. Another way of stating this is as follows.

DEFINITION

A **polynomial in x** is an expression of the form

$$a_nx^n + a_{n-1}x^{n-1} + \cdots + a_1x + a_0$$

where n is a nonnegative integer and each coefficient a_i is a real number.

In the preceding definition, each of the expressions a_kx^k in the sum is called a **term** of the polynomial. If a coefficient a_i is zero, we usually delete the term a_ix^i. The coefficient a_n of the highest power of x is the **leading coefficient** of the polynomial and, if $a_n \neq 0$, we say that the polynomial has **degree n.** By definition, two polynomials are **equal** if and only if they have the same degree and corresponding coefficients are equal. If all the coefficients of a polynomial are zero, it is called the **zero polynomial** and is denoted by 0. It is customary not to assign a degree to the zero polynomial.

If some of the coefficients are negative, then for convenience we often use minus signs between appropriate terms. To illustrate, instead of $3x^2 + (-5)x + (-7)$, we write $3x^2 - 5x - 7$ for this polynomial of degree 2. Polynomials in other variables may also be considered. For example, $\frac{2}{5}z^2 - 3z^7 + 8 - \sqrt{5}z^4$ is a polynomial in z of degree 7. We ordinarily arrange the terms in order of decreasing powers of the variable; thus, for this polynomial we write $-3z^7 - \sqrt{5}z^4 + \frac{2}{5}z^2 + 8$.

According to the definition of degree, if c is a nonzero real number, then c is a polynomial of degree 0. Such polynomials (together with the zero polynomial) are called **constant polynomials.**

A polynomial in x may be thought of as an algebraic expression obtained by employing only additions, subtractions, and multiplications involving x. In particular, the expressions

$$\frac{1}{x} + 3x, \qquad \frac{x-5}{x^2+2}, \qquad 3x^2 + \sqrt{x} - 2$$

are not polynomials since they involve divisions by variables or contain roots of variables.

In more advanced courses, coefficients of polynomials are often chosen from some mathematical system other than the set of real numbers. However, in this text, unless mentioned otherwise, *the terminology "polynomial" will always refer to polynomials with real coefficients.*

Since polynomials, and the monomials that make up polynomials, are symbols representing real numbers, all of the properties in Section 1.1 can be applied. Thus, if additions, multiplications, and subtractions are carried out with polynomials, we may simplify the result by using various properties of real numbers.

EXAMPLE 1 Find the sum of $x^3 + 2x^2 - 5x + 7$ and $4x^3 - 5x^2 + 3$.

Solution Rearranging terms and using the Distributive Properties of real numbers gives us

$$\begin{aligned}(x^3 + 2x^2 - 5x + 7) &+ (4x^3 - 5x^2 + 3)\\ &= x^3 + 4x^3 + 2x^2 - 5x^2 - 5x + 7 + 3\\ &= (1 + 4)x^3 + (2 - 5)x^2 + (-5)x + (7 + 3)\\ &= 5x^3 - 3x^2 - 5x + 10.\end{aligned}$$

■

Example 1 illustrates the fact that the sum of any two polynomials in x can be obtained by adding coefficients of like powers of x. To keep track of the coefficients it is sometimes convenient to use the following scheme for finding the sum.

$$\begin{array}{r} x^3 + 2x^2 - 5x + 7\\ 4x^3 - 5x^2 + 3\\ \hline 5x^3 - 3x^2 - 5x + 10\end{array}$$

The difference of two polynomials is found by subtracting coefficients of like powers, as indicated by the next example.

EXAMPLE 2 Subtract $4x^3 - 5x^2 + 3$ from $x^3 + 2x^2 - 5x + 7$.

Solution

$$\begin{aligned}(x^3 + 2x^2 - 5x + 7) &- (4x^3 - 5x^2 + 3)\\ &= x^3 + 2x^2 - 5x + 7 - 4x^3 + 5x^2 - 3\\ &= x^3 - 4x^3 + 2x^2 + 5x^2 - 5x + 7 - 3\\ &= (1 - 4)x^3 + (2 + 5)x^2 - 5x + (7 - 3)\\ &= -3x^3 + 7x^2 - 5x + 4\end{aligned}$$

■

The intermediate steps in the previous solution were used for completeness. They may be omitted after you become proficient with such manipulations. In order to multiply two polynomials, we use the Distributive

Properties together with the Laws of Exponents and combine like terms, as illustrated in the next example.

EXAMPLE 3 Find the product of $x^2 + 5x - 4$ and $2x^3 + 3x - 1$.

Solution

$$\begin{aligned}(x^2 + 5x - 4)(2x^3 + 3x - 1) \\ &= x^2(2x^3 + 3x - 1) + 5x(2x^3 + 3x - 1) - 4(2x^3 + 3x - 1) \\ &= 2x^5 + 3x^3 - x^2 + 10x^4 + 15x^2 - 5x - 8x^3 - 12x + 4 \\ &= 2x^5 + 10x^4 + (3 - 8)x^3 + (-1 + 15)x^2 + (-5 - 12)x + 4 \\ &= 2x^5 + 10x^4 - 5x^3 + 14x^2 - 17x + 4\end{aligned}$$

For convenience, this procedure may be arranged as follows:

$$\begin{array}{rrrrrr} 2x^3 & + \; 3x & - 1 & & & \\ x^2 & + \; 5x & - 4 & & & \\ \hline 2x^5 & & + 3x^3 & - \; x^2 & & \\ & 10x^4 & & + 15x^2 & - \; 5x & \\ & & - 8x^3 & & - 12x & + 4 \\ \hline 2x^5 & + 10x^4 & - 5x^3 & + 14x^2 & - 17x & + 4 \end{array}$$

■

We may also consider polynomials in more than one variable. For example, a polynomial in *two* variables x and y is a sum of terms, each of the form ax^my^k for some real number a and nonnegative integers m and k. An example is

$$3x^4y + 2x^3y^5 + 7x^2 - 4xy + 8y - 5.$$

In like manner, we may consider polynomials in three variables x, y, z or, for that matter, in *any* number of variables. Addition, subtraction, and multiplication are performed using properties of real numbers. A simple illustration is given in the next example.

EXAMPLE 4 Find the product of $x^2 + xy + y^2$ and $x - y$.

Solution

$$\begin{aligned}(x^2 + xy + y^2)(x - y) &= (x^2 + xy + y^2)x - (x^2 + xy + y^2)y \\ &= x^3 + x^2y + xy^2 - x^2y - xy^2 - y^3 \\ &= x^3 - y^3\end{aligned}$$

■

Division by a monomial is relatively easy, as seen in the next example.

EXAMPLE 5 Divide $6x^2y^3 + 4x^3y^2 - 10xy$ by $2xy$.

Solution Using the properties of quotients and the Laws of Exponents, we obtain

$$\begin{aligned}\frac{6x^2y^3 + 4x^3y^2 - 10xy}{2xy} &= \frac{6x^2y^3}{2xy} + \frac{4x^3y^2}{2xy} - \frac{10xy}{2xy}\\ &= 3xy^2 + 2x^2y - 5.\end{aligned}$$ ■

Certain products occur so frequently in algebra that they deserve special attention. We list some of these next, using letters to represent real numbers. The reader should check the validity of each formula by actually carrying out the multiplications.

PRODUCT FORMULAS

(i) $(x + y)(x - y) = x^2 - y^2$

(ii) $(x + y)^2 = x^2 + 2xy + y^2$

(iii) $(x - y)^2 = x^2 - 2xy + y^2$

(iv) $(x + y)^3 = x^3 + 3x^2y + 3xy^2 + y^3$

(v) $(x - y)^3 = x^3 - 3x^2y + 3xy^2 - y^3$

Since the symbols x and y used in these formulas represent real numbers, they may be replaced by algebraic expressions, as illustrated in the next example.

EXAMPLE 6 Find the following products.

(a) $(2r^2 - \sqrt{s})(2r^2 + \sqrt{s})$ (b) $\left(\sqrt{c} + \frac{1}{\sqrt{c}}\right)^2$ (c) $(2a - 5b)^3$

Solution

(a) Using Product Formula (i) with $x = 2r^2$ and $y = \sqrt{s}$,

$$\begin{aligned}(2r^2 - \sqrt{s})(2r^2 + \sqrt{s}) &= (2r^2)^2 - (\sqrt{s})^2\\ &= 4r^4 - s.\end{aligned}$$

(b) Using Product Formula (ii) with $x = \sqrt{c}$ and $y = 1/\sqrt{c}$,

$$\begin{aligned}\left(\sqrt{c} + \frac{1}{\sqrt{c}}\right)^2 &= (\sqrt{c})^2 + 2\sqrt{c}\cdot\frac{1}{\sqrt{c}} + \left(\frac{1}{\sqrt{c}}\right)^2\\ &= c + 2 + \frac{1}{c}.\end{aligned}$$

(c) Applying Product Formula (v) with $x = 2a$ and $y = 5b$,

$$\begin{aligned}(2a - 5b)^3 &= (2a)^3 - 3(2a)^2(5b) + 3(2a)(5b)^2 - (5b)^3\\ &= 8a^3 - 60a^2b + 150ab^2 - 125b^3.\end{aligned}$$ ■

If a polynomial is written as a product of other polynomials, then each polynomial in the product is called a **factor** of the original polynomial. For example, since $x^2 - 9 = (x + 3)(x - 3)$, we see that $x + 3$ and $x - 3$ are factors of $x^2 - 9$. Any polynomial has, as a factor, *every* nonzero real number c. As an illustration, given $3x^2 - 5x + 2$ and any nonzero real number c, we can write

$$3x^2 - 5x + 2 = c\left(\frac{3}{c}x^2 - \frac{5}{c}x + \frac{2}{c}\right).$$

A factor c of this type is called a **trivial factor.** We shall be interested primarily in **nontrivial factors** of polynomials; that is, factors that contain polynomials of degree greater than zero.

An integer $a > 1$ is prime if it cannot be written as a product of two positive integers greater than 1. In similar fashion, if S denotes a set of numbers, then a polynomial with coefficients in S is said to be **prime,** or **irreducible** over S, if it cannot be written as a product of two polynomials of positive degree with coefficients in S. Note that a polynomial may be irreducible over one set S but not over another. For example, $x^2 - 2$ is irreducible over the rational numbers since it cannot be expressed as a product of two polynomials of positive degree that have *rational* coefficients. If we allow the factors to have *real* coefficients, then $x^2 - 2$ is not prime, since

$$x^2 - 2 = (x + \sqrt{2})(x - \sqrt{2}).$$

Similarly, $x^2 + 1$ is irreducible over the real numbers but, as we shall see in Section 3.4, not over the complex numbers. It can be shown that every polynomial $ax + b$ of degree 1 is irreducible.

EXAMPLE 7 Express the polynomial $6x^2 - 7x - 3$ as a product of irreducible polynomials.

Solution If we write $6x^2 - 7x - 3 = (ax + b)(cx + d)$, then the product of a and c is 6; the product of b and d is -3; and $ad + bc = -7$. Trying various possibilities, we arrive at the factorization

$$6x^2 - 7x - 3 = (2x - 3)(3x + 1).$$ ■

It is usually difficult to factor polynomials of degree greater than 2. In simple cases the following formulas may be useful. Each can be verified by multiplication.

FACTORING FORMULAS

$a^2 - b^2 = (a + b)(a - b)$	(Difference of two squares)
$a^3 - b^3 = (a - b)(a^2 + ab + b^2)$	(Difference of two cubes)
$a^3 + b^3 = (a + b)(a^2 - ab + b^2)$	(Sum of two cubes)

EXAMPLE 8 Express each of the following as a product of polynomials that are irreducible over the set of integers.

(a) $16x^4 - 81$ (b) $x^3 - 8$ (c) $3x^3 + 2x^2 - 12x - 8$

Solution

(a) We apply the difference-of-two-squares formula twice, as follows:

$$\begin{aligned} 16x^4 - 81 &= (4x^2)^2 - 9^2 = (4x^2 + 9)(4x^2 - 9) \\ &= (4x^2 + 9)[(2x)^2 - 3^2] = (4x^2 + 9)(2x + 3)(2x - 3). \end{aligned}$$

(b) Using the difference-of-two-cubes formula, with $a = x$ and $b = 2$, we obtain

$$\begin{aligned} x^3 - 8 &= x^3 - 2^3 \\ &= (x - 2)(x^2 + 2x + 4). \end{aligned}$$

By trying all possibilities we can show that $x^2 + 2x + 4$ cannot be expressed as a product of two first-degree polynomials with integer coefficients.

(c) We first employ the Distributive Property and then the difference-of-two-squares formula, as follows:

$$\begin{aligned} 3x^3 + 2x^2 - 12x - 8 &= x^2(3x + 2) - 4(3x + 2) \\ &= (x^2 - 4)(3x + 2) \\ &= (x + 2)(x - 2)(3x + 2). \end{aligned}$$

The technique used in part (c) of Example 8 is called **factorization by grouping.**

Quotients of algebraic expressions are called **fractional expressions.** As a special case, a quotient of two polynomials is called a **rational expression.** Some examples are

$$\frac{x^2 - 5x + 1}{x^3 + 7}, \quad \frac{z^2x^4 - 3yz}{5z}, \quad \text{and} \quad \frac{1}{4xy}.$$

Many problems in mathematics involve combining fractional expressions and then simplifying the result. Since fractional expressions are quotients that contain symbols representing real numbers, the Properties of Quotients listed in Section 1.1 may be used. Of course, the letters a, b, c, and d will now be replaced by polynomials. Of particular importance in simplification problems is the formula

$$\frac{ad}{bd} = \frac{a}{b}$$

which is obtained by multiplying the numerator and denominator of the given fraction by $1/d$. This rule is sometimes phrased "a common factor in the numerator and denominator may be *canceled* from the quotient." To use this technique in problems involving rational expressions, we factor both the numerator and denominator into prime factors and then cancel common factors that occur in the numerator and denominator. We refer to the resulting expression as being *simplified*, or *reduced to lowest terms*.

EXAMPLE 9 Simplify $\dfrac{3x^2 - 5x - 2}{x^2 - 4}$.

Solution Factoring the numerator and denominator and canceling common factors gives us

$$\frac{3x^2 - 5x - 2}{x^2 - 4} = \frac{(3x + 1)(x - 2)}{(x + 2)(x - 2)} = \frac{3x + 1}{x + 2}.$$ ■

In the preceding example we canceled the common factor $x - 2$; that is, we divided numerator and denominator by $x - 2$. This simplification is valid only if $x - 2 \neq 0$, that is $x \neq 2$. However, 2 is not in the domain of x since it leads to a zero denominator when substituted in the original expression. Hence, our manipulations are valid. We shall always assume such restrictions when simplifying rational expressions.

EXAMPLE 10 Simplify $\dfrac{2 - x - 3x^2}{6x^2 - x - 2}$.

Solution

$$\frac{2 - x - 3x^2}{6x^2 - x - 2} = \frac{(1 + x)(2 - 3x)}{(2x + 1)(3x - 2)} = \frac{-(1 + x)}{2x + 1}$$

The fact that $(2 - 3x) = -(3x - 2)$ accounts for the minus sign in the final answer since $(2 - 3x)/(3x - 2) = -(3x - 2)/(3x - 2) = -1$. Another method is to change the form of the numerator as follows:

$$\frac{2 - x - 3x^2}{6x^2 - x - 2} = \frac{-(3x^2 + x - 2)}{6x^2 - x - 2}$$

$$= -\frac{(3x - 2)(x + 1)}{(3x - 2)(2x + 1)} = -\frac{x + 1}{2x + 1}$$ ■

Multiplication and division of fractional expressions are performed using rules for quotients and then simplifying, as illustrated in the next example.

EXAMPLE 11 Perform the indicated operations and simplify.

(a) $\dfrac{x^2-6x+9}{x^2-1}\cdot\dfrac{2x-2}{x-3}$ (b) $\dfrac{x+2}{2x-3}\div\dfrac{x^2-4}{2x^2-3x}$

Solution

(a)
$$\frac{x^2-6x+9}{x^2-1}\cdot\frac{2x-2}{x-3}=\frac{(x-3)^2}{(x+1)(x-1)}\cdot\frac{2(x-1)}{x-3}$$
$$=\frac{2(x-3)^2(x-1)}{(x+1)(x-1)(x-3)}=\frac{2(x-3)}{x+1}$$

(b)
$$\frac{x+2}{2x-3}\div\frac{x^2-4}{2x^2-3x}=\frac{x+2}{2x-3}\cdot\frac{2x^2-3x}{x^2-4}$$
$$=\frac{(x+2)x(2x-3)}{(2x-3)(x+2)(x-2)}=\frac{x}{x-2}$$ ■

When adding or subtracting two rational expressions, we usually find a common denominator and use the following properties of real numbers:

$$\frac{a}{d}+\frac{c}{d}=\frac{a+c}{d};\qquad \frac{a}{d}-\frac{c}{d}=\frac{a-c}{d}.$$

If the expressions have different denominators a common denominator may be introduced by multiplying numerator and denominator of each of the fractions by suitable polynomials. It is usually desirable to use the **least common denominator (lcd)** of the two fractions. The lcd can be found by obtaining the prime fractorization for each denominator and then forming the product of the different prime factors, using the *highest* exponent that appears with each prime factor. Let us begin with a numerical example of this technique.

EXAMPLE 12 Express as a rational number in lowest terms:

$$\frac{7}{24}+\frac{5}{18}.$$

Solution The prime factorizations of the denominators are $24=(2^3)(3)$ and $18=(2)(3^2)$. To find the lcd we form the product of the different prime factors, using the highest exponent associated with each factor. This gives us

$(2^3)(3^2)$, or 72. We now change each fraction to an equal fraction with denominator 72 and add, as follows:

$$\frac{7}{24}+\frac{5}{18}=\frac{7}{24}\cdot\frac{3}{3}+\frac{5}{18}\cdot\frac{4}{4}$$
$$=\frac{21}{72}+\frac{20}{72}=\frac{41}{72}.$$

If we use the Addition Property of Quotients, stated on page 4, we obtain

$$\frac{7}{24}+\frac{5}{18}=\frac{(7)(18)+(24)(5)}{(24)(18)}$$
$$=\frac{126+120}{432}=\frac{246}{432}=\frac{41}{72}.$$

■

The method for finding the lcd for rational expressions is analogous to the process illustrated in the solution of Example 12. The only difference is that we use factorizations of polynomials instead of integers.

EXAMPLE 13 Change to a rational expression in lowest terms:

$$\frac{6}{x(3x-2)}+\frac{5}{3x-2}-\frac{2}{x^2}$$

Solution The denominators are already in factored form. Evidently, the lcd is $x^2(3x-2)$. In order to obtain three fractions having that denominator, we multiply numerator and denominator of the first fraction by x, of the second by x^2, and of the third by $3x-2$. This gives us

$$\frac{6}{x(3x-2)}+\frac{5}{3x-2}-\frac{2}{x^2}=\frac{6x}{x^2(3x-2)}+\frac{5x^2}{x^2(3x-2)}-\frac{2(3x-2)}{x^2(3x-2)}$$
$$=\frac{6x+5x^2-(6x-4)}{x^2(3x-2)}=\frac{5x^2+4}{x^2(3x-2)}.$$

■

EXAMPLE 14 Change to a rational expression:

$$\frac{2x+5}{x^2+6x+9}+\frac{x}{x^2-9}+\frac{1}{x-3}$$

Solution We begin by factoring denominators:

$$\frac{2x+5}{x^2+6x+9}+\frac{x}{x^2-9}+\frac{1}{x-3}=\frac{2x+5}{(x+3)^2}+\frac{x}{(x+3)(x-3)}+\frac{1}{x-3}$$

Since the lcd is $(x + 3)^2(x - 3)$, we multiply numerator and denominator of the first fraction by $x - 3$, of the second by $x + 3$, and of the third by $(x + 3)^2$, and add:

$$\frac{(2x + 5)(x - 3)}{(x + 3)^2(x - 3)} + \frac{x(x + 3)}{(x + 3)^2(x - 3)} + \frac{(x + 3)^2}{(x + 3)^2(x - 3)}$$

$$= \frac{(2x^2 - x - 15) + (x^2 + 3x) + (x^2 + 6x + 9)}{(x + 3)^2(x - 3)}$$

$$= \frac{4x^2 + 8x - 6}{(x + 3)^2(x - 3)} = \frac{2(2x^2 + 4x - 3)}{(x + 3)^2(x - 3)}$$ ■

The concept of *derivative* is used extensively in calculus. For certain calculus problems involving derivatives it is necessary to simplify expressions of the types given in the next two examples. (Also see Exercises 79–96.)

EXAMPLE 15 Simplify $\dfrac{\dfrac{1}{(x + h)^2} - \dfrac{1}{x^2}}{h}$.

Solution

$$\frac{\dfrac{1}{(x + h)^2} - \dfrac{1}{x^2}}{h} = \frac{\dfrac{x^2 - (x + h)^2}{(x + h)^2x^2}}{h}$$

$$= \frac{x^2 - (x^2 + 2xh + h^2)}{(x + h)^2x^2h}$$

$$= \frac{x^2 - x^2 - 2xh - h^2}{(x + h)^2x^2h}$$

$$= \frac{-h(2x + h)}{(x + h)^2x^2h}$$

$$= -\frac{2x + h}{(x + h)^2x^2}$$ ■

EXAMPLE 16 Simplify $\dfrac{3x^2(2x + 5)^{1/2} - x^3(\frac{1}{2})(2x + 5)^{-1/2}(2)}{[(2x + 5)^{1/2}]^2}$.

Solution

$$\frac{3x^2(2x+5)^{1/2} - x^3(\tfrac{1}{2})(2x+5)^{-1/2}(2)}{[(2x+5)^{1/2}]^2} = \frac{3x^2(2x+5)^{1/2} - \dfrac{x^3}{(2x+5)^{1/2}}}{2x+5}$$

$$= \frac{\dfrac{3x^2(2x+5) - x^3}{(2x+5)^{1/2}}}{2x+5}$$

$$= \frac{6x^3 + 15x^2 - x^3}{(2x+5)^{1/2}(2x+5)}$$

$$= \frac{5x^3 + 15x^2}{(2x+5)^{3/2}}$$

$$= \frac{5x^2(x+3)}{(2x+5)^{3/2}}$$

EXERCISES 1.3

In Exercises 1–8 perform the indicated operations and find the degree of the resulting polynomial.

1 $(2x^3 + 4x^2 + 3x + 7) + (x^4 - 5x^3 + x - 2)$

2 $(2x^5 - 3x^2 + 2) + (-2x^4 + x^2 - 4x + 5)$

3 $(x^3 + 5x^2 - 7x + 2) - (x^3 + 5x^2 + x + 2)$

4 $(2x^4 - 3x^2 + 1) - (3x^4 + x + 1)$

5 $(2x^2 - 5x + 7)(3x - 5)$

6 $(3x^3 - 2x^2 + 1)(8x - 1)$

7 $(2x^3 - x + 5)(x^2 + x + 2)$

8 $(7x^4 + x^2 - 1)(7x^4 - x^3 + 4x)$

Find the products in Exercises 9–22.

9 $(5x^2 + 2y)(3x^2 - 7y)$ 10 $(x + 9y^2)(3x - 4y^2)$

11 $(6t - 5v)(6t + 5v)$ 12 $(8u + 3)^2$

13 $(3r + 10s)^2$ 14 $(4v - 3w)(4v + 3w)$

15 $(4x^2 - 5y^2)^2$ 16 $(10p^2 + 7q^2)^2$

17 $(x - 2y)^3$ 18 $(4x - y)^3$

19 $(3r + 4s)^3$ 20 $(2a + 5b)^3$

21 $(x^2 + y^2)^3$ 22 $(u^2 - 3v)^3$

In Exercises 23–44 express each polynomial as a product of polynomials that are irreducible over the set of integers.

23 $3x^2 + 10x - 8$ 24 $10x^2 + 29x - 21$

25 $12x^3 + 21x^2 - 45x$ 26 $6x^4 - 5x^3 - 6x^2$

27 $25x^2 - 9$ 28 $49 - 36x^2$

29 $27x^3 - 1$ 30 $64x^3 + 125$

31 $x^4 - 13x^2 + 36$ 32 $3x^4 - 11x^2 - 4$

33 $x^2 + 2x + 1$ 34 $x^2 - 6x + 9$

35 $x^2 + x + 1$ 36 $x^2 + 16$

37 $x^3 + x^2 + x + 1$ 38 $2x^3 - x^2 - 2x + 1$

39 $x^8 - 10x^4 + 9$ 40 $x^6 - 64$

41 $x^6 - 1$ 42 $(x^2 - 25)^3$

43 $4x^3 - 8x^2 - 9x + 18$ 44 $x^6 - x^4$

Simplify the expressions in Exercises 45–96.

45 $\dfrac{6x^2 + 7x - 10}{6x^2 + 13x - 15}$ 46 $\dfrac{10x^2 + 29x - 21}{5x^2 - 23x + 12}$

47 $\dfrac{12y^2 + 3y}{20y^2 + 9y + 1}$ 48 $\dfrac{4z^2 + 12z + 9}{2z^2 + 3z}$

49 $\dfrac{6 - 7a - 5a^2}{10a^2 - a - 3}$

50 $\dfrac{6y - 5y^2}{25y^2 - 36}$

51 $\dfrac{4x^3 - 9x}{10x^4 + 11x^3 - 6x^2}$

52 $\dfrac{16x^4 + 8x^3 + x^2}{4x^3 + 25x^2 + 6x}$

53 $\dfrac{6r}{3r - 1} - \dfrac{4r}{2r + 5}$

54 $\dfrac{3s}{s^2 + 1} - \dfrac{6}{2s - 1}$

55 $\dfrac{2x + 1}{2x - 1} - \dfrac{x - 1}{x + 1}$

56 $\dfrac{3u + 2}{u - 4} + \dfrac{4u + 1}{5u + 2}$

57 $\dfrac{9t - 6}{8t^3 - 27} \cdot \dfrac{4t^2 - 9}{12t^2 + 10t - 12}$

58 $\dfrac{a^2 + 4a + 3}{3a^2 + a - 2} \cdot \dfrac{3a^2 - 2a}{2a^2 + 13a + 21}$

59 $\dfrac{5a^2 + 12a + 4}{a^4 - 16} \div \dfrac{25a^2 + 20a + 4}{a^2 - 2a}$

60 $\dfrac{x^3 - 8}{x^2 - 4} \div \dfrac{x}{x^3 + 8}$

61 $\dfrac{2}{3x + 1} - \dfrac{9}{(3x + 1)^2}$

62 $\dfrac{4}{(5x - 2)^2} + \dfrac{x}{5x - 2}$

63 $\dfrac{1}{c} - \dfrac{c + 2}{c^2} + \dfrac{3}{c^3}$

64 $\dfrac{6}{3t} + \dfrac{t + 5}{t^3} + \dfrac{1 - 2t^2}{t^4}$

65 $\dfrac{5}{x - 1} + \dfrac{8}{(x - 1)^2} - \dfrac{3}{(x - 1)^3}$

66 $\dfrac{8}{x} + \dfrac{3}{2x - 4} + \dfrac{7x}{x^2 - 4}$

67 $\dfrac{2}{x} + \dfrac{7}{x^2} + \dfrac{5}{2x - 3} + \dfrac{1}{(2x - 3)^2}$

68 $\dfrac{4}{x} + \dfrac{3x^2 + 5}{x^3} - \dfrac{6}{2x + 1}$

69 $\dfrac{p^4 + 3p^3 - 8p - 24}{p^3 - 2p^2 - 9p + 18}$

70 $\dfrac{2ac + bc - 6ad - 3bd}{6ac + 2ad + 3bc + bd}$

71 $\dfrac{5}{7x - 3} - \dfrac{2}{2x + 1} + \dfrac{4x}{14x^2 + x - 3}$

72 $2 + \dfrac{3}{x} + \dfrac{7x}{3x + 10}$

73 $\dfrac{\dfrac{a}{b} - \dfrac{b}{a}}{\dfrac{1}{a} + \dfrac{1}{b}}$

74 $\dfrac{\dfrac{1}{x + 1} - 5}{\dfrac{1}{x} - x}$

75 $\dfrac{\dfrac{x}{y^2} - \dfrac{y}{x^2}}{\dfrac{1}{y^2} - \dfrac{1}{x^2}}$

76 $\dfrac{\dfrac{r}{s} + \dfrac{s}{r}}{\dfrac{r^2}{s^2} - \dfrac{s^2}{r^2}}$

77 $\dfrac{\dfrac{5}{x + 1} + \dfrac{2x}{x + 3}}{\dfrac{x}{x + 1} + \dfrac{7}{x + 3}}$

78 $\dfrac{\dfrac{3}{w} - \dfrac{6}{2w + 1}}{\dfrac{5}{w} + \dfrac{8}{2w + 1}}$

79 $\dfrac{(x + h)^2 - 7(x + h) - (x^2 - 7x)}{h}$

80 $\dfrac{(x + h)^3 + 4(x + h) - (x^3 + 4x)}{h}$

81 $\dfrac{\dfrac{1}{(x + h)^3} - \dfrac{1}{x^3}}{h}$

82 $\dfrac{\dfrac{1}{x + h} - \dfrac{1}{x}}{h}$

83 $\dfrac{\dfrac{1}{2x + 2h + 3} - \dfrac{1}{2x + 3}}{h}$

84 $\dfrac{\dfrac{7}{5x + 5h - 2} - \dfrac{7}{5x - 2}}{h}$

85 $(2x^2 - 3x + 1)(4)(3x + 2)^3(3) + (3x + 2)^4(4x - 3)$

86 $(6x - 5)^3(2)(x^2 + 4)(2x) + (x^2 + 4)^2(3)(6x - 5)^2(6)$

87 $(x^2 - 4)^{1/2}(3)(2x + 1)^2(2) + (2x + 1)^3(\tfrac{1}{2})(x^2 - 4)^{-1/2}(2x)$

88 $(3x + 2)^{1/3}(2)(4x - 5)(4) + (4x - 5)^2(\tfrac{1}{3})(3x + 2)^{-2/3}(3)$

89 $(3x + 1)^6(\tfrac{1}{2})(2x - 5)^{-1/2}(2) + (2x - 5)^{1/2}(6)(3x + 1)^5(3)$

90 $(x^2 + 9)^4(-\tfrac{1}{3})(x + 6)^{-4/3} + (x + 6)^{-1/3}(4)(x^2 + 9)^3(2x)$

91 $\dfrac{(6x + 1)^3(27x^2 + 2) - (9x^3 + 2x)(3)(6x + 1)^2(6)}{(6x + 1)^6}$

92 $\dfrac{(x^2 - 1)^4(2x) - x^2(4)(x^2 - 1)^3(2x)}{(x^2 - 1)^8}$

93 $\dfrac{(x^2+4)^{1/3}(3)-(3x)(\frac{1}{3})(x^2+4)^{-2/3}(2x)}{[(x^2+4)^{1/3}]^2}$

94 $\dfrac{(1-x^2)^{1/2}(2x)-x^2(\frac{1}{2})(1-x^2)^{-1/2}(-2x)}{[(1-x^2)^{1/2}]^2}$

95 $\dfrac{(4x^2+9)^{1/2}(2)-(2x+3)(\frac{1}{2})(4x^2+9)^{-1/2}(8x)}{[(4x^2+9)^{1/2}]^2}$

96 $\dfrac{(3x+2)^{1/2}(\frac{1}{3})(2x+3)^{-2/3}(2)-(2x+3)^{1/3}(\frac{1}{2})(3x+2)^{-1/2}(3)}{[(3x+2)^{1/2}]^2}$

Test

SECTION 1.4

EQUATIONS AND INEQUALITIES

If x is a variable, then expressions such as

$$x+3=0, \quad x^2-5=4x, \quad \text{or} \quad (x^2-9)\sqrt[3]{x+1}=0$$

are called **equations** in x. If certain numbers are substituted for x, true statements are obtained, whereas substitutions of other numbers produce false statements. For example, the equation $x+3=0$ leads to a false statement for every value of x except -3. If 2 is substituted for x in the equation $x^2-5=4x$, we obtain $4-5=8$, or $-1=8$, a false statement. However, if we let $x=5$, then $(5)^2-5=4\cdot 5$, or $20=20$, which is true. Given an equation in x, if a true statement is obtained when x is replaced by some real number a from the domain of x, then a is called a **solution** or a **root** of the equation. We also say that a **satisfies** the equation. To **solve** an equation means to find all the solutions.

If every number in the domain of the variable is a solution of an equation, the equation is called an **identity.** An example of an identity is

$$\frac{1}{x^2-4}=\frac{1}{(x+2)(x-2)},$$

since this equation is true for every number in the domain of x. An equation is called a **conditional equation** if there are numbers that are *not* solutions.

The solutions of an equation depend on the system of numbers under consideration. For example, if we demand that solutions be rational numbers, then the equation $x^2=2$ has no solutions since there is no rational number whose square is 2. However, if we allow *real* numbers, then the solutions are $-\sqrt{2}$ and $\sqrt{2}$. Similarly, the equation $x^2=-1$ has no real solutions; however, we shall see later that this equation has solutions if *complex* numbers are allowed.

Two equations are said to be **equivalent** if they have exactly the same solutions. For example, the equations

$$x=3, \quad x-1=2, \quad 5x=15, \quad \text{and} \quad 2x+1=7$$

are all equivalent.

One method of solving an equation is to replace it by a list of equivalent equations, each in some sense simpler than the preceding one, and terminating the list with an equation for which the solutions are obvious. This is often accomplished by using various properties of real numbers. For example, we may add the same expression to both sides of an equation without changing the solutions. Similarly, the same expression may be subtracted from both sides of an equation. We can also multiply or divide both sides of an equation by an expression that represents a nonzero real number. The following example illustrates these remarks.

EXAMPLE 1 Solve the equation $6x - 7 = 2x + 5$.

Solution The equations in the following list are equivalent. (Why?)

$$6x - 7 = 2x + 5$$

$$(6x - 7) + 7 = (2x + 5) + 7$$

$$6x = 2x + 12$$

$$6x - 2x = (2x + 12) - 2x$$

$$4x = 12$$

$$\frac{1}{4}(4x) = \frac{1}{4}(12)$$

$$x = 3$$

Since the last equation has exactly one solution, 3, it follows that 3 is the only solution of the original equation, $6x - 7 = 2x + 5$. ■

In the process of finding solutions of equations, errors may be introduced because of incorrect manipulations or mistakes in arithmetic. Thus, it is advisable to check answers by substitution in the original equation. To apply a check in Example 1, we substitute 3 for x in the equation $6x - 7 = 2x + 5$, obtaining

$$6(3) - 7 = 2(3) + 5$$

$$18 - 7 = 6 + 5$$

$$11 = 11,$$

which is a true statement.

We sometimes say that an equation such as that given in Example 1 *has the solution* $x = 3$. As another illustration, the equation $x^2 = 4$ is said to have solutions $x = 2$ and $x = -2$.

If we multiply both sides of an equation by an expression that equals zero for some value of x, then the equation obtained may not be equivalent to the previous one. The following example illustrates a situation in which this may happen.

EXAMPLE 2 Solve the equation $\dfrac{3x}{x-2} = 1 + \dfrac{6}{x-2}$.

Solution Multiplying both sides by $x - 2$ and simplifying leads to

$$\left(\frac{3x}{x-2}\right)(x-2) = 1(x-2) + \left(\frac{6}{x-2}\right)(x-2)$$

$$3x = (x-2) + 6$$

$$2x = 4$$

$$x = 2.$$

Let us check to see whether 2 is a solution of the original equation. Substituting 2 for x, we obtain

$$\frac{3(2)}{2-2} = 1 + \frac{6}{2-2}, \quad \text{or} \quad \frac{6}{0} = 1 + \frac{6}{0}.$$

Since division by 0 is not permissible, 2 is not a solution. Actually, the given equation has no solutions, for we have shown that if the equation is true for some value of x, then that value must be 2. However, as we have seen, 2 is not a solution. ■

The preceding example indicates that *it is essential to check answers that are obtained after multiplying both sides of an equation by an expression that contains variables.*

Some equations may be solved by using the fact that if a product of real numbers is zero, then at least one of the factors is zero. This so-called **method of factoring** is illustrated in the next two examples.

EXAMPLE 3 Solve the equation $3x^2 + x - 10 = 0$.

Solution The equation may be written in the form

$$(3x-5)(x+2) = 0,$$

and hence, either $\quad 3x - 5 = 0 \quad \text{or} \quad x + 2 = 0.$

The solutions of these equations are

$$x = \frac{5}{3} \quad \text{and} \quad x = -2.$$

This shows that the only possible solutions of $3x^2 + x - 10 = 0$ are $\frac{5}{3}$ or -2. The fact that both numbers actually are roots may be verified by direct substitution in the given equation. ■

EXAMPLE 4 Solve $x^3 + 2x^2 - x - 2 = 0$.

Solution The left side may be factored by grouping, as follows:

$$x^3 + 2x^2 - x - 2 = 0$$
$$x^2(x + 2) - (x + 2) = 0$$
$$(x^2 - 1)(x + 2) = 0$$
$$(x + 1)(x - 1)(x + 2) = 0$$

Setting each factor equal to 0 gives us

$$x + 1 = 0, \qquad x - 1 = 0, \qquad x + 2 = 0,$$

or equivalently, $\quad x = -1, \qquad x = 1, \qquad x = -2.$

That these three numbers are solutions may be checked by substitution in the original equation. ■

It is sometimes necessary to consider inequalities that involve variables. As a first illustration, let us consider

$$x^2 - 3 < 2x + 4.$$

If certain numbers such as 4 or 5 are substituted for x, we obtain the false statements $13 < 12$ or $22 < 14$, respectively. Other numbers, such as 1 or 2, produce the true statements $-2 < 6$ or $1 < 8$. In general, if we are given an inequality involving one variable x, and if a true statement is obtained when x is replaced by a real number a, then a is called a **solution** of the inequality. Thus 1 and 2 are solutions of the inequality $x^2 - 3 < 2x + 4$, whereas 4 and 5 are not solutions. To **solve** an inequality means to find all solutions. We say that two inequalities are **equivalent** if they have exactly the same solutions.

A standard method for solving an inequality is to replace it with a list of equivalent inequalities, terminating the list with an inequality for which the solutions are obvious. Among the main tools used in applying this method are the following properties, where a, b, and c denote real numbers.

PROPERTIES OF >

(i) If $a > b$ and $b > c$, then $a > c$.

(ii) If $a > b$, then $a + c > b + c$.

(iii) If $a > b$ and $c > 0$, then $ac > bc$.

(iv) If $a > b$ and $c < 0$, then $ac < bc$.

We can supply proofs for all four properties by making use of the fact that both the sum and product of any two positive real numbers are positive. Thus to prove (i) we first note that if $a > b$ and $b > c$, then $a - b$ and $b - c$ are both positive. Consequently, the sum $(a - b) + (b - c)$ is positive. Since the sum reduces to $a - c$, we see that $a - c$ is positive, which means that $a > c$.

To establish (ii) we again note that if $a > b$, then $a - b$ is positive. Since $(a + c) - (b + c) = a - b$, it follows that $(a + c) - (b + c)$ is positive; that is, $a + c > b + c$.

To prove (iii) observe that if $a > b$ and $c > 0$, then $a - b$ and c are both positive and hence so is the product $(a - b)c$. Consequently, $ac - bc$ is positive; that is, $ac > bc$.

Finally, to prove (iv) we first note that if $c < 0$, then $0 - c$, or $-c$, is positive. In addition, if $a > b$, then $a - b$ is positive and hence the product $(a - b)(-c)$ is positive. However, $(a - b)(-c) = -ac + bc$ and hence $bc - ac$ is positive. This means that $bc > ac$, or $ac < bc$.

Similar results are true for the symbol $<$. Specifically, the following properties can be proved.

PROPERTIES OF $<$

(i) If $a < b$ and $b < c$, then $a < c$.

(ii) If $a < b$, then $a + c < b + c$.

(iii) If $a < b$ and $c > 0$, then $ac < bc$.

(iv) If $a < b$ and $c < 0$, then $ac > bc$.

If x represents a real number, then adding the same expression in x to both sides of an inequality leads to an equivalent inequality. We may multiply both sides of an inequality by an expression containing x if we are certain that the expression is positive for all values of x under consideration. To illustrate, multiplication by $x^4 + 3x^2 + 5$ would be permissible since this expression is always positive. If we multiply both sides of an inequality by an expression that is always negative, such as $-7 - x^2$, then the inequality sign is reversed.

EXAMPLE 5 Solve the inequality $2x + 5 > 4x - 3$.

Solution The following inequalities are equivalent. (Why?)

$$2x + 5 > 4x - 3$$
$$(2x + 5) - 5 > (4x - 3) - 5$$
$$2x > 4x - 8$$
$$2x - 4x > (4x - 8) - 4x$$

$$-2x > -8$$

$$\left(-\frac{1}{2}\right)(-2x) < \left(-\frac{1}{2}\right)(-8)$$

$$x < 4$$

Hence, the solutions of $2x + 5 > 4x - 3$ consist of all real numbers x such that $x < 4$. ■

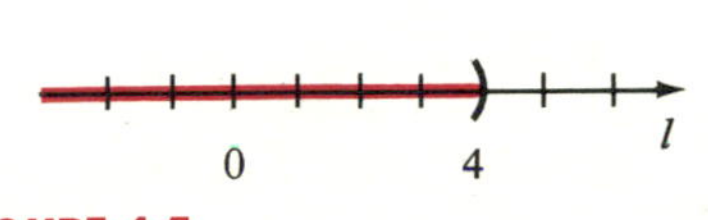

FIGURE 1.5

It is often convenient to give graphical interpretations for solutions of inequalities. By the **graph** of a set of real numbers we mean the collection of all points on a coordinate line l that correspond to the numbers. To **sketch a graph** we darken (or color) an appropriate portion of l. The graph corresponding to the solutions of Example 5 consists of all points to the left of the point with coordinate 4 and is sketched in Figure 1.5, where it is understood that the colored portion extends indefinitely to the left. The parenthesis in the figure indicates that the point corresponding to 4 is not part of the graph. If we wish to include such an *endpoint* of a graph, we use a bracket instead of a parenthesis. This will be illustrated in Example 7.

If $a < b$, the symbol (a, b) is sometimes used for all real numbers between a and b. This set is called an **open interval.**

DEFINITION OF OPEN INTERVAL

$$(a, b) = \{x : a < x < b\}$$

The symbol to the right of the equal sign is translated "the set of all x such that $a < x < b$." The numbers a and b are called the **endpoints** of the interval. The graph of (a, b) consists of all points on a coordinate line that lie between the points corresponding to a and b. In Figure 1.6 we have sketched the graph of a general open interval (a, b) and also the special open intervals $(-1, 3)$ and $(2, 4)$. The parentheses in the figure indicate that the endpoints of the intervals are not to be included. For convenience, we shall use the terms *open interval* and *graph of an open interval* interchangeably.

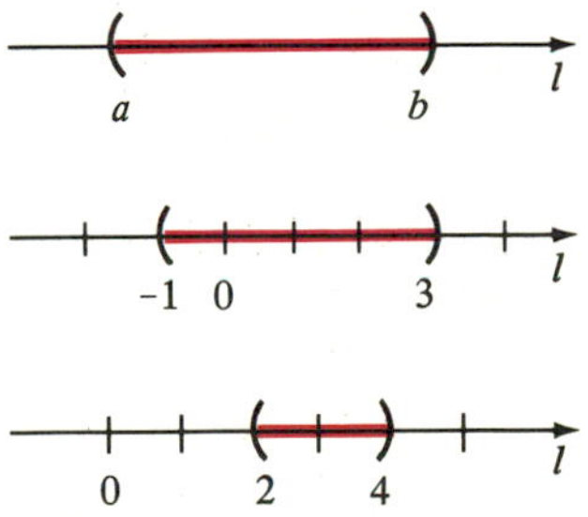

The solutions of an inequality may often be described in terms of intervals, as illustrated in the next example.

FIGURE 1.6 Open intervals (a, b), $(-1, 3)$, and $(2, 4)$.

EXAMPLE 6 Solve the inequality $-6 < 2x - 4 < 2$ and represent the solutions graphically.

Solution A real number x is a solution of the given inequality if and only if it is a solution of *both* of the inequalities

$$-6 < 2x - 4 \quad \text{and} \quad 2x - 4 < 2.$$

The first inequality is equivalent to each of the following:

$$-6 < 2x - 4$$
$$-6 + 4 < (2x - 4) + 4$$
$$-2 < 2x$$
$$\tfrac{1}{2}(-2) < \tfrac{1}{2}(2x)$$
$$-1 < x.$$

The second inequality is equivalent to each of the following:

$$2x - 4 < 2$$
$$2x < 6$$
$$x < 3.$$

Thus, x is a solution of the given inequality if and only if *both*

$$-1 < x \quad \text{and} \quad x < 3,$$

that is,

$$-1 < x < 3.$$

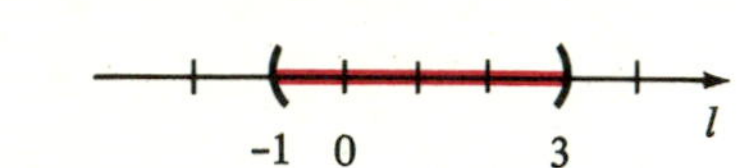

FIGURE 1.7

Hence, the solutions are all numbers in the open interval $(-1, 3)$. The graph is sketched in Figure 1.7.

An alternative (and shorter) method is to work with both inequalities simultaneously, as follows:

$$-6 < 2x - 4 < 2$$
$$-6 + 4 < 2x < 2 + 4$$
$$-2 < 2x < 6$$
$$-1 < x < 3.$$

The last inequality agrees with the first solution. ■

If we wish to include an endpoint of an interval, a bracket is used instead of a parenthesis. If $a < b$, then **closed intervals,** denoted by $[a, b]$, and **half-open intervals,** denoted by $[a, b)$ or $(a, b]$, are defined as follows.

DEFINITION OF CLOSED AND HALF-OPEN INTERVALS

$$[a, b] = \{x : a \le x \le b\}$$
$$[a, b) = \{x : a \le x < b\}$$
$$(a, b] = \{x : a < x \le b\}$$

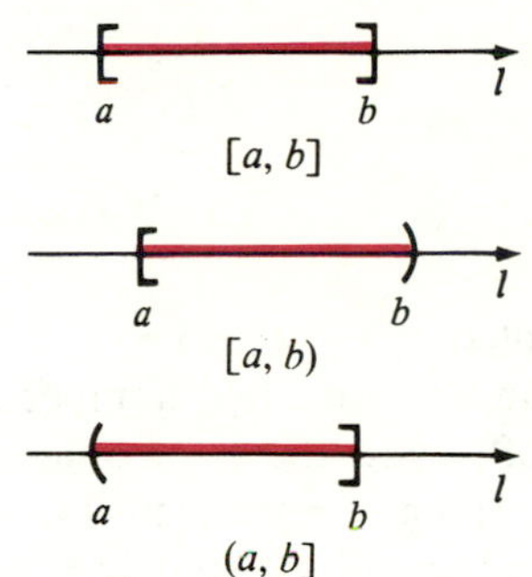

FIGURE 1.8

Typical graphs are sketched in Figure 1.8. A bracket indicates that the corresponding endpoint is part of the graph.

EXAMPLE 7 Solve the inequality

$$-5 \leq \frac{4 - 3x}{2} < 1$$

and sketch the graph corresponding to the solutions.

Solution A number x is a solution if and only if it satisfies both of the inequalities

$$-5 \leq \frac{4 - 3x}{2} \quad \text{and} \quad \frac{4 - 3x}{2} < 1.$$

We can either work with each inequality separately or proceed as in the alternative solution of Example 6, as follows:

$$-5 \leq \frac{4 - 3x}{2} < 1$$

$$-10 \leq 4 - 3x < 2$$

$$-10 - 4 \leq -3x < 2 - 4$$

$$-14 \leq -3x < -2$$

$$(-\tfrac{1}{3})(-14) \geq (-\tfrac{1}{3})(-3x) > (-\tfrac{1}{3})(-2)$$

$$\tfrac{14}{3} \geq x > \tfrac{2}{3}$$

$$\tfrac{2}{3} < x \leq \tfrac{14}{3}.$$

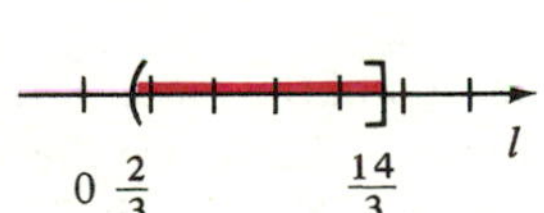

FIGURE 1.9

Hence, the solutions of the inequality consist of all numbers in the interval $(\frac{2}{3}, \frac{14}{3}]$. The graph is sketched in Figure 1.9. ■

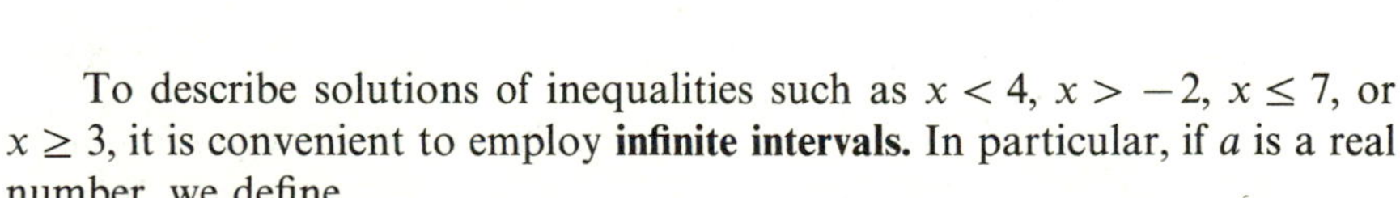

To describe solutions of inequalities such as $x < 4$, $x > -2$, $x \leq 7$, or $x \geq 3$, it is convenient to employ **infinite intervals.** In particular, if a is a real number, we define

$$(-\infty, a) = \{x : x < a\}.$$

Thus $(-\infty, a)$ denotes the set of all real numbers less than a. The symbol ∞ (**infinity**) is merely a notational device and is not to be interpreted as representing a real number. To illustrate, the solutions in Example 5 are the numbers in the infinite interval $(-\infty, 4)$. The graph of $(-\infty, 4)$ is sketched in Figure 1.5. If we wish to include the point corresponding to a we write

$$(-\infty, a] = \{x : x \leq a\}.$$

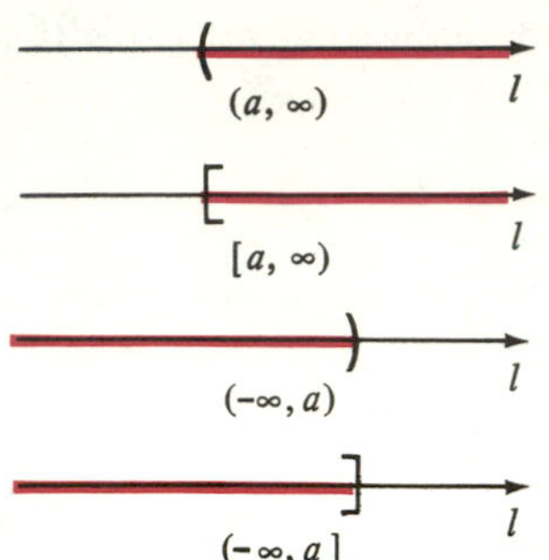

FIGURE 1.10

Other types of infinite intervals are defined by

$$(a, \infty) = \{x : x > a\}$$
$$[a, \infty) = \{x : x \geq a\}.$$

The set $\mathbb{R}$ of real numbers is sometimes denoted by $(-\infty, \infty)$.

Typical graphs of infinite intervals for an arbitrary real number a are sketched in Figure 1.10, where the absence of a parentheses or bracket, on the left for the graphs of $(-\infty, a)$ and $(-\infty, a]$, and on the right for (a, ∞) and $[a, \infty)$, indicates that the colored portions extend indefinitely.

EXAMPLE 8 Solve $\dfrac{1}{x - 2} > 0$ and sketch the graph corresponding to the solutions.

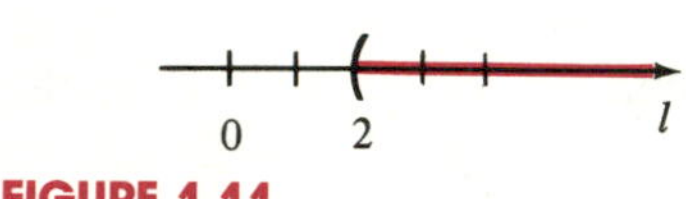

FIGURE 1.11

Solution Since the numerator is positive, the fraction is positive if and only if $x - 2 > 0$, or equivalently, if $x > 2$. Hence, the solutions are all numbers in the infinite interval $(2, \infty)$. The graph is sketched in Figure 1.11. ■

The next two examples illustrate how equations and inequalities may be used to solve certain applied problems.

EXAMPLE 9 A box with a square base and no top is to be made from a square piece of tin by cutting out 3-inch squares from each corner and folding up the sides. If the box is to hold 48 cubic inches, what size piece of tin should be used?

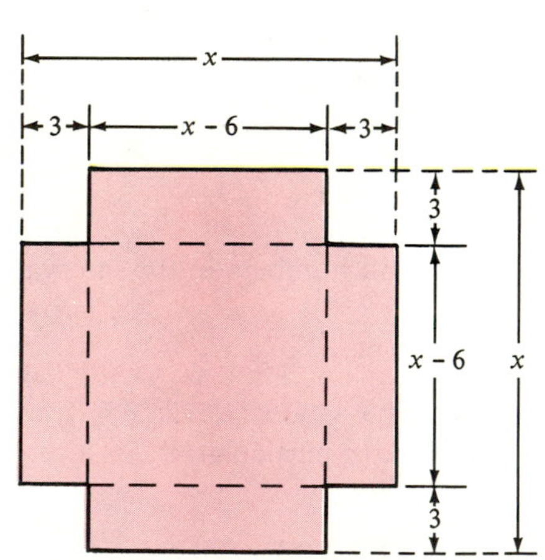

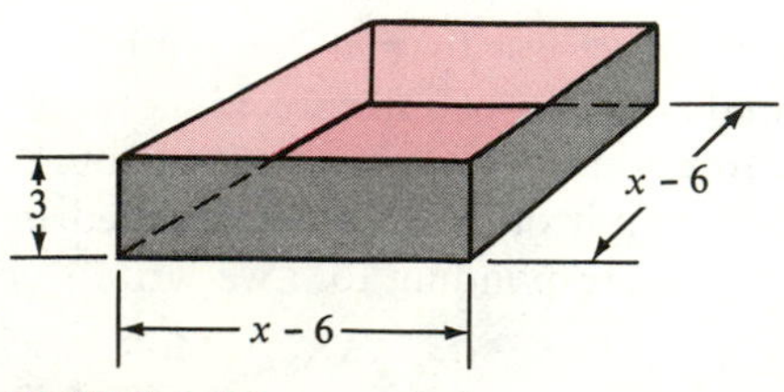

FIGURE 1.12

Solution If we let x denote the length of the side of the piece of tin, then the length of the base of the box is $x - 6$ (see Figure 1.12). Since the area of the base is $(x - 6)^2$ and the height is 3, the volume of the box is $3(x - 6)^2$. Moreover, since the box is to hold 48 cubic inches,

$$3(x - 6)^2 = 48.$$

The solutions of this equation may be found as follows:

$$(x - 6)^2 = 16$$
$$x^2 - 12x + 36 = 16$$
$$x^2 - 12x + 20 = 0$$
$$(x - 10)(x - 2) = 0$$

Consequently, either $x = 10$ or $x = 2$.

Let us now check each of these numbers. Referring to Figure 1.12, we see that 2 is unacceptable since no box is possible in this case. (Why?) However, if we begin with a 10-inch square of tin, cut out 3-inch corners, and fold, we obtain a box having dimensions 4 inches, 4 inches, and 3 inches. The box has the desired volume of 48 cubic inches. Thus, 10 inches is the answer to the problem. ■

As illustrated in Example 9, even though an equation is formulated correctly, it is possible, because of the physical nature of a given problem, to arrive at meaningless solutions. These solutions should be discarded. For example, we would not accept the answer -7 years for the age of an individual, nor $\sqrt{50}$ for the number of automobiles in a parking lot.

EXAMPLE 10 The temperature readings on the Fahrenheit and Celsius scales are related by the equation $C = \frac{5}{9}(F - 32)$. What values of F correspond to $30 \le C \le 40$?

Solution The following inequalities are equivalent:

$$30 \le C \le 40$$
$$30 \le \tfrac{5}{9}(F - 32) \le 40$$
$$\tfrac{9}{5}(30) \le F - 32 \le \tfrac{9}{5}(40)$$
$$54 \le F - 32 \le 72$$
$$86 \le F \le 104$$

Thus, a Celsius temperature range from 30° C to 40° C is the same as a Fahrenheit range from 86° F to 104° F. ■

EXERCISES 1.4

Solve the equations in Exercises 1–18.

1 $5x + 3 = 7x - 2$

2 $8x - 5 = 6x + 4$

3 $(4x - 3)(2x + 3) - 8x(x - 4) = 0$

4 $(2x + 3)(3x - 2) = 6x^2 + 1$

5 $6x^2 + 11x - 10 = 0$

6 $15x^2 + x - 6 = 0$

7 $4y^2 + 29y + 30 = 0$

8 $8z^2 + 19z - 27 = 0$

9 $3x^2 = 5x$

10 $x^2 = 4x^3$

11 $4x^3 + 12x^2 - 9x - 27 = 0$

12 $2x^3 + 5x^2 - 8x - 20 = 0$

13 $\dfrac{9x}{3x - 1} = 2 + \dfrac{3}{3x - 1}$

14 $\dfrac{2x}{2x + 3} + \dfrac{6}{4x + 6} = 5$

15 $\dfrac{7}{y^2 - 4} - \dfrac{4}{y + 2} = \dfrac{5}{y - 2}$

16 $\dfrac{-3}{4x - 2} + \dfrac{2}{2x - 1} = \dfrac{7}{2}$

17 $\sqrt{7 - 5x} = 8$

18 $\sqrt[3]{6 - s^2} + 5 = 0$

In Exercises 19–28 express the inequality in interval notation and sketch a graph of the interval.

19 $2 < x < 5$

20 $-1 < x < 2$

21 $-1 < x \leq 3$

22 $-2 \geq x \geq -3$

23 $4 \geq x \geq 1$

24 $0 < x \leq 4$

25 $x > -1$

26 $x < 1$

27 $x \leq 2$

28 $x \geq -3$

Express the intervals in Exercises 29–34 as inequalities in the variable x.

29 $(-1, 7)$

30 $[5, 12]$

31 $(8, 9]$

32 $[-5, 0)$

33 $(5, \infty)$

34 $(-\infty, 7]$

Solve the inequalities in Exercises 35–54 and express the solutions in terms of intervals.

35 $5x - 6 > 11$

36 $3x - 5 < 10$

37 $2 - 7x \leq 16$

38 $7 - 2x \geq -3$

39 $3x + 1 < 5x - 4$

40 $6x - 5 > 9x + 1$

41 $4 - \frac{1}{2}x \geq -7 + \frac{1}{4}x$

42 $\frac{1}{3}x - 4 \leq \frac{1}{4}x - 6$

43 $-4 < 3x + 5 < 8$

44 $6 > 2x - 6 > 4$

45 $3 \geq \dfrac{7 - x}{2} \geq 1$

46 $-2 \leq \dfrac{5 - 3x}{4} \leq \dfrac{1}{2}$

47 $0 < 2 - \frac{3}{4}x \leq \frac{1}{2}$

48 $-3 < \frac{1}{2}x - 4 \leq 0$

49 $(3x + 1)(2x - 4) < (6x - 2)(x + 5)$

50 $(x + 2)(x - 5) < (x - 1)^2$

51 $(x + 2)^2 > x(x - 2)$

52 $2x(8x - 1) \geq (4x + 1)(4x - 3)$

53 $\dfrac{5}{3x + 7} > 0$

54 $\dfrac{1}{6 - 2x} < 0$

55 According to Hooke's Law, the force F (in pounds) required to stretch a certain spring x inches beyond its natural length is given by the formula $F = (4.5)x$. (See the following figure.) If $10 \leq F \leq 18$, what are the corresponding values for x?

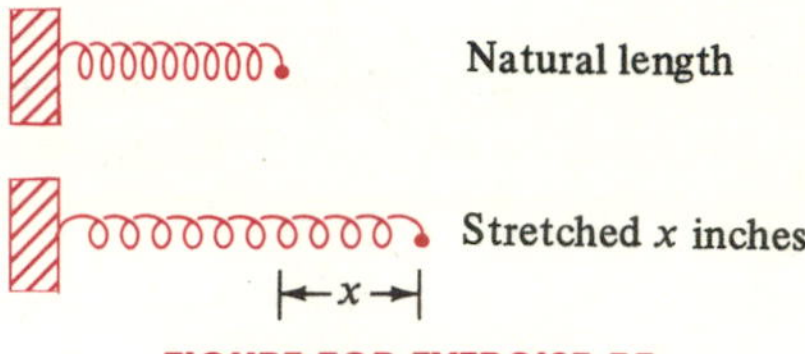

FIGURE FOR EXERCISE 55

56 In the study of electricity, Ohm's Law states that if R denotes the resistance of an object (in ohms), E the potential difference across the object (in volts), and I the current that flows through it (in amperes), then $R = E/I$ (see figure). If the voltage is 110, what values of the resistance will result in a current that does not exceed 10 amperes?

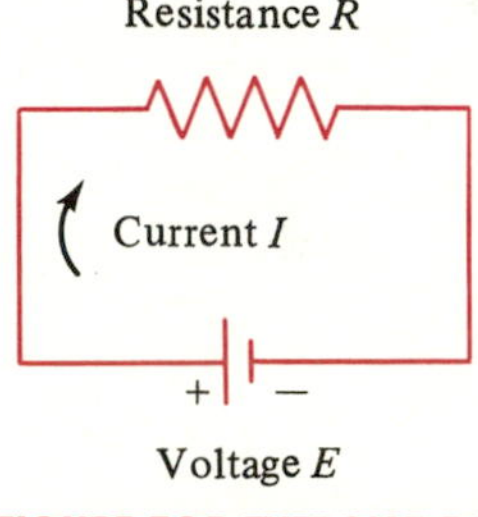

FIGURE FOR EXERCISE 56

57 A student in an algebra course has test scores of 75, 82, 71, and 84. What score on the next test will raise the student's average to 80?

58 Going into a final exam, a student has test scores of 72, 80, 65, 78, and 60. If the final exam counts as 1/3 of the final grade, what score must the student receive to finish with an average of 76?

59 Two boys who are 224 meters apart start walking toward each other at the same time. If they walk at rates of 1.5 and 2 meters per second, respectively, when will they meet? How far will each have walked? (*Hint:* distance = (rate) × (time).)

60 A projectile is fired horizontally at a target, and the sound of its impact is heard 1.5 seconds later. If the speed of the projectile is 3300 ft/sec and the speed of sound is 1100 ft/sec, how far away is the target?

61 After playing 100 games, a major league baseball team has a record of 0.650. If it wins only 50% of its games for the remainder of the season, when will its record be 0.600?

62 A rectangle is twice as long as it is wide. If the length and width are decreased by 2 cm and 3 cm, respectively, the area is decreased by 30 cm^2. Find the original dimensions.

63 Twenty liters of a solution contains 20% of a certain chemical. How much water should be added so that the resulting solution contains 15% of the chemical?

64 A chemist has two acid solutions, the first containing 20% acid and the second 35% acid. How many ml of each should be mixed to obtain 50 ml of a solution containing 30% acid?

65 A salesperson purchased an automobile that was advertised as averaging 25 miles per gallon in the city and 40 miles per gallon on the highway. On a certain sales trip it required 51 gallons of gasoline to cover 1800 miles. Assuming that the advertised mileage estimates were correct, how many miles were driven in the city?

66 A farmer plans to use 180 feet of fencing to enclose a rectangular region, using part of a straight river bank instead of fencing as one side of the rectangle. Find the area of the region if the length of the side parallel to the river bank is

(a) twice the length of an adjacent side.

(b) one-half the length of an adjacent side.

(c) the same as the length of an adjacent side.

67 A boy can row a boat at a rate of 5 mph in still water. If he rows upstream for 15 minutes, and then rows downstream, returning to his starting point in another 12 minutes, find:

(a) the rate of the current.

(b) the total distance traveled.

68 Ohm's Law states that $R = E/I$. (See Exercise 56.) In a certain circuit $E = 110$ and $R = 50$. If E and R are to be changed by the same numerical amount, what change in them will cause I to double?

69 An airplane flying north at 200 mph passed over a point on the ground at 2:00 P.M. Another airplane at the same altitude passed over the point at 2:30 P.M., flying east at 400 mph (see figure).

(a) If t denotes the time (in hours) after 2:30 P.M., express the distance d between the planes in terms of t.

(b) At what time were the airplanes 500 miles apart?

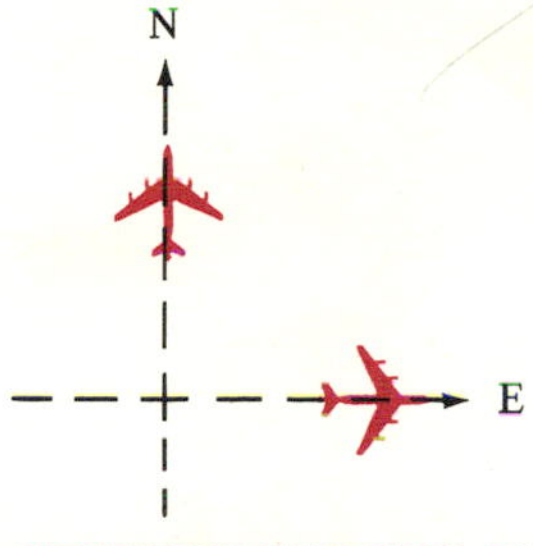

FIGURE FOR EXERCISES 69

70 A box with an open top is to be constructed by cutting out 3-inch squares from a rectangular sheet of tin whose length is twice its width. What size sheet will produce a box having a volume of 60 in.^{3}?

SECTION 1.5
MORE ON INEQUALITIES

Among the most important inequalities that occur in areas of mathematics such as calculus are those involving absolute values. As we discussed in Section 1.1, if a represents a real number, then $|a|$ is the distance on a coordinate line between the origin and the point corresponding to a. It follows that the solutions of an inequality of the form $|a| < 5$ consist of all real numbers a such that $-5 < a < 5$; that is, the numbers in the open interval $(-5, 5)$. In general, if b is any positive real number, the following properties can be proved.

PROPERTIES OF ABSOLUTE VALUES ($b > 0$)

(i) $|a| < b$ if and only if $-b < a < b$

(ii) $|a| > b$ if and only if either $a > b$ or $a < -b$

(iii) $|a| = b$ if and only if $a = b$ or $a = -b$.

Properties (ii) and (iii) are also true if $b = 0$. It follows that if $b \geq 0$, then

$$|a| \leq b \quad \text{if and only if} \quad -b \leq a \leq b$$

and

$$|a| \geq b \quad \text{if and only if} \quad a \geq b \quad \text{or} \quad a \leq -b.$$

EXAMPLE 1 Solve the following inequalities.

(a) $|x| < 4$ (b) $|x| > 4$ (c) $|x| = 4$

Solution

(a) Using property (i) with $a = x$ and $b = 4$, we see that

$$|x| < 4 \quad \text{if and only if} \quad -4 < x < 4.$$

Hence, the solutions are all real numbers in the open interval $(-4, 4)$.

(b) By property (ii), $|x| > 4$ means that either $x > 4$ or $x < -4$. Thus, the solutions consist of all real numbers in the two infinite intervals $(-\infty, -4)$ and $(4, \infty)$.

(c) By property (iii)

$$|x| = 4 \quad \text{if and only if} \quad x = 4 \text{ or } x = -4.$$ ■

For solutions of the type obtained in (b) of Example 1 it is sometimes convenient to use the union symbol, $\cup$, and write

$$(-\infty, -4) \cup (4, \infty)$$

to denote all real numbers x such that either x is in $(-\infty, -4)$ or x is in $(4, \infty)$.

EXAMPLE 2 Solve the inequality $|x - 3| < 0.1$.

Solution By property (i), with $a = x - 3$ and $b = 0.1$, the inequality is equivalent to

$$-0.1 < x - 3 < 0.1$$

and hence to

$$-0.1 + 3 < (x - 3) + 3 < 0.1 + 3$$

or to

$$2.9 < x < 3.1.$$

Consequently, the solutions are the real numbers in the interval (2.9, 3.1). ■

In the study of calculus it is necessary to consider inequalities of the type given in the next example.

EXAMPLE 3 If a and δ denote real numbers and $\delta > 0$, solve the inequality

$$|x - a| < \delta$$

and represent the solutions graphically.

Solution We may proceed as in Example 2. Thus,

$$-\delta < x - a < \delta$$

$$a - \delta < a + (x - a) < a + \delta$$

$$a - \delta < x < a + \delta.$$

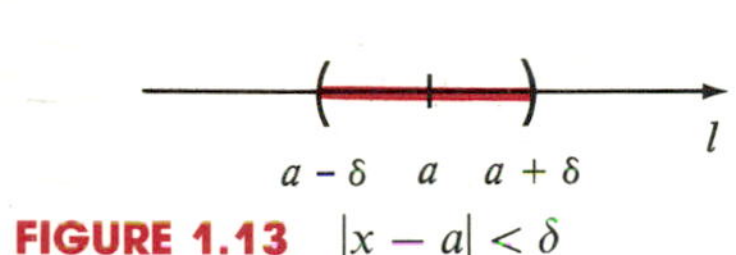

FIGURE 1.13 $|x - a| < \delta$

The solutions of the inequality consist of all real numbers in the interval $(a - \delta, a + \delta)$. A typical graph is sketched in Figure 1.13. ■

Note that Example 2 is the special case of Example 3 with $a = 3$ and $\delta = 0.1$.

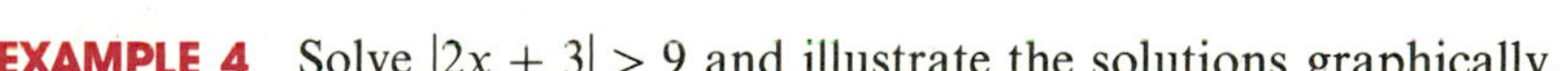

EXAMPLE 4 Solve $|2x + 3| > 9$ and illustrate the solutions graphically.

Solution By property (ii) of absolute values with $a = 2x + 3$ and $b = 9$, the solutions of the inequality are the solutions of the following *two* inequalities:

$$\text{(i)}\quad 2x + 3 > 9 \qquad \text{(ii)}\quad 2x + 3 < -9.$$

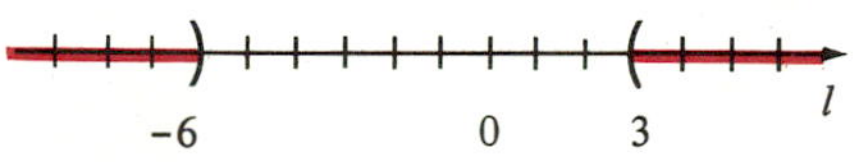

Inequality (i) is equivalent to $2x > 6$, or $x > 3$. This gives us the infinite interval $(3, \infty)$. Inequality (ii) is equivalent to $2x < -12$, or $x < -6$, which leads to the interval $(-\infty, -6)$. Consequently, the solutions of $|2x + 3| > 9$ consist of the numbers in the union $(-\infty, -6) \cup (3, \infty)$. The graph is sketched in Figure 1.14. ■

-6 0 3 l

FIGURE 1.14

Before proceeding, let us make an important remark about notation. The solutions in Example 4 consist of all real numbers x such that either $x > 3$ or $x < -6$. Some students employ the *incorrect* form $3 < x < -6$ to denote these inequalities. The notation $a < b < c$ should be used if and only if b satisfies *both* of the conditions $a < b$ and $b < c$. In particular, there is no real number x such that $3 < x < -6$ since it is impossible for x to satisfy $3 < x$ and $x < -6$ simultaneously.

Most of the inequalities considered up to this time were equivalent to inequalities containing only first-degree polynomials. To solve inequalities involving polynomials of higher degree we shall make use of an important result. This result, which will be discussed further in Section 3.1, implies that if real numbers c and d are *successive* solutions of a (polynomial) equation $a_nx^n + \cdots + a_1x + a_0 = 0$, that is, *if $c < d$ and if there are no other solutions between c and d, then either all values of the polynomial are positive, or all values are negative, when x is in the interval* (c, d). Thus, if we choose *any* number k, such that $c < k < d$, and if the value of the polynomial is positive for $x = k$, then the polynomial is positive for *every* x in (c, d). Similarly, if the polynomial is negative for $x = k$, then it is negative throughout (c, d). We shall call the value of the polynomial at $x = k$ a **test value** for the interval (c, d). Test values may also be used on infinite intervals of the form $(-\infty, a)$ or (a, ∞), provided the polynomial equation has no solutions on these intervals. The use of test values is demonstrated in the following examples.

EXAMPLE 5 Solve $2x^2 - x - 3 < 0$.

Solution The following inequalities are equivalent:

$$2x^2 - x - 3 < 0$$

$$(x + 1)(2x - 3) < 0$$

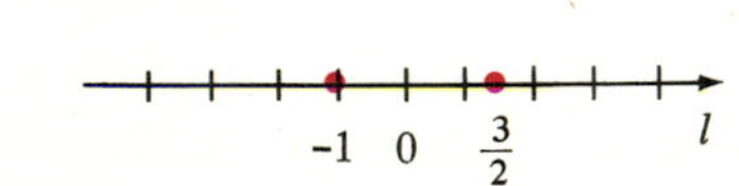

FIGURE 1.15

We see from the factored form that the equation $2x^2 - x - 3 = 0$ has solutions -1 and $\frac{3}{2}$. For reference, let us plot the corresponding points on a real axis, as in Figure 1.15. These points divide the axis into three parts and determine the following intervals:

$$(-\infty, -1), \qquad (-1, \tfrac{3}{2}), \qquad (\tfrac{3}{2}, \infty).$$

According to the discussion preceding this example, the sign of the polynomial $2x^2 - x - 3$ in each interval can be determined by finding a suitable test value.

If we choose -2 in $(-\infty, -1)$, then the polynomial $2x^2 - x - 3$ has the value

$$2(-2)^2 - (-2) - 3 = 8 + 2 - 3 = 7.$$

Since 7 is positive, it follows that $2x^2 - x - 3 > 0$ for every x in $(-\infty, -1)$.

If we choose 0 in $(-1, \frac{3}{2})$, then the polynomial has the value

$$2(0)^2 - (0) - 3 = -3.$$

Since -3 is negative, $2x^2 - x - 3 < 0$ for every x in $(-1, \frac{3}{2})$.

Finally, choosing 2 in $(\frac{3}{2}, \infty)$, we obtain

$$2(2)^2 - 2 - 3 = 8 - 2 - 3 = 3,$$

and since 3 is positive, $2x^2 - x - 3 > 0$ throughout $(\frac{3}{2}, \infty)$. It is convenient to tabulate these facts as follows.

Interval	$(-\infty, -1)$	$(-1, \frac{3}{2})$	$(\frac{3}{2}, \infty)$
k	-2	0	2
Test value of $2x^2 - x - 3$ at k	7	-3	3
Sign of $2x^2 - x - 3$ in interval	+	−	+

Hence, the solutions of the inequality $2x^2 - x - 3 < 0$ are the real numbers in the interval $(-1, \frac{3}{2})$. ■

EXAMPLE 6 Solve the inequality $x^2 - 7x + 10 > 0$ and represent the solutions graphically.

Solution The inequality may be written in the form

$$(x - 2)(x - 5) > 0.$$

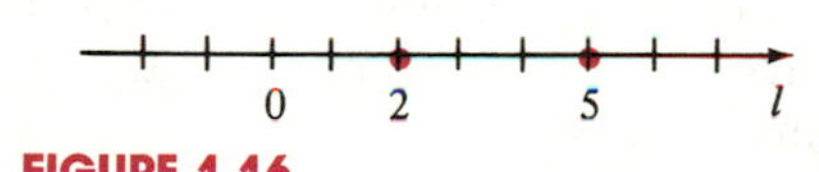

FIGURE 1.16

Points corresponding to the solutions 2 and 5 of $x^2 - 7x + 10 = 0$ are plotted in Figure 1.16. Referring to the figure, we obtain the following intervals:

$$(-\infty, 2), \qquad (2, 5), \qquad (5, \infty).$$

We next use test values to determine the sign of $x^2 - 7x + 10$ in each interval. The following table summarizes results. (*Check each entry.*)

Interval	$(-\infty, 2)$	$(2, 5)$	$(5, \infty)$
k	0	3	6
Test value of $x^2 - 7x + 10$ at k	$0^2 - 7(0) + 10 = 10$	$3^2 - 7(3) + 10 = -2$	$6^2 - 7(6) + 10 = 4$
Sign of $x^2 - 7x + 10$ in interval	+	−	+

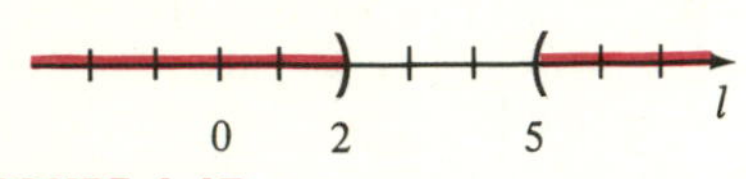

FIGURE 1.17

Referring to the table, we see that $x^2 - 7x + 10 > 0$ if x is in either $(-\infty, 2)$ or $(5, \infty)$. Thus, the solutions of the inequality are given by $(-\infty, 2) \cup (5, \infty)$. The graph corresponding to the solutions is sketched in Figure 1.17. ■

EXAMPLE 7 Solve $\dfrac{x+1}{x+3} \le 2$ and represent the solutions graphically.

Solution The following inequalities are equivalent:

$$\frac{x+1}{x+3} \le 2$$

$$\frac{x+1}{x+3} - 2 \le 0$$

$$\frac{x+1-2x-6}{x+3} \le 0$$

$$\frac{-x-5}{x+3} \le 0$$

$$\frac{x+5}{x+3} \ge 0$$

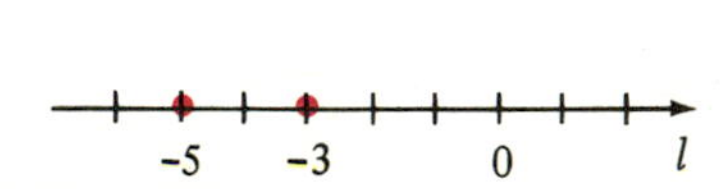

FIGURE 1.18

The numerator and denominator of $(x+5)/(x+3)$ equal zero at $x = -5$ and $x = -3$, respectively. For reference, we plot these points as in Figure 1.18. Note that -5 is a solution of $(x+5)/(x+3) \ge 0$; however, -3 is *not* a solution since a zero denominator occurs if -3 is substituted for x. The points in the figure determine the following intervals:

$$(-\infty, -5), \qquad (-5, -3), \qquad (-3, \infty).$$

Since $(x+5)/(x+3)$ is a quotient of two polynomials, it is always positive or always negative throughout each interval. As before, we may use test values to determine the sign in each interval. The following table summarizes results.

Interval	$(-\infty, -5)$	$(-5, -3)$	$(-3, \infty)$
k	-6	-4	0
Test value of $(x+5)/(x+3)$ at k	$\frac{1}{3}$	-1	$\frac{5}{3}$
Sign of $(x+5)/(x+3)$ in interval	$+$	$-$	$+$

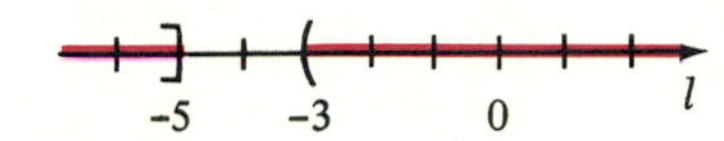

FIGURE 1.19

Referring to the table, we see that the solutions of $(x + 5)/(x + 3) > 0$ are given by $(-\infty, -5) \cup (-3, \infty)$. Hence, the solutions of $(x + 5)/(x + 3) \geq 0$ are given by $(-\infty, -5] \cup (-3, \infty)$. The graph is sketched in Figure 1.19. ■

EXAMPLE 8 Solve the inequality $(x + 2)(x - 1)(x - 5) > 0$ and represent the solutions graphically.

FIGURE 1.20

Solution The expression $(x + 2)(x - 1)(x - 5)$ is zero at -2, 1, and 5. The corresponding points are plotted on a real axis in Figure 1.20. These points determine the four intervals:

$$(-\infty, -2), \quad (-2, 1), \quad (1, 5), \quad (5, \infty).$$

We next use test values, as shown in the following table.

Interval	$(-\infty, -2)$	$(-2, 1)$	$(1, 5)$	$(5, \infty)$
k	-3	0	2	6
Test value of $(x + 2)(x - 1)(x - 5)$	$(-1)(-4)(-8) = -32$	$(2)(-1)(-5) = 10$	$(4)(1)(-3) = -12$	$(8)(5)(1) = 40$
Sign of $(x + 2)(x - 1)(x - 5)$ in interval	$-$	$+$	$-$	$+$

Referring to the table, we see that the solutions of $(x + 2)(x - 1)(x - 5) > 0$ are given by $(-2, 1) \cup (5, \infty)$. The graph is sketched in Figure 1.21.

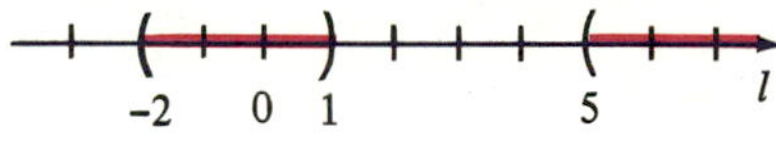

FIGURE 1.21

■

EXERCISES 1.5

Solve the inequalities in Exercises 1–44 and express the solutions in terms of intervals.

1 $|x| < 2$

2 $|x| < 25$

3 $|x| > 6$

4 $|x| > \frac{1}{2}$

5 $|-x| \leq 10$

6 $|x| \geq 7$

7 $|x - 10| < 0.05$

8 $|x - 5| < 0.001$

9 $\left|\dfrac{x + 4}{3}\right| \leq 2$

10 $\left|\dfrac{x + 1}{5}\right| \leq 1$

11 $|5 - 3x| < 7$

12 $|7x + 4| < 10$

13 $|2x + 4| > 8$

14 $|6x - 7| \geq 4$

15 $\dfrac{3}{|5 - 2x|} \leq 1$

16 $\left|\dfrac{4 - x}{2}\right| > 9$

17 $x^2 - x - 6 < 0$

18 $x^2 + 6x + 5 < 0$

19 $x(2x + 3) > 5$

20 $x(3x - 1) > 4$

21 $x^2 \leq 10x$

22 $4x^2 \geq x$

23 $\dfrac{5}{x - 4} < 0$

24 $\dfrac{-3}{x + 5} > 0$

25 $\dfrac{-4}{2x+7} < 0$

26 $\dfrac{x+3}{x-8} \le 0$

27 $\dfrac{2x+1}{10-3x} < 0$

28 $\dfrac{9-4x}{x+1} \ge 0$

29 $\dfrac{6}{x^2-9} < 0$

30 $\dfrac{10}{25x^2-16} > 0$

31 $\dfrac{7x}{x^2-16} \ge 0$

32 $\dfrac{x-2}{x^2-9} < 0$

33 $\dfrac{x^2-25}{4x^2-9} \le 0$

34 $\dfrac{10}{x^4-16} > 0$

35 $\dfrac{x+4}{2x-1} < 3$

36 $\dfrac{1}{x-2} > \dfrac{3}{x+1}$

37 $x^3 < x$

38 $x^4 + 3x^2 > 4$

39 $x^3 - x^2 - 4x + 4 \ge 0$

40 $(x^2 - 4x + 4)(3x - 7) < 0$

41 $\dfrac{x^2-x-2}{x^2-4x+3} \ge 0$

42 $\dfrac{x^2+2x-3}{x^2-4x} \le 0$

43 $0 < |x-a| < \delta$ where $\delta > 0$

44 $|x-a| \ge \delta$ where $\delta > 0$

45 If a projectile is fired straight upward from level ground with an initial velocity of 72 ft/sec, its altitude s (in feet) after t seconds given by $s = -16t^2 + 72t$. During what time interval will the projectile be at least 32 feet above the ground?

46 The period T (sec) of a simple pendulum of length l (cm) is given by $T = 2\pi\sqrt{l/g}$, where g is a physical constant. If, under certain conditions, $g = 980$ and $98 \le l \le 100$, what is the corresponding range for T?

SECTION 1.6
REVIEW

Define or discuss each of the following.

1 Commutative Properties of real numbers
2 Associative Properties
3 Distributive Properties
4 Integers
5 Prime number
6 Rational and irrational numbers
7 Coordinate line
8 A number a is greater than a number b.
9 A number a is less than a number b.
10 Absolute value of a real number
11 The distance between points on a coordinate line
12 Exponent
13 Laws of Exponents
14 Principal nth root
15 Radical notation
16 Rationalizing a denominator
17 Rational exponents
18 Scientific form for real numbers
19 Variable
20 Domain of a variable
21 Algebraic expression
22 Polynomial
23 Degree of a polynomial
24 Prime polynomial
25 Irreducible polynomial
26 Rational expression
27 Solution of an equation
28 Identity
29 Conditional equation
30 Equivalent equations

31 Solution of an inequality
32 Properties of inequalities
33 Equivalent inequalities
34 The graph of a set of real numbers
35 Open interval
36 Closed interval
37 Half-open interval
38 Infinite interval

EXERCISES 1.6

1 Express each of the following as a rational number with least positive numerator.

(a) $\left(\frac{2}{3}\right)\left(-\frac{5}{8}\right)$ (b) $\frac{3}{4}+\frac{6}{5}$

(c) $\frac{5}{8}-\frac{6}{7}$ (d) $\frac{3}{4}\div\frac{6}{5}$

2 Replace □ with either <, >, or =.

(a) -0.1 □ -0.01 (b) $\sqrt{9}$ □ -3

(c) $\frac{1}{6}$ □ 0.166

3 Express in terms of inequalities:

(a) x is negative.

(b) a is between $\frac{1}{2}$ and $\frac{1}{3}$.

(c) The absolute value of x is not greater than 4.

4 Rewrite without using the absolute value symbol:

(a) $|-7|$ (b) $\frac{|-5|}{-5}$

(c) $|3^{-1}-2^{-1}|$

5 If points A, B, and C on a coordinate line have coordinates -8, 4, and -3, respectively, find the following:

(a) $d(A, C)$ (b) $d(C, A)$

(c) $d(B, C)$

6 Determine whether the following are identities.

(a) $(x+y)^2 = x^2 + y^2$

(b) $\frac{1}{\sqrt{x+y}} = \frac{1}{\sqrt{x}} + \frac{1}{\sqrt{y}}$

(c) $\frac{1}{\sqrt{c}-\sqrt{d}} = \frac{\sqrt{c}+\sqrt{d}}{c-d}$

Simplify the expressions in Exercises 7–28.

7 $(3a^2b)^2(2ab^3)$

8 $\frac{6r^3y^2}{2r^5y}$

9 $\frac{(3x^2y^{-3})^{-2}}{x^{-5}y}$

10 $\left(\frac{a^{2/3}b^{3/2}}{a^2b}\right)^6$

11 $(-2p^2q)^3(p/4q^2)^2$

12 $c^{-4/3}c^{3/2}c^{1/6}$

13 $\left(\frac{xy^{-1}}{\sqrt{z}}\right)^4 \div \left(\frac{x^{1/3}y^2}{z}\right)^3$

14 $\left(\frac{-64x^3}{z^6y^9}\right)^{2/3}$

15 $((a^{2/3}b^{-2})^3)^{-1}$

16 $\frac{(3u^2v^5w^{-4})^3}{(2uv^{-3}w^2)^4}$

17 $\frac{r^{-1}+s^{-1}}{(rs)^{-1}}$

18 $(u+v)^3(u+v)^{-2}$

19 $\sqrt[3]{(x^4y^{-1})^6}$

20 $\sqrt[3]{8x^5y^3z^4}$

21 $\frac{1}{\sqrt[3]{4}}$

22 $\sqrt{\frac{a^2b^3}{c}}$

23 $\sqrt[3]{4x^2y}\sqrt[3]{2x^5y^2}$

24 $\sqrt[4]{(-4a^3b^2c)^2}$

25 $\frac{1}{\sqrt{t}}\left(\frac{1}{\sqrt{t}}-1\right)$

26 $\sqrt{\sqrt[3]{(c^3d^6)^4}}$

27 $\frac{\sqrt{12x^4y}}{\sqrt{3x^2y^5}}$

28 $\sqrt[3]{(a+2b)^3}$

In Exercises 29 and 30 rationalize the denominator.

29 $\frac{1-\sqrt{x}}{1+\sqrt{x}}$

30 $\frac{1}{\sqrt{a}+\sqrt{a-2}}$

31 Express in scientific form:

(a) 93,700,000,000 (b) 0.00000402

32 Express as a decimal:

(a) 6.8×10^7 (b) 7.3×10^{-4}

Perform the operations indicated in Exercises 33–36 and find the degree of the resulting polynomial.

33 $(2x^3 - x^2 + 5x - 7) - (3x^3 + 2x^2 - x + 2)$

34 $(x^4 + 3x^2 - 7) + (x^3 - 2x^2 + x + 1)$

35 $(x^2 - 3x + 4)(x - 5)$

36 $(x^2 - 2)(x^3 - 4x^2 + 3)$

Find the products in Exercises 37 and 38.

37 $(2x^2 - 3y)^2$

38 $(2x + y^2)^3$

Express the polynomials in Exercises 39–44 as products of polynomials that are irreducible over the set of integers.

39 $8x^2 + 10x - 3$

40 $9x^4 + 12x^2 + 4$

41 $3x^4 - 2x^3 + 12x^2 - 8x$

42 $64x^3 + 8$

43 $2x^4 - 32$

44 $3x^4 + 6x^3 - 24x^2$

Simplify the expressions in Exercises 45–54.

45 $\dfrac{6x^2 - 7x - 5}{4x^2 + 4x + 1}$

46 $\dfrac{r^3 - t^3}{r^2 - t^2}$

47 $\dfrac{6x^2 - 5x - 6}{x^2 - 4} \div \dfrac{2x^2 - 3x}{x + 2}$

48 $\dfrac{2}{4x - 5} - \dfrac{5}{10x + 1}$

49 $\dfrac{7}{x + 2} + \dfrac{3x}{(x + 2)^2} - \dfrac{5}{x}$

50 $\dfrac{x + x^{-2}}{1 + x^{-2}}$

51 $\dfrac{1}{x} - \dfrac{2}{x^2 + x} - \dfrac{3}{x + 3}$

52 $(a^{-1} + b^{-1})^{-1}$

53 $(x^2 + 1)^{3/2}(4)(x + 5)^3 + (x + 5)^4(\frac{3}{2})(x^2 + 1)^{1/2}(2x)$

54 $\dfrac{(4 - x^2)(\frac{1}{3})(6x + 1)^{-2/3}(6) - (6x + 1)^{1/3}(-2x)}{(4 - x^2)^2}$

Solve the equations and inequalities in Exercises 55–74.

55 $\dfrac{3x + 1}{5x + 7} = \dfrac{6x + 11}{10x - 3}$

56 $2 - \dfrac{1}{x} = 1 + \dfrac{4}{x}$

57 $12x^2 - 5x - 2 = 0$

58 $2x^2 + 5x - 12 = 0$

59 $(2x - 1)(x - 1) = 21$

60 $20x^3 + 8x^2 - 35x - 14 = 0$

61 $\sqrt[3]{4x - 5} - 2 = 0$

62 $10 - 7x < 4 + 2x$

63 $-\dfrac{1}{2} < \dfrac{2x + 3}{5} < \dfrac{3}{2}$

64 $(3x - 1)(10x + 4) \geq (6x - 5)(5x - 7)$

65 $\dfrac{6}{10x + 3} < 0$

66 $|4x + 7| < 21$

67 $|16 - 3x| \geq 5$

68 $2 < |x - 6| < 4$

69 $10x^2 + 11x - 6 > 0$

70 $x^2 - 3x \leq 10$

71 $\dfrac{3}{2x + 3} < \dfrac{1}{x - 2}$

72 $\dfrac{x + 1}{x^2 - 25} \leq 0$

73 $x^3 > x^2$

74 $(x^2 - x)(x^2 - 5x + 6) < 0$

75 An airplane flew with the wind for 30 minutes and returned the same distance in 45 minutes. If the cruising speed of the airplane is 320 mph, find the speed of the wind.

76 A merchant wishes to mix peanuts costing $1.50 per pound with cashews costing $4.00 per pound, obtaining 50 pounds of a mixture costing $2.40 per pound. How many pounds of each should be used?

77 A chemist has 80 ml of a solution containing 25% acid. How many milliliters should be removed and replaced by pure acid to obtain a solution containing 40% acid?

78 A man puts a fence around a rectangular field and then subdivides the field into three smaller rectangular plots by placing two fences parallel to one of the sides. If the area of the field is 31,250 square yards, and 1,000 yards of fencing were used, find the dimensions of the field.

79 The width of a page in a book is 2 inches smaller than its length. The printed area is 72 square inches, with 1-inch margins at the top and bottom and $\frac{1}{2}$-inch margins at each side. Find the dimensions of the page.

80 A sales representative for a company estimates that gasoline consumption for her automobile averages 28 miles per gallon on the highway and 22 miles per gallon in the city. On a recent trip she covered 627 miles and used 24 gallons of gasoline. How much of the trip was spent driving in the city?

2

FUNCTIONS

One of the most useful concepts in mathematics is that of function. Indeed, it is safe to say that without the notion of function, little progress could be made in mathematics or in any area of science. In the first two sections of this chapter we consider coordinate systems and graphs in two dimensions. The remainder of the chapter contains a discussion of functions and their graphs.

SECTION 2.1
COORDINATE SYSTEMS IN TWO DIMENSIONS

In Section 1.1 we discussed how coordinates may be assigned to points on a line. Coordinate systems can also be introduced in planes by means of *ordered pairs*. The term **ordered pair** refers to two real numbers, where one is designated as the "first" number and the other as the "second." The symbol (a, b) is used to denote the ordered pair consisting of the real numbers a and b where a is first and b is second. There are many uses for ordered pairs. In Section 1.4 they were used to denote open intervals. In this section they will represent .points in a plane. Although ordered pairs are employed in different situations, there is little chance that we will confuse them, since it should always be clear from the discussion whether the symbol (a, b) represents an interval, a point, or some other mathematical object. We consider two ordered pairs (a, b) and (c, d) equal, and write

$$(a, b) = (c, d) \quad \text{if and only if} \quad a = c \text{ and } b = d.$$

This implies, in particular, that $(a, b) \neq (b, a)$ if $a \neq b$.

A **rectangular,** or **Cartesian,*** **coordinate system** may be introduced in a plane by considering two perpendicular coordinate lines in the plane that intersect in the origin O on each line. Unless specified otherwise, the same unit of length is chosen on each line. Usually one of the lines is horizontal with positive direction to the right, and the other line is vertical with positive direction upward, as indicated by the arrowheads in Figure 2.1. The two lines are called **coordinate axes** and the point O is called the **origin.** The horizontal line is usually referred to as the ***x*-axis** and the vertical line as the ***y*-axis,** and they are labeled x and y, respectively. The plane is then called a **coordinate plane** or, with the preceding notation for coordinate axes, an ***xy*-plane.** In certain applications different labels such as d, t, etc., are used for the coordinate lines. The coordinate axes divide the plane into four parts called the **first, second, third,** and **fourth quadrants** and labeled I, II, III, and IV, respectively, as shown in Figure 2.1.

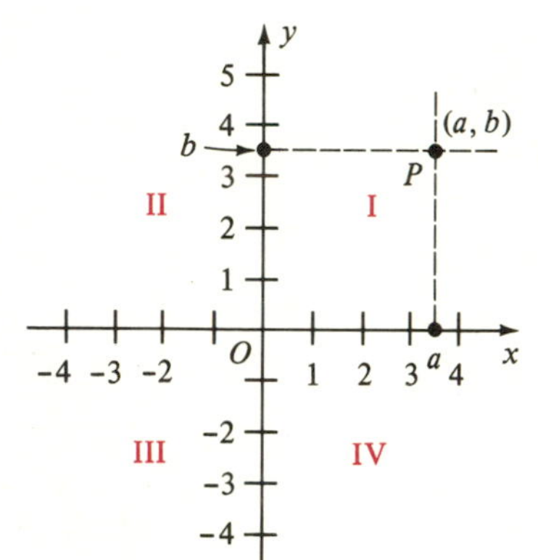

FIGURE 2.1

Each point P in an xy-plane may be assigned a unique ordered pair. If vertical and horizontal lines through P intersect the x- and y-axes at points with coordinates a and b, respectively (see Figure 2.1), then P is assigned the ordered pair (a, b). The number a is called the ***x*-coordinate** (or **abscissa**) of P, and b is called the ***y*-coordinate** (or **ordinate**) of P. We sometimes say that P *has coordinates* (a, b). Conversely, every order pair (a, b) determines a point P in the xy-plane with coordinates a and b. Specifically, P is the point of intersection of lines perpendicular to the x-axis and y-axis at the points having coordinates a and b, respectively. This establishes a one-to-one correspondence between the set of all points in the xy-plane and the set of all ordered pairs. It is sometimes convenient to refer to the *point* (a, b), meaning the point with x-coordinate a and y-coordinate b. The symbol $P(a, b)$ will denote the point P with coordinates (a, b). To **plot a point** $P(a, b)$ means to locate, in a coordinate plane, the point P with coordinates (a, b). This point is represented by a dot in the appropriate position, as illustrated in Figure 2.2.

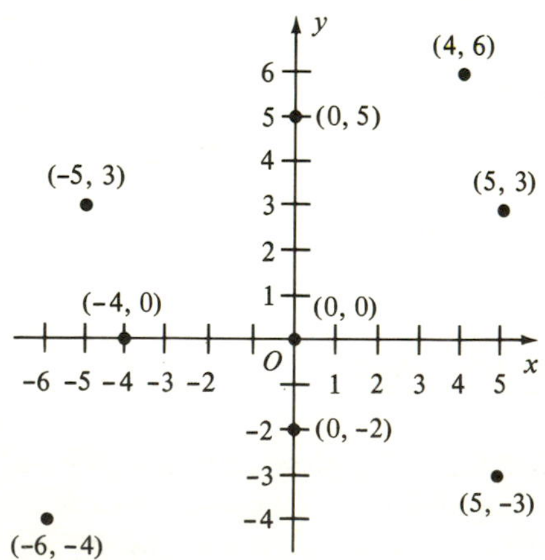

FIGURE 2.2

The next statement provides a formula for finding the distance between two points in a coordinate plane.

* The term "Cartesian" is used in honor of the French mathematician and philosopher René Descartes (1596–1650), who was one of the first to employ such coordinate systems.

DISTANCE FORMULA

The distance $d(P_1,P_2)$ between any two points $P_1(x_1, y_1)$ and $P_2(x_2, y_2)$ in a coordinate plane is

$$d(P_1, P_2) = \sqrt{(x_2 - x_1)^2 + (y_2 - y_1)^2}.$$

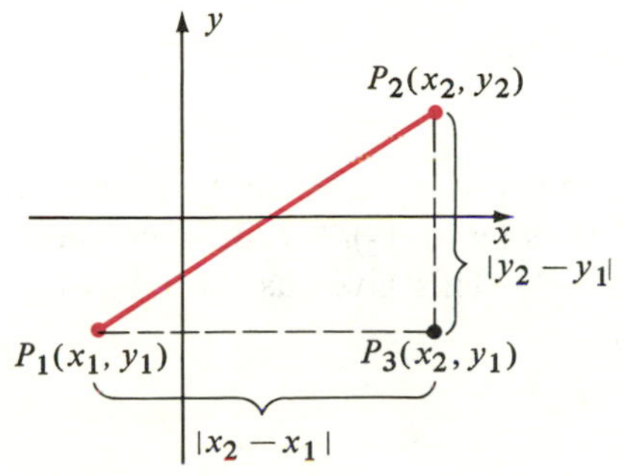

FIGURE 2.3

Proof If $x_1 \neq x_2$ and $y_1 \neq y_2$, then, as illustrated in Figure 2.3, the points P_1, P_2, and $P_3(x_2, y_1)$ are vertices of a right triangle. By the Pythagorean Theorem,

$$[d(P_1, P_2)]^2 = [d(P_1, P_3)]^2 + [d(P_3, P_2)]^2.$$

Using the fact that $d(P_1, P_3) = |x_2 - x_1|$ and $d(P_3, P_2) = |y_2 - y_1|$ leads to the desired formula.

If $y_1 = y_2$, the points P_1 and P_2 lie on the same horizontal line and

$$d(P_1, P_2) = |x_2 - x_1| = \sqrt{(x_2 - x_1)^2}.$$

Similarly, if $x_1 = x_2$, the points are on the same vertical line and

$$d(P_1, P_2) = |y_2 - y_1| = \sqrt{(y_2 - y_1)^2}.$$

These are special cases of the Distance Formula.

Although we referred to Figure 2.3, the argument used in our proof is independent of the positions of the points P_1 and P_2. □

In applying the Distance Formula, note that $d(P_1, P_2) = d(P_2, P_1)$ and, hence, the order in which we subtract the x-coordinates and the y-coordinates of the points is immaterial.

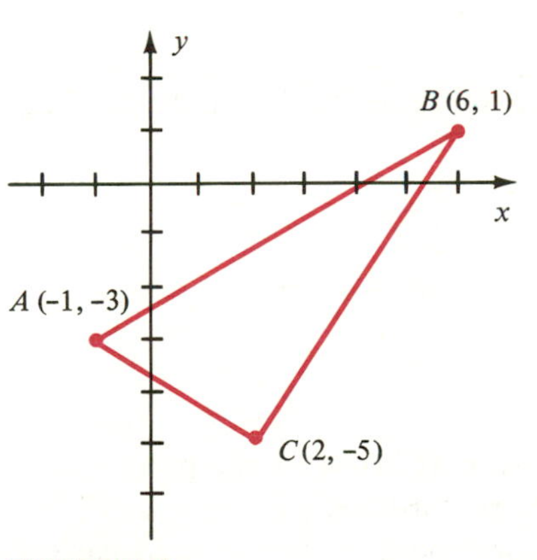

FIGURE 2.4

EXAMPLE 1 Plot the points $A(-1, -3)$, $B(6, 1)$, and $C(2, -5)$. Prove that the triangle with vertices A, B, and C is a right triangle, and find its area.

Solution The points and the triangle are shown in Figure 2.4. From plane geometry, a triangle is a right triangle if and only if the sum of the squares of two of its sides is equal to the square of the remaining side. Using the Distance Formula,

$$d(A, B) = \sqrt{(-1 - 6)^2 + (-3 - 1)^2} = \sqrt{49 + 16} = \sqrt{65}$$
$$d(B, C) = \sqrt{(6 - 2)^2 + (1 + 5)^2} = \sqrt{16 + 36} = \sqrt{52}$$
$$d(A, C) = \sqrt{(-1 - 2)^2 + (-3 + 5)^2} = \sqrt{9 + 4} = \sqrt{13}.$$

Since $[d(A, B)]^2 = [d(B, C)]^2 + [d(A, C)]^2$, the triangle is a right triangle with hypotenuse AB. The area is $\frac{1}{2}\sqrt{52}\sqrt{13} = 13$ square units. ■

It is easy to obtain a formula for the midpoint of a line segment. Let $P_1(x_1, y_1)$ and $P_2(x_2, y_2)$ be two points in a coordinate plane and let M be the midpoint of the segment P_1P_2. The lines through P_1 and P_2 parallel to the y-axis intersect the x-axis at $A_1(x_1, 0)$ and $A_2(x_2, 0)$ and, from plane

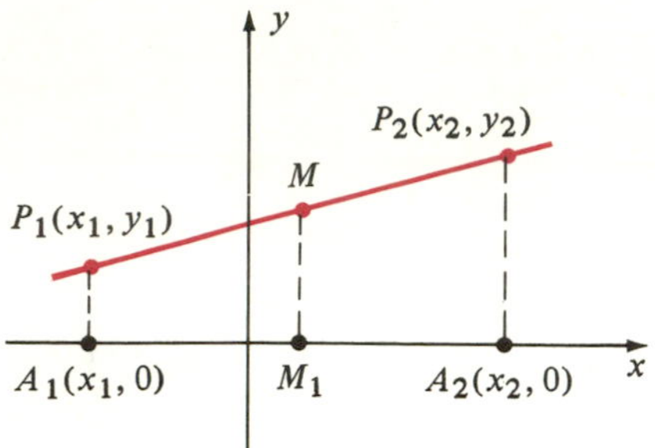

FIGURE 2.5

geometry, the line through M parallel to the y-axis bisects the segment A_1A_2 (see Figure 2.5). If $x_1 < x_2$, then $x_2 - x_1 > 0$, and hence $d(A_1, A_2) = x_2 - x_1$. Since M_1 is halfway from A_1 to A_2, the x-coordinate of M_1 is

$$\begin{aligned} x_1 + \tfrac{1}{2}(x_2 - x_1) &= x_1 + \tfrac{1}{2}x_2 - \tfrac{1}{2}x_1 \\ &= \tfrac{1}{2}x_1 + \tfrac{1}{2}x_2 \\ &= \frac{x_1 + x_2}{2}. \end{aligned}$$

It follows that the x-coordinate of M is also $(x_1 + x_2)/2$. It can be shown in similar fashion that the y-coordinate of M is $(y_1 + y_2)/2$. Moreover, these formulas hold for all positions of P_1 and P_2. This gives us the following result.

MIDPOINT FORMULA

The midpoint of the line segment from $P_1(x_1, y_1)$ to $P_2(x_2, y_2)$ is

$$\left(\frac{x_1 + x_2}{2}, \frac{y_1 + y_2}{2}\right).$$

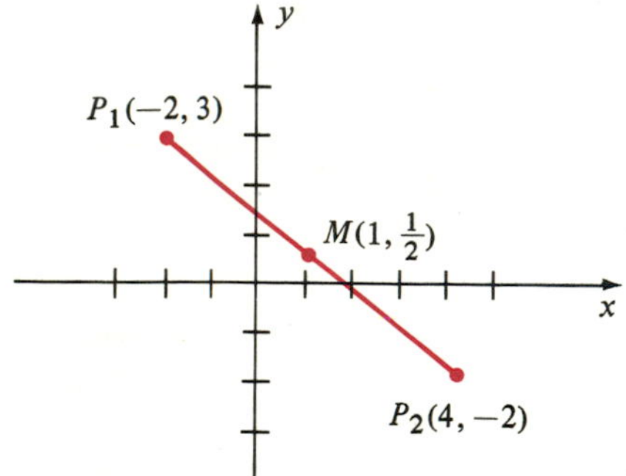

FIGURE 2.6

EXAMPLE 2 Find the midpoint M of the line segment from $P_1(-2, 3)$ to $P_2(4, -2)$. Plot the points P_1, P_2, M and verify that $d(P_1, M) = d(P_2, M)$.

Solution Applying the Midpoint Formula, we find that the coordinates of M are

$$\left(\frac{-2 + 4}{2}, \frac{3 + (-2)}{2}\right) \quad \text{or} \quad \left(1, \frac{1}{2}\right).$$

The three points P_1, P_2, and M are plotted in Figure 2.6. Using the Distance Formula, we obtain

$$\begin{aligned} d(P_1, M) &= \sqrt{(-2 - 1)^2 + (3 - \tfrac{1}{2})^2} = \sqrt{9 + \tfrac{25}{4}} \\ d(P_2, M) &= \sqrt{(4 - 1)^2 + (-2 - \tfrac{1}{2})^2} = \sqrt{9 + \tfrac{25}{4}}. \end{aligned}$$

Hence $d(P_1, M) = d(P_2, M)$. ■

If W is a set of ordered pairs, then we may consider the point $P(x, y)$ in a coordinate plane that corresponds to the ordered pair (x, y) in W. The **graph** of W is the set of all points that correspond to the ordered pairs in W. The phrase "sketch the graph of W" means to illustrate the significant features of the graph geometrically on a coordinate plane.

EXAMPLE 3 Sketch the graph of $W = \{(x, y): |x| \leq 2, |y| \leq 1\}$.

Solution The notation describing W may be translated "the set of all

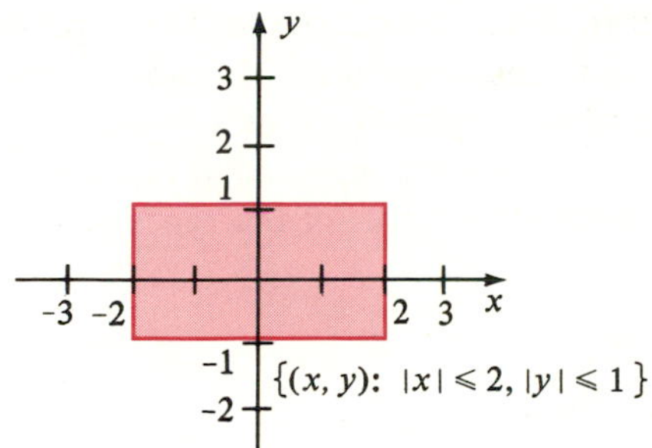

FIGURE 2.7

ordered pairs (x, y) such that $|x| \leq 2$ and $|y| \leq 1$." These inequalities are equivalent to $-2 \leq x \leq 2$ and $-1 \leq y \leq 1$. Hence the graph of W consists of all points within and on the boundary of the rectangular region shown in Figure 2.7. ■

EXAMPLE 4 Sketch the graph of $W = \{(x, y): y = 2x - 1\}$.

Solution We begin by finding points with coordinates of the form (x, y), where the ordered pair (x, y) is in W. It is convenient to list these coordinates in the following tabular form, where for each real number x the corresponding value for y is $2x - 1$.

x	-2	-1	0	1	2	3
y	-5	-3	-1	1	3	5

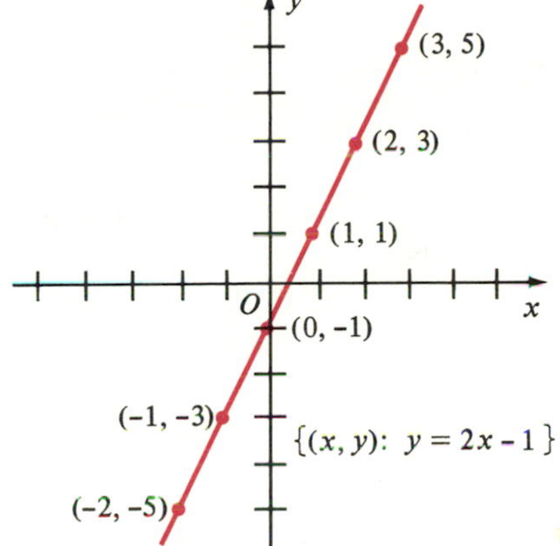

FIGURE 2.8

After plotting the points with these coordinates, it appears that they lie on a line and we sketch the graph (see Figure 2.8). Ordinarily, the few points we have plotted would not be enough to illustrate the graph; however, in this elementary case we can be reasonably sure that the graph is a line. In Section 2.3 we will prove that our conjecture is correct. ■

The x-coordinates of points at which a graph intersects the x-axis are called the **x-intercepts** of the graph. Similarly, the y-coordinates of points at which a graph intercepts the y-axis are called the **y-intercepts.** The graph in Figure 2.8 has one x-intercept, $\frac{1}{2}$, and one y-intercept, -1.

It is impossible to sketch the entire graph in Example 4 since x may be assigned values that are numerically as large as desired. Nevertheless, we often call a drawing of the type given in Figure 2.8 *the graph of W* or *a sketch of the graph*. It is understood that the drawing is only a device for visualizing the actual graph and the line does not terminate as shown in the figure. In general, the sketch of a graph should illustrate enough of the graph so that the remaining parts are evident.

The graph in Example 4 is determined by the equation $y = 2x - 1$ in the sense that for every real number x, the equation can be used to find a number y such that (x, y) is in W. Given an equation in x and y, we say that an ordered pair (a, b) is a **solution** of the equation if equality is obtained when a is substituted for x and b for y. Two equations in x and y are said to be **equivalent** if they have exactly the same solutions. The solutions of an equation in x and y determine a set S of ordered pairs, and we define the **graph of the equation** as the graph of S. Note that the graph of the equation $y = 2x - 1$ is the same as the graph of W (see Figure 2.8).

For some of the equations in this chapter the technique we will use for sketching the graph consists of plotting a sufficient number of points until some pattern emerges, and then sketching the graph accordingly. This is obviously a crude (and often inaccurate) way to arrive at the graph;

however, it is a method often employed in elementary courses. As we progress through this text, techniques will be introduced that will enable us to sketch accurate graphs without plotting many points. To give accurate descriptions of graphs when complicated expressions are involved, it is usually necessary to employ more advanced mathematical tools of the types introduced in the study of calculus.

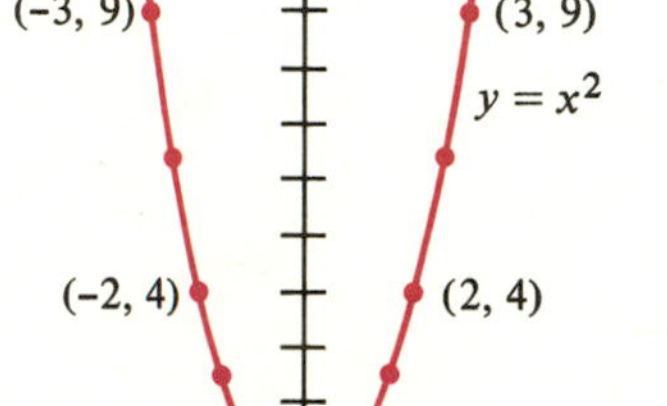

FIGURE 2.9

EXAMPLE 5 Sketch the graph of the equation $y = x^2$.

Solution To obtain the graph, we must plot more points than in the previous example. Increasing successive x-coordinates by $\frac{1}{2}$, we obtain the following table.

x	-3	$-\frac{5}{2}$	-2	$-\frac{3}{2}$	-1	$-\frac{1}{2}$	0	$\frac{1}{2}$	1	$\frac{3}{2}$	2	$\frac{5}{2}$	3
y	9	$\frac{25}{4}$	4	$\frac{9}{4}$	1	$\frac{1}{4}$	0	$\frac{1}{4}$	1	$\frac{9}{4}$	4	$\frac{25}{4}$	9

Larger numerical values of x produce even larger values of y. For example, the points (4, 16), (5, 25), and (6, 36) are on the graph, as are $(-4, 16)$, $(-5, 25)$, and $(-6, 36)$. Plotting the points given by the table and drawing a smooth curve through these points gives us the sketch in Figure 2.9, where several points are labeled. ■

The graph in Example 5 is called a **parabola.** The lowest point (0, 0) is called the **vertex** of the parabola and we say that the parabola **opens upward.** If the graph were inverted, as would be the case for $y = -x^2$, then the parabola **opens downward.** The y-axis is called the **axis of the parabola.** In general, the graph of *any* equation of the form $y = ax^2$, where $a \neq 0$, is a parabola with vertex (0, 0). Parabolas may also open to the right or to the left (see Example 6). The definition of parabola and a detailed discussion of its properties may be found in Chapter 9.

If the coordinate plane in Figure 2.9 is folded along the y-axis, then the graph that lies in the left half of the plane coincides with that in the right half. We say that **the graph is symmetric with respect to the y-axis.** As in (i) of Figure 2.10, a graph is symmetric with respect to the y-axis provided that

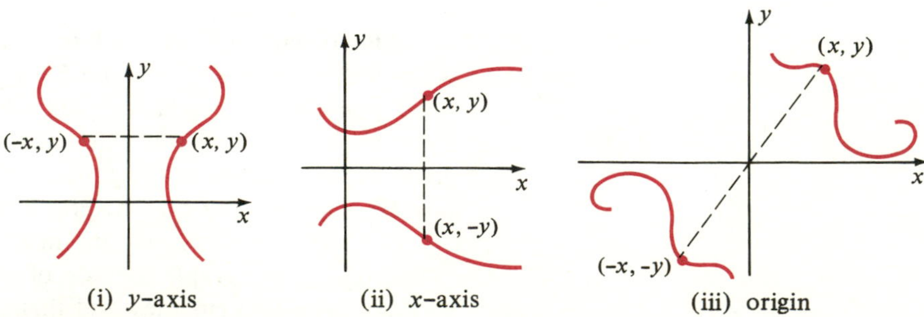

FIGURE 2.10 Symmetries

the point $(-x, y)$ is on the graph whenever (x, y) is on the graph. Similarly, as in (ii) of Figure 2.10, **a graph is symmetric with respect to the x-axis** if, whenever a point (x, y) is on the graph, then $(x, -y)$ is also on the graph. In this case, if we fold the coordinate plane along the x-axis, the part of the graph that lies above the x-axis will coincide with the part that lies below. Certain graphs possess another type of symmetry, called **symmetry with respect to the origin.** In this situation, whenever a point (x, y) is on the graph, then $(-x, -y)$ is also on the graph, as illustrated in (iii) of Figure 2.10.

The following tests are useful for investigating these three types of symmetry for graphs of equations in x and y.

TESTS FOR SYMMETRY

(i) The graph of an equation is symmetric with respect to the y-axis if substitution of $-x$ for x leads to an equivalent equation.

(ii) The graph of an equation is symmetric with respect to the x-axis if substitution of $-y$ for y leads to an equivalent equation.

(iii) The graph of an equation is symmetric with respect to the origin if the simultaneous substitution of $-x$ for x and $-y$ for y leads to an equivalent equation.

If, in the equation of Example 5, we substitute $-x$ for x, we obtain $y = (-x)^2$, which is equivalent to $y = x^2$. Hence, by Test (i), the graph is symmetric with respect to the y-axis.

If symmetry with respect to an axis exists, then it is sufficient to determine the graph in half of the coordinate plane, since the remainder of the graph is a mirror image, or reflection, of that half.

EXAMPLE 6 Sketch the graph of the equation $y^2 = x$.

Solution Since substitution of $-y$ for y does not change the equation, the graph is symmetric with respect to the x-axis. (See Symmetry Test (ii).) It is sufficient, therefore, to plot points with nonnegative y-coordinates and then reflect through the x-axis. Since $y^2 = x$, the y-coordinates of points above the x-axis are given by $y = \sqrt{x}$. Coordinates of some points on the graph are listed in the following table.

x	0	1	2	3	4	9
y	0	1	$\sqrt{2}$	$\sqrt{3}$	2	3

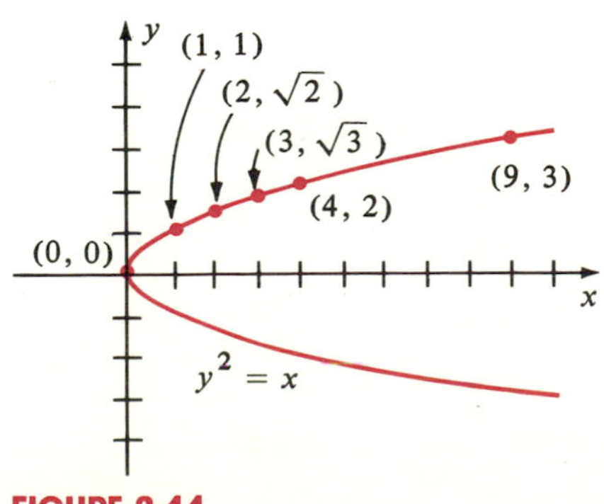

FIGURE 2.11

A portion of the graph is sketched in Figure 2.11. The graph is a parabola that opens to the right, with its vertex at the origin. In this case the x-axis is the axis of the parabola. ■

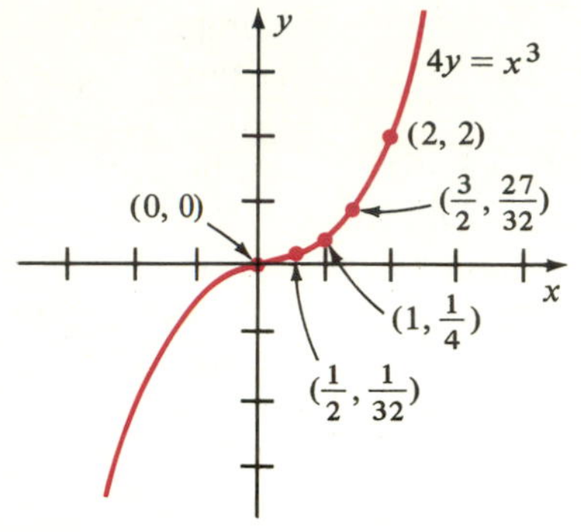

FIGURE 2.12

EXAMPLE 7 Sketch the graph of the equation $4y = x^3$.

Solution If we substitute $-x$ for x and $-y$ for y, then

$$4(-y) = (-x)^3 \quad \text{or} \quad -4y = -x^3.$$

Multiplying both sides by -1, we see that the last equation has the same solutions as the given equation $4y = x^3$. Hence, from Symmetry Test (iii), the graph is symmetric with respect to the origin. The following table lists some points on the graph.

x	0	$\frac{1}{2}$	1	$\frac{3}{2}$	2	$\frac{5}{2}$
y	0	$\frac{1}{32}$	$\frac{1}{4}$	$\frac{27}{32}$	2	$\frac{125}{32}$

By symmetry (or substitution) we see that the points $(-1, -\frac{1}{4})$, $(-2, -2)$, etc., are on the graph. Plotting points leads to the graph in Figure 2.12. ■

If $C(h, k)$ is a point in a coordinate plane, then a circle with center C and radius $r > 0$ may be defined as the collection of all points in the plane that are r units from C. As shown in (i) of Figure 2.13, a point $P(x, y)$ is on the circle if and only if $d(C, P) = r$ or, by the Distance Formula, if and only if

$$\sqrt{(x-h)^2 + (y-k)^2} = r.$$

The following equivalent equation is called the **equation of a circle of radius *r* and center (*h*, *k*).**

EQUATION OF A CIRCLE

$$(x-h)^2 + (y-k)^2 = r^2, \qquad r > 0$$

If $h = 0$ and $k = 0$, this equation reduces to $x^2 + y^2 = r^2$, which is an equation of a circle of radius r with center at the origin (see (ii) of Figure 2.13). If $r = 1$, the graph is called a **unit circle.**

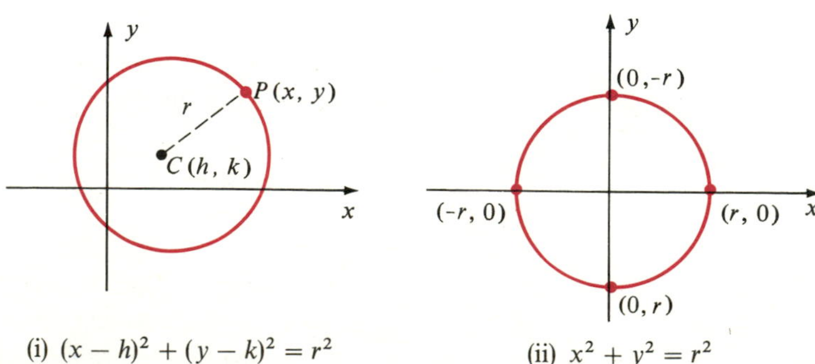

FIGURE 2.13

EXAMPLE 8 Find an equation of the circle that has center $C(-2, 3)$ and contains the point $D(4, 5)$.

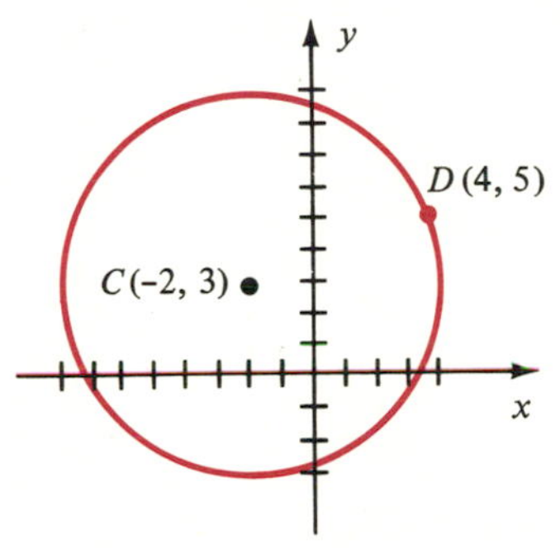

FIGURE 2.14

Solution The circle is illustrated in Figure 2.14. Since D is on the circle, the radius r is $d(C, D)$. By the Distance Formula,

$$r = \sqrt{(-2-4)^2 + (3-5)^2} = \sqrt{36+4} = \sqrt{40}.$$

Using the equation of a circle with $h = -2$, $k = 3$, and $r = \sqrt{40}$, we obtain

$$(x+2)^2 + (y-3)^2 = 40,$$

or

$$x^2 + y^2 + 4x - 6y - 27 = 0.$$

■

Squaring terms of $(x-h)^2 + (y-k)^2 = r^2$ and simplifying leads to an equation of the form

$$x^2 + y^2 + ax + by + c = 0$$

where a, b, and c are real numbers. Conversely, if we begin with the last equation it is always possible, by *completing the squares* in x and y, to obtain an equation of the form

$$(x-h)^2 + (y-k)^2 = d.$$

The method will be illustrated in Example 9. If $d > 0$, the graph is a circle with center (h, k) and radius $r = \sqrt{d}$. If $d = 0$, then, since $(x-h)^2 \geq 0$ and $(y-k)^2 \geq 0$, the only solution of the equation is (h, k), and hence the graph consists of only one point. Finally, if $d < 0$, the equation has no real solutions and there is no graph.

EXAMPLE 9 Find the center and radius of the circle with equation

$$x^2 + y^2 - 4x + 6y - 3 = 0.$$

Solution We begin by arranging the equation as follows:

$$(x^2 - 4x) + (y^2 + 6y) = 3.$$

Next we complete the squares by adding appropriate numbers within the parentheses. Of course, to obtain equivalent equations, we must add the numbers to *both* sides of the equation. To complete the square for an expression of the form $x^2 + ax$, we add the square of half the coefficient of x, that is, $(a/2)^2$, to both sides of the equation. Similarly, for $y^2 + by$, we add $(b/2)^2$ to both sides. In this example $a = -4$, $b = 6$, $(a/2)^2 = (-2)^2 = 4$, and $(b/2)^2 = 3^2 = 9$. This leads to

$$(x^2 - 4x + 4) + (y^2 + 6y + 9) = 3 + 4 + 9$$

or $$(x-2)^2+(y+3)^2=16.$$

It follows that the center is $(2, -3)$ and the radius is 4. ■

EXERCISES 2.1

In Exercises 1–6 find (a) the distance $d(A, B)$ between the points A and B, and (b) the midpoint of the segment AB.

1 $A(6, -2)$, $B(2, 1)$ **2** $A(-4, -1)$, $B(2, 3)$

3 $A(0, -7)$, $B(-1, -2)$ **4** $A(4, 5)$, $B(4, -4)$

5 $A(-3, -2)$, $B(-8, -2)$ **6** $A(11, -7)$, $B(-9, 0)$

In Exercises 7 and 8 prove that the triangle with vertices A, B, and C is a right triangle, and find its area.

7 $A(-3, 4)$, $B(2, -1)$, $C(9, 6)$

8 $A(7, 2)$, $B(-4, 0)$, $C(4, 6)$

9 Prove that the following points are vertices of a parallelogram: $A(-4, -1)$, $B(0, -2)$, $C(6, 1)$, $D(2, 2)$.

10 Given $A(-4, -3)$ and $B(6, 1)$, find a formula that expresses the fact that $P(x, y)$ is on the perpendicular bisector of AB.

11 For what values of a is the distance between $(a, 3)$ and $(5, 2a)$ greater than $\sqrt{26}$?

12 Given the points $A(-2, 0)$ and $B(2, 0)$, find a formula that contains no radicals and expresses the fact that the sum of the distances from $P(x, y)$ to A and to B, respectively, is 5.

13 Prove that the midpoint of the hypotenuse of any right triangle is equidistant from the vertices. (*Hint:* Label the vertices of the triangle $O(0, 0)$, $A(a, 0)$, and $B(0, b)$.)

14 Prove that the diagonals of any parallelogram bisect each other. (*Hint:* Label three of the vertices of the parallelogram $O(0, 0)$, $A(a, b)$, and $C(c, 0)$.)

In Exercises 15–20 sketch the graph of the set W.

15 $W = \{(x, y): x = 4\}$

16 $W = \{(x, y): y = -3\}$

17 $W = \{(x, y): xy < 0\}$

18 $W = \{(x, y): xy = 0\}$

19 $W = \{(x, y): |x| < 2, |y| > 1\}$

20 $W = \{(x, y): |x| > 1, |y| \le 2\}$

In Exercises 21–42 sketch the graph of the equation, and test for symmetry.

21 $y = 3x + 1$ **22** $y = 4x - 3$

23 $y = -2x + 3$ **24** $y = 2 - 3x$

25 $y = 2x^2 - 1$ **26** $y = -x^2 + 2$

27 $4y = x^2$ **28** $3y + x^2 = 0$

29 $y = -\frac{1}{2}x^3$ **30** $y = \frac{1}{2}x^3$

31 $y = x^3 - 2$ **32** $y = 2 - x^3$

33 $y = \sqrt{x}$ **34** $y = \sqrt{x} - 1$

35 $y = \sqrt{-x}$ **36** $y = \sqrt{x-1}$

37 $x^2 + y^2 = 16$ **38** $4x^2 + 4y^2 = 25$

39 $y = \sqrt{9 - x^2}$ **40** $y = -\sqrt{4 - x^2}$

41 $x = -\sqrt{9 - y^2}$ **42** $x = \sqrt{4 - y^2}$

In Exercises 43–50 find an equation of a circle that satisfies the stated conditions.

43 Center $C(3, -2)$, radius 4

44 Center $C(-5, 2)$, radius 5

45 Center at the origin, passing through $P(-3, 5)$

46 Center $C(-4, 6)$, passing through $P(1, 2)$

47 Center $C(-4, 2)$, tangent to the x-axis

48 Center $C(3, -5)$, tangent to the y-axis

49 Endpoints of a diameter $A(4, -3)$ and $B(-2, 7)$

50 Tangent to both axes, center in the first quadrant, radius 2

In Exercises 51–56 find the center and radius of the circle with the given equation.

51 $x^2 + y^2 + 4x - 6y + 4 = 0$

52 $x^2 + y^2 - 10x + 2y + 22 = 0$

53 $x^2 + y^2 + 6x = 0$

54 $x^2 + y^2 + x + y - 1 = 0$

55 $2x^2 + 2y^2 - x + y - 3 = 0$

56 $9x^2 + 9y^2 - 6x + 12y - 31 = 0$

CALCULATOR EXERCISES 2.1

Sketch the graphs of each equation, after approximating the coordinates of a sufficient number of points.

1 $y = x^{2/3}$

2 $y = x^{3/2}$

3 $y = x^{5/2}$

4 $y = 2x^{3/5} - 1$

5 $y = \sqrt[5]{x} - 1$

6 $y = \sqrt[4]{x + 1}$

7 $x^{2/3} + y^{2/3} = 1$

8 $y = x/(x^2 + 1)$

SECTION 2.2
LINES

The following concept is fundamental to the study of lines. All lines referred to are considered to be in some fixed coordinate plane.

DEFINITION Let l be a line that is not parallel to the y-axis, and let $P_1(x_1, y_1)$ and $P_2(x_2, y_2)$ be distinct points on l. The **slope m** of l is

$$m = \frac{y_2 - y_1}{x_2 - x_1}.$$

If l is parallel to the y-axis, then the slope is not defined.

Typical points P_1 and P_2 on a line l are shown in Figure 2.15. The numerator $y_2 - y_1$ in the formula for m measures the vertical change in direction

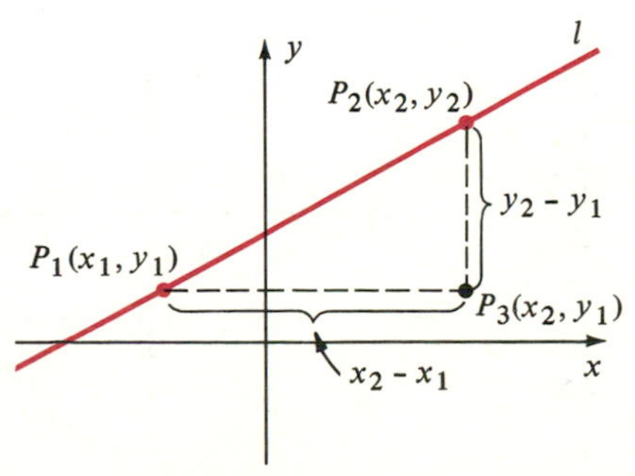

(i) Positive slope

(ii) Negative slope

FIGURE 2.15

in proceeding from P_1 to P_2 and may be positive, negative, or zero. The denominator $x_2 - x_1$ measures the amount of horizontal change in going from P_1 to P_2, and it may be positive or negative, but never zero, because l is not parallel to the y-axis.

In finding the slope of a line, it is immaterial which point is labeled P_1 and which is labeled P_2 since

$$\frac{y_2 - y_1}{x_2 - x_1} = \frac{y_1 - y_2}{x_1 - x_2}.$$

Consequently, we may assume that the points are labeled so that $x_1 < x_2$, as in Figure 2.15. In this event $x_2 - x_1 > 0$, and hence the slope is positive, negative, or zero, depending on whether $y_2 > y_1$, $y_2 < y_1$, or $y_2 = y_1$. The slope of the line shown in (i) of Figure 2.15 is positive, whereas the slope of the line shown in (ii) of the figure is negative.

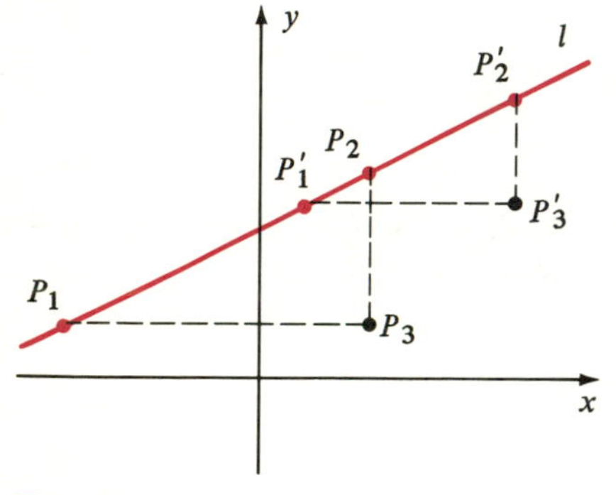

FIGURE 2.16

A **horizontal line** is a line that is parallel to the x-axis. Note that a line is horizontal if and only if its slope is 0. A **vertical line** is a line that is parallel to the y-axis. The slope of a vertical line is undefined.

It is important to note that the definition of slope is independent of the two points that are chosen on l, for if other points $P'_1(x'_1, y'_1)$ and $P'_2(x'_2, y'_2)$ are used, then as in Figure 2.16, the triangle with vertices P'_1, P'_2, and $P'_3(x'_2, y'_1)$ is similar to the triangle with vertices P_1, P_2, and $P_3(x_2, y_1)$. Since the ratios of corresponding sides are equal, it follows that

$$\frac{y_2 - y_1}{x_2 - x_1} = \frac{y'_2 - y'_1}{x'_2 - x'_1}.$$

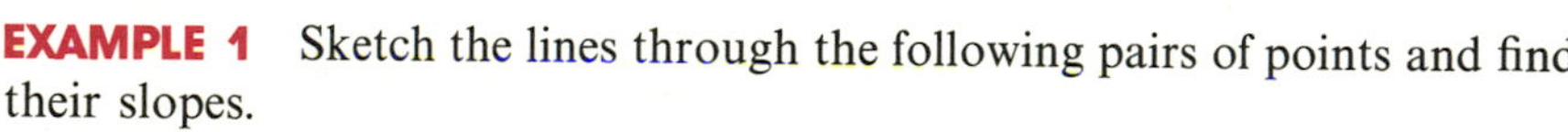

EXAMPLE 1 Sketch the lines through the following pairs of points and find their slopes.

(a) $A(-1, 4)$ and $B(3, 2)$ (b) $A(2, 5)$ and $B(-2, -1)$

(c) $A(4, 3)$ and $B(-2, 3)$ (d) $A(4, -1)$ and $B(4, 4)$.

Solution The lines are sketched in Figure 2.17.

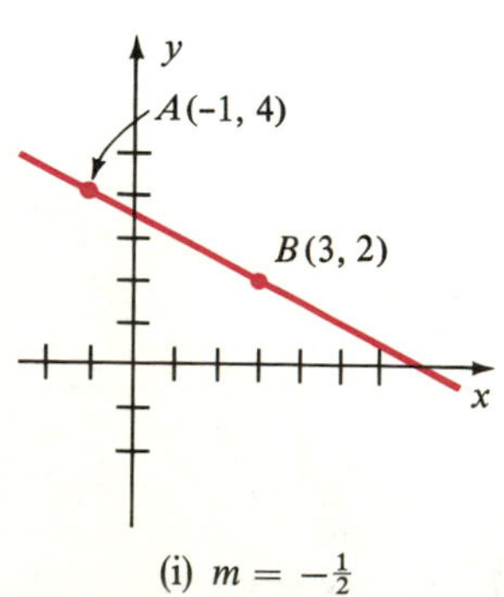

(i) $m = -\frac{1}{2}$

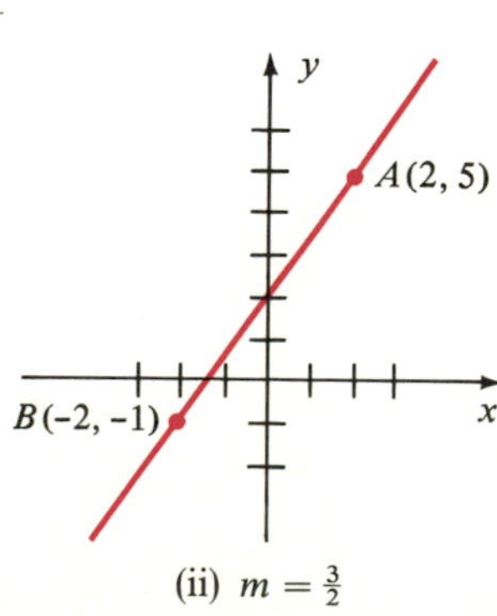

(ii) $m = \frac{3}{2}$

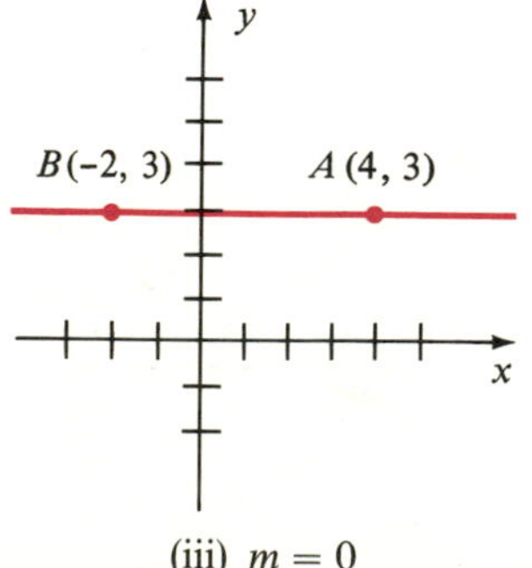

(iii) $m = 0$

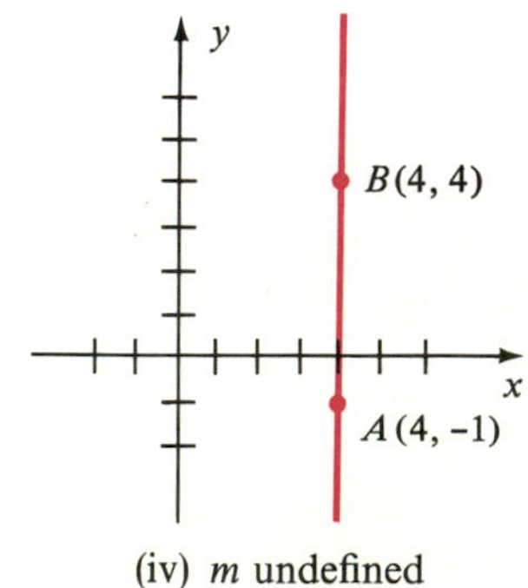

(iv) m undefined

FIGURE 2.17

Using the definition of slope gives us

(a) $$m = \frac{2-4}{3-(-1)} = \frac{-2}{4} = -\frac{1}{2}$$

(b) $$m = \frac{5-(-1)}{2-(-2)} = \frac{6}{4} = \frac{3}{2}$$

(c) $$m = \frac{3-3}{-2-4} = \frac{0}{-6} = 0$$

(d) The slope is undefined since the line is vertical. This is also seen by noting that, if the formula for m is used, the denominator is zero. ■

EXAMPLE 2 Construct a line through $P(2, 1)$ that has slope (a) $\frac{5}{3}$; (b) $-\frac{5}{3}$.

Solution If the slope of a line is a/b and b is positive, then for every change of b units in the horizontal direction, the line rises or falls $|a|$ units, depending on whether a is positive or negative, respectively. If $P(2, 1)$ is on the line and $m = \frac{5}{3}$, we can obtain another point on the line by starting at P and moving 3 units to the right and 5 units upward. This gives us the point $Q(5, 6)$, and the line is determined (see (i) of Figure 2.18). Similarly, if $m = -\frac{5}{3}$, we move 3 units to the right and 5 units downward, obtaining $Q(5, -4)$ as in (ii) of Figure 2.18.

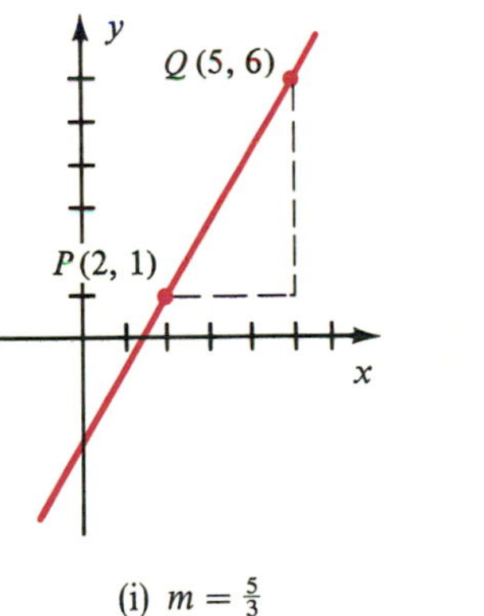

(i) $m = \frac{5}{3}$

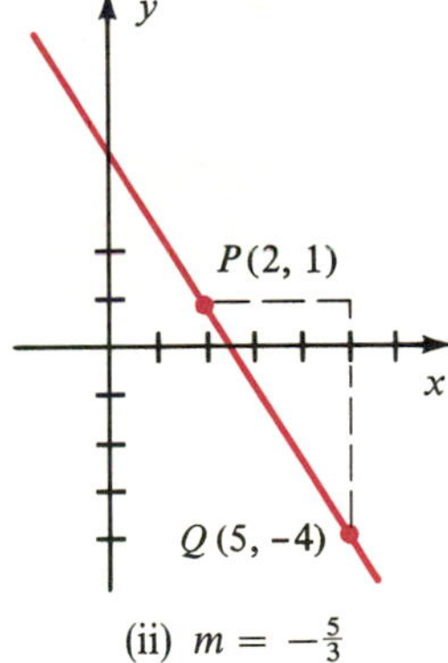

(ii) $m = -\frac{5}{3}$

FIGURE 2.18 ■

THEOREM

(i) The graph of the equation $x = a$ is a vertical line that has x-intercept a.

(ii) The graph of the equation $y = b$ is a horizontal line that has y-intercept b.

Proof The equation $x = a$, where a is a real number, may be considered as an equation in two variables x and y, since we can write it in the

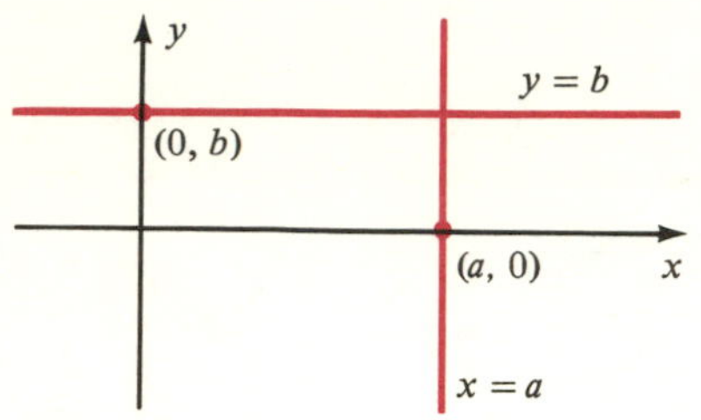

FIGURE 2.19

form

$$x + (0)y = a.$$

Some typical solutions of the equation are $(a, -2)$, $(a, 1)$, and $(a, 3)$. Evidently, all solutions of the equation consist of pairs of the form (a, y), where y may have any value and a is fixed. It follows that the graph of $x = a$ is a line parallel to the y-axis with x-intercept a, as illustrated in Figure 2.19. This proves (i). Part (ii) is proved in similar fashion. □

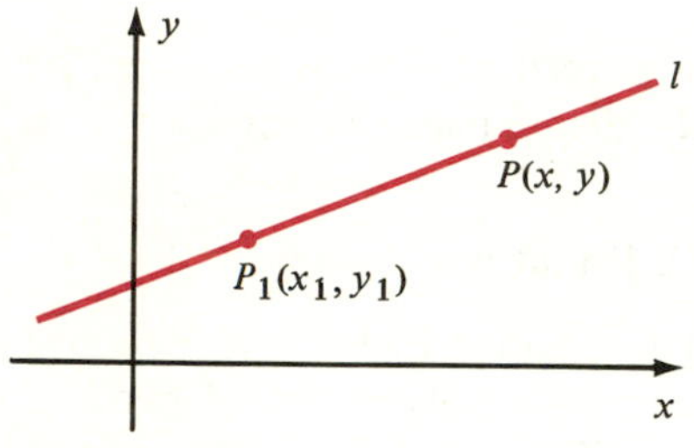

FIGURE 2.20

Let us next find an equation of a line l through a point $P_1(x_1, y_1)$ with slope m (only one such line exists). If $P(x, y)$ is any point with $x \neq x_1$ (see Figure 2.20), then P is on l if and only if the slope of the line through P_1 and P is m; that is,

$$\frac{y - y_1}{x - x_1} = m.$$

This equation may be written in the form

$$y - y_1 = m(x - x_1).$$

Note that (x_1, y_1) is also a solution of the last equation and hence the points on l are precisely the points that correspond to the solutions. This equation for l is referred to as the **Point-Slope Form.** Our discussion may be summarized as follows.

POINT-SLOPE FORM FOR THE EQUATION OF A LINE

An equation for the line through the point $P(x_1, y_1)$ with slope m is

$$y - y_1 = m(x - x_1).$$

EXAMPLE 3 Find an equation of the line through the points $A(1, 7)$ and $B(-3, 2)$.

Solution The slope m of the line is

$$m = \frac{7 - 2}{1 - (-3)} = \frac{5}{4}.$$

We may use the coordinates of either A or B for (x_1, y_1) in the Point-Slope Form. Using $A(1, 7)$ gives us

$$y - 7 = \tfrac{5}{4}(x - 1),$$

which is equivalent to

$$4y - 28 = 5x - 5 \quad \text{or} \quad 5x - 4y + 23 = 0. \qquad ■$$

The Point-Slope Form may be rewritten as $y = mx - mx_1 + y_1$, which is of the form

$$y = mx + b$$

with $b = -mx_1 + y_1$. The real number b is the y-intercept of the graph, as we may see by setting $x = 0$. Since the equation $y = mx + b$ displays the slope m and y-intercept b of l, it is called the **Slope-Intercept Form** for the equation of a line. Conversely, if we start with $y = mx + b$, we may write

$$y - b = m(x - 0).$$

Comparing this last equation with the Point-Slope Form, we see that the graph is a line that has slope m and passes through the point $(0, b)$. This gives us the next result.

SLOPE-INTERCEPT FORM FOR THE EQUATION OF A LINE

The graph of the equation $y = mx + b$ is a line having slope m and y-intercept b.

We have shown that every line is the graph of an equation of the form

$$ax + by + c = 0$$

where a, b, and c are real numbers, and a and b are not both zero. We call such an equation a **linear equation** in x and y. Let us show, conversely, that the graph of $ax + by + c = 0$, where a and b are not both zero, is always a line. On the one hand, if $b \neq 0$, we may solve for y, obtaining

$$y = \left(-\frac{a}{b}\right)x + \left(-\frac{c}{b}\right)$$

which, by the Slope-Intercept Form, is an equation of a line with slope $-a/b$ and y-intercept $-c/b$. On the other hand, if $b = 0$ but $a \neq 0$, then we may solve for x, obtaining $x = -c/a$, which is the equation of a vertical line with x-intercept $-c/a$. This establishes the following important theorem.

THEOREM

The graph of a linear equation $ax + by + c = 0$ is a line and, conversely, every line is the graph of a linear equation.

For simplicity, we shall use the terminology *the line* $ax + by + c = 0$ instead of the more accurate phrase *the line with equation* $ax + by + c = 0$.

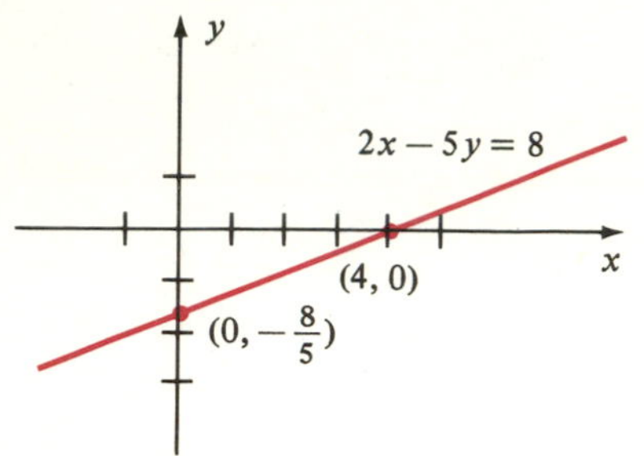

FIGURE 2.21

EXAMPLE 4 Sketch the graph of $2x - 5y = 8$.

Solution From the preceding theorem we know that the graph is a line, and hence it is sufficient to find two points on the graph. Let us find the x- and y-intercepts. Substituting $y = 0$ in the given equation, we obtain the x-intercept 4. Substituting $x = 0$, we see that y-intercept is $-\frac{8}{5}$. This leads to the graph in Figure 2.21.

Another method of solution is to express the given equation in Slope-Intercept Form. To do this we begin by isolating the term involving y on one side of the equal sign, obtaining

$$5y = 2x - 8.$$

Next, dividing both sides by 5 gives us

$$y = \frac{2}{5}x + \left(\frac{-8}{5}\right),$$

which is in the form $y = mx + b$. Hence, the slope is $m = \frac{2}{5}$, and the y-intercept is $b = -\frac{8}{5}$. We may then sketch a line through the point $(0, -\frac{8}{5})$ with slope $\frac{2}{5}$. ■

The following theorem can be proved.

THEOREM Two nonvertical lines are parallel if and only if they have the same slope.

We shall use this fact in the next example.

EXAMPLE 5 Find an equation of a line that passes through the point $(5, -7)$ and is parallel to the line $6x + 3y - 4 = 0$.

Solution Let us express the given equation in Slope-Intercept Form. We begin by writing

$$3y = -6x + 4$$

and then divide both sides by 3, obtaining

$$y = -2x + \tfrac{4}{3}.$$

The last equation is in Slope-Intercept Form with $m = -2$, and hence the slope is -2. Since parallel lines have the same slope, the required line also has slope -2. Applying the Point-Slope Form gives us

$$y + 7 = -2(x - 5).$$

This is equivalent to

$$y + 7 = -2x + 10 \quad \text{or} \quad 2x + y - 3 = 0.$$

The next result specifies conditions for perpendicular lines.

THEOREM

Two lines with slopes m_1 and m_2 are perpendicular if and only if $m_1 m_2 = -1$.

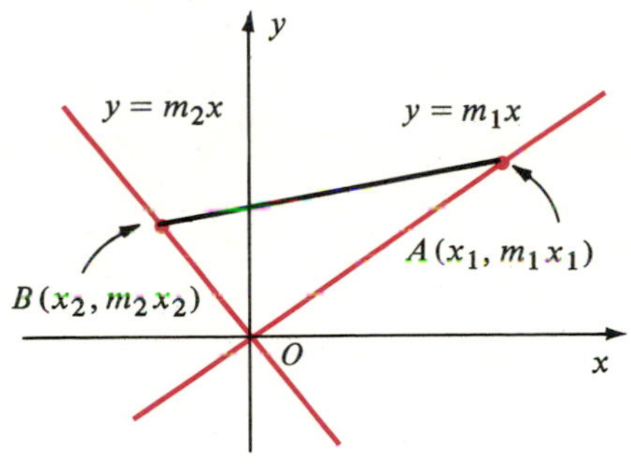

FIGURE 2.22

Proof For simplicity, let us consider the special case of two lines that intersect at the origin O, as illustrated in Figure 2.22. In this case equations of the lines are $y = m_1x$ and $y = m_2x$. If, as in the figure, we choose points $A(x_1, m_1x_1)$ and $B(x_2, m_2x_2)$ different from O on the lines, then the lines are perpendicular if and only if angle AOB is a right angle. By the Pythagorean Theorem, angle AOB is a right angle if and only if

$$[d(A, B)]^2 = [d(O, B)]^2 + [d(O, A)]^2$$

or, by the Distance Formula,

$$(m_2x_2 - m_1x_1)^2 + (x_2 - x_1)^2 = (m_2x_2)^2 + x_2^2 + (m_1x_1)^2 + x_1^2.$$

Squaring the indicated terms and simplifying gives us

$$-2m_1m_2x_1x_2 - 2x_1x_2 = 0.$$

Dividing both sides by $-2x_1x_2$, we see that the lines are perpendicular if and only if $m_1m_2 + 1 = 0$, or $m_1m_2 = -1$.

The same type of proof may be given if the lines intersect at *any* point (a, b). □

A convenient way to remember the conditions for perpendicularity is to note that m_1 and m_2 must be *negative reciprocals* of one another, that is, $m_1 = -1/m_2$ and $m_2 = -1/m_1$.

EXAMPLE 6 Find an equation of a line that passes through the point $(5, -7)$ and is perpendicular to the line $6x + 3y - 4 = 0$.

Solution The line $6x + 3y - 4 = 0$ was considered in Example 5, and we found that its slope is -2. Hence the slope of the required line is the negative reciprocal $-[1/(-2)]$, or $\frac{1}{2}$. Applying the Point-Slope Form gives us

$$y + 7 = \tfrac{1}{2}(x - 5),$$

which is equivalent to

$$2y + 14 = x - 5 \quad \text{or} \quad x - 2y - 19 = 0.$$ ■

EXAMPLE 7 Find an equation for the perpendicular bisector of the line segment from $A(1, 7)$ to $B(-3, 2)$.

Solution By the Midpoint Formula, the midpoint M of the segment AB is $(-1, \frac{9}{2})$. Since the slope of AB is $\frac{5}{4}$ (see Example 3), it follows from the preceding theorem that the slope of the perpendicular bisector is $-\frac{4}{5}$. Applying the Point-Slope Form,

$$y - \frac{9}{2} = -\frac{4}{5}(x + 1).$$

Multiplying both sides by 10 and simplifying leads to $8x + 10y - 37 = 0$. ■

EXERCISES 2.2

In Exercises 1–4 plot the points A and B and find the slope of the line through A and B.

1 $A(-4, 6)$, $B(-1, 18)$ **2** $A(6, -2)$, $B(-3, 5)$

3 $A(-1, -3)$, $B(-1, 2)$ **4** $A(-3, 4)$, $B(2, 4)$

5 Show that $A(-3, 1)$, $B(5, 3)$, $C(3, 0)$, and $D(-5, -2)$ are vertices of a parallelogram.

6 Show that $A(2, 3)$, $B(5, -1)$, $C(0, -6)$, and $D(-6, 2)$ are vertices of a trapezoid.

7 Prove that the points $A(6, 15)$, $B(11, 12)$, $C(-1, -8)$, and $D(-6, -5)$ are vertices of a rectangle.

8 Prove that the points $A(1, 4)$, $B(6, -4)$, and $C(-15, -6)$ are vertices of a right triangle.

9 If three consecutive vertices of a parallelogram are $A(-1, -3)$, $B(4, 2)$, and $C(-7, 5)$, find the fourth vertex.

10 Let $A(x_1, y_1)$, $B(x_2, y_2)$, $C(x_3, y_3)$, and $D(x_4, y_4)$ denote the vertices of an arbitrary quadrilateral. Prove that the line segments joining midpoints of adjacent sides form a parallelogram.

In Exercises 11–20 find an equation for the line satisfying the given conditions.

11 Through $A(2, -6)$, slope $\frac{1}{2}$

12 Slope -3, y-intercept 5

13 Through $A(-5, -7)$, $B(3, -4)$

14 x-intercept -4, y-intercept 8

15 Through $A(8, -2)$, y-intercept -3

16 Slope 6, x-intercept -2

17 Through $A(10, -6)$, parallel to (a) the y-axis; (b) the x-axis.

18 Through $A(-5, 1)$, perpendicular to (a) the y-axis; (b) the x-axis.

19 Through $A(7, -3)$, perpendicular to the line with equation $2x - 5y = 8$.

20 Through $(-\frac{3}{4}, -\frac{1}{2})$, parallel to the line with equation $x + 3y = 1$.

21 Given $A(3, -1)$ and $B(-2, 6)$, find an equation for the perpendicular bisector of the line segment AB.

22 Find an equation for the line that bisects the second and fourth quadrants.

23 Find equations for the altitudes of the triangle with vertices $A(-3, 2)$, $B(5, 4)$, $C(3, -8)$, and find the point at which the altitudes intersect.

24 Find equations for the medians of the triangle in Exercise 23, and find their point of intersection.

In Exercises 25–34 use the Slope-Intercept Form to find the slope and y-intercept of the line with the given equation, and sketch the graph of each equation.

25 $3x - 4y + 8 = 0$ **26** $2y - 5x = 1$

27 $x + 2y = 0$ **28** $8x = 1 - 4y$

29 $y = 4$ **30** $x + 2 = \frac{1}{2}y$

31 $5x + 4y = 20$ **32** $y = 0$

33 $x = 3y + 7$ **34** $x - y = 0$

35 Find a real number k such that the point $P(-1, 2)$ is on the line $kx + 2y - 7 = 0$.

36 Find a real number k such that the line $5x + ky - 3 = 0$ has y-intercept -5.

37 If a line l has nonzero x- and y-intercepts a and b, respectively, prove that an equation for l is $(x/a) + (y/b) = 1$. (This is called the **intercept form** for the equation of a line.) Express the equation $4x - 2y = 6$ in intercept form.

38 Prove that an equation of the line through $P_1(x_1, y_1)$ and $P_2(x_2, y_2)$ is

$$(y - y_1)(x_2 - x_1) = (y_2 - y_1)(x - x_1).$$

(This is called the **two-point form** for the equation of a line.) Use the two-point form to find an equation of the line through $A(7, -1)$ and $B(4, 6)$.

39 Find all values of r such that the slope of the line through the points $(r, 4)$ and $(1, 3 - 2r)$ is less than 5.

40 Find all values of t such that the slope of the line through $(t, 3t + 1)$ and $(1 - 2t, t)$ is greater than 4.

41 Six years ago a house was purchased for \$59,000. This year it is appraised at \$95,000. Assuming that the value increased by the same amount each year, find an equation that specifies the value at any time after the purchase date. When was the house worth \$73,000?

42 Charles' Law for gases states that if the pressure remains constant, then the relationship between the volume V that a gas occupies and its temperature T in degrees Celsius is given by $V = V_0(1 + \frac{1}{273}T)$.

(a) What is the significance of V_0?

(b) What increase in temperature is needed to increase the volume from V_0 to $2V_0$?

(c) Sketch the graph of the equation on a TV-plane for the case $V_0 = 100$ and $T \geq 273$.

SECTION 2.3
DEFINITION OF FUNCTION

The notion of **correspondence** is encountered frequently in everyday life. For example, to each book in a library there corresponds the number of pages in the book. As another example, to each human being there corresponds a birth date. To cite a third example, if the temperature of the air is recorded throughout a day, then at each instant of time there is a corresponding temperature. These examples of correspondence involve two sets X and Y. In our first example, X denotes the set of books in a library and Y the set of positive integers. For each book x in X there corresponds a positive integer y, namely the number of pages in the book.

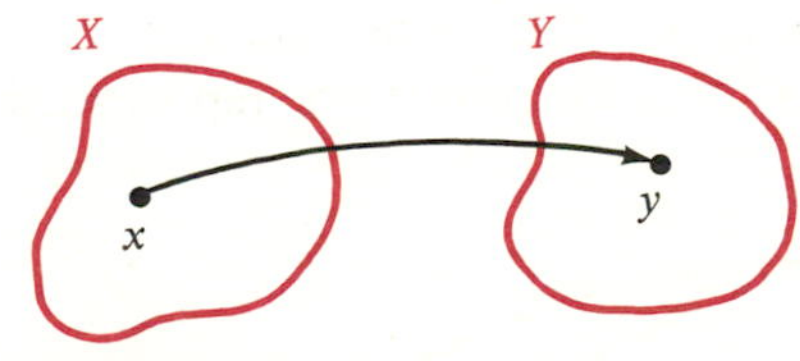

FIGURE 2.23

We sometimes represent correspondences by diagrams of the type shown in Figure 2.23, where the sets X and Y are represented by points within regions in a plane. The curved arrow indicates that the element y of Y corresponds to the element x of X. We have pictured X and Y as different sets. However, X and Y may have elements in common. As a matter of fact, we often have $X = Y$.

Our examples indicate that *to each x in X there corresponds one and only one y in Y;* that is, *y is unique* for a given x. However, the same element of Y may correspond to different elements of X. For example, two different books may have the same number of pages, two different people may have the same birthday, and so on.

In most of our work X and Y will be sets of numbers. To illustrate, let X and Y both denote the set $\mathbb{R}$ of real numbers, and to each real number x let us assign its square x^2. Thus, to 3 we assign 9, to -5 we assign 25, and to $\sqrt{2}$ the number 2. This gives us a correspondence from $\mathbb{R}$ to $\mathbb{R}$.

Each of the preceding examples of a correspondence is a *function*.

DEFINITION

> A **function** f from a set X to a set Y is a correspondence that assigns to each element x of X a unique element y of Y. The element y is called the **image** of x under f and is denoted by $f(x)$. The set X is called the **domain** of the function. The **range** of the function consists of all images of elements of X.

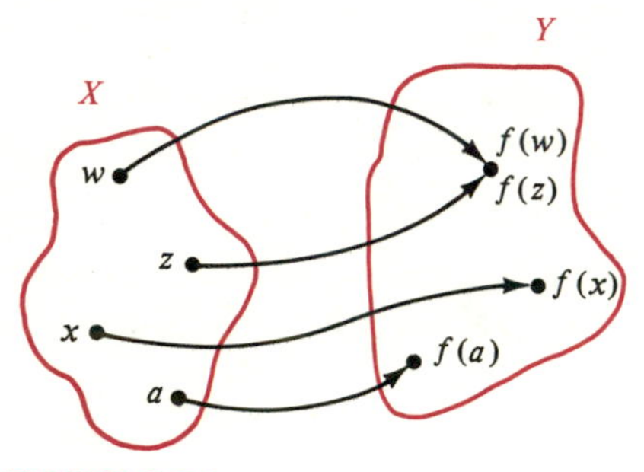

FIGURE 2.24

The notation $f(x)$ for the element of Y that corresponds to x is read "f of x." We also call $f(x)$ the **value** of f at x. Using the pictorial representation given earlier, we may now sketch a diagram as in Figure 2.24. The curved arrows indicate that the elements $f(x)$, $f(w)$, $f(z)$, and $f(a)$ of Y correspond to the elements x, w, z, and a of X. Let us repeat the important fact that *to each x in X there is assigned precisely one image $f(x)$ in Y;* however, different elements of X, such as w and z in Figure 2.24, may have the same image in Y.

Occasionally, one of the notations

$$X \xrightarrow{f} Y \quad \text{or} \quad f: X \to Y$$

is used to signify that f is a function from X to Y. This notation is often read as "f **maps** X into Y." The analogous statement for elements is *f maps x into* $f(x)$ and is denoted by

$$f: x \to f(x).$$

Beginning students are sometimes confused by the symbols f and $f(x)$. Remember that f is used to represent the function. It is neither in X nor in Y. However, $f(x)$ is an element of Y, namely the element that f assigns to x.

If the sets X and Y in the definition of function are intervals, or other sets of real numbers, then instead of using points within regions to represent elements, we may use two coordinate lines l and l'. This technique is illustrated in Figure 2.25, where two images for a function f are represented graphically.

FIGURE 2.25

Two functions f and g from X to Y are said to be **equal,** and we write

$$f = g \quad \text{if and only if} \quad f(x) = g(x)$$

for every x in X. For example, if $g(x) = (\frac{1}{2})(2x^2 - 6) + 3$ and $f(x) = x^2$ for all x in $\mathbb{R}$, then $g = f$.

EXAMPLE 1 Let f be the function with domain $\mathbb{R}$ such that $f(x) = x^2$ for every x in $\mathbb{R}$. Find $f(-6)$, $f(\sqrt{3})$, and $f(a)$ where a is any real number. What is the range of f?

Solution Values of f (or images under f) may be found by substituting for x in the equation $f(x) = x^2$. Thus,

$$f(-6) = (-6)^2 = 36, \quad f(\sqrt{3}) = (\sqrt{3})^2 = 3, \quad \text{and} \quad f(a) = a^2.$$

If T denotes the range of f, then, by definition, T consists of all numbers of the form $f(a)$ where a is in $\mathbb{R}$. Hence, T is the set of all squares a^2 where a is a real number. Since the square of every real number is nonnegative, T is contained in the set of all nonnegative real numbers. Moreover, every nonnegative real number c is an image under f since $f(\sqrt{c}) = (\sqrt{c})^2 = c$. Hence, the range of f is the set of all nonnegative real numbers. ■

It is usually difficult to find the range of a function by using an algebraic approach as in the preceding example. In the next section we shall see that the range can often be obtained from a graph.

If a function is defined as in Example 1, the symbols used for the function and variable are immaterial; that is, expressions such as $f(x) = x^2$, $f(s) = s^2$, $g(t) = t^2$, $k(r) = r^2$, etc., all define the same function. This is true because if a is any number in the domain, then the same image a^2 is obtained no matter which expression is employed.

In the remainder of our work, unless specified otherwise, the phase "f is a function" will mean that the domain and range are sets of real numbers. If a function is defined by means of some expression, as in Example 1, and the domain X is not stated explicitly, then X is considered to be the totality of real numbers for which the given expression is meaningful. To illustrate, if $f(x) = \sqrt{x - 2}$, then the domain is assumed to be the set of real numbers x such that $\sqrt{x - 2}$ is real; that is, $x - 2 \geq 0$, or $x \geq 2$. Thus, the domain is the infinite interval $[2, \infty)$. If x is in the domain, we sometimes say that f **is defined at** x, or that $f(x)$ **exists.** If a subset S is contained in the domain we often say that f **is defined on** S. The terminology f **is undefined at** x means that x is not in the domain of f.

EXAMPLE 2 If $g(x) = \dfrac{\sqrt{4 + x}}{1 - x}$, find

(a) the domain of g.

(b) $g(5)$, $g(-2)$, $g(-a)$, $-g(a)$.

Solution

(a) The fractional expression is a real number if and only if the radicand is nonnegative and the denominator is different from 0. Thus, $g(x)$ exists if and only if

$$4 + x \geq 0 \quad \text{and} \quad 1 - x \neq 0,$$

or equivalently,

$$x \geq -4 \quad \text{and} \quad x \neq 1.$$

This may be expressed in terms of intervals as $[-4, 1) \cup (1, \infty)$.

(b) To find the desired values of g, we substitute for x as follows:

$$g(5) = \frac{\sqrt{4+5}}{1-5} = \frac{\sqrt{9}}{-4} = -\frac{3}{4}$$

$$g(-2) = \frac{\sqrt{4+(-2)}}{1-(-2)} = \frac{\sqrt{2}}{3}$$

$$g(-a) = \frac{\sqrt{4+(-a)}}{1-(-a)} = \frac{\sqrt{4-a}}{1+a}$$

$$-g(a) = -\frac{\sqrt{4+a}}{1-a} = \frac{\sqrt{4+a}}{a-1}$$

■

Certain branches of mathematics such as calculus, involve manipulations of the type given in the next example.

EXAMPLE 3 Suppose $f(x) = x^2 + 3x - 2$. If a and h are real numbers, and $h \neq 0$, find

$$\frac{f(a+h) - f(a)}{h}.$$

Solution Since

$$f(a+h) = (a+h)^2 + 3(a+h) - 2 = (a^2 + 2ah + h^2) + (3a + 3h) - 2$$

and

$$f(a) = a^2 + 3a - 2,$$

we see that

$$\frac{f(a+h) - f(a)}{h} = \frac{[(a^2 + 2ah + h^2) + (3a + 3h) - 2] - (a^2 + 3a - 2)}{h}$$

$$= \frac{2ah + h^2 + 3h}{h}$$

$$= 2a + h + 3.$$

■

Many formulas that occur in mathematics and the sciences determine functions. As an illustration, the formula $A = \pi r^2$ for the area A of a circle of radius r assigns to each positive real number r a unique value of A. This determines a function f where $f(r) = \pi r^2$, and we may write $A = f(r)$. The letter r, which represents an arbitrary number from the domain of f, is often called an **independent variable.** The letter A, which represents a number from the range of f, is called a **dependent variable,** since its value depends on the number assigned to r. If two variables r and A are related in this manner, it is customary to use the phrase "A is a function of r." To cite another example, if an automobile travels at a uniform rate of 50 miles per hour, then the distance d (miles) traveled in time t (hours) is given by $d = 50t$, and hence, the distance d is a function of time t.

EXAMPLE 4 Express the radius of a circle as a function of its area.

Solution If A and r denote the area and radius, then

$$A = \pi r^2 \quad \text{and} \quad r^2 = \frac{A}{\pi}.$$

Since both A and r are positive,

$$r = \sqrt{\frac{A}{\pi}}.$$

This shows that r is a function of A, since to each value of A there corresponds a unique value $\sqrt{A/\pi}$ of r. ■

We have seen that different elements in the domain of a function may have the same image. If images are always different, then, as in the next definition, the function is called *one-to-one*.

DEFINITION

A function f from X to Y is a **one-to-one function** if, whenever $a \neq b$ in X, then $f(a) \neq f(b)$ in Y.

If f is one-to-one, then each $f(x)$ in the range is the image of *precisely one* x in X. The function illustrated in Figure 2.24 is not one-to-one since two different elements w and z of X have the same image in Y. A one-to-one function is often called a **one-to-one correspondence.** To illustrate, the association between real numbers and points on a coordinate line is an example of a one-to-one correspondence.

EXAMPLE 5

(a) If $f(x) = 3x + 2$, prove that f is one-to-one.

(b) If $g(x) = x^2 + 5$, prove that g is not one-to-one.

Solution

(a) If $a \neq b$, then $3a \neq 3b$ and hence $3a + 2 \neq 3b + 2$, or $f(a) \neq f(b)$. Thus, f is one-to-one.

(b) The function g is not one-to-one since different numbers in the domain may have the same image. For example, although $-1 \neq 1$, both $g(-1)$ and $g(1)$ are equal to 6. ■

If f is a function from X to X and if $f(x) = x$ for every x, that is, every element x maps into itself, then f is called the **identity function** on X. A function f is a **constant function** if there is some (fixed) element c such that $f(x) = c$ for every x in the domain. When a constant function is represented by a diagram of the type shown in Figure 2.24, *every* arrow from X terminates at the same point in Y.

The types of functions described in the next definition occur frequently.

DEFINITION

A function f with domain X is

(i) **even** if $f(-x) = f(x)$ for every x in X.

(ii) **odd** if $f(-x) = -f(x)$ for every x in X.

EXAMPLE 6

(a) If $f(x) = 3x^4 - 2x^2 + 5$, show that f is an even function.

(b) If $g(x) = 2x^5 - 7x^3 + 4x$, show that f is an odd function.

Solution If x is any real number, then

(a)
$$\begin{aligned} f(-x) &= 3(-x)^4 - 2(-x^2) + 5 \\ &= 3x^4 - 2x^2 + 5 = f(x) \end{aligned}$$

and hence f is even.

(b)
$$\begin{aligned} g(-x) &= 2(-x)^5 - 7(-x)^3 + 4(-x) \\ &= -2x^5 + 7x^3 - 4x \\ &= -(2x^5 - 7x^3 + 4x) = -g(x) \end{aligned}$$

Thus, g is odd. ■

The concept of ordered pair can be used to obtain an alternative approach to functions. We first observe that a function f from X to Y determines the following set W of ordered pairs:

$$W = \{(x, f(x)) : x \text{ is in } X\}.$$

Thus, W is the totality of ordered pairs for which the first number is in X and the second number is the image of the first. In Example 3, where $f(x) = x^2 + 3x - 2$, W consists of all pairs of the form $(x, x^2 + 3x - 2)$ where x is any real number. It is important to note that for each x there is exactly one ordered pair (x, y) in W that has x in the first position.

Conversely, if we begin with a set W of ordered pairs such that each x in X appears exactly once in the first position of an ordered pair, and numbers from Y appear in the second position, then W determines a function from X to Y. Specifically, for any x in X there is a unique pair (x, y) in W and, by letting y correspond to x, we obtain a function from X to Y.

It follows from the preceding discussion that the statement given below could also be used as a definition of function; however, we prefer to think of it as an alternative approach to this concept.

ALTERNATIVE DEFINITION OF FUNCTION

A function with domain X is a set W of ordered pairs such that for each x in X there is exactly one ordered pair (x, y) in W that has x in the first position.

EXERCISES 2.3

1 If $f(x) = 2x^2 - 3x + 4$, find $f(1)$, $f(-1)$, $f(0)$, and $f(2)$.

2 If $f(x) = x^3 + 5x^2 - 1$, find $f(2)$, $f(-2)$, $f(0)$, and $f(-1)$.

3 If $f(x) = \sqrt{x-1} + 2x$, find $f(1)$, $f(3)$, $f(5)$, and $f(10)$.

4 If $f(x) = \dfrac{x}{x-2}$, find $f(1)$, $f(3)$, $f(-2)$, and $f(0)$.

In Exercises 5–8 find each of the following, where a, b, and h are real numbers:

(a) $f(a)$ (b) $f(-a)$

(c) $-f(a)$ (d) $f(a+h)$

(e) $f(a) + f(h)$

(f) $\dfrac{f(a+h) - f(a)}{h}$ provided $h \neq 0$

5 $f(x) = 5x - 2$

6 $f(x) = 3 - 4x$

7 $f(x) = 2x^2 - x + 3$

8 $f(x) = x^3 - 2x$

In Exercises 9–12 find the following:

(a) $g(1/a)$ (b) $1/g(a)$ (c) $g(a^2)$

(d) $(g(a))^2$ (e) $g(\sqrt{a})$ (f) $\sqrt{g(a)}$

9 $g(x) = 3x^2$

10 $g(x) = 3x - 8$

11 $g(x) = \dfrac{2x}{x^2 + 1}$

12 $g(x) = \dfrac{x^2}{x + 1}$

In Exercises 13–20 find the domain of the function f.

13 $f(x) = \sqrt{3x - 5}$

14 $f(x) = \sqrt{7 - 2x}$

15 $f(x) = \sqrt{4 - x^2}$

16 $f(x) = \sqrt{x^2 - 9}$

17 $f(x) = \dfrac{x + 1}{x^3 - 9x}$

18 $f(x) = \dfrac{4x + 7}{6x^2 + 13x - 5}$

19 $f(x) = \dfrac{\sqrt{x}}{2x^2 - 11x + 12}$

20 $f(x) = \dfrac{x^3 - 1}{x^2 - 1}$

In Exercises 21–26 find the number that maps into 4. If $a > 0$, what number maps into a? Find the range of f.

21 $f(x) = 7x - 5$ **22** $f(x) = 3x$

23 $f(x) = \sqrt{x - 3}$ **24** $f(x) = 1/x$

25 $f(x) = x^3$ **26** $f(x) = \sqrt[3]{x - 4}$

In Exercises 27–34 determine if the function f is one-to-one.

27 $f(x) = 2x + 9$ **28** $f(x) = 1/(7x + 9)$

29 $f(x) = 5 - 3x^2$ **30** $f(x) = 2x^2 - x - 3$

31 $f(x) = \sqrt{x}$ **32** $f(x) = x^3$

33 $f(x) = |x|$ **34** $f(x) = 4$

In Exercises 35–44 determine whether f is even, odd, or neither even nor odd.

35 $f(x) = 3x^3 - 4x$ **36** $f(x) = 7x^6 - x^4 + 7$

37 $f(x) = 9 - 5x^2$ **38** $f(x) = 2x^5 - 4x^3$

39 $f(x) = 2$ **40** $f(x) = 2x^3 + x^2$

41 $f(x) = 2x^2 - 3x + 4$ **42** $f(x) = \sqrt{x^2 + 1}$

43 $f(x) = \sqrt[3]{x^3 - 4}$ **44** $f(x) = |x| + 5$

45 If $f(x)$ is a polynomial, and if the coefficients of all odd powers of x are 0, show that f is an even function.

46 If $f(x)$ is a polynomial, and if the coefficients of all even powers of x are 0, show that f is an odd function.

47 Find a formula that expresses the radius r of a circle as a function of its circumference C. If the circumference of *any* circle is increased by 12 inches, determine how much the radius increases.

48 Find a formula that expresses the volume of a cube as a function of its surface area. Find the volume if the surface area is 36 square inches.

49 Express the perimeter P of a square as a function of its area A.

50 Find a formula that expresses the area A of an equilateral triangle as a function of the length s of a side.

51 An open box is to be made from a rectangular piece of cardboard having dimensions 20 inches × 30 inches by cutting out identical squares of area x^2 from each corner and turning up the sides (see figure). Express the volume V of the box as a function of x.

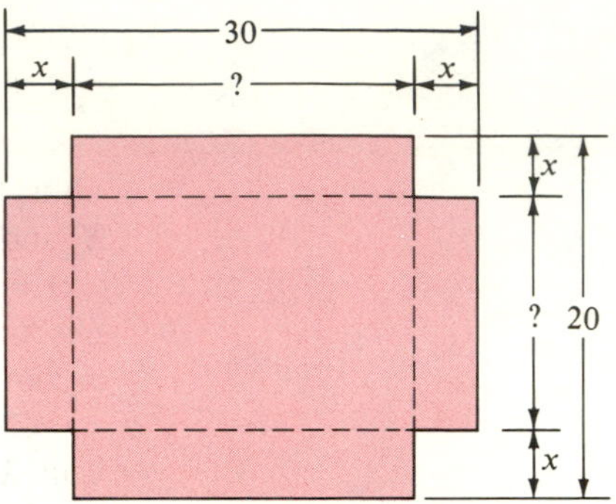

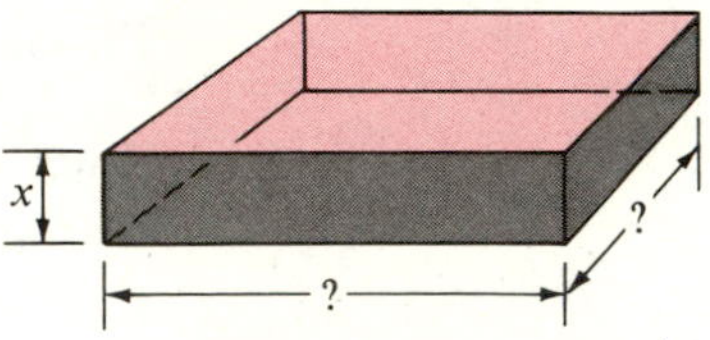

FIGURE FOR EXERCISE 51

52 Express the surface area S of a sphere as a function of its volume V.

53 A hot-air balloon is released at 1:00 P.M. and rises vertically at a rate of 2 meters per second. An observation point is situated 100 meters from a point on the ground directly below the balloon (see figure). If t denotes the time (in seconds) after 1:00 P.M., express the distance d between the balloon and the observation point as a function of t.

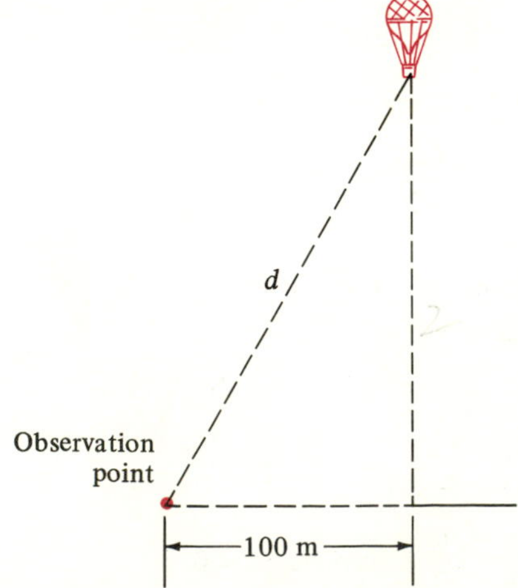

FIGURE FOR EXERCISE 53

54 Two ships leave port at 9:00 A.M., one sailing south at a rate of 16 mph and the other west at a rate of 20 mph (see figure). If t denotes the time (in hours) after 9:00 A.M., express the distance d between the ships as a function of t.

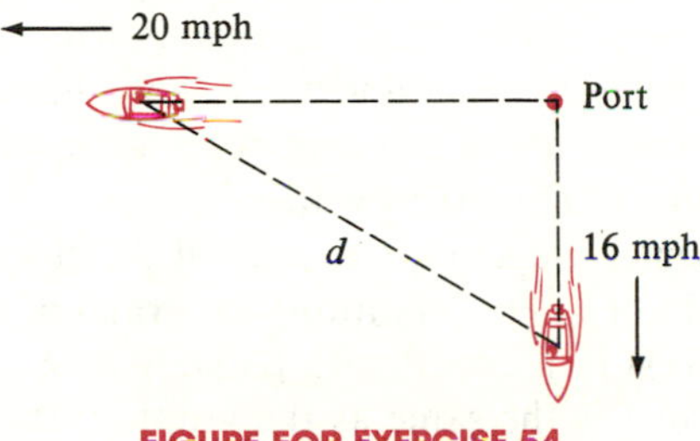

FIGURE FOR EXERCISE 54

55 A company sells running shoes to dealers at a rate of \$20 per pair if less than 50 pairs are ordered. If a dealer orders 50 or more pairs (up to 600), the price per pair is reduced at a rate of 2 cents times the number ordered. Let A denote the amount of money received when x pairs are ordered. Express A as a function of x.

56 A travel agency offers a tour at a cost of \$30 per person if not more than 60 people take the tour. If more than 60 take the tour, the cost is reduced 10 cents for each person in excess of 60. Let A denote the total cost if x people go on the tour. Express A as a function of x.

57 A steel storage tank for propane gas is to be constructed in the shape of a right circular cylinder of altitude 10 feet with a hemisphere attached to each end (see figure). The radius x is yet to be determined.

(a) Express the volume V of the tank as a function of x.

(b) Express the surface area S of the tank as a function of x.

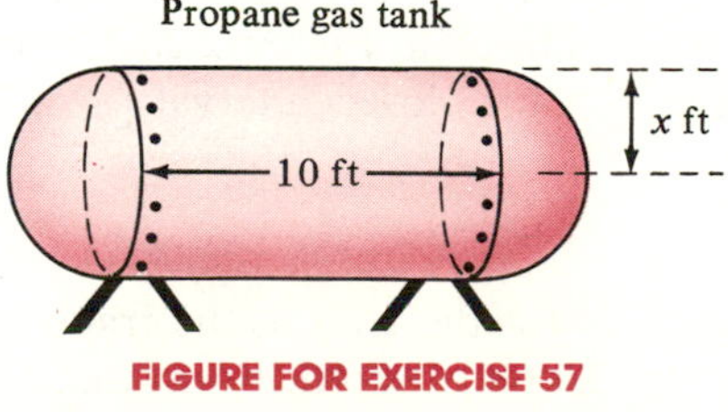

FIGURE FOR EXERCISE 57

58 A man in a rowboat, 2 miles from the nearest point A on a straight shoreline, wishes to reach a house located at a point B, 6 miles further downshore (see figure). He plans to row to a point P that is between A and B and is x miles from the house, and then walk the remainder of the distance. Suppose he can row at a rate of 3 mph and can walk at a rate of 5 mph. If T is the total time required to reach the house, express T as a function of x.

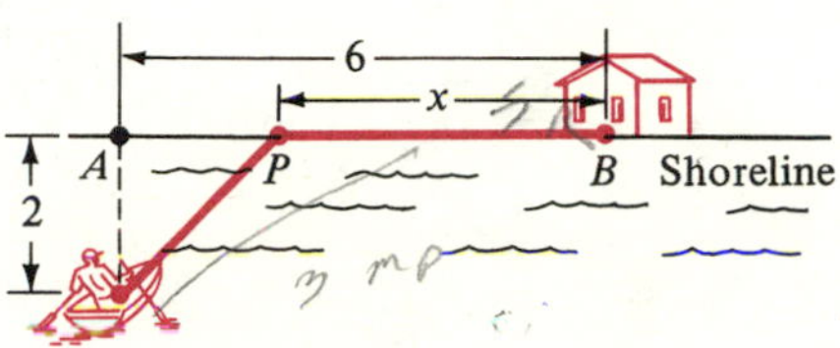

FIGURE FOR EXERCISE 58

59 A right circular cylinder of radius r and height h is inscribed in a cone of altitude 12 and base radius 4, as illustrated in the figure.

(a) Express h as a function of r. (*Hint:* Use similar triangles.)

(b) Express the volume V of the cylinder as a function of r.

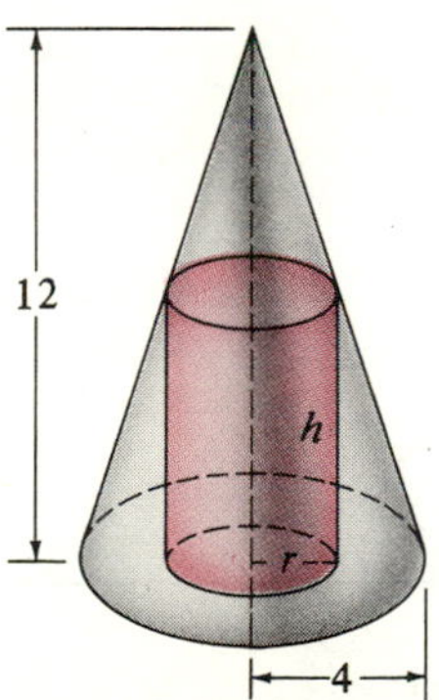

FIGURE FOR EXERCISE 59

60 Suppose a point $P(x, y)$ is moving along the parabola $y = x^2$. Let d denote the distance from the point $A(3, 1)$ to P. Express d as a function of x.

SECTION 2.4
GRAPHS OF FUNCTIONS

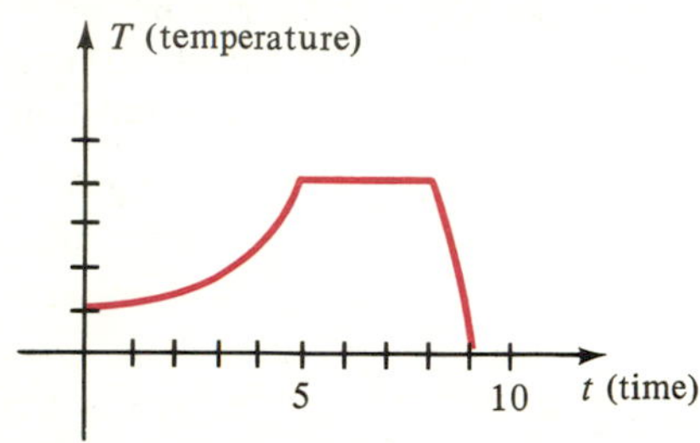

FIGURE 2.26

Graphs are often used to describe the variation of physical quantities. For example, a scientist may use a graph as in Figure 2.26 to indicate the temperature T of a certain solution at various times t during an experiment. The sketch shows that the temperature increased gradually from time $t = 0$ to time $t = 5$, did not change between $t = 5$ and $t = 8$, and then decreased rapidly from $t = 8$ to $t = 9$. A visual aid of this type reveals the variation of T more clearly than a long table of numerical values.

If f is a function, we frequently use a graph to exhibit the change in $f(x)$ as x varies through the domain of f. By definition, the **graph of a function** f is the set of all points $(x, f(x))$ in a coordinate plane, where x is in the domain of f. Thus, the graph of f is the same as the graph of the equation $y = f(x)$. If $P(a, b)$ is on the graph, then the y-coordinate b is the functional value $f(a)$, as illustrated in Figure 2.27. The figure also exhibits the domain of f (the set of possible values of x) and the range of f (the corresponding values of y). Although we have pictured the domain and range as closed intervals, they may be infinite intervals or other sets of real numbers.

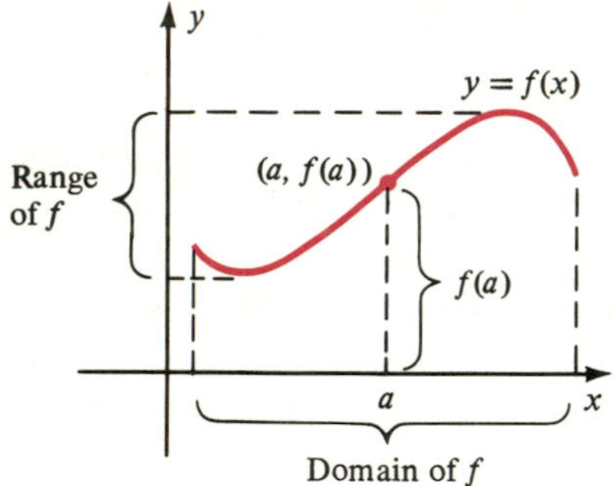

FIGURE 2.27

It is important to note that since there is a unique value $f(a)$ for each a in the domain, only *one* point on the graph has x-coordinate a. Thus, *every vertical line intersects the graph of a function in at most one point.* Consequently, for graphs of functions it is impossible to obtain a figure, such as a circle, in which some vertical lines intersect the graph in more than one point.

The x-intercepts of the graph of a function f are the solutions of the equation $f(x) = 0$. These numbers are called the **zeros** of the function. The y-intercept of the graph is $f(0)$, if it exists.

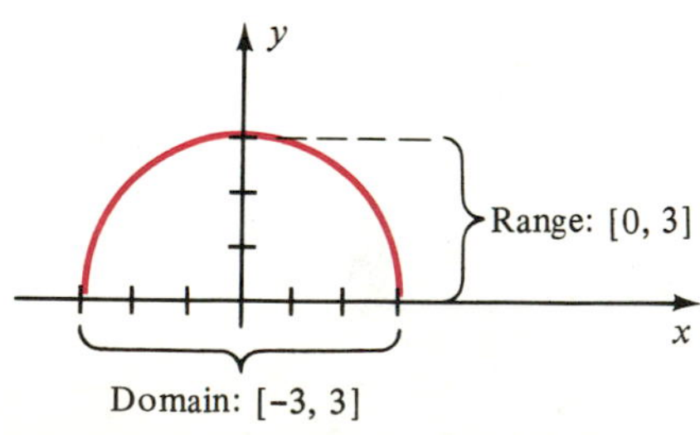

FIGURE 2.28 $f(x) = \sqrt{9 - x^2}$

EXAMPLE 1 If $f(x) = \sqrt{9 - x^2}$, sketch the graph of f and find the domain and range of f.

Solution The graph of f is the same as the graph of the equation $y = \sqrt{9 - x^2}$. We know from Section 2.1 that the graph of $x^2 + y^2 = 9$ is a circle of radius 3 with center at the origin (see Figure 2.13). Solving the equation $x^2 + y^2 = 9$ for y gives us $y = \pm\sqrt{9 - x^2}$. It follows that the graph of f is the *upper half* of the circle, as illustrated in Figure 2.28. Referring to the figure, we see that the domain of f is the closed interval $[-3, 3]$ and the range of f is the interval $[0, 3]$. ■

If $f(x) = ax + b$, then the graph of f is the same as the graph of the equation $y = ax + b$. It follows from the Slope-Intercept Form that the graph is a line with slope a and y-intercept b. For this reason, any function f such that $f(x) = ax + b$ is called a **linear function.**

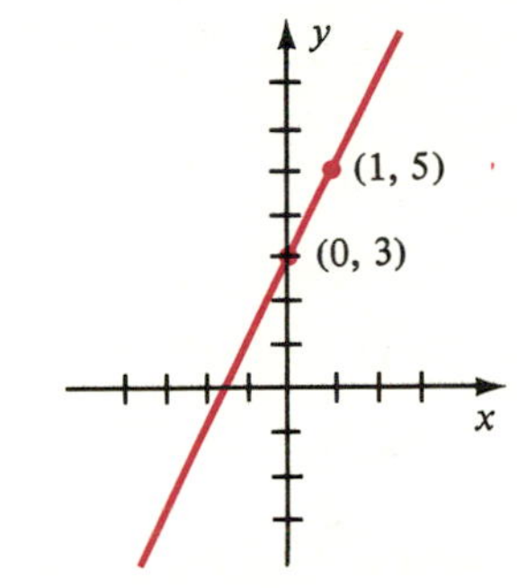

FIGURE 2.29 $f(x) = 2x + 3$

EXAMPLE 2 If $f(x) = 2x + 3$, sketch the graph of f and find the domain and range of f.

Solution The graph is the same as the graph of the equation $y = 2x + 3$, and, hence, by the Slope-Intercept Form, is a line of slope 2 and y-intercept 3. (See Figure 2.29.) Since the values of x and y may be any real numbers, both the domain and range of f are $\mathbb{R}$. ■

If f is the linear function in Example 2, and if $x_1 < x_2$, then $2x_1 + 3 < 2x_2 + 3$; that is, $f(x_1) < f(x_2)$. This means that, as x-coordinates of points increase, y-coordinates also increase. The function f is then said to be *increasing*. If f is increasing, then the graph rises as x increases. For certain functions $f(x_1) > f(x_2)$ whenever $x_1 < x_2$. In this case the graph of f falls as x increases and the function is called a *decreasing* function. In general, we shall speak of functions that increase or decrease on certain intervals, as in the following definition. Functions that neither increase nor decrease are called *constant*.

DEFINITION

Let a function f be defined on an interval I and let x_1, x_2 be numbers in I.

(i) f is **increasing** on I if $f(x_1) < f(x_2)$ whenever $x_1 < x_2$.

(ii) f is **decreasing** on I if $f(x_1) > f(x_2)$ whenever $x_1 < x_2$.

(iii) f is **constant** on I if $f(x_1) = f(x_2)$ for every x_1 and x_2.

Geometric interpretations of the preceding definition are given in Figure 2.30, where the interval I is not indicated. If f is constant on I, then the graph is part of a horizontal line. (Why?)

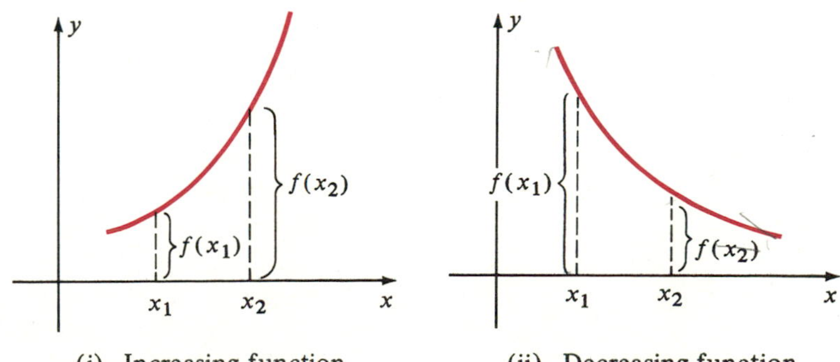

FIGURE 2.30

We shall use the phrases "f is increasing" and "$f(x)$ is increasing" interchangeably. This will also be done for the term "decreasing."

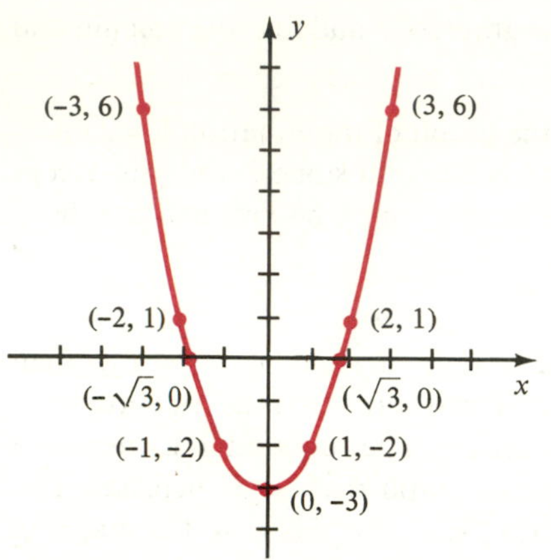

FIGURE 2.31 $f(x) = x^2 - 3$

EXAMPLE 3 If $f(x) = x^2 - 3$, sketch the graph of f, and find the domain and range of f.

Solution Coordinates $(x, f(x))$ of some points on the graph of f are listed in the following table:

x	-3	-2	-1	0	1	2	3
$f(x)$	6	1	-2	-3	-2	1	6

The x-intercepts are the solutions of the equation $f(x) = 0$, that is, of $x^2 - 3 = 0$. Thus the x-intercepts are $\pm\sqrt{3}$. The y-intercept is $f(0) = -3$. Plotting the points given by the table and using the x-intercepts leads to the sketch in Figure 2.31.

Since x can have any value, the domain of f is $\mathbb{R}$. The range of f consists of all real numbers y such that $y \geq -3$, and hence is the interval $[-3, \infty)$. ■

To sketch the graph in Figure 2.31, we assumed that f is decreasing on the interval $(\infty, 0]$ and increasing on the interval $[0, \infty)$. These facts can be proved using properties of inequalities; however, for more complicated functions it is often necessary to use techniques studied in calculus. In this text we shall usually rely on sufficient point plotting to determine where functions increase or decrease.

Referring to Figure 2.31, we see that $f(x) = x^2 - 3$ takes on its least value at $x = 0$. This smallest value, -3, is called the **minimum value** of f. The corresponding point $(0, -3)$ is the lowest point on the graph. Clearly, $f(x)$ does not attain a **maximum value,** that is, a *largest* value.

The solution to Example 3 could have been shortened by observing that since $(-x)^2 - 3 = x^2 - 3$, the graph of $y = x^2 - 3$ is symmetric with respect to the y-axis. This fact also follows from (i) of the next theorem.

THEOREM ON SYMMETRY

(i) The graph of an even function is symmetric with respect to the y-axis.

(ii) The graph of an odd function is symmetric with respect to the origin.

Proof If f is even, then $f(-x) = f(x)$, and hence the equation $y = f(x)$ is not changed if $-x$ is substituted for x. Statement (i) now follows from Symmetry Test (i) of Section 2.1. The proof of (ii) is left to the reader. □

EXAMPLE 4 Sketch the graph of f if $f(x) = |x|$.

Solution If $x \geq 0$, then $f(x) = x$ and hence the part of the graph to the

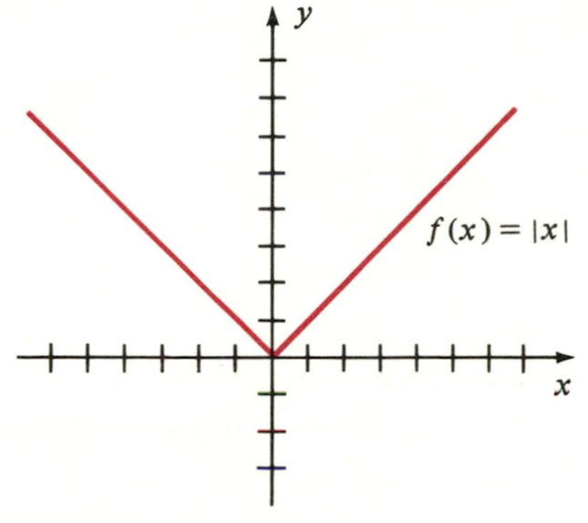

FIGURE 2.32 $f(x) = |x|$

right of the y-axis is identical to the graph of $y = x$, which is a line through the origin with slope 1. Since $|-x| = |x|$, we see that f is an even function and, hence, by the preceding theorem, the graph is symmetric with respect to the y-axis. Plotting several points and using symmetry leads to the sketch in Figure 2.32. As in Example 3, the function decreases on $(-\infty, 0]$ and increases on $[0, \infty)$.

EXAMPLE 5 Sketch the graph of f if $f(x) = \sqrt{x - 1}$.

Solution The domain of f consists of all real numbers x such that $x \geq 1$. The following table lists some points $(x, f(x))$ on the graph.

x	1	2	3	4	5	6
$f(x)$	0	1	$\sqrt{2}$	$\sqrt{3}$	2	$\sqrt{5}$

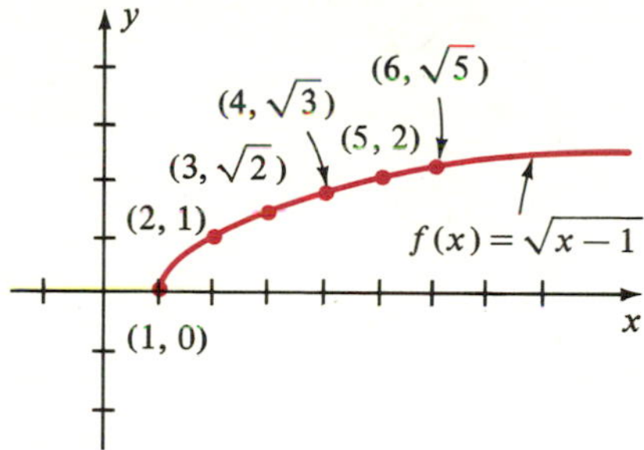

FIGURE 2.33 $f(x) = \sqrt{x - 1}$

Plotting points leads to the sketch shown in Figure 2.33. The function is increasing throughout its domain. The x-intercept is 1 and there is no y-intercept. ■

EXAMPLE 6 Describe the graph of a constant function with domain $\mathbb{R}$.

Solution By definition f is a constant function if there is some (fixed) real number c such that $f(x) = c$ for every x. Thus, the graph of f is the same as the graph of the equation $y = c$ and hence is a horizontal line with y-intercept c. ■

EXAMPLE 7 Given $f(x) = x^2 + c$, sketch the graph of f if

(a) $c = 4$; (b) $c = -2$.

Solution We shall sketch both graphs on the same coordinate axes. The graph of $y = x^2$ was sketched in Figure 2.9, and, for reference, is represented by dashes in Figure 2.34. To find the graph of $f(x) = x^2 + 4$ we may simply increase the y-coordinate of each point on the graph of $y = x^2$ by 4, as shown in the figure. (Why?) This amounts to *shifting* the graph of $y = x^2$ *upward* 4 units. For $c = -2$ we decrease y-coordinates by 2 and, hence, this graph may be obtained by shifting the graph of $y = x^2$ downward 2 units. Each graph is a parabola symmetric with respect to the y-axis. To determine the correct position of each graph, it is advisable to plot several points on each. ■

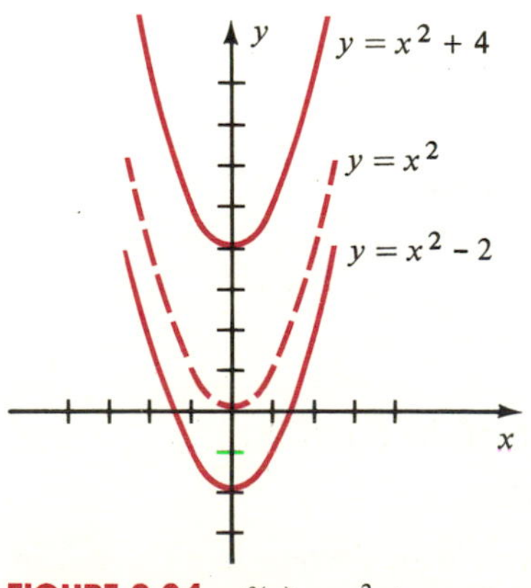

FIGURE 2.34 $f(x) = x^2 + c$

The graphs in the preceding example illustrate **vertical shifts** of the graph of $y = x^2$ and are special cases of the following general rules.

VERTICAL SHIFTS OF GRAPHS ($c > 0$)

To obtain the graph of:	shift the graph of $y = f(x)$:
$y = f(x) - c$	c units downward
$y = f(x) + c$	c units upward

Similar rules can be stated for **horizontal shifts.** Specifically, if $c > 0$, consider the graphs of $y = f(x)$ and $y = f(x - c)$ sketched on the same coordinate axes, as illustrated in Figure 2.35.

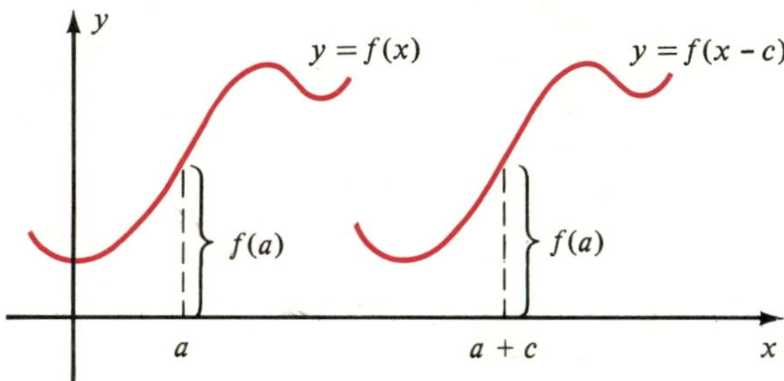

FIGURE 2.35

Since $f(a) = f(a + c - c)$, we see that the point with x-coordinate a on the graph of $y = f(x)$ has the same y-coordinate as the point with x-coordinate $a + c$ on the graph of $y = f(x - c)$. This implies that the latter graph can be obtained by shifting the graph of $y = f(x)$ to the right c units. Similarly, the graph of $y = f(x + c)$ can be obtained by shifting the graph of f to the left c units. These rules are listed for reference in the next box.

HORIZONTAL SHIFTS OF GRAPHS ($c > 0$)

To obtain the graph of	shift the graph of $y = f(x)$:
$y = f(x - c)$	c units to the right
$y = f(x + c)$	c units to the left

EXAMPLE 8 Sketch the graph of f for each of the following:

(a) $f(x) = (x - 4)^2$ (b) $f(x) = (x + 2)^2$

Solution The graph of $y = x^2$ is sketched, with dashes, in Figure 2.36. According to the rules for horizontal shifts, shifting this graph to the right

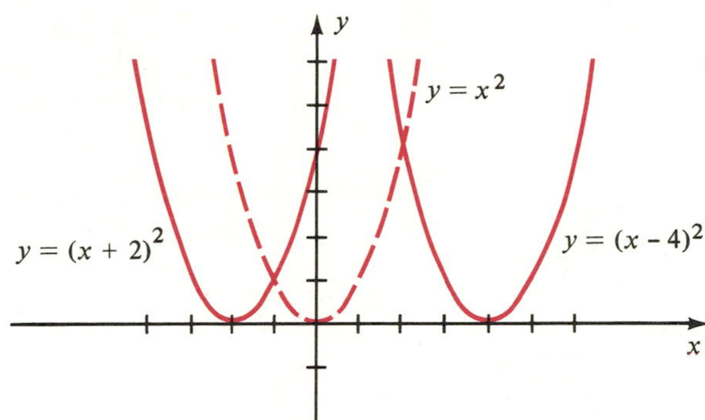

FIGURE 2.36

4 units gives us the graph of $y = (x - 4)^2$, whereas shifting to the left 2 units leads to the graph of $y = (x + 2)^2$. Students who are not convinced of the validity of this technique are urged to plot several points on each graph.

To obtain the graph of $y = cf(x)$, we may *multiply* the y-coordinates of points on the graph of $y = f(x)$ by c. For example, if $y = 2f(x)$, we double y-coordinates, or if $y = \frac{1}{2}f(x)$, we multiply each y-coordinate by $\frac{1}{2}$. If $c > 0$ (and $c \neq 1$) we shall refer to this procedure as **stretching** the graph of $y = f(x)$. ■

EXAMPLE 9 Sketch the graph of (a) $y = 4x^2$; (b) $y = \frac{1}{4}x^2$.

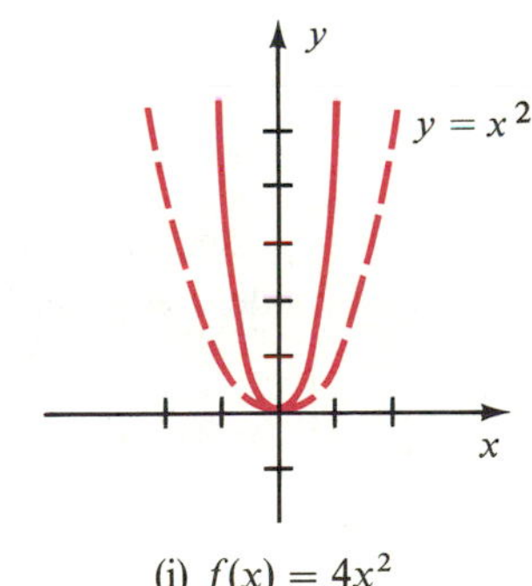

(i) $f(x) = 4x^2$

Solution

(a) To sketch the graph of $y = 4x^2$ we may refer to the graph of $y = x^2$ (see dashes in Figure 2.37) and multiply the y-coordinate of each point by 4. This gives us a narrower parabola that is sharper at the vertex, as illustrated in (i) of the figure. To obtain the correct shape, several points such as (0, 0), $(\frac{1}{2}, 1)$, and (1, 4) should be plotted.

(b) The graph of $y = \frac{1}{4}x^2$ may be sketched by multiplying y-coordinates of points on the graph of $y = x^2$ by $\frac{1}{4}$. The graph is a wider parabola that is flatter at the vertex, as shown in (ii) of Figure 2.37.

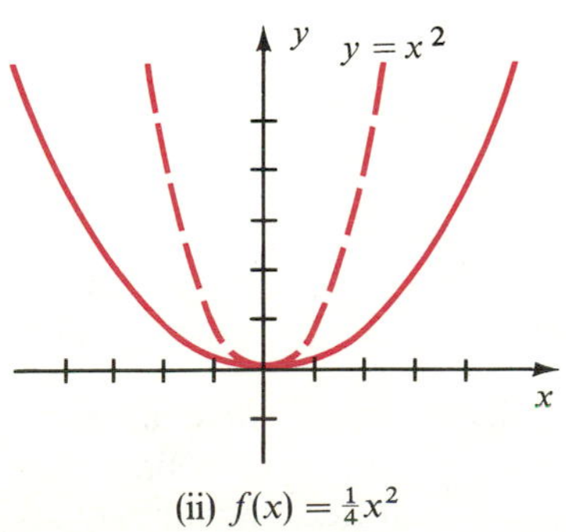

(ii) $f(x) = \frac{1}{4}x^2$

FIGURE 2.37

The graph of $y = -f(x)$ may be obtained by multiplying the y-coordinate of each point on the graph of $y = f(x)$ by -1. In particular, every point (a, b) on the graph of $y = f(x)$ that lies above the x-axis determines a point $(a, -b)$ below the x-axis. Similarly, if (c, d) lies below the x-axis (that is, $d < 0$), then $(c, -d)$ lies above the x-axis. The graph of $y = -f(x)$ is called a **reflection** of the graph of $y = f(x)$ through the x-axis. ■

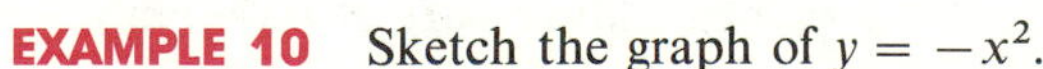

EXAMPLE 10 Sketch the graph of $y = -x^2$.

Solution The graph may be found by plotting points; however, since the graph of $y = x^2$ is well known, we sketch it with dashes, as in Figure 2.38, and then multiply y-coordinates of points by -1. This gives us the reflection through the x-axis indicated in the figure. ■

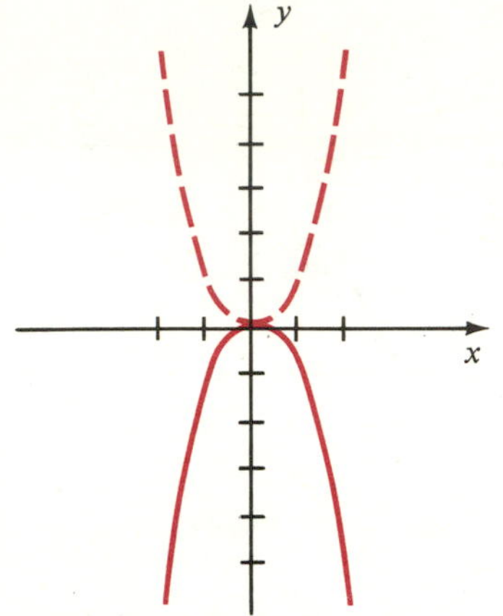

FIGURE 2.38 $f(x) = -x^2$

Sometimes functions are described in terms of more than one expression, as in the next examples.

EXAMPLE 11 Sketch the graph of the function f that is defined as follows:

$$f(x) = \begin{cases} 2x + 3 & \text{if } x < 0 \\ x^2 & \text{if } 0 \le x < 2 \\ 1 & \text{if } x \ge 2 \end{cases}$$

Solution If $x < 0$, then $f(x) = 2x + 3$. This means that if x is negative, the expression $2x + 3$ should be used to find functional values. Consequently, if $x < 0$, then the graph of f coincides with the line $y = 2x + 3$ and we sketch that portion of the graph to the left of the y-axis as indicated in Figure 2.39.

If $0 \le x < 2$, we use x^2 to find functional values of f, and therefore this part of the graph of f coincides with the graph of the equation $y = x^2$. We then sketch the corresponding part of the graph of f between $x = 0$ and $x = 2$, as indicated in Figure 2.39.

Finally, if $x \ge 2$, the graph of f coincides with the constant function having values equal to 1. That part of the graph is the horizontal half line illustrated in Figure 2.39. ■

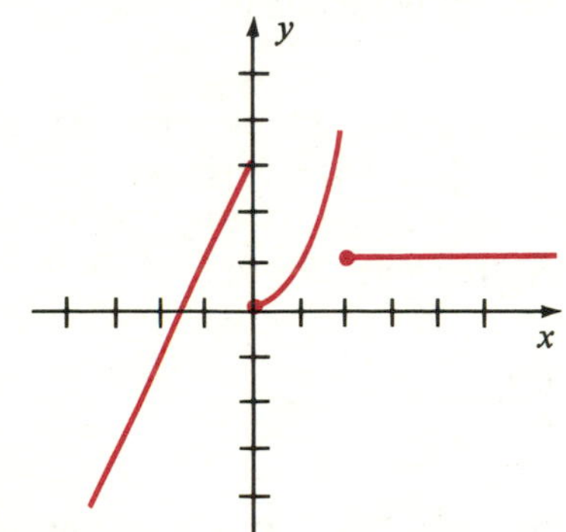

FIGURE 2.39

EXAMPLE 12 If x is any real number, then there exist consecutive integers n and $n + 1$ such that $n \le x < n + 1$. Let f be the function defined as follows: If $n \le x < n + 1$, then $f(x) = n$. Sketch the graph of f.

Solution The x- and y-coordinates of some points on the graph may be listed as follows:

Values of x	$f(x)$
...	...
$-2 \le x < -1$	-2
$-1 \le x < 0$	-1
$0 \le x < 1$	0
$1 \le x < 2$	1
$2 \le x < 3$	2
...	...

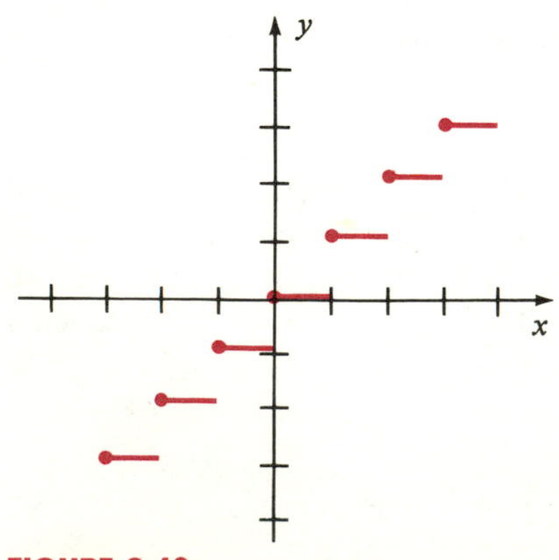

FIGURE 2.40

Since f is a constant function whenever x is between successive integers, the corresponding part of the graph is a segment of a horizontal line. Part of the graph is sketched in Figure 2.40. The graph continues indefinitely to the right and to the left. ■

The symbol $[\![x]\!]$ is often used to denote the largest integer n such that $n \leq x$. For example $[\![1.6]\!] = 1$, $[\![\sqrt{5}]\!] = 2$, $[\![\pi]\!] = 3$, and $[\![-3.5]\!] = -4$. Using this notation, the function f of Example 12 may be defined by $f(x) = [\![x]\!]$. It is customary to refer to f as the **greatest integer function.**

EXERCISES 2.4

In Exercises 1–20 sketch the graph of f, find the domain and range of f, and describe the intervals in which f is increasing, is decreasing, or is constant.

1 $f(x) = 5x$

2 $f(x) = -3x$

3 $f(x) = -4x + 2$

4 $f(x) = 4x - 2$

5 $f(x) = 3 - x^2$

6 $f(x) = 4x^2 + 1$

7 $f(x) = 2x^2 - 4$

8 $f(x) = \frac{1}{100}x^2$

9 $f(x) = \sqrt{x + 4}$

10 $f(x) = \sqrt{4 - x}$

11 $f(x) = \sqrt{x} + 2$

12 $f(x) = 2 - \sqrt{x}$

13 $f(x) = 4/x$

14 $f(x) = 1/x^2$

15 $f(x) = |x - 2|$

16 $f(x) = |x + 2|$

17 $f(x) = |x| - 2$

18 $f(x) = |x| + 2$

19 $f(x) = \dfrac{x}{|x|}$

20 $f(x) = x + |x|$

In Exercises 21–30 sketch, on the same coordinate axes, the graphs of f for the stated values of c. (Make use of vertical shifts, horizontal shifts, stretching, or reflecting.)

21 $f(x) = 3x + c$; $c = 0, c = 2, c = -1$

22 $f(x) = -2x + c$; $c = 0, c = 1, c = -3$

23 $f(x) = x^3 + c$; $c = 0, c = 1, c = -2$

24 $f(x) = -x^3 + c$; $c = 0, c = 2, c = -1$

25 $f(x) = \sqrt{4 - x^2} + c$; $c = 0, c = 4, c = -3$

26 $f(x) = c - |x|$; $c = 0, c = 5, c = -2$

27 $f(x) = 3(x - c)$; $c = 0, c = 2, c = 3$

28 $f(x) = -2(x - c)^2$; $c = 0, c = 3, c = 1$

29 $f(x) = (x + c)^3$; $c = 0, c = 2, c = -2$

30 $f(x) = c\sqrt{9 - x^2}$; $c = 0, c = 2, c = 3$

31 The graph of a function f with domain $0 \leq x \leq 4$ is shown in the accompanying figure. Sketch the graph of each of the following:

(a) $y = f(x + 2)$ (b) $y = f(x - 2)$

(c) $y = f(x) + 2$ (d) $y = f(x) - 2$

(e) $y = 2f(x)$ (f) $y = \frac{1}{2}f(x)$

(g) $y = -2f(x)$ (h) $y = f(x - 3) + 1$

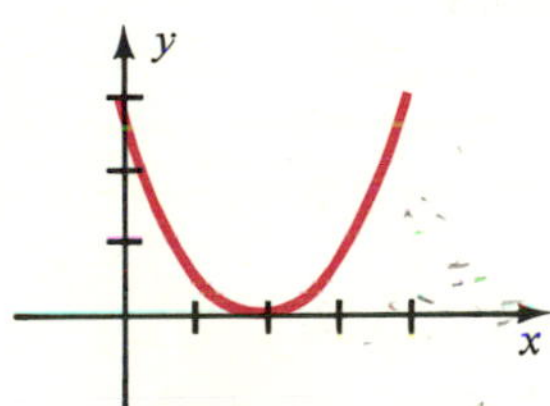

FIGURE FOR EXERCISE 31

32 The graph of a function f with domain $0 \leq x \leq 4$ is shown in the accompanying figure. Sketch the graph of each of the following:

(a) $y = f(x - 1)$ (b) $y = f(x + 3)$

(c) $y = f(x) - 2$ (d) $y = f(x) + 1$

(e) $y = 3f(x)$ (f) $y = \frac{1}{2}f(x)$

(g) $y = f(x + 2) - 2$ (h) $y = f(2x)$

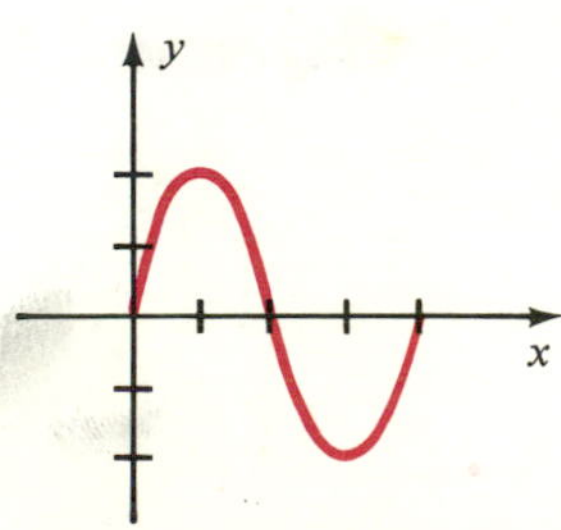

FIGURE FOR EXERCISE 32

In Exercises 33–40 sketch the graph of f.

33 $f(x) = \begin{cases} 2 & \text{if } x < 0 \\ -1 & \text{if } x \geq 0 \end{cases}$

34 $f(x) = \begin{cases} -1 & \text{if } x \text{ is an integer} \\ 1 & \text{if } x \text{ is not an integer} \end{cases}$

35 $f(x) = \begin{cases} 3 & \text{if } x < -3 \\ -x & \text{if } -3 \leq x \leq 3 \\ -3 & \text{if } x > 3 \end{cases}$

36 $f(x) = \begin{cases} x & \text{if } x < 0 \\ -2 & \text{if } 0 \leq x < 1 \\ x^2 & \text{if } x \geq 1 \end{cases}$

37 $f(x) = \begin{cases} x^2 & \text{if } x \leq -1 \\ x^3 & \text{if } |x| < 1 \\ 2x & \text{if } x \geq 1 \end{cases}$

38 $f(x) = \begin{cases} x & \text{if } x \leq 1 \\ -x^2 & \text{if } 1 < x < 2 \\ x & \text{if } x \geq 2 \end{cases}$

39 $f(x) = \begin{cases} \dfrac{x^2 - 4}{x - 2} & \text{if } x \neq 2 \\ 3 & \text{if } x = 2 \end{cases}$

40 $f(x) = \begin{cases} \dfrac{x^2 - 1}{1 - x} & \text{if } x \neq 1 \\ 2 & \text{if } x = 1 \end{cases}$

41 If $[\![x]\!]$ denotes values of the greatest integer function, sketch the graph of f in each of the following:

(a) $f(x) = [\![x - 3]\!]$ (b) $f(x) = 2[\![x]\!]$

(c) $f(x) = -[\![x]\!]$

42 Explain why the graph of the equation $y^2 = x$ is not the graph of a function.

43 Sketch the graph of f if $f(x) = x$ for every x.

44 Define a function f whose graph is the upper half of a circle with center at the origin and radius 1.

45 Define a function f whose graph is the lower half of a circle with center at the origin and radius 1.

46 If a function f is increasing throughout its domain, prove that f is one-to-one.

47 If a function f is decreasing throughout its domain, prove that f is one-to-one.

48 Prove that a function f is one-to-one if and only if every horizontal line intersects the graph of f in at most one point.

SECTION 2.5

QUADRATIC FUNCTIONS

Functions of the following type occur frequently throughout mathematics and the sciences.

DEFINITION

> A function f is a **quadratic function** if
>
> $$f(x) = ax^2 + bx + c$$
>
> where a, b, and c are real numbers and $a \neq 0$.

If $b = c = 0$ in the definition, then $f(x) = ax^2$ and the graph is a parabola with vertex at the origin, opening upward if $a > 0$ or downward if $a < 0$ (see

Figures 2.37 and 2.38). If $b = 0$ and $c \neq 0$, then

$$f(x) = ax^2 + c$$

and from the discussion of vertical shifts in Section 2.4, the graph is a parabola with vertex at the point $(0, c)$ on the y-axis. Some typical graphs are illustrated in Figure 2.34. Another is given in the next example.

EXAMPLE 1 Sketch the graph of f if $f(x) = -\frac{1}{2}x^2 + 4$.

Solution The graph of $f(x) = -\frac{1}{2}x^2$ is similar to the graph of $f(x) = -x^2$, sketched in Figure 2.38, but is somewhat wider. The graph of $f(x) = -\frac{1}{2}x^2 + 4$ may be found by shifting the graph of $f(x) = -\frac{1}{2}x^2$ upward 4 units. Coordinates of several points on the graph are listed in the following table.

x	0	1	2	$\sqrt{8}$	3
$f(x)$	4	$\frac{7}{2}$	2	0	$-\frac{1}{2}$

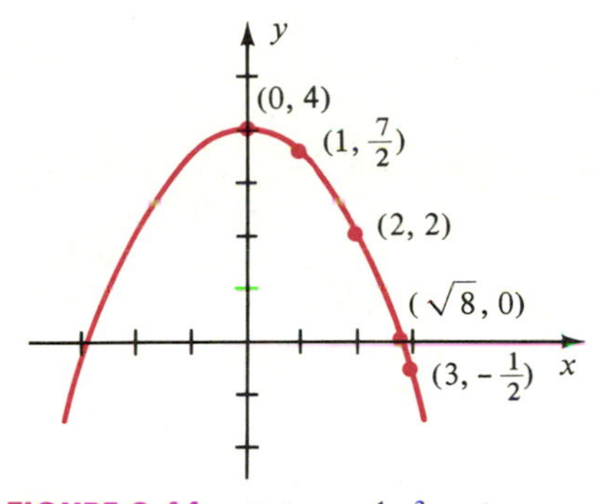

FIGURE 2.41 $f(x) = -\frac{1}{2}x^2 + 4$

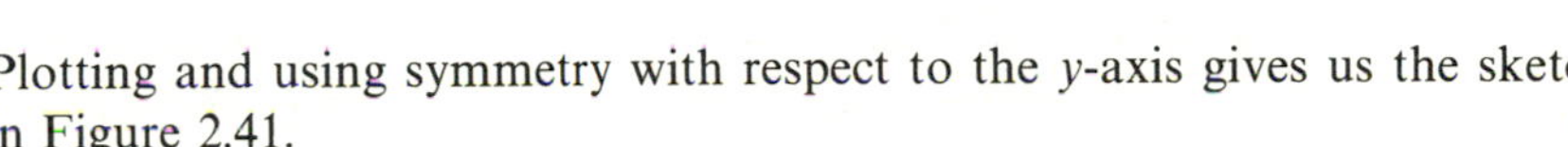

Plotting and using symmetry with respect to the y-axis gives us the sketch in Figure 2.41. ■

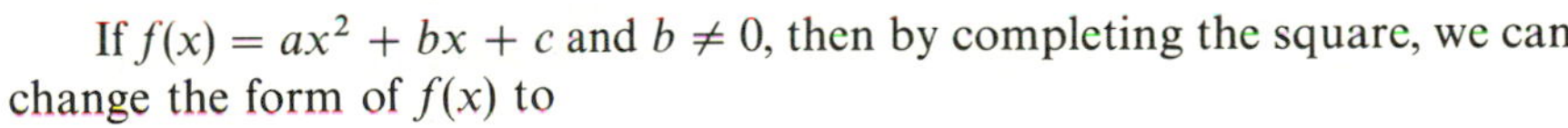

If $f(x) = ax^2 + bx + c$ and $b \neq 0$, then by completing the square, we can change the form of $f(x)$ to

$$f(x) = a(x - h)^2 + k$$

where h and k are real numbers. This technique is illustrated in the next example. As we shall see, the graph of f can be readily obtained from this new form.

EXAMPLE 2 If $f(x) = 3x^2 + 24x + 50$, express $f(x)$ in the form $a(x - h)^2 + k$.

Solution Before we complete the square *it is essential to remove the coefficient of x^2 from the first two terms* of $f(x)$ as follows:

$$\begin{aligned} f(x) &= 3x^2 + 24x + 50 \\ &= 3(x^2 + 8x \qquad) + 50. \end{aligned}$$

We may now complete the square for the expression within parentheses by adding the square of one-half the coefficient of x, that is, $(\frac{8}{2})^2$, or 16. However, if we add 16 to the expression within parentheses, then, because of the factor 3, we are actually adding 48 to $f(x)$. Hence, we must compensate by

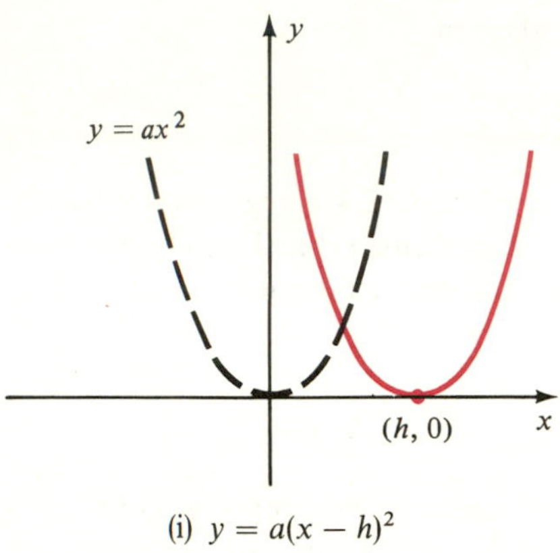

(i) $y = a(x - h)^2$

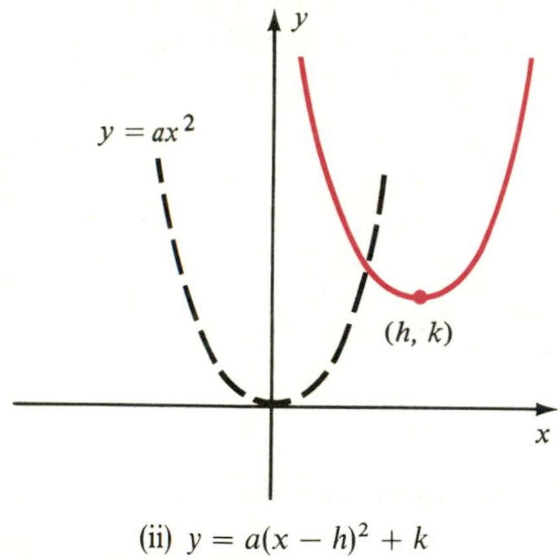

(ii) $y = a(x - h)^2 + k$

FIGURE 2.42

subtracting 48:

$$\begin{aligned} f(x) &= 3(x^2 + 8x \qquad) + 50 \\ &= 3(x^2 + 8x + 16) + 50 - 48 \\ &= 3(x + 4)^2 + 2, \end{aligned}$$

which has the desired form with $a = 3$, $h = -4$, and $k = 2$ ■

If $f(x) = ax^2 + bx + c$, then by completing the square as in Example 2, we see that the graph of f is the same as the graph of an equation of the form

$$y = a(x - h)^2 + k.$$

Applying the discussion of horizontal shifts in Section 2.4, we can find the graph of $y = a(x - h)^2$ by shifting the graph of $y = ax^2$ either to the left or right, depending on the sign of h. Thus, $y = a(x - h)^2$ is an equation of a parabola that has vertex $(h, 0)$ and a vertical axis. A typical graph is sketched in (i) of Figure 2.42 for the case in which both a and h are positive. Since the graph of $y = a(x - h)^2 + k$ can be obtained from that of $y = a(x - h)^2$ by a *vertical* shift of $|k|$ units, it follows that *the graph of a quadratic function f is a parabola* that has vertex (h, k) and a vertical axis. The sketch in (ii) of Figure 2.42 illustrates one possible graph. Observe that since (h, k) is the lowest (or highest) point on the parabola, $f(x)$ has its minimum (or maximum) value at $x = h$. This value is $f(h) = k$.

In the preceding discussion we derived the following general equation of a parabola that has vertex (h, k) and a vertical axis.

STANDARD EQUATION OF A PARABOLA (VERTICAL AXIS)

$$y - k = a(x - h)^2$$

A parabola with a vertical axis opens upward if $a > 0$ or downward if $a < 0$.

EXAMPLE 3 Sketch the graph of $f(x) = 2x^2 - 6x + 4$ and find the minimum value of $f(x)$.

Solution The graph of f is the same as the graph of $y = 2x^2 - 6x + 4$. From the preceding discussion we know that the graph is a parabola, and we begin by completing the square:

$$\begin{aligned} y &= 2x^2 - 6x + 4 \\ &= 2(x^2 - 3x \qquad) + 4 \\ &= 2(x^2 - 3x + \tfrac{9}{4}) + (4 - \tfrac{9}{2}) \\ &= 2(x - \tfrac{3}{2})^2 - \tfrac{1}{2}. \end{aligned}$$

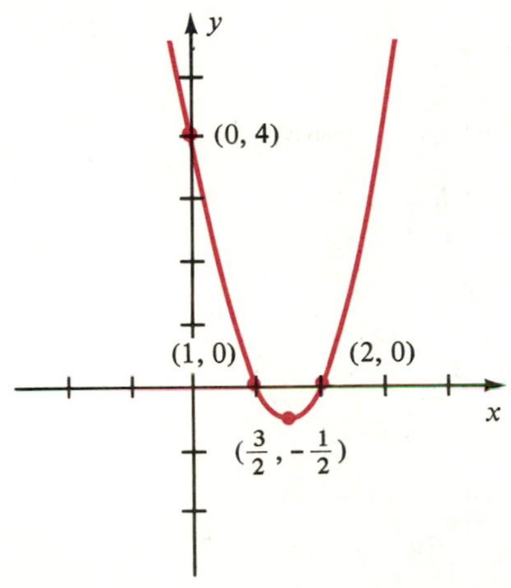

FIGURE 2.43 $f(x) = 2x^2 - 6x + 4$

If we write the last equation as

$$y + \tfrac{1}{2} = 2(x - \tfrac{3}{2})^2$$

and compare it with the standard equation of a parabola, we see that $h = \frac{3}{2}$ and $k = -\frac{1}{2}$. Consequently, the vertex (h, k) of the parabola is $(\frac{3}{2}, -\frac{1}{2})$. Since $a = 2 > 0$, the parabola opens upward. The y-intercept is $f(0) = 4$. To find the x-intercepts we solve $2x^2 - 6x + 4 = 0$, or the equivalent equation $(2x - 2)(x - 2) = 0$, obtaining $x = 1$ and $x = 2$. Plotting the vertex together with the x- and y-intercepts provides enough points for a reasonably accurate sketch (see Figure 2.43). The minimum value of f occurs at the vertex, and hence is $f(\frac{3}{2}) = -\frac{1}{2}$. ■

EXAMPLE 4 Sketch the graph of f if $f(x) = 8 - 2x - x^2$, and find the maximum value of $f(x)$.

Solution We know from the discussion in this section that the graph is a parabola. Let us set $y = f(x)$ and find the vertex by completing the square:

$$\begin{aligned} y &= -x^2 - 2x + 8 \\ &= -(x^2 + 2x) + 8 \\ &= -(x^2 + 2x + 1) + 8 + 1 \\ &= -(x + 1)^2 + 9. \end{aligned}$$

If we now write

$$y - 9 = -(x + 1)^2$$

and compare with the standard equation of a parabola, we see that $h = -1$, $k = 9$, and hence the vertex is $(-1, 9)$. Since $a = -1 < 0$, the parabola opens downward. To find the x-intercepts we solve the equation $8 - 2x - x^2 = 0$, or equivalently, $x^2 + 2x - 8 = 0$. Factoring gives us $(x + 4)(x - 2) = 0$, and hence the intercepts are $x = -4$ and $x = 2$. The y-intercept is 8. Using this information gives us the sketch in Figure 2.44. The maximum value of f occurs at the vertex and hence is $f(-1) = 9$. ■

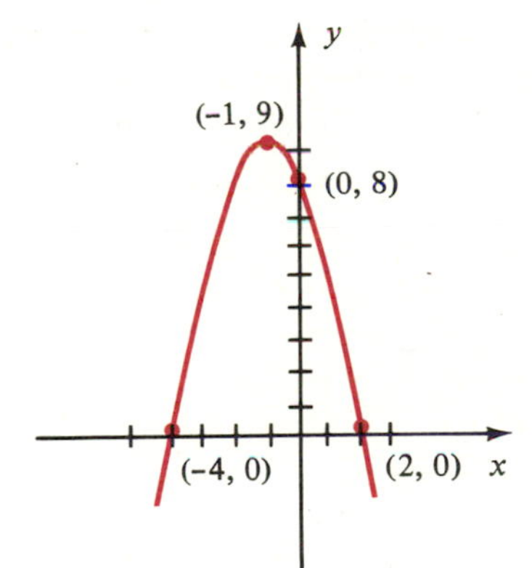

FIGURE 2.44 $y = 8 - 2x - x^2$

The x-intercepts of the graph of $f(x) = ax^2 + bx + c$ are the solutions of $ax^2 + bx + c = 0$. An equation of this type is called a **quadratic equation.** In the preceding examples we found solutions by factoring. Since factorizations are not always obvious, let us derive a general formula for solving any equation of the form

$$ax^2 + bx + c = 0$$

where $a \neq 0$. Dividing both sides by a gives us

$$x^2 + \frac{b}{a}x + \frac{c}{a} = 0$$

or equivalently,
$$x^2 + \frac{b}{a}x = -\frac{c}{a}.$$

We next complete the square by adding $(b/2a)^2$ to both sides:
$$x^2 + \frac{b}{a}x + \left(\frac{b}{2a}\right)^2 = \left(\frac{b}{2a}\right)^2 - \frac{c}{a}.$$

We may now write
$$\left(x + \frac{b}{2a}\right)^2 = \frac{b^2 - 4ac}{4a^2}.$$

If $b^2 - 4ac \geq 0$, then
$$x + \frac{b}{2a} = \pm\sqrt{\frac{b^2 - 4ac}{4a^2}}$$

and using the fact that $\sqrt{4a^2} = |2a|$,
$$x = -\frac{b}{2a} \pm \sqrt{\frac{b^2 - 4ac}{4a^2}} = -\frac{b}{2a} \pm \frac{\sqrt{b^2 - 4ac}}{2a}.$$

Thus, if the quadratic equation $ax^2 + bx + c = 0$ has solutions, then they are given by the two numbers in the last formula. Moreover, it can be shown by direct substitution that the two numbers do satisfy the equation. The case $b^2 - 4ac < 0$ is discussed in Section 3.4, where it is shown that the same formula holds if complex numbers are considered. The following statement summarizes our work.

QUADRATIC FORMULA

If $a \neq 0$, then the solutions of $ax^2 + bx + c = 0$ are
$$x = \frac{-b \pm \sqrt{b^2 - 4ac}}{2a}.$$

If $b^2 - 4ac > 0$, the equation has two real and unequal solutions and the graph has two x-intercepts. If $b^2 - 4ac = 0$, the equation has one (double) solution and the graph is tangent to the x-axis. If $b^2 - 4ac < 0$, the equation has no real solutions, and hence, the graph has no x-intercepts. We have illustrated the three cases in Figure 2.45 if $a > 0$. A similar situation occurs if $a < 0$, but in this case the parabolas open downward.

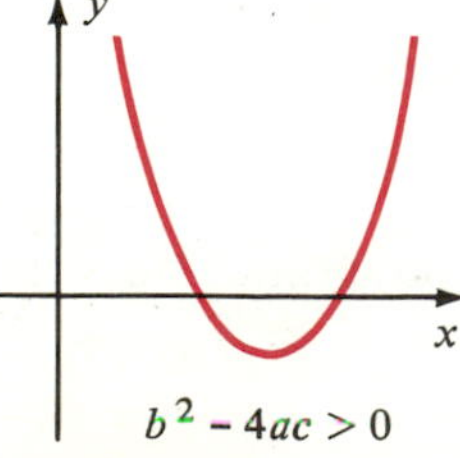

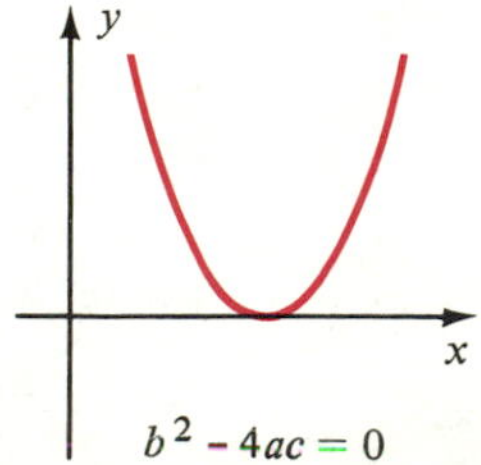

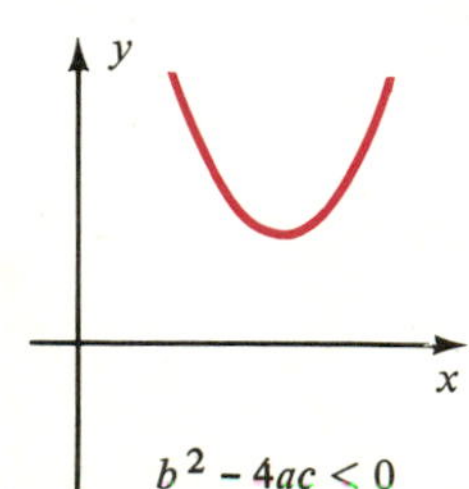

FIGURE 2.45 $f(x) = ax^2 + bx + c$

EXAMPLE 5 Solve the equation $2x^2 - 6x + 3 = 0$.

Solution Letting $a = 2$, $b = -6$, and $c = 3$ in the Quadratic Formula,

$$x = \frac{6 \pm \sqrt{(-6)^2 - 4(2)(3)}}{2(2)} = \frac{6 \pm \sqrt{12}}{4}$$

$$= \frac{6 \pm 2\sqrt{3}}{4} = \frac{2(3 \pm \sqrt{3})}{4}$$

$$= \frac{3 \pm \sqrt{3}}{2}.$$

Hence, the solutions are $x = (3 \pm \sqrt{3})/2$. ■

We can solve certain applied problems by finding maximum or minimum values of quadratic functions. The next example is one illustration.

EXAMPLE 6 A long rectangular sheet of metal, 12 inches wide, is to be made into a rain gutter by turning up two sides so that they are perpendicular to the sheet. How many inches should be turned up to give the gutter its greatest capacity?

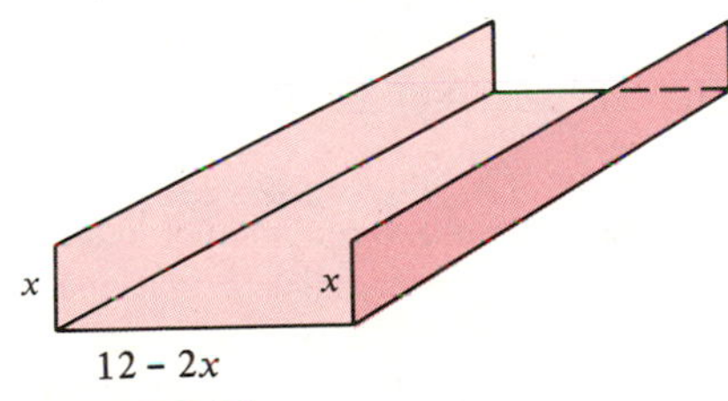

FIGURE 2.46

Solution The gutter is illustrated in Figure 2.46, where, if x denotes the number of inches turned up on each side, then the width of the base of the gutter is $12 - 2x$ inches. (Why?) The capacity will be greatest when the area of the rectangle with sides of length x and $12 - 2x$ has its greatest value. Letting $f(x)$ denote this area, we have

$$f(x) = x(12 - 2x) = 12x - 2x^2 = -2x^2 + 12x.$$

We may find the maximum value of $f(x)$ by completing the square:

$$\begin{aligned} f(x) &= -2(x^2 - 6x) \\ &= -2(x^2 - 6x + 9) + 18 \\ &= -2(x - 3)^2 + 18. \end{aligned}$$

The maximum value of $f(x)$ occurs at $x = 3$. (Why?) Hence, 3 inches should be turned up to achieve maximum capacity. ■

EXERCISES 2.5

1 Given $f(x) = ax^2 + 2$, sketch the graph of f for each value of a.

(a) $a = 2$ (b) $a = 5$

(c) $a = \frac{1}{2}$ (d) $a = -3$

2 Given $f(x) = 4x^2 + c$, sketch the graph of f for each value of c.

(a) $c = 2$ (b) $c = 5$

(c) $c = \frac{1}{2}$ (d) $c = -3$

In Exercises 3–6 sketch the graph of f and find the vertex.

3 $f(x) = 4x^2 - 9$

4 $f(x) = 2x^2 + 3$

5 $f(x) = 9 - 4x^2$

6 $f(x) = 16 - 9x^2$

In Exercises 7–14 use the Quadratic Formula to find the zeros of f.

7 $f(x) = 4x^2 - 11x - 3$

8 $f(x) = 25x^2 + 10x + 1$

9 $f(x) = 9x^2 - 12x + 4$

10 $f(x) = 10x^2 + x - 21$

11 $f(x) = 2x^2 + x - 5$

12 $f(x) = 2x^2 + 4x - 3$

13 $f(x) = \dfrac{2x + 1}{x - 3} - \dfrac{x - 1}{x + 2}$

14 $f(x) = 9x^2 + 13x$

In Exercises 15–20 use the technique of completing the square to express $f(x)$ in the form $a(x - h)^2 + k$, and find the maximum or minimum value of $f(x)$.

15 $f(x) = 2x^2 - 16x + 23$

16 $f(x) = 3x^2 - 12x + 7$

17 $f(x) = -5x^2 - 10x + 3$

18 $f(x) = -2x^2 - 12x - 12$

19 $f(x) = 4x^2 - 4x + 2$

20 $f(x) = 9x^2 + 12x - 8$

In Exercises 21–30 sketch the graph of f and find the maximum or minimum value of $f(x)$.

21 $f(x) = x^2 + 5x + 4$

22 $f(x) = x^2 - 6x$

23 $f(x) = 8x - 12 - x^2$

24 $f(x) = 10 + 3x - x^2$

25 $f(x) = x^2 + x + 3$

26 $f(x) = x^2 + 2x + 5$

27 $f(x) = 3x^2 - 12x + 16$

28 $f(x) = 2x^2 + 12x + 13$

29 $f(x) = -5x^2 - 10x - 4$

30 $f(x) = -3x^2 + 24x - 42$

31 If $f(x) = x^2 + kx - 5 + 3k$, find two different values for k such that f has only one zero.

32 If $f(x) = 3x^2 + kx - 4k$, find all values of k such that f has no real zeros.

33 If one zero of $f(x) = kx^2 + 3x + k$ is -2, find the other zero.

34 Find a real number k such that the graph of the equation $y = x^2 + kx + 3$ contains the point $P(-1, -2)$.

35 Find two real numbers whose difference is 40 and whose product is a minimum. (*Hint:* Let x and $x - 40$ denote the numbers.)

36 Find two positive real numbers whose sum is 40 and whose product is a maximum.

37 A piece of wire 24 inches long is bent into the shape of a rectangle. Prove that the enclosed area will be greatest if the rectangle is a square.

38 A person standing on the top of a building projects an object directly upward with a velocity of 144 ft/sec. Its height $s(t)$ in feet above the ground after t seconds is given by $s(t) = -16t^2 + 144t + 100$. What is its maximum height? What is the height of the building?

39 A man wishes to put a fence around a rectangular field and then subdivide the field into three smaller rectangular plots by placing two fences parallel to one of the sides (see figure). If he can afford only 1000 yards of fencing, what dimensions will give the maximum rectangular area?

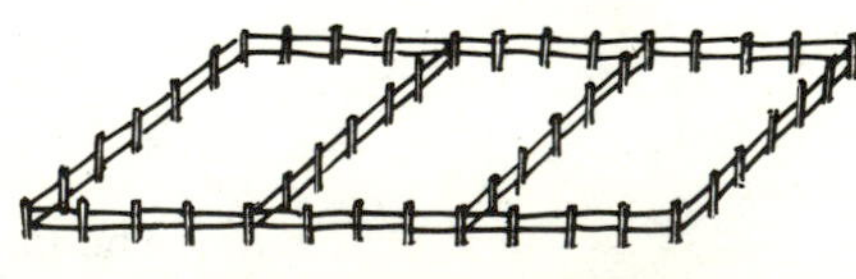

FIGURE FOR EXERCISE 39

40 A company sells running shoes to dealers at a rate of \$20 per pair if less than 50 pairs are ordered. If a dealer orders 50 or more pairs (up to 600), the price per pair is reduced at a rate of 2 cents times the number ordered. What size order will produce the maximum amount of money for the company?

SECTION 2.6

COMPOSITE AND INVERSE FUNCTIONS

Suppose X, Y, and Z are sets of real numbers, and let f be a function from X to Y and g a function from Y to Z. Using arrow notation we have

$$X \xrightarrow{f} Y \xrightarrow{g} Z.$$

This is read "f maps X into Y and g maps Y into Z." We shall use f and g to define a function from X to Z. For every x in X, the number $f(x)$ is in Y. Since the domain of g is Y, we may then find the image of $f(x)$ under g. Of course, this element of Z is written as $g(f(x))$. By associating $g(f(x))$ with x, we obtain a function from X to Z called the *composite function* of g by f. This is illustrated pictorially in Figure 2.47, where the dashes indicate the correspondence we have defined from X to Z.

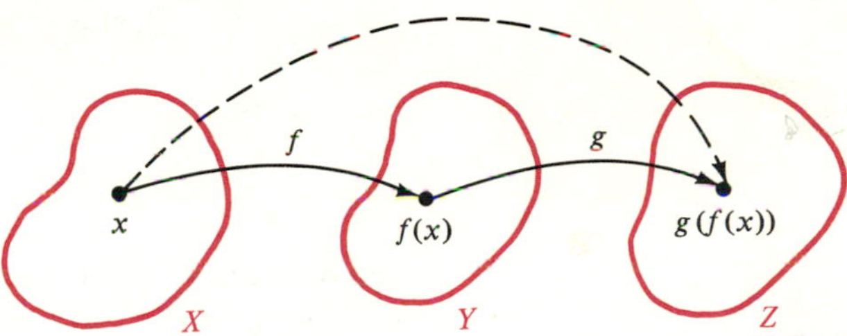

FIGURE 2.47

We sometimes use an operational symbol $\circ$ and denote a composite function by $g \circ f$. The following definition summarizes our discussion.

DEFINITION

> If f is a function from X to Y and g is a function from Y to Z, then the **composite function $g \circ f$** is the function from X to Z defined by
>
> $$(g \circ f)(x) = g(f(x))$$
>
> for every x in X.

EXAMPLE 1 If $f(x) = x^3$ and $g(x) = 5x^2 + 2x + 1$, find $(g \circ f)(x)$.

Solution By definition,

$$(g \circ f)(x) = g(f(x)) = g(x^3).$$

Since $g(x^3)$ means that x^3 should be substituted for x in the expression for $g(x)$, we have

$$g(x^3) = 5(x^3)^2 + 2(x^3) + 1.$$

Consequently, $$(g \circ f)(x) = 5x^6 + 2x^3 + 1.$$ ■

When we apply the definition of $g \circ f$, it is not essential that the domain of g be all of Y, but merely that it *contain* the range of f. In certain cases we may wish to restrict x to some subset of X so that $f(x)$ is in the domain of g, as illustrated in the next example.

EXAMPLE 2 If the functions f and g are defined by $f(x) = x - 2$ and $g(x) = 5x + \sqrt{x}$, find $(g \circ f)(x)$ and the domain of $g \circ f$.

Solution Formal substitutions give us the following:

$$\begin{aligned}(g \circ f)(x) &= g(f(x)) && \text{(definition of } g \circ f) \\ &= g(x - 2) && \text{(definition of } f) \\ &= 5(x - 2) + \sqrt{x - 2} && \text{(definition of } g) \\ &= 5x - 10 + \sqrt{x - 2} && \text{(simplifying)}\end{aligned}$$

The domain X of f is the set of all real numbers; however, the last equality implies that $(g \circ f)(x)$ is a real number only if $x \geq 2$. Thus, the domain of the composite function $g \circ f$ is the interval $[2, \infty)$. ■

Given f and g, it may also be possible to find $f(g(x))$. In this case we first obtain the image of x under g and then apply f to $g(x)$. This gives us a function called the *composite function of f by g* and denoted by $f \circ g$.

DEFINITION OF $f \circ g$

$$(f \circ g)(x) = f(g(x))$$

EXAMPLE 3 If $f(x) = x^2 - 1$ and $g(x) = 3x + 5$, find $(f \circ g)(x)$ and $(g \circ f)(x)$.

Solution

$$\begin{aligned}(f \circ g)(x) &= f(g(x)) && \text{(definition of } f \circ g) \\ &= f(3x + 5) && \text{(definition of } g) \\ &= (3x + 5)^2 - 1 && \text{(definition of } f) \\ &= 9x^2 + 30x + 24 && \text{(simplifying)}\end{aligned}$$

Similarly,

$$\begin{aligned}(g \circ f)(x) &= g(f(x)) && \text{(definition of } g \circ f\text{)}\\ &= g(x^2 - 1) && \text{(definition of } f\text{)}\\ &= 3(x^2 - 1) + 5 && \text{(definition of } g\text{)}\\ &= 3x^2 + 2 && \text{(simplifying).}\end{aligned}$$

We see from Example 3 that $f(g(x))$ and $g(f(x))$ are not always the same, that is, $f \circ g \neq g \circ f$. In certain cases it may happen that equality *does* occur. Of major importance is the case when $f(g(x))$ and $g(f(x))$ are not only identical, but both are equal to x. Of course, f and g must be very special functions for this to happen. In the following discussion we indicate the manner in which these functions will be restricted.

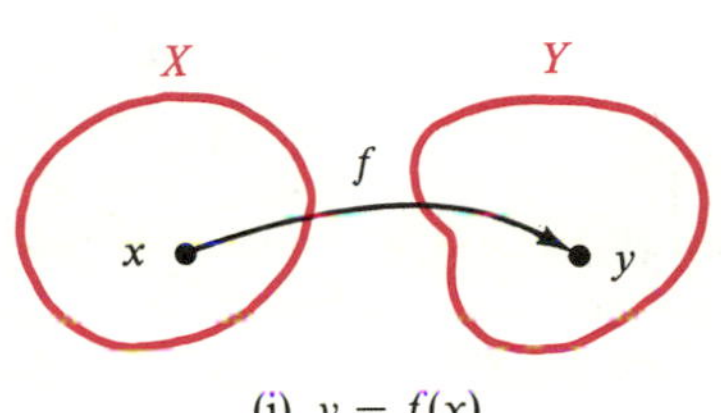

(i) $y = f(x)$

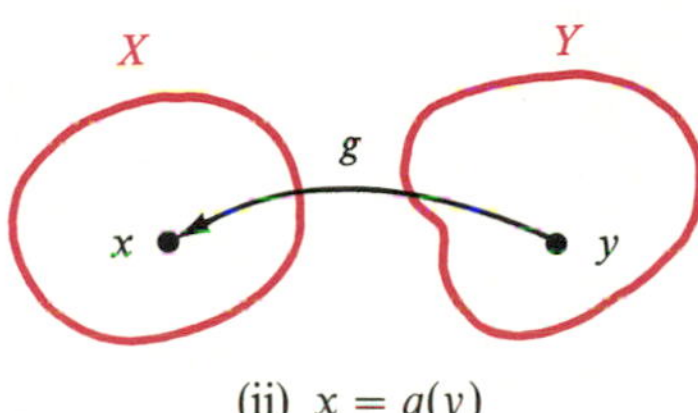

(ii) $x = g(y)$

FIGURE 2.48

Suppose f is a *one-to-one function* with domain X and range Y. This implies that each element of Y is the image of precisely one element of X. Another way of phrasing this is to say that *each element of Y can be written in one and only one way in the form $f(x)$*, where x is in X. We may then define a function g from Y to X by demanding that

$$g(f(x)) = x \quad \text{for every } x \text{ in } X.$$

This amounts to *reversing* the correspondence given by f. If f is represented geometrically by drawing arrows, as in (i) of Figure 2.48, then g can be represented by simply *reversing* these arrows, as illustrated in (ii) of the figure. It follows that g is a one-to-one function with domain Y and range X. (See Exercise 38.)

As illustrated in Figure 2.48, if $f(x) = y$, then $x = g(y)$. This means that

$$f(g(y)) = y \quad \text{for every } y \text{ in } Y.$$

Since the notation used for the variable is immaterial, we may write

$$f(g(x)) = x \quad \text{for every } x \text{ in } Y.$$

The functions f and g are called *inverse functions* of one another, according to the following definition.

DEFINITION

Let f be a one-to-one function with domain X and range Y. A function g with domain Y and range X is called the **inverse function of f** if

$$g(f(x)) = x \quad \text{for every } x \text{ in } X$$

and $$f(g(x)) = x \quad \text{for every } x \text{ in } Y.$$

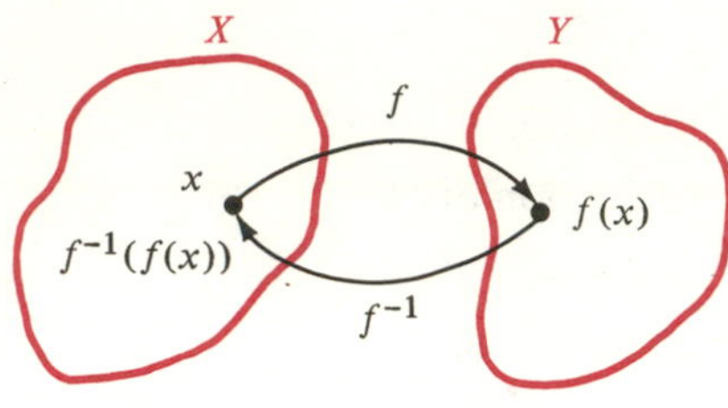

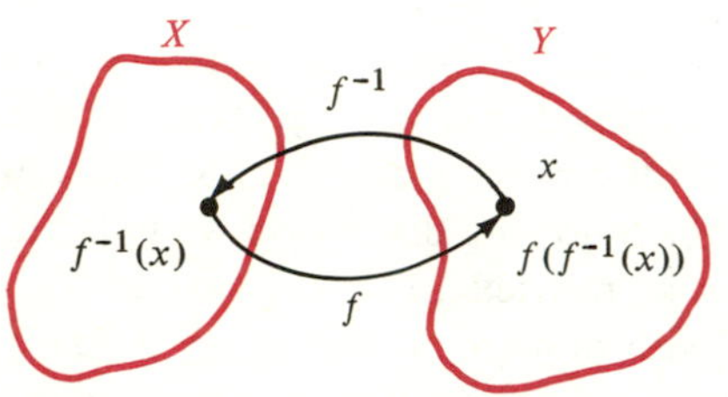

FIGURE 2.49

The symbol f^{-1} is often used to denote the inverse function of f. Employing this notation,

$$f^{-1}(f(x)) = x \quad \text{for every } x \text{ in } X,$$
$$f(f^{-1}(x)) = x \quad \text{for every } x \text{ in } Y.$$

The symbol -1 used here should not be mistaken for an exponent; that is, $f^{-1}(x)$ does not mean $1/f(x)$. The reciprocal $1/f(x)$ may be denoted by $[f(x)]^{-1}$.

Although the preceding definition seems abstract, it is merely a formal way of stating that f^{-1} reverses the correspondence given by f, and that f reverses the correspondence given by f^{-1}. Geometric interpretations for the last two identities are shown in Figure 2.49.

Inverse functions are very important in the study of trigonometry. In Chapter 4 we shall discuss two other special types of inverse functions. It is important to remember that to define the inverse of a function f, *it is absolutely essential that f be one-to-one.* The most common examples of one-to-one functions are those that are either increasing or decreasing on their domains, for, in this case, if $a \neq b$ in the domain, then $f(a) \neq f(b)$ in the range.

The diagrams in Figure 2.48 contain a hint for finding the inverse of a one-to-one function f in certain cases. If possible, *solve the equation $y = f(x)$ for x in terms of y*, obtaining an equation of the form $x = g(y)$. If the two conditions $f(g(x)) = x$ and $g(f(x)) = x$ are true for all x in the domains of f and g, then g is the required inverse function f^{-1}. The success of this method depends on the nature of the equation $y = f(x)$. The next two examples illustrate the method if f is either a linear or quadratic function.

EXAMPLE 4 If $f(x) = 3x - 5$, find the inverse function of f.

Solution The graph of the linear function f is a line of slope 3 (Why?), and hence f is increasing for all x. It follows that f is a one-to-one function with domain and range $\mathbb{R}$, and hence the inverse function g exists. If we let

$$y = 3x - 5$$

and then solve for x in terms of y, we obtain

$$x = \frac{y + 5}{3}.$$

Letting $$g(y) = \frac{y + 5}{3}$$

gives us a function g that reverses the correspondence determined by f. Since the symbol used for the independent variable is immaterial, we may replace y by x in the expression for g, obtaining

$$g(x) = \frac{x + 5}{3}.$$

To verify that g is actually the inverse function of f, we must verify that the two conditions $f(g(x)) = x$ and $g(f(x)) = x$ are fulfilled. Thus,

$$\begin{aligned} f(g(x)) &= f\left(\frac{x+5}{3}\right) && \text{(definition of } g) \\ &= 3\left(\frac{x+5}{3}\right) - 5 && \text{(definition of } f) \\ &= x && \text{(simplifying).} \end{aligned}$$

Also,

$$\begin{aligned} g(f(x)) &= g(3x-5) && \text{(definition of } f) \\ &= \frac{(3x-5)+5}{3} && \text{(definition of } g) \\ &= x && \text{(simplifying).} \end{aligned}$$

This proves that g is the inverse function of f. Using the f^{-1} notation,

$$f^{-1}(x) = \frac{x+5}{3}.$$

■

EXAMPLE 5 Find the inverse function of f if the domain X is the interval $[0, \infty)$ and $f(x) = x^2 - 3$ for all x in X.

Solution The domain has been restricted so that f is increasing and hence is one-to-one. The range of f is the interval $[-3, \infty)$. As in Example 4 we begin by considering the equation

$$y = x^2 - 3.$$

Solving for x gives us $\quad x = \pm\sqrt{y+3}.$

Since x is nonnegative we reject $x = -\sqrt{y+3}$ and, as in the preceding example, we let

$$g(y) = \sqrt{y+3} \quad \text{or, equivalently,} \quad g(x) = \sqrt{x+3}.$$

Since $\quad f(g(x)) = f(\sqrt{x+3}) = (\sqrt{x+3})^2 - 3 = (x+3) - 3 = x$

and $\quad g(f(x)) = g(x^2-3) = \sqrt{(x^2-3)+3} = x,$

we see that $\quad f^{-1}(x) = \sqrt{x+3} \quad \text{where } x \geq -3.$ ■

An interesting relationship exists between the graphs of a function f and its inverse function f^{-1}. We first note that f maps a into b if and only if f^{-1} maps b into a; that is, $b = f(a)$ means the same thing as $a = f^{-1}(b)$. These equations imply that the point (a, b) is on the graph of f if and only if the point (b, a) is on the graph of f^{-1}. As an illustration, in Example 5

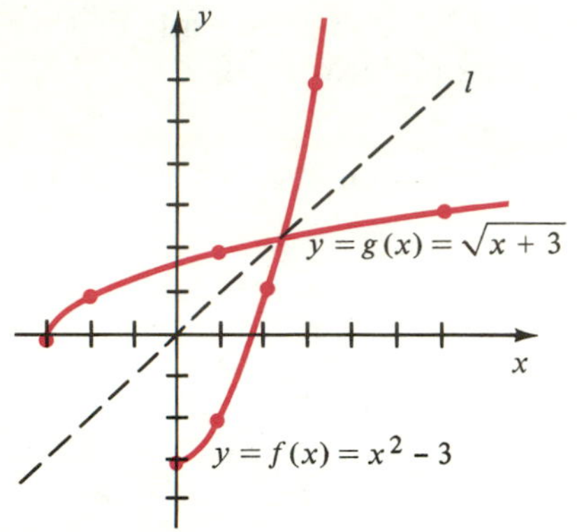

FIGURE 2.50

we found that the functions f and f^{-1} given by

$$f(x) = x^2 - 3 \quad \text{and} \quad f^{-1}(x) = \sqrt{x+3}$$

are inverse functions of one another, provided that x is suitably restricted. Some points on the graph of f are $(0, -3)$, $(1, -2)$, $(2, 1)$, and $(3, 6)$. Corresponding points on the graph of f^{-1} are $(-3, 0)$, $(-2, 1)$, $(1, 2)$, and $(6, 3)$. The graphs of f and f^{-1} are sketched on the same coordinate axes in Figure 2.50. If the page is folded along the line l that bisects quadrants I and III (as indicated by the dashes in the figure), then the graphs of f and f^{-1} coincide. Note that an equation for l is $y = x$. The two graphs are said to be *reflections* of one another through the line l (or *symmetric* with respect to l). This is typical of the graph of every function f that has an inverse function f^{-1}.

EXERCISES 2.6

In Exercises 1–18 find $(f \circ g)(x)$ and $(g \circ f)(x)$.

1 $f(x) = 3x + 2$, $g(x) = 2x - 1$

2 $f(x) = x + 4$, $g(x) = 5x - 3$

3 $f(x) = 4x^2 - 5$, $g(x) = 3x$

4 $f(x) = 7x + 1$, $g(x) = 2x^2$

5 $f(x) = 3x^2 + 2x$, $g(x) = 2x - 1$

6 $f(x) = 5x + 7$, $g(x) = 4 - x^2$

7 $f(x) = x - 1$, $g(x) = x^3$

8 $f(x) = x^3 + 5x$, $g(x) = 4x$

9 $f(x) = x^2 + 9x$, $g(x) = \sqrt{x+9}$

10 $f(x) = \sqrt[3]{x^2+1}$, $g(x) = x^3 + 1$

11 $f(x) = \dfrac{1}{2x-5}$, $g(x) = \dfrac{1}{x^2}$

12 $f(x) = \dfrac{x}{x+1}$, $g(x) = x - 1$

13 $f(x) = |x|$, $g(x) = -5$

14 $f(x) = 7$, $g(x) = 10$

15 $f(x) = x^2$, $g(x) = 1/x^2$

16 $f(x) = \dfrac{1}{x+1}$, $g(x) = x + 1$

17 $f(x) = 2x - 3$, $g(x) = \dfrac{x+3}{2}$

18 $f(x) = x^3 - 1$, $g(x) = \sqrt[3]{x+1}$

In Exercises 19–22 prove that f and g are inverse functions of one another and sketch the graphs of f and g on the same coordinate plane.

19 $f(x) = 7x + 5$; $g(x) = (x - 5)/7$

20 $f(x) = x^2 - 1, x \geq 0$; $g(x) = \sqrt{x+1}, x \geq -1$

21 $f(x) = \sqrt{2x-4}, x \geq 2$; $g(x) = \frac{1}{2}(x^2 + 4), x \geq 0$

22 $f(x) = x^3 + 1$; $g(x) = \sqrt[3]{x-1}$

In Exercises 23–36 find the inverse function of f.

23 $f(x) = 4x - 3$

24 $f(x) = 9 - 7x$

25 $f(x) = \dfrac{1}{2x+5}$, $x > -\frac{5}{2}$

26 $f(x) = \dfrac{1}{3x-1}$, $x > \frac{1}{3}$

27 $f(x) = 9 - x^2$, $x \geq 0$

28 $f(x) = 4x^2 + 1$, $x \geq 0$

29 $f(x) = 5x^3 - 2$

30 $f(x) = 7 - 2x^3$

31 $f(x) = \sqrt{3x - 5},\ x \geq \frac{5}{3}$

32 $f(x) = \sqrt{4 - x^2},\ 0 \leq x \leq 2$

33 $f(x) = \sqrt[3]{x} + 8$

34 $f(x) = (x^3 + 1)^5$

35 $f(x) = x$

36 $f(x) = -x$

37 (a) Prove that the linear function defined by $f(x) = ax + b$, where $a \neq 0$, has an inverse function, and find $f^{-1}(x)$.

(b) Does a contant function have an inverse? Explain.

38 If f is a one-to-one function with domain X and range Y, prove that f^{-1} is a one-to-one function with domain Y and range X.

39 Prove that a one-to-one function can have at most one inverse function.

40 If $f(x) = ax + b$ and $g(x) = cx + d$, find conditions on c and d in terms of a and b that will guarantee that $f \circ g = g \circ f$. Discuss the case in which $a = b = 1$.

41 If f is a linear function and g is a quadratic function, show that $f \circ g$ and $g \circ f$ are quadratic functions.

42 If both f and g are quadratic functions, show that $(f \circ g)(x)$ is a polynomial of degree 4.

SECTION 2.7 REVIEW

Define or discuss each of the following.

1 Ordered pair
2 Rectangular coordinate system in a plane
3 Coordinate axes
4 Quadrants
5 Coordinates of a point
6 Distance formula
7 Midpoint formula
8 Graph of an equation in x and y
9 Tests for symmetry
10 Equation of a circle
11 Unit circle
12 Slope of a line
13 Point-Slope Form
14 Slope-Intercept Form
15 Linear equation in x and y
16 Function
17 Domain and range of a function
18 One-to-one function
19 Identity function
20 Constant function
21 Even function
22 Odd function
23 Graph of a function
24 Linear function
25 Increasing function
26 Decreasing function
27 Vertical shifts of graphs
28 Horizontal shifts of graphs
29 Stretching of graphs
30 Reflections of graphs
31 Quadratic function
32 Graph of a quadratic function
33 Quadratic formula
34 Composite function of two functions
35 Inverse function

EXERCISES 2.7

1 Plot the points $A(3, 1)$, $B(-5, -3)$, and $C(4, -1)$, and prove that they are vertices of a right triangle. What is the area of the triangle?

2 Given points $P(-5, 9)$ and $Q(-8, -7)$, find (a) the midpoint of the segment PQ, and (b) a point T such that Q is the midpoint of PT.

3 Describe the set of all points (x, y) in a coordinate plane such that $y/x < 0$.

4 Find the slope of the line through $C(11, -5)$ and $D(-8, 6)$.

5 Prove that the points $A(-3, 1)$, $B(1, -1)$, $C(4, 1)$, and $D(3, 5)$ are vertices of a trapezoid.

6 Find an equation of the circle that has center $C(7, -4)$ and passes through the point $Q(-3, 3)$.

7 Find an equation of the circle that has center $C(-5, -1)$ and is tangent to the line $x = 4$.

8 Express the equation $8x + 3y - 24 = 0$ in Slope-Intercept Form.

9 Find an equation of the line through $A(\frac{1}{2}, -\frac{1}{3})$ that is (a) parallel to the line $6x + 2y + 5 = 0$; (b) perpendicular to the line $6x + 2y + 5 = 0$.

10 Find an equation of the line that has x-intercept -3 and passes through the center of the circle that has equation $x^2 + y^2 - 4x + 10y + 26 = 0$.

Sketch the graphs of the equations in Exercises 11–21.

11 $2y + 5x - 8 = 0$ **12** $x = 3y + 4$

13 $x + 5 = 0$ **14** $2y - 7 = 0$

15 $y = \sqrt{1 - x}$ **16** $3x - 7y^2 = 0$

17 $9y^2 + 2x = 0$ **18** $y^2 = 16 - x^2$

19 $x^2 + y^2 + 4x - 16y + 64 = 0$

20 $y = -4x^2 + 8x - 6$

21 $y = x^2 + 6x + 16$

22 Find the domain and range of f if:

(a) $f(x) = \sqrt{3x - 4}$

(b) $f(x) = 1/(x + 3)^2$

23 If $f(x) = x/\sqrt{x + 3}$ find (a)–(g):

(a) $f(1)$ (b) $f(-1)$

(c) $f(0)$ (d) $f(-x)$

(e) $-f(x)$ (f) $f(x^2)$

(g) $(f(x))^2$

In Exercises 24–30 sketch the graph of f. Find the domain, the range, and the intervals in which f is increasing or decreasing.

24 $f(x) = |x + 3|$ **25** $f(x) = \dfrac{1 - 3x}{2}$

26 $f(x) = \sqrt{2 - x}$ **27** $f(x) = 1 - \sqrt{x + 1}$

28 $f(x) = 9 - x^2$ **29** $f(x) = 1000$

30 $f(x) = \begin{cases} x^2 & \text{if } x < 0 \\ 3x & \text{if } 0 \le x < 2 \\ 6 & \text{if } x \ge 2 \end{cases}$

31 Sketch the graphs of the following equations, making use of shifting, stretching, or reflecting.

(a) $y = \sqrt{x}$ (b) $y = \sqrt{x + 4}$

(c) $y = \sqrt{x} + 4$ (d) $y = 4\sqrt{x}$

(e) $y = \frac{1}{4}\sqrt{x}$ (f) $y = -\sqrt{x}$

32 Determine whether f is even, odd, or neither even nor odd.

(a) $f(x) = \sqrt[3]{x^3 + 4x}$

(b) $f(x) = \sqrt[3]{3x^2 - x^3}$

(c) $f(x) = \sqrt[3]{x^4 + 3x^2 + 5}$

In Exercises 33 and 34 find the zeros of f.

33 $f(x) = 2x^2 - 4x - 5$ **34** $f(x) = 5x^2 + 9x + 3$

In Exercises 35 and 36 find the maximum or minimum value of $f(x)$ by completing the square.

35 $f(x) = 5x^2 + 30x + 49$

36 $f(x) = -3x^2 + 30x - 82$

37 The interior of a half-mile racetrack consists of a rectangle with semicircles at two opposite ends. Find the dimensions that will maximize the area of the rectangle.

38 At 1:00 P.M. ship A is 30 miles due south of ship B and is sailing north at a rate of 15 miles per hour. If ship B is sailing west at a rate of 10 miles per hour, find the time at which the distance between the ships is minimal (see the following figure).

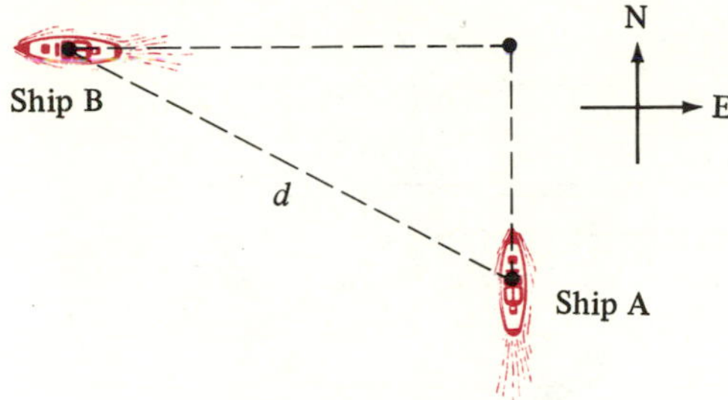

FIGURE FOR EXERCISE 38

In Exercises 39 and 40 find $(f \circ g)(x)$ and $(g \circ f)(x)$.

39 $f(x) = 2x^2 - 5x + 1, \; g(x) = 3x + 2$

40 $f(x) = \sqrt{3x + 2}, \; g(x) = 1/x^2$

In Exercises 41 and 42 find $f^{-1}(x)$ and sketch the graphs of f and f^{-1} on the same coordinate plane.

41 $f(x) = 10 - 15x$

42 $f(x) = 9 - 2x^2, \; x \le 0$

43 If the altitude and radius of a right circular cylinder are equal, express the volume V as a function of the circumference C of the base.

44 A company plans to manufacture a container having the shape of a right circular cylinder, open at the top, and having a capacity of 24π in.^{3}. If the cost of the material for the bottom is 30¢ per in.^{2} and that for the curved sides is 10¢ per in.^{2}, express the total cost C of the material as a function of the radius r of the base of the container.

3

POLYNOMIAL AND RATIONAL FUNCTIONS

Polynomial functions are the most basic functions in algebra—they are defined only in terms of additions and multiplications. Techniques for sketching their graphs are discussed in the first section of this chapter. We then turn our attention to the operation of division and study methods for finding zeros of polynomial functions. Finally, we consider quotients of polynomial functions, that is, *rational functions.*

SECTION 3.1
GRAPHS OF POLYNOMIAL FUNCTIONS

Among the most important functions in mathematics are polynomial functions.

DEFINITION

A function f is a **polynomial function** if

$$f(x) = a_nx^n + a_{n-1}x^{n-1} + \cdots + a_1x + a_0$$

where the coefficients $a_0, a_1, \ldots, a_n$ are real numbers and the exponents are nonnegative integers.

According to our work in Section 1.3, f is a polynomial function if $f(x)$ is a polynomial, and we say that f has **degree n** if $f(x)$ has degree n. The domain of a polynomial function is $\mathbb{R}$. If the degree is odd, then the range is also $\mathbb{R}$; however, if the degree is even, then the range is an infinite interval of the form $(-\infty, a]$ or $[a, \infty)$. These facts are illustrated by the graphs in this section.

Recall that if $f(c) = 0$, then c is called a **zero** of f, or of $f(x)$. We also call c a **solution,** or **root,** of the equation $f(x) = 0$. The zeros of f are the x-intercepts of the graph of f.

If a polynomial function has degree 0, then $f(x) = a$ for some nonzero real number a, and the graph is a horizontal line. Polynomial functions of degrees 1 and 2 are linear and quadratic functions, respectively. We discussed their graphs (lines and parabolas) in Chapter 2. In this section we shall concentrate on graphs of polynomial functions of degree greater than 2.

If, in the preceding definition, all the coefficients except a_n are zero, then we may write

$$f(x) = ax^n \qquad \text{where } a = a_n \neq 0.$$

In this case, if $n = 1$, the graph of f is a line that passes through the origin, whereas if $n = 2$, the graph is a parabola with vertex at the origin. Several illustrations with $n = 3$ are given in the next example.

EXAMPLE 1 Sketch the graph of f if (a) $f(x) = \frac{1}{2}x^3$; (b) $f(x) = -\frac{1}{2}x^3$.

Solution

(a) The following table lists several points on the graph of $f(x) = \frac{1}{2}x^3$.

x	0	$\frac{1}{2}$	1	$\frac{3}{2}$	2	$\frac{5}{2}$
$f(x)$	0	$\frac{1}{16}$	$\frac{1}{2}$	$\frac{27}{16}$	4	$\frac{125}{16} \approx 7.8$

Since f is an odd function, the graph of f is symmetric with respect to the origin (see Section 2.3), and hence the points $(-\frac{1}{2}, -\frac{1}{16})$, $(-1, -\frac{1}{2})$, etc., are also on the graph. The graph is sketched in (i) of Figure 3.1.

(b) If $f(x) = -\frac{1}{2}x^3$, the graph can be obtained from that in part (a) by multiplying all y-coordinates by -1. Reflecting the graph in part (a) through the x-axis, we obtain the sketch shown in (ii) of Figure 3.1. ■

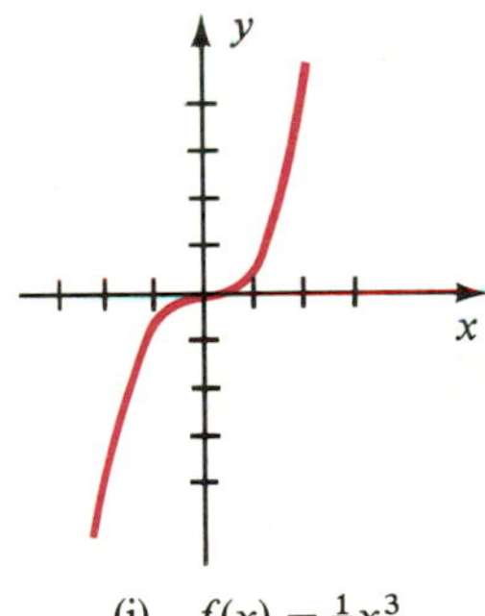

(i) $f(x) = \frac{1}{2}x^3$

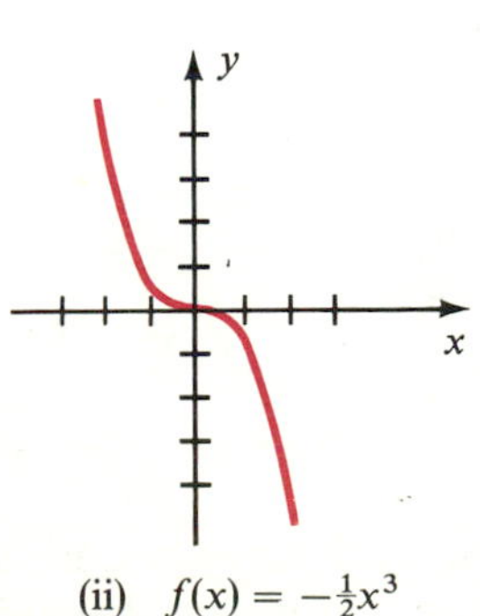

(ii) $f(x) = -\frac{1}{2}x^3$

FIGURE 3.1

In general, if $f(x) = ax^3$, then increasing the absolute value of the coefficient a results in a graph that rises or falls more sharply. For example, if $f(x) = 10x^3$, then $f(1) = 10$, $f(2) = 80$, and $f(-2) = -80$. The effect of using a larger exponent and holding a fixed as, for example, in $f(x) = \frac{1}{2}x^5$, also leads to a graph that rises or falls more rapidly.

If n is an *even* integer, then the graph of $f(x) = ax^n$ is symmetric with respect to the y-axis. (Why?) Two illustrations (with $a = 1$) are given in (i) and (ii) of Figure 3.2. Note that as the exponent increases, the graph becomes flatter at the origin. It also rises (or falls) more rapidly if we let x increase through positive (or negative) values. Using a negative coefficient, such as -1, inverts the graph.

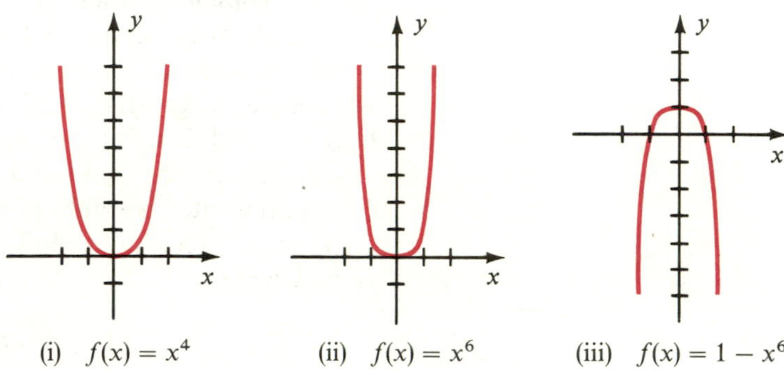

(i) $f(x) = x^4$ (ii) $f(x) = x^6$ (iii) $f(x) = 1 - x^6$

FIGURE 3.2

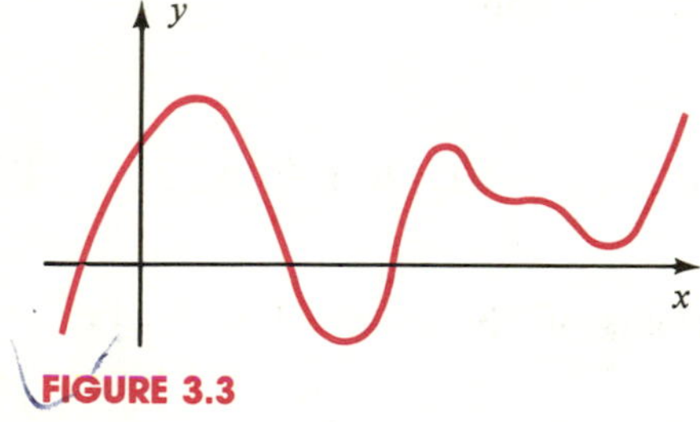

FIGURE 3.3

The graph of $f(x) = ax^n + c$ where $c \neq 0$ can be found by a vertical shift of the graph of $f(x) = ax^n$. For example, if $f(x) = 1 - x^6$, we obtain the sketch shown in (iii) of Figure 3.2.

A complete analysis of graphs of polynomial functions of degree greater than 2 requires methods that are used in calculus. As the degree increases, the graphs usually become more complicated. However, they always have a smooth appearance with a number of "hills" and "valleys," as illustrated in Figure 3.3. The number of hills may be large if the degree is large, but it may also be small, as illustrated in Figures 3.1 and 3.2.

A crude method for obtaining a rough sketch of the graph of a polynomial function is to plot many points and fit a curve to the resulting configuration; however, this is usually an extremely tedious procedure. The method involves use of a property of polynomial functions called **continuity.** Continuity, which is studied extensively in calculus, implies that a small change in x produces a small change in the functional value $f(x)$. The next theorem specifies another important property of polynomial functions. The proof requires advanced mathematical methods.

INTERMEDIATE VALUE THEOREM FOR POLYNOMIAL FUNCTIONS

If f is a polynomial function and $f(a) \neq f(b)$, where $a < b$, then f takes on every value between $f(a)$ and $f(b)$ in the interval $[a, b]$.

The Intermediate Value Theorem states that if w is any number between $f(a)$ and $f(b)$, then there is a number c between a and b such that $f(c) = w$. If the graph of the polynomial function f is regarded as extending con-

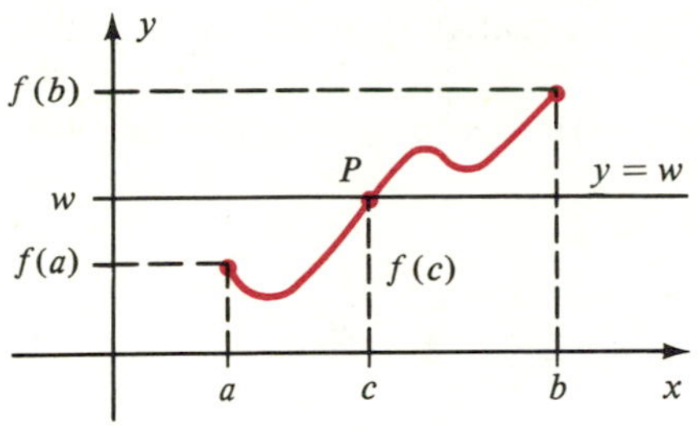

FIGURE 3.4

tinuously from the point $(a, f(a))$ to the point $(b, f(b))$, as illustrated in Figure 3.4, then for any number w between $f(a)$ and $f(b)$ it appears that a horizontal line with y-intercept w should intersect the graph in at least one point P. The x-coordinate c of P is a number such that $f(c) = w$.

A corollary to the Intermediate Value Theorem is that if $f(a)$ and $f(b)$ have opposite signs, then there is at least one number c between a and b such that $f(c) = 0$; that is, f has a zero at c. Geometrically, this implies that if the point $(a, f(a))$ on the graph of a polynomial function lies below the x-axis, and the point $(b, f(b))$ lies above the x-axis, or vice versa, then the graph crosses the x-axis at least once between the points $(a, 0)$ and $(b, 0)$. A by-product of this fact is that if c and d are *successive* zeros of $f(x)$, that is, there are no other zeros between c and d, then $f(x)$ *does not change sign on the interval* (c, d). Thus, if we choose any number k such that $c < k < d$, and if $f(k)$ is positive, then $f(x)$ is positive throughout (c, d). Similarly, if $f(k)$ is negative, then $f(x)$ is negative throughout (c, d). We shall call $f(k)$ a **test value** for $f(x)$ on the interval (c, d). Test values may also be used on infinite intervals of the form $(-\infty, a)$ or (a, ∞), provided that $f(x)$ has no zeros on these intervals. The use of test values in graphing is similar to that used for inequalities in Section 1.5.

EXAMPLE 2 If $f(x) = x^3 + x^2 - 4x - 4$, determine all values of x such that $f(x) > 0$ and all values of x such that $f(x) < 0$. Use this information as an aid in sketching the graph of f.

Solution The geometric significance of the stated inequalities is that the graph lies above the x-axis for values of x such that $f(x) > 0$, whereas the graph lies below the x-axis if $f(x) < 0$. We may factor $f(x)$ by grouping terms as follows:

$$\begin{aligned} f(x) &= (x^3 + x^2) - (4x + 4) \\ &= x^2(x + 1) - 4(x + 1) \\ &= (x^2 - 4)(x + 1) \\ &= (x + 2)(x - 2)(x + 1) \end{aligned}$$

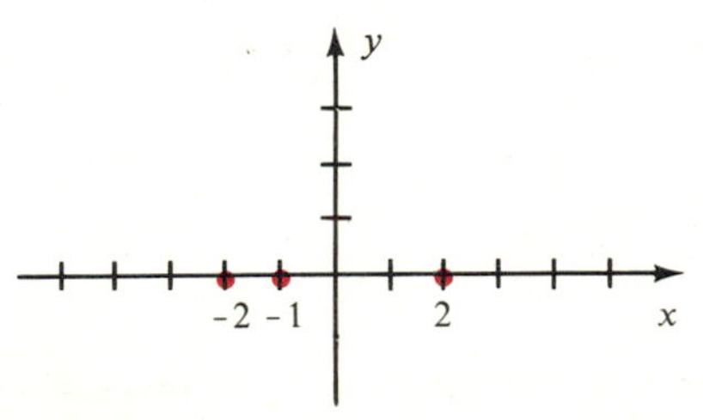

FIGURE 3.5

From the last equation it follows that the zeros of $f(x)$ are -2, -1, and 2. The corresponding points on the graph (see Figure 3.5) divide the x-axis into four parts, and we consider the open intervals

$$(-\infty, -2), \quad (-2, -1), \quad (-1, 2), \quad \text{and} \quad (2, \infty).$$

The sign of $f(x)$ in each of these intervals can be determined by finding a suitable test value. Thus, if we choose -3 in $(-\infty, -2)$, then

$$\begin{aligned} f(-3) &= (-3)^3 + (-3)^2 - 4(-3) - 4 \\ &= -27 + 9 + 12 - 4 = -10. \end{aligned}$$

Since the test value $f(-3)$ is negative, $f(x)$ is negative throughout the interval $(-\infty, -2)$.

If we choose $-\frac{3}{2}$ in the interval $(-2, -1)$, then the corresponding test value is

$$\begin{aligned} f(-\tfrac{3}{2}) &= (-\tfrac{3}{2})^3 + (-\tfrac{3}{2})^2 - 4(-\tfrac{3}{2}) - 4 \\ &= -\tfrac{27}{8} + \tfrac{9}{4} + 6 - 4 = \tfrac{7}{8}. \end{aligned}$$

Since $f(-\frac{3}{2})$ is positive, $f(x)$ is positive throughout $(-2, -1)$. The following table summarizes these facts and lists suitable test values for the remaining two intervals.

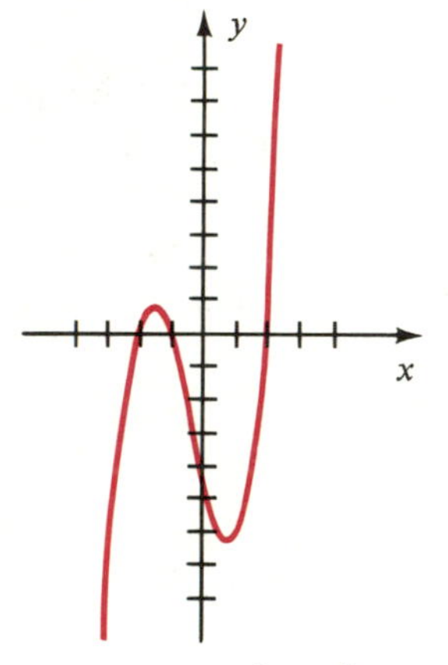

FIGURE 3.6 $f(x) = x^3 + x^2 - 4x - 4$

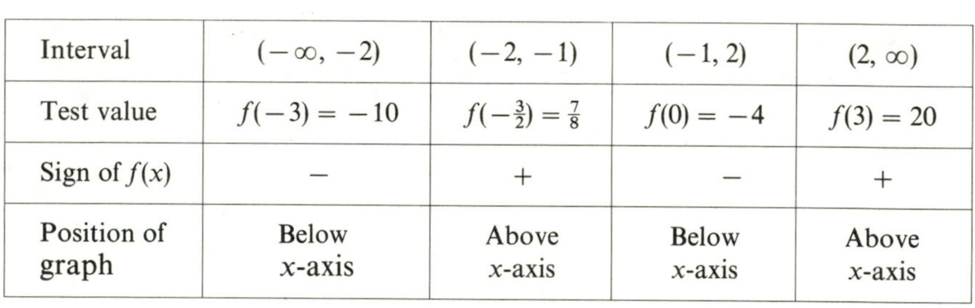

Interval	$(-\infty, -2)$	$(-2, -1)$	$(-1, 2)$	$(2, \infty)$
Test value	$f(-3) = -10$	$f(-\frac{3}{2}) = \frac{7}{8}$	$f(0) = -4$	$f(3) = 20$
Sign of $f(x)$	$-$	$+$	$-$	$+$
Position of graph	Below x-axis	Above x-axis	Below x-axis	Above x-axis

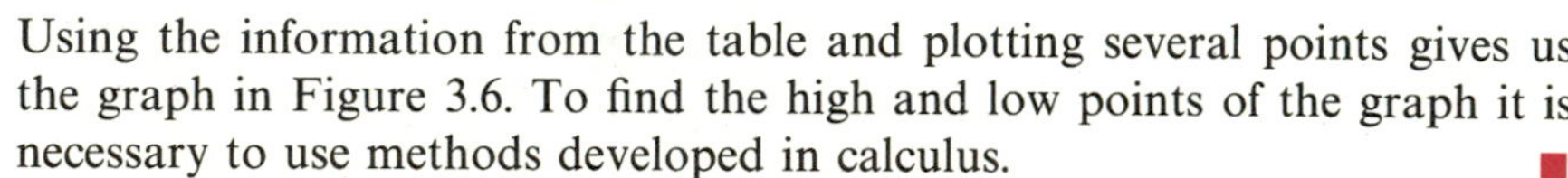

Using the information from the table and plotting several points gives us the graph in Figure 3.6. To find the high and low points of the graph it is necessary to use methods developed in calculus. ■

The graph of every polynomial function of degree 3 has an S-shaped appearance similar to that shown in Figure 3.6, or it has an inverted version of that graph if the coefficient of x^3 is negative. However, sometimes the graph may have only one x-intercept or the S shape may be elongated, as in Figure 3.1.

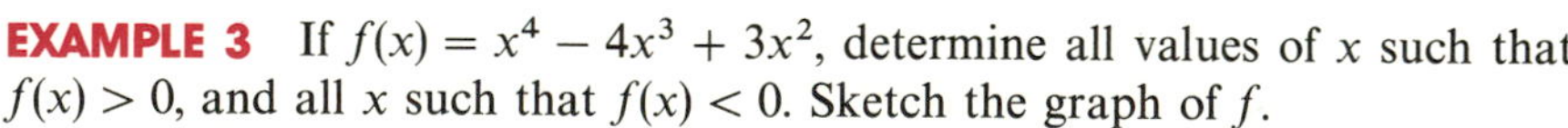

EXAMPLE 3 If $f(x) = x^4 - 4x^3 + 3x^2$, determine all values of x such that $f(x) > 0$, and all x such that $f(x) < 0$. Sketch the graph of f.

Solution We begin by factoring $f(x)$:

$$\begin{aligned} f(x) &= x^2(x^2 - 4x + 3) \\ &= x^2(x - 1)(x - 3) \end{aligned}$$

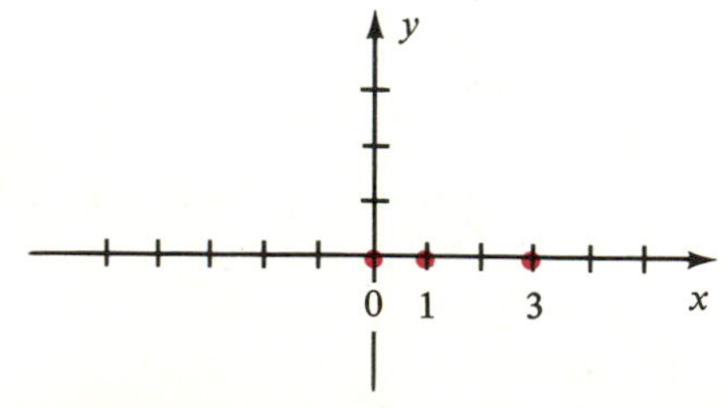

FIGURE 3.7

The zeros of $f(x)$, that is, the x-intercepts of the graph, are, in increasing order, 0, 1, and 3. The corresponding points on the graph are shown in Figure 3.7.

We next determine the sign of $f(x)$ in each of the intervals $(-\infty, 0)$, $(0, 1)$, $(1, 3)$, and $(3, \infty)$ by finding suitable test values, as in the preceding example. The following table summarizes our results. The reader should carefully check each entry.

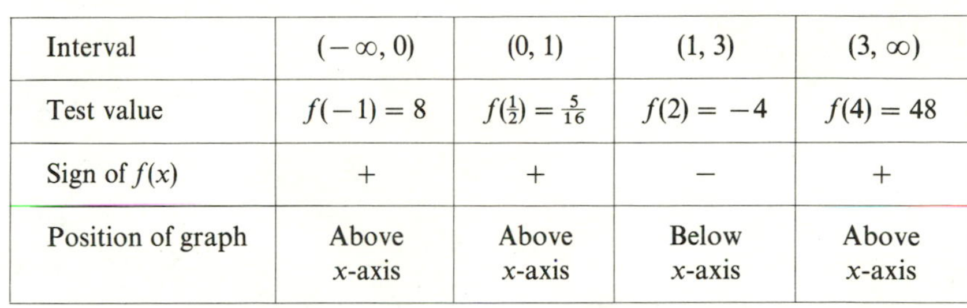

Interval	$(-\infty, 0)$	$(0, 1)$	$(1, 3)$	$(3, \infty)$
Test value	$f(-1) = 8$	$f(\frac{1}{2}) = \frac{5}{16}$	$f(2) = -4$	$f(4) = 48$
Sign of $f(x)$	$+$	$+$	$-$	$+$
Position of graph	Above x-axis	Above x-axis	Below x-axis	Above x-axis

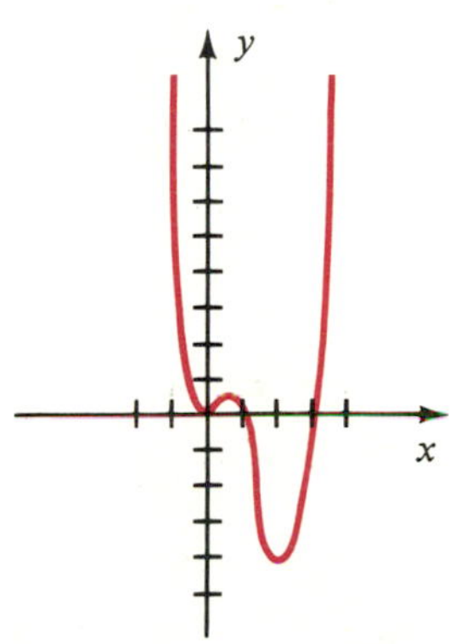

FIGURE 3.8 $f(x) = x^4 - 4x^3 + 3x^2$

Making use of the information in the table and plotting several points gives us the sketch in Figure 3.8. ■

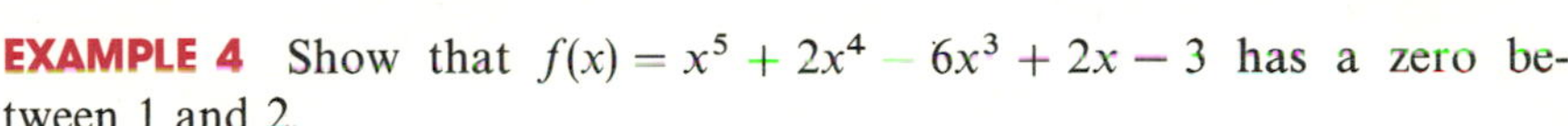

EXAMPLE 4 Show that $f(x) = x^5 + 2x^4 - 6x^3 + 2x - 3$ has a zero between 1 and 2.

Solution Substitution for x gives us

$$f(1) = 1 + 2 - 6 + 2 - 3 = -4$$
$$f(2) = 32 + 32 - 48 + 4 - 3 = 17.$$

Since $f(1)$ and $f(2)$ have opposite signs, it follows from the Intermediate Value Theorem that $f(c) = 0$ for some real number c between 1 and 2. ■

The preceding example illustrates a scheme for locating zeros of polynomials. By using a method of *successive approximation*, each such zero can be approximated to any degree of accuracy (see, for example, Calculator Exercises 1–10). Additional techniques for locating zeros of polynomials will be introduced later in this chapter.

EXERCISES 3.1

1 If $f(x) = ax^3 + 2$, sketch the graph of f for each value of a.

(a) $a = 2$ (b) $a = 4$

(c) $a = \frac{1}{4}$ (d) $a = -2$

2 If $f(x) = 2x^3 + c$, sketch the graph of f for each value of c.

(a) $c = 2$ (b) $c = 4$

(c) $c = \frac{1}{4}$ (d) $c = -2$

In Exercises 3–18 determine all x such that $f(x) > 0$, and all x such that $f(x) < 0$. Sketch the graph of f.

3 $f(x) = \frac{1}{2}x^3 - 4$ **4** $f(x) = -\frac{1}{4}x^3 - 16$

5 $f(x) = \frac{1}{8}x^4 + 2$ **6** $f(x) = 1 - x^5$

7 $f(x) = x^3 - 9x$ **8** $f(x) = 16x - x^3$

9 $f(x) = -x^3 - x^2 + 2x$ **10** $f(x) = x^3 + x^2 - 12x$

11 $f(x) = (x + 4)(x - 1)(x - 5)$

12 $f(x) = (x + 2)(x - 3)(x - 4)$

13 $f(x) = x^4 - 16$

14 $f(x) = 16 - x^4$

15 $f(x) = -x^4 - 3x^2 + 4$

16 $f(x) = x^4 - 7x^2 - 18$

17 $f(x) = x(x - 2)(x + 1)(x + 3)$

18 $f(x) = x(x + 1)^2(x - 3)(x - 5)$

19 If $f(x) = 3x^3 - kx^2 + x - 5k$, find a number k such that the graph of f contains the point $(-1, 4)$.

20 If one zero of $f(x) = x^3 - 2x^2 - 16x + 16k$ is 2, find two other zeros.

In Exercises 21–26 show that f has a zero between a and b.

21 $f(x) = x^3 - 4x^2 + 3x - 2;\quad a = 3, b = 4$

22 $f(x) = 2x^3 + 5x^2 - 3;\quad a = -3, b = -2$

23 $f(x) = -x^4 + 3x^3 - 2x + 1;\quad a = 2, b = 3$

24 $f(x) = 2x^4 + 3x - 2;\quad a = \frac{1}{2}, b = \frac{3}{4}$

25 $f(x) = x^5 + x^3 + x^2 + x + 1;\quad a = -\frac{1}{2}, b = -1$

26 $f(x) = x^5 - 3x^4 - 2x^3 + 3x^2 - 9x - 6;$
$a = 3, b = 4$

CALCULATOR EXERCISES 3.1

If a zero of a polynomial lies between two integers, then a method of *successive approximation* may be used to approximate the zero to any degree of accuracy. One such method is illustrated in Exercise 1.

1 If $f(x) = x^3 - 3x + 1$,

(a) show that f has a zero between 1 and 2.

(b) by increasing values of x by tenths, show that f has a zero between 1.5 and 1.6.

(c) by increasing values of x by hundredths, show that f has a zero between 1.53 and 1.54.

In (c) of Exercise 1, a zero of $x^3 - 3x + 1$ is said to be *isolated between successive hundredths* in the interval $[1, 2]$. In Exercises 2–6 isolate a zero of $f(x)$ between successive hundredths in the interval $[a, b]$.

2 $f(x) = x^3 + 5x - 3;\quad a = 0, b = 1$

3 $f(x) = 2x^3 - 4x^2 - 3x + 1;\quad a = 2, b = 3$

4 $f(x) = x^4 - 4x^3 + 3x^2 - 8x + 2;\quad a = 3, b = 4$

5 $f(x) = x^5 - 2x^2 + 4;\quad a = -2, b = -1$

6 $f(x) = x^4 - 2x^3 + 10x - 25;\quad a = 2, b = 3$

In Exercises 7–10 isolate a zero of $f(x)$ between successive thousandths in the interval $[a, b]$.

7 $f(x) = x^4 + 2x^3 - 5x^2 + 1;\quad a = 1, b = 2$

8 $f(x) = x^4 - 5x^2 + 2x - 5;\quad a = 2, b = 3$

9 $f(x) = x^5 + x^2 - 9x - 3;\quad a = -2, b = -1$

10 $f(x) = x^5 + 3x^4 - x^3 + 2x^2 + 6x - 2;$
$a = -4, b = -3$

SECTION 3.2
PROPERTIES OF DIVISION

In the following discussion, symbols such as $f(x)$ and $g(x)$ will be used to denote polynomials in x. If a polynomial $g(x)$ is a factor of a polynomial $f(x)$, then $f(x)$ is said to be **divisible** by $g(x)$. For example, $x^2 - 25$ is divisible by $x - 5$ and by $x + 5$. A division process may also be introduced if $g(x)$ is *not* a factor of $f(x)$. The process is similar to that used for integers. For example, the integer 24 has positive factors 1, 2, 3, 4, 6, 8, 12, and 24; that is, 24 is *divisible* by those numbers. A different situation exists if we divide 24 by nonfactors such as 5, 7, 15, and 32. In such cases *long division* is used, a process that yields a *quotient* and *remainder*. Let us consider the process from elementary arithmetic where, for example, the work involved in dividing 4126 by 23 can be arranged as follows:

$$
\begin{array}{r|r}
 & 179 \\ \cline{2-2}
23 & 4126 \\
 & 23 \\ \cline{2-2}
 & 182 \\
 & 161 \\ \cline{2-2}
 & 216 \\
 & 207 \\ \cline{2-2}
 & 9
\end{array}
$$

The number 179 is called the *quotient* and 9 is the *remainder*. To complete the terminology, 23 is called the *divisor* and 4126, the *dividend*. The remainder should always be less than the divisor for, otherwise, the quotient can be increased. The preceding result is sometimes written as

$$\frac{4126}{23} = 179 + \frac{9}{23}.$$

Multiplying by 23 gives us

$$4126 = (23)(179) + 9.$$

This form is very useful for theoretical purposes and can be generalized to arbitrary integers. Specifically, it can be shown that if a and b are integers with $b > 0$, then there exist unique integers q and r such that

$$a = bq + r$$

where $0 \leq r < b$. The integer q is called the **quotient** and r is the **remainder** in the division of a by b.

A similar discussion can be given for polynomials. For example, the polynomial $x^4 - 16$ is divisible by $x^2 - 4$, by $x^2 + 4$, by $x + 2$, and by $x - 2$; but $x^2 + 3x + 1$ is not a factor of $x^4 - 16$. However, by long division, we can write

$$
\begin{array}{r|l}
 & x^2 - 3x + 8 \\ \cline{2-2}
x^2 + 3x + 1 & x^4 \qquad\qquad\qquad\qquad\qquad - 16 \\
 & x^4 + 3x^3 + x^2 \\ \cline{2-2}
 & \quad - 3x^3 - x^2 \\
 & \quad - 3x^3 - 9x^2 - 3x \\ \cline{2-2}
 & \qquad\qquad 8x^2 + 3x - 16 \\
 & \qquad\qquad 8x^2 + 24x + 8 \\ \cline{2-2}
 & \qquad\qquad\quad - 21x - 24
\end{array}
$$

This arrangement displays the quotient $x^2 - 3x + 8$ and the remainder $-21x - 24$. Note that the process ends when we arrive at a polynomial (the remainder) that either is 0 or has smaller degree than the divisor. As with integers, the result of this division is often written

$$\frac{x^4 - 16}{x^2 + 3x + 1} = (x^2 - 3x + 8) + \left(\frac{-21x - 24}{x^2 + 3x + 1}\right).$$

Multiplying both sides of the equation by $x^2 + 3x + 1$, we obtain

$$x^4 - 16 = (x^2 + 3x + 1)(x^2 - 3x + 8) + (-21x - 24).$$

This equation has the same general form $a = bq + r$ that was given for integers.

The preceding example illustrates the following theorem, which we state without proof.

DIVISION ALGORITHM FOR POLYNOMIALS

If $f(x)$ and $g(x)$ are polynomials and if $g(x) \neq 0$, then there exist unique polynomials $q(x)$ and $r(x)$ such that

$$f(x) = g(x)q(x) + r(x)$$

where either $r(x) = 0$ or the degree of $r(x)$ is less than the degree of $g(x)$. The polynomial $q(x)$ is called the **quotient** and $r(x)$ is the **remainder** in the division of $f(x)$ by $g(x)$.

An interesting special case occurs if $f(x)$ is divided by a linear polynomial of the form $x - c$ where c is a real number. If $x - c$ is a factor of $f(x)$, then

$$f(x) = (x - c)q(x)$$

for some polynomial $q(x)$; that is, the remainder $r(x)$ is 0. If $x - c$ is not a factor of $f(x)$, then the degree of the remainder $r(x)$ is less than the degree of the divisor $x - c$, and hence $r(x)$ must have degree 0. This, in turn, means that the remainder is a nonzero number. Consequently, in all cases we have

$$f(x) = (x - c)q(x) + d$$

where d is some real number (possibly $d = 0$). If c is substituted for x in the equation $f(x) = (x - c)q(x) + d$, we obtain

$$f(c) = (c - c)q(c) + d,$$

which reduces to $f(c) = d$; that is,

$$f(x) = (x - c)q(x) + f(c).$$

We have proved the following theorem.

REMAINDER THEOREM

> If a polynomial $f(x)$ is divided by $x - c$, then the remainder is $f(c)$.

EXAMPLE 1 If $f(x) = x^3 - 3x^2 + x + 5$, use the Remainder Theorem to find $f(2)$.

Solution According to the Remainder Theorem, $f(2)$ is the remainder when $f(x)$ is divided by $x - 2$. By long division,

$$\begin{array}{r|l} & x^2 - x - 1 \\ \hline x - 2 & x^3 - 3x^2 + x + 5 \\ & x^3 - 2x^2 \\ \hline & \quad -x^2 + x \\ & \quad -x^2 + 2x \\ \hline & \qquad -x + 5 \\ & \qquad -x + 2 \\ \hline & \qquad\quad 3 \end{array}$$

Hence, $f(2) = 3$. To check our work by direct substitution, we have $f(2) = 2^3 - 3(2)^2 + 2 + 5 = 3$. ■

FACTOR THEOREM

> A polynomial $f(x)$ has a factor $x - c$ if and only if $f(c) = 0$.

Proof By the Remainder Theorem, $f(x) = (x - c)q(x) + f(c)$ for some quotient $q(x)$. If $f(c) = 0$, then $f(x) = (x - c)q(x)$; that is, $x - c$ is a factor of $f(x)$. Conversely, if $x - c$ is a factor, then the remainder upon division of $f(x)$ by $x - c$ must be 0, and hence, by the Remainder Theorem, $f(c) = 0$. □

The Factor Theorem is useful for finding factors of polynomials, as illustrated in the next example.

EXAMPLE 2 Show that $x - 2$ is a factor of the polynomial

$$f(x) = x^3 - 4x^2 + 3x + 2.$$

Solution Since $f(2) = 8 - 16 + 6 + 2 = 0$, it follows from the Factor Theorem that $x - 2$ is a factor of $f(x)$. Of course, another method of solution would be to divide $f(x)$ by $x - 2$ and show that the remainder is 0. The quotient in the division would be another factor of $f(x)$. ■

EXAMPLE 3 Find a polynomial $f(x)$ of degree 3 that has zeros 2, -1, and 3.

Solution By the Factor Theorem, $f(x)$ has factors $x - 2$, $x + 1$, and $x - 3$. We may then write

$$f(x) = a(x - 2)(x + 1)(x - 3)$$

where any nonzero value may be assigned to a. If we let $a = 1$ and multiply, we obtain

$$f(x) = x^3 - 4x^2 + x + 6.$$

■

EXERCISES 3.2

In Exercises 1–6 find the quotient $q(x)$ and the remainder $r(x)$ if $f(x)$ is divided by $g(x)$.

1 $f(x) = x^4 + 3x^3 - 2x + 5$, $g(x) = x^2 + 2x - 4$

2 $f(x) = 4x^3 - x^2 + x - 3$, $g(x) = x^2 - 5x$

3 $f(x) = 5x^3 - 2x$, $g(x) = 2x^2 + 1$

4 $f(x) = 3x^4 - x^3 - x^2 + 3x + 4$, $g(x) = 2x^3 - x + 4$

5 $f(x) = 7x^3 - 5x + 2$, $g(x) = 2x^4 - 3x^2 + 9$

6 $f(x) = 10x - 4$, $g(x) = 8x^2 - 5x + 17$

In Exercises 7–12 use the Remainder Theorem to find $f(c)$.

7 $f(x) = 2x^3 - x^2 - 5x + 3$, $c = 4$

8 $f(x) = 4x^3 - 3x^2 + 7x + 10$, $c = 3$

9 $f(x) = x^4 + 5x^3 - x^2 + 5$, $c = -2$

10 $f(x) = x^4 - 7x^2 + 2x - 8$, $c = -3$

11 $f(x) = x^6 - 3x^4 + 4$, $c = \sqrt{2}$

12 $f(x) = x^5 - x^4 + x^3 - x^2 + x - 1$, $c = -1$

13 Determine k so that $f(x) = x^3 + kx^2 - kx + 10$ is divisible by $x + 3$.

14 Determine all values of k such that $f(x) = k^2x^3 - 4kx - 3$ is divisible by $x - 1$.

15 Use the Factor Theorem to show that $x - 2$ is a factor of $f(x) = x^4 - 3x^3 - 2x^2 + 5x + 6$.

16 Show that $x + 2$ is a factor of $f(x) = x^{12} - 4096$.

17 Prove that $f(x) = 3x^4 + x^2 + 5$ has no factor of the form $x - c$ where c is a real number.

18 Find the remainder if the polynomial $3x^{100} + 5x^{85} - 4x^{38} + 2x^{17} - 6$ is divided by $x + 1$.

19 Use the Factor Theorem to prove that $x - y$ is a factor of $x^n - y^n$ for all positive integers n. If n is even, show that $x + y$ is also a factor of $x^n - y^n$.

20 If n is an odd positive integer, prove that $x + y$ is a factor of $x^n + y^n$.

SECTION 3.3
SYNTHETIC DIVISION

To apply the Remainder Theorem we must divide by polynomials of the form $x - c$. The process, referred to as *synthetic division*, simplifies the work if divisors are of the form $x - c$. We shall illustrate the process by means of examples.

If the polynomial $3x^4 - 8x^3 + 9x + 5$ is divided by $x - 2$, the following result is obtained:

$$
\begin{array}{r|rrrrr}
 & 3x^3 & -\,2x^2 & -\,4x & +\,1 & \\ \hline
x-2 & 3x^4 & -\,8x^3 & +\,0x^2 & +\,9x & +\,5 \\
 & 3x^4 & -\,6x^3 & & & \\ \cline{2-3}
 & & -2x^3 & +\,0x^2 & & \\
 & & -2x^3 & +\,4x^2 & & \\ \cline{3-4}
 & & & -4x^2 & +\,9x & \\
 & & & -4x^2 & +\,8x & \\ \cline{4-5}
 & & & & x & +\,5 \\
 & & & & x & -\,2 \\ \cline{5-6}
 & & & & & 7
\end{array}
$$

Note that the term $0x^2$ has been inserted in the dividend so that *all* powers of x are accounted for. Since this division involves a great deal of labor for so simple a problem, we look for a means of simplifying the notation. If we arrange the terms that involve like powers of x in vertical columns, we can see that the repeated expressions $3x^4$, $-2x^3$, $-4x^2$, and x may be deleted without much chance of confusion. Also, it is apparently unnecessary to "bring down" the terms $0x^2$, $9x$, and 5 from the dividend as we did in the previous arrangement. With the elimination of those repetitions, our work takes on this form:

$$
\begin{array}{r|rrrrr}
 & 3x^3 & -\,2x^2 & -\,4x & +\,1 & \\ \hline
x-2 & 3x^4 & -\,8x^3 & +\,0x^2 & +\,9x & +\,5 \\
 & & -6x^3 & & & \\ \cline{2-3}
 & & -2x^3 & & & \\
 & & & 4x^2 & & \\ \cline{3-4}
 & & & -4x^2 & & \\
 & & & & 8x & \\ \cline{4-5}
 & & & & x & \\
 & & & & & -2 \\ \cline{5-6}
 & & & & & 7
\end{array}
$$

If we take care to keep like powers of x aligned vertically, and if we account for missing terms by supplying zero coefficients, then we can save some labor by omitting the symbol x. Doing this, the preceding display may may be written as:

$$
\begin{array}{r|rrrrr}
 & 3 & -2 & -4 & 1 & \\ \hline
1-2 & 3 & -8 & 0 & 9 & 5 \\
 & & -6 & & & \\ \cline{2-3}
 & & -2 & & & \\
 & & & 4 & & \\ \cline{3-4}
 & & & -4 & & \\
 & & & & 8 & \\ \cline{4-5}
 & & & & 1 & \\
 & & & & & -2 \\ \cline{5-6}
 & & & & & 7
\end{array}
$$

Since the divisor is polynomial of the form $x - c$, the two coefficients in the far left position are always $1 - c$. With this in mind, we shall discard the coefficient 1. Moreover, to make our notation more compact, let us move the numbers up in the following way:

$$\begin{array}{r|rrrrr}
 & 3 & -2 & -4 & 1 & \\ \cline{2-6}
-2 & 3 & -8 & 0 & 9 & 5 \\
 & & -6 & 4 & 8 & -2 \\ \cline{2-6}
 & & -2 & -4 & 1 & 7
\end{array}$$

If we now insert the leading coefficient, 3, in the first position of the last row, the first four numbers of that row are the coefficients 3, -2, -4, and 1 of the quotient, and the final number, 7, is the remainder. Since it is unnecessary to write the coefficients of the quotient two times, we discard the first row in our scheme:

$$\begin{array}{r|rrrrr}
-2 & 3 & -8 & 0 & 9 & 5 \\
 & & -6 & 4 & 8 & -2 \\ \cline{2-6}
 & 3 & -2 & -4 & 1 & \boxed{7}
\end{array}$$

The rule at the top has also been deleted, since we no longer need it, and we have singled out the remainder by color.

There is a simple way of interpreting the last display. Note that every number in the second row can be obtained by multiplying the number in the third row of the *preceding* column by -2. Moreover, each number in the third row can be found by subtracting the number above it in the second row from the corresponding number in the first row. This suggests a method for carrying out the procedure without actually thinking of the division process: After arranging the terms of the polynomial in decreasing powers of x, we write the coefficients in a row, supplying 0 for any missing term. Next we write $-c$ (in our case, -2) to the left of this row. We then bring down the leading coefficient, 3, to the third row, and multiply that number by -2 to obtain the first number, -6, in the second row. We subtract -6 from -8 to obtain the second number, -2, in the third row, and then we multiply by $-c$ (in our case, -2) to obtain the second number, 4, in the second row. Again we subtract to get the third number, -4, in the third row. This process is continued until the final number in the third row (the remainder) is obtained.

It is possible to avoid subtractions if we use the number c in place of $-c$ in the far left position of the first row. In this event the signs of the elements in the second row are changed, and hence to find elements of the third row, we *add* the number above it in the second row to the corresponding number in the first row. With this change our example is written as follows:

$$\begin{array}{r|rrrrr}
2 & 3 & -8 & 0 & 9 & 5 \\
 & & 6 & -4 & -8 & 2 \\ \cline{2-6}
 & 3 & -2 & -4 & 1 & \boxed{7}
\end{array}$$

This scheme is called the process of **synthetic division.**

EXAMPLE 1 Use synthetic division to find the quotient and remainder if $2x^4 + 5x^3 - 2x - 8$ is divided by $x + 3$.

Solution Since the divisor is $x + 3$, the c in the expression $x - c$ is -3. Hence, the synthetic division takes this form:

-3	2	5	0	-2	-8
		-6	3	-9	33
	2	-1	3	-11	25

The first four numbers in the third row are the coefficients of the quotient, and the last number is the remainder. Hence, the quotient is the polynomial $2x^3 - x^2 + 3x - 11$ and the remainder is 25. ■

Synthetic division can be used to find values of polynomial functions, as illustrated in the next example.

EXAMPLE 2 If $f(x) = 3x^5 - 38x^3 + 5x^2 - 1$, use synthetic division to find $f(4)$.

Solution By the Remainder Theorem, $f(4)$ is the remainder when $f(x)$ is divided by $x - 4$. Dividing synthetically, we obtain

4	3	0	-38	5	0	-1
		12	48	40	180	720
	3	12	10	45	180	719

Consequently, $f(4) = 719$. ■

Synthetic division may be employed to help find zeros of polynomials. By the method illustrated in the preceding example, $f(c) = 0$ if and only if the remainder in the synthetic division by $x - c$ is 0.

EXAMPLE 3 Show that -11 is a zero of the polynomial

$$f(x) = x^3 + 8x^2 - 29x + 44$$

Solution Dividing synthetically by $x + 11$ gives us

-11	1	8	-29	44
		-11	33	-44
	1	-3	4	0

Thus, $f(-11) = 0$. ■

Example 3 shows that the number -11 is a solution of the equation $x^3 + 8x^2 - 29x + 44 = 0$. In Section 3.6 we shall use synthetic division to find rational solutions of equations.

EXERCISES 3.3

In Exercises 1–10 use synthetic division to find the quotient and remainder if the first polynomial is divided by the second.

1 $2x^3 - 3x^2 + 4x - 5,\ x - 2$

2 $3x^3 - 4x^2 - x + 8,\ x + 4$

3 $x^3 - 8x - 5,\ x + 3$

4 $5x^3 - 6x^2 + 15,\ x - 4$

5 $3x^5 + 6x^2 + 7,\ x + 2$

6 $-2x^4 + 10x - 3,\ x - 3$

7 $4x^4 - 5x^2 + 1,\ x - \frac{1}{2}$

8 $9x^3 - 6x^2 + 3x - 4,\ x - \frac{1}{3}$

9 $x^n - 1,\ x - 1$, where n is any positive integer

10 $x^n + 1,\ x + 1$, where n is any positive integer

Use synthetic division to solve Exercises 11–16.

11 If $f(x) = x^4 - 4x^3 + x^2 - 3x - 5$, find $f(2)$ and $f(-2)$.

12 If $f(x) = 0.3x^3 + 0.04x - 0.034$, find $f(0.2)$ and $f(-0.2)$.

13 If $f(x) = x^6 - x^5 + x^4 - x^3 + x^2 - x + 1$, find $f(4)$.

14 If $f(x) = 8x^5 - 3x^2 + 7$, find $f(\frac{1}{2})$.

15 If $f(x) = x^2 + 3x - 5$, find $f(2 + \sqrt{3})$.

16 If $f(x) = x^3 - 3x^2 - 8$, find $f(1 + \sqrt{2})$.

In Exercises 17–20 use synthetic division to show that c is a zero of $f(x)$.

17 $f(x) = 3x^4 + 8x^3 - 2x^2 - 10x + 4,\ c = -2$

18 $f(x) = 4x^3 - 9x^2 - 8x - 3,\ c = 3$

19 $f(x) = 4x^3 - 6x^2 + 8x - 3,\ c = \frac{1}{2}$

20 $f(x) = 27x^4 - 9x^3 + 3x^2 + 6x + 1,\ c = -\frac{1}{3}$

SECTION 3.4
COMPLEX NUMBERS

The zeros of a polynomial function f, or equivalently, the solutions of the equation $f(x) = 0$, are not always real numbers. For example, since the square of a real number cannot be negative, the equation $x^2 = -5$ has no solutions if x is restricted to $\mathbb{R}$. It is possible to construct a system of numbers in which such equations have solutions: the system of *complex numbers*, which is defined in this section.

Consider the problem of inventing a new number system $\mathbb{C}$ that contains the set $\mathbb{R}$ of real numbers and can be used to solve equations such as $x^2 = -5$. We wish to define operations in $\mathbb{C}$ called *addition* and *multiplication* in such a way that if we restrict the elements to the subset $\mathbb{R}$, then the operations have the same properties as addition and multiplication of real numbers. Since we are extending the notions of addition and multiplication to the set $\mathbb{C}$, we shall continue to use the symbols $+$ and $\cdot$ for those operations.

To gain some insight into the construction of $\mathbb{C}$, let us begin by taking an intuitive approach. If we want equations of the form $x^2 = -k$ to have solutions when k is a positive real number, then in particular when $k = 1$, $\mathbb{C}$ must contain some element i such that $i^2 = -1$. If b is in $\mathbb{R}$, then b is also in $\mathbb{C}$, and hence bi must be in $\mathbb{C}$. Moreover, if a is in $\mathbb{R}$ and if $\mathbb{C}$ is to be closed relative to addition, then $a + bi$ is in $\mathbb{C}$. Thus, $\mathbb{C}$ must contain elements of the form $a + bi$, $c + di$, etc., where a, b, c, and d are real numbers and $i^2 = -1$. If we want the Commutative, Associative, and Distributive Properties to be valid, then these elements must add as follows:

$$(a + bi) + (c + di) = (a + c) + (bi + di)$$

or

(A) $$(a + bi) + (c + di) = (a + c) + (b + d)i$$

Similarly, if $i^2 = -1$, the following manipulations should be valid:

$$\begin{aligned}(a + bi)(c + di) &= (a + bi)c + (a + bi)(di)\\ &= ac + (bi)c + a(di) + (bi)(di)\\ &= ac + (bc)i + (ad)i + (bd)(i^2)\\ &= ac + (bd)(-1) + (ad)i + (bc)i\\ &= (ac - bd) + (ad + bc)i\end{aligned}$$

To summarize, the following rule for multiplication must hold in $\mathbb{C}$:

(M) $$(a + bi)(c + di) = (ac - bd) + (ad + bc)i$$

The preceding discussion indicates the manner in which elements behave in a system of the required type. Moreover, our discussion provides a key to the actual construction of $\mathbb{C}$. Thus, we begin by *defining* a **complex number** as any symbol of the form $a + bi$ where a and b are real numbers. The real number a is called the **real part** of the complex number and b is called the **imaginary part.** At the outset, the letter i is given no specific meaning and the $+$ sign that appears in $a + bi$ is not interpreted as the symbol for addition, but only as part of the notation for a complex number. As above, $\mathbb{C}$ will denote the set of all complex numbers. Two complex numbers $a + bi$ and $c + di$ are said to be **equal,** and we write

$$a + bi = c + di \quad \text{if and only if} \quad a = c \text{ and } b = d.$$

Next we *define* **addition** and **multiplication** of complex numbers by means of formulas (A) and (M). Observe that the + sign is used in three different ways in (A): First, by our previous remarks, it is part of the symbol for a complex number. Second, it is used to denote addition of the complex numbers $a + bi$ and $c + di$. Third, it is the addition sign for real numbers, as in the expressions $a + c$ and $b + d$. The need for remembering this threefold use of + will disappear after we agree on the notational conventions that follow.

Let us consider the subset $\mathbb{R}'$ of $\mathbb{C}$ consisting of all complex numbers of the form $a + 0i$, where a is a real number. By associating a with $a + 0i$, we obtain a one-to-one correspondence between the set $\mathbb{R}$ and $\mathbb{R}'$. Applying (A) and (M) to the elements of $\mathbb{R}'$ (by letting $b = d = 0$), we obtain

$$(a + 0i) + (c + 0i) = (a + c) + 0i$$

$$(a + 0i)(c + 0i) = ac + 0i.$$

Thus, to add (or multiply) two elements of $\mathbb{R}'$, we may add (or multiply) the real parts, *disregarding* the imaginary parts. Hence, as far as properties of addition and multiplication are concerned, the only difference between $\mathbb{R}$ and $\mathbb{R}'$ is the notation for the elements. Accordingly, we shall use the symbol a in place of $a + 0i$. For example, an element such as 3 of $\mathbb{R}$ (or $\mathbb{C}$) is considered the same as the element $3 + 0i$ of $\mathbb{C}$. It is also convenient to abbreviate the complex number $0 + bi$ by the symbol bi. Applying (A) gives us

$$(a + 0i) + (0 + bi) = (a + 0) + (0 + b)i = a + bi.$$

This indicates that $a + bi$ may be thought of as the sum of two complex numbers a and bi (that is, $a + 0i$ and $0 + bi$). With these agreements on notation, all the + signs in (A) may be regarded as addition of complex numbers.

EXAMPLE 1 Write each expression in the form $a + bi$ where a and b are real numbers.

(a) $(3 + 4i) + (2 + 5i)$ (b) $(3 + 4i)(2 + 5i)$

Solution

(a) Applying (A),

$$(3 + 4i) + (2 + 5i) = (3 + 2) + (4 + 5)i = 5 + 9i.$$

(b) Using (M),

$$(3 + 4i)(2 + 5i) = (3 \cdot 2 - 4 \cdot 5) + (3 \cdot 5 + 4 \cdot 2)i = -14 + 23i.$$ ■

It is not difficult to show that the operations of addition and multiplication of complex numbers are both commutative and associative. The Distributive Property is also true for complex numbers. The identity element relative to addition is 0 (or, equivalently, $0 + 0i$) since

$$\begin{aligned}(a + bi) + 0 &= (a + bi) + (0 + 0i)\\ &= (a + 0) + (b + 0)i\\ &= a + bi.\end{aligned}$$

As in $\mathbb{R}$, we refer to 0 as **zero.** It follows from (M) that the *product* of zero and any complex number is zero. We may also use (M) to prove that 1 (that is, $1 + 0i$) is the identity element relative to multiplication.

If $(-a) + (-b)i$ is added to $a + bi$, we obtain 0. This implies that $(-a) + (-b)i$ is the additive inverse of $a + bi$; that is,

$$-(a + bi) = (-a) + (-b)i.$$

Subtraction of complex numbers is defined using additive inverses:

$$(a + bi) - (c + di) = (a + bi) + [-(c + di)].$$

Since $-(c + di) = (-c) + (-d)i$, it follows that

$$(a + bi) - (c + di) = (a - c) + (b - d)i.$$

The special case where $b = c = 0$ gives us

$$(a + 0i) - (0 + di) = (a - 0) + (0 - d)i,$$

which may be written in the form

$$a - di = a + (-d)i.$$

If c, d, and k are real numbers, then, by (M) and our agreement on notation,

$$k(c + di) = (k + 0i)(c + di) = (kc - 0d) + (kd + 0c)i,$$

that is,

$$k(c + di) = kc + (kd)i.$$

One illustration of this formula is

$$3(5 + 2i) = 15 + 6i.$$

The special case where $k = -1$ gives us

$$(-1)(c + di) = (-c) + (-d)i = -(c + di).$$

Hence, as in $\mathbb{R}$, the additive inverse of a complex number may be found by multiplying the complex number by -1.

The complex number $0 + 1i$ will be denoted by i. Observe that

$$b(0 + 1i) = (b \cdot 0) + (b \cdot 1)i = 0 + bi = bi.$$

Thus, the symbol bi that has been used throughout this section may be regarded as the *product* of b and i.

Finally, using (M) with $a = c = 0$ and $b = d = 1$, we obtain

$$\begin{aligned} i^2 &= (0 + 1i)(0 + 1i) \\ &= (0 \cdot 0 - 1 \cdot 1) + (0 \cdot 1 + 1 \cdot 0)i \\ &= -1 + 0i. \end{aligned}$$

This gives us the following important rule:

$$i^2 = -1.$$

If we collect all of the formulas and remarks made in this section, it becomes evident that, when working with complex numbers, *we may treat all symbols as though they represented real numbers with exactly one exception: wherever the symbol i^2 appears, it may be replaced by* -1. Consequently, manipulations can be carried out without referring to (A) or (M), which is what we have worked toward from the very beginning of our discussion! We shall use this technique in the solution of the next example. If, as in Example 2, we are asked to write an expression in the form $a + bi$, we shall also accept the form $a - di$, since we have seen that it equals $a + (-d)i$.

EXAMPLE 2 Write each expression in the form $a + bi$.

(a) $4(2 + 5i) - (3 - 4i)$ (b) $(4 - 3i)(2 + i)$

(c) $i(3 - 2i)^2$ (d) i^{51}

Solution

(a) $4(2 + 5i) - (3 - 4i) = 8 + 20i - 3 + 4i = 5 + 24i$

(b) $(4 - 3i)(2 + i) = 8 - 6i + 4i - 3i^2 = 11 - 2i$

(c) $i(3 - 2i)^2 = i(9 - 12i + 4i^2) = i(5 - 12i) = 5i - 12i^2 = 12 + 5i$

(d) Taking successive powers of i, we obtain $i^1 = i$, $i^2 = -1$, $i^3 = -i$, $i^4 = 1$, and then the cycle starts over: $i^5 = i$, $i^6 = i^2 = -1$, etc. In particular, $i^{51} = i^{48}i^3 = (i^4)^{12}i^3 = (1)^{12}i^3 = i^3 = -i$. ■

The complex number $a - bi$ is called the **conjugate** of the complex number $a + bi$. Since

$$(a + bi)(a - bi) = a^2 + b^2,$$

we see that the product of a complex number and its conjugate is a real number. Conjugates are useful for finding the multiplicative inverse

$1/(a + bi)$ of $a + bi$, or for simplifying the quotient $(a + bi)/(c + di)$ of two complex numbers, as illustrated in the next example.

EXAMPLE 3 Express each fraction in the form $a + bi$.

(a) $\dfrac{1}{9 + 2i}$ (b) $\dfrac{7 - i}{3 - 5i}$

Solution We multiply numerator and denominator by the conjugate of the denominator, as follows:

(a) $$\frac{1}{9 + 2i} = \frac{1}{9 + 2i} \cdot \frac{9 - 2i}{9 - 2i} = \frac{9 - 2i}{81 + 4} = \frac{9}{85} - \frac{2}{85}i$$

(b) $$\frac{7 - i}{3 - 5i} = \frac{7 - i}{3 - 5i} \cdot \frac{3 + 5i}{3 + 5i} = \frac{21 - 3i + 35i - 5i^2}{9 - 25i^2}$$

$$= \frac{26 + 32i}{34} = \frac{13}{17} + \frac{16}{17}i$$ ■

It is easy to see that if p is any positive real number, then the equation $x^2 = -p$ has solutions in $\mathbb{C}$. As a matter of fact, one solution is $\sqrt{p}i$, since

$$(\sqrt{p}i)^2 = (\sqrt{p})^2 i^2 = p(-1) = -p.$$

Similarly, $-\sqrt{p}i$ is also a solution. Moreover, $\sqrt{p}i$ and $-\sqrt{p}i$ are the only solutions, for if a complex number z is a solution, then $z^2 + p = 0$, and hence

$$(z + \sqrt{p}i)(z - \sqrt{p}i) = 0.$$

This implies that either $z = -\sqrt{p}i$ or $z = \sqrt{p}i$.

The next definition is motivated by the fact that $(\sqrt{p}i)^2 = -p$.

DEFINITION

If p is a positive real number, then the **principal square root** of $-p$ is denoted by $\sqrt{-p}$ and is defined to be the complex number $\sqrt{p}i$.

Some examples of principal square roots are

$$\sqrt{-9} = \sqrt{9}i = 3i, \quad \sqrt{-5} = \sqrt{5}i, \quad \sqrt{-1} = \sqrt{1}i = i.$$

Care must be taken in using the radical sign if the radicand is negative. For example, the formula $\sqrt{a}\sqrt{b} = \sqrt{ab}$, which holds for positive real numbers, is not true when both a and b are negative. To illustrate,

$$\sqrt{-3}\sqrt{-3} = (\sqrt{3}i)(\sqrt{3}i) = (\sqrt{3})^2 i^2 = 3(-1) = -3,$$

whereas $$\sqrt{(-3)(-3)} = \sqrt{9} = 3.$$

Hence, $$\sqrt{-3}\sqrt{-3} \neq \sqrt{(-3)(-3)}.$$

However, if only *one* of a or b is negative, then $\sqrt{a}\sqrt{b} = \sqrt{ab}$. In general, we shall not apply laws of radicals if radicands are negative. Instead, we shall change the form of radicals before performing any operations, as illustrated in the next example.

EXAMPLE 4 Express $(5 - \sqrt{-9})(-1 + \sqrt{-4})$ in the form $a + bi$.

Solution

$$\begin{aligned}(5 - \sqrt{-9})(-1 + \sqrt{-4}) &= (5 - 3i)(-1 + 2i)\\ &= -5 + 3i + 10i - 6i^2\\ &= -5 + 13i + 6 = 1 + 13i\end{aligned}$$ ■

In Section 2.5 we proved that if a, b, and c are real numbers such that $b^2 - 4ac \geq 0$ and if $a \neq 0$, then the solutions of the quadratic equation $ax^2 + bx + c = 0$ are

$$\frac{-b + \sqrt{b^2 - 4ac}}{2a} \quad \text{and} \quad \frac{-b - \sqrt{b^2 - 4ac}}{2a}.$$

We may now extend this fact to the case where $b^2 - 4ac < 0$. Indeed, the same manipulations used to obtain the Quadratic Formula, together with the developments in this section, show that if $b^2 - 4ac < 0$, then the solutions of $ax^2 + bx + c = 0$ are the two *complex* numbers given above. Notice that the solutions are conjugates of one another.

EXAMPLE 5 Find the solutions of the equation $5x^2 + 2x + 1 = 0$.

Solution By the Quadratic Formula,

$$x = \frac{-2 \pm \sqrt{4 - 20}}{10} = \frac{-2 \pm \sqrt{-16}}{10} = \frac{-2 \pm 4i}{10}.$$

Dividing numerator and denominator by 2, we see that the solutions of the equation are $-\frac{1}{5} + \frac{2}{5}i$ and $-\frac{1}{5} - \frac{2}{5}i$. ■

EXAMPLE 6 Find the solutions of the equation $x^3 - 1 = 0$.

Solution The equation $x^3 - 1 = 0$ may be written as

$$(x - 1)(x^2 + x + 1) = 0.$$

Setting each factor equal to zero and solving the resulting equations, we obtain the solutions

$$1, \quad \frac{-1 \pm \sqrt{1 - 4}}{2},$$

which may be written as

$$1, \quad -\frac{1}{2}+\frac{\sqrt{3}}{2}i, \quad -\frac{1}{2}-\frac{\sqrt{3}}{2}i.$$

The three solutions of $x^3 - 1 = 0$ are called the **cube roots of unity.** If n is any positive integer, then the equation $x^n - 1 = 0$ has n distinct complex solutions, called the ***n*th roots of unity.** We shall investigate these in more detail in Section 6.10.

The solutions of quadratic equations with complex coefficients are also given by the Quadratic Formula. Since, in this case, $b^2 - 4ac$ may be complex, the solution of such an equation may involve finding the square root of a complex number. We shall not discuss the general theory of roots of complex numbers at this time.

EXERCISES 3.4

In Exercises 1–34 write the expression in the form $a + bi$.

1 $(3 + 2i) + (-5 + 4i)$

2 $(8 - 5i) + (2 - 3i)$

3 $(-4 + 5i) + (2 - i)$

4 $(5 + 7i) + (-8 - 4i)$

5 $(16 + 10i) - (9 + 15i)$

6 $(2 - 6i) - (7 + 2i)$

7 $7 - (3 - 7i)$

8 $-9 + (5 + 9i)$

9 $5i - (6 + 2i)$

10 $(10 + 7i) - 12i$

11 $(4 + 3i)(-1 + 2i)$

12 $(3 - 6i)(2 + i)$

13 $(-7 + i)(-3 + i)$

14 $(5 + 2i)(5 - 2i)$

15 $(3 + 4i)(3 - 4i)$

16 $7i(13 + 8i)$

17 $-9i(4 - 8i)$

18 $(6i)(-2i)$

19 $4(8 - 11i)$

20 $-3(-6 + 12i)$

21 $(-3i)(5i)$

22 $(1 - i)(1 + i)$

23 $\dfrac{1}{3 + 2i}$

24 $\dfrac{1}{5 + 8i}$

25 $\dfrac{7}{5 - 6i}$

26 $\dfrac{-3}{2 - 5i}$

27 $\dfrac{4 - 3i}{2 + 4i}$

28 $\dfrac{4 + 3i}{-1 + 2i}$

29 $\dfrac{6 + 4i}{1 - 5i}$

30 $\dfrac{7 - 6i}{-5 - i}$

31 $\dfrac{21 - 7i}{i}$

32 $\dfrac{10 + 9i}{-3i}$

33 $\dfrac{2 - 3i}{1 + i} + \dfrac{7 + 4i}{3 + 5i}$

34 $\dfrac{6 - 2i}{3 + i} - \dfrac{3 - 7i}{i}$

Find the solutions of the equations in Exercises 35–48.

35 $x^2 - 3x + 10 = 0$

36 $x^2 - 5x + 20 = 0$

37 $x^2 + 2x + 5 = 0$

38 $x^2 + 3x + 6 = 0$

39 $4x^2 + x + 3 = 0$

40 $-3x^2 + x - 5 = 0$

41 $x^3 - 125 = 0$

42 $x^3 + 27 = 0$

43 $x^6 - 64 = 0$

44 $x^4 = 81$

45 $4x^4 + 25x^2 + 36 = 0$

46 $27x^4 + 21x^2 + 4 = 0$

47 $x^3 + 3x^2 + 4x = 0$

48 $8x^3 - 12x^2 + 2x - 3 = 0$

49 If $z = a + bi$ is any complex number, its conjugate is often denoted by $\bar{z}$, that is, $\bar{z} = a - bi$. Prove each of the following.

(a) $\overline{z + w} = \bar{z} + \bar{w}$

(b) $\overline{z \cdot w} = \bar{z} \cdot \bar{w}$

(c) $\overline{z^2} = (\bar{z})^2$, $\overline{z^3} = (\bar{z})^3$, and $\overline{z^4} = (\bar{z})^4$

(d) $\bar{z} = z$ if and only if z is real.

50 Refer to Exercise 49. Prove that $\overline{z - w} = \bar{z} - \bar{w}$ and $\bar{\bar{z}} = z$.

SECTION 3.5

THE ZEROS OF A POLYNOMIAL

We can extend the Factor and Remainder Theorems, which we discussed in Section 3.2, to the system of complex numbers. Thus, a complex number c is a zero of a polynomial $f(x)$, that is, $f(c) = 0$, if and only if $x - c$ is a factor of $f(x)$. Except in special cases, zeros of polynomials are very difficult to find. For example, if $f(x) = x^5 - 3x^4 + 4x^3 + 4x - 10$, the function has no obvious zeros. Moreover, there is no device, such as the Quadratic Formula, that can be used to find the zeros. In spite of the practical difficulty of determining zeros of polynomials, it is possible to make some headway concerning the *theory* of such zeros. The results in this section form the basis for work in what is known as *The Theory of Equations.*

FUNDAMENTAL THEOREM OF ALGEBRA

If $f(x)$ is a polynomial of positive degree, and has complex coefficients, then $f(x)$ has at least one complex zero.

The usual proof of this theorem requires results from the field of mathematics called *functions of a complex variable.* In turn, a prerequisite for studying this field is a strong background in calculus. The first proof of the Fundamental Theorem of Algebra was given by the German mathematician Carl Friedrich Gauss (1777–1855), who is considered by many to be the greatest mathematician of all time.

As a special case of the Fundamental Theorem, if all the coefficients of $f(x)$ are real, then $f(x)$ has at least one complex zero. We should also remark that if $a + bi$ is a complex zero of a polynomial, it may happen that $b = 0$, in which case we refer to the number as a **real zero.** If the Fundamental Theorem is combined with the Factor Theorem, the following useful corollary is obtained.

COROLLARY

Every polynomial of positive degree has a factor of the form $x - c$ where c is a complex number.

The corollary enables us, at least in theory, to express every polynomial $f(x)$ of positive degree as a product of polynomials of degree 1. If $f(x)$ has a degree $n > 0$, then applying the corollary gives us

$$f(x) = (x - c_1)f_1(x)$$

where c_1 is a complex number and $f_1(x)$ is a polynomial of degree $n - 1$.

If $n - 1 > 0$, we may apply the corollary again, obtaining

$$f_1(x) = (x - c_2)f_2(x)$$

where c_2 is a complex number and $f_2(x)$ is a polynomial of degree $n - 2$. Hence,

$$f(x) = (x - c_1)(x - c_2)f_2(x).$$

Continuing this process, after n steps we arrive at a polynomial $f_n(x)$ of degree 0. Thus, $f_n(x) = a$ for some nonzero number a and we may write

$$f(x) = a(x - c_1)(x - c_2) \cdots (x - c_n)$$

where each complex number c_j is a zero of $f(x)$. Evidently, the leading coefficient of the polynomial on the right in the last equation is a. It follows that a is the leading coefficient of $f(x)$. We have proved the following theorem.

THEOREM

If $f(x)$ is a polynomial of degree $n > 0$, then there exist n complex numbers $c_1, c_2, \ldots, c_n$ such that

$$f(x) = a(x - c_1)(x - c_2) \cdots (x - c_n)$$

where a is the leading coefficient of $f(x)$. Each number c_j is a zero of $f(x)$.

COROLLARY

A polynomial of degree $n > 0$ has at most n different complex zeros.

Proof We shall give an indirect proof. Suppose $f(x)$ has *more* than n different complex zeros. Let us choose $n + 1$ of these zeros and label them $c_1, c_2, \ldots, c_n$, and c. We may use the c_j to obtain the factorization indicated in the statement of the theorem. Substituting c for x and using the fact that $f(c) = 0$, we obtain

$$0 = a(c - c_1)(c - c_2) \cdots (c - c_n).$$

However, each factor on the right side is different from zero because $c \neq c_j$ for every j. Since the product of nonzero numbers cannot equal zero, we have a contradiction. □

EXAMPLE 1 Find a polynomial $f(x)$ of degree 3 with zeros 2, -1, and 3 that has the value 5 at $x = 1$.

Solution By the Factor Theorem $f(x)$ has factors $x - 2$, $x + 1$, and $x - 3$. No other factors of degree 1 exist, since, by the Factor Theorem, another linear factor $x - c$ would produce a fourth zero of $f(x)$ in violation of the last corollary. Hence, $f(x)$ has the form

$$f(x) = a(x - 2)(x + 1)(x - 3)$$

for some number a. If $f(x)$ has the value 5 at $x = 1$, then $f(1) = 5$, that is,

$$a(1 - 2)(1 + 1)(1 - 3) = 5 \quad \text{or} \quad 4a = 5.$$

Consequently, $a = \frac{5}{4}$ and

$$f(x) = \tfrac{5}{4}(x - 2)(x + 1)(x - 3).$$

Multiplying the four factors we obtain

$$f(x) = \tfrac{5}{4}x^3 - 5x^2 + \tfrac{5}{4}x + \tfrac{15}{2}.$$ ■

The numbers $c_1, c_2, \ldots, c_n$ in the theorem on page 129 are not necessarily all different. To illustrate, the polynomial $f(x) = x^3 + x^2 - 5x + 3$ has the factorization

$$f(x) = (x + 3)(x - 1)(x - 1).$$

If a factor $x - c$ occurs m times in the factorization, then c is called a **zero of multiplicity *m*** of $f(x)$, or a **root of multiplicity *m*** of the equation $f(x) = 0$. In the preceding illustration, 1 is a zero of multiplicity 2 and -3 is a zero of multiplicity 1.

As another illustration, if

$$f(x) = (x - 2)(x - 4)^3(x + 1)^2$$

then $f(x)$ has degree 6 and possesses three distinct zeros 2, 4, and -1, where 2 has multiplicity 1, 4 has multiplicity 3, and -1 has multiplicity 2.

If $f(x)$ is a polynomial of degree n and $f(x) = a(x - c_1)(x - c_2) \cdots (x - c_n)$, then the n complex numbers $c_1, c_2, \ldots, c_n$ are zeros of $f(x)$. If a zero of multiplicity m is counted as m zeros, this tells us that $f(x)$ has at least n zeros (not necessarily all different). Combining this with the fact that $f(x)$ has at most n zeros gives us the next result.

THEOREM

If $f(x)$ is a polynomial of degree $n > 0$ and if a zero of multiplicity m is counted m times, then $f(x)$ has precisely n zeros.

EXAMPLE 2 Express $f(x) = x^5 - 4x^4 + 13x^3$ as a product of linear factors, and list the five zeros of $f(x)$.

Solution We begin by writing

$$f(x) = x^3(x^2 - 4x + 13).$$

By the Quadratic Formula, the zeros of the polynomial $x^2 - 4x + 13$ are

$$\frac{4 \pm \sqrt{16 - 52}}{2} = \frac{4 \pm \sqrt{-36}}{2} = \frac{4 \pm 6i}{2} = 2 \pm 3i.$$

Hence, by the Factor Theorem, $x^2 - 4x + 13$ has factors $x - (2 + 3i)$ and $x - (2 - 3i)$, and we obtain the desired factorization,

$$f(x) = x \cdot x \cdot x \cdot (x - 2 - 3i)(x - 2 + 3i).$$

Since $x - 0$ occurs as a factor three times, the number 0 is zero of multiplicity three, and the five zeros of $f(x)$ are 0, 0, 0, $2 + 3i$, and $2 - 3i$. ■

The next theorem may be used to obtain information about the zeros of a polynomial $f(x)$ with real coefficients. In the statement of the theorem it is assumed that the terms of $f(x)$ are arranged in order of decreasing powers of x, and that terms with zero coefficients are deleted. It is also assumed that the **constant term,** that is, the term that does not contain x, is different from 0. We say there is a **variation of sign** in $f(x)$ if two consecutive coefficients have opposite signs. To illustrate, the polynomial

$$f(x) = 2x^5 - 7x^4 + 3x^2 + 6x - 5$$

has three variations in sign, since there is one variation from $2x^5$ to $-7x^4$, a second from $-7x^4$ to $3x^2$, and a third from $6x$ to -5.

The theorem also refers to the variations of sign in $f(-x)$. Using the previous illustration, note that

$$\begin{aligned} f(-x) &= 2(-x)^5 - 7(-x)^4 + 3(-x)^2 + 6(-x) - 5 \\ &= -2x^5 - 7x^4 + 3x^2 - 6x - 5. \end{aligned}$$

Hence, there are two variations of sign in $f(-x)$: one from $-7x^4$ to $3x^2$, and a second from $3x^2$ to $-6x$.

DESCARTES' RULE OF SIGNS

Let $f(x)$ be a polynomial with real cofficients and nonzero constant term.

(i) The number of positive real solutions of the equation $f(x) = 0$ either is equal to the number of variations of sign in $f(x)$ or is less than that number by an even integer.

(ii) The number of negative real solutions of the equation $f(x) = 0$ either is equal to the number of variations of sign in $f(-x)$ or is less than that number by an even integer.

The proof of Descartes' Rule is rather technical and will not be given in this text.

EXAMPLE 3 Discuss the number of possible positive and negative real solutions, and nonreal complex solutions of the equation

$$2x^5 - 7x^4 + 3x^2 + 6x - 5 = 0.$$

Solution The polynomial $f(x)$ on the left side of the equation is the same as the one given in the illustration preceding the statement of Descartes' Rule. Since there are three variations of sign in $f(x)$, the equation has either three positive real solutions or one positive real solution.

Since $f(-x) = -2x^5 - 7x^4 + 3x^2 - 6x - 5$ has two variations of sign, the given equation has either two negative solutions or no negative solutions. The solutions that are not real numbers are complex numbers of the form $a + bi$ where a and b are real with $b \neq 0$. The following table summarizes the various possibilities that can occur for solutions of the equation.

Number of positive real solutions	1	1	3	3
Number of negative real solutions	0	2	0	2
Number of nonreal, complex solutions	4	2	2	0
Total number of solutions	5	5	5	5

Descartes' Rule stipulates that the constant term of the polynomial is different from 0. If the constant term is 0, as in the equation

$$x^4 - 3x^3 + 2x^2 - 5x = 0,$$

then we may write

$$x(x^3 - 3x^2 + 2x - 5) = 0.$$

In this case $x = 0$ is a solution and Descartes' Rule may be applied to $x^3 - 3x^2 + 2x - 5 = 0$. ■

EXAMPLE 4 Discuss the nature of the roots of the equation

$$3x^5 + 4x^3 + 2x - 5 = 0.$$

Solution The polynomial $f(x)$ on the left side of the equation has one variation of sign and hence, by (i) of Descartes' Rule, the equation has precisely one positive real root. Since

$$\begin{aligned} f(-x) &= 3(-x)^5 + 4(-x)^3 + 2(-x) - 5 \\ &= -3x^5 - 4x^3 - 2x - 5, \end{aligned}$$

$f(-x)$ has no variations of sign and hence, by (ii) of Descartes' Rule, there are no negative real roots. Thus, the equation has one real root and four nonreal, complex roots. Moreover, since $f(0) = -5$ and $f(1) = 4$, the real root is between 0 and 1. (Why?) ■

When applying Descartes' Rule, we count roots of multiplicity k as k roots. For example, given $x^2 - 2x + 1 = 0$, the polynomial $x^2 - 2x + 1$ has two variations of sign and hence the equation has either two positive real roots or none. The factored form of the equation is $(x - 1)^2 = 0$, and hence 1 is a root of multiplicity 2.

We shall conclude this section with a discussion of *bounds* for the real solutions of an equation $f(x) = 0$ where $f(x)$ is a polynomial with real coefficients. By definition, a real number b is an **upper bound** for the solutions if no solution is greater than b. A real number a is a **lower bound** for the solutions if no solution is less than a. Thus, if r is a real solution of $f(x) = 0$, then $a \leq r \leq b$. Note that upper and lower bounds are not unique, since any number greater than b is also an upper bound, and any number less than a is a lower bound.

We may use synthetic division to find upper and lower bounds for the solutions of $f(x) = 0$. Recall that if $f(x)$ is divided synthetically by $x - c$, then the third row that appears in the division process consists of the coefficients of the quotient $q(x)$ together with the remainder $f(c)$. The following theorem indicates how this third row may be used to find upper and lower bounds for the real solutions.

BOUNDS FOR REAL ZEROS OF POLYNOMIALS

Suppose that $f(x)$ is a polynomial with real coefficients and positive leading coefficient, and that $f(x)$ is divided synthetically by $x - c$.

(i) If $c > 0$ and if all numbers in the third row of the division process are either positive or zero, then c is an upper bound for the real solutions of the equation $f(x) = 0$.

(ii) If $c < 0$ and if the numbers in the third row of the division process are alternately positive and negative (where a 0 in the third row is considered to be either positive or negative), then c is a lower bound for the real solutions of the equation $f(x) = 0$.

A general proof of this theorem can be patterned after the solution given in the next example.

EXAMPLE 5 Find upper and lower bounds for the real solutions of the equation

$$2x^3 + 5x^2 - 8x - 7 = 0.$$

Solution If we divide synthetically by $x - 1$ and $x - 2$, we obtain

$$\begin{array}{r|rrrr} 1 & 2 & 5 & -8 & -7 \\ & & 2 & 7 & -1 \\ \hline & 2 & 7 & -1 & -8 \end{array} \qquad \begin{array}{r|rrrr} 2 & 2 & 5 & -8 & -7 \\ & & 4 & 18 & 20 \\ \hline & 2 & 9 & 10 & 13 \end{array}$$

Since all numbers in the third row of the synthetic division by $x - 2$ are positive, it follows from (i) of the preceding theorem that 2 is an upper bound for the real solutions of the equation. This fact is also evident if we express the division by $x - 2$ in the Division Algorithm form

$$2x^3 + 5x^2 - 8x - 7 = (x - 2)(2x^2 + 9x + 10) + 13.$$

If $x > 2$, then it is obvious that the right side of the equation is positive, and hence is not zero. Consequently, $2x^3 + 5x^2 - 8x - 7$ is not zero if $x > 2$.

After some trial-and-error attempts using $x - (-1)$ and $x - (-2)$, we find that synthetic division by $x - (-3)$ and $x - (-4)$ gives us

$$\begin{array}{r|rrrr} -3 & 2 & 5 & -8 & -7 \\ & & -6 & 3 & 15 \\ \hline & 2 & -1 & -5 & 8 \end{array} \qquad \begin{array}{r|rrrr} -4 & 2 & 5 & -8 & -7 \\ & & -8 & 12 & -16 \\ \hline & 2 & -3 & 4 & -23 \end{array}$$

Since the numbers in the third row of the synthetic division by $x - (-4)$ are alternately positive and negative, it follows from (ii) of the preceding theorem that -4 is a lower bound for the real solutions. This can also be proved by expressing the division by $x + 4$ in the form

$$2x^3 + 5x^2 - 8x - 7 = (x + 4)(2x^2 - 3x + 4) - 23.$$

If $x < -4$, then the right side of this equation is negative (Why?), and therefore is not zero. Hence, $2x^3 + 5x^2 - 8x - 7$ is not zero if $x < -4$. ■

EXERCISES 3.5

In Exercises 1–6 find a polynomial $f(x)$ of degree 3 with the indicated zeros and satisfying the given conditions.

1 $5, -2, -3;\quad f(2) = 4$

2 $4, 1, -6;\quad f(5) = 2$

3 $2, 3, 1;\quad f(0) = 12$

4 $\sqrt{2}, \pi, 0;\quad f(0) = 0$

5 $2 + i, 2 - i, -4;\quad f(1) = 3$

6 $1 + 2i, 1 - 2i, 5;\quad f(-2) = 1$

7 Find a polynomial of degree 4 such that both -2 and 3 are zeros of multiplicity 2.

8 Find a polynomial of degree 5 such that -2 is a zero of multiplicity 3 and 4 is a zero of multiplicity 2.

9 Find a polynomial $f(x)$ of degree 8 such that 2 is a zero of multiplicity 3, 0 is a zero of multiplicity 5, and $f(3) = 54$.

10 Find a polynomial $f(x)$ of degree 7 such that 1 is a zero of multiplicity 2, -1 is a zero of multiplicity 2, 0 is a zero of multiplicity 3, and $f(2) = 36$.

In Exercises 11–18 find the zeros of the polynomials and state the multiplicity of each zero.

11 $f(x) = (x + 4)^3(3x - 4)$

12 $f(x) = (x - 5)^2(4x + 7)^3$

13 $f(x) = 2x^5 - 8x^4 - 10x^3$

14 $f(x) = (4x^2 - 5)^2$

15 $f(x) = (9x^2 - 25)^4(x^2 + 16)$

16 $f(x) = (2x^2 + 13x - 7)^3$

17 $f(x) = (x^2 + x - 2)^2(x^2 - 4)$

18 $f(x) = 4x^6 + x^4$

19 Show that -3 is a zero of multiplicity 2 of the polynomial $f(x) = x^4 + 7x^3 + 13x^2 - 3x - 18$, and express $f(x)$ as a product of linear factors.

20 Show that 4 is a zero of multiplicity 2 of the polynomial $f(x) = x^4 - 9x^3 + 22x^2 - 32$, and express $f(x)$ as a product of linear factors.

21 Show that 1 is a zero of multiplicity 5 of $f(x) = x^6 - 4x^5 + 5x^4 - 5x^2 + 4x - 1$, and express $f(x)$ as a product of linear factors.

22 Show that -1 is a zero of multiplicity 4 of $f(x) = x^5 + x^4 - 6x^3 - 14x^2 - 11x - 3$, and express $f(x)$ as a product of linear factors.

In Exercises 23–30 use Descartes' Rule of Signs to determine the number of possible positive, negative, and nonreal, complex solutions of the equation.

23 $4x^3 - 6x^2 + x - 3 = 0$

24 $5x^3 - 6x - 4 = 0$

25 $4x^3 + 2x^2 + 1 = 0$

26 $3x^3 - 4x^2 + 3x + 7 = 0$

27 $3x^4 + 2x^3 - 4x + 2 = 0$

28 $2x^4 - x^3 + x^2 - 3x + 4 = 0$

29 $x^5 + 4x^4 + 3x^3 - 4x + 2 = 0$

30 $2x^6 + 5x^5 + 2x^2 - 3x + 4 = 0$

In Exercises 31–36 find the smallest and largest integers that are upper and lower bounds, respectively, for the real solutions of the equation.

31 $x^3 - 4x^2 - 5x + 7 = 0$

32 $2x^3 - 5x^2 + 4x - 8 = 0$

33 $x^4 - x^3 - 2x^2 + 3x + 6 = 0$

34 $2x^4 - 9x^3 - 8x - 10 = 0$

35 $2x^5 - 13x^3 + 2x - 5 = 0$

36 $3x^5 + 2x^4 - x^3 - 8x^2 - 7 = 0$

37 Let $f(x)$ and $g(x)$ be polynomials of degree not greater than n, where n is a positive integer. Show that if $f(x)$ and $g(x)$ are equal in value for more than n distinct values of x, then $f(x)$ and $g(x)$ are identical; that is, coefficients of like powers are the same. (*Hint:* Write

$$f(x) = a_nx^n + a_{n-1}x^{n-1} + \cdots + a_1x + a_0$$
$$g(x) = b_nx^n + b_{n-1}x^{n-1} + \cdots + b_1x + b_0$$

and consider

$$h(x) = f(x) - g(x) = (a_n - b_n)x^n + \cdots + (a_0 - b_0).$$

Then show that $h(x)$ has more than n distinct zeros and conclude that $a_i = b_i$ for all i.)

SECTION 3.6

COMPLEX AND RATIONAL ZEROS OF POLYNOMIALS

Example 2 of the preceding section illustrates an interesting fact about polynomials with real coefficients: The two complex zeros $2 + 3i$ and $2 - 3i$ of $x^5 - 4x^4 + 13x^3$ are conjugates of one another. The relationship is not accidental, since the following general result is true.

THEOREM If $f(x)$ is a polynomial of degree $n > 0$, and has real coefficients, and if z is a complex zero of $f(x)$, then the conjugate $\bar{z}$ of z is also a zero of $f(x)$.

Proof We may write

$$f(x) = a_nx^n + a_{n-1}x^{n-1} + \cdots + a_1x + a_0$$

where the a_i are real numbers and $a_n \neq 0$. If $f(z) = 0$, then

$$a_n z^n + a_{n-1} z^{n-1} + \cdots + a_1 z + a_0 = 0.$$

If two complex numbers are equal, then so are their conjugates. Consequently, the conjugate of the left side of the last equation equals the conjugate of the right side; that is,

$$\overline{a_n z^n + a_{n-1} z^{n-1} + \cdots + a_1 z + a_0} = \overline{0} = 0$$

where the fact that $\overline{0} = 0$ follows from $\overline{0} = \overline{0 + 0i} = 0 - 0i = 0$.

It is not difficult to prove that if z and w are complex numbers, then $\overline{z + w} = \overline{z} + \overline{w}$. More generally, the conjugate of *any* sum of complex numbers is the sum of the conjugates. Consequently,

$$\overline{a_n z^n} + \overline{a_{n-1} z^{n-1}} + \cdots + \overline{a_1 z} + \overline{a_0} = 0.$$

It can also be shown that if z and w are complex numbers, then $\overline{z \cdot w} = \overline{z} \cdot \overline{w}$, $\overline{z^n} = \overline{z}^n$ for every positive integer n, and $\overline{z} = z$ if and only if z is real (cf. Exercise 49 of Section 3.4). Thus, for every k,

$$\overline{a_j z^k} = \overline{a_j} \cdot \overline{z^k} = \overline{a_j} \cdot \overline{z}^k = a_j \overline{z}^k$$

and therefore

$$a_n \overline{z}^n + a_{n-1} \overline{z}^{n-1} + \cdots + a_1 \overline{z} + a_0 = 0.$$

The last equation states that $f(\overline{z}) = 0$, which completes the proof. □

EXAMPLE 1 Find a polynomial $f(x)$ of degree 4 that has real coefficients and zeros $2 + i$ and $-3i$.

Solution By the preceding theorem, $f(x)$ must also have zeros $2 - i$ and $3i$, and hence by the Factor Theorem, $f(x)$ has factors $x - (2 + i)$, $x - (2 - i)$, $x - (-3i)$, and $x - (3i)$. Multiplying those factors gives us a polynomial of the required type:

$$\begin{aligned} f(x) &= [x - (2 + i)][x - (2 - i)](x - 3i)(x + 3i) \\ &= (x^2 - 4x + 5)(x^2 + 9) \\ &= x^4 - 4x^3 + 14x^2 - 36x + 45. \end{aligned}$$

■

If a polynomial with real coefficients is factored as on page 129, some of the factors $x - c_k$ may have a complex coefficient c_k. However, it is always possible to obtain a factorization into polynomials with real coefficients, as stated in the next theorem.

THEOREM

Every polynomial with real coefficients and positive degree n can be expressed as a product of linear and quadratic polynomials with real coefficients, where the quadratic factors have no real zeros.

Proof Since $f(x)$ has precisely n complex zeros $c_1, c_2, \ldots, c_n$, we may write

$$f(x) = a(x - c_1)(x - c_2) \cdots (x - c_n)$$

where a is the leading coefficient of $f(x)$. Of course, some of the c_k may be real. In such cases we obtain the linear factors referred to in the statement of the theorem. If a zero c_k is not real, then by the preceding theorem the conjugate $\overline{c_k}$ is also a zero of $f(x)$ and hence must be one of the numbers $c_1, c_2, \ldots, c_n$. This implies that both $x - c_k$ and $x - \overline{c_k}$ appear in the factorization of $f(x)$ If those factors are multiplied, we obtain

$$(x - c_k)(x - \overline{c_k}) = x^2 - (c_k + \overline{c_k})x + c_k\overline{c_k},$$

which has *real* coefficients, since $c_k + \overline{c_k}$ and $c_k\overline{c_k}$ are real numbers. (Check this fact.) Thus, the complex zeros of $f(x)$ and their conjugates give rise to quadratic polynomials that are irreducible over $\mathbb{R}$. This completes the proof. □

EXAMPLE 2 Express $x^4 - 2x^2 - 3$ as a product (a) of linear polynomials, and (b) of linear and quadratic polynomials with real coefficients that are irreducible over $\mathbb{R}$.

Solution

(a) We can find the zeros of the given polynomial by solving the equation $x^4 - 2x^2 - 3 = 0$, which may be regarded as quadratic in x^2. Solving for x^2 by means of the Quadratic Formula, we obtain

$$x^2 = \frac{2 \pm \sqrt{4 + 12}}{2} = \frac{2 \pm 4}{2}$$

or $x^2 = 3$ and $x^2 = -1$. Hence, the zeros are $\sqrt{3}$, $-\sqrt{3}$, i, and $-i$, and we obtain the factorization

$$x^4 - 2x^2 - 3 = (x - \sqrt{3})(x + \sqrt{3})(x - i)(x + i).$$

(b) Multiplying the last two factors in the preceding factorization gives us

$$x^4 - 2x^2 - 3 = (x - \sqrt{3})(x + \sqrt{3})(x^2 + 1),$$

which is of the form stated in the preceding theorem.

The solution of this example could also have been obtained by factoring the original expression without first finding the zeros. Thus,

$$\begin{aligned} x^4 - 2x^2 - 3 &= (x^2 - 3)(x^2 + 1) \\ &= (x + \sqrt{3})(x - \sqrt{3})(x + i)(x - i). \end{aligned}$$ ■

We have already pointed out that it is generally very difficult to find the zeros of a polynomial of high degree. However, if all the coefficients are integers or rational numbers, we can employ a method for finding the *rational* zeros, if they exist. The method is a consequence of the following theorem.

THEOREM ON RATIONAL ZEROS

Suppose that $f(x) = a_nx^n + a_{n-1}x^{n-1} + \cdots + a_1x + a_0$ is a polynomial with integer coefficients. If c/d is a rational zero of $f(x)$, where c and d have no common prime factors and $c > 0$, then c is a factor of a_0 and d is a factor of a_n.

Proof Let us show that c is a factor of a_0. If $c = 1$, the theorem follows at once, since 1 is a factor of *any* number. Now suppose that $c \neq 1$. In this case $c/d \neq 1$, for if $c/d = 1$, we obtain $c = d$, and since c and d have no prime factor in common, this implies that $c = d = 1$, a contradiction. Hence, in the following discussion we have $c \neq 1$ and $c \neq d$.

Since $f(c/d) = 0$,

$$a_n(c^n/d^n) + a_{n-1}(c^{n-1}/d^{n-1}) + \cdots + a_1(c/d) + a_0 = 0.$$

Multiplying by d^n and then adding $-a_0d^n$ to both sides, we obtain

$$a_nc^n + a_{n-1}c^{n-1}d + \cdots + a_1cd^{n-1} = -a_0d^n$$

or

$$c(a_nc^{n-1} + a_{n-1}c^{n-2}d + \cdots + a_1d^{n-1}) = -a_0d^n.$$

This shows that c is a factor of the integer a_0d^n. If c is factored into primes, such as $c = p_1p_2 \cdots p_k$, then each prime p_j is also a factor of a_0d^n. However, by hypothesis, none of the p_j is a factor of d. This implies that each p_j is a factor of a_0, that is, c is a factor of a_0. A similar argument may be used to prove that d is a factor of a_n. □

The technique of using the preceding theorem for finding rational solutions of equations with integral coefficients is illustrated in the following example.

EXAMPLE 3 Find all rational solutions of the equation

$$3x^4 + 14x^3 + 14x^2 - 8x - 8 = 0.$$

Solution The problem is equivalent to finding the rational zeros of the polynomial on the left side of the equation. According to the preceding theorem, if c/d is a rational zero and $c > 0$, then c is a divisor of -8 and d is a divisor of 3. Hence, the possible choices for c are 1, 2, 4, and 8, and the choices for d are ± 1 and ± 3. Consequently, any rational roots are included among the numbers ± 1, ± 2, ± 4, ± 8, $\pm \frac{1}{3}$, $\pm \frac{2}{3}$, $\pm \frac{4}{3}$, and $\pm \frac{8}{3}$. We can reduce the number of possibilities by finding upper and lower bounds for the real solutions; however, we shall not do so here. It is necessary to check to see which of the numbers, if any, are zeros. Synthetic division is the appropriate method for this task. After perhaps many trial-and-error attempts, we obtain:

$$\begin{array}{r|rrrrr} -2 & 3 & 14 & 14 & -8 & -8 \\ & & -6 & -16 & 4 & 8 \\ \hline & 3 & 8 & -2 & -4 & \boxed{0} \end{array}$$

This result shows that -2 is a zero. Moreover, the synthetic division provides the coefficients of the quotient in the division of the polynomial by $x + 2$. Hence, we have the following factorization of the given polynomial:

$$(x + 2)(3x^3 + 8x^2 - 2x - 4).$$

The remaining solutions of the equation must be zeros of the second factor, and therefore we may use the latter polynomial to check for solutions. Again proceeding by trial and error, we ultimately find that synthetic division by $x + \frac{2}{3}$ gives us the following result:

$$\begin{array}{r|rrrr} -\frac{2}{3} & 3 & 8 & -2 & -4 \\ & & -2 & -4 & 4 \\ \hline & 3 & 6 & -6 & \boxed{0} \end{array}$$

Therefore, $-\frac{2}{3}$ is a zero.

The remaining zeros are solutions of the equation $3x^2 + 6x - 6 = 0$, or equivalently, $x^2 + 2x - 2 = 0$. By the Quadratic Formula, this equation has solutions

$$\frac{-2 \pm \sqrt{4 - 4(-2)}}{2} = \frac{-2 \pm \sqrt{12}}{2} = \frac{-2 \pm 2\sqrt{3}}{2}.$$

Hence, the given polynomial has two rational roots, -2 and $-\frac{2}{3}$, and two irrational roots, $-1 + \sqrt{3}$ and $-1 - \sqrt{3}$. ■

The Theorem on Rational Zeros may be applied to equations with rational coefficients. We merely multiply both sides of the equation by the least common denominator of all the coefficients to obtain an equation with integral coefficients, and then proceed as in Example 3.

EXAMPLE 4 Find all rational solutions of the equation

$$\tfrac{2}{3}x^4 + \tfrac{1}{2}x^3 - \tfrac{5}{4}x^2 - x - \tfrac{1}{6} = 0.$$

Solution Multiplying both sides of the equation by 12, we obtain the equivalent equation

$$8x^4 + 6x^3 - 15x^2 - 12x - 2 = 0.$$

If c/d is a rational solution, then the choices for c are 1 and 2 and the choices for d are ± 1, ± 2, ± 4, and ± 8. Hence, the only possible rational roots are ± 1, ± 2, $\pm \frac{1}{2}$, $\pm \frac{1}{4}$, and $\pm \frac{1}{8}$. Trying various possibilities we obtain, by synthetic division,

$$\begin{array}{r|rrrrr} -\frac{1}{2} & 8 & 6 & -15 & -12 & -2 \\ & & -4 & -1 & 8 & 2 \\ \hline & 8 & 2 & -16 & -4 & \boxed{0} \end{array}$$

Hence, $-\frac{1}{2}$ is a solution. Using synthetic division on the coefficients of the quotient gives us

$$\begin{array}{r|rrrr} -\frac{1}{4} & 8 & 2 & -16 & -4 \\ & & -2 & 0 & 4 \\ \hline & 8 & 0 & -16 & \boxed{0} \end{array}$$

Consequently, $-\frac{1}{4}$ is a solution. The last synthetic division gave us the quotient $8x^2 - 16$. Setting this equal to zero and solving, we obtain $x^2 = 2$, or $x = \pm\sqrt{2}$. Thus, the given equation has rational solutions $-\frac{1}{2}$, $-\frac{1}{4}$ and irrational solutions $\sqrt{2}$, $-\sqrt{2}$. ■

The discussion in this section gives no practical information about finding the irrational zeros of polynomials. The examples we have worked are not typical of problems encountered in applications. Indeed, it is not unusual for a polynomial with rational coefficients to have *no* rational zeros. Except in the simplest cases, the best that we can accomplish is to find decimal approximations to the irrational zeros. We can approximate irrational zeros to any degree of accuracy by methods such as those described in Calculator Exercises 3.1. A standard way to approximate zeros is to use a calculus technique called Newton's Method. In practice, computers have taken over the task of approximating irrational solutions of equations.

EXERCISES 3.6

In Exercises 1–8 find a polynomial with real coefficients that has the given zero (or zeros) and degree.

1 $4 + i$; degree 2

2 $3 - 3i$; degree 2

3 $4, 3 - 2i$; degree 3

4 $-2, 1 + 5i$; degree 3

5 $1 + 4i, 2 - i$; degree 4

6 $3, 0, 8 - 7i$; degree 4

7 $5i, 1 + i, 0$; degree 5

8 $i, 2i, 3i$; degree 6

9 Does there exist a polynomial of degree 3 with real coefficients that has zeros 1, -1, and i? Justify your answer.

10 The complex number i is a zero of the polynomial $f(x) = x^3 - ix^2 + 2ix + 2$; however, the conjugate $-i$ of i is not a zero. Why doesn't this contradict the first theorem of this section?

In Exercises 11–28 find all solutions of the given equations.

11 $x^3 - x^2 - 10x - 8 = 0$

12 $x^3 + x^2 - 14x - 24 = 0$

13 $2x^3 - 3x^2 - 17x + 30 = 0$

14 $12x^3 + 8x^2 - 3x - 2 = 0$

15 $x^4 + 3x^3 - 30x^2 - 6x + 56 = 0$

16 $3x^5 - 10x^4 - 6x^3 + 24x^2 + 11x - 6 = 0$

17 $2x^3 - 7x^2 - 10x + 24 = 0$

18 $2x^3 - 3x^2 - 8x - 3 = 0$

19 $6x^3 + 19x^2 + x - 6 = 0$

20 $6x^3 + 5x^2 - 17x - 6 = 0$

21 $8x^3 + 18x^2 + 45x + 27 = 0$

22 $3x^3 - x^2 + 11x - 20 = 0$

23 $x^4 - x^3 - 9x^2 - 3x - 36 = 0$

24 $3x^4 + 16x^3 + 28x^2 + 31x + 30 = 0$

25 $9x^4 + 15x^3 - 20x^2 - 20x + 16 = 0$

26 $15x^4 + 4x^3 + 11x^2 + 4x - 4 = 0$

27 $4x^5 + 12x^4 - 41x^3 - 99x^2 + 10x + 24 = 0$

28 $4x^5 + 24x^4 - 13x^3 - 174x^2 + 9x + 270 = 0$

Show that the equations in Exercises 29–36 have no rational roots.

29 $x^3 + 3x^2 - 4x + 6 = 0$

30 $3x^3 - 4x^2 + 7x + 5 = 0$

31 $x^5 - 3x^3 + 4x^2 + x - 2 = 0$

32 $2x^5 + 3x^3 + 7 = 0$

33 $3x^4 + 7x^3 + 3x^2 - 8x - 10 = 0$

34 $2x^4 - 3x^3 + 6x^2 - 24x + 5 = 0$

35 $8x^4 + 16x^3 - 26x^2 - 12x + 15 = 0$

36 $5x^5 + 2x^4 + x^3 - 10x^2 - 4x - 2 = 0$

37 If n is an odd positive integer, prove that a polynomial of degree n with real coefficients has at least one real zero.

38 Show that the first theorem of this section is not necessarily true if $f(x)$ has complex coefficients.

39 Complete the proof of the Theorem on Rational Zeros by showing that d is a factor of a_n.

40 If a polynomial of the form

$$x^n + a_{n-1}x^{n-1} + \cdots + a_1x + a_0,$$

where each a_i is an integer, has a rational root r, show that r is an integer and is a factor of a_0.

SECTION 3.7

RATIONAL FUNCTIONS

A function f is a **rational function** if, for all x in its domain,

$$f(x) = \frac{g(x)}{h(x)}$$

where $g(x)$ and $h(x)$ are polynomials. Of major importance are the zeros of the numerator and denominator. *Throughout this section we shall assume that $g(x)$ and $h(x)$ have no common factors* and hence, no common zeros. Note that $f(c) = 0$ if and only if $g(c) = 0$, and hence the zeros of the numerator $g(x)$ are the zeros of $f(x)$. However, if c is a zero of the denominator $h(x)$, then $f(c) = g(c)/0$ is undefined, and hence the behavior of $f(x)$ requires special attention when x is near c.

EXAMPLE 1 Sketch the graph of f if $f(x) = \dfrac{1}{x-2}$.

Solution Since the numerator 1 is never zero, $f(x)$ has no zeros. Hence, the graph has no x-intercepts. Since the denominator $x - 2$ has the zero 2, $f(2)$ is undefined. If x is very close to 2, and $x > 2$, then $f(x)$ is very large. For example,

$$f(2.1) = \frac{1}{2.1-2} = \frac{1}{0.1} = 10$$

$$f(2.01) = \frac{1}{2.01-2} = \frac{1}{0.01} = 100$$

$$f(2.001) = \frac{1}{2.001-2} = \frac{1}{0.001} = 1{,}000.$$

If x is close to 2, and $x < 2$, then $|f(x)|$ is large, but $f(x)$ is negative. Thus,

$$f(1.9) = \frac{1}{1.9-2} = \frac{1}{-0.1} = -10$$

$$f(1.99) = \frac{1}{1.99-2} = \frac{1}{-0.01} = -100$$

$$f(1.999) = \frac{1}{1.999-2} = \frac{1}{-0.001} = -1{,}000.$$

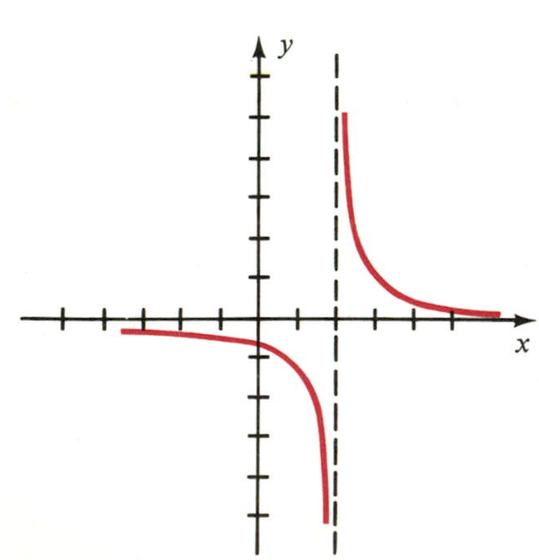

FIGURE 3.9 $f(x) = \dfrac{1}{x-2}$

Several other values of $f(x)$ are displayed in the following table:

x	-8	-1	0	1	3	4	12
$f(x)$	$-\frac{1}{10}$	$-\frac{1}{3}$	$-\frac{1}{2}$	-1	1	$\frac{1}{2}$	$\frac{1}{10}$

Observe that as $|x|$ increases, $f(x)$ approaches zero. Plotting points and taking into account what happens near $x = 2$, we arrive at the sketch in Figure 3.9. ■

In Example 1, $f(x)$ can be made as large as we desire by choosing x sufficiently close to 2 (and $x > 2$). This fact will be denoted symbolically by

$$f(x) \to \infty \quad \text{as} \quad x \to 2^{+}$$

This notation is read $f(x)$ *increases without bound as x approaches* 2 *from the right.* The phrase "$f(x)$ becomes positively infinite" is sometimes used in place of "$f(x)$ increases without bound." It is important to remember that the symbol ∞ (read "infinity") does not represent a real number, but is used merely as an abbreviation for certain types of functional behavior. In like manner, the notation

$$f(x) \to -\infty \quad \text{as} \quad x \to 2^-$$

is read $f(x)$ *decreases without bound as x approaches 2 from the left.* The phrase "$f(x)$ becomes negatively infinite" is sometimes used instead of "$f(x)$ decreases without bound."

Let us extend the preceding discussion to arbitrary functions and to the case where x approaches *any* real number a. In particular, the symbol $x \to a^+$ will signify that x approaches a from the *right*, that is, through values *greater* than a. The symbol $x \to a^-$ will mean that x approaches a from the *left*, that is, through values *less* than a. Some illustrations of the manner in which a function may increase or decrease without bound, together with the notation used, are shown in Figure 3.10. In the figure, a is pictured as positive, but we can also have $a < 0$.

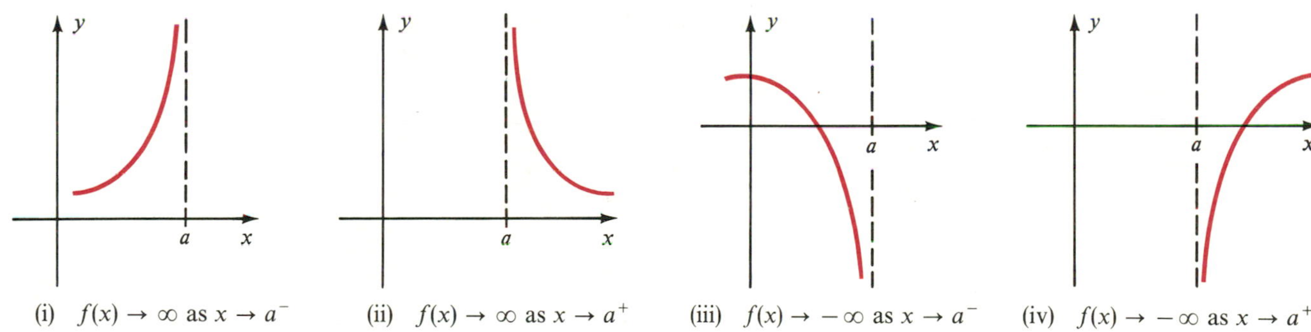

(i) $f(x) \to \infty$ as $x \to a^-$ (ii) $f(x) \to \infty$ as $x \to a^+$ (iii) $f(x) \to -\infty$ as $x \to a^-$ (iv) $f(x) \to -\infty$ as $x \to a^+$

FIGURE 3.10

DEFINITION

The line $x = a$ is a **vertical asymptote** for the graph of a function f if

$$f(x) \to \infty \quad \text{or} \quad f(x) \to -\infty$$

as x approaches a from either the left or right.

In Figure 3.10 the dashed lines represent vertical asymptotes. In Figure 3.9 the line $x = 2$ is a vertical asymptote for the graph of $f(x) = 1/(x - 2)$.

Vertical asymptotes are a common characteristic of graphs of rational functions. Indeed, *if the number a is a zero of the denominator* $h(x)$, *then the graph of* $f(x) = g(x)/h(x)$ *has the vertical asymptote* $x = a$.

We are also interested in values of $f(x)$ when $|x|$ is large. As an illustration, if, for the function given in Example 1, we assign very large values to x, then $f(x)$ is close to 0. Thus,

$$f(1{,}002) = \frac{1}{1{,}000} = 0.001 \quad \text{and} \quad f(1{,}000{,}002) = \frac{1}{1{,}000{,}000} = 0.000001.$$

Moreover, we can make $f(x)$ as close to 0 as we desire by choosing x sufficiently large. This is expressed symbolically by

$$f(x) \to 0 \quad \text{as} \quad x \to \infty,$$

which is read $f(x)$ *approaches* 0 *as x increases without bound* (or *as x becomes positively infinite*).

Similarly, in Example 1, we write

$$f(x) \to 0 \quad \text{as} \quad x \to -\infty,$$

which is read $f(x)$ *approaches* 0 *as x decreases without bound* (or *as x becomes negatively infinite*).

In Example 1 the line $y = 0$, that is, the x-axis, is called a *horizontal asymptote* for the graph. In general, we have the following definition. The notation should be self-evident.

DEFINITION

The line $y = b$ is a **horizontal asymptote** for the graph of a function f if

$$f(x) \to b \quad \text{as} \quad x \to \infty \quad \text{or as} \quad x \to -\infty.$$

Graphs of rational functions often have horizontal asymptotes. Some typical cases (for $x \to \infty$) are illustrated in Figure 3.11. The manner in which the graph "approaches" the line $y = b$ may vary, depending on the nature of the function. Similar sketches may be made for the case $x \to -\infty$. Note that, as in the third sketch, the graph of f may cross a horizontal asymptote.

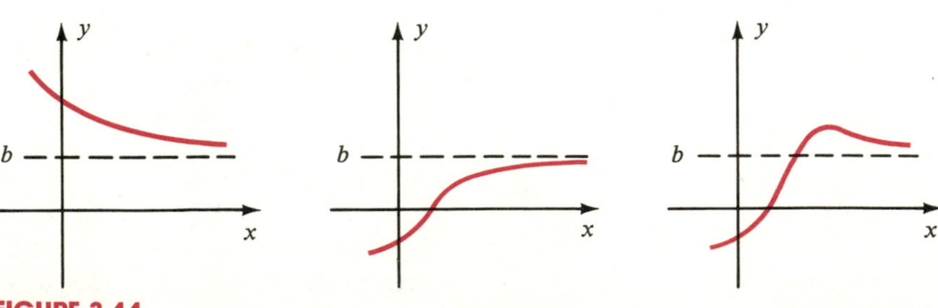

FIGURE 3.11

We shall next list some guidelines for sketching the graph of a rational function. Their use will be illustrated in Examples 2, 3, and 4.

Guidelines for sketching the graph of $f(x) = \dfrac{g(x)}{h(x)}$

where $g(x)$ and $h(x)$ are polynomials that have no common factor

STEP 1 Find the real zeros of the numerator $g(x)$, and use them to plot the points corresponding to the x-intercepts.

STEP 2 Find the real zeros of the denominator $h(x)$. For each zero a, the line $x = a$ is a vertical asymptote. Represent $x = a$ with dashes.

STEP 3 Find the sign of $f(x)$ in each of the intervals determined by the zeros of $g(x)$ and $h(x)$. Use these signs to determine whether the graph lies above or below the x-axis in each interval.

STEP 4 If $x = a$ is a vertical asymptote, use the information in Step 3 to determine whether $f(x) \to \infty$ or $f(x) \to -\infty$ for each case:

$$\text{(i)}\quad x \to a^-; \qquad \text{(ii)}\quad x \to a^+.$$

Make note of this by sketching a portion of the graph on each side of $x = a$.

STEP 5 Use the information in Step 3 to determine the manner in which the graph intersects the x-axis.

STEP 6 Determine the behavior of $f(x)$ as $x \to \infty$ or $x \to -\infty$. If $f(x) \to b$, then the line $y = b$ is a horizontal asymptote. If $b \neq 0$, represent $y = b$ with dashes.

STEP 7 Sketch the graph, using the information found in the preceding steps and plotting points wherever necessary.

EXAMPLE 2 Sketch the graph of f if $f(x) = \dfrac{x - 1}{x^2 - x - 6}$.

Solution It is convenient to factor the denominator as follows:

$$f(x) = \frac{x - 1}{(x + 2)(x - 3)}.$$

We shall obtain the graph by following the steps listed in the guidelines.

Step 1: The numerator $x - 1$ has the zero 1, and hence we plot the point $(1, 0)$ on the graph, as shown in Figure 3.12.

Step 2: The denominator has zeros -2 and 3. Hence, the lines $x = -2$ and $x = 3$ are vertical asymptotes, and we represent them with dashes, as in Figure 3.12.

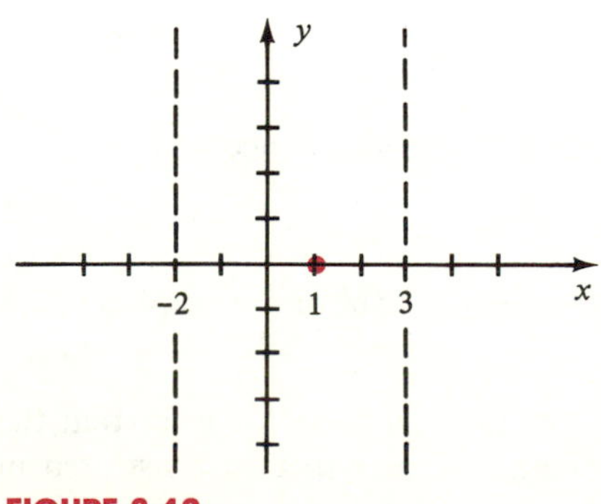

FIGURE 3.12

Step 3: The zeros -2, 1, and 3 of the numerator and denominator of $f(x)$

determine the following intervals:

$$(-\infty, -2), \quad (-2, 1), \quad (1, 3), \quad \text{and} \quad (3, \infty).$$

Since $f(x)$ is a quotient of two polynomials, it follows from our work in Section 3.1 that $f(x)$ is always positive or always negative throughout each interval. Using test values to determine the sign of $f(x)$, we arrive at the following table.

Interval	$(-\infty, -2)$	$(-2, 1)$	$(1, 3)$	$(3, \infty)$
Test value	$f(-3) = -\frac{2}{3}$	$f(0) = \frac{1}{6}$	$f(2) = -\frac{1}{4}$	$f(4) = \frac{1}{2}$
Sign of $f(x)$	$-$	$+$	$-$	$+$
Position of graph	Below x-axis	Above x-axis	Below x-axis	Above x-axis

Step 4: We shall use the fourth row of the table in Step 3 to investigate the behavior of $f(x)$ near each vertical asymptote.

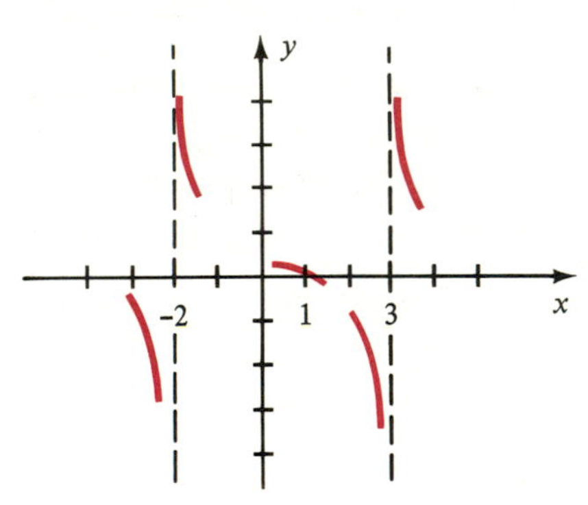

FIGURE 3.13

(a) Consider the vertical asymptote $x = -2$. Since the graph lies *below* the x-axis throughout the interval $(-\infty, -2)$, it follows that

$$f(x) \to -\infty \quad \text{as} \quad x \to -2^-.$$

Since the graph lies *above* the x-axis throughout the interval $(-2, 1)$, it follows that

$$f(x) \to \infty \quad \text{as} \quad x \to -2^+.$$

We note these facts in Figure 3.13 by sketching portions of the graph on each side of the line $x = -2$.

(b) Consider the vertical asymptote $x = 3$. The graph lies *below* the x-axis throughout the interval $(1, 3)$, and hence

$$f(x) \to -\infty \quad \text{as} \quad x \to 3^-.$$

The graph lies *above* the x-axis throughout $(3, \infty)$, and hence

$$f(x) \to \infty \quad \text{as} \quad x \to 3^+.$$

We note these facts in Figure 3.13 by sketching portions of the graph on each side of $x = 3$.

Step 5: Referring to the fourth row of the table in Step 3, we see that the graph crosses the x-axis at $(1, 0)$ in a manner similar to that illustrated in Figure 3.13.

Step 6: To determine what is true if $x \to \infty$ or $x \to -\infty$, we divide numerator and denominator of $f(x)$ by x^2, obtaining

$$f(x) = \frac{(1/x) - (1/x^2)}{1 - (1/x) - (6/x^2)}.$$

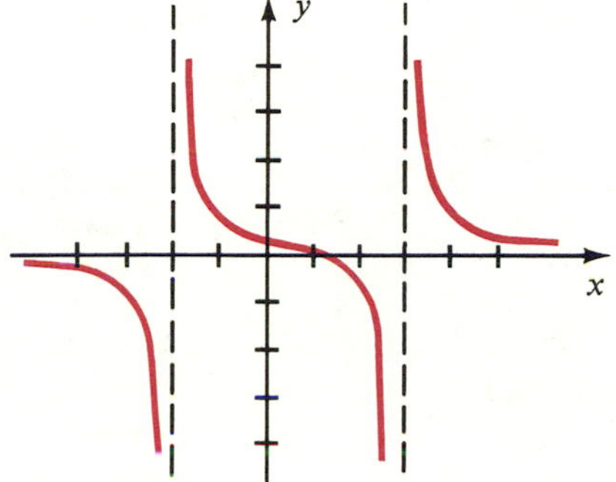

FIGURE 3.14 $f(x) = \dfrac{x - 1}{x^2 - x - 6}$

If $|x|$ is very large, then the expressions in parentheses are close to 0, and hence

$$f(x) \approx \frac{0 - 0}{1 - 0 - 0} = \frac{0}{1} = 0.$$

This indicates that

$$f(x) \to 0 \quad \text{as} \quad x \to \infty \qquad \text{and} \qquad f(x) \to 0 \quad \text{as} \quad x \to -\infty.$$

Thus, the line $y = 0$ (the x-axis) is a horizontal asymptote for the graph.

Step 7: Using the information found in Steps 4, 5, and 6, and plotting several points, we obtain the sketch in Figure 3.14. ■

The technique used in Step 6 of Example 2 can be generalized. Specifically, if we wish to investigate $f(x) = g(x)/h(x)$ as $x \to \infty$ or as $x \to -\infty$, and if the degree of $g(x)$ is not greater than the degree of $h(x)$, then we divide numerator and denominator by x^m, where m is the degree of $h(x)$. If the degree n of $g(x)$ is greater than the degree m of $h(x)$, it can be shown that $f(x)$ either increases or decreases without bound as $x \to \infty$ or as $x \to -\infty$. If $n \leq m$, then the graph has a horizontal asymptote (cf. Exercise 25).

EXAMPLE 3 Sketch the graph of f if $f(x) = \dfrac{x^2}{x^2 - x - 2}$.

Solution Factoring the denominator gives us

$$f(x) = \frac{x^2}{(x + 1)(x - 2)}.$$

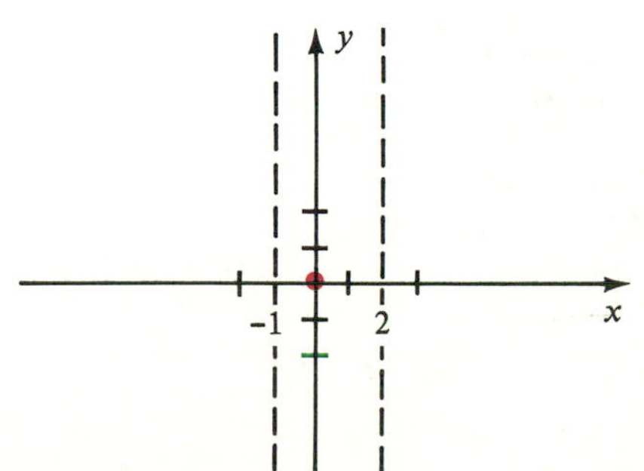

FIGURE 3.15

We shall again follow the guidelines listed earlier.

Step 1: The numerator x^2 has 0 as a zero, and hence the graph intersects the x-axis at (0, 0), as shown in Figure 3.15.

Step 2: Since the denominator has zeros -1 and 2, the lines $x = -1$ and $x = 2$ are vertical asymptotes, and we represent them with dashes, as in Figure 3.15.

Step 3: The intervals determined by the zeros in Steps 1 and 2 are

$$(-\infty, -1), \quad (-1, 0), \quad (0, 2), \quad \text{and} \quad (2, \infty).$$

Following the procedure used in Step 3 of Example 2, we arrive at the following table.

Interval	$(-\infty, -1)$	$(-1, 0)$	$(0, 2)$	$(2, \infty)$
Test value	$f(-2) = 1$	$f(-\frac{1}{2}) = -\frac{1}{5}$	$f(1) = -\frac{1}{2}$	$f(3) = \frac{9}{4}$
Sign of $f(x)$	+	−	−	+
Position of graph	Above x-axis	Below x-axis	Below x-axis	Above x-axis

Step 4: We refer to the fourth row of the table in Step 3 and proceed as follows:

(a) Consider the vertical asymptote $x = -1$. Since the graph is above the x-axis in $(-\infty, -1)$, it follows that

$$f(x) \to \infty \quad \text{as} \quad x \to -1^-.$$

Since the graph is below the x-axis in $(-1, 0)$, we have

$$f(x) \to -\infty \quad \text{as} \quad x \to -1^+.$$

(b) Consider the vertical asymptote $x = 2$. The graph is below the x-axis in the interval $(0, 2)$, and hence

$$f(x) \to -\infty \quad \text{as} \quad x \to 2^-.$$

The graph is above the x-axis in $(2, \infty)$, and hence

$$f(x) \to \infty \quad \text{as} \quad x \to 2^+.$$

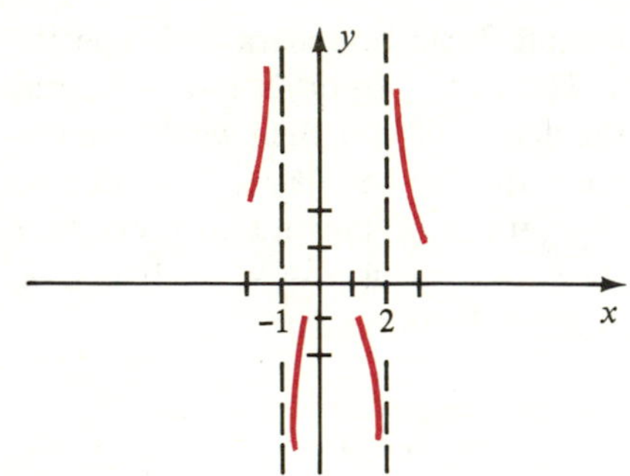

FIGURE 3.16

These facts are noted in Figure 3.16.

Step 5: Referring to the table in Step 3, we see that the graph lies below the x-axis in both of the intervals $(-1, 0)$ and $(0, 2)$. Consequently, the graph intersects, but does not cross, the x-axis at $(0, 0)$.

Step 6: To determine what happens as $x \to \infty$ or as $x \to -\infty$, we use the method discussed prior to this example. Since the degree of the denominator, $x^2 - x - 2$, is 2, we divide numerator and denominator of $f(x)$ by x^2, obtaining

$$f(x) = \frac{x^2/x^2}{(x^2 - x - 2)/x^2} = \frac{1}{1 - (1/x) - (2/x^2)}.$$

If we assign very large values to $|x|$, the expressions $1/x$ and $2/x^2$ are very close to 0, and hence

$$f(x) \approx \frac{1}{1 - 0 - 0} = 1.$$

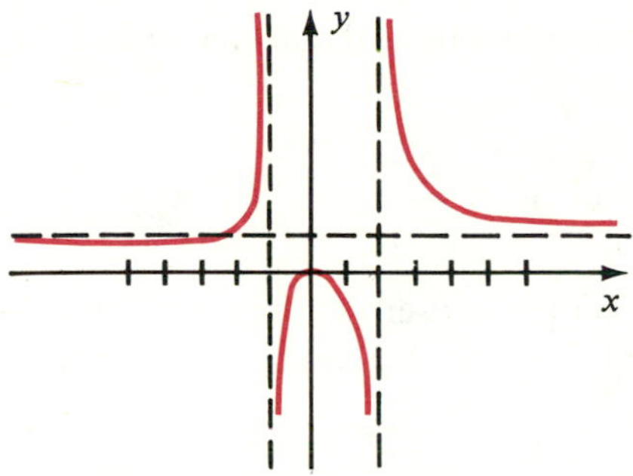

FIGURE 3.17 $f(x) = \dfrac{x^2}{x^2 - x - 2}$

This indicates that

$$f(x) \to 1 \quad \text{as} \quad x \to \infty \qquad \text{and} \qquad f(x) \to 1 \quad \text{as} \quad x \to -\infty.$$

Hence, the line $y = 1$ is a horizontal asymptote for the graph, and we sketch the asymptote with dashes, as in Figure 3.17.

Step 7: Using the information found in Steps 4, 5, and 6, and plotting several points, we obtain the graph, as shown in Figure 3.17. The graph intersects the horizontal asymptote at $x = -2$; this may be verified by solving the equation $x^2/(x^2 - x - 2) = 1$. The fact that the graph lies below the horizontal asymptote if $x < -2$ and above it if $-2 < x < -1$ may be verified by plotting points. ■

EXAMPLE 4 Sketch the graph of f if $f(x) = \dfrac{2x^4}{x^4 + 1}$.

Solution In this solution we shall not formally write down each step in the guidelines. Note that since $f(-x) = f(x)$, the function is even, and hence the graph is symmetric with respect to the y-axis.

The graph intersects the x-axis at (0, 0). Since the denominator of $f(x)$ has no real zeros, the graph has no vertical asymptotes. To examine what happens to $f(x)$ as $x \to \infty$ or as $x \to -\infty$, we divide numerator and denominator by x^4, obtaining

$$f(x) = \frac{2}{1 + (1/x^4)}.$$

Since $\qquad \dfrac{1}{x^4} \to 0 \quad \text{as} \quad x \to \infty$

it follows that $\qquad f(x) \to \dfrac{2}{1 + 0} = 2 \quad \text{as} \quad x \to \infty.$

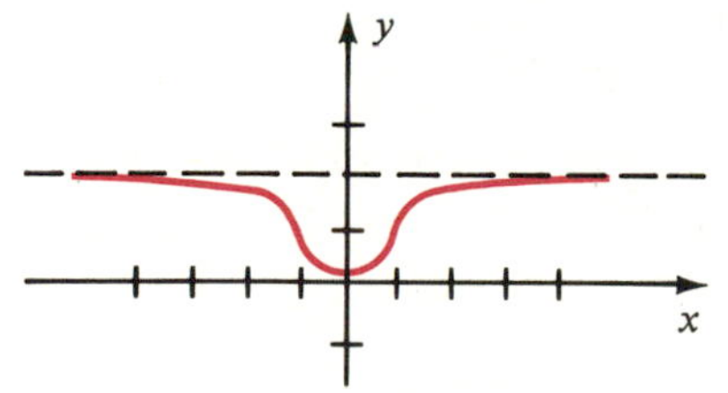

FIGURE 3.18 $f(x) = \dfrac{2x^4}{x^4 + 1}$

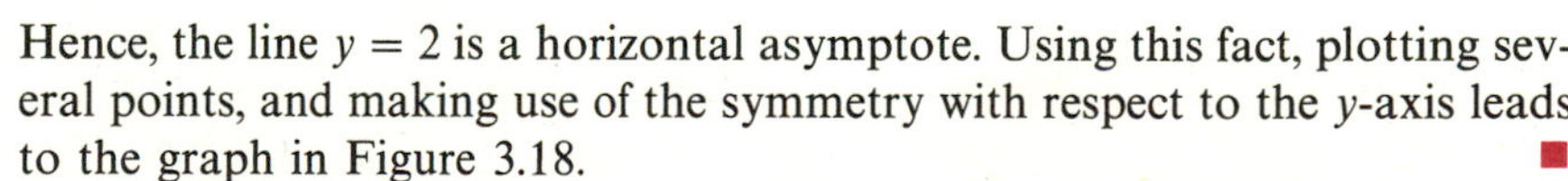

Hence, the line $y = 2$ is a horizontal asymptote. Using this fact, plotting several points, and making use of the symmetry with respect to the y-axis leads to the graph in Figure 3.18. ■

If $f(x) = g(x)/h(x)$ for polynomials $g(x)$ and $h(x)$, and *if the degree of* $g(x)$ *is one greater than the degree of* $h(x)$, then the graph of f has an **oblique asymptote** $y = ax + b$; that is, the graph approaches this line as $x \to \infty$ or as $x \to -\infty$. To find the oblique asymptote we may use division to express $f(x)$ in the form

$$f(x) = \frac{g(x)}{h(x)} = (ax + b) + \frac{r(x)}{h(x)}$$

where either $r(x) = 0$ or the degree of $r(x)$ is less than the degree of $h(x)$. It follows that

$$\frac{r(x)}{h(x)} \to 0 \quad \text{as} \quad x \to \infty \quad \text{or as} \quad x \to -\infty.$$

(See (ii) of Exercise 25.) Consequently, $f(x)$ gets closer and closer to $ax + b$ as $|x|$ increases without bound. The next example illustrates a special case of this procedure.

EXAMPLE 5 Find all the asymptotes and sketch the graph of f if

$$f(x) = \frac{x^2 - 9}{2x - 4}.$$

Solution A vertical asymptote occurs if $2x - 4 = 0$, that is, if $x = 2$. (Why?) Moreover, since the degree of the numerator $x^2 - 9$ is one greater than the degree of the denominator, $2x - 4$, the graph has an oblique asymptote. Dividing, we obtain

$$\begin{array}{r|l} & \tfrac{1}{2}x + 1 \\ \hline 2x - 4 & x^2 \qquad\quad - 9 \\ & x^2 - 2x \\ \hline & \qquad 2x - 9 \\ & \qquad 2x - 4 \\ \hline & \qquad\quad -5 \end{array}$$

Therefore,
$$\frac{x^2 - 9}{2x - 4} = \left(\frac{1}{2}x + 1\right) - \frac{5}{2x - 4}.$$

As we indicated in the discussion preceding this example, the line $y = \frac{1}{2}x + 1$ is an oblique asymptote. This line and the vertical asymptote $x = 2$ are sketched (with dashes) in (i) of Figure 3.19.

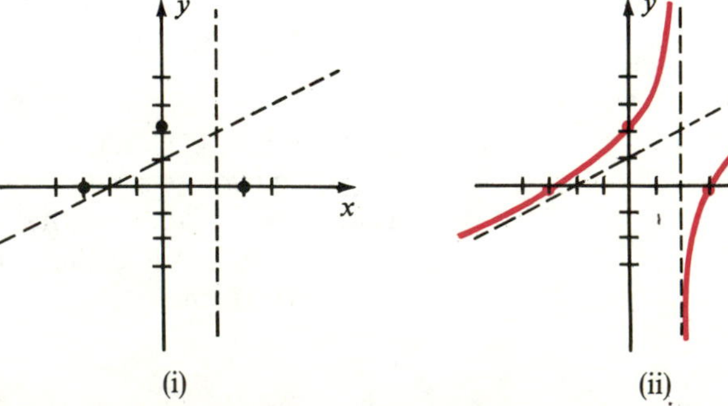

FIGURE 3.19

The x-intercepts of the graph are the solutions of the equation $x^2 - 9 = 0$, and hence are 3 and -3. The y-intercept is $f(0) = \frac{9}{4}$. The corresponding points are plotted in (i) of Figure 3.19. It is now easy to show that the graph has the shape indicated in (ii) of Figure 3.19. ■

Graphs of rational functions may become increasingly complicated as the degrees of the polynomials in the numerator and denominator increase. Techniques developed in calculus must be employed for a thorough treatment of such graphs.

EXERCISES 3.7

Sketch the graph of f in Exercises 1–20.

1 $f(x) = \dfrac{1}{x+2}$

2 $f(x) = \dfrac{1}{x+3}$

3 $f(x) = \dfrac{-2}{x+4}$

4 $f(x) = \dfrac{-3}{x-1}$

5 $f(x) = \dfrac{x}{x-5}$

6 $f(x) = \dfrac{x}{3x+2}$

7 $f(x) = \dfrac{4}{(x-1)^2}$

8 $f(x) = \dfrac{-1}{(x+2)^2}$

9 $f(x) = \dfrac{1}{x^2-4}$

10 $f(x) = \dfrac{2}{x^2+x-2}$

11 $f(x) = \dfrac{5x}{4-x^2}$

12 $f(x) = \dfrac{x^2}{x^2-4}$

13 $f(x) = \dfrac{x^2}{x^2-7x+10}$

14 $f(x) = \dfrac{x}{x^2-x-6}$

15 $f(x) = \dfrac{3x+2}{x}$

16 $f(x) = \dfrac{x^2-4}{x^2}$

17 $f(x) = \dfrac{4}{x^2+4}$

18 $f(x) = \dfrac{3x}{x^2+1}$

19 $f(x) = \dfrac{1}{x^3+x^2-6x}$

20 $f(x) = \dfrac{x^2-x}{16-x^2}$

In Exercises 21–24 find the vertical and oblique asymptotes, and sketch the graph of f.

21 $f(x) = \dfrac{x^2-x-6}{x+1}$

22 $f(x) = \dfrac{2x^2-x-3}{x-2}$

23 $f(x) = \dfrac{8-x^3}{2x^2}$

24 $f(x) = \dfrac{x^3+1}{x^2-9}$

25 Let $f(x) = \dfrac{a_nx^n + a_{n-1}x^{n-1} + \cdots + a_1x + a_0}{b_mx^m + b_{m-1}x^{m-1} + \cdots + b_1x + b_0}$

where $a_n \neq 0$ and $b_m \neq 0$. Give intuitive arguments for the following.

(a) If $n = m$, then $y = a_n/b_m$ is a horizontal asymptote for the graph of f.

(b) If $n < m$, then $y = 0$ is a horizontal asymptote for the graph of f.

(c) If $n > m$, then the graph has no horizontal asymptotes.

SECTION 3.8
PARTIAL FRACTIONS

It is easy to verify that

$$\frac{2}{x^2-1}=\frac{1}{x-1}+\frac{-1}{x+1}.$$

The expression on the right side of this equation is called *the partial fraction decomposition* of $2/(x^2-1)$. In general, if f is *any* rational function, it is theoretically possible to express $f(x)$ as a sum of quotients of polynomials whose denominators involve powers of polynomials of degree not greater than two. This fact has important applications in the study of calculus.

If $g(x)$ and $h(x)$ are polynomials *and the degree of* $g(x)$ *is less than the degree of* $h(x)$, then it follows from a theorem in algebra that

$$\frac{g(x)}{h(x)}=F_1+F_2+\cdots+F_k$$

where each F_i has one of the forms

$$\frac{A}{(px+q)^m} \quad \text{or} \quad \frac{Cx+D}{(ax^2+bx+c)^n}$$

for some nonnegative integers m and n, and where ax^2+bx+c is **irreducible** in the sense that this quadratic polynomial has no real zeros, that is, $b^2-4ac<0$. The sum $F_1+F_2+\cdots+F_k$ is called the **partial fraction decomposition** of $g(x)/h(x)$ and each F_i is called a **partial fraction.** We shall not prove this algebraic result but will, instead, give rules for obtaining the decomposition.

To find the partial fraction decomposition of $g(x)/h(x)$ *it is essential* that $g(x)$ have lower degree than $h(x)$. If this is not the case, then long division should be employed to arrive at such an expression. For example, given

$$\frac{x^3-6x^2+5x-3}{x^2-1}$$

we obtain
$$\frac{x^3-6x^2+5x-3}{x^2-1}=x-6+\frac{6x-9}{x^2-1}.$$

The partial fraction decomposition is then found for $(6x-9)/(x^2-1)$.

To obtain the decomposition of $g(x)/h(x)$, we begin by expressing the denominator $h(x)$ as a product of factors of the form $px+q$, or of irreducible quadratic factors ax^2+bx+c. Repeated factors are then collected so that $h(x)$ is a product of *different* factors of the form $(px+q)^m$ or $(ax^2+bx+c)^n$ where m and n are nonnegative integers and ax^2+bx+c is irreducible. We then apply the following rules.

PARTIAL FRACTION DECOMPOSITIONS

Rule 1. For each factor of $h(x)$ of the form $(px + q)^m$ where $m \geq 1$, the partial fraction decomposition of $g(x)/h(x)$ contains a sum of m partial fractions of the form

$$\frac{A_1}{px + q} + \frac{A_2}{(px + q)^2} + \cdots + \frac{A_m}{(px + q)^m}$$

where each A_i is a real number.

Rule 2. For each factor of $h(x)$ of the form $(ax^2 + bx + c)^n$ where $n \geq 1$ and $b^2 - 4ac < 0$, the partial fraction decomposition of $g(x)/h(x)$ contains a sum of n partial fractions of the form

$$\frac{A_1x + B_1}{ax^2 + bx + c} + \frac{A_2x + B_2}{(ax^2 + bx + c)^2} + \cdots + \frac{A_nx + B_n}{(ax^2 + bx + c)^n}$$

where, for each i, A_i and B_i are real numbers.

EXAMPLE 1 Find the partial fraction decomposition of

$$\frac{4x^2 + 13x - 9}{x^3 + 2x^2 - 3x}.$$

Solution The denominator of the quotient has the factored form $x(x + 3)(x - 1)$. Each of the linear factors is handled according to Rule 1 with $m = 1$. Thus, for the factor x there corresponds a partial fraction of the form A/x. Similarly, for the factors $x + 3$ and $x - 1$ there correspond partial fractions $B/(x + 3)$ and $C/(x - 1)$, respectively. The partial fraction decomposition has the form

$$\frac{4x^2 + 13x - 9}{x(x + 3)(x - 1)} = \frac{A}{x} + \frac{B}{x + 3} + \frac{C}{x - 1}.$$

Multiplying by the lowest common denominator, $x(x + 3)(x - 1)$, gives us

$$4x^2 + 13x - 9 = A(x + 3)(x - 1) + Bx(x - 1) + Cx(x + 3).$$

In a case such as this, in which the factors are all linear and nonrepeated, the values for A, B, and C can be found by substituting values for x that make the various factors zero. If we let $x = 0$ in the last equation, then $-9 = -3A$, or $A = 3$. Letting $x = 1$ gives us $8 = 4C$, or $C = 2$. Finally, if $x = -3$, then $-12 = 12B$, or $B = -1$. The partial fraction decomposition is, therefore,

$$\frac{4x^2 + 13x - 9}{x(x + 3)(x - 1)} = \frac{3}{x} + \frac{-1}{x + 3} + \frac{2}{x - 1}.$$

■

EXAMPLE 2 Find the partial fraction decomposition of

$$\frac{x^2 + 10x - 36}{x(x-3)^2}.$$

Solution By Rule 1 with $m = 1$, there is a partial fraction A/x corresponding to the factor x. Next, applying Rule 1 with $m = 2$, the factor $(x-3)^2$ gives rise to a sum of two partial fractions $B/(x-3) + C/(x-3)^2$. Thus, the partial fraction decomposition has the form

$$\frac{x^2 + 10x - 36}{x(x-3)^2} = \frac{A}{x} + \frac{B}{x-3} + \frac{C}{(x-3)^2}.$$

Multiplying both sides by $x(x-3)^2$ gives us

$$(*) \qquad x^2 + 10x - 36 = A(x-3)^2 + Bx(x-3) + Cx.$$

(The symbol $(*)$ is used for later identification of this equation.) Two of the unknown constants may be determined easily. Letting $x = 3$ in $(*)$ gives us

$$9 + 30 - 36 = A(0) + B(0) + 3C.$$

This reduces to $3 = 3C$, or $C = 1$. In like manner, letting $x = 0$, we get

$$0 + 0 - 36 = A(-3)^2 + B(0) + C(0).$$

Hence, $-36 = 9A$, or $A = -4$. The remaining constant can be found by comparing coefficients. If the right side of $(*)$ is expanded and like powers of x are collected, we obtain

$$x^2 + 10x - 36 = (A + B)x^2 + (-6A - 3B + C)x + 9A.$$

The coefficient $A + B$ of x^2 on the right must equal the coefficient of x^2 on the left, that is, $A + B = 1$. Since $A = -4$, it follows that $B = 1 - A = 1 - (-4) = 5$. The partial fraction decomposition is therefore

$$\frac{x^2 + 10x - 36}{x(x-3)^2} = \frac{-4}{x} + \frac{5}{x-3} + \frac{1}{(x-3)^2}.$$

■

EXAMPLE 3 Find the partial fraction decomposition of

$$\frac{3x^3 - 18x^2 + 29x - 4}{(x+1)(x-2)^3}.$$

Solution By Rule 1, there is a partial fraction of the form $A/(x+1)$ corresponding to the factor $x + 1$ in the denominator. For the factor $(x-2)^3$ we apply Rule 1 with $m = 3$, obtaining a sum of three partial fractions $B/(x-2)$,

$C/(x-2)^2$, and $D/(x-2)^3$. Consequently, the decomposition has the form

$$\frac{3x^3 - 18x^2 + 29x - 4}{(x+1)(x-2)^3} = \frac{A}{x+1} + \frac{B}{x-2} + \frac{C}{(x-2)^2} + \frac{D}{(x-2)^3}.$$

Multiplying both sides by $(x+1)(x-2)^3$ gives us

$$(*) \quad \begin{aligned} &3x^3 - 18x^2 + 29x - 4 \\ &\quad = A(x-2)^3 + B(x+1)(x-2)^2 + C(x+1)(x-2) + D(x+1). \end{aligned}$$

Two of the unknown constants may be determined easily. If we let $x = 2$ in $(*)$, then

$$24 - 72 + 58 - 4 = 3D, \quad 6 = 3D, \quad \text{and} \quad D = 2.$$

Similarly, letting $x = -1$ in $(*)$ gives us

$$-3 - 18 - 29 - 4 = -27A, \quad -54 = -27A, \quad \text{and} \quad A = 2.$$

The remaining constants may be found by comparing coefficients. If the right side of $(*)$ is expanded and like powers of x are collected, we see that the coefficient of x^3 is $A + B$. This must be equal to the coefficient of x^3 on the left, that is,

$$A + B = 3.$$

Since $A = 2$, it follows that $B = 3 - A = 3 - 2 = 1$. Finally, we compare the constant terms in $(*)$ by letting $x = 0$. This gives us

$$-4 = -8A + 4B - 2C + D.$$

Substituting the values we have found for A, B, and D leads to

$$-4 = -16 + 4 - 2C + 2,$$

which has the solution $C = -3$. The partial fraction decomposition is therefore

$$\frac{3x^3 - 18x^2 + 29x - 4}{(x+1)(x-2)^3} = \frac{2}{x+1} + \frac{1}{x-2} + \frac{-3}{(x-2)^2} + \frac{2}{(x-2)^3}. \quad \blacksquare$$

EXAMPLE 4 Find the partial fraction decomposition of

$$\frac{x^2 - x - 21}{2x^3 - x^2 + 8x - 4}.$$

Solution The denominator may be factored by grouping, as follows:

$$2x^3 - x^2 + 8x - 4 = x^2(2x-1) + 4(2x-1) = (x^2+4)(2x-1).$$

Applying Rule 2 to the irreducible quadratic factor $x^2 + 4$, we see that one of the partial fractions has the form $(Ax + B)/(x^2 + 4)$. By Rule 1, there is also a partial fraction $C/(2x - 1)$ corresponding to factor $2x - 1$. Consequently,

$$\frac{x^2 - x - 21}{2x^3 - x^2 + 8x - 4} = \frac{Ax + B}{x^2 + 4} + \frac{C}{2x - 1}.$$

As in previous examples, this leads to

$$(*) \qquad x^2 - x - 21 = (Ax + B)(2x - 1) + C(x^2 + 4).$$

Substituting $x = \frac{1}{2}$, we obtain $\frac{1}{4} - \frac{1}{2} - 21 = \frac{17}{4}C$, which has the solution $C = -5$. The remaining constants may be found by comparing coefficients. Rearranging the right side of (∗) gives us

$$x^2 - x - 21 = (2A + C)x^2 + (-A + 2B)x - B + 4C.$$

Comparing the coefficients of x^2, we see that $2A + C = 1$. Since $C = -5$, it follows that $2A = 6$, or $A = 3$. Similarly, comparing the constant terms, $-B + 4C = -21$ and hence $-B - 20 = -21$, or $B = 1$. Thus, the partial fraction decomposition is

$$\frac{x^2 - x - 21}{2x^3 - x^2 + 8x - 4} = \frac{3x + 1}{x^2 + 4} + \frac{-5}{2x - 1}.$$ ■

EXAMPLE 5 Find the partial fraction decomposition of

$$\frac{5x^3 - 3x^2 + 7x - 3}{(x^2 + 1)^2}.$$

Solution Applying Rule 2 with $n = 2$,

$$\frac{5x^3 - 3x^2 + 7x - 3}{(x^2 + 1)^2} = \frac{Ax + B}{x^2 + 1} + \frac{Cx + D}{(x^2 + 1)^2}.$$

Multiplying both sides by $(x^2 + 1)^2$ gives us

$$5x^3 - 3x^2 + 7x - 3 = (Ax + B)(x^2 + 1) + Cx + D$$

or $$5x^3 - 3x^2 + 7x - 3 = Ax^3 + Bx^2 + (A + C)x + (B + D).$$

Comparing the coefficients of x^3 and x^2, we obtain $A = 5$ and $B = -3$. From the coefficients of x we see that $A + C = 7$, or $C = 7 - A = 7 - 5 = 2$. Finally, the constant terms give us $B + D = -3$, or $D = -3 - B = -3 - (-3) = 0$. Therefore,

$$\frac{5x^3 - 3x^2 + 7x - 3}{(x^2 + 1)^2} = \frac{5x - 3}{x^2 + 1} + \frac{2x}{(x^2 + 1)^2}.$$ ■

EXERCISES 3.8

Find the partial fraction decompositions in Exercises 1–26.

1 $\dfrac{8x - 1}{(x - 2)(x + 3)}$

2 $\dfrac{x - 29}{(x - 4)(x + 1)}$

3 $\dfrac{x + 34}{x^2 - 4x - 12}$

4 $\dfrac{5x - 12}{x^2 - 4x}$

5 $\dfrac{4x^2 - 15x - 1}{(x - 1)(x + 2)(x - 3)}$

6 $\dfrac{x^2 + 19x + 20}{x(x + 2)(x - 5)}$

7 $\dfrac{4x^2 - 5x - 15}{x^3 - 4x^2 - 5x}$

8 $\dfrac{37 - 11x}{(x + 1)(x^2 - 5x + 6)}$

9 $\dfrac{2x + 3}{(x - 1)^2}$

10 $\dfrac{5x^2 - 4}{x^2(x + 2)}$

11 $\dfrac{19x^2 + 50x - 25}{3x^3 - 5x^2}$

12 $\dfrac{10 - x}{x^2 + 10x + 25}$

13 $\dfrac{x^2 - 6}{(x + 2)^2(2x - 1)}$

14 $\dfrac{2x^2 + x}{(x - 1)^2(x + 1)^2}$

15 $\dfrac{3x^3 + 11x^2 + 16x + 5}{x(x + 1)^3}$

16 $\dfrac{4x^3 + 3x^2 + 5x - 2}{x^3(x + 2)}$

17 $\dfrac{x^2 + x - 6}{(x^2 + 1)(x - 1)}$

18 $\dfrac{x^2 - x - 21}{(x^2 + 4)(2x - 1)}$

19 $\dfrac{9x^2 - 3x + 8}{x^3 + 2x}$

20 $\dfrac{2x^3 + 2x^2 + 4x - 3}{x^4 + x^2}$

21 $\dfrac{4x^3 - x^2 + 4x + 2}{(x^2 + 1)^2}$

22 $\dfrac{3x^3 + 13x - 1}{(x^2 + 4)^2}$

23 $\dfrac{2x^4 - 2x^3 + 6x^2 - 5x + 1}{x^3 - x^2 + x - 1}$

24 $\dfrac{x^3}{x^3 - 3x^2 + 9x - 27}$

25 $\dfrac{4x^3 + 4x^2 - 4x + 2}{2x^2 - x + 1}$

26 $\dfrac{x^5 - 5x^4 + 7x^3 - x^2 - 4x + 12}{x^3 - 3x^2}$

SECTION 3.9
REVIEW

Define or discuss each of the following.

1 Polynomial function
2 Intermediate Value Theorem
3 Graph of a polynomial function
4 Division Algorithm for Polynomials
5 Remainder Theorem
6 Factor Theorem
7 Synthetic division
8 Complex numbers
9 Conjugate of a complex number
10 Fundamental Theorem of Algebra
11 Multiplicity of a zero of a polynomial
12 The number of zeros of a polynomial
13 Descartes' Rule of Signs
14 Upper and lower bounds for solutions of an equation
15 Rational zeros of a polynomial function
16 Rational function
17 Graph of a rational function
18 Vertical and horizontal asymptotes
19 Oblique asymptotes
20 Partial fractions

EXERCISES 3.9

In Exercises 1–6 sketch the graph of f.

1 $f(x) = (x + 2)^3$

2 $f(x) = x^6 - 32$

3 $f(x) = x^3 + 2x^2 - 8x$

4 $f(x) = 2x^2 + x^3 - x^4$

5 $f(x) = \frac{1}{4}(x + 2)(x - 1)^2(3 - x)$

6 $f(x) = \frac{1}{15}(x^5 - 20x^3 + 64x)$

In Exercises 7–12 find all asymptotes and sketch the graph of f.

7 $f(x) = \dfrac{-2}{(x + 1)^2}$

8 $f(x) = \dfrac{1}{(x - 1)^3}$

9 $f(x) = \dfrac{3x^2}{16 - x^2}$

10 $f(x) = \dfrac{x}{(x + 5)(x^2 - 5x + 4)}$

11 $f(x) = \dfrac{x^2 + 2x - 8}{x + 3}$

12 $f(x) = \dfrac{x^4 - 16}{x^3}$

In Exercises 13–16 find the quotient and remainder if $f(x)$ is divided by $g(x)$.

13 $f(x) = 3x^5 - 4x^3 + x + 5,\ g(x) = x^3 - 2x + 7$

14 $f(x) = 7x^2 + 3x - 10,\ g(x) = x^3 - x^2 + 10$

15 $f(x) = 9x + 4,\ g(x) = 2x - 5$

16 $f(x) = 4x^3 - x^2 + 2x - 1,\ g(x) = x^2$

17 If $f(x) = -4x^4 + 3x^3 - 5x^2 + 7x - 10$, use the Remainder Theorem to find $f(-2)$.

18 Use the Remainder Theorem to prove that $x - 3$ is a factor of $2x^4 - 5x^3 - 4x^2 + 9$.

In Exercises 19 and 20 use synthetic division to find the quotient and remainder if $f(x)$ is divided by $g(x)$.

19 $f(x) = 6x^5 - 4x^2 + 8,\ g(x) = x + 2$

20 $f(x) = 2x^3 + 5x^2 - 2x + 1,\ g(x) = x - \sqrt{2}$

Express the complex numbers in Exercises 21–26 in the form $a + bi$ where a and b are real numbers.

21 $(7 + 5i) + (-8 + 3i)$

22 $(4 + 2i)(-5 + 4i)$

23 $(3 + 8i)^2$

24 $\dfrac{1}{9 - 2i}$

25 $\dfrac{6 - 3i}{2 + 7i}$

26 $\dfrac{20 - 8i}{4i}$

In Exercises 27 and 28 find polynomials with real coefficients that have the indicated zeros and degrees, and that satisfy the given conditions.

27 $-3 + 5i,\ -1;$ degree 3; $f(1) = 4$

28 $1 - i,\ 3,\ 0;$ degree 4; $f(2) = -1$

29 Find a polynomial of degree 7 such that -3 is a zero of multiplicity 2 and 0 is a zero of multiplicity 5.

30 Show that 2 is a zero of multiplicity 3 of the polynomial $x^5 - 4x^4 - 3x^3 + 34x^2 - 52x + 24$ and express this polynomial as a product of linear factors.

In Exercises 31 and 32 find the zeros of the polynomials and state the multiplicity of each zero.

31 $(x^2 - 2x + 1)^2(x^2 + 2x - 3)$

32 $x^6 + 2x^4 + x^2$

In Exercises 33 and 34 (a) use Descartes' Rule of Signs to determine the number of positive, negative, and nonreal, complex solutions; and (b) find the smallest and largest integers that are upper and lower bounds, respectively, of the real solutions.

33 $2x^4 - 4x^3 + 2x^2 - 5x - 7 = 0$

34 $x^5 - 4x^3 + 6x^2 + x + 4 = 0$

35 Prove that $7x^6 + 2x^4 + 3x^2 + 10$ has no real zeros.

Find all solutions of the equations in Exercises 36–38.

36 $x^4 + 9x^3 + 31x^2 + 49x + 30 = 0$

37 $16x^3 - 20x^2 - 8x + 3 = 0$

38 $x^4 - 7x^2 + 6 = 0$

In Exercises 39 and 40 find the partial fraction decompositions.

39 $\dfrac{4x^2 + 54x + 134}{(x + 3)(x^2 + 4x - 5)}$

40 $\dfrac{x^2 + 14x - 13}{x^3 + 5x^2 + 4x + 20}$

EXPONENTIAL AND LOGARITHMIC FUNCTIONS

Exponential and logarithmic functions are very important in mathematics and the sciences. In this chapter we shall define these functions and study some of their properties.

SECTION 4.1 EXPONENTIAL FUNCTIONS

In Section 1.2 we defined a^r for any positive real number a and any *rational* number r. It is also possible to define a^x for every *real* number x in such a way that the Laws of Exponents remain valid. To illustrate, given a^π, we could use the nonterminating decimal representation 3.14159 . . . for π and consider the numbers a^3, $a^{3.1}$, $a^{3.14}$, $a^{3.141}$, $a^{3.1415}$, ... We might expect that each successive power gets closer to a^π. This is precisely what happens if a^x is properly defined. However, the definition requires more advanced concepts than are available to us and, consequently, it is better to leave it for a mathematics course such as calculus. Although definitions and proofs are omitted in this chapter, we shall assume that the Laws of Exponents are valid for all *real* exponents.

Before continuing the discussion of real exponents, we shall establish several results about rational exponents. Let us first show that if $a > 1$, then $a^r > 1$ for every positive rational number r. If $a > 1$, then, multiplying both sides by a, we obtain $a^2 > a$, and hence $a^2 > a > 1$. Multiplying by a again we obtain $a^3 > a^2 > a > 1$. Continuing, we see that $a^n > 1$ for every positive integer n. (A complete proof requires the method of mathematical induction, which is discussed in Section 8.1.) Similarly, if $0 < a < 1$, it follows that $a^n < 1$ for every positive integer n. Next, suppose that $a > 1$ and $r = p/q$ for positive integers p and q. If $a^{p/q} \leq 1$, then from the previous discussion, $(a^{p/q})^q \leq 1$, or $a^p \leq 1$, which contradicts the fact that $a^n > 1$ for every positive integer n. Consequently, $a^r > 1$ for every positive rational number r. We may use this fact to prove the following theorem.

THEOREM If $a > 1$ and if r and s are rational numbers such that $r < s$, then $a^r < a^s$.

Proof If $r < s$, then $s - r$ is a positive rational number, and from the previous discussion, $1 < a^{s-r}$. Multiplying both sides by a^r, we obtain

$$a^r < (a^{s-r})a^r \quad \text{or} \quad a^r < a^s,$$

which is what we wished to prove. □

The preceding theorem may be extended to *real* exponents r and s.

Since to each real number x there corresponds a unique real number a^x, we can define a function as follows.

DEFINITION Let $a > 0$. The **exponential function f with base a** is defined by

$$f(x) = a^x$$

for every real number x.

If $a > 1$, and if x_1 and x_2 are real numbers such that $x_1 < x_2$, then $a^{x_1} < a^{x_2}$, that is, $f(x_1) < f(x_2)$. This means that if $a > 1$, then the exponential function f with base a is increasing throughout its domain. It can also be shown that if $0 < a < 1$, then f is decreasing.

EXAMPLE 1 Sketch the graph of f if $f(x) = 2^x$.

Solution Coordinates of some points on the graph are listed in the following table.

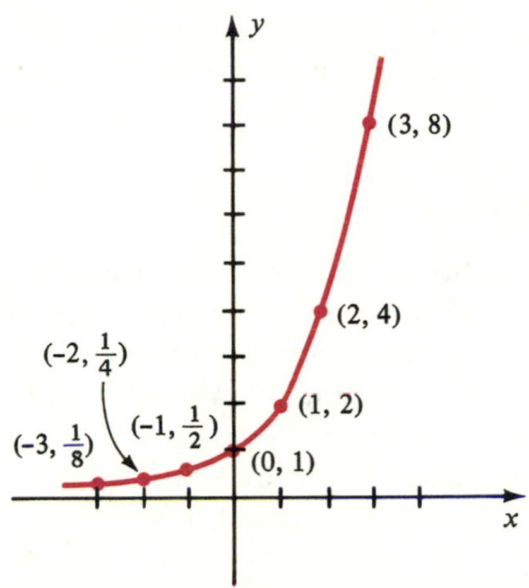

FIGURE 4.1 $f(x) = 2^x$

x	-3	-2	-1	0	1	2	3	4
2^x	$\frac{1}{8}$	$\frac{1}{4}$	$\frac{1}{2}$	1	2	4	8	16

Plotting points and using the fact that f is increasing gives us the sketch in Figure 4.1. ■

The graph in Figure 4.1 is typical of the exponential function with base a if $a > 1$. Since $a^0 = 1$, the y-intercept is always 1. Observe that as x decreases through negative values, the graph approaches the x-axis but never intersects it, since $a^x > 0$ for all x. This means that the x-axis is a *horizontal asymptote* for the graph. As x increases through positive values, the graph rises very rapidly. Indeed, if we begin with $x = 0$ and consider successive unit changes in x, then the corresponding changes in y are 1, 2, 4, 8, 16, 32, 64, etc. This type of variation is characteristic of the **exponential law of growth.** In this case f is called a **growth function.** At the end of this section we shall give several practical applications of growth functions.

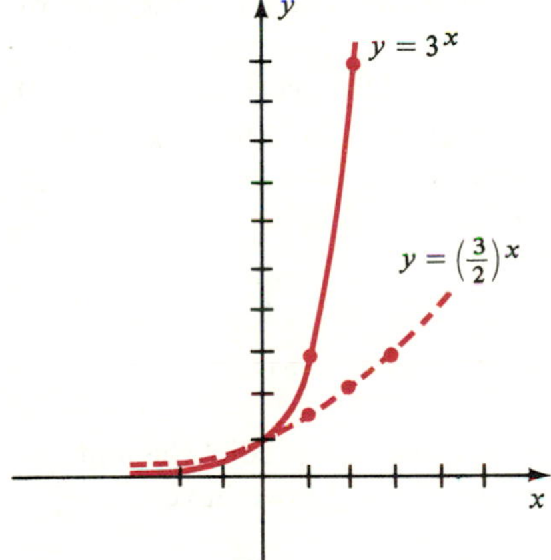

FIGURE 4.2

EXAMPLE 2 If $f(x) = (\frac{3}{2})^x$ and $g(x) = 3^x$, sketch the graphs of f and g on the same coordinate plane.

Solution The following table displays coordinates of several points on the graphs.

x	-2	-1	0	1	2	3	4
$(\frac{3}{2})^x$	$\frac{4}{9}$	$\frac{2}{3}$	1	$\frac{3}{2}$	$\frac{9}{4}$	$\frac{27}{8}$	$\frac{81}{16}$
3^x	$\frac{1}{9}$	$\frac{1}{3}$	1	3	9	27	81

Plotting these points we obtain Figure 4.2, in which dashes have been used for the graph of f to distinguish it from the graph of g. ■

Example 2 illustrates the fact that if $1 < a < b$, then $a^x < b^x$ for positive values of x and $b^x < a^x$ for negative values of x. In particular, since $\frac{3}{2} < 2 < 3$, the graph of $y = 2^x$ in Example 1 lies between the graphs of f and g in Example 2.

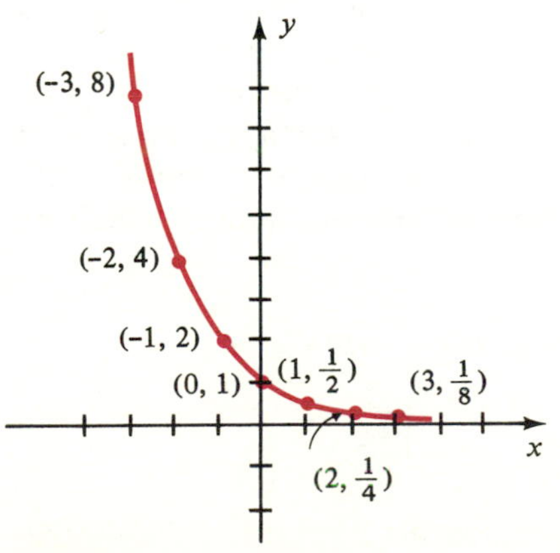

FIGURE 4.3 $y = (\frac{1}{2})^x = 2^{-x}$

EXAMPLE 3 Sketch the graph of the equation $y = (\frac{1}{2})^x$.

Solution Some points on the graph may be obtained from the following table:

x	-3	-2	-1	0	1	2	3
$(\frac{1}{2})^x$	8	4	2	1	$\frac{1}{2}$	$\frac{1}{4}$	$\frac{1}{8}$

The graph is sketched in Figure 4.3. Since $(\frac{1}{2})^x = 2^{-x}$, the graph is the same as the graph of the equation $y = 2^{-x}$.

In advanced mathematics and applications it is often necessary to consider functions such that $f(x) = a^p$, where p is some expression in x. The next example illustrates the case $p = -x^2$.

EXAMPLE 4 Sketch the graph of f if $f(x) = 2^{-x^2}$.

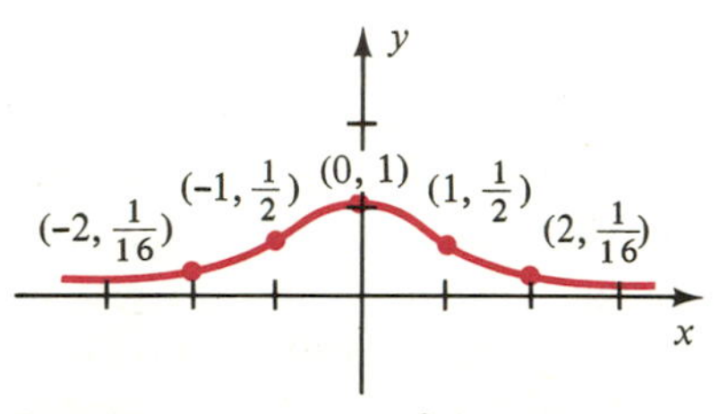

FIGURE 4.4 $y = 2^{-x^2}$

Solution Since $f(x) = 1/2^{x^2}$, it follows that if x increases numerically, the corresponding point $(x, f(x))$ on the graph approaches the x-axis. Thus, the x-axis is a horizontal asymptote for the graph. The maximum value of $f(x)$ occurs at $x = 0$. Since f is an even function, the graph is symmetric with respect to the y-axis. Several points on the graph are $(0, 1)$, $(1, \frac{1}{2})$, and $(2, \frac{1}{16})$. Plotting and using symmetry gives us the sketch in Figure 4.4. Functions similar to f arise in the study of the branch of mathematics called *probability*. (See Calculator Exercise 14.) ■

APPLICATION: Bacterial growth

The variations of many physical entities can be described by means of exponential functions. One of the most common examples occurs in the growth of certain populations. As an illustration, it might be observed experimentally that the number of bacteria in a culture doubles every hour. If 1,000 bacteria are present at the start of the experiment, then the experimenter would obtain the readings listed below, where t is the time in hours and $f(t)$ is the bacteria count at time t.

t	0	1	2	3	4
$f(t)$	1,000	2,000	4,000	8,000	16,000

It appears that $f(t) = (1{,}000)2^t$. With this formula we can predict the number of bacteria present at any time t. For example, at $t = 1.5$ we have

$$f(t) = (1{,}000)2^{3/2} = 1{,}000\sqrt{2^3} = 1{,}000\sqrt{8} \approx 2{,}828.$$

APPLICATION: Radioactive decay

Certain physical quantities *decrease* exponentially. In such cases the base a of the exponential function is between 0 and 1. One of the most common examples is the decay of a radioactive substance. As an illustration, the polonium isotope ^{210}Po has a half-life of approximately 140 days; that is, given any amount, one-half of it will disintegrate in 140 days. If 20 mg of ^{210}Po is present initially, then the following table indicates the amount remaining after various intervals of time.

t (days)	0	140	280	420	560
Amount remaining (mg)	20	10	5	2.5	1.25

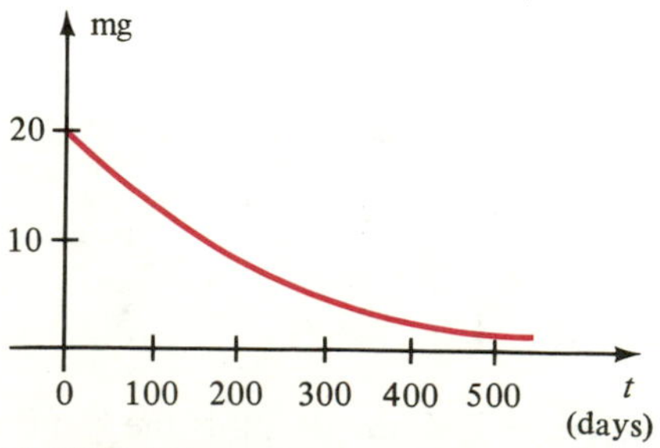

FIGURE 4.5 Decay of polonium

The sketch in Figure 4.5 illustrates the exponential nature of the disintegration.

An electrical condenser can also be used to illustrate exponential decay. If the condenser is allowed to discharge, the initial rate of discharge is relatively high, but it then tapers off as in the preceding example of radioactive decay.

APPLICATION: Compound Interest

Compound interest provides a good illustration of exponential growth, If a sum of money P, called the **principal,** is invested at a *simple* interest rate r, then the interest at the end of one interest period is the product Pr when r is expressed as a decimal. For example, if $P = \$1{,}000$ and the interest rate is 9% per year, then $r = 0.09$, and the interest at the end of one year is \$1,000(0.09), or \$90.

If the interest is reinvested at the end of this period, then the new principal is

$$P + Pr \quad \text{or} \quad P(1 + r).$$

Note that to find the new principal we multiply the original principal by $(1 + r)$. In the preceding illustration the new principal is \$1,000(1.09), or \$1,090.

After another time period has elapsed the new principal may be found by multiplying $P(1 + r)$ by $(1 + r)$. Thus, the principal after two time periods is $P(1 + r)^2$. If we to continue to reinvest, the principal after three periods is $P(1 + r)^3$; after four it is $P(1 + r)^4$; and in general, the amount A invested after k time periods is

$$A = P(1 + r)^k.$$

Interest accumulated by means of this formula is called **compound interest.** Note that A is expressed in terms of an exponential function whose base is $1 + r$ and whose exponent is k. The time period may vary and may be measured in years, months, weeks, days, or any other suitable unit of time. When applying the formula for A, remember that r is the interest rate per time period. For example, if the rate is stated as 9% *per year compounded monthly*, then the rate per month is $\frac{9}{12}\%$, or equivalently, 0.75%. Thus, $r = 0.0075$ and k is the number of months.

Generally, suppose that r is the interest rate and that interest is compounded n times per year: The interest rate per time period is r/n. If the principal P is invested for t years, then the number of interest periods is nt, and the amount A after t years is given by the following formula.

COMPOUND INTEREST FORMULA

$$A = P\left(1 + \frac{r}{n}\right)^{nt}$$

EXAMPLE 5 Suppose that \$1,000 is invested at an interest rate of 9% compounded monthly. Find the new amount of principal after 5 years; after 10 years; after 15 years.

Solution Applying the Compound Interest Formula with $r = 0.09$, $n = 12$, and $P = \$1000$, the amount after t years is

$$A = 1{,}000\left(1 + \frac{0.09}{12}\right)^{12t} = 1{,}000\,(1.0075)^{12t}.$$

Substituting $t = 5$, 10, and 15, and using a calculator, we obtain the following amounts:

$$\begin{aligned} &\text{After 5 years:} && A = 1{,}000\,(1.0075)^{60} = \$1{,}565.68\\ &\text{After 10 years:} && A = 1{,}000\,(1.0075)^{120} = \$2{,}451.36\\ &\text{After 15 years:} && A = 1{,}000\,(1.0075)^{180} = \$3{,}838.04 \end{aligned}$$

The exponential nature of the increase is indicated by the fact that during the first five years the growth in the investment is \$565.68; during the second five-year period the growth is \$885.68; and during the last five-year period, it is \$1,368.68. ■

The next example illustrates what happens if the rate and total time invested are fixed, but the *time period* for compounding interest is varied.

EXAMPLE 6 Suppose \$1,000 is invested at a compound interest rate of 9%. Find the new amount of principal after one year if the interest is compounded monthly; weekly; daily; hourly; each minute.

Solution Letting $P = \$1{,}000$, $t = 1$, and $r = 0.09$ in the Compound Interest Formula,

$$A = 1{,}000\left(1 + \frac{0.09}{n}\right)^{n}$$

for n interest periods per year. To find the desired amounts, we let

$$n = 12,\ 52,\ 365,\ 8{,}760,\ 525{,}600.$$

We have assumed there are 365 days in a year and, hence, $(365)(24) = 8{,}760$ hours, and $(8{,}760)(60) = 525{,}600$ minutes. Using the Compound Interest Formula (and a calculator), we obtain the following table.

Time period for compounding interest	Amount of principal after one year
Month	$1{,}000\left(1+\frac{0.09}{12}\right)^{12} = \$1{,}093.81$
Week	$1{,}000\left(1+\frac{0.09}{52}\right)^{52} = \$1{,}094.09$
Day	$1{,}000\left(1+\frac{0.09}{365}\right)^{365} = \$1{,}094.16$
Hour	$1{,}000\left(1+\frac{0.09}{8{,}760}\right)^{8{,}760} = \$1{,}094.17$
Minute	$1{,}000\left(1+\frac{0.09}{525{,}600}\right)^{525{,}600} = \$1{,}094.17$

Note that, in the preceding example, after a certain time period is reached, the number of interest periods per year has little effect on the final amount. If interest had been compounded each *second*, the result would still be \$1,094.17. Thus, the amount approaches a fixed value as n increases.

If we let $P = 1$, $r = 1$, and $t = 1$ in the Compound Interest Formula, we obtain

$$A = \left(1 + \frac{1}{n}\right)^n.$$

The expression on the right of the equation occurs in the study of calculus. In Example 6, in a similar situation, as n increased, A approached a limiting value. The same phenomenon occurs here, as illustrated by the following table, which was obtained using a calculator.

n	Approximation to $\left(1+\frac{1}{n}\right)^n$
1	2.0000000
10	2.5937425
100	2.7048138
1,000	2.7169238
10,000	2.7181459
100,000	2.7182546
1,000,000	2.7182818
10,000,000	2.7182818

It can be proved that as n increases, $(1 + 1/n)^n$ gets closer and closer to a certain irrational number, denoted by e. As indicated by the values in the table, e can be assigned the following decimal approximation.

APPROXIMATION TO *e*

$$e \approx 2.71828$$

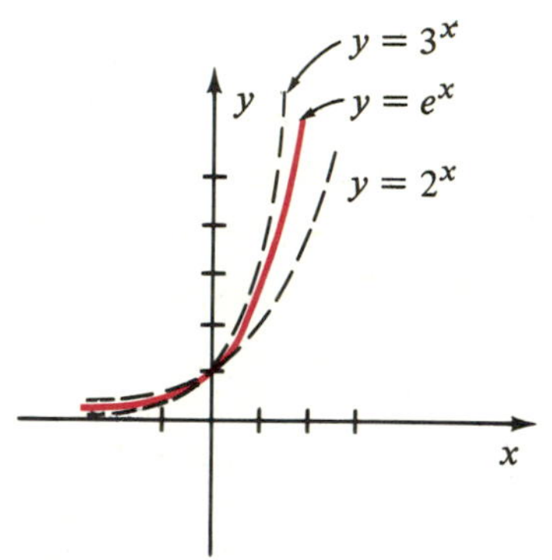

FIGURE 4.6

The number e arises naturally in the investigation of many physical phenomena. For this reason, the function f defined by $f(x) = e^x$ is called the **natural exponential function.** It is one of the most important functions that occurs in advanced mathematics and applications. Since $2 < e < 3$, the graph of $y = e^x$ lies "between" the graphs of $y = 2^x$ and $y = 3^x$, as shown in Figure 4.6.

A brief table of values of e^x is given in Table 2 of Appendix II. Many calculators have a $\boxed{y^x}$ key that can be used to approximate values of exponential functions. In this case, approximations to e^x can be found by calculating $(2.71828)^x$. Another approximating technique—using inverse functions—is described at the end of Section 4.3.

APPLICATION: Radiotherapy

One of the many fields in which exponential functions with base e play an important role is *radiotherapy*, the treatment of tumors by radiation. Of major interest is the fraction of a tumor population that survives a treatment. This *surviving fraction* depends not only on the energy and nature of the radiation, but also on the depth, size, and characteristics of the tumor itself. The exposure to radiation may be thought of as a number of potentially damaging events, where only one "hit" is required to kill a tumor cell. Suppose that each cell has exactly one "target" that must be hit. If k denotes the average target size of a tumor cell, and if x is the number of damaging events (the *dose*), then the surviving fraction $f(x)$ is

$$f(x) = e^{-kx}.$$

This is called the *one-target–one-hit surviving fraction.*

Suppose next that each cell has n targets and that hitting any one of the targets results in the death of a cell. In this case, the *n-target–one-hit surviving fraction* is

$$f(x) = 1 - (1 - e^{-kx})^n.$$

The graph of f may be analyzed to determine what effect increasing the dosage x will have on decreasing the surviving fraction of tumor cells. Note that $f(0) = 1$; that is, if there is no dose, then all cells survive. As a special case, if $k = 1$ and $n = 2$, then

$$\begin{aligned} f(x) &= 1 - (1 - e^{-x})^2 = 1 - (1 - 2e^{-x} + e^{-2x}) \\ &= 2e^{-x} - e^{-2x}. \end{aligned}$$

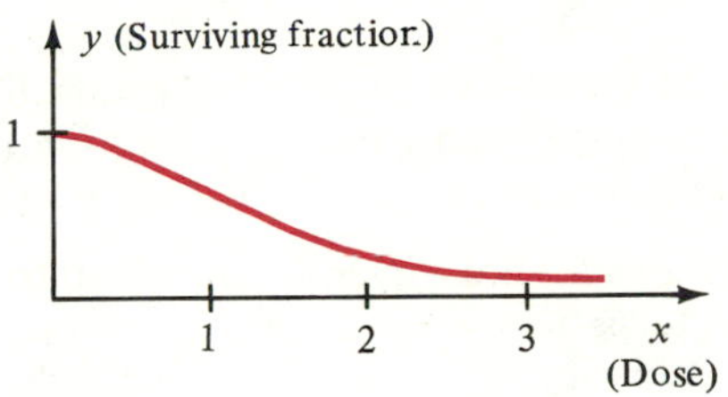

FIGURE 4.7 Surviving fraction of tumor cells after a radiation treatment

A complete analysis of the graph of f requires methods of calculus. It can be shown that the graph has the shape indicated in Figure 4.7. The "shoulder" on the curve near the point (0, 1) represents the threshold nature of the treatment; that is, a small dose results in very little tumor elimination. Note that for a large x, an increase in dosage has little effect on the surviving fraction. To determine the ideal dose that should be administered to a patient, specialists in radiation therapy must also take into account the number of healthy cells that are killed during a treatment.

In Calculator Exercise 21 you are asked to verify the graph in Figure 4.7.

Problems of the type illustrated in the next example occur in the study of calculus (cf. Exercises 25–28).

EXAMPLE 7 Find the zeros of f if $f(x) = x^2(-2e^{-2x}) + 2xe^{-2x}$.

Solution We may factor $f(x)$ as follows:

$$\begin{aligned} f(x) &= 2xe^{-2x} - 2x^2e^{-2x} \\ &= 2xe^{-2x}(1 - x). \end{aligned}$$

To find the zeros of f, we must solve the equation $f(x) = 0$. Since $e^{-2x} > 0$ for all x, it follows that $f(x) = 0$ if and only if $x = 0$ or $1 - x = 0$. Thus, the zeros of f are 0 and 1. ■

EXERCISES 4.1

In Exercises 1–24 sketch the graph of the function f.

1 $f(x) = 4^x$

2 $f(x) = 5^x$

3 $f(x) = 10^x$

4 $f(x) = 8^x$

5 $f(x) = 3^{-x}$

6 $f(x) = 4^{-x}$

7 $f(x) = -2^x$

8 $f(x) = -3^x$

9 $f(x) = 4 - 2^{-x}$

10 $f(x) = 2 + 3^{-x}$

11 $f(x) = (\frac{2}{3})^x$

12 $f(x) = (\frac{3}{4})^{-x}$

13 $f(x) = (\frac{5}{2})^{-x}$

14 $f(x) = (\frac{4}{3})^x$

15 $f(x) = 2^{|x|}$

16 $f(x) = 2^{-|x|}$

17 $f(x) = 2^{x+3}$

18 $f(x) = 3^{x+2}$

19 $f(x) = 2^{3-x}$

20 $f(x) = 3^{-2-x}$

21 $f(x) = 3^{1-x^2}$

22 $f(x) = 2^{-(x+1)^2}$

23 $f(x) = 3^x + 3^{-x}$

24 $f(x) = 3^x - 3^{-x}$

In Exercises 25–28 find the zeros of f.

25 $f(x) = xe^x + e^x$

26 $f(x) = -x^2e^{-x} + 2xe^{-x}$

27 $f(x) = x^3(4e^{4x}) + 3x^2e^{4x}$

28 $f(x) = x^2(2e^{2x}) + 2xe^{2x} + e^{2x} + 2xe^{2x}$

Simplify the expressions in Exercises 29–30.

29 $\dfrac{(e^x + e^{-x})(e^x + e^{-x}) - (e^x - e^{-x})(e^x - e^{-x})}{(e^x + e^{-x})^2}$

30 $\dfrac{(2^x - 2^{-x})^2 - (2^x + 2^{-x})^2}{(2^x + 2^{-x})^2}$

31 If $f(x) = 2^x$ and $g(x) = x^2$, illustrate the difference in the rates of growth of f and g for $x \geq 0$ by sketching the graphs of both functions on the same coordinate plane.

32 Repeat Exercise 31 if $f(x) = 2^{-x}$ and $g(x) = x^{-2}$.

33 The number of bacteria in a certain culture increased from 600 to 1800 between 7:00 A.M. and 9:00 A.M. Assuming exponential growth and using methods of calculus, it can be shown that the number $f(t)$ of bacteria t hours after 7:00 A.M. was given by $f(t) = 600(3)^{t/2}$.

(a) Estimate the number of bacteria in the culture at 8:00 A.M.; 10:00 A.M.; 11:00 A.M.

(b) Sketch the graph of f from $t = 0$ to $t = 4$.

34 According to Newton's Law of Cooling, the rate at which an object cools is directly proportional to the difference in temperature between the object and the surrounding medium. If a certain object cools from 125° to 100° in 30 minutes when surrounded by air at a temperature of 75°, then its temperature $f(t)$ after t hours of cooling is given by $f(t) = 50(2)^{-2t} + 75$.

(a) If $t = 0$ corresponds to 1:00 P.M. approximate, to the nearest tenth of a degree, the temperature at 2:00 P.M.; 3:30 P.M.; 4:00 P.M.

(b) Sketch the graph of f from $t = 0$ to $t = 4$.

35 The radioactive isotope ^{210}Bi has a half-life of 5 days; that is, the number of radioactive particles will decrease to one-half the number in 5 days. If there are 100 mg of ^{210}Bi present at $t = 0$, then the amount $f(t)$ remaining after t days is given by $f(t) = 100(2)^{-t/5}$.

(a) How much ^{210}Bi remains after 5 days? 10 days? 12.5 days?

(b) Sketch the graph of f from $t = 0$ to $t = 30$.

36 The number of bacteria in a certain culture at time t is given by $Q(t) = 2(3^t)$, where t is measured in hours and $Q(t)$, in thousands. What is the initial number of bacteria? What is the number after 10 minutes? after 30 minutes? after 1 hour?

37 The half-life of radium is 1,600 years; that is, given any quantity, one-half of it will disintegrate in 1,600 years. If the initial amount is q_0 milligrams, then the quantity $q(t)$ remaining after t years is given by $q(t) = q_0 2^{kt}$. Find k.

38 If 10 grams of salt are added to a quantity of water, then the amount $q(t)$ that is undissolved after t minutes is given by $q(t) = 10(\frac{4}{5})^t$. Sketch a graph that shows the value $q(t)$ at any time from $t = 0$ to $t = 10$.

39 If \$1,000, is invested at a rate of 12% per year compounded monthly, what is the principal after 1 month? 2 months? 6 months? 1 year?

40 If a savings fund pays interest at a rate of 10% compounded semiannually, how much money invested now will amount to \$5,000 after one year?

41 If a certain make of automobile is purchased for C dollars, then its trade-in value $v(t)$ at the end of t years is given by $v(t) = 0.78C(0.85)^{t-1}$. If the original cost is \$10,000, calculate, to the nearest dollar, the value after (a) 1 year; (b) 4 years; (c) 7 years.

42 If the value of real estate increases at a rate of 10% per year, then after t years the value V of a house purchased for P dollars is given by $V = P(1.1)^t$. If a house was purchased for \$80,000 in 1983, what will it be worth in 1987?

43 Why was $a < 0$ ruled out in the discussion of a^x?

44 Prove that if $0 < a < 1$ and r and s are rational numbers such that $r < s$, then $a^r > a^s$.

45 How does the graph of $y = a^x$ compare with the graph of $y = -a^x$?

46 If $a > 1$, how does the graph of $y = a^x$ compare with the graph of $y = a^{-x}$?

CALCULATOR EXERCISES 4.1

1 Approximate $3^{\sqrt{2}}$ and compare this number with the successive approximations

$$3^{1.4},\quad 3^{1.41},\quad 3^{1.414},\quad 3^{1.4142}.$$

2 Approximate 2^{π} and compare this number with the successive approximations

$$2^{3.1},\quad 2^{3.14},\quad 2^{3.141},\quad 2^{3.1415}.$$

In Exercises 3 and 4 use decimal approximations to replace the symbol □ by <, >, or =.

3 $(\sqrt{2})^{\sqrt{3}} \square (\sqrt{3})^{\sqrt{2}}$

4 $\pi^{22/7} \square (22/7)^{\pi}$

In Exercises 5 and 6 demonstrate that the given equality is true by expressing each side as a decimal. (Do not use Laws of Exponents.)

5 (a) $5^{\sqrt{2}}5^{\sqrt{3}} = 5^{\sqrt{2}+\sqrt{3}}$
(b) $5^{\sqrt{3}}/5^{\sqrt{2}} = 5^{\sqrt{3}-\sqrt{2}}$

6 $(5^{\sqrt{3}})^{\sqrt{2}} = 5^{\sqrt{6}}$

In Exercises 7 and 8 sketch the graph corresponding to $-3 \leq x \leq 3$ by choosing values of x at intervals of length 0.5; that is, $x = -3, -2.5, -2, -1.5$, etc.

7 $y = (1.8)^x$

8 $y = (2.3)^{-x^2}$

Solve Exercises 9–12 by using the Compound Interest Formula.

9 If \$1,000 is invested at an interest rate of 6% per year compounded quarterly, find the principal at the end of (a) one year; (b) two years; (c) five years; (d) ten years.

10 Rework Exercise 9 if the interest rate is 6% per year compounded monthly.

11 If \$10,000 is invested at a rate of 9% per year compounded semiannually, how long will it take for the principal to exceed (a) \$15,000? (b) \$20,000? (c) \$30,000?

12 A certain department store requires its credit card customers to pay interest at the rate of 18% per year compounded monthly on any unpaid bills. If a man buys a television set for \$500 on credit and then makes no payments for one year, how much does he owe at the end of the year?

13 Sketch the graph of f if $f(x)$ is:

(a) e^x

(b) e^{-x}

(c) $\dfrac{e^x + e^{-x}}{2}$

(d) $\dfrac{2}{e^x + e^{-x}}$

(e) $\dfrac{e^x - e^{-x}}{2}$

(f) $\dfrac{2}{e^x - e^{-x}}$

14 In statistics the **normal distribution function** is defined by

$$f(x) = \frac{1}{\sigma\sqrt{2\pi}} e^{(-1/2)[(x-\mu)/\sigma]^2}$$

for real numbers μ and σ with $\sigma > 0$. (μ is called the *mean* and σ is the *variance* of the distribution.) Sketch the graph of f for the case $\sigma = 1$ and $\mu = 0$.

15 Sketch the graph of $y = x(2^x)$.

16 If $f(x) = x^2(5^{\sqrt{x}})$, find $f(1.01) - f(1)$.

17 Under certain conditions the atmospheric pressure p (in inches) at altitude h feet is given by

$$p = 29e^{-0.000034h}.$$

What is the pressure at an altitude of 40,000 feet?

18 Starting with c milligrams of the polonium isotope ^{210}Po, the amount remaining after t days may be approximated by $A = ce^{-0.00495t}$. If the initial amount is 50 milligrams, find, to the nearest hundredth, the amount remaining after (a) 30 days; (b) 180 days; (c) 365 days.

19 Growths of physical quantities are often more stable than that given by $f(t) = ke^{at}$, where t is time. In biology the formula

$$G(t) = ke^{-(Ae^{-Bt})}$$

for positive constants A, B, and k is used to estimate certain populations. The graph of G is called a **Gompertz growth curve.** Sketch the graph of G for the special case $k = 10$, $A = \frac{1}{2}$, $B = 1$, and $t \geq 0$, and describe what happens as t increases without bound.

20 The graph of

$$f(x) = a + b(1 - e^{-cx})$$

for positive constants a, b, and c may be used to describe certain learning processes. To illustrate, suppose a manufacturer estimates that a new employee can produce 3 items the first day on the job. As the employee becomes more proficient, more items per day can be produced until a certain maximum production is reached. Suppose that after n days on the job, the number $f(n)$ of items produced is approximated by the formula

$$f(n) = 3 + 20(1 - e^{-0.1n}).$$

(a) Estimate the number of items produced on the fifth day; the ninth day; the twenty-fourth day; the thirtieth day.

(b) Sketch the graph of f from $n = 0$ to $n = 30$. (Graphs of this type are called **learning curves,** and are used frequently in education and psychology).

(c) What happens as n increases without bound?

21 Verify the graph sketched in Figure 4.7.

22 If a drug is injected into the blood stream, its concentration t minutes later is given by

$$C(t) = \frac{k}{a-b}(e^{-bt} - e^{-at})$$

for positive constants a, b, and k with $a > b$. Sketch the graph of f if $a = 3$, $b = 2$, and $k = 1$. What can be said about the concentration after a long period of time? (Use properties of exponential functions to justify your answer.)

SECTION 4.2
LOGARITHMS

Throughout this section and the next, it is assumed that a is any positive real number and $a \neq 1$. Let us begin by examining the graph of f if $f(x) = a^x$ (see Figure 4.8 for the case $a > 1$). It appears that every positive real number u is the y-coordinate of some point on the graph; that is, there is a number v such that $u = a^v$. Indeed, v is the x-coordinate of the point where the line $y = u$ intersects the graph. Moreover, since f is increasing throughout its domain, the number u can occur as a y-coordinate only once. A similar situation exists if $0 < a < 1$. These facts are summarized in the next theorem.

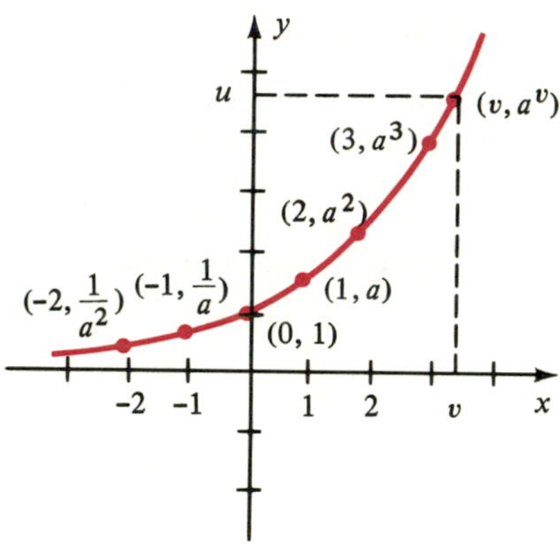

FIGURE 4.8 $f(x) = a^x$, $a > 1$

THEOREM

For each positive real number u there is a unique real number v such that

$$a^v = u.$$

A rigorous proof of this theorem requires concepts studied in calculus.

DEFINITION

If u is any positive real number, then the exponent v such that $a^v = u$ is called the **logarithm of u with base a** and denoted by $\log_a u$.

A convenient way to memorize this definition is as follows:

$$v = \log_a u \quad \text{if and only if} \quad a^v = u.$$

As illustrations,

$$3 = \log_2 8 \quad \text{since} \quad 2^3 = 8$$
$$-2 = \log_5 \tfrac{1}{25} \quad \text{since} \quad 5^{-2} = \tfrac{1}{25}$$
$$4 = \log_{10} 10{,}000 \quad \text{since} \quad 10^4 = 10{,}000.$$

THEOREM

(i) $a^{\log_a x} = x$ for every $x > 0$.

(ii) $\log_a a^x = x$ for every x.

(iii) $\log_a 1 = 0$.

Proof Formula (i) is true since, by definition, $\log_a x$ is the exponent v such that $a^v = x$. To prove (ii) we begin with $a^x = a^x$ and use the definition of logarithm with $v = x$ and $u = a^x$. Finally, to prove (iii) we note that $a^0 = 1$ and apply the definition of logarithm. □

EXAMPLE 1 Find s in each of the following.

(a) $s = \log_4 2$ (b) $\log_5 s = 2$ (c) $\log_s 8 = 3$

Solution

(a) If $s = \log_4 2$, then $4^s = 2$ and hence $s = \frac{1}{2}$.

(b) If $\log_5 s = 2$, then $5^2 = s$ and hence $s = 25$.

(c) If $\log_s 8 = 3$, then $s^3 = 8$ and hence $s = \sqrt[3]{8} = 2$. ■

EXAMPLE 2 Solve the equation $\log_4 (5 + x) = 3$.

Solution If $\log_4 (5 + x) = 3$, then

$$5 + x = 4^3 \quad \text{or} \quad 5 + x = 64.$$

Hence, the solution is $x = 59$. ■

The following laws are fundamental for all work with logarithms of positive real numbers u and w.

LAWS OF LOGARITHMS

(i) $\log_a (uw) = \log_a u + \log_a w$

(ii) $\log_a (u/w) = \log_a u - \log_a w$

(iii) $\log_a (u^c) = c \log_a u$ for every real number c.

Proof To prove (i) we begin by letting

$$r = \log_a u \quad \text{and} \quad s = \log_a w.$$

Applying the definition of logarithm, $a^r = u$ and $a^s = w$. Consequently,

$$a^r a^s = uw$$

and hence,

$$a^{r+s} = uw.$$

By the definition of logarithm, the last equation is equivalent to

$$r + s = \log_a (uw).$$

Substituting for r and s from the first step in the proof gives us

$$\log_a u + \log_a w = \log_a uw.$$

This completes the proof of (i).

To prove (ii) we begin as in the proof of (i), but divide a^r by a^s, obtaining

$$\frac{a^r}{a^s} = \frac{u}{w}, \quad \text{or} \quad a^{r-s} = \frac{u}{w}.$$

The last equation is equivalent to

$$r - s = \log_a (u/w).$$

Substituting for r and s gives us (ii).

Finally, if c is any real number, then

$$(a^r)^c = u^c, \quad \text{or} \quad a^{cr} = u^c.$$

According to the definition of logarithm, the last equality implies that

$$cr = \log_a u^c.$$

Substituting for r, we obtain

$$c \log_a u = \log_a u^c.$$

This proves (iii) of the Laws of Logarithms. □

The following examples illustrate uses of the Laws of Logarithms.

EXAMPLE 3 If $\log_a 3 = 0.4771$ and $\log_a 2 = 0.3010$, find the following.

(a) $\log_a 6$ (b) $\log_a \frac{3}{2}$ (c) $\log_a \sqrt{2}$ (d) $\dfrac{\log_a 3}{\log_a 2}$

Solution

(a) Since $6 = 2 \cdot 3$, we may use Law (i) to obtain

$$\begin{aligned}\log_a 6 &= \log_a (2 \cdot 3) = \log_a 2 + \log_a 3\\ &= 0.4771 + 0.3010 = 0.7781.\end{aligned}$$

(b) By Law (ii),

$$\log_a \tfrac{3}{2} = \log_a 3 - \log_a 2$$
$$= 0.4771 - 0.3010 = 0.1761.$$

(c) Using Law (iii),

$$\log_a \sqrt{2} = \log_a 2^{1/2} = \tfrac{1}{2} \log_a 2$$
$$= \tfrac{1}{2}(0.3010) = 0.1505.$$

(d) There is no law of logarithms that allows us to simplify $(\log_a 3)/(\log_a 2)$. Consequently, we *divide* 0.4771 by 0.3010, obtaining the approximation 1.585. It is important to notice the difference between this problem and part (b). ■

EXAMPLE 4 Solve the following equations:

(a) $\log_5 (2x + 3) = \log_5 11 + \log_5 3$

(b) $\log_4 (x + 6) - \log_4 10 = \log_4 (x - 1) - \log_4 2$

Solution

(a) Using Law (i), the equation may be written as

$$\log_5 (2x + 3) = \log_5 (11 \cdot 3) = \log_5 33.$$

Consequently, $2x + 3 = 33$, or $2x = 30$, and the solution is $x = 15$.

(b) The equation is equivalent to

$$\log_4 (x + 6) - \log_4 (x - 1) = \log_4 10 - \log_4 2.$$

Applying Law (ii),

$$\log_4 \left(\frac{x+6}{x-1}\right) = \log_4 \frac{10}{2} = \log_4 5$$

and hence, $$\frac{x+6}{x-1} = 5.$$

The last equation implies that

$$x + 6 = 5x - 5, \quad \text{or} \quad 4x = 11.$$

Thus, the solution is $x = \frac{11}{4}$. ■

Extraneous solutions sometimes occur in the process of solving logarithmic equations, as illustrated in the next example.

EXAMPLE 5 Solve the equation $2 \log_7 x = \log_7 36$.

Solution Applying Law (iii), we obtain $2 \log_7 x = \log_7 x^2$, and substitution in the given equation leads to

$$\log_7 x^2 = \log_7 36.$$

Consequently, $x^2 = 36$ and, hence, either $x = 6$ or $x = -6$. However, $x = -6$ is not a solution of the original equation since x must be positive in order for $\log_7 x$ to exist. Thus, the only solution is $x = 6$.

The preceding difficulty could have been avoided by writing

$$\log_7 x = \tfrac{1}{2} \log_7 36 = \log_7 36^{1/2} = \log_7 6$$

and, therefore, $x = 6$. ■

It is sometimes necessary to *change the base* of a logarithm by expressing $\log_b u$ in terms of $\log_a u$, for some positive real number $b \neq 1$. We begin with the equivalent equations

$$v = \log_b u \quad \text{and} \quad b^v = u.$$

Taking the logarithm, base a, of both sides of the second equation gives us

$$\log_a b^v = \log_a u.$$

Applying Law (iii), $\qquad v \log_a b = \log_a u.$

Solving for v (that is, $\log_b u$), we obtain formula (i) in the next box.

CHANGE OF BASE FORMULAS

$$\text{(i)}\quad \log_b u = \frac{\log_a u}{\log_a b} \qquad \text{(ii)}\quad \log_b a = \frac{1}{\log_a b}$$

To obtain formula (ii) we let $u = a$ in (i) and use the fact that $\log_a a = 1$.

The Laws of Logarithms are often used as in the following example.

EXAMPLE 6 Express $\log_a \dfrac{x^3\sqrt{y}}{z^2}$ in terms of the logarithms of x, y, and z.

Solution Using the three Laws of Logarithms,

$$\begin{aligned}\log_a \frac{x^3\sqrt{y}}{z^2} &= \log_a (x^3\sqrt{y}) - \log_a z^2 \\ &= \log_a x^3 + \log_a \sqrt{y} - \log_a z^2 \\ &= 3 \log_a x + \tfrac{1}{2} \log_a y - 2 \log_a z.\end{aligned}$$

■

The procedure in Example 6 can also be reversed, as illustrated in the next example.

EXAMPLE 7 Express in terms of one logarithm:

$$\tfrac{1}{3}\log_a (x^2 - 1) - \log_a y - 4\log_a z.$$

Solution Using the Laws of Logarithms we have

$$\begin{aligned}\tfrac{1}{3}\log_a (x^2 - 1) - \log_a y - 4\log_a z &= \log_a (x^2 - 1)^{1/3} - \log_a y - \log_a z^4\\ &= \log_a \sqrt[3]{x^2 - 1} - (\log_a y + \log_a z^4)\\ &= \log_a \sqrt[3]{x^2 - 1} - \log_a yz^4\\ &= \log_a \frac{\sqrt[3]{x^2 - 1}}{yz^4}\end{aligned}$$

As a final remark, note that there is no general law for expressing $\log_a (u + w)$ in terms of simpler logarithms. It is evident that it does not always equal $\log_a u + \log_a w$, since this sum equals $\log_a (uw)$.

EXERCISES 4.2

Change the equations in Exercises 1–8 to logarithmic form.

1 $4^3 = 64$ **2** $3^5 = 243$

3 $2^7 = 128$ **4** $5^3 = 125$

5 $10^{-3} = 0.001$ **6** $10^{-2} = 0.01$

7 $t^r = s$ **8** $v^w = u$

Change the equations in Exercises 9–16 to exponential form.

9 $\log_{10} 1000 = 3$ **10** $\log_3 81 = 4$

11 $\log_3 \frac{1}{243} = -5$ **12** $\log_4 \frac{1}{64} = -3$

13 $\log_7 1 = 0$ **14** $\log_9 1 = 0$

15 $\log_t r = p$ **16** $\log_v w = q$

Find the numbers in Exercises 17–30.

17 $\log_4 \frac{1}{16}$ **18** $\log_2 32$

19 $\log_{10} 100$ **20** $\log_8 64$

21 $10^{\log_{10} 5}$ **22** $\log_{10} 0.0001$

23 $\log_7 \sqrt[3]{7}$ **24** $10^{2\log_{10} 3}$

25 $\log_{10} \frac{1}{10}$ **26** $\log_{10} 100{,}000$

27 $\log_{1/2} 8$ **28** $\log_3 \frac{1}{81}$

29 $3^{4\log_3 2}$ **30** $\log_6 \sqrt[5]{6}$

If $\log_a 7 = 0.8$ and $\log_a 3 = 0.5$, change the expressions in Exercises 31–40 to decimal form.

31 $\log_a \frac{7}{3}$ **32** $\log_a \frac{3}{7}$

33 $\log_a 7^3$ **34** $\log_a 3^7$

35 $\log_a (1/\sqrt{3})$ **36** $\log_a \sqrt[4]{3}$

37 $\log_a 63$ **38** $\dfrac{\log_a 7}{\log_a 3}$

39 $\log_7 a$ **40** $(\log_3 a)(\log_a 3)$

Find the solutions of the equations in Exercises 41–58.

41 $\log_3 (x - 4) = 2$

42 $\log_2 (x - 5) = 4$

43 $\log_9 x = \frac{3}{2}$

44 $\log_4 x = -\frac{3}{2}$

45 $\log_5 x^2 = -2$

46 $\log_{10} x^2 = -4$

47 $\log_2 (x^2 - 5x + 14) = 3$

48 $\log_2 (x^2 - 10x + 18) = 1$

49 $\log_x 7 = 3$

50 $5^{\log_5 x} = 10$

51 $\log_6 (2x - 3) = \log_6 12 - \log_6 3$

52 $2 \log_3 x = 3 \log_3 5$

53 $\log_2 x - \log_2 (x + 1) = 3 \log_2 4$

54 $\log_5 x + \log_5 (x + 6) = \frac{1}{2} \log_5 9$

55 $\log_{10} x^2 = \log_{10} x$

56 $\log_x 10 = 10$

57 $\frac{1}{2} \log_5 (x - 2) = 3 \log_5 2 - \frac{3}{2} \log_5 (x - 2)$

58 $\log_{10} 5^x = \log_3 1$

In Exercises 59–66 express the logarithm in terms of logarithms of x, y, and z.

59 $\log_a \dfrac{x^2 y}{z^3}$

60 $\log_a \dfrac{x^3 y^2}{z^5}$

61 $\log_a \dfrac{\sqrt{x} z^2}{y^4}$

62 $\log_a x \sqrt[3]{\dfrac{y^2}{z^4}}$

63 $\log_a \sqrt[3]{\dfrac{x^2}{yz^5}}$

64 $\log_a \dfrac{\sqrt{x} y^6}{\sqrt[3]{z^2}}$

65 $\log_a \sqrt{x\sqrt{yz^3}}$

66 $\log_a \sqrt[3]{x^2 y \sqrt{z}}$

In Exercises 67–70 write the expression as one logarithm.

67 $2 \log_a x + \frac{1}{3} \log_a (x - 2) - 5 \log_a (2x + 3)$

68 $5 \log_a x - \frac{1}{2} \log_a (3x - 4) + 3 \log_a (5x + 1)$

69 $\log_a (y^2 x^3) - 2 \log_a x \sqrt[3]{y} + 3 \log_a \left(\dfrac{x}{y}\right)$

70 $2 \log_a \dfrac{y^3}{x} - 3 \log_a y + \frac{1}{2} \log_a x^4 y^2$

CALCULATOR EXERCISES 4.2

The [log] key on a calculator is used to find logarithms with base 10. In Exercises 1–6 demonstrate that the given statement is true by expressing each side as a decimal (do not use Laws of Logarithms).

1 $\log_{10} (2 \cdot 5) = \log_{10} 2 + \log_{10} 5$

2 $\log_{10} \frac{5}{2} = \log_{10} 5 - \log_{10} 2$

3 $\log_{10} (2^5) = 5 \log_{10} 2$

4 $\log_{10} (2 + 5) \neq \log_{10} 2 + \log_{10} 5$

5 $10^{\log_{10} 5} = 5$

6 $\log_{10} 1 = 0$

If $a = 10$ in the Change of Base Formula (i), then

$$\log_b u = \frac{\log_{10} u}{\log_{10} b}.$$

Use this formula to find the following to four-decimal-place accuracy.

7 $\log_3 8$

8 $\log_9 7$

9 $\log_{17} 412$

10 $\log_\pi \sqrt{5}$

SECTION 4.3
LOGARITHMIC FUNCTIONS

In the following definition we use the concept of logarithm to introduce a new function whose domain is the set of positive real numbers.

DEFINITION

Let $a > 0$ and $a \neq 1$. The **logarithmic function with base *a*** is defined by

$$f(x) = \log_a x$$

for every $x > 0$.

The graph of f is the same as the graph of the equation $y = \log_a x$, which, by the definition of logarithm, is equivalent to

$$x = a^y.$$

To find some pairs that are solutions of the preceding equation, we may substitute for y and find the corresponding values of x, as shown in the following table.

y	-3	-2	-1	0	1	2	3
x	$\frac{1}{a^3}$	$\frac{1}{a^2}$	$\frac{1}{a}$	1	a	a^2	a^3

If $a > 1$, we obtain the sketch in Figure 4.9. In this case, f is an increasing function throughout its domain. If $0 < a < 1$, then the graph has the general shape shown in Figure 4.10, and hence f is a decreasing function. Note that for every a under consideration, no part of the graph is to the left of the y-axis. The graph has no y-intercept, and the x-intercept is 1.

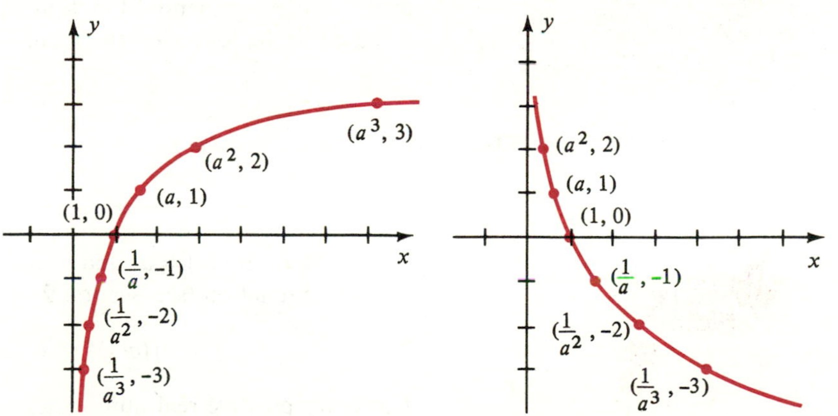

FIGURE 4.9 $f(x) = \log_a x,\ a > 1$

FIGURE 4.10 $f(x) = \log_a x,\ 0 < a < 1$

Functions defined by expressions of the form $\log_a p$, for some expression p that involves x, often occur in mathematics and applications. Functions of this type are classified as members of the logarithmic family; however, the graphs may differ from those sketched in Figures 4.9 and 4.10, as illustrated in the following examples.

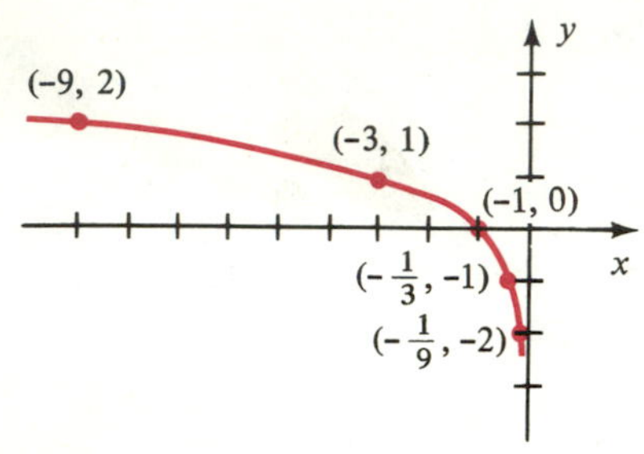

FIGURE 4.11 $f(x) = \log_3(-x)$

EXAMPLE 1 Sketch the graph of f if $f(x) = \log_3(-x)$, $x < 0$.

Solution If $x < 0$, then $-x > 0$, and hence $\log_3(-x)$ is defined. We wish to sketch the graph of the equation $y = \log_3(-x)$, or equivalently, $3^y = -x$. The following table displays coordinates of some points on the graph sketched in Figure 4.11.

y	-2	-1	0	1	2
x	$-\frac{1}{9}$	$-\frac{1}{3}$	-1	-3	-9

■

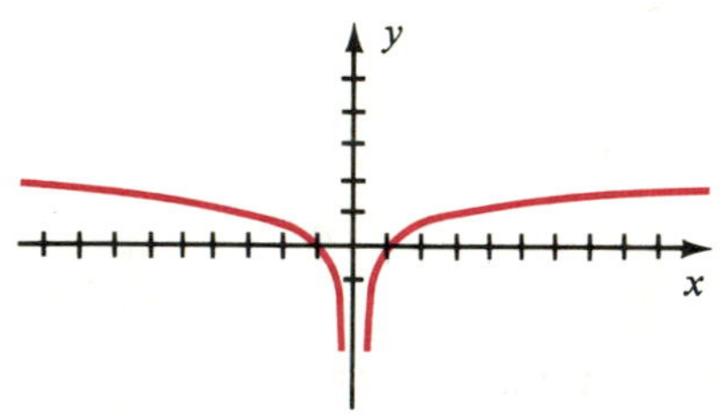

FIGURE 4.12 $f(x) = \log_3|x|$

EXAMPLE 2 Sketch the graph of the equation $y = \log_3|x|$, $x \neq 0$.

Solution Since $|x| > 0$ for all $x \neq 0$, the graph includes points corresponding to negative values of x as well as to positive values. If $x > 0$, then $|x| = x$ and hence to the right of the y-axis the graph coincides with the graph of $y = \log_3 x$, or equivalently, $x = 3^y$. If $x < 0$, then $|x| = -x$ and the graph is the same as that of $y = \log_3(-x)$ (see Example 1). The graph is sketched in Figure 4.12. Note that the graph is symmetric with respect to the y-axis. ■

There is a close relationship between exponential and logarithmic functions. This is to be expected, since the logarithmic function was defined in terms of the exponential function. A precise description of the relationship is stated in the following theorem.

THEOREM

> The exponential and logarithmic functions with base a are inverse functions of one another.

Proof If $f(x) = a^x$ and $g(x) = \log_a x$, then, according to the definition of inverse function (see Section 2.4), we must show that

$$f(g(x)) = x \quad \text{and} \quad g(f(x)) = x.$$

For every positive real number x,

$$f(g(x)) = a^{g(x)} = a^{\log_a x} = x.$$

For every real number x,

$$g(f(x)) = \log_a f(x) = \log_a a^x = x.$$

This proves the theorem. □

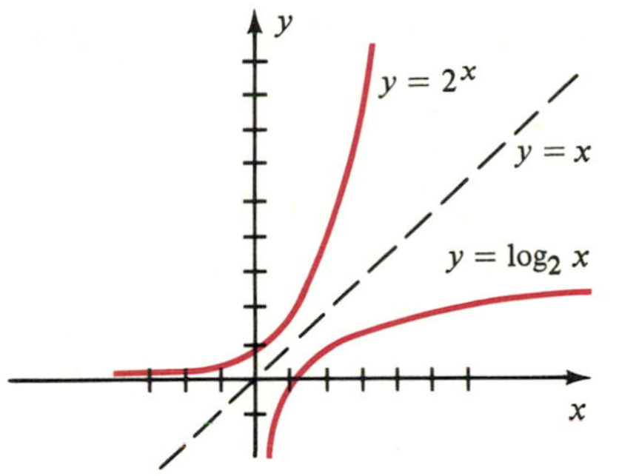

FIGURE 4.13

EXAMPLE 3 Use the preceding theorem to obtain the graph of $y = \log_2 x$ from that of $y = 2^x$.

Solution If we let $f(x) = 2^x$ and $g(x) = \log_2 x$, then by the theorem, f and g are inverse functions of one another. It follows from the discussion in Section 2.6 that their graphs are symmetric with respect to the line $y = x$. The graph of $y = 2^x$ was sketched in Figure 4.1 and is resketched in Figure 4.13. If we reflect this graph through the line $y = x$, we obtain the graph of $y = \log_2 x$, as shown in the figure. ■

Logarithmic functions occur frequently in applications. Indeed, if two variables u and v are related such that u is an exponential function of v, then v is a logarithmic function of u. As a specific example, if \$1.00 is invested at a rate of 9% per year compounded monthly, then from the Compound Interest Formula given in Section 4.1, the amount A after t years is

$$A = (1.0075)^{12t}.$$

The logarithmic form of this equation is

$$12t = \log_{1.0075} A, \quad \text{or} \quad t = \tfrac{1}{12}\log_{1.0075} A.$$

Thus, the number of years required for \$1.00 to grow to an amount A is related logarithmically to A.

Another example we considered in Section 4.1 was that of population growth. In particular, the equation for the number N of bacteria in a certain culture after t hours was

$$N = (1000)2^t \quad \text{or} \quad 2^t = \frac{N}{1000}.$$

Changing to logarithmic form, we obtain

$$t = \log_2 \frac{N}{1000}.$$

Hence, the time t is a logarithmic function of N.

In Section 4.1 we defined the natural exponential function f by means of the equation $f(x) = e^x$. The logarithmic function with base e is called the **natural logarithmic function.** We use **ln *x*** as an abbreviation for $\log_e x$ and refer to it as the **natural logarithm of *x*.** Thus *the natural logarithmic and natural exponential functions are inverse functions of one another*. Since $e \approx 3$, the graph of $y = \ln x$ is similar in appearance to the graph of $y = \log_3 x$. The Laws of Logarithms for natural logarithms are written as follows.

LAWS OF NATURAL LOGARITHMS

(i) $\ln (uv) = \ln u + \ln v$

(ii) $\ln \dfrac{u}{v} = \ln u - \ln v$

(iii) $\ln u^c = c \ln u$

Substituting e for a in the theorem on page 171 gives us the next result.

THEOREM

(i) $e^{\ln x} = x$ for every $x > 0$.

(ii) $\ln e^x = x$ for every x.

(iii) $\ln 1 = 0$

EXAMPLE 4 According to Newton's Law of Cooling, the rate at which an object cools is directly proportional to the difference in temperature between the object and its surrounding medium. Newton's Law can be used to show that under certain conditions the temperature T of an object at time t is given by

$$T = 75e^{-2t}.$$

Express t as a function of T.

Solution The equation $T = 75e^{-2t}$ may be written as

$$e^{-2t} = \frac{T}{75}.$$

Using logarithms with base e yields

$$-2t = \log_e \frac{T}{75} = \ln \frac{T}{75}.$$

Consequently,

$$t = -\frac{1}{2} \ln \frac{T}{75} \quad \text{or} \quad t = -\frac{1}{2} [\ln T - \ln 75].$$ ■

Calculators that have a key labeled [ln x] can be used to approximate values of the natural logarithmic function. A short table of values of $\ln x$ is given in Table 3 of Appendix II.

Most scientific calculators do not have a key for finding e^x directly, but we can use the fact that the natural exponential function is the inverse of the natural logarithmic function to approximate such values. Thus, if we denote the inverse function of ln by $\ln^{-1}$, then $e^x = \ln^{-1} x$ for all x. On a typical calculator, $\ln^{-1} x$ is obtained by entering x and then pressing, successively, [INV] and [ln x]. This technique for approximating e^x is illustrated in the next example.

EXAMPLE 5 Use a calculator to approximate (a) $e^{4.7}$; (b) e.

Solution

(a) To find $e^{4.7}$ we proceed as follows:

Enter: 4.7

Press [INV] [ln x]: 109.94717

The approximation listed in Table 2 in Appendix II is $e^{4.70} \approx 109.95$.

(b) Using the fact that $e = e^1$,

Enter: 1

Press [INV] [ln x]: 2.7182818 ■

EXERCISES 4.3

Sketch the graph of f in Exercises 1–10.

1 (a) $f(x) = \log_2 x$
(b) $f(x) = \log_4 x$

2 (a) $f(x) = \log_5 x$
(b) $f(x) = \log_{10} x$

3 (a) $f(x) = \log_3 x$
(b) $f(x) = \log_3 (x - 2)$

4 (a) $f(x) = \log_2 (-x)$
(b) $f(x) = \log_2 (3 - x)$

5 (a) $f(x) = \log_3 (3x)$
(b) $f(x) = 3 \log_3 x$

6 (a) $f(x) = \log_2 (x^2)$
(b) $f(x) = \log_2 (x^3)$

7 (a) $f(x) = \log_2 \sqrt{x}$
(b) $f(x) = \log_2 \sqrt[3]{x}$

8 (a) $f(x) = \log_2 |x|$
(b) $f(x) = |\log_2 x|$

9 (a) $f(x) = \log_3 (1/x)$
(b) $f(x) = 1/(\log_3 x)$

10 (a) $f(x) = \log_3 (2 + x)$
(b) $f(x) = 2 + \log_3 x$

11 What is the geometric relationship between the graphs of $y = \log_a x$ and $y = a^x$? Why does this relationship exist?

12 Describe the relationship of the graph of $y = \ln x$ to the graphs of $y = \log_2 x$ and $y = \log_3 x$.

13 Starting with q_0 milligrams of pure radium, the amount q remaining after t years is

$$q = q_0(2)^{-t/1600}.$$

Use logarithms with base 2 to solve for t in terms of q and q_0.

14 The number of bacteria in a certain culture at time t is given by $N = 10^4(3)^t$. Use logarithms with base 3 to solve for t in terms of N.

15 An electrical condenser with initial charge Q_0 is allowed to discharge. After t seconds the charge Q is

$$Q = Q_0 e^{kt}$$

where k is a constant of proportionality. Use natural logarithms to solve for t in terms of Q_0, Q, and k.

16 Under certain conditions the atmospheric pressure p at altitude h is given by

$$p = 29e^{-0.000034h}$$

Use natural logarithms to solve for h as a function of p.

17 The loudness of sound, as experienced by the human ear, is based upon intensity levels. A formula used for finding the intensity level α that corresponds to a sound intensity I, in decibels, is

$$\alpha = 10 \log_{10} \frac{I}{I_0}$$

where I_0 is a special value of I agreed to be the weakest sound that can be detected by the ear under certain conditions. Find α if:

(a) I is 10 times as great as I_0;

(b) I is 1,000 times as great as I_0;

(c) I is 10,000 times as great as I_0. (This is the intensity level of the average voice.)

18 The current I at time t in a certain electrical circuit is given by $I = 20e^{-Rt/L}$, where R and L denote the resistance and inductance, respectively. Use natural logarithms to solve for t in terms of the remaining variables.

19 Using the Richter scale, the magnitude R of an earthquake of intensity I may be found by means of the formula $R = \log_{10}(I/I_0)$, where I_0 is a certain minimum intensity. Find R if an earthquake has intensity:
(a) 100 times that of I_0;
(b) 10,000 times that of I_0;
(c) 100,000 times that of I_0.

20 Refer to Exercise 19. The magnitude of the San Francisco earthquake of 1906 was approximately 8 on the Richter scale. What is the corresponding intensity in terms of I_0?

CALCULATOR EXERCISES 4.3

1 Sketch the graph of

(a) $y = \ln x$ (b) $y = \ln|x|$
(c) $y = |\ln x|$

2 Sketch the graph of f if $f(x) = x^2 \ln x$.

3 Demonstrate that $\ln e^x = x$ and $e^{\ln x} = x$ for the following values of x:

(a) 1 (b) 3.4
(c) 0.056 (d) 8.143

4 Refer to Calculator Exercise 18 of Section 4.1. For what value of t will the amount remaining be 15 milligrams?

5 Letting $a = e$ in the Change of Base Formula

$$\log_b u = \frac{\log_a u}{\log_a b} \quad \text{leads to} \quad \log_b u = \frac{\ln u}{\ln b}.$$

Use this formula to approximate

(a) $\log_3 4$ (b) $\log_5 (1.64)$
(c) $\log_7 \sqrt{2}$ (d) $\log_{10} 382$
(e) $\log_{10} \frac{47}{62}$

6 The $\boxed{\log}$ key on a calculator refers to the logarithmic function with base 10. Use a calculator as an aid to sketching the graph of f if $f(x) = \log_{10} x$.

SECTION 4.4
COMMON LOGARITHMS

Before electronic calculators were invented, logarithms with base 10 were used for complicated numerical computations involving products, quotients, and powers of real numbers. Base 10 was employed because it is well suited for numbers that are expressed in decimal form. Logarithms with base 10 are called **common logarithms.** The symbol $\log x$ is used as an abbreviation for $\log_{10} x$. Thus, we have the following definition.

DEFINITION OF COMMON LOGARITHMS

$$\log x = \log_{10} x \quad \text{for every } x > 0.$$

Since inexpensive calculators are now available, we have little need for logarithms as a tool for computational work. However, base 10 does occur in certain applications (cf. Exercises 19–26), and hence scientific calculators have a key labeled "log" that can be used to find common logarithms. We have already discussed the key, usually labeled "ln x," that is used for approximating natural logarithms.

In the examples of this section, a calculator was used to approximate logarithms of numbers. Appendix II contains tables of common and natural logarithms that may be used if a calculator either is not available or is inoperative. The use of the table of common logarithms is explained in Appendix I.

EXAMPLE 1 Use a calculator to approximate log 436; log 0.0436, ln 436, and ln 0.0436.

Solution Entering the indicated numbers and pressing the appropriate keys, we obtain the following approximations:

$$\log 436 \approx 2.6394865$$

$$\log 0.0436 \approx -1.3605135$$

$$\ln 436 \approx 6.0776422$$

$$\ln 0.0436 \approx -3.1326981$$

■

To solve certain problems we must find x when given either $\log x$ or $\ln x$. One way to accomplish this is by using the inverse function key [INV]. If we first press [INV] and then press [log], we obtain the *inverse logarithmic function* $\log^{-1}$ (from the theorem on page 178, $\log^{-1}$ is the exponential function with base 10). Since $\log^{-1}(\log x) = x$, we can obtain x by entering $\log x$ and then pressing, successively, [INV] and [log]. Similarly, given $\ln x$, we can find x by entering $\ln x$ and pressing [INV] [ln x]. This procedure is illustrated in the next example.

EXAMPLE 2 Approximate x to three decimal places:

(a) $\log x = 1.7959$ (b) $\ln x = 4.7$

Solution

(a) Given $\log x = 1.7959$,

Enter: 1.7959

Press [INV] [log]: 62.502876

Thus, $x \approx 62.503$.

(b) Given $\ln x = 4.7$,

Enter: 4.7

Press [INV] [ln x]: 109.94717

Hence, $x \approx 109.95$. ■

Example 2, part (a) could alternatively be solved by noting that:

$$\text{if} \quad \log x = 1.7959, \quad \text{then} \quad x = 10^{1.7959}.$$

The number x can then be approximated using a $\boxed{y^x}$ key. Similarly, part (b) of the example could be solved by using the fact that natural logarithms have base e. Specifically,

$$\text{if} \quad \ln x = 4.7, \quad \text{then} \quad x = e^{4.7}.$$

Table 2 in Appendix II or a calculator could then be used to approximate x (cf. Example 5 of Section 4.3).

The variables in certain equations appear as exponents or logarithms, as illustrated in the following examples.

EXAMPLE 3 Solve the equation $3^x = 21$.

Solution Taking the common logarithm of both sides and using (iii) of the Laws of Logarithms, we obtain

$$\begin{aligned} \log (3^x) &= \log 21 \\ x \log 3 &= \log 21 \\ x &= \frac{\log 21}{\log 3}. \end{aligned}$$

If an approximation is desired, we may use a calculator or Table 1 in Appendix II to obtain

$$x \approx \frac{1.3222}{0.4771} \approx 2.77$$

A partial check on the solution is to note that, since $3^2 = 9$ and $3^3 = 27$, the number x such that $3^x = 21$ should lie between 2 and 3, somewhat closer to 3 than to 2.

We could also have solved for x by using natural logarithms. In this case, a calculator yields

$$\begin{aligned} x = \frac{\ln 21}{\ln 3} &\approx \frac{3.0445224}{1.0986123} \\ &\approx 2.7712437 \approx 2.77 \end{aligned}$$

■

EXAMPLE 4 Solve $5^{2x+1} = 6^{x-2}$.

Solution If we take the common logarithm of both sides and use (iii) of the Laws of Logarithms, we obtain

$$(2x + 1) \log 5 = (x - 2) \log 6.$$

We may now solve for x:

$$\begin{aligned} 2x \log 5 + \log 5 &= x \log 6 - 2 \log 6 \\ 2x \log 5 - x \log 6 &= -\log 5 - 2 \log 6 \\ x(2 \log 5 - \log 6) &= -(\log 5 + \log 6^2) \\ x &= \frac{-(\log 5 + \log 36)}{2 \log 5 - \log 6} \\ &= \frac{-\log (5 \cdot 36)}{\log 5^2 - \log 6}. \end{aligned}$$

Thus $$x = -\frac{\log 180}{\log (25/6)}.$$

If an approximation to the solution is desired, we could proceed as in Example 1:

$$x \approx -\frac{2.2553}{.6198} \approx -3.64$$

Natural logarithms could also have been used:

$$x = -\frac{\ln 180}{\ln (25/6)} \approx \frac{5.1929569}{1.4271164} \approx -3.64$$ ■

EXAMPLE 5 Solve the equation $\log (5x - 1) - \log (x - 3) = 2$.

Solution The equation may be written

$$\log \frac{5x - 1}{x - 3} = 2.$$

Using the definition of logarithm with $a = 10$ gives us

$$\frac{5x - 1}{x - 3} = 10^2.$$

Consequently,

$$5x - 1 = 10^2(x - 3) = 100x - 300, \quad \text{or} \quad 299 = 95x.$$

Hence, $$x = \frac{299}{95}.$$

We leave it to the reader to check that this is the solution of the equation. ■

EXAMPLE 6 Solve the equation $\dfrac{5^x - 5^{-x}}{2} = 3.$

Solution We multiply both sides of the equation, first by 2 and then by 5^x, obtaining

$$5^x - 5^{-x} = 6$$
$$5^{2x} - 1 = 6(5^x),$$

which may be written as

$$(5^x)^2 - 6(5^x) - 1 = 0.$$

Letting $u = 5^x$ gives us the quadratic equation

$$u^2 - 6u - 1 = 0$$

in the variable u. Applying the Quadratic Formula,

$$u = \frac{6 \pm \sqrt{36 + 4}}{2} = 3 \pm \sqrt{10},$$

that is, $5^x = 3 \pm \sqrt{10}$. Since 5^x is never negative, the number $3 - \sqrt{10}$ must be discarded; therefore,

$$5^x = 3 + \sqrt{10}.$$

Taking the common logarithm of both sides and using (iii) of the Laws of Logarithms,

$$x \log 5 = \log (3 + \sqrt{10}) \quad \text{or} \quad x = \frac{\log (3 + \sqrt{10})}{\log 5}.$$

To obtain an approximate solution we may write $3 + \sqrt{10} \approx 6.16$ and use Table 1 or a calculator. This gives us

$$x \approx \frac{\log 6.16}{\log 5} \approx \frac{0.7896}{0.6990} \approx 1.13.$$

Natural logarithms could also have been used to obtain

$$x = \frac{\ln (3 + \sqrt{10})}{\ln 5}.$$

■

EXERCISES 4.4

In Exercises 1–6 approximate, to four decimal places, (a) the common logarithms of the numbers; (b) the natural logarithms of the numbers.

1 347; 0.00347; 3.47

2 86.2; 8,620; 0.862

3 0.54; 540; 540,000

4 208; 2.08; 20,800

5 60.2; 0.0000602; 602

6 5; 0.5; 0.0005

In Exercises 7–18 approximate x to three significant figures.

7 $\log x = 3.6274$

8 $\log x = 1.8965$

9 $\log x = 0.9469$

10 $\log x = 4.9680$

11 $\log x = -1.6253$

12 $\log x = -2.2118$

13 $\ln x = 2.3$

14 $\ln x = 3.7$

15 $\ln x = 0.05$

16 $\ln x = 0.95$

17 $\ln x = -1.6$

18 $\ln x = -5$

19 Chemists use a number denoted by "pH" to describe quantitatively the acidity or basicity of solutions. By definition,

$$\text{pH} = -\log [\text{H}^+]$$

where $[\text{H}^+]$ is the hydrogen ion concentration in moles per liter. Approximate the pH of each substance:

(a) vinegar: $[\text{H}^+] \approx 6.3 \times 10^{-3}$

(b) carrots: $[\text{H}^+] \approx 1.0 \times 10^{-5}$

(c) sea water: $[\text{H}^+] \approx 5.0 \times 10^{-9}$

20 Approximate the hydrogen ion concentration $[\text{H}^+]$ in each of the following substances. (Refer to Exercise 19 and use a calculator or the closest entry in Table 1 to approximate $[\text{H}^+]$.)

(a) apples: pH ≈ 3.0 (b) beer: pH ≈ 4.2

(c) milk: pH ≈ 6.6

21 A solution is considered acidic if $[\text{H}^+] > 10^{-7}$ or basic if $[\text{H}^+] < 10^{-7}$. What are the corresponding inequalities involving pH?

22 Many solutions have a pH between 1 and 14. What is the corresponding range of the hydrogen ion content $[\text{H}^+]$?

23 The sound intensity level formula considered in Exercise 17 of Section 4.3 may be written

$$\alpha = 10 \log \frac{I}{I_0}.$$

Find α if I is 285,000 times as great as I_0.

24 A sound intensity level of 140 decibels produces pain in the average human ear. Approximately how many times greater than I_0 must I be in order for α to reach this level? (Refer to Exercise 23 and use a calculator or the nearest entry in Table 1 to find the approximation.)

25 The Richter scale formula given in Exercise 19 of Section 4.3 may be written as $R = \log (I/I_0)$. An earthquake of magnitude 4.5 on this scale may cause damage to buildings. Find the intensity (in terms of I_0) of such an earthquake.

26 Refer to Exercise 25. The largest recorded magnitudes of earthquakes have been between 8 and 9 on the Richter scale. Find the corresponding intensities in terms of I_0.

Find the solutions of the equations in Exercises 27–46 without using a calculator or a table of values.

27 $10^x = 7$

28 $5^x = 8$

29 $4^x = 3$

30 $10^x = 6$

31 $3^{4-x} = 5$

32 $(\frac{1}{3})^x = 100$

33 $3^{x+4} = 2^{1-3x}$

34 $4^{2x+3} = 5^{x-2}$

35 $2^{-x} = 8$

36 $2^{-x^2} = 5$

37 $\log x = 1 - \log (x - 3)$

38 $\log (5x + 1) = 2 + \log (2x - 3)$

39 $\log (x^2 + 4) - \log (x + 2) = 3 + \log (x - 2)$

40 $\log (x - 4) - \log (3x - 10) = \log (1/x)$

41 $\log (x^2) = (\log x)^2$

42 $\log \sqrt{x} = \sqrt{\log x}$

43 $\log (\log x) = 2$

44 $\log \sqrt{x^3 - 9} = 2$

45 $x^{\sqrt{\log x}} = 10^8$

46 $\log (x^3) = (\log x)^3$

In Exercises 47–50 solve for x in terms of y.

47 $y = \dfrac{10^x + 10^{-x}}{2}$

48 $y = \dfrac{10^x - 10^{-x}}{2}$

49 $y = \dfrac{10^x - 10^{-x}}{10^x + 10^{-x}}$

50 $y = \dfrac{1}{10^x - 10^{-x}}$

In Exercises 51–54 use natural logarithms to solve for x in terms of y.

51 $y = \dfrac{e^x - e^{-x}}{2}$

52 $y = \dfrac{e^x + e^{-x}}{2}$

53 $y = \dfrac{e^x - e^{-x}}{e^x + e^{-x}}$

54 $y = \dfrac{e^x + e^{-x}}{e^x - e^{-x}}$

55 The current I in a certain electrical circuit at time t is given by

$$I = \frac{E}{R}(1 - e^{-Rt/L})$$

where E, R, and L denote the electromotive force, the resistance, and the inductance, respectively (see figure). Use natural logarithms to solve for t in terms of the remaining symbols.

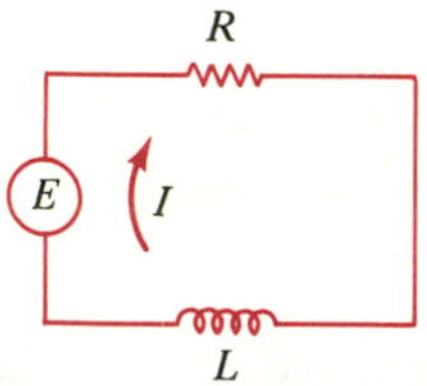

FIGURE FOR EXERCISE 55

56 Solve the Compound Interest Formula

$$A = P\left(1 + \frac{r}{n}\right)^{nt}$$

for t in terms of the other symbols.

57 In Exercise 17 of Section 4.3 we considered the sound intensity level formula

$$\alpha = 10 \log \frac{I}{I_0}.$$

(a) Solve for I in terms of α and I_0.

(b) Show that a one-decibel rise in the intensity level corresponds to a 26% increase in the intensity.

58 The formula $\text{pH} = -\log [\text{H}^+]$ was introduced in Exercise 19. Solve for $[\text{H}^+]$ in terms of pH.

59 If a beam of light that has intensity k is projected vertically downward into water, then its intensity $I(x)$ at a depth of x meters is $I(x) = ke^{-1.4x}$. At what depth is the intensity (a) one-half its value at the surface? (b) one-tenth its value?

60 The current $I(t)$ at time t in a certain electrical circuit is given by $I(t) = I_0 e^{-Rt/L}$, where R and L denote the resistance and inductance, respectively, and I_0 is the current at time $t = 0$. At what time is the current $(1/100)I_0$?

CALCULATOR EXERCISES 4.4

Use natural logarithms to approximate the solutions of the equations in Exercises 1–4.

1 $5^x = 7$

2 $4^{-x} = 10$

3 $e^{-x^2} = 0.163$

4 $e^{-2x+1} = 5^x$

5 How long will it take a sum of money to double if it is invested at a rate of 6% per year compounded monthly?

6 If \$10,000 is invested at an interest rate of 8% per year compounded quarterly, when will the principal exceed \$20,000?

7 A certain radioactive substance decays according to the formula $q(t) = q_0 e^{-0.0063t}$, where q_0 is the initial amount of the substance and t is the time in days. Use natural logarithms to approximate its half-life, that is, the number of days it takes for half of the substance to decay.

8 The air pressure $p(h)$, in lb/in.^{2}, at an altitude of h feet above sea level may be approximated by

$$p(h) = 14.7e^{-0.0000385h}.$$

At approximately what altitude h is the air pressure (a) 10 lb/in.^{2}? (b) one-half its value at sea level?

9 The graph of $G(t) = 10e^{-(1/2)e^{-t}}$ is a Gompertz growth curve. (See Calculator Exercise 19 of Section 4.1.) Approximate, to the nearest thousandth, the value of t such that $G(t) = 8$.

10 The formula $f(n) = 3 + 20(1 - e^{-0.1n})$ may be used to estimate certain learning processes (see Calculator Exercise 20 of Section 4.1). Approximate, to the nearest hundredth, the value of n such that $f(n) = 13$.

SECTION 4.5
REVIEW

Define or discuss the following.

1 The exponential function with base a
2 The natural exponential function
3 The logarithm of u with base a
4 Laws of Logarithms
5 Logarithmic functions
6 The natural logarithmic function
7 Common logarithms

EXERCISES 4.5

Find the numbers in Exercises 1–10 without the aid of a calculator or table.

1 $\log_2 \frac{1}{16}$
2 $\log_5 \sqrt[3]{5}$
3 $6^{\log_6 4}$
4 $10^{3 \log 2}$
5 $\log 1{,}000{,}000$
6 $\ln e$
7 $\log_4 2$
8 $\log_\pi 1$
9 $e^{\ln 5}$
10 $\log \log 10^{10}$

In Exercises 11–20 sketch the graph of f.

11 $f(x) = 3^{x+2}$
12 $f(x) = (\frac{3}{5})^x$
13 $f(x) = (\frac{3}{2})^{-x}$
14 $f(x) = 3^{-2x}$
15 $f(x) = 3^{-x^2}$
16 $f(x) = 1 - 3^{-x}$
17 $f(x) = \log_6 x$
18 $f(x) = \log_3 (x^2)$
19 $f(x) = 2 \log_3 x$
20 $f(x) = \log_2 (x + 4)$

Find the solutions of the equations in Exercises 21–30 without using a calculator or table.

21 $\log_8 (x - 5) = \frac{2}{3}$
22 $\log_4 (x + 1) = 2 + \log_4 (3x - 2)$
23 $2 \log_3 (x + 3) - \log_3 (x + 1) = 3 \log_3 2$
24 $\log \sqrt[4]{x + 1} = \frac{1}{2}$
25 $2^{5-x} = 6$
26 $3^{x^2} = 7$
27 $2^{5x+3} = 3^{2x+1}$
28 $e^{\ln (x+1)} = 3$
29 $x^2(-2xe^{-x^2}) + 2xe^{-x^2} = 0$
30 $\ln x = 1 + \ln (x + 1)$

31 Express $\log x^4 \sqrt[3]{y^2/z}$ in terms of logarithms of x, y, and z.

32 Express $\log (x^2/y^3) + 4 \log y - 6 \log \sqrt{xy}$ as one logarithm.

Solve the equations in Exercises 33 and 34 for x in terms of y.

33 $y = \dfrac{10^x + 10^{-x}}{10^x - 10^{-x}}$
34 $y = \dfrac{1}{10^x + 10^{-x}}$

In Exercises 35–40 approximate x using (a) a calculator; (b) tables.

35 $\log x = 1.8938$
36 $\log x = -2.4260$
37 $\ln x = 1.8$
38 $\ln x = -0.75$
39 $x = \ln 6.6$
40 $x = \log 84.7$

5

THE TRIGONOMETRIC FUNCTIONS

Trigonometry originated over 2,000 years ago when the Greeks needed precise methods for measuring angles and sides of triangles. We shall discuss angles in Section 5.1, but from a modern point of view. Trigonometric functions of angles and of real numbers are defined in Sections 5.2 and 5.3, respectively. Section 5.4 contains a discussion of how values of trigonometric functions may be found. Graphs involving the trigonometric functions are considered in Sections 5.5–5.7. We conclude the chapter with some applications of trigonometry to problems about right triangles and harmonic motion.

SECTION 5.1

ANGLES

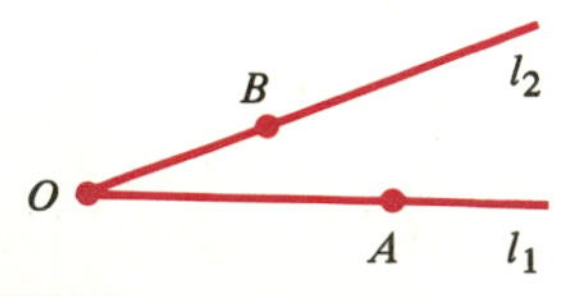

FIGURE 5.1

In geometry an angle is often regarded as the set of all points on two rays or half-lines, l_1 and l_2, having the same initial point O. If A and B are points on l_1 and l_2, respectively (see Figure 5.1), then we may refer to **angle** ***AOB***. The same is true for finite line segments with a common endpoint. For trigonometric purposes it is convenient to regard angle AOB as generated by starting with the fixed ray l_1 with endpoint O and rotating it about

O, in a plane, to a position specified by ray l_2. We call l_1 the **initial side,** l_2 the **terminal side,** and O the **vertex** of the angle. The amount or direction of rotation is not restricted in any way. Thus, we might let l_1 make several revolutions in either direction about O before coming to the position l_2, as illustrated by the curved arrows in Figure 5.2. It is important to observe that there are many different angles that have the same initial and terminal sides. Any two such angles are called **coterminal.**

FIGURE 5.2

If a rectangular coordinate system is introduced, then the **standard position** of an angle is obtained by taking the vertex at the origin and letting the initial side l_1 coincide with the positive x-axis. If l_1 is rotated in a counterclockwise direction to the terminal position l_2, then the angle is considered **positive,** whereas if l_1 is rotated in a clockwise direction, the angle is **negative.** We often denote angles by lowercase Greek letters and specify the direction of rotation by means of a circular arc or spiral with an arrow attached. Figure 5.3 contains sketches of two positive angles α and β, and a negative angle γ. If the terminal side of an angle is in a certain quadrant, we speak of the *angle* as being in that quadrant. In Figure 5.3, α is in quadrant III, β is in quadrant I, and γ is in quadrant II. If the terminal side coincides with a coordinate axis, then the angle is referred to as a **quadrantal angle.**

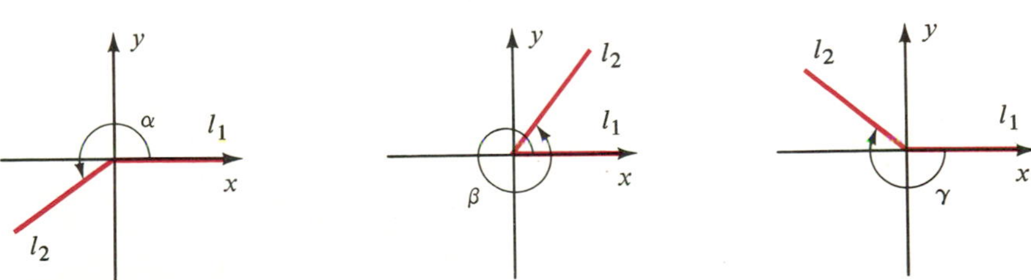

FIGURE 5.3

One of the units of measurement for angles is the **degree.** If the angle is placed in standard position on a rectangular coordinate system, then an angle of 1 degree is, by definition, the measure of the angle formed by $\frac{1}{360}$ of a complete revolution in the counterclockwise direction. The symbol $^\circ$ is used to denote the degrees in the measure of an angle. In Figure 5.4 several angles measured in degrees are sketched in standard position on a rectangular coordinate system. As shown in the figure, the notation $\theta = 60°$ is employed to specify an angle θ whose measure is 60°. We also use phrases such as "an angle of 60°," or "a 60° angle" instead of the more precise (but cumbersome) "an angle having degree measure 60°."

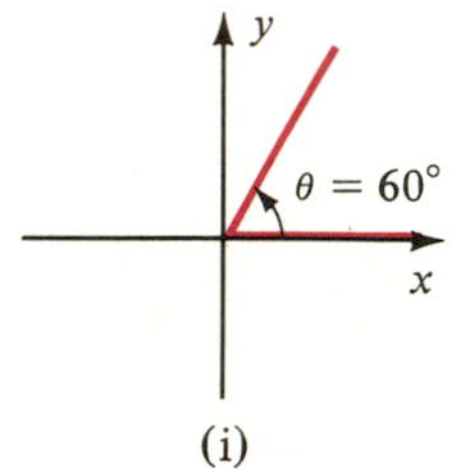

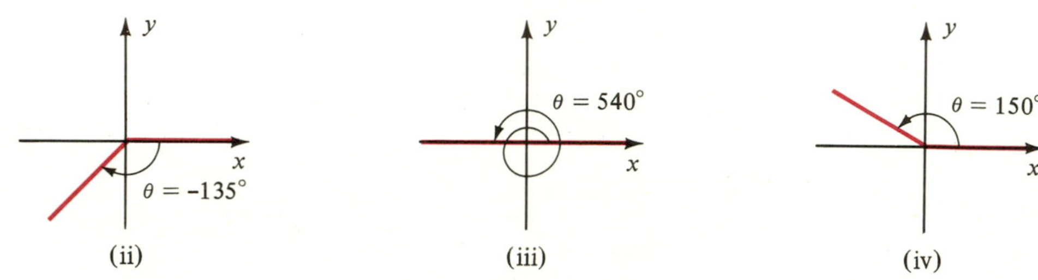

FIGURE 5.4

A 90° angle is called a **right angle.** An angle θ is **acute** if $0° < \theta < 90°$, or **obtuse** if $90° < \theta < 180°$. Two acute angles are **complementary** if their sum is 90°. For example, 20° and 70° are complementary angles. In general, if θ is acute then θ and $90° - \theta$ are complementary. Two positive angles are **supplementary** if their sum is 180°.

If smaller measurements than those afforded by the degree are required, we can use tenths, hundredths, or thousandths of degrees. Another method of measurement is to divide each degree into 60 equal parts, called **minutes** (denoted by ′), and each minute into 60 equal parts, called **seconds** (denoted by ″). Thus 1′ is $\frac{1}{60}$ of 1°, and 1″ is $\frac{1}{60}$ of 1′. A notation such as $\theta = 73°56'18''$ refers to an angle θ of measure 73 degrees, 56 minutes, and 18 seconds.

EXAMPLE 1 If $\theta = 60°$, find two positive angles and two negative angles that are coterminal with θ.

Solution The angle θ is sketched in (i) of Figure 5.4. To find positive coterminal angles, we may add either 360° or 720° to θ, since this amounts to one or two additional complete revolutions. This gives us

$$60° + 360° = 420° \quad \text{and} \quad 60° + 720° = 780°.$$

To find negative coterminal angles we add $-360°$ or $-720°$, obtaining

$$60° + (-360°) = -300° \quad \text{and} \quad 60° + (-720°) = -660°.$$ ■

EXAMPLE 2 Find the angle that is complementary to θ if:

(a) $\theta = 25°43'37''$ (b) $\theta = 73.26°$

Solution The desired angle is $90° - \theta$. We may arrange our work as follows:

(a)
$$\begin{array}{rl} 90^\circ & = 89^\circ59'60'' \\ \theta & = 25^\circ43'37'' \\ \hline 90^\circ - \theta & = 64^\circ16'23'' \end{array}$$

(b)
$$\begin{array}{rl} 90^\circ & = 90.00^\circ \\ \theta & = 73.26^\circ \\ \hline 90^\circ - \theta & = 16.74^\circ \end{array}$$

■

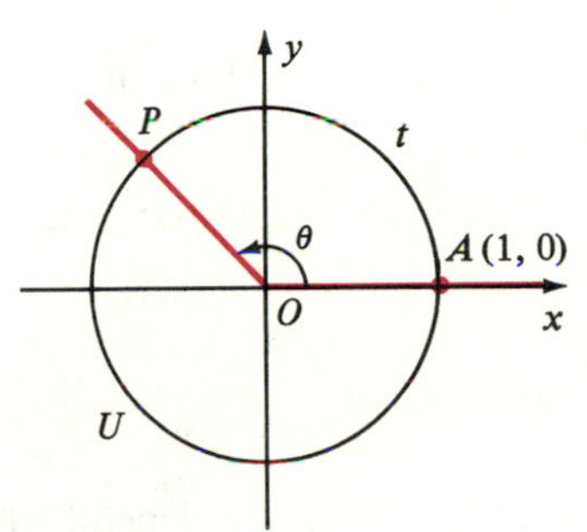

FIGURE 5.5

In calculus and its applications, *radian measure* is usually employed to specify the magnitude of an angle. To define this unit of measurement, consider a unit circle U—a circle with radius 1 and with the center at the origin O of a rectangular coordinate system—and let θ be an angle in standard position (see Figure 5.5). We regard θ as generated by rotating the positive x-axis about O. As the x-axis rotates to its terminal position, its point of intersection with U travels a certain distance t before arriving at its final position P, as illustrated in Figure 5.5. We shall consider t positive for a counterclockwise rotation and negative for a clockwise rotation; that is, t is positive or negative, depending on whether θ is positive or negative, respectively. The **radian measure** of θ is defined as the number t. We refer to θ as *an angle of t radians* and write $\theta = t$ or $\theta = t$ *radians*. If $0^\circ \leq \theta \leq 360^\circ$, we say that the arc $\widehat{AP}$ of length t in Figure 5.5 **subtends** θ, or that θ is *subtended by* the arc. In particular, if $\theta = 1$, then θ is an angle that is subtended by an arc of unit length on the unit circle U. The notation $\theta = -7.5$ means that θ is an angle generated by a clockwise rotation in which the point of intersection of the terminal side of θ with the unit circle U travels 7.5 units. Several angles, measured in radians, are sketched in Figure 5.6.

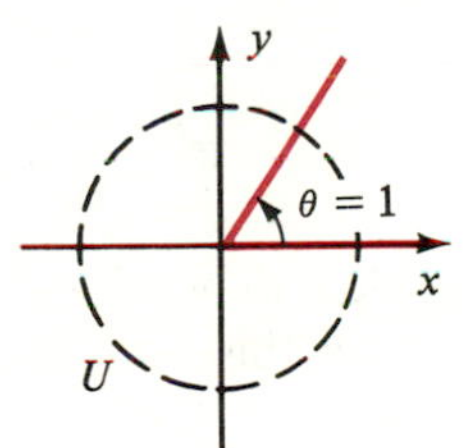

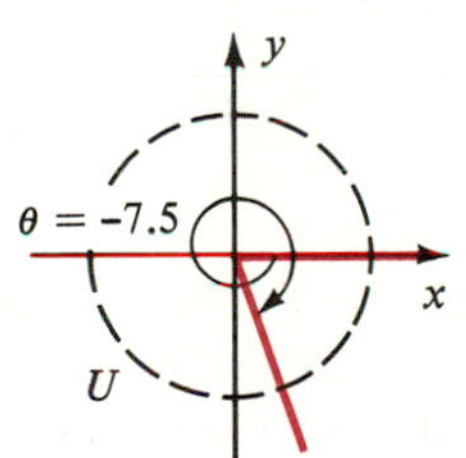

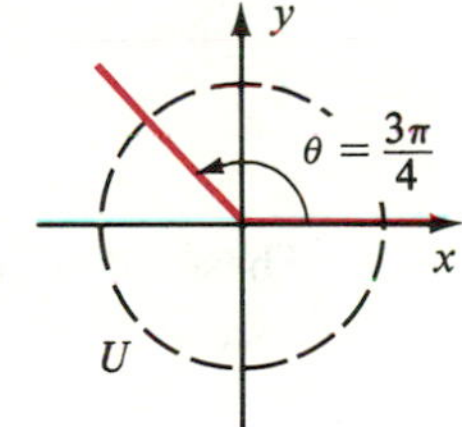

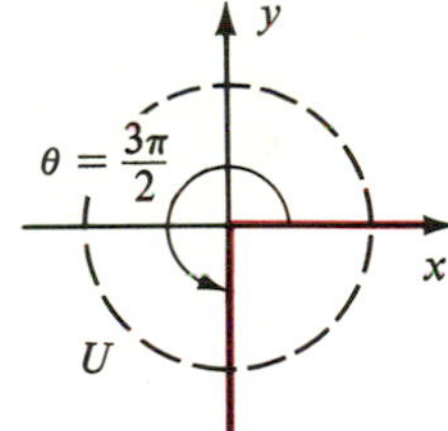

FIGURE 5.6

The radian measure of an angle can be found by using a circle of *any* radius. In the following discussion, the terminology **central angle** of a circle refers to an angle whose vertex is at the center of the circle. Suppose that θ is a central angle of a circle of radius r and that θ is subtended by an arc of length s, where $0 \leq s \leq 2\pi r$. To find the radian measure of θ, let us place θ in standard position on a rectangular coordinate system and superimpose a unit circle U, as shown in Figure 5.7. If t is the length of the arc on U that subtends θ, then, by definition, we may write $\theta = t$. From plane geometry, the ratio of the arcs in Figure 5.7 is the same as the ratio of the radii; that is,

$$\frac{t}{s} = \frac{1}{r} \quad \text{or} \quad t = \frac{s}{r}.$$

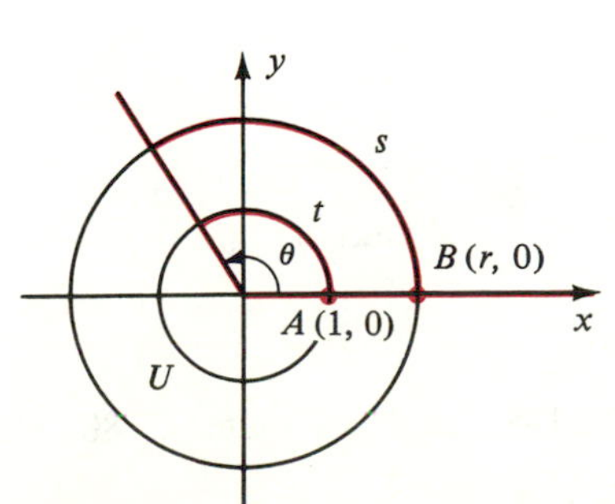

FIGURE 5.7

This gives us the following result.

THEOREM

If a central angle θ of a circle of radius r is subtended by an arc of length s, then the radian measure of θ is

$$\theta = \frac{s}{r}.$$

EXAMPLE 3 A central angle θ is subtended by an arc 10 cm long on a circle of radius 4 cm. Find the radian measure of θ.

Solution Substituting in the formula $\theta = s/r$ gives us the radian measure:

$$\theta = \frac{10}{4} = 2.5$$

■

The formula $\theta = s/r$ for radian measure of an angle is independent of the size of the circle. For example, if the radius of the circle is $r = 4$ cm and an arc of length 8 cm subtends a central angle θ, then the radian measure of θ is

$$\theta = \frac{8 \text{ cm}}{4 \text{ cm}} = 2.$$

If the radius of the circle is 5 km and the arc is 10 km, then

$$\theta = \frac{10 \text{ km}}{5 \text{ km}} = 2.$$

These calculations indicate that the radian measure of an angle is dimensionless, and hence may be regarded as a real number. Indeed, for this reason, we usually employ the notation $\theta = t$ instead of $\theta = t$ radians.

It is not difficult to transform angular measure from one system to another. If we consider the angle θ in standard position, generated by one-half of a complete counterclockwise rotation, then $\theta = 180°$. Using the formula $\theta = s/r$, we see that the radian measure of θ is π. This gives us the following relationships.

RELATIONSHIPS BETWEEN DEGREES AND RADIANS

$$180° = \pi \text{ radians}, \qquad 1° = \frac{\pi}{180} \text{ radians}, \qquad 1 \text{ radian} = \left(\frac{180}{\pi}\right)°.$$

If we use the approximation $\pi \approx 3.14159$ and calculate $\pi/180$ and $180/\pi$, we obtain

$$1° \approx 0.01745 \text{ radians} \quad \text{and} \quad 1 \text{ radian} \approx 57.296°.$$

The following theorem is a direct consequence of the preceding formulas.

THEOREM

(i) To change radian measure to degrees, multiply by $180/\pi$.
(ii) To change degree measure to radians, multiply by $\pi/180$.

If an angle has radian measure 5, we shall write $\theta = 5$ instead of $\theta = 5$ *radians*. The same is true for any other angles. There should be no confusion as to whether radian or degree measure is being used, since if θ has degree measure $5°$, we write $\theta = 5°$, and *not* $\theta = 5$.

EXAMPLE 4

(a) Find the radian measure of θ if $\theta = 150°$ and if $\theta = 225°$.

(b) Find the degree measure of θ if $\theta = 7\pi/4$ and if $\theta = \pi/3$.

Solution

(a) By the preceding theorem, we can find the number of radians in $150°$ by multiplying 150 by $\pi/180$. Thus,

$$150° = 150\left(\frac{\pi}{180}\right) = \frac{5\pi}{6}.$$

Similarly,

$$225° = 225\left(\frac{\pi}{180}\right) = \frac{5\pi}{4}.$$

(b) According to the preceding theorem, to find the number of degrees in $7\pi/4$ radians, we multiply by $180/\pi$, obtaining

$$\frac{7\pi}{4} = \frac{7\pi}{4}\left(\frac{180}{\pi}\right) = 315°.$$

Similarly,

$$\frac{\pi}{3} = \frac{\pi}{3}\left(\frac{180}{\pi}\right) = 60°.$$

■

EXAMPLE 5 If the measure of an angle θ is 3 radians, find the approximate measure of θ in terms of degrees, minutes, and seconds.

Solution Since 1 radian $\approx 57.296°$,

$$3 \text{ radians} \approx 3(57.296°) = 171.888° = 171° + 0.888°.$$

Since there are $60'$ in each degree, the number of minutes in $0.888°$ is $60(0.888)$, or $53.28'$. Hence,

$$3 \text{ radians} \approx 171°53.28'.$$

Finally, $0.28' = (0.28)60'' \approx 17''$. Therefore,

$$3 \text{ radians} \approx 171°53'17''.$$

Some calculators have keys that can be used to convert the degree measure of an angle to radians, and vice versa. Refer to the user's guide of the particular calculator for information. Since entries are made in the decimal system, angles that are expressed in terms of degrees and minutes must be converted to decimal form before they are entered into the calculator. For example,

$$67°30' = 67.5° \quad \text{since} \quad 30' = (30/60)° = 0.5°$$

$$123°45' = 123.75° \quad \text{since} \quad 45' = (45/60)° = 0.75°$$

$$8°13' \approx 8.21667° \quad \text{since} \quad 13' = (13/60)° \approx 0.21667°.$$

If seconds are involved, another appropriate decimal must be calculated. To illustrate, since there are 3,600 seconds in each degree, $18'' = (18/3600)° = 0.005'$ and hence,

$$67°30'18'' = 67.505°.$$

EXERCISES 5.1

In Exercises 1–12 place the angle with the indicated measure in standard position on a rectangular coordinate system, and find the measure of two positive angles and two negative angles that are coterminal with the given angle.

1 $120°$ **2** $240°$ **3** $135°$

4 $315°$ **5** $-30°$ **6** $-150°$

7 $620°$ **8** $570°$ **9** $5\pi/6$

10 $2\pi/3$ **11** $-\pi/4$ **12** $-5\pi/4$

In Exercises 13–24 find the radian measure that corresponds to the degree measure.

13 $150°$ **14** $120°$ **15** $-60°$

16 $-135°$ **17** $225°$ **18** $210°$

19 $450°$ **20** $630°$ **21** $72°$

22 $54°$ **23** $100°$ **24** $95°$

In Exercises 25–36 find the degree measure that corresponds to the radian measure.

25 $2\pi/3$ **26** $5\pi/6$ **27** $11\pi/6$

28 $4\pi/3$ **29** $3\pi/4$ **30** $11\pi/4$

31 $-7\pi/2$ **32** $-5\pi/2$ **33** 7π

34 9π **35** $\pi/9$ **36** $\pi/16$

In Exercises 37–40 find the approximate measure of θ in terms of degrees, minutes, and seconds.

37 $\theta = 2$ **38** $\theta = 1.5$

39 $\theta = 5$ **40** $\theta = 4$

41 A central angle θ is subtended by an arc 7 cm long on a circle of radius 4 cm. Approximate the measure of θ in (a) radians; (b) degrees.

42 A central angle θ is subtended by an arc 3 feet long on a circle of radius 20 inches. Approximate the measure of θ in (a) radians; (b) degrees.

43 Approximate the length of an arc that subtends a central angle of measure $50°$ on a circle of diameter 16 meters.

44 Approximate the length of an arc that subtends a central angle of 2.2 radians on a circle of diameter 120 cm.

45 If a central angle θ of measure $20°$ is subtended by a circular arc of length 3 km, find the radius of the circle.

46 If a central angle θ of radian measure 4 is subtended by a circular arc 10 cm long, find the radius of the circle.

47 The distance between two points A and B on the earth is measured along a circle having center C at the center of the earth and radius equal to the distance from C to the surface. If the diameter of the earth is approximately 8,000 miles, approximate the distance between A and B if angle ACB has measure (a) 60°, (b) 45°, (c) 30°, (d) 10°, and (e) 1°.

48 Refer to Exercise 47. If angle ACB has measure 1′, then the distance between A and B is called a **nautical mile.** Approximate the number of ordinary (**statute**) miles in a nautical mile.

49 Refer to Exercise 47. If two points A and B are 500 miles apart, find angle ACB in both degree measure and radian measure.

50 The **angular speed** of a wheel that is rotating at a constant rate is the angle generated by a line segment from the center of the wheel to the circumference, per unit of time. If a wheel of diameter 3 feet is rotating at a rate of 2,400 rpm (revolutions per minute), find the angular speed.

CALCULATOR EXERCISES 5.1

In Exercises 1–6 approximate, to four decimal places, the radian measure that corresponds to the given degree measure of θ, or the degree measure that corresponds to the given radian measure.

1 $\theta = 73°24'$

2 $\theta = 261°37'$

3 $\theta = 482°16'43''$

4 $\theta = \sqrt{317} + \pi$

5 $\theta = \ln(13.6521)$

6 $\theta = e^{\sqrt{\pi}}$

7–9 The diameter of the earth at the equator is estimated as 7,926.41 miles. Rework Exercises 47–49 of this section using that measurement, assuming that points A and B are on the equator.

10 Refer to Exercise 50 of this section. If a wheel of diameter 1.683 meters is rotating at a rate of 2,875 rpm, find the angular speed.

SECTION 5.2
TRIGONOMETRIC FUNCTIONS OF ANGLES

We shall introduce the trigonometric functions in the manner that they arose historically, as ratios of sides of a right triangle. This will lead, in a natural way, to the definition in terms of arbitrary angles. We will discuss trigonometric functions of real numbers in Section 5.3.

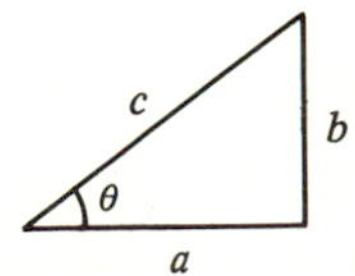

FIGURE 5.8

If θ is any acute angle, we may consider a right triangle having θ as one of its angles, as illustrated in Figure 5.8. The following six ratios can be obtained using the lengths a, b, c of the three sides of the triangle:

$$\frac{b}{c}, \quad \frac{a}{c}, \quad \frac{b}{a}, \quad \frac{c}{b}, \quad \frac{c}{a}, \quad \frac{a}{b}.$$

Moreover, these ratios depend only on θ, and not on the size of the triangle. Suppose that θ is also an angle of a different right triangle, as shown in

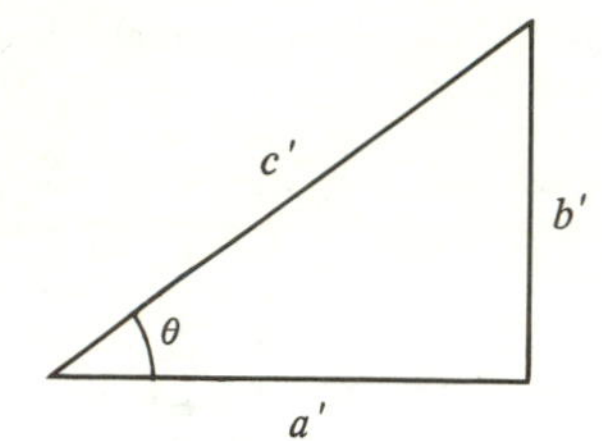

FIGURE 5.9

Figure 5.9. Since the two triangles have equal angles, they are similar, and therefore ratios of corresponding sides are proportional. For example,

$$\frac{b}{c} = \frac{b'}{c'}, \quad \frac{a}{c} = \frac{a'}{c'}, \quad \text{and} \quad \frac{b}{a} = \frac{b'}{a'}.$$

Thus, for each θ, the six ratios are uniquely determined and, hence, are functions of θ. They are called **trigonometric functions.** It is customary to designate them as the **sine, cosine, tangent, cosecant, secant,** and **cotangent** functions, abbreviated **sin, cos, tan, csc, sec,** and **cot,** respectively. The symbol sin (θ), or sin θ, is used for the ratio b/c, which the sine function associates with θ. Values of the other five functions are denoted in like manner. To summarize, if θ is an acute angle of a right triangle, as in Figure 5.8, then by definition

$$\sin\theta = \frac{b}{c} \qquad \csc\theta = \frac{c}{b}$$

$$\cos\theta = \frac{a}{c} \qquad \sec\theta = \frac{c}{a}$$

$$\tan\theta = \frac{b}{a} \qquad \cot\theta = \frac{a}{b}$$

Each of these six trigonometric functions has for its domain the set of all acute angles. Later in this section we will extend the domains to larger sets of angles.

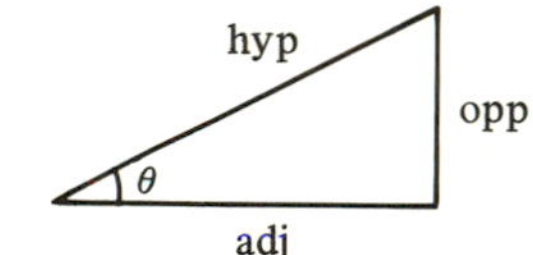

FIGURE 5.10

If θ is the angle in Figure 5.8, we refer to the sides of the triangle of lengths a, b, and c as the **adjacent side, opposite side,** and **hypotenuse,** respectively. For convenience we shall use **adj, opp,** and **hyp** to denote the lengths of these sides. We may then represent the triangle as in Figure 5.10. Using this notation, the trigonometric functions may be expressed as follows.

TRIGONOMETRIC FUNCTIONS OF ACUTE ANGLES

$$\sin\theta = \frac{\text{opp}}{\text{hyp}} \qquad \csc\theta = \frac{\text{hyp}}{\text{opp}}$$

$$\cos\theta = \frac{\text{adj}}{\text{hyp}} \qquad \sec\theta = \frac{\text{hyp}}{\text{adj}}$$

$$\tan\theta = \frac{\text{opp}}{\text{adj}} \qquad \cot\theta = \frac{\text{adj}}{\text{opp}}$$

It is better to memorize these formulas than those in the preceding definition, since they can be applied to any right triangle without attaching the labels a, b, c to the sides.

EXAMPLE 1 Find the values of the trigonometric functions that correspond to each θ:

(a) $\theta = 60°$ (b) $\theta = 30°$ (c) $\theta = 45°$

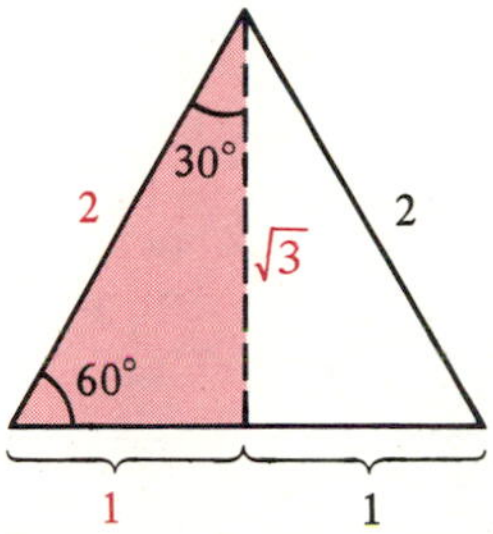

FIGURE 5.11

Solution Let us consider an equilateral triangle having sides of length 2. The median from one vertex to the opposite side bisects the angle at that vertex, as illustrated by the dashes in Figure 5.11. By the Pythagorean Theorem the side opposite 60° in the colored right triangle is $\sqrt{3}$. Using the preceding formulas, we obtain the values for (a) and (b) as follows:

(a)
$$\sin 60° = \frac{\sqrt{3}}{2} \qquad \csc 60° = \frac{2}{\sqrt{3}} = \frac{2\sqrt{3}}{3}$$
$$\cos 60° = \frac{1}{2} \qquad \sec 60° = \frac{2}{1} = 2$$
$$\tan 60° = \frac{\sqrt{3}}{1} = \sqrt{3} \qquad \cot 60° = \frac{1}{\sqrt{3}} = \frac{\sqrt{3}}{3}$$

(b)
$$\sin 30° = \frac{1}{2} \qquad \csc 30° = \frac{2}{1} = 2$$
$$\cos 30° = \frac{\sqrt{3}}{2} \qquad \sec 30° = \frac{2}{\sqrt{3}} = \frac{2\sqrt{3}}{3}$$
$$\tan 30° = \frac{1}{\sqrt{3}} = \frac{\sqrt{3}}{3} \qquad \cot 30° = \frac{\sqrt{3}}{1} = \sqrt{3}$$

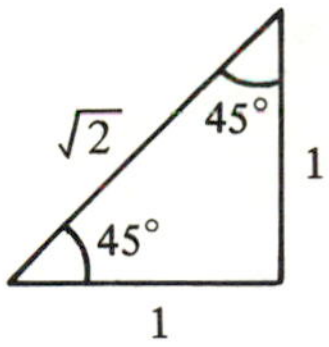

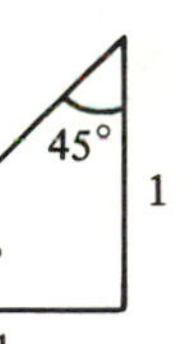

FIGURE 5.12

(c) To find the functional values for $\theta = 45°$, let us consider an isosceles right triangle whose two equal sides have length 1, as illustrated of Figure 5.12. By the Pythagorean Theorem, the length of the hypotenuse is $\sqrt{2}$. Hence,

$$\sin 45° = \frac{1}{\sqrt{2}} = \frac{\sqrt{2}}{2} = \cos 45° \qquad \tan 45° = \frac{1}{1} = 1$$
$$\csc 45° = \frac{\sqrt{2}}{1} = \sqrt{2} = \sec 45° \qquad \cot 45° = \frac{1}{1} = 1$$

■

For reference, the values found in Example 1 are listed in the following table, together with the radian measures of the angles. Two reasons for stressing these values are that (1) they are exact, and (2) they occur frequently in work involving trigonometry. Because of the importance of these special values, it is a good idea either to memorize the table or to be able to find the values quickly by using triangles as in Example 1.

SPECIAL VALUES OF THE TRIGONOMETRIC FUNCTIONS

θ (radians)	θ (degrees)	$\sin\theta$	$\cos\theta$	$\tan\theta$	$\cot\theta$	$\csc\theta$	$\sec\theta$
$\frac{\pi}{6}$	30°	$\frac{1}{2}$	$\frac{\sqrt{3}}{2}$	$\frac{\sqrt{3}}{3}$	$\sqrt{3}$	2	$\frac{2\sqrt{3}}{3}$
$\frac{\pi}{4}$	45°	$\frac{\sqrt{2}}{2}$	$\frac{\sqrt{2}}{2}$	1	1	$\sqrt{2}$	$\sqrt{2}$
$\frac{\pi}{3}$	60°	$\frac{\sqrt{3}}{2}$	$\frac{1}{2}$	$\sqrt{3}$	$\frac{\sqrt{3}}{3}$	$\frac{2\sqrt{3}}{3}$	2

Functional values corresponding to *any* angle will be considered in Section 5.4.

The next example illustrates a practical use for trigonometric functions of acute angles. Additional applications involving right triangles will be considered in Section 5.8.

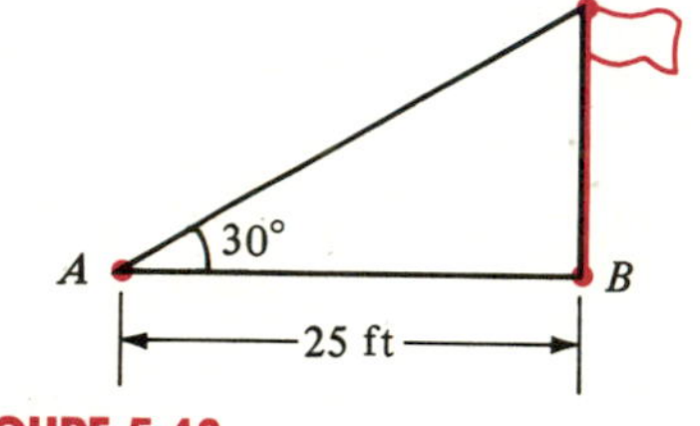

FIGURE 5.13

EXAMPLE 2 An engineer notes that at a point A, located on level ground a distance 25.0 feet from the base B of a flagpole, the angle between the ground and the top of the pole is 30° (see Figure 5.13). Approximate the height of the pole to the nearest tenth of a foot.

Solution If the height of the pole is h, then referring to Figure 5.13, we see that

$$\tan 30^\circ = \frac{h}{25} \quad \text{or} \quad h = 25 \tan 30^\circ.$$

Using the value of tan 30° from Example 1,

$$h = 25\left(\frac{\sqrt{3}}{3}\right) \approx 25(0.577) \approx 14.4 \text{ feet.}$$

■

Before extending the definition to arbitrary angles we shall discuss several important relationships that exist among the trigonometric functions. The formulas listed in the next box are, without doubt, the most important identities in trigonometry, because they may be used to simplify and to unify many different aspects of the subject. Since the formulas are true for every acute angle θ and are part of the foundation for work in trigonometry, they are called the *Fundamental Identities.*

Three of the Fundamental Identities involve squares such as $(\sin\theta)^2$ and $(\cos\theta)^2$. In general, if n is an integer and $n \neq -1$, then powers such as $(\cos\theta)^n$ are written in the form $\cos^n\theta$. The symbols $\sin^{-1}$ and $\cos^{-1}$ are reserved for inverse trigonometric functions, which will be discussed in the next chapter.

With this agreement on notation we have, for example,

$$\cos^2 \theta = (\cos \theta)^2 = (\cos \theta)(\cos \theta)$$

$$\tan^3 \theta = (\tan \theta)^3 = (\tan \theta)(\tan \theta)(\tan \theta)$$

$$\sec^4 \theta = (\sec \theta)^4 = (\sec \theta)(\sec \theta)(\sec \theta)(\sec \theta).$$

Let us first list all the fundamental identities and then discuss the proofs. The following formulas are true for every acute angle θ.

THE FUNDAMENTAL IDENTITIES

$$\csc \theta = \frac{1}{\sin \theta} \qquad \tan \theta = \frac{\sin \theta}{\cos \theta} \qquad \sin^2 \theta + \cos^2 \theta = 1$$

$$\sec \theta = \frac{1}{\cos \theta} \qquad \cot \theta = \frac{\cos \theta}{\sin \theta} \qquad 1 + \tan^2 \theta = \sec^2 \theta$$

$$\cot \theta = \frac{1}{\tan \theta} \qquad 1 + \cot^2 \theta = \csc^2 \theta$$

The proofs of the Fundamental Identities follow directly from the definition of the trigonometric functions of θ given at the beginning of this section. For example,

$$\sec \theta = \frac{c}{a} = \frac{1}{a/c} = \frac{1}{\cos \theta}$$

$$\cot \theta = \frac{a}{b} = \frac{1}{b/a} = \frac{1}{\tan \theta}$$

$$\tan \theta = \frac{b}{a} = \frac{b/c}{a/c} = \frac{\sin \theta}{\cos \theta}$$

The formulas $\sec \theta = 1/\cos \theta$ and $\cot \theta = \cos \theta/\sin \theta$ are proved in like manner.

If we apply the Pythagorean Theorem to the triangle in Figure 5.8, we obtain

$$b^2 + a^2 = c^2.$$

Dividing both sides by c^2 leads to

$$\left(\frac{b}{c}\right)^2 + \left(\frac{a}{c}\right)^2 = 1, \quad \text{that is,} \quad \sin^2 \theta + \cos^2 \theta = 1.$$

Dividing both sides of the last equation by $\cos^2\theta$ gives us

$$\frac{\sin^2\theta}{\cos^2\theta}+\frac{\cos^2\theta}{\cos^2\theta}=\frac{1}{\cos^2\theta}\quad\text{or}\quad\left(\frac{\sin\theta}{\cos\theta}\right)^2+1=\left(\frac{1}{\cos\theta}\right)^2$$

Since $\sin\theta/\cos\theta=\tan\theta$ and $1/\cos\theta=\sec\theta$, this implies that

$$\tan^2\theta+1=\sec^2\theta.$$

The proof that $1+\cot^2\theta=\csc^2\theta$ is left as an exercise.

EXAMPLE 3 If θ is an acute angle and $\sin\theta=\frac{2}{3}$, use the Fundamental Identities to find the values of $\cos\theta$ and $\tan\theta$.

Solution Since $\sin^2\theta+\cos^2\theta=1$, we may write $\cos^2\theta=1-\sin^2\theta$, and hence

$$\cos\theta=\sqrt{1-\sin^2\theta}=\sqrt{1-(\tfrac{2}{3})^2}=\sqrt{\frac{5}{9}}=\frac{\sqrt{5}}{3}.$$

Next, $$\tan\theta=\frac{\sin\theta}{\cos\theta}=\frac{2/3}{\sqrt{5}/3}=\frac{2}{\sqrt{5}}=\frac{2\sqrt{5}}{5}.$$ ■

EXAMPLE 4 Express $\tan\theta$ in terms of $\cos\theta$, where θ is any acute angle.

Solution We begin with $\tan\theta=\sin\theta/\cos\theta$. Since $\sin^2\theta=1-\cos^2\theta$ and θ is acute, we may write

$$\sin\theta=\sqrt{1-\cos^2\theta}.$$

Hence, $$\tan\theta=\frac{\sin\theta}{\cos\theta}=\frac{\sqrt{1-\cos^2\theta}}{\cos\theta}.$$ ■

Many applied problems require the use of angles that are not acute. To employ trigonometry as a tool for solving such problems, it is necessary to extend the definition of the trigonometric functions. This extension may be made by using the standard position of an angle θ on a rectangular coordinate system. If θ is acute, then we have a situation similar to that illustrated in Figure 5.14, where we have chosen a point $P(a, b)$ on the terminal side of θ and where $c=\sqrt{a^2+b^2}$. Evidently,

$$\sin\theta=\frac{b}{c},\quad\cos\theta=\frac{a}{c},\quad\tan\theta=\frac{b}{a}.$$

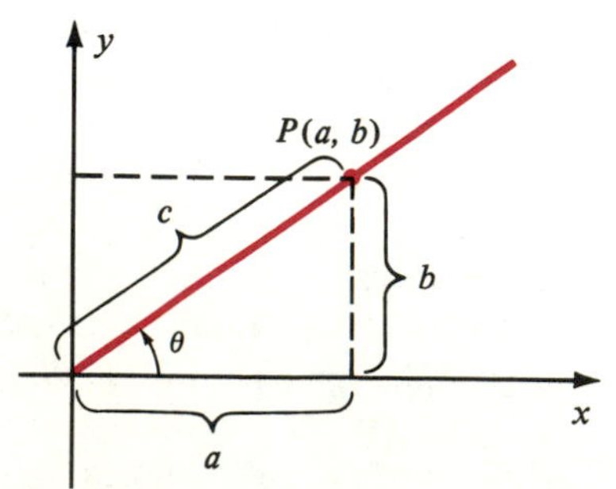

FIGURE 5.14

We now wish to consider angles of the type illustrated in Figure 5.15 (or any *other* angle, either positive, negative, or zero).

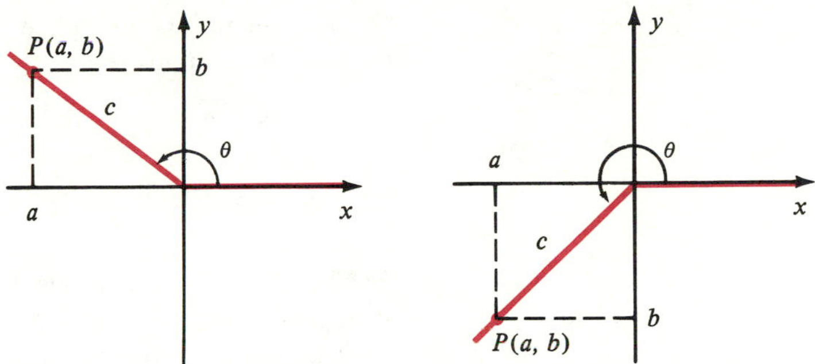

FIGURE 5.15

Our objective is to define the trigonometric functions so that their values agree with those given previously whenever the angle is acute. The next definition accomplishes this in a simple manner.

TRIGONOMETRIC FUNCTIONS OF ANY ANGLE

Let θ be an angle in standard position on a rectangular coordinate system and let $P(a, b)$ be any point other than O on the terminal side of θ. If $d(O, P) = c$, then

$$\sin\theta = \frac{b}{c} \qquad \csc\theta = \frac{c}{b} \quad (\text{if } b \neq 0)$$

$$\cos\theta = \frac{a}{c} \qquad \sec\theta = \frac{c}{a} \quad (\text{if } a \neq 0)$$

$$\tan\theta = \frac{b}{a} \quad (\text{if } a \neq 0) \qquad \cot\theta = \frac{a}{b} \quad (\text{if } b \neq 0)$$

It can be shown, by using similar triangles, that the formulas in this definition are independent of the point $P(a, b)$ that is chosen on the terminal side of θ. Moreover, the numbers a and b may be positive, negative, or zero, depending on the quadrant containing θ, or on whether the terminal side is on a coordinate axis.

The domains of the sine and cosine functions consist of all angles θ. Note, however, that $\tan\theta$ and $\sec\theta$ are undefined if $a = 0$, that is, if the terminal side of θ is on the y-axis. Thus the domain of both the tangent and secant functions consists of all angles *except* those having radian measure $(\pi/2) + n\pi$ where n is any integer. Some special cases are $\pm\pi/2$, $\pm 3\pi/2$, and $\pm 5\pi/2$. In like manner, the domain of both the cotangent and cosecant functions consists of all angles except those whose terminal sides are on the x-axis; that is, angles of radian measure $n\pi$ where n is any integer. (Why?)

For all points $P(a, b)$ in the preceding definition we have $|a| \leq c$ and $|b| \leq c$. This implies that

$$|\sin \theta| \leq 1, \qquad |\cos \theta| \leq 1, \qquad |\csc \theta| \geq 1, \qquad |\sec \theta| \geq 1$$

for every θ in the domains of these functions.

EXAMPLE 5 If θ is an angle in standard position on a rectangular coordinate system and if the point $P(-15, 8)$ is on the terminal side of θ, find the values of the trigonometric functions of θ.

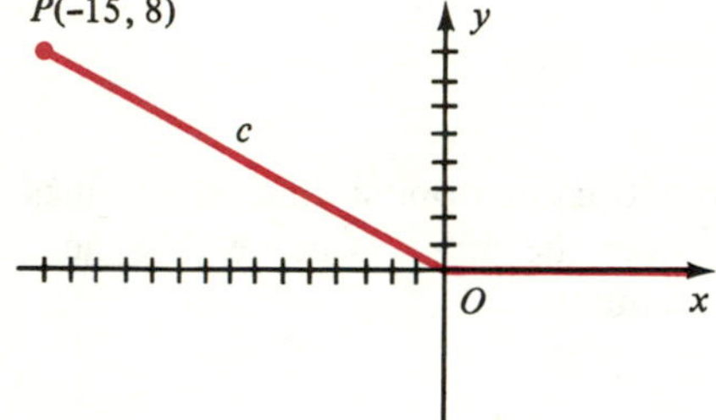

FIGURE 5.16

Solution The point $P(-15, 8)$ is plotted in Figure 15.16. Note that the terminal side of θ is in quadrant II. By the Distance Formula, the distance c from the origin O to any point $P(a, b)$ is

$$c = d(O, P) = \sqrt{(a - 0)^2 + (b - 0)^2} = \sqrt{a^2 + b^2}.$$

Hence, for $P(-15, 8)$, we have

$$c = \sqrt{(-15)^2 + 8^2} = \sqrt{225 + 64} = \sqrt{289} = 17.$$

Applying the definition with $a = -15$, $b = 8$, and $c = 17$,

$$\sin \theta = \frac{8}{17} \qquad \csc \theta = \frac{17}{8}$$

$$\cos \theta = -\frac{15}{17} \qquad \sec \theta = -\frac{17}{15}$$

$$\tan \theta = -\frac{8}{15} \qquad \cot \theta = -\frac{15}{8}$$

■

EXAMPLE 6 An angle θ is in standard position on a rectangular coordinate system. If its terminal side is in quadrant III and lies on the line $y = 3x$, find the values of the trigonometric functions of θ.

Solution The graph of $y = 3x$ is sketched in Figure 5.17, together with the (colored) initial and terminal sides of θ.

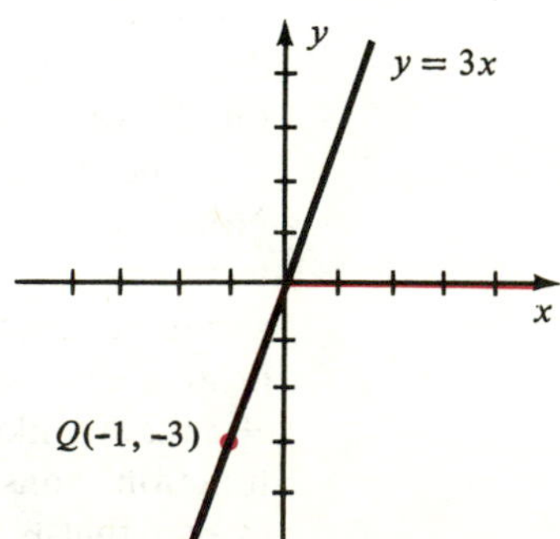

FIGURE 5.17

We begin by choosing a convenient point, say $Q(-1, -3)$ on the terminal side. The distance c from the origin to Q is

$$c = d(O, Q) = \sqrt{(-1)^2 + (-3)^2} = \sqrt{10}.$$

Applying the definition with $a = -1$, $b = -3$, and $c = \sqrt{10}$ gives us

$$\sin\theta = \frac{-3}{\sqrt{10}} = \frac{-3\sqrt{10}}{10} \qquad \csc\theta = \frac{\sqrt{10}}{-3} = -\frac{\sqrt{10}}{3}$$

$$\cos\theta = \frac{-1}{\sqrt{10}} = \frac{-\sqrt{10}}{10} \qquad \sec\theta = \frac{\sqrt{10}}{-1} = -\sqrt{10}$$

$$\tan\theta = \frac{-3}{-1} = 3 \qquad \cot\theta = \frac{-1}{-3} = \frac{1}{3}$$

■

The definition of the trigonometric functions of any angle may be applied if the terminal side of θ is on a coordinate axis, as illustrated in the next example.

EXAMPLE 7 Find the values of the trigonometric functions of θ if $\theta = 3\pi/2$.

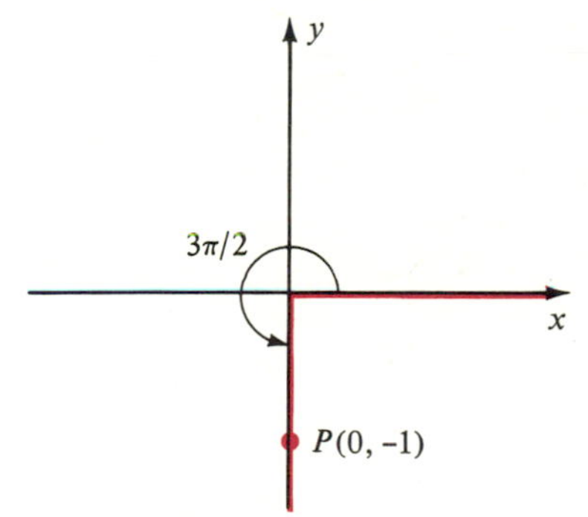

FIGURE 5.18

Solution Note that $3\pi/2 = 270°$. (Why?) Placing θ in standard position, the terminal side of θ coincides with the negative y-axis, as shown in Figure 5.18. To use the definition we may choose any point P on the terminal side of θ. For simplicity we consider $P(0, -1)$. In this case $c = 1$, $a = 0$, $b = -1$, and hence

$$\sin\frac{3\pi}{2} = \frac{-1}{1} = -1 \qquad \csc\frac{3\pi}{2} = \frac{1}{-1} = -1$$

$$\cos\frac{3\pi}{2} = \frac{0}{1} = 0 \qquad \cot\frac{3\pi}{2} = \frac{0}{-1} = 0$$

The tangent and secant functions are undefined, since the meaningless expressions $\tan\theta = (-1)/0$ and $\sec\theta = 1/0$ arise when we substitute in the appropriate formulas. ■

Let us determine the signs associated with the trigonometric functions. If θ is in quadrant II and $P(a, b)$ is a point on the terminal side, then a is negative, b is positive and hence, by definition, $\sin\theta$ and $\csc\theta$ are positive, whereas the other four functions are negative. The reader should check the remaining quadrants. The following table indicates the signs in all four quadrants.

SIGNS OF THE TRIGONOMETRIC FUNCTIONS

Quadrant containing θ	Positive functions	Negative functions
I	All	None
II	sin, csc	cos, sec, tan, cot
III	tan, cot	sin, csc, cos, sec
IV	cos, sec	sin, csc, tan, cot

EXAMPLE 8 Find the quadrant containing θ if $\sin\theta < 0$ and $\cos\theta > 0$.

Solution Referring to the above table we see that $\sin\theta < 0$ if θ is in quadrants III or IV, and $\cos\theta > 0$ if θ is in quadrants I or IV. Hence, for both conditions to be satisfied, θ must be in quadrant IV. ■

The Fundamental Identities (see page 201) that were established for acute angles are also true for trigonometric functions of any angle. Indeed, the proofs are identical to those given earlier.

EXERCISES 5.2

In Exercises 1–10 find the values of the trigonometric functions of θ if θ is the angle of the pictured right triangle.

1

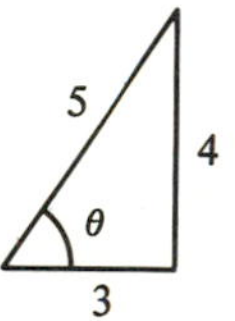

2

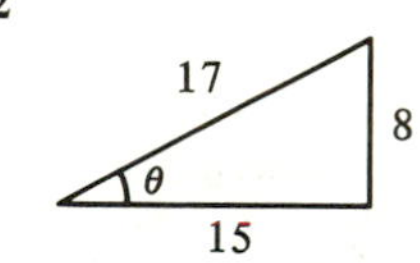

3

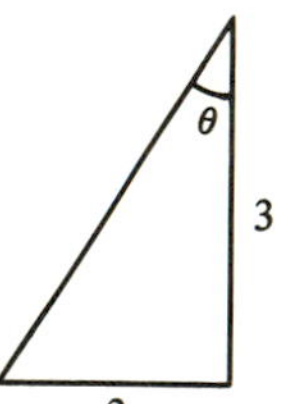

4

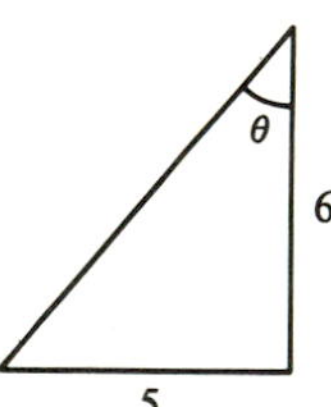

5

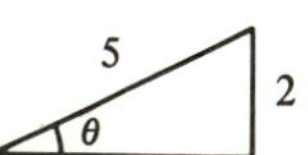

6

3 θ 1

7

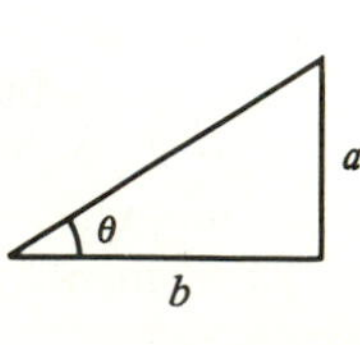

8

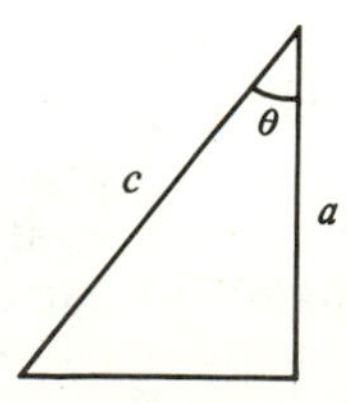

9

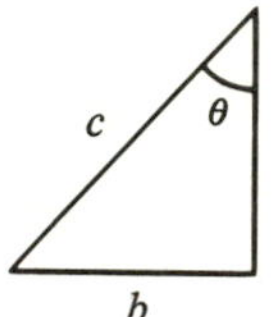

10

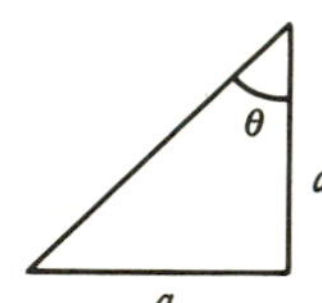

In Exercises 11–18 find the values of the other trigonometric functions for the acute angle θ.

11 $\sin\theta = \frac{3}{5}$ **12** $\cos\theta = \frac{8}{17}$

13 $\sec\theta = \frac{6}{5}$ **14** $\csc\theta = 4$

15 $\cos\theta = \sqrt{2}/2$ **16** $\sin\theta = \sqrt{3}/2$

17 $\sin\theta = b/\sqrt{a^2 + b^2}$, where $a^2 + b^2 \neq 0$

18 $\cos\theta = x$, where $0 < x < 1$

In Exercises 19–24 write the first expression in terms of the second.

19 $\cot\theta,\ \sin\theta$ **20** $\tan\theta,\ \sin\theta$

21 $\sec\theta,\ \sin\theta$ **22** $\csc\theta,\ \cos\theta$

23 $\tan\theta,\ \sec\theta$ **24** $\cot\theta,\ \csc\theta$

Prove the identities in Exercises 25 and 26.

25 $1 + \cot^2\theta = \csc^2\theta$ **26** $\cot\theta = \dfrac{\cos\theta}{\sin\theta}$

In Exercises 27–44 find the values of the six trigonometric functions of θ if θ is in standard position and satisfies the given condition.

27 The point $P(4, -3)$ is on the terminal side of θ.

28 The point $P(-8, -15)$ is on the terminal side of θ.

29 The point $P(-2, -5)$ is on the terminal side of θ.

30 The point $P(-1, 2)$ is on the terminal side of θ.

31 The terminal side of θ is in quadrant II and lies on the line $y = -4x$.

32 The terminal side of θ is in quadrant IV and lies on the line $3y + 5x = 0$.

33 The terminal side of θ is in quadrant III and is perpendicular to the line $7y + 2x + 2 = 0$.

34 The terminal side of θ is in quadrant II and is parallel to the line through points $A(1, 4)$ and $B(3, -2)$.

35 The terminal side of θ is in quadrant I and lies on a line having slope $\frac{4}{3}$.

36 The terminal side of θ bisects the third quadrant.

37 $\theta = 90°$ **38** $\theta = 0°$

39 $\theta = 180°$ **40** $\theta = -270°$

41 $\theta = 2\pi$ **42** $\theta = 5\pi/2$

43 $\theta = 7\pi/2$ **44** $\theta = 3\pi$

In Exercises 45–54 find the quadrant containing θ if the given conditions are true.

45 $\cos\theta > 0$ and $\sin\theta < 0$

46 $\tan\theta < 0$ and $\cos\theta > 0$

47 $\sin\theta < 0$ and $\cot\theta > 0$

48 $\sec\theta > 0$ and $\tan\theta < 0$

49 $\csc\theta > 0$ and $\sec\theta < 0$

50 $\csc\theta > 0$ and $\cot\theta < 0$

51 $\sec\theta < 0$ and $\tan\theta > 0$

52 $\sin\theta < 0$ and $\sec\theta > 0$

53 $\cos\theta > 0$ and $\tan\theta > 0$

54 $\cos\theta < 0$ and $\csc\theta < 0$

55 Is there an angle θ such that $7\sin\theta = 9$? Explain.

56 Is there an angle θ such that $3\csc\theta = 1$? Explain.

SECTION 5.3
THE CIRCULAR FUNCTIONS

In Section 2.3 a function was defined as a correspondence that assigns to each element of a set X (the domain) a unique element of a set Y (the range). For each trigonometric function discussed in the preceding section, the domain is a set of angles and the range is a set of real numbers. For example, given the angle 45°, the tangent function assigns the real number tan 45°, or 1. In calculus and in many applications, the domains of functions consist of real numbers. It is therefore necessary to take a different point of view, and to regard the domains of trigonometric functions as subsets of $\mathbb{R}$. Perhaps the simplest way to accomplish this is to use the following definition.

TRIGONOMETRIC FUNCTIONS OF REAL NUMBERS

If t is a real number, then the value of any trigonometric function at t is its value at an angle of t radians.

As an illustration, if we regard $\pi/3$ as a real number (not necessarily the radian measure of an angle), then using the preceding definition and the table on page 200,

$$\sin\frac{\pi}{3}=\frac{\sqrt{3}}{2},\quad \cos\frac{\pi}{3}=\frac{1}{2},\quad \text{and}\quad \tan\frac{\pi}{3}=\sqrt{3}.$$

In Section 5.4 we shall consider values of trigonometric functions that correspond to *any* real number.

According to our definition of trigonometric functions of real numbers, a notation such as sin 2 may be interpreted as *either* the sine of the real number 2 *or* the sine of an angle of 2 radians. As in Section 5.2, if degree measure is used we shall write sin 2°. Of course, $\sin 2 \neq \sin 2°$.

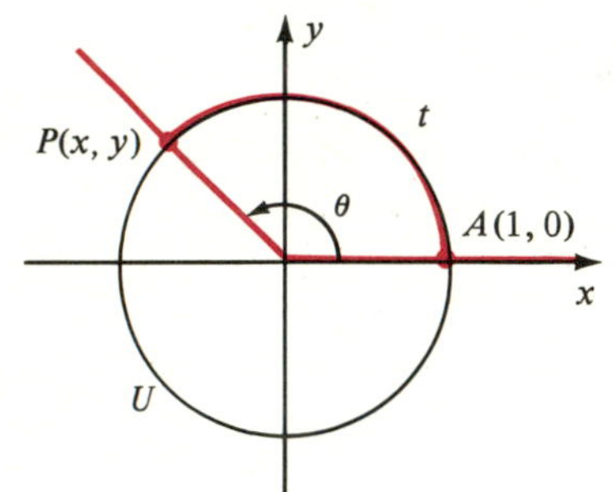

FIGURE 5.19

The concept of trigonometric functions as functions of real numbers may be interpreted geometrically by referring to a unit circle U with center at the origin of a coordinate plane. Thus, let t be any real number and let θ be the angle of radian measure t. Place θ in standard position and let $P(x, y)$ be the point at which the terminal side of θ intersects U. If $0 \leq \theta \leq 2\pi$ we have a situation similar to that shown in Figure 5.19, which pictures θ as a second-quadrant angle. As in Section 5.1, the number t may be regarded as either the radian measure of θ, or as the length of the arc $\widehat{AP}$ on U.

If $t > 2\pi$, then the x-axis must make more than one revolution to generate θ. In this event t is the distance traveled by the point of intersection of the x-axis with U before arriving at its final position $P(x, y)$. If $t < 0$, then $-t$ equals that distance.

The preceding discussion indicates how we may associate with each real number t, a unique point $P(x, y)$ on U. We shall call $P(x, y)$ **the point on the unit circle U that corresponds to t.** Note that by the definition of trigonometric functions of real numbers, together with that for any angle (see page 203), we may write

$$\sin t = \sin\theta = \frac{y}{1} = y.$$

The same procedure for the remaining five functions gives us the following formulas.

TRIGONOMETRIC FUNCTIONS IN TERMS OF A UNIT CIRCLE

If t is a real number and $P(x, y)$ is the point on the unit circle U that corresponds to t, then

$$\sin t = y \qquad \csc t = \frac{1}{y}\quad (\text{if } y \neq 0)$$

$$\cos t = x \qquad \sec t = \frac{1}{x}\quad (\text{if } x \neq 0)$$

$$\tan t = \frac{y}{x}\quad (\text{if } x \neq 0) \qquad \cot t = \frac{x}{y}\quad (\text{if } y \neq 0).$$

Since these formulas express values in terms of the coordinates of a point $P(x, y)$ on a unit circle, the trigonometric functions are sometimes referred to as the **circular functions.**

EXAMPLE 1 Use the unit circle U to find the values of the sine, cosine, and tangent functions at t.

(a) $t = 0$ (b) $t = \pi/4$ (c) $t = \pi/2$

Solution

(a) The point on U that corresponds to $t = 0$ is $P(1, 0)$. Thus, we let $x = 1$ and $y = 0$ in the unit circle formulas, obtaining

$$\sin 0 = 0, \quad \cos 0 = 1 \quad \text{and} \quad \tan 0 = \frac{0}{1} = 0.$$

(b) If $t = \pi/4$, then $P(x, y)$ lies on the line $y = x$ that bisects the first quadrant. Since an equation for U is $x^2 + y^2 = 1$, and since $P(x, x)$ is on U, we have

$$x^2 + x^2 = 1, \quad \text{or} \quad 2x^2 = 1.$$

Solving for x gives us

$$x = 1/\sqrt{2} = \sqrt{2}/2.$$

Thus P is the point $(\sqrt{2}/2, \sqrt{2}/2)$, and we obtain

$$\sin\frac{\pi}{4} = \frac{\sqrt{2}}{2}, \quad \cos\frac{\pi}{4} = \frac{\sqrt{2}}{2}, \quad \text{and} \quad \tan\frac{\pi}{4} = \frac{\sqrt{2}/2}{\sqrt{2}/2} = 1.$$

(c) The point on U corresponding to $t = \pi/2$ is $P(0, 1)$. Letting $x = 0$ and $y = 1$ in the unit circle formulas, we see that $\sin(\pi/2) = 1$ and $\cos(\pi/2) = 0$. Since $y/x = 1/0$, it follows that $\tan(\pi/2)$ is undefined. ■

The domains of trigonometric functions of angles were discussed in Section 5.2. If we wish to express the domains in terms of real numbers, we could refer to the radian measures of angles. An alternative method is to use the unit circle approach discussed in this section. Thus, if $P(x, y)$ is the point on U corresponding to the real number t, then since $\sin t = x$ and $\cos t = y$, the sine and cosine functions always exist and, therefore, each has domain $\mathbb{R}$. In the definitions of the tangent and secant functions, x appears in the denominator and hence we must exclude values of t for which x is 0; that is, the values of t corresponding to the points $(0, 1)$ and $(0, -1)$ on the unit circle U. It follows that the domain of the tangent and secant functions consists of all numbers *except* those of the form $(\pi/2) + n\pi$ where n is an integer. In particular, we exclude $\pm\pi/2$, $\pm 3\pi/2$, $\pm 5\pi/2$, etc.

Since $\cot t = x/y$ and $\csc t = 1/y$, the domain of the cotangent and cosecant functions is the set of all real numbers except those numbers t for which the y-coordinate of P is 0. These include the numbers, 0, $\pm\pi$, $\pm 2\pi$, $\pm 3\pi$, and, in general, all numbers of the form $n\pi$ where n is an integer. In the future, when we work with tan t, cot t, sec t, and csc t, we will always assume that t is in the appropriate domain, even though this fact will not always be mentioned explicitly.

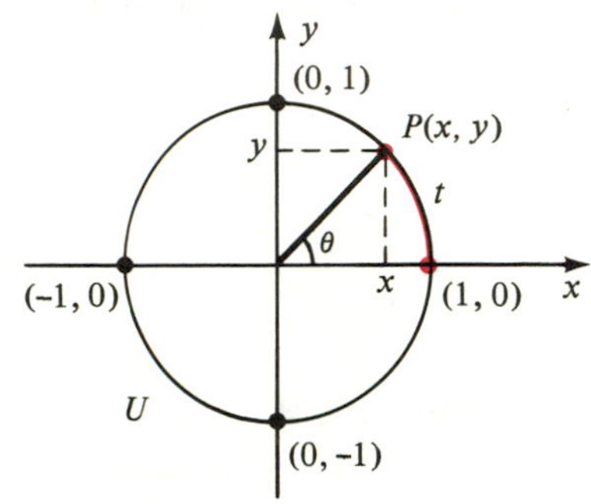

FIGURE 5.20

The unit circle approach can be used to make some general observations about the variation of the trigonometric functions. For now, let us concentrate on the sine function. Referring to the unit circle U in Figure 5.20, we see that if t varies from 0 to $\pi/2$, then the coordinates of $P(x, y)$ vary from $(1, 0)$ to $(0, 1)$. In particular, the y-coordinate—that is, the value of the sine function—increases from 0 to 1. Moreover, this function takes on *all* values between 0 and 1. If we let t vary from $\pi/2$ to π, then the coordinates of $P(x, y)$ vary from $(0, 1)$ to $(-1, 0)$, and hence the sine function (that is, the y-coordinate of P) decreases from 1 to 0. In similar fashion, we see that if t varies from π to $3\pi/2$, then sin t decreases from 0 to -1; and as t varies from $3\pi/2$ to 2π, sin t increases from -1 to 0. The following table partially indicates this variation of sin t in the interval $[0, 2\pi]$. For a more complete description we would have to insert many more values of t and sin t.

t	0	$\frac{\pi}{2}$	π	$\frac{3\pi}{2}$	2π
$\sin t$	0	1	0	-1	0

If we let t vary through the interval $[2\pi, 4\pi]$, then $P(x, y)$ traces the circle again and the identical pattern for sin t is repeated, that is,

$$\sin (t + 2\pi) = \sin t$$

for every t in $[0, 2\pi]$. The same is true for other intervals of length 2π. Indeed, for every integer n,

$$\sin (t + 2\pi n) = \sin t.$$

According to the next definition, this repetitive variation is specified by saying that the sine function is *periodic*.

DEFINITION

A function f is **periodic** if there exists a positive real number k such that

$$f(t + k) = f(t)$$

for every t in the domain of f. The least such positive real number k, if it exists, is called the **period** of f.

The period of the sine function is 2π (see Exercise 15).

The variation of the cosine function in the interval $[0, 2\pi]$ can be determined by observing the variation of the x-coordinate of the point P on U as t varies from 0 to 2π. Again referring to Figure 5.20, we see that the cosine function (that is, x) decreases from 1 to 0 in the interval $[0, \pi/2]$, decreases from 0 to -1 in $[\pi/2, \pi]$, increases from -1 to 0 in $[\pi, 3\pi/2]$, and increases from 0 to 1 in $[3\pi/2, 2\pi]$. The pattern is then repeated in successive intervals of length 2π. It follows that the cosine function is periodic with period 2π.

The preceding discussion suggests that the graphs of the sine and cosine functions may have the appearance of those in Figures 5.21 and 5.22, respectively. In the figures, we have plotted several points of the form $(t, \sin t)$ and $(t, \cos t)$ using the functional values mentioned earlier. Since each function has period 2π, each graph repeats itself every 2π units along the horizontal axis. These sketches should be considered as very rough approximations, since we have not yet developed techniques for obtaining many intermediate points. Note, however, that we could use Example 1 of the preceding section to get the points $(\pi/4, \sin \pi/4)$ and $(\pi/4, \cos \pi/4)$, that is, $(\pi/4, \sqrt{2}/2)$. Similarly, the point $(\pi/6, 1/2)$ is on the sine graph, and $(\pi/6, \sqrt{3}/2)$ is on the cosine graph. We will study the graphs of the trigonometric functions more carefully in Section 5.5.

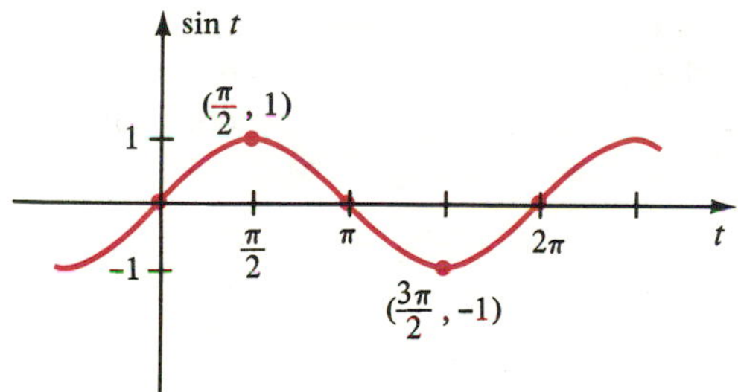

FIGURE 5.21 Sine function

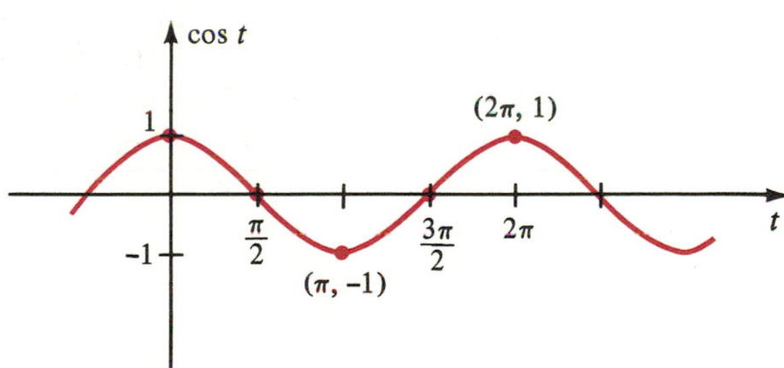

FIGURE 5.22 Cosine function

The secant and cosecant functions are also periodic with period 2π (see Exercises 17 and 18). The variations of these functions will be discussed in Section 5.5. At that time we shall see that the tangent and cotangent functions are periodic with period π.

Let us briefly discuss the range of each trigonometric function. Since the pair (x, y) in the definition of the trigonometric functions gives the coordinates of a point on the unit circle U, we have

$$-1 \leq x \leq 1 \quad \text{and} \quad -1 \leq y \leq 1.$$

Since $\sin t = y$ and $\cos t = x$, it follows that the range of both the sine and cosine functions is the set of all numbers in the closed interval $[-1, 1]$.

If x is a nonzero real number such that $-1 \leq x \leq 1$, then $|x| \leq 1$. Multiplying both sides of the last inequality by $1/|x|$, we obtain $1 \leq 1/|x|$. Since

$\sec t = 1/x$, we see that $1 \leq |\sec t|$ for every real number t, and therefore the range of the secant function consists of all real numbers having absolute value greater than or equal to 1. The same is true for the range of the cosecant function.

It can be shown that $\tan t$ and $\cot t$ take on all real values (see Exercises 19 and 20). That will also become apparent in Section 5.5, when we study the graphs of the trigonometric functions.

We shall conclude this section by obtaining formulas for trigonometric functions of $-t$, where t is any real number or the measure of an angle. Since a minus sign is involved, we shall call them *Formulas for Negatives.*

FORMULAS FOR NEGATIVES

$$\sin(-t) = -\sin t$$
$$\cos(-t) = \cos t$$
$$\tan(-t) = -\tan t$$

To establish the formulas consider the unit circle in Figure 5.23. If we let t vary from 0 to 2π, then the point $P(x, y)$ traces the unit circle U once in the counterclockwise direction, whereas the point $Q(x, -y)$ corresponding to $-t$ traces U once in the clockwise direction. Using the formulas for trigonometric functions in terms of a unit circle (see page 208):

$$\sin(-t) = -y = -\sin t$$

$$\cos(-t) = x = \cos t$$

$$\tan(-t) = \frac{-y}{x} = -\frac{y}{x} = -\tan t,$$

which is what we wished to prove.

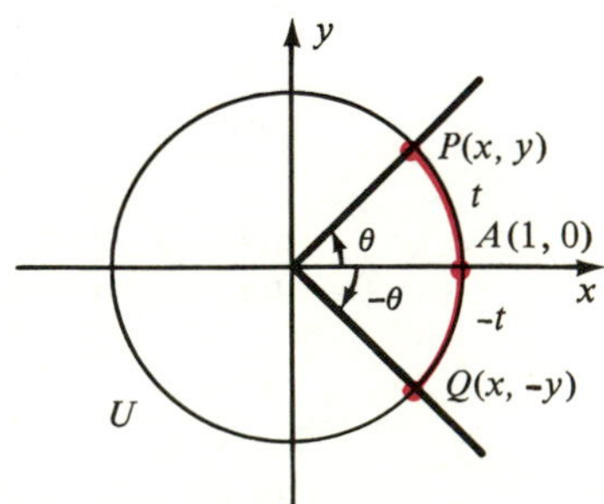

FIGURE 5.23

EXAMPLE 2 Use the Formulas for Negatives to find $\sin(-\pi/4)$, $\cos(-30°)$, and $\tan(-\pi/3)$.

Solution Using the formulas and the table on page 200,

$$\sin(-\pi/4) = -\sin(\pi/4) = -\sqrt{2}/2$$
$$\cos(-30°) = \cos 30° = \sqrt{3}/2$$
$$\tan(-\pi/3) = -\tan(\pi/3) = -\sqrt{3}.$$

EXERCISES 5.3

In Exercises 1–10 find (a) the point on the unit circle that corresponds to the given real number t, and (b) the values of the trigonometric functions at t.

1 π **2** $-3\pi/2$

3 $7\pi/2$ **4** 5π

5 $3\pi/4$ **6** $5\pi/4$

7 $\pi/6$ **8** $\pi/3$

9 $5\pi/6$ **10** $4\pi/3$

In Exercises 11 and 12 use the Formulas for Negatives to find the numbers.

11 (a) $\sin(-\pi/6)$
(b) $\cos(-\pi/4)$
(c) $\tan(-\pi/6)$

12 (a) $\tan(-\pi/4)$
(b) $\cos(-\pi/3)$
(c) $\sin(-\pi/3)$

13 Prove that (a)–(c) are true for every t in the domain of the function.

(a) $\csc(-t) = -\csc t$

(b) $\sec(-t) = \sec t$

(c) $\cot(-t) = -\cot t$

14 Prove that (a) the cosine function is even; (b) the sine and tangent functions are odd.

15 Prove that the sine function has period 2π. (*Hint:* Assume that there is a positive real number k less than 2π such that $\sin(t + k) = \sin t$ for all t. Arrive at a contradiction by letting $t = 0$.)

16 Prove that the cosine function has period 2π.

17 Prove that the cosecant function has period 2π.

18 Prove that the secant function has period 2π.

19 Prove that the range of the tangent function is $\mathbb{R}$ by showing that if a is any real number, then there is a point P on U corresponding to t such that $\tan t = a$. (*Hint:* If $P = (x, y)$ is on U, consider the equation $\tan t = y/x = a$, where $x^2 + y^2 = 1$.)

20 Prove that the range of the cotangent function is $\mathbb{R}$.

SECTION 5.4
VALUES OF THE TRIGONOMETRIC FUNCTIONS

In previous sections we calculated several special values of the trigonometric functions. Let us now turn to the problem of finding *all* values. Since the sine function has period 2π, it is sufficient to know the values of $\sin t$ for $0 \le t \le 2\pi$, because these values are repeated in every t-interval of length 2π.

The same is true for the other trigonometric functions. As a matter of fact, the values of any trigonometric function can be determined if its values for $0 \le t \le \pi/2$ are known. To prove this fact we shall make use of the following concepts.

DEFINITION

Let t be a real number, let θ be an angle of radian measure t, and let P be the point on the unit circle U that corresponds to t.

(i) The **reference number** associated with t is the shortest arc length t' on U between P and the x-axis.

(ii) The **reference angle** associated with θ is the acute angle θ' that the terminal side of θ makes with the positive or negative x-axis.

If P is a point on the unit circle U with center at the origin O, and if the line segment OP is the terminal side of an angle θ of radian measure t, then the reference number t' and reference angle θ' are as indicated in Figure 5.24. Note that $0 \le t' \le \pi/2$ and $0 \le \theta' \le \pi/2$ for all values of t and θ.

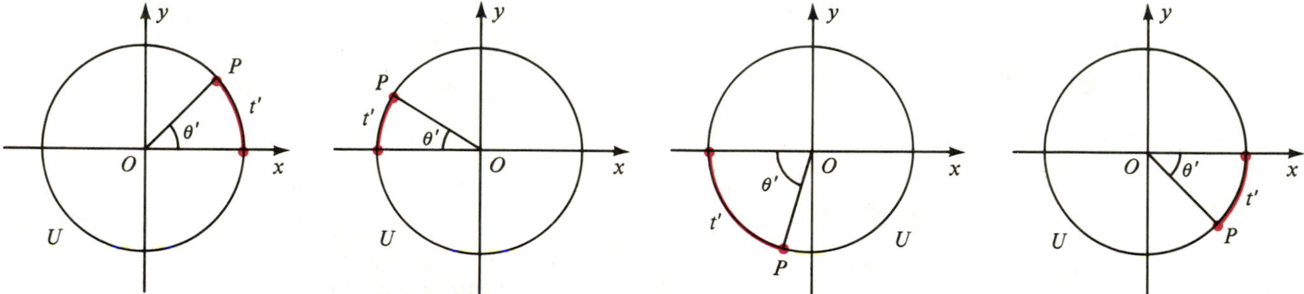

FIGURE 5.24 Reference numbers and angles

Using the definition of trigonometric functions of real numbers in Section 5.3, we see that

$$\sin \theta' = \sin t', \qquad \cos \theta' = \cos t', \qquad \tan \theta' = \tan t'.$$

Similar formulas are true for the other trigonometric functions. In the following example we shall use the approximation $\pi \approx 3.1416$ for estimating reference numbers.

EXAMPLE 1 Approximate the reference number t' if

(a) $t = 2$ (b) $t = 4$ (c) $t = 7\pi/4$ (d) $t = -2\pi/3$.

Solution Let $P(t)$ denote the point on the unit circle U that corresponds to t. Each point $P(t)$ in (a)–(d) and the magnitude of its reference number t', are indicated in Figure 5.25.

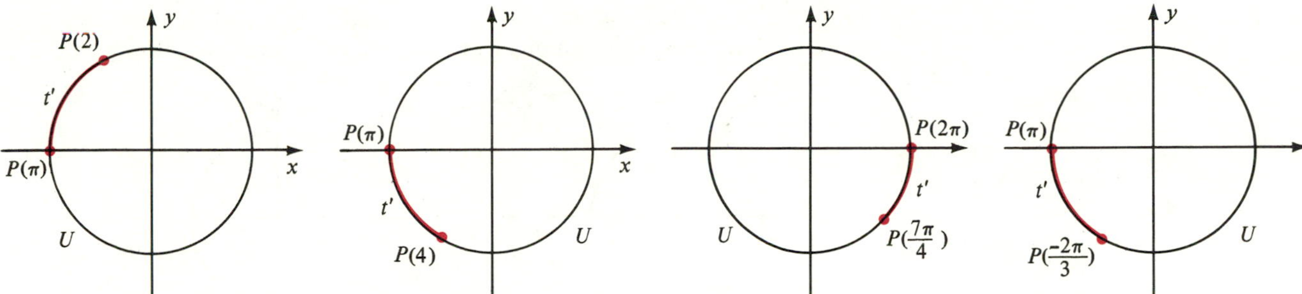

FIGURE 5.25

Referring to the figure, we see that

(a) $t' = \pi - 2 \approx 3.1416 - 2 = 1.1416$

(b) $t' = 4 - \pi \approx 4 - 3.1416 = 0.8584$

(c) $t' = 2\pi - (7\pi/4) = \pi/4$

(d) $t' = \pi - (2\pi/3) = \pi/3$ ■

EXAMPLE 2 Sketch the reference angle θ' for each value of θ, and express the measure of θ' in terms of both radians and degrees.

(a) $\theta = 5\pi/6$ (b) $\theta = 315°$ (c) $\theta = -240°$

Solution Each angle θ and its reference angle θ' are sketched in Figure 5.26.

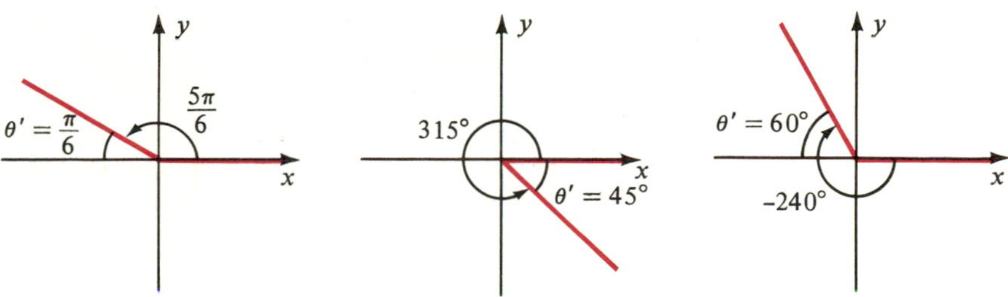

FIGURE 5.26

Evidently,

(a) $\theta' = \pi/6 = 30°$ (b) $\theta' = 45° = \pi/4$ (c) $\theta' = 60° = \pi/3$ ■

Let us now demonstrate how reference numbers or reference angles can be used to find values of the trigonometric functions. Let $P(x, y)$ be the point on the unit circle U corresponding to t. If P is not on a coordinate axis and t' is the reference number for t, then $0 < t' < \pi/2$. Consider the point $A(1, 0)$

and let $P'(x', y')$ be the point on U in quadrant I such that $\widehat{AP'} = t'$. Illustrations in which $P(x, y)$ lies in quadrants II, III, or IV are given in Figure 5.27.

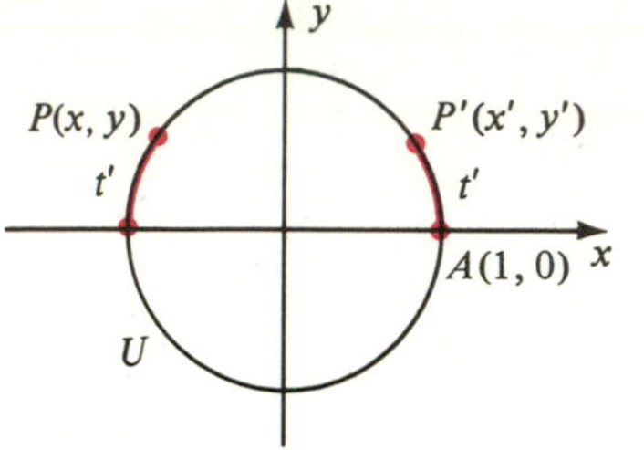

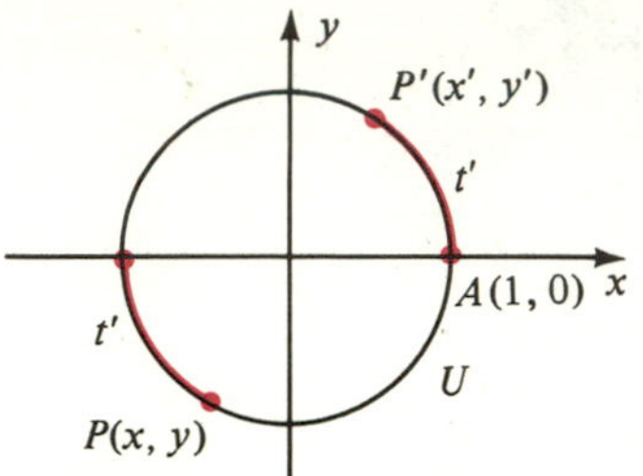

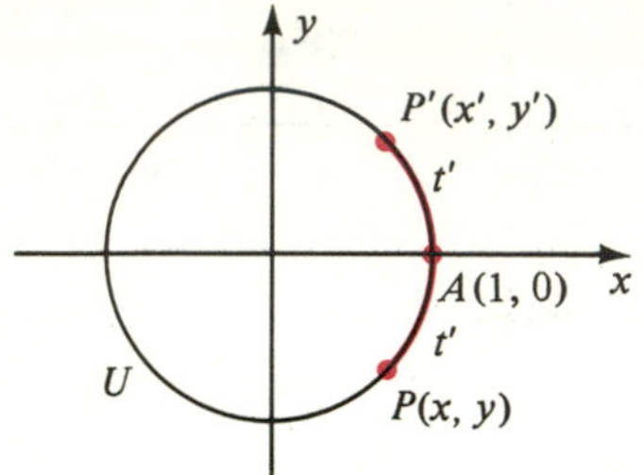

FIGURE 5.27

We see that in all cases

$$x' = |x| \quad \text{and} \quad y' = |y|$$

and hence we have

$$|\cos t| = |x| = x' = \cos t'$$

$$|\sin t| = |y| = y' = \sin t'.$$

It is easy to show that the absolute value of *every* trigonometric function at t is the same as its value at t'. For example,

$$|\tan t| = \left|\frac{y}{x}\right| = \frac{|y|}{|x|} = \frac{y'}{x'} = \tan t'.$$

In terms of angles, if θ' is the reference angle of θ, then

$$|\sin \theta| = \sin \theta' \quad \text{and} \quad |\cos \theta| = \cos \theta'.$$

Similar results apply to other trigonometric functions. We may state the following rules.

RULES FOR FINDING VALUES OF TRIGONOMETRIC FUNCTIONS

(i) To find the value of a trigonometric function at a real number t, determine its value for the reference number t' associated with t and prefix the appropriate sign.

(ii) To find the value of a trigonometric function at an angle θ, determine its value for the reference angle θ' associated with θ and prefix the appropriate sign.

The "appropriate sign" can be determined from the table of signs on page 206.

EXAMPLE 3 Find $\sin(7\pi/4)$ and $\cos(-7\pi/6)$.

Solution We can interpret $7\pi/4$ and $-7\pi/6$ either as real numbers or as radian measures of angles. If we let $P(t)$ denote the point on the unit circle

U that corresponds to t, then we have the situation shown in Figure 5.28, where $\pi/4$ is the reference number (angle) for $7\pi/4$, and $\pi/6$ is the reference number (angle) for $-7\pi/6$. Using the rules preceding this example,

$$\sin\frac{7\pi}{4} = -\sin\frac{\pi}{4} = -\frac{\sqrt{2}}{2}$$

$$\cos\frac{-7\pi}{6} = -\cos\frac{\pi}{6} = -\frac{\sqrt{3}}{2}.$$

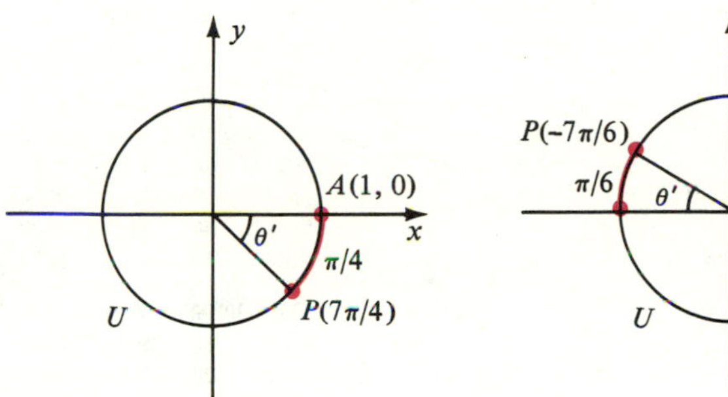

FIGURE 5.28 ■

EXAMPLE 4 Find the following functional values:

(a) sin 150° (b) tan 315° (c) sec (−240°)

Solution The angles and their reference angles are shown in Figure 5.29.

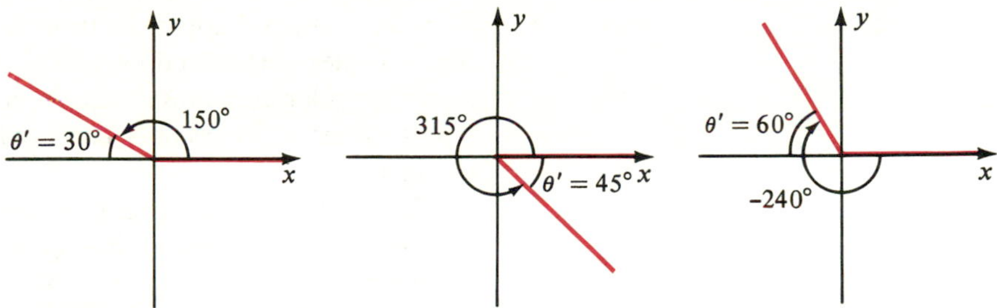

FIGURE 5.29

Using the rules for finding functional values,

(a) $\sin 150° = \sin 30° = \frac{1}{2}$

(b) $\tan 315° = -\tan 45° = -1$

(c) $\sec(-240°) = -\sec 60° = -2$ ■

By employing advanced techniques it is possible to approximate, to any degree of accuracy, all the values of the trigonometric functions in the t-interval $[0, \pi/2]$ or, equivalently, in the degree interval $[0°, 90°]$. Table 4, part of which is reproduced here, gives four-decimal-place approximations to such values.

t	t degrees	sin t	cos t	tan t	cot t	sec t	csc t		
.4887	**28°00′**	.4695	.8829	.5317	1.881	1.133	2.130	**62°00′**	1.0821
.4916	10	.4720	.8816	.5354	1.868	1.134	2.118	50	1.0792
.4945	20	.4746	.8802	.5392	1.855	1.136	2.107	40	1.0763
.4974	30	.4772	.8788	.5430	1.842	1.138	2.096	30	1.0734
.5003	40	.4797	.8774	.5467	1.829	1.140	2.085	20	1.0705
.5032	50	.4823	.8760	.5505	1.816	1.142	2.074	10	1.0676
.5061	**29°00′**	.4848	.8746	.5543	1.804	1.143	2.063	**61°00′**	1.0647
.5091	10	.4874	.8732	.5581	1.792	1.145	2.052	50	1.0617
.5120	20	.4899	.8718	.5619	1.780	1.147	2.041	40	1.0588
.5149	30	.4924	.8704	.5658	1.767	1.149	2.031	30	1.0559
.5178	40	.4950	.8689	.5696	1.756	1.151	2.202	20	1.0530
.5207	50	.4975	.8675	.5735	1.744	1.153	2.010	10	1.0501
.5236	**30°00′**	.5000	.8660	.5774	1.732	1.155	2.000	**60°00′**	1.0472
.5265	10	.5025	.8646	.5812	1.720	1.157	1.990	50	1.0443
.5294	20	.5050	.8631	.5851	1.709	1.159	1.980	40	1.0414
.5323	30	.5075	.8616	.5890	1.698	1.161	1.970	30	1.0385
.5352	40	.5100	.8601	.5930	1.686	1.163	1.961	20	1.0356
.5381	50	.5125	.8587	.5969	1.675	1.165	1.951	10	1.0327
		cos t	sin t	cot t	tan t	csc t	sec t	t degrees	t

The domain of t in Table 4 is from 0 to 1.5708. The last number is a four-decimal-place approximation to $\pi/2$. The table is arranged so that functional values corresponding to angles in degree measure may be found directly. Angular measures are given in 10′ intervals from 0° to 90°. The reason that t varies at intervals of approximately 0.0029 is because $10' \approx 0.0029$ radians.

To find values of trigonometric functions if $0 \leq t \leq 0.7854 \approx \pi/4$ or $0° \leq \theta \leq 45°$, the labels at the *top* of the columns in Table 4 should be used. For example,

$$\sin(0.5003) \approx 0.4797 \qquad \tan 28°30' \approx 0.5430$$
$$\cos(0.4945) \approx 0.8802 \qquad \sec 29°00' \approx 1.143$$
$$\cot(0.5120) \approx 1.780 \qquad \csc 29°40' \approx 2.020.$$

However, if $0.7854 \leq t \leq 1.5708$, or if $45° \leq \theta \leq 90°$, then the labels at the *bottom* of the columns should be employed. For example,

$$\sin(1.0705) \approx 0.8774 \qquad \csc 62°00' \approx 1.133$$
$$\cos(1.0530) \approx 0.4950 \qquad \cot 61°10' \approx 0.5505$$
$$\tan(1.0821) \approx 1.881 \qquad \sec 60°30' \approx 2.031.$$

The reason that the table can be so arranged follows from the fact, to be proved later, that

$$\sin t = \cos\left(\frac{\pi}{2} - t\right), \qquad \cot t = \tan\left(\frac{\pi}{2} - t\right), \qquad \csc t = \sec\left(\frac{\pi}{2} - t\right)$$

or equivalently,

$$\sin \theta = \cos (90^\circ - \theta), \qquad \cot \theta = \tan (90^\circ - \theta), \qquad \csc \theta = \sec (90^\circ - \theta).$$

In particular,

$$\sin 29^\circ = \cos (90^\circ - 29^\circ) = \cos 61^\circ$$

$$\cot 28^\circ 20' = \tan (90^\circ - 28^\circ 20') = \tan 61^\circ 40',$$

as shown in the table.

If it is necessary to find functional values when t lies *between* numbers given in the table, the method of linear interpolation, discussed in Appendix I, may be employed. Similarly, given a value such as $\sin t = 0.6371$, we may refer to the body of Table 4 and use linear interpolation, if necessary, to obtain an approximation to t. If t is measured in degrees we round off to the nearest minute (see Appendix I for examples).

Table 5 can be used to find functional values of t if t is a two-decimal-place approximation of a real number or the radian measure of an angle, and $0 \le t \le 1.57$. For example, referring to Table 5, we see that

$$\sin (0.95) \approx 0.8134$$

$$\cos (1.48) \approx 0.0907$$

$$\cot (0.50) \approx 1.830.$$

Scientific calculators have keys, labeled [sin], [cos], and [tan], that can be used to approximate values of these functions. The values of csc, sec, and cot may then be found by means of the reciprocal key. *Before using a calculator to find functional values that correspond to a real number (or the radian measure of an angle), be sure that the calculator is in the radian mode. For values corresponding to degree measure, the degree mode should be selected.* Readers should consult their user's manuals for specific instructions.

If a calculator is employed to approximate functional values, it is unnecessary to use reference numbers or angles. As an illustration, to find $\sin 210^\circ$, place the calculator in the degree mode, enter the number 210 and press the [sin] key, obtaining $\sin 210^\circ = -0.5$, which is the exact value. Using the same procedure for 240° we obtain, on a typical calculator,

$$\sin 240^\circ \approx -0.8660254,$$

which is a seven-decimal-place approximation. If the *exact* value of $\sin 240^\circ$ is desired, a calculator should not be used. In this case we find the reference angle 60° of 240° and use the rules stated on page 216 together with a knowledge of special angles to obtain

$$\sin 240^\circ = -\sin 60^\circ = -\frac{\sqrt{3}}{2}.$$

If minutes are involved, as in finding tan 67°29′, then before using the [tan] key, *the degree measure of the angle should be expressed in decimal form.* Since

$$29' = \left(\frac{29}{60}\right)^\circ \approx 0.4833333,$$

we may write

$$\tan 67^\circ 29' \approx \tan 67.4833333^\circ \approx 2.4122286.$$

EXAMPLE 5 Approximate cot 837.4°.

Solution Place the calculator in degree mode and use the reciprocal relationship $\cot\theta = 1/\tan\theta$ as follows:

Enter:	837.4	(degree value of θ)
Press [tan]:	−1.9291956	(value of $\tan\theta$)
Press [1/x]:	−0.5183508	(value of $1/\tan\theta$)

Hence cot 837.4° ≈ −0.5183508. ■

Inverse functions that correspond to trigonometric functions are defined in Section 6.7, where it is shown that there is a function denoted by $\sin^{-1}$ such that

$$\sin^{-1}(\sin t) = t \quad \text{provided} \quad -\pi/2 \le t \le \pi/2.$$

This fact can be used to find a value for t when given $\sin t$. Note that this value is restricted to the t-interval $[-\pi/2, \pi/2]$. The corresponding degree measure is between −90° and 90°. The reason for this restriction is discussed in Section 6.7, where *inverse trigonometric functions* are treated in detail.

To use $\sin^{-1}$ with a calculator, we first press [INV] (for inverse) and then [sin]. This technique is analogous to that in Section 4.4, where given $\log x$, we pressed [INV] followed by [log] to approximate x.

EXAMPLE 6 If $\sin t = 0.5$, use a calculator to approximate

(a) the degree measure of t; (b) the real number t.

Solution

(a) We place the calculator in degree mode and proceed as follows:

Enter: 0.5 (the value of sin t)
Press [INV] [sin]: 30 (the degree measure of t)

(b) We place the calculator in radian mode and do the following:

Enter: 0.5 (the value of sin t)
Press [INV] [sin]: 0.5235988 (the number t)

The last number is a decimal approximation for an angle of $\pi/6$ radians (or 30°). ■

In Section 6.7 we will define functions denoted by $\cos^{-1}$ and $\tan^{-1}$ such that

$$\cos^{-1}(\cos t) = t \quad \text{if} \quad 0 \le t \le \pi$$

and

$$\tan^{-1}(\tan t) = t \quad \text{if} \quad -\pi/2 < t < \pi/2.$$

It is important to note the restrictions on t.

EXAMPLE 7 If $\cos t = -0.5$, use a calculator to approximate

(a) the degree measure of t; (b) the real number t.

Solution

(a) Using the degree mode,

Enter: −0.5 (the value of cos t)
Press [INV] [cos]: 120 (the degree measure of t)

(b) Using the radian mode,

Enter: −0.5 (the value of cos t)
Press [INV] [cos]: 2.0943951 (the number t)

The last decimal is an approximation to the number $2\pi/3$. ■

The tangent function is treated in like manner.

EXERCISES 5.4

In Exercises 1–6 find the reference number t' if t has the given value.

1 (a) $3\pi/4$ (b) $4\pi/3$ (c) $-\pi/6$

2 (a) $5\pi/6$ (b) $2\pi/3$ (c) $-3\pi/4$

3 (a) $9\pi/4$ (b) $7\pi/6$ (c) $-2\pi/3$

4 (a) $8\pi/3$ (b) $7\pi/4$ (c) $-7\pi/6$

5 (a) 1.5 (b) 5

6 (a) 3.5 (b) −4

In Exercises 7–10 find the reference angle θ' if θ has the given measure.

7 (a) 240°
(b) 340°
(c) -110°

8 (a) 165°
(b) 275°
(c) -202°

9 (a) $130^\circ 40'$
(b) -405°
(c) $-260^\circ 35'$

10 (a) $335^\circ 20'$
(b) -620°
(c) $-185^\circ 40'$

In Exercises 11–16 find the exact values without the use of tables or a calculator.

11 (a) $\sin 2\pi/3$
(b) $\sin 4\pi/3$

12 (a) $\cos 5\pi/6$
(b) $\cos 7\pi/6$

13 (a) $\tan(-5\pi/4)$
(b) $\cot 315^\circ$

14 (a) $\sin 210^\circ$
(b) $\csc(-150^\circ)$

15 (a) $\csc 300^\circ$
(b) $\sec(-120^\circ)$

16 (a) $\tan(-135^\circ)$
(b) $\sec 225^\circ$

In Exercises 17–30 use Table 4 to approximate the numbers, and compare your answers with those found by means of a calculator.

17 $\sin(0.2676)$

18 $\cos(0.3258)$

19 $\tan(0.9948)$

20 $\sec(0.8988)$

21 $\csc(0.7738)$

22 $\cot(0.7883)$

23 $\cos 38^\circ 30'$

24 $\sin 73^\circ 20'$

25 $\cot 9^\circ 10'$

26 $\csc 43^\circ 40'$

27 $\sec 168^\circ 50'$

28 $\tan 207^\circ 10'$

29 $\sin 342^\circ 20'$

30 $\cos 432^\circ 40'$

In Exercises 31–42 use a calculator to approximate the numbers to four decimal places.

31 $\sin(0.48)$

32 $\cos(0.82)$

33 $\tan 2$

34 $\cot 6$

35 $\sec \frac{1}{4}$

36 $\csc(1.54)$

37 $\cos 97^\circ 43'$

38 $\sin 22^\circ 34'$

39 $\cot 62^\circ 27'$

40 $\tan 57^\circ 16'$

41 $\csc 16^\circ 51'$

42 $\sec 9^\circ 12'$

In Exercises 43–50 use a calculator to approximate, to two decimal places, (a) the degree measure of t, and (b) the real number t.

43 $\cos t = 0.8620$

44 $\sin t = 0.6612$

45 $\tan t = 3.7$

46 $\cos t = 0.8$

47 $\sin t = 0.4217$

48 $\tan t = 4.91$

49 $\sec t = 4.246$

50 $\csc t = 11$

SECTION 5.5

GRAPHS OF THE TRIGONOMETRIC FUNCTIONS

In preceding sections we used notations such as $\sin t$ or $\sin \theta$ to denote values of the sine function at t or θ, respectively. Since we now wish to sketch the graph of the sine function on an xy-coordinate system, we shall consider equations of the form $y = \sin x$, where x denotes a real number (or the radian measure of an angle). The variable x used here should not be confused with that employed in the unit circle approach to trigonometric functions on page 208, where x denoted the x-coordinate of a point P on the unit circle U. Similar remarks hold for other graphs involving the trigonometric functions.

It is not difficult to sketch the graph of the sine function, that is, the equation $y = \sin x$. Since $-1 \le \sin x \le 1$ for every real number x, the graph lies between the horizontal lines $y = 1$ and $y = -1$. Moreover, since the sine function is periodic with period 2π, it is sufficient to determine the graph for $0 \le x \le 2\pi$, because the same pattern is repeated in intervals of length 2π along the x-axis. In Section 5.3 our discussion of the variation of $\sin x$ in the interval $[0, 2\pi]$ concentrated on the y-coordinate of a point P, as P traversed the unit circle U once in the counterclockwise direction (see page 210). The manner in which $\sin x$ varies in $[0, 2\pi]$ may be briefly summarized as follows:

Variation of x	Variation of $\sin x$
0 to $\pi/2$	0 to 1
$\pi/2$ to π	1 to 0
π to $3\pi/2$	0 to -1
$3\pi/2$ to 2π	-1 to 0

Using these facts, we conjectured that the graph of the sine function might look like that in Figure 5.21. This guess may be further justified by calculating some intermediate values of $\sin x$, as indicated in the following table.

x	0	$\frac{\pi}{4}$	$\frac{\pi}{2}$	$\frac{3\pi}{4}$	π	$\frac{5\pi}{4}$	$\frac{3\pi}{2}$	$\frac{7\pi}{4}$	2π
$y = \sin x$	0	$\frac{\sqrt{2}}{2}$	1	$\frac{\sqrt{2}}{2}$	0	$\frac{-\sqrt{2}}{2}$	-1	$\frac{-\sqrt{2}}{2}$	0

To obtain a rough sketch we may plot the points (x, y) given by this table, draw a smooth curve through them, and extend the configuration to the right and left in periodic fashion. This gives us the portion of the graph shown in Figure 5.30. Of course, the graph does not terminate, but continues indefinitely to the right and left. For greater accuracy, additional points could be plotted using, for example, $\sin \pi/6 = \frac{1}{2}$, $\sin \pi/3 = \sqrt{3}/2 \approx 0.86$, etc. We could also use Tables 4 and 5 or a calculator to obtain many additional points on the graph. We refer to the part of the graph corresponding to the interval $[0, 2\pi]$ as a **sine wave.**

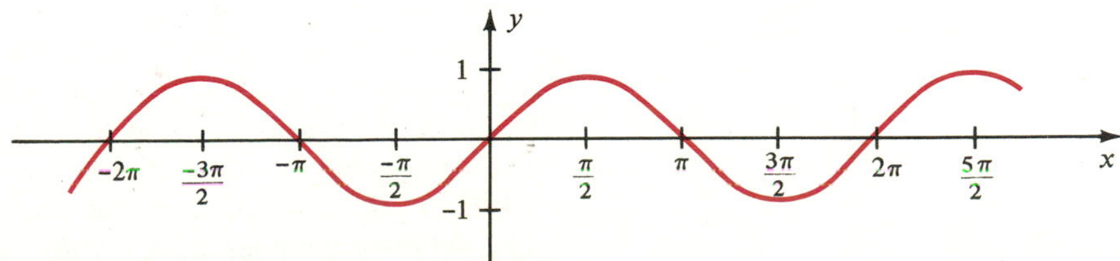

FIGURE 5.30 $y = \sin x$

The graph of the cosine function can be found in similar fashion, by studying the x-coordinate of a point P, as P travels around the unit circle U once in the counterclockwise direction. We may briefly summarize the behavior of $\cos x$ in the interval $[0, 2\pi]$ as follows:

Variation of x	Variation of $\cos x$
0 to $\pi/2$	1 to 0
$\pi/2$ to π	0 to -1
π to $3\pi/2$	-1 to 0
$3\pi/2$ to 2π	0 to 1

Using this information, together with the fact that the cosine function has period 2π, and plotting points as we did for the sine function, leads to the sketch in Figure 5.31. Note that the graph of $y = \cos x$ can be obtained by shifting the graph of $y = \sin x$ to the left a distance $\pi/2$.

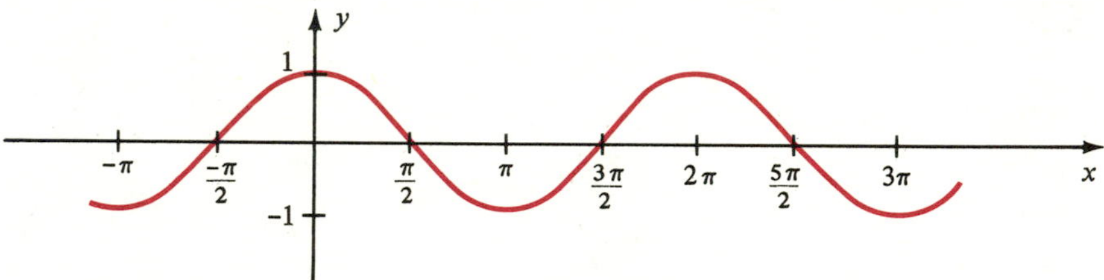

FIGURE 5.31 $y = \cos x$

Several values of the tangent function are listed in the following table.

x	$-\frac{\pi}{3}$	$-\frac{\pi}{4}$	$-\frac{\pi}{6}$	0	$\frac{\pi}{6}$	$\frac{\pi}{4}$	$\frac{\pi}{3}$
$\tan x$	$-\sqrt{3}$	-1	$\frac{-\sqrt{3}}{3}$	0	$\frac{\sqrt{3}}{3}$	1	$\sqrt{3}$

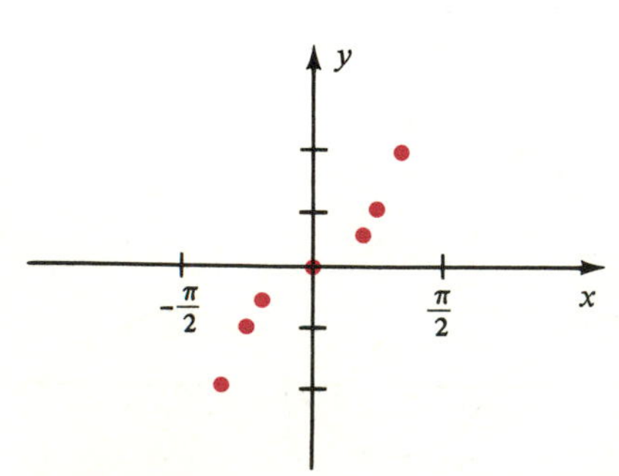

FIGURE 5.32

The corresponding points are plotted in Figure 5.32. The values of $\tan x$ near $x = \pi/2$ demand special consideration. As x increases through positive values toward $\pi/2$, the point $P = (a, b)$ on the unit circle U that corresponds to x approaches the point $(0, 1)$. Since, by definition, $\tan x = b/a$, it follows that if b approaches 1 and a approaches 0 (through positive values), then $\tan x$ takes on large positive values. First recall that $\pi/2 \approx \frac{1}{2}(3.1416) = 1.5708$. The following values of $\tan x$ for x close to $\pi/2$ were obtained using a calculator:

$$\begin{aligned}
\tan(1.5700) &\approx 1{,}255.8\\
\tan(1.5703) &\approx 2{,}014.8\\
\tan(1.5706) &\approx 5{,}093.5\\
\tan(1.5707) &\approx 10{,}381.3\\
\tan(1.57079) &\approx 158{,}057.9
\end{aligned}$$

Notice how rapidly $\tan x$ increases as x gets close to $\pi/2$. Indeed, $\tan x$ can be made arbitrarily large by choosing x sufficiently close to $\pi/2$. Using the terminology introduced in Section 3.7, we say that $\tan x$ *increases without bound as x approaches $\pi/2$ through values less than $\pi/2$*, and we write

$$\tan x \to \infty \quad \text{as} \quad x \to \frac{\pi}{2}^{-}.$$

Similarly, as x approaches $-\pi/2$ through values greater than $-\pi/2$, $\tan x$ *decreases without bound*, that is,

$$\tan x \to -\infty \quad \text{as} \quad x \to -\frac{\pi}{2}^{+}$$

This variation in the open interval $(-\pi/2, \pi/2)$ is illustrated in Figure 5.33. The lines $x = \pi/2$ and $x = -\pi/2$ are vertical asymptotes for the graph. It is not difficult to show that the same pattern is repeated in the open intervals $(\pi/2, 3\pi/2)$, $(3\pi/2, 5\pi/2)$, and in similar intervals of length π, as shown in Figure 5.33. Thus, *the tangent function has period π.*

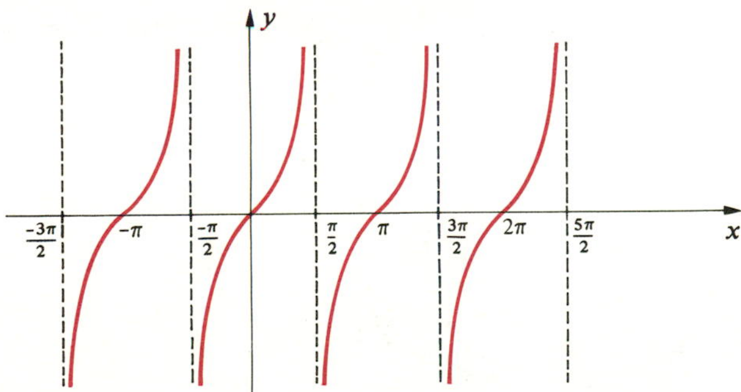

FIGURE 5.33 $y = \tan x$

The graphs of the remaining three trigonometric functions can now be easily obtained. For example, since $\csc x = 1/\sin x$, we may find the y-coordinate of a point on the graph of the cosecant function by taking the reciprocal of the corresponding y-coordinate of the sine graph. This is possible except for $x = n\pi$, where n is an integer, for in this case $\sin x = 0$ and hence $1/\sin x$ is undefined. As an aid to sketching the graph of the cosecant function it is convenient to sketch the graph of the sine function with dashes (see Figure 5.34) and then take reciprocals of y-coordinates to obtain points on the cosecant graph. Notice the manner in which the cosecant function increases or decreases without bound as x approaches $n\pi$, where n is an integer. The graph has vertical asymptotes as indicated in the figure.

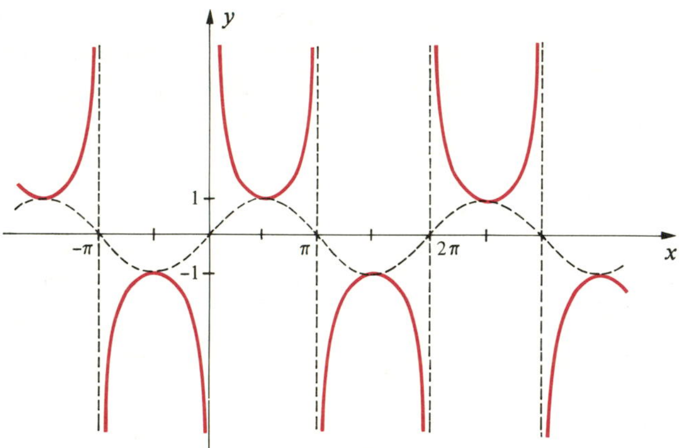

FIGURE 5.34 $y = \csc x$

Since $\sec x = 1/\cos x$ and $\cot x = 1/\tan x$, the graphs of the secant and cotangent functions may be obtained in similar fashion (see Figures 5.35 and 5.36). Their verifications are left as exercises.

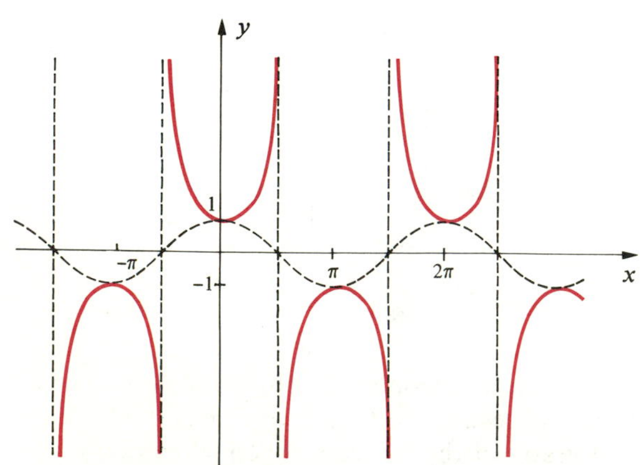

FIGURE 5.35 $y = \sec x$

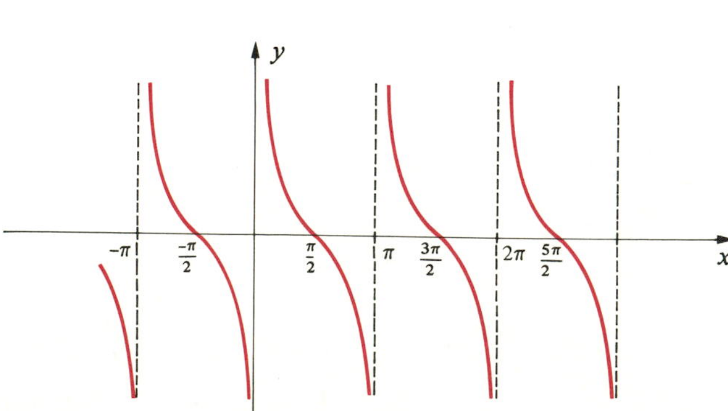

FIGURE 5.36 $y = \cot x$

EXAMPLE 1 Sketch the graph of the function f if $f(x) = 2 \sin x$.

Solution Although the graph could be obtained by plotting points, note that for each x-coordinate x_1, the y-coordinate $f(x_1)$ is always twice that of the corresponding y-coordinate on the sine graph. A simple graphical technique is to sketch the graph of $y = \sin x$ with dashes and then double each y-coordinate to find points on the graph of $y = 2 \sin x$ (see Figure 5.37). This amounts to stretching the graph of a function, as discussed in Section 2.5.

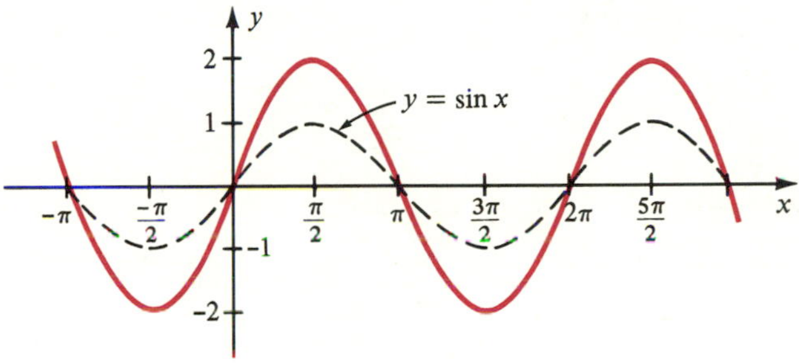

FIGURE 5.37 $y = 2 \sin x$ ■

EXERCISES 5.5

1 (a) Sketch the graph of the cosine function after plotting a sufficient number of points.

(b) Sketch the graph of the secant function by taking reciprocals in part (a).

(c) Describe the intervals between -2π and 2π in which the secant function is increasing.

(d) Describe the intervals between -2π and 2π in which the secant function is decreasing.

(e) With the aid of Table 4 or a calculator, discuss the variation of $\sec x$ as x approaches $\pi/2$ through values less than $\pi/2$ and through values greater than $\pi/2$.

(f) Discuss the symmetry of the graph of the secant function.

2 (a) Sketch the graph of the cotangent function after plotting a sufficient number of points.

(b) In what intervals is the cotangent function increasing?

(c) In what intervals is the cotangent function decreasing?

(d) With the aid of Table 4 or a calculator, discuss the variation of $\cot x$ as x approaches 0 through values greater than 0 and through values less than 0.

(e) Discuss the symmetry of the graph of the cotangent function.

3 Practice sketching graphs of the sine, cosine, and tangent functions, taking different units of length on the horizontal and vertical axes. Continue this practice until you reach the stage at which, if you were awakened from a sound sleep in the middle of the night and asked to sketch one of these graphs, you could do so in less than thirty seconds.

4 Repeat Exercise 3 for the cosecant, secant, and cotangent functions.

In Exercises 5–14 use the method illustrated in the Example of this section to sketch the graph of f.

5 $f(x) = 4 \sin x$

6 $f(x) = 3 \sin x$

7 $f(x) = \frac{1}{2} \sin x$

8 $f(x) = \frac{1}{4} \sin x$

9 $f(x) = 2 \cos x$

10 $f(x) = 4 \cos x$

11 $f(x) = \frac{1}{3} \cos x$

12 $f(x) = \frac{1}{2} \cos x$

13 $f(x) = -\sin x$

14 $f(x) = -\cos x$

CALCULATOR EXERCISES 5.5

If f is a function, the notation

$$f(x) \to L \quad \text{as} \quad x \to a^+$$

is sometimes used to express the fact that $f(x)$ gets very close to L as x gets close to a (through values greater than a). Use a calculator to support Exercises 1–6 by substituting the following values for x: $x = 0.1$, $x = 0.01$, $x = 0.001$, $x = 0.0001$.

1 $\dfrac{\sin x}{x} \to 1 \quad \text{as} \quad x \to 0^+$

2 $\dfrac{\tan x}{x} \to 1 \quad \text{as} \quad x \to 0^+$

3 $\dfrac{1 - \cos x}{x} \to 0 \quad \text{as} \quad x \to 0^+$

4 $\dfrac{\sin x}{1 + \cos x} \to 0 \quad \text{as} \quad x \to 0^+$

5 $x \cot x \to 1 \quad \text{as} \quad x \to 0^+$

6 $\dfrac{x + \tan x}{\sin x} \to 2 \quad \text{as} \quad x \to 0^+$

SECTION 5.6
TRIGONOMETRIC GRAPHS

In this section we shall consider graphs of equations of the form

$$y = a \sin (bx + c)$$

for real numbers a, b and c. We shall also discuss graphs of similar equations that involve different trigonometric functions. To sketch a graph we could begin by plotting many points; however, it is generally easier to use information about the graphs of the trigonometric functions discussed in the preceding section. Let us consider the special case in which $c = 0$, $b = 1$, and $a > 0$. Thus, we wish to sketch the graph of the equation

$$y = a \sin x.$$

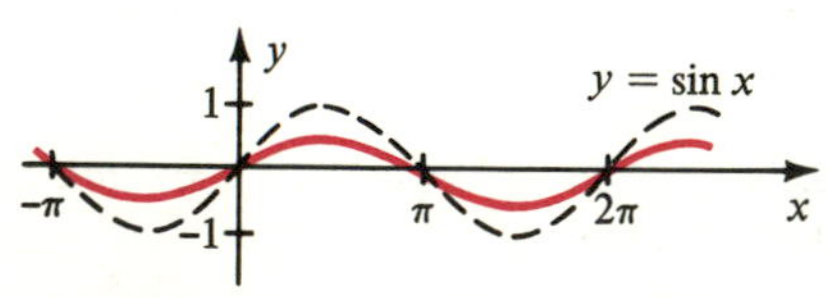

FIGURE 5.38 $y = \frac{1}{2} \sin x$

As in our discussion of stretching of graphs (see Section 2.5), we may find the y-coordinate of a point on the graph by multiplying the corresponding y-coordinate on the graph of $y = \sin x$ by a. Thus, if $y = 2 \sin x$, we multiply by 2; if $y = \frac{1}{2} \sin x$, we multiply by $\frac{1}{2}$; etc. The graph of $y = 2 \sin x$ is sketched in Figure 5.37. The graph of $y = \frac{1}{2} \sin x$ is sketched in Figure 5.38, where for comparison we have indicated the graph of $y = \sin x$ with dashes.

EXAMPLE 1 Sketch the graph of the equation $y = 3 \cos x$.

Solution The graph of the equation may be obtained from the graph of $y = \cos x$ by multiplying y-coordinates of points by 3. We first sketch

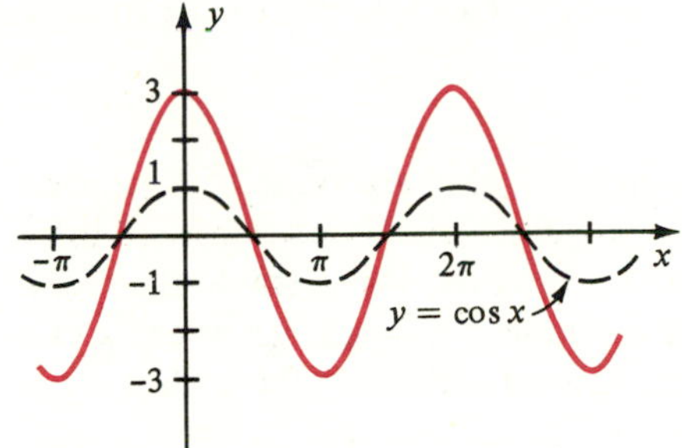

FIGURE 5.39 $y = 3 \cos x$

$y = \cos x$ with dashes, and then we triple the y-coordinates. This gives us the sketch in Figure 5.39. ■

If $a < 0$, then the y-coordinates of points on the graph of $y = a \sin x$ are negatives of the corresponding y-coordinates of points on the graph of $y = |a| \sin x$, as illustrated in the next example.

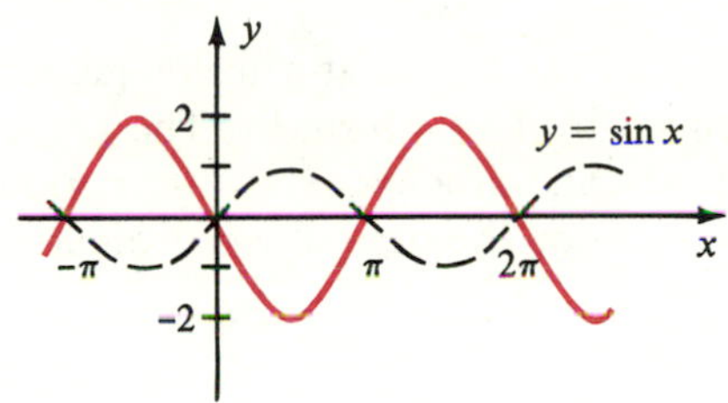

FIGURE 5.40 $y = -2 \sin x$

EXAMPLE 2 Sketch the graph of $y = -2 \sin x$.

Solution We first sketch the graph of $y = \sin x$ with dashes as a guide, and then we multiply each y-coordinate by -2. This gives us the sketch shown in Figure 5.40. The sketch may also be obtained by multiplying the y-coordinates of the graph of $y = 2 \sin x$ (see Figure 5.37) by -1. Thus, the graph of $y = -2 \sin x$ is a *reflection through the x-axis* of the graph of $y = 2 \sin x$. ■

If $f(x) = a \sin (bx + c)$, then the **amplitude of** f (or of its graph) is defined as the largest y-coordinate of points on the graph. If $a > 0$, the largest y-coordinate occurs if $\sin (bx + c) = 1$, whereas if $a < 0$, we must take $\sin (bx + c) = -1$. In either case the amplitude is $|a|$. As illustrations, if $f(x) = \frac{1}{2} \sin x$, then from Figure 5.38 we see that the amplitude is $\frac{1}{2}$. If $f(x) = -2 \sin x$, then from Figure 5.40 we see that the amplitude is 2. The same is true if f is given by $f(x) = a \cos (bx + c)$.

The next theorem provides some information about periods.

THEOREM

If $f(x) = \sin bx$ and $b \neq 0$, then the period of f is $2\pi/|b|$.

Proof If $b > 0$, then as bx varies from 0 to 2π, we obtain exactly one sine wave for the graph of f. The conclusion of the theorem follows from the fact that bx ranges from 0 to 2π if and only if x ranges from 0 to $2\pi/b$. A similar proof may be given if $b < 0$. □

EXAMPLE 3 Find the period and sketch the graph of each function f:

(a) $f(x) = \sin 2x$ (b) $f(x) = \sin \frac{1}{2}x$

Solution Both functions have the form given in the preceding theorem. Hence, in part (a) the period is $2\pi/2 = \pi$, which means that there is exactly one sine wave of amplitude 1 corresponding to the interval $[0, \pi]$. The graph is sketched in Figure 5.41, where for convenience we have used different scales on the x- and y-axes. For part (b) the period is $2\pi/(1/2) = 4\pi$ and

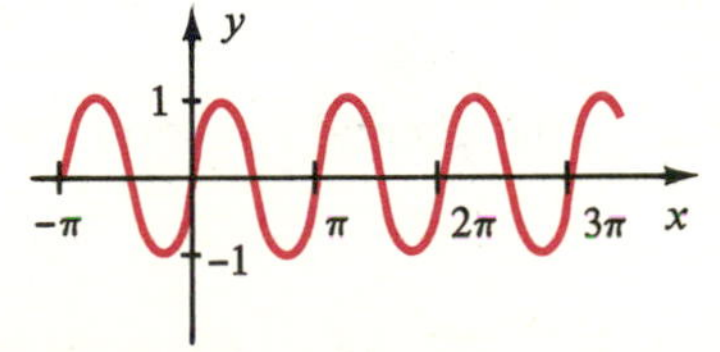

FIGURE 5.41 $y = \sin 2x$

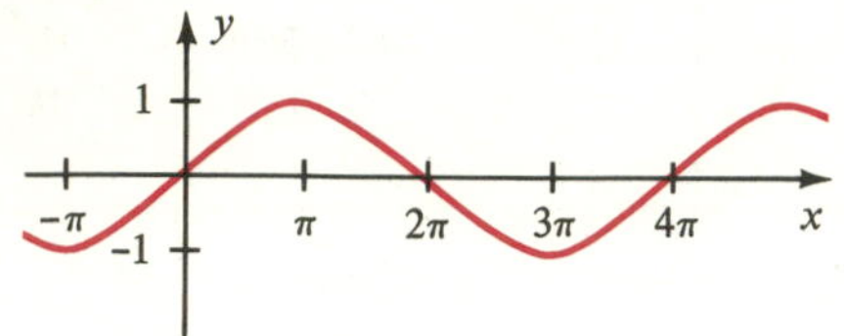

FIGURE 5.42 $y = \sin \frac{1}{2}x$

hence there is one sine wave of amplitude 1 corresponding to the interval $[0, 4\pi]$, as illustrated in Figure 5.42. ■

Given $f(x) = \sin bx$, where $b > 0$, then, roughly speaking, if b is large, $2\pi/b$ is small and the sine waves are close together. As a matter of fact, there are b sine waves in an interval of 2π units. However, if b is small, then $2\pi/b$ is large and the waves are shallow. For example, if $y = \sin \frac{1}{10}x$, then $\frac{1}{10}$ of a sine wave occurs in the interval $[0, 2\pi]$, and an interval 20π units long is required for one complete wave.

If $b < 0$, we can use the fact that $\sin(-x) = -\sin x$ to obtain the graph. To illustrate, the graph of $y = \sin(-2x)$ is the same as the graph of $y = -\sin 2x$.

By combining the preceding remarks we can arrive at a technique for sketching the graph of a function f defined by $f(x) = a \sin bx$. The graph has the basic sine wave pattern. However, the amplitude is $|a|$ and the period is $2\pi/|b|$. If $a < 0$ or $b < 0$, we make adjustments on the signs of y-coordinates, as discussed earlier.

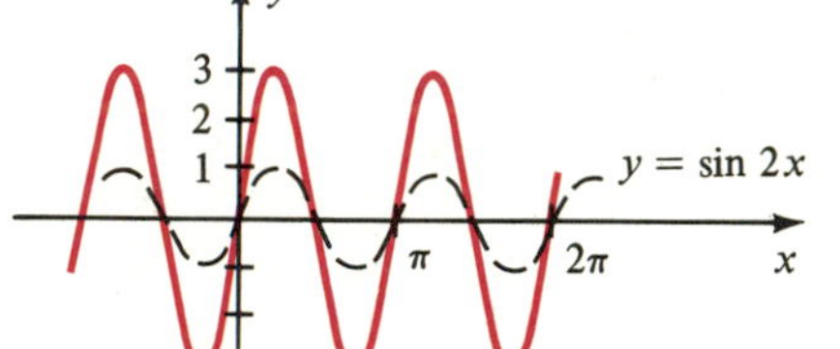

FIGURE 5.43 $y = 3 \sin 2x$

EXAMPLE 4 Sketch the graph of f if $f(x) = 3 \sin 2x$.

Solution From the preceding discussion we see that the amplitude of f is 3 and the period is $2\pi/2 = \pi$. The graph is readily obtained by first sketching the graph of $y = \sin 2x$ with dashes, and then multiplying the y-coordinates of each point by 3. This leads to the sketch in Figure 5.43. ■

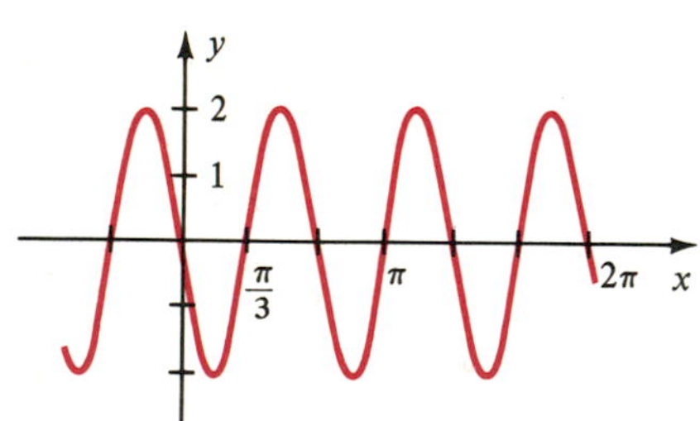

FIGURE 5.44 $y = -2 \sin 3x$

EXAMPLE 5 Sketch the graph of f if $f(x) = 2 \sin(-3x)$.

Solution Since $f(x) = -2 \sin 3x$, we see that the amplitude is 2 and the period is $2\pi/3$. Thus, there is one sine wave in every interval of length $2\pi/3$. The minus sign indicates a reflection through the x-axis. If we consider the interval $[0, 2\pi/3]$ and sketch a sine wave of amplitude 2 (reflected through the x-axis), the shape of the graph is apparent. The configuration given in the interval $[0, 2\pi/3]$ is carried along periodically, as illustrated in Figure 5.44. ■

Similar discussions can be given if f is defined by either $f(x) = a \cos bx$ or $f(x) = a \tan bx$. In the latter case the period is $\pi/|b|$ since the tangent function has period π. There is no largest y-coordinate for points on the graph of the tangent function, and hence we do not refer to its amplitude; however, we may still use the process of multiplying y-coordinates by a to obtain points on the graph of $y = a \tan bx$.

Let us conclude this section by considering the general situation

$$f(x) = a \sin(bx + c).$$

We have already observed that the amplitude is $|a|$. We also know that one sine wave for the graph of $y = \sin(bx + c)$ is obtained if $bx + c$ varies from

0 to 2π, that is, if bx ranges from $-c$ to $2\pi - c$. In turn, the variation of bx is obtained by letting x range from $-c/b$ to $(2\pi - c)/b$. If $-c/b > 0$, this amounts graphically to *shifting* the graph of $y = a \sin bx$ to the right $-c/b$ units. If $-c/b < 0$, the shift is to the left. The number $-c/b$ is sometimes called the **phase shift** associated with the function. Similar remarks can be made for the other functions. It is unnecessary to remember a general formula for finding the phase shift. In any specific problem the interval that contains one complete sine wave can be found by solving the two equations

$$bx + c = 0 \quad \text{and} \quad bx + c = 2\pi$$

for x, as illustrated in the next example.

EXAMPLE 6 Sketch the graph of $f(x) = 3 \sin\left(2x - \dfrac{\pi}{2}\right)$.

Solution The equation is of the form discussed in this section, with $a = 3$, $b = 2$, and $c = -\pi/2$. It follows from our discussion that the graph has the sine wave pattern with amplitude 3 and period $2\pi/2 = \pi$. To obtain an interval containing exactly one sine wave, we let $2x - (\pi/2)$ range from 0 to 2π. The endpoints of the interval can be found by solving the two equations

$$2x - \frac{\pi}{2} = 0 \quad \text{and} \quad 2x - \frac{\pi}{2} = 2\pi.$$

This gives us

$$x = \frac{\pi}{4} \quad \text{and} \quad x = \frac{5\pi}{4}.$$

Thus, one sine wave of amplitude 3 will occur in the interval $[\pi/4, 5\pi/4]$. Sketching that wave and then repeating it to the right and left gives us the graph in Figure 5.45. ■

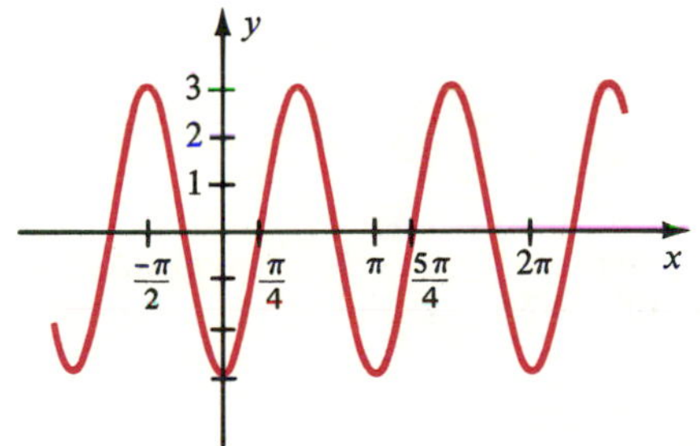

FIGURE 5.45 $y = 3 \sin\left(2x - \dfrac{\pi}{2}\right)$

EXERCISES 5.6

1 Without plotting many points, sketch the graph and determine the amplitude and period of each function f.

(a) $f(x) = 4 \sin x$ (b) $f(x) = \sin 4x$

(c) $f(x) = \frac{1}{4} \sin x$ (d) $f(x) = \sin \frac{1}{4}x$

(e) $f(x) = 2 \sin \frac{1}{4}x$ (f) $f(x) = \frac{1}{2} \sin 4x$

(g) $f(x) = -4 \sin x$ (h) $f(x) = \sin(-4x)$

2 Sketch the graphs of the functions that involve the cosine and are analogous to the functions defined in Exercise 1.

3 Without plotting many points, sketch the graph of each function f defined as follows, determining the amplitude and period in each case:

(a) $f(x) = 3 \cos x$ (b) $f(x) = \cos 3x$

(c) $f(x) = \frac{1}{3} \cos x$ (d) $f(x) = \cos \frac{1}{3}x$

(e) $f(x) = 2 \cos \frac{1}{3}x$ (f) $f(x) = \frac{1}{3} \cos 2x$

(g) $f(x) = -3 \cos x$ (h) $f(x) = \cos(-3x)$

4 Sketch the graphs of the functions that involve the sine and are analogous to the functions defined in Exercise 3.

Sketch the graphs of the equations in Exercises 5–44.

5 $y = \sin\left(x - \dfrac{\pi}{2}\right)$ 6 $y = \cos\left(x + \dfrac{\pi}{4}\right)$

7 $y = 3\sin\left(x - \frac{\pi}{2}\right)$

8 $y = 4\cos\left(x + \frac{\pi}{4}\right)$

9 $y = \cos\left(x + \frac{\pi}{3}\right)$

10 $y = \sin\left(x - \frac{\pi}{3}\right)$

11 $y = 4\cos\left(x + \frac{\pi}{3}\right)$

12 $y = 5\sin\left(x - \frac{\pi}{3}\right)$

13 $y = \sin(3x + \pi)$

14 $y = \cos(3x - \pi)$

15 $y = -2\sin(3x + \pi)$

16 $y = 6\cos(3x - \pi)$

17 $y = 5\sin\left(3x - \frac{\pi}{2}\right)$

18 $y = -4\cos\left(2x + \frac{\pi}{3}\right)$

19 $y = 6\sin \pi x$

20 $y = 3\cos \frac{1}{2}\pi x$

21 $y = 2\cos \frac{\pi}{2}x$

22 $y = 4\sin 3\pi x$

23 $y = \frac{1}{2}\sin 2\pi x$

24 $y = \frac{1}{2}\cos \frac{\pi}{2}x$

25 $y = \frac{1}{2}\sec x$

26 $y = \frac{1}{4}\csc x$

27 $y = -2\csc x$

28 $y = -3\sec x$

29 $y = \sec 2x$

30 $y = \csc 3x$

31 $y = \csc\left(x - \frac{\pi}{4}\right)$

32 $y = \sec\left(x + \frac{\pi}{3}\right)$

33 $y = \tan \frac{1}{2}x$

34 $y = \cot 2x$

35 $y = \tan\left(x + \frac{\pi}{2}\right)$

36 $y = \cot\left(x - \frac{\pi}{4}\right)$

37 $y = \frac{1}{2}\cot x$

38 $y = -2\tan x$

39 $y = \tan(-x)$

40 $y = -\cot x$

41 $y = 2\sin\left(x - \frac{\pi}{2}\right)$

42 $y = \cos\left(x - \frac{\pi}{2}\right)$

43 $y = \sin\left(x + \frac{\pi}{2}\right)$

44 $y = \cos\left(x + \frac{\pi}{2}\right)$

CALCULATOR EXERCISES 5.6

1–6 The graphs in Examples 1–6 of this section were obtained *qualitatively*, in the sense that we determined the general shape by referring to other graphs and not by plotting points. Verify them *quantitatively*, by calculating the coordinates of many points on the graphs.

SECTION 5.7
ADDITIONAL GRAPHICAL TECHNIQUES

Mathematical applications often involve functions that are defined in terms of sums and products of expressions, such as

$$f(x) = \sin 2x + \cos x \quad \text{or} \quad f(x) = 2^{-x}\sin x.$$

When we work with sums, the graphical technique known as **addition of y-coordinates** is useful. The method applies not only to trigonometric expressions, but to arbitrary expressions as well. If f is a sum of two functions g and h that have the same domain X, then

$$f(x) = g(x) + h(x)$$

for every x in X. A sketch of the graph of f may be obtained from the graphs of g and h. We begin by sketching the graphs of the equations $y = g(x)$ and $y = h(x)$ on the same coordinate axes, as illustrated by the dashes in Figure 5.46.

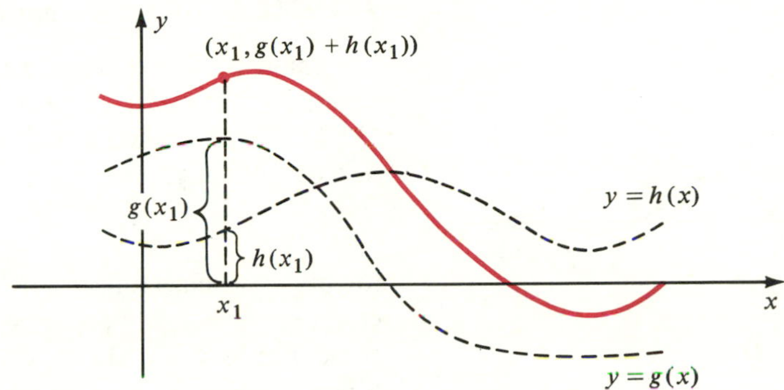

FIGURE 5.46 $y = g(x) + h(x)$

Since $f(x_1) = g(x_1) + h(x_1)$ for every x_1 in X, the y-coordinate of the point on the graph of $y = g(x) + h(x)$ with x-coordinate x_1 is the *sum* of the corresponding y-coordinates of points on the graphs of g and h. If we draw a vertical line at the point with coordinates $(x_1, 0)$, then the y-coordinates $g(x_1)$ and $h(x_1)$ may be added geometrically by means of a compass or ruler, as is illustrated in Figure 5.46. If either $g(x_1)$ or $h(x_1)$ is negative, then a *subtraction* of y-coordinates may be employed. By using this technique a sufficient number of times, we obtain a sketch of the graph of f.

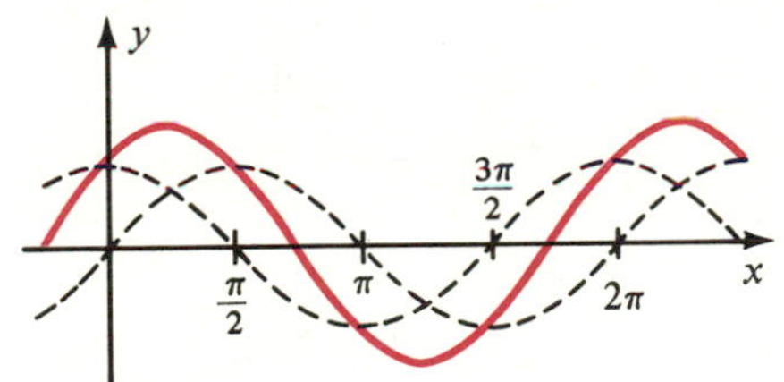

FIGURE 5.47 $y = \cos x + \sin x$

EXAMPLE 1 If $f(x) = \cos x + \sin x$, use addition of y-coordinates to sketch the graph of f.

Solution We begin by sketching (with dashes) the graphs of the equations $y = \cos x$ and $y = \sin x$. Next, for various numbers x_1, we add y-coordinates geometrically. After a sufficient number of y-coordinates are added and a pattern emerges, we draw a smooth curve through the points, as indicated by the sketch in Figure 5.47. As a check, it would be worthwhile to plot some points on the graph by substituting numbers for x; however, we shall leave such verifications to the reader. It can be seen from the graph that the function f is periodic with period 2π. ■

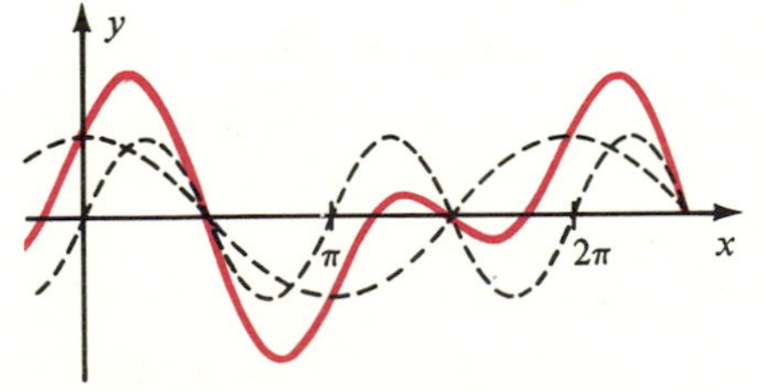

FIGURE 5.48 $y = \cos x + \sin 2x$

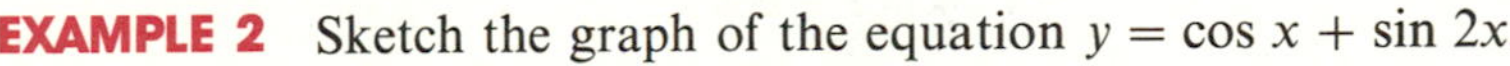

EXAMPLE 2 Sketch the graph of the equation $y = \cos x + \sin 2x$.

Solution We sketch (with dashes) the graphs of the equations $y = \cos x$ and $y = \sin 2x$ on the same coordinate axes and use the method of addition of y-coordinates. The graph is illustrated in Figure 5.48. Evidently, f is periodic with period 2π. ■

The next example illustrates the fact that if f is the product of two functions g and h, then the graphs of g and h can often be used to supply information about the graph of f.

EXAMPLE 3 Sketch the graph of f if $f(x) = 2^{-x} \sin x$.

Solution The function f is the product of two functions g and h, defined by $g(x) = 2^{-x}$ and $h(x) = \sin x$. Using properties of absolute values, $|f(x)| = |2^{-x}|\,|\sin x|$. Since $|\sin x| \le 1$ and $2^{-x} > 0$, it follows that $|f(x)| \le 2^{-x}$ for all x. Consequently,

$$-2^{-x} \le f(x) \le 2^{-x}$$

for all x, which implies that the graph of f lies between the graphs of the equations $y = -2^{-x}$ and $y = 2^{-x}$. The graph of f will coincide with one of these graphs if $|\sin x| = 1$, that is, if $x = (\pi/2) + n\pi$, where n is an integer. Since $2^{-x} > 0$, the x-intercepts on the graph of f occur at $\sin x = 0$, that is, at $x = n\pi$. With this information we obtain the sketch shown in Figure 5.49. ■

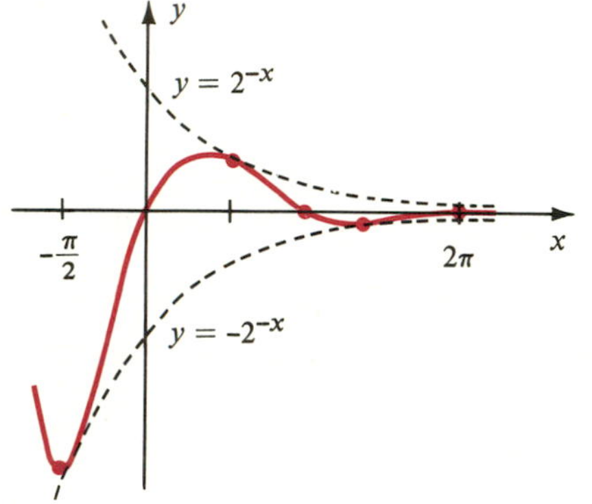

FIGURE 5.49 $y = 2^{-x} \sin x$

The graph of $y = 2^{-x} \sin x$, sketched in Figure 5.49, is called a **damped sine wave,** and 2^{-x} is called the **damping factor.** By using different damping factors, we may obtain other compressed or expanded variations of sine waves. The analysis of such graphs is important in physics and engineering.

EXERCISES 5.7

Use addition of y-coordinates to sketch the graphs of the equations in Exercises 1–20.

1 $y = \cos x + 3 \sin x$

2 $y = \sin x + 3 \cos x$

3 $y = 2 \cos x + 3 \sin x$

4 $y = 2 \sin x + 3 \cos x$

5 $y = \sin x + \cos 2x$

6 $y = 2 \cos x + \sin 2x$

7 $y = \cos x - \sin x$

8 $y = 2 \sin x - \cos x$

9 $y = 2 \cos x - \frac{1}{2} \sin 2x$

10 $y = 2 \cos x + \cos \frac{1}{2}x$

11 $y = 1 + \sin x$

12 $y = 2 + \tan x$

13 $y = \frac{1}{2}x + \sin x$

14 $y = 1 + \cos x$

15 $y = 2^x + \sin x$

16 $y = \csc x - 1$

17 $y = 2 + \sec x$

18 $y = x - \sin x$

19 $y = \cos x - 1$

20 $y = |x + 1| + \cos 2x$

In Exercises 21–30 sketch the graph of f.

21 $f(x) = 2^{-x} \cos x$

22 $f(x) = 2^x \sin x$

23 $f(x) = |x| \sin x$

24 $f(x) = (1 + x^2) \cos x$

25 $f(x) = \frac{1}{2}x^2 \sin x$

26 $f(x) = \cot x \tan x$

27 $f(x) = |\sin x|$

28 $f(x) = |\cos x| + 1$

29 $f(x) = \sin |x|$

30 $f(x) = \cos |x + 1|$

CALCULATOR EXERCISES 5.7

1 Use a calculator to help sketch the graph of the equation $y = (\sin x)/x$ on the interval $(0, 2\pi]$.

2 Verify some of the graphs in this section quantitatively, by calculating the coordinates of many points.

SECTION 5.8

APPLICATIONS INVOLVING RIGHT TRIANGLES

Trigonometry was developed to help solve problems involving measurements of angles and lengths of sides of triangles. Problems of that type are no longer the most important applications; however, trigonometric techniques are still used to answer questions about triangles that arise in physical situations. We shall restrict our discussion in this section to right triangles. Triangles that do not contain a right angle will be considered in the next chapter.

In this section and Chapter 6 we shall often use the following notation. The vertices of a triangle will be denoted by A, B, and C. The angles at A, B, and C will be denoted by α, β, and γ, respectively, and the lengths of the sides opposite these angles by a, b, and c, respectively. The triangle itself will be referred to as *triangle ABC*. To *solve* a triangle means to find all of its parts, that is, the lengths of the three sides and the measures of the three angles. If the triangle is a right triangle and if one of the acute angles and a side are known, or if two sides are given, then the formulas in Section 5.2 that express the trigonometric functions as ratios of sides of a triangle may be used to find the remaining parts.

EXAMPLE 1 If, in triangle ABC, $\gamma = 90°$, $\alpha = 34°$, and $b = 10.5$, approximate the remaining parts of the triangle.

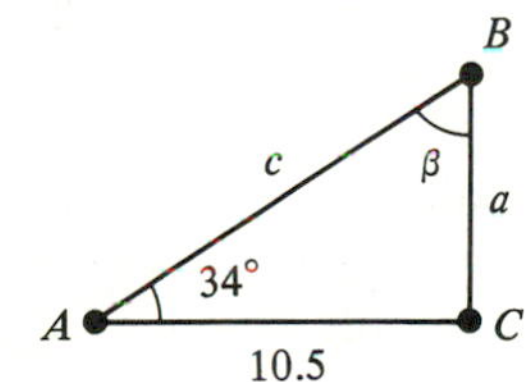

FIGURE 5.50

Solution Since the sum of the angles is 180° it follows that $\beta = 56°$. Referring to Figure 5.50 and using the fact that $\tan \alpha = \text{opp/adj}$,

$$\tan 34° = \frac{a}{10.5} \quad \text{or} \quad a = (10.5) \tan 34°.$$

Using a calculator or Table 4, we obtain

$$a \approx (10.5)(0.6745) \approx 7.1.$$

Side c can be found using either the cosine or secant functions. Since $\cos \alpha = \text{adj/hyp}$,

$$\cos 34° = \frac{10.5}{c} \quad \text{or} \quad c = \frac{10.5}{\cos 34°}.$$

Using Table 4 or a calculator, we obtain

$$c \approx \frac{10.5}{0.8290} \approx 12.7.$$

The division involved in finding c could have been avoided through the use of the secant, which equals hyp/adj. Thus,

$$\sec 34° = \frac{c}{10.5} \quad \text{or} \quad c = (10.5) \sec 34°.$$

Using Table 4 or a calculator gives us

$$c \approx (10.5)(1.2062) = 12.6651 \approx 12.7.$$

■

As illustrated in Example 1, when working with triangles, we will round off answers. One reason for doing so is that in applications, the lengths of sides of triangles and measures of angles are usually found by mechanical devices, and hence are only approximations to the exact values. Consequently, a number such as 10.5 in Example 1 is assumed to have been rounded off to the nearest tenth. We cannot expect more accuracy in the calculated values for the remaining sides and, therefore, they should also be rounded off to the nearest tenth.

In some problems a number with many digits, such as 36.4635, may be given for the side of a triangle. If a calculator is used, the number may be entered as usual; however, if Table 4 is employed, a number of this type should be rounded off before we begin any calculations. If a number x is written in the scientific form $x = c \cdot 10^k$, where $1 \leq c < 10$, then before using Table 4, c should be rounded off to three decimal places. Another way of saying this is that x should be rounded off to four **significant figures.** Some examples will help to clarify the procedure. If $x = 36.4635$, we round off to 36.46. The number 684,279 should be rounded off to 684,300. For a decimal such as 0.096202 we use 0.09620. The reason for rounding off is that the values of the trigonometric functions in Table 4 have been rounded off to four significant figures, and hence we cannot expect more than four-figure accuracy in our computations.

As a final remark on approximations, answers should also be rounded off when we are finding angles. In general, we shall use the following rules: If the sides of a triangle are known to four significant figures, then measures of angles obtained from Table 4 or a calculator should be rounded off to the nearest minute; if the sides are known to three significant figures, then calculated measures of angles should be rounded off to the nearest multiple of ten minutes; and if the sides are known to only two significant figures, then calculations should be rounded off to the nearest degree. To justify these rules, we would have to make a much deeper analysis of problems involving approximate data.

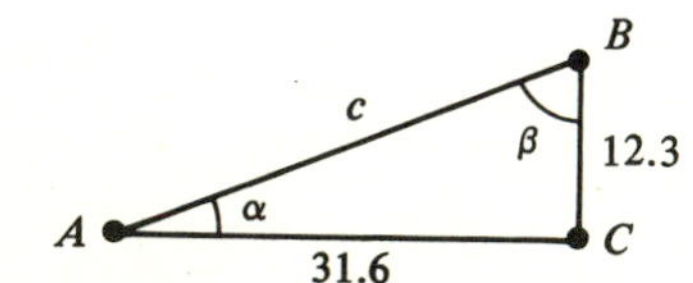

FIGURE 5.51

EXAMPLE 2 If, in triangle ABC, $\gamma = 90°$, $a = 12.3$, and $b = 31.6$, find the remaining parts.

Solution Referring to the triangle illustrated in Figure 5.51, we have

$$\tan \alpha = \frac{12.3}{31.6} \approx 0.3892.$$

By the rule stated in the preceding paragraph, α should be rounded off to the nearest multiple of 10′. From Table 4 we see that $\alpha \approx 21°20'$.

If a calculator is used to approximate α, we may proceed as follows:

Enter:	0.3892	(the value of $\tan\alpha$)
Press [INV] [tan]:	$21.265988 \approx 21°16' \approx 21°20'$	(the approximation to α)

Since $\alpha \approx 21°20'$, $\quad \beta \approx 90° - 21°20' = 68°40'.$

Again referring to Figure 5.51,

$$\sec\alpha = \frac{c}{31.6} \quad \text{or} \quad c = (31.6)\sec\alpha$$

and $\quad c \approx (31.6)\sec 21°20' \approx (31.6)(1.0736) \approx 33.9.$

Side c can also be found using the cosine. Referring to Figure 5.51, we see that

$$\cos\alpha = \frac{31.6}{c}$$

and hence, $\quad c = \dfrac{31.6}{\cos 21°20'} \approx \dfrac{31.6}{0.9315} \approx 33.9.$ ■

Right triangles are useful in solving various types of applied problems. The following examples give two illustrations; others are given in the exercises.

EXAMPLE 3 From a point on level ground 135 feet from the base of a tower, the angle of elevation of the top of the tower is 57°20′. Approximate the height of the tower.

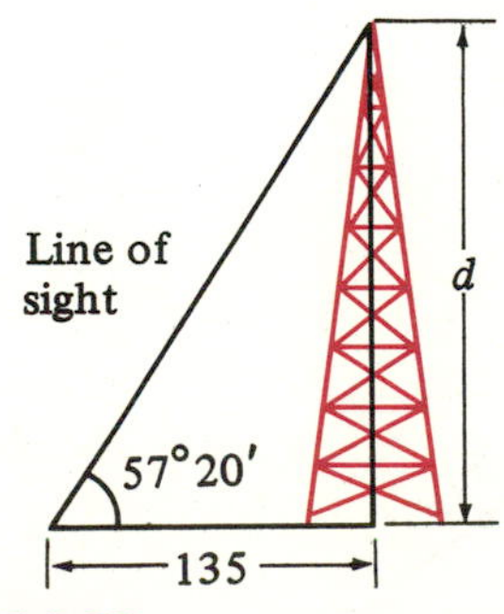

FIGURE 5.52

Solution The **angle of elevation** is the angle that the line of sight makes with the horizontal. If we let d denote the height of the tower, then the given facts may be represented by the triangle in Figure 5.52. Referring to the figure, we see that

$$\tan 57°20' = \frac{d}{135} \quad \text{or} \quad d = (135)\tan 57°20'.$$

Using a table or calculator,

$$d \approx (135)(1.560) \approx 210.6 \approx 211 \text{ feet.}$$ ■

EXAMPLE 4 From the top of a building that overlooks the ocean, a man observes a boat sailing directly toward him. If the man is 100 feet above sea level and if the angle of depression of the boat changes from 25° to 40° during the period of observation, find the approximate distance the boat travels during that time.

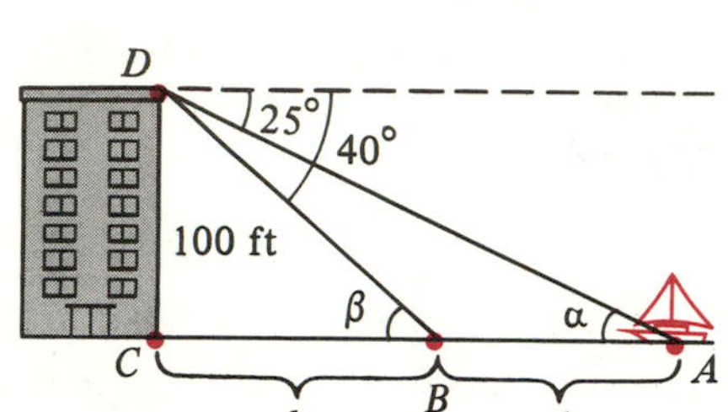

FIGURE 5.53

Solution The **angle of depression** is the angle between the line of sight and the horizontal. Let A and B be the positions of the boat that correspond to the 25° and 40° angles, respectively. Suppose that the man is at point D, and C is the point 100 feet directly below him. Let d denote the distance the boat travels and let k denote the distance from B to C. This gives us the drawing in Figure 5.53, where α and β denote angles DAC and DBC, respectively. It follows that $\alpha = 25°$ and $\beta = 40°$. (Why?)

From triangle BCD,

$$\cot \beta = \frac{k}{100} \quad \text{or} \quad k = 100 \cot \beta.$$

From triangle DAC,

$$\cot \alpha = \frac{d + k}{100} \quad \text{or} \quad d + k = 100 \cot \alpha.$$

Consequently,

$$\begin{aligned} d &= 100 \cot \alpha - k = 100 \cot \alpha - 100 \cot \beta \\ &= 100(\cot \alpha - \cot \beta) = 100(\cot 25° - \cot 40°) \\ &\approx 100(2.145 - 1.192) = 100(0.953) = 95.3 \end{aligned}$$

Hence, $d \approx 95$ feet. ■

In certain navigation and surveying problems, the **direction,** or **bearing,** from a point P to a point Q is often specified by stating the acute angle that the half-line from P through Q varies to the east or west of the north-south line. Figure 5.54 illustrates four such lines. The north-south and east-west

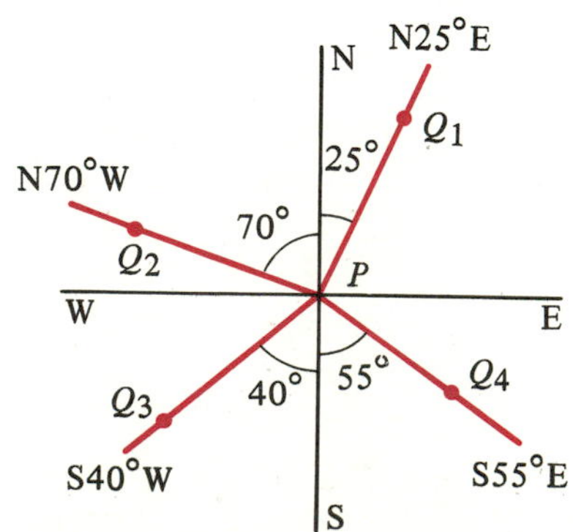

FIGURE 5.54

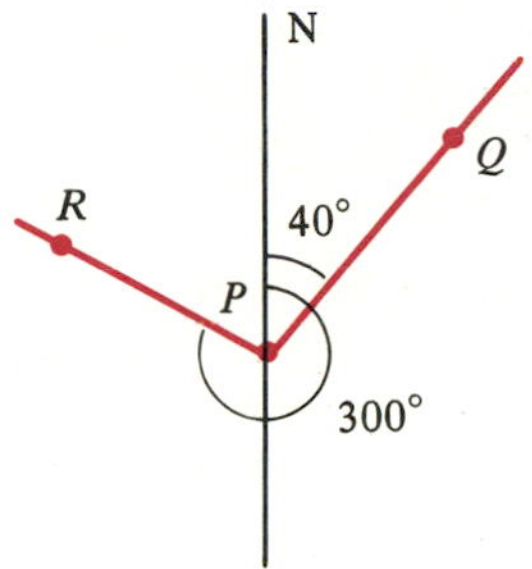

FIGURE 5.55

lines are labeled NS and WE, respectively. The bearing from P to Q_1 is 25° east of north and is denoted by N25°E. We also refer to the **direction** N25°E, meaning the direction from P to Q_1. The bearings from P to Q_2, Q_3, and Q_4 are represented in a similar manner in the figure.

In air navigation, directions and bearings are specified by measuring from the north in a clockwise direction. In this case, a positive measure is assigned to the angle instead of the negative measure to which we are accustomed for clockwise rotations. Thus, referring to Figure 5.55, we see that the direction of PQ is 40°, whereas the direction of PR is 300°.

We shall use these notations in Exercises 35–37.

EXERCISES 5.8

Given the indicated parts of triangle ABC with $\gamma = 90°$, approximate the remaining parts of the triangles in Exercises 1–16.

1 $\alpha = 30°,\ b = 20$

2 $\beta = 45°,\ b = 35$

3 $\beta = 52°,\ a = 15$

4 $\alpha = 37°,\ b = 24$

5 $\alpha = 17°40',\ a = 4.50$

6 $\beta = 64°20',\ a = 20.1$

7 $\beta = 71°51',\ b = 240.0$

8 $\alpha = 31°10',\ a = 510$

9 $a = 25,\ b = 45$

10 $a = 31,\ b = 9.0$

11 $c = 5.8,\ b = 2.1$

12 $a = 0.42,\ c = 0.68$

13 $\alpha = 37°46',\ b = 512.0$

14 $\beta = 10°17',\ b = 68.40$

15 $a = 614,\ c = 806$

16 $c = 37.4,\ b = 21.6$

17 Approximate the angle of elevation of the sun if a boy 5.0 feet tall casts a shadow 4.0 feet long on level ground.

18 From a point 15 meters above level ground an observer measures the angle of depression of an object on the ground as 68°. Approximate the distance from the object to the point on the ground directly beneath the observer.

19 The string on a kite is taut and makes an angle of 54°20′ with the horizontal. Find the approximate height of the kite above level ground if 85.0 meters of string are out and the end of the string is held 1.50 meters above the ground.

20 The side of a regular pentagon is 24.0 cm long. Approximate the radius of the circumscribed circle.

21 From a point P on level ground the angle of elevation of the top of a tower is 26°50′. From a point 25.0 meters closer to the tower and on the same line with P and the base of the tower, the angle of elevation of the top is 53°30′. Approximate the height of the tower.

22 A ladder 20 feet long leans against the side of a building. If the angle between the ladder and the building is 22°, approximate the distance from the bottom of the ladder to the building. If the distance from the bottom of the ladder to the building is increased by 3.0 feet, approximately how far does the top of the ladder move down the building?

23 As a weather balloon rises vertically, its angle of elevation from a point P on the level ground 110 km from the point Q directly underneath the balloon changes from 19°20′ to 31°50′. Approximately how far does the balloon rise during this period?

24 From a point A that is 8.20 meters above level ground, the angle of elevation of the top of a building is 31°20′ and the angle of depression of the base of the building is 12°50′. Approximate the height of the building.

25 To find the distance d between two points P and Q on opposite shores of a lake, a surveyor locates a point R that is 50.0 meters from P such that RP is perpendicular to PQ. Next, using a transit, the surveyor measures angle PRQ as 72°40′. What is d?

26 A guy wire is attached to the top of a radio antenna and to a point on horizontal ground that is 40.0 meters from the base of the antenna. If the wire makes an angle of 58°20′ with the ground, approximate the length of the wire.

27 An octagon is inscribed in a circle of radius 12.0 cm. Approximate the perimeter of the octagon.

28 A builder wishes to construct a ramp 24 feet long that rises to a height of 5.0 feet above level ground. Approximate the angle that the ramp should make with the horizontal.

29 A rocket is fired at sea level and climbs at a constant angle of 75° through a distance of 10,000 feet. Approximate its altitude to the nearest foot.

30 A CB antenna is located on the top of a garage that is 16 feet tall. From a point on level ground that is 100 feet from a point directly below the antenna, the antenna subtends an angle of 12°. Approximate the length of the antenna.

31 An airplane flying at an altitude of 10,000 feet passes directly over a fixed object on the ground. One minute later the angle of depression of the object is 42°. Approximate the speed of the airplane to the nearest mph.

32 A motorist is traveling along a level highway at a speed of 60 km/h directly toward a distant mountain. She observes that between 1:00 P.M. and 1:10 P.M. the angle of elevation of the top of the mountain changes from 10° to 70°. Approximate the height of the mountain.

33 An airplane pilot wishes to make his approach to an airstrip at an angle of 10° with the horizontal. If he is flying at an altitude of 5,000 feet, approximately how far from the airstrip should he begin his descent? (Give answer to the nearest 100 feet.)

34 In order to measure the height h of a cloud cover, a spotlight is directed vertically upward from the ground. From a point on level ground that is d meters from the spotlight, the angle of elevation θ of the light image on the clouds is then measured. Find a formula that expresses h in terms of d and θ. As a special case, approximate h if $d = 1{,}000$ meters and $\theta = 59°0'$.

35 A ship leaves port at 1:00 P.M. and sails in the direction N34°W at a rate of 24 mph. Another ship leaves port at 1:30 P.M. and sails in the direction N56°E at a rate of 18 mph. Approximately how far apart are the ships at 3:00 P.M.?

36 From an observation point A a forest ranger sights a fire in the direction S35°50′W. From a point B, 5 miles due west of A, another ranger sights the same fire in the direction S54°10′E. Approximate, to the nearest mile, the the distance of the fire from A.

37 An airplane flying at a speed of 360 mph flies from a point A in the direction 137° for 30 minutes and then flies in the direction 227° for 45 minutes. Approximate, to the nearest mile, the distance from the airplane to A.

38 Generalize Exercise 19 to the case where the angle is α, the number of meters of string out is d, and the end of the string is held c meters above the ground. Express the height h of the kite in terms of α, d, and c.

39 Generalize Exercise 23 to the case where the distance from P to Q is d km and the angle of elevation changes from α to β.

40 Generalize Exercise 24 to the case where point A is d meters above ground and the angles of elevation and depression are α and β, respectively. Express the height h of the building in terms of d, α, and β.

CALCULATOR EXERCISES 5.8

Given the indicated parts of triangle ABC with $\gamma = 90°$, approximate the remaining parts of each triangle in Exercises 1–10. Round off answers to four significant figures.

1 $\alpha = 41.27°$, $a = 314.6$

2 $\beta = 24.96°$, $b = 209.3$

3 $\beta = 37.06°$, $a = 0.4613$

4 $\alpha = 17.69°$, $b = 1.307$

5 $\beta = 2.71°$, $b = 7149$

6 $\alpha = 84.07°$, $a = 0.1024$

7 $a = 46.87$, $b = 13.12$

8 $a = 6.948$, $b = 8.371$

9 $b = 2{,}462$, $c = 5{,}074$

10 $a = 88.12$, $c = 94.06$

SECTION 5.9

HARMONIC MOTION

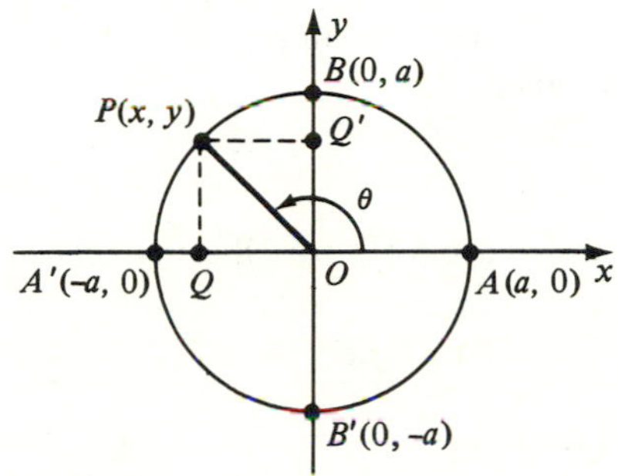

FIGURE 5.56

Trigonometric functions are important in the investigation of vibratory or oscillatory motion, such as the motion of a particle in a vibrating guitar string or in a spring that has been compressed or elongated, and then released to oscillate back and forth. The fundamental type of particle displacement inherent in these illustrations is termed *harmonic motion.* As an aid to introducing this concept, let us consider a point P moving at a constant rate around a circle of radius a with center at the origin O of a rectangular coordinate system. Suppose that the initial position of P is $A(a, 0)$ and θ is the angle generated by the ray OP after t units of time (see Figure 5.56).

The **angular speed** ω of OP is, by definition, the rate at which the measure of θ changes per unit time. To state that P moves around the circle at a constant rate is equivalent to stating that the angular speed ω is constant. If ω is constant, then $\theta = \omega t$. To illustrate, if $\omega = \pi/6$ radians per second, then $\theta = (\pi/6)t$. In this case, at $t = 1$ second, $\theta = (\pi/6)(1) = \pi/6$, and P is one-third of the way from $A(a, 0)$ to $B(0, a)$. At $t = 2$ seconds, $\theta = (\pi/6)(2) = \pi/3$, and P is two-thirds of the way from A to B. At $t = 6$ seconds, $\theta = (\pi/6)(6) = \pi$, and P is at $A'(-a, 0)$, and so on.

If the coordinates of P are (x, y) then $\sin\theta = y/a$ and $\cos\theta = x/a$. Multiplying both sides of these equations by a, and using the fact that $\theta = \omega t$, gives us

$$x = a\cos\omega t, \qquad y = a\sin\omega t.$$

The last two equations specify the position (x, y) of P at any time t. Let us next consider the point $Q(x, 0)$, which is called the **projection of P on the x-axis.** The position of Q is given by $x = a\cos\omega t$. As P moves around the circle several times, the point Q oscillates back and forth between $A(a, 0)$ and $A'(-a, 0)$. Similarly, the point $Q'(0, y)$ is called the **projection of P on the y-axis,** and its position is given by $y = a\sin\omega t$. As P moves around the circle, Q' oscillates between $B'(0, -a)$ and $B(0, a)$. The motions of Q and Q' are of the type described in the next definition.

DEFINITION

> A point that moves on a coordinate line such that its distance d from the origin at time t is given by either $d = a\cos\omega t$ or $d = a\sin\omega t$, where a and ω are real numbers, is said to be in **simple harmonic motion.**

In the preceding definition, the **amplitude** of the motion is the maximum displacement $|a|$ of the point from the origin. The **period** is the time, $2\pi/\omega$,

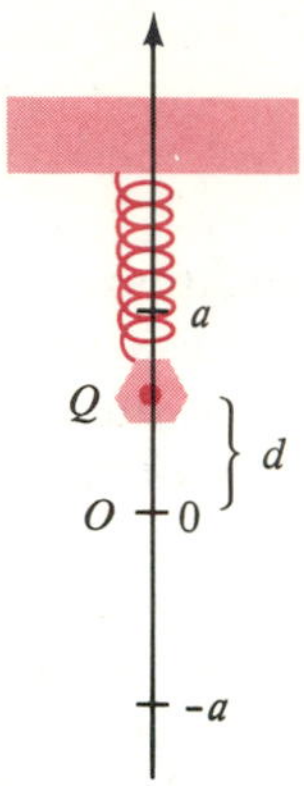

FIGURE 5.57

required for one complete oscillation. The **frequency** $\omega/2\pi$ is the number of oscillations per unit of time.

A physical interpretation of simple harmonic motion can be obtained by considering a spring that has an attached weight that is oscillating vertically relative to a coordinate line, as illustrated in Figure 5.57. The number d represents the coordinate of a fixed point Q in the weight, and we assume that the amplitude a of the motion is constant. In this case no frictional force is retarding the motion. If friction is present, then the amplitude decreases with time, and the motion is said to be *damped.*

EXAMPLE 1 Suppose the weight shown in Figure 5.57 is oscillating according to the law

$$d = 10 \cos\left(\frac{\pi}{6} t\right)$$

where t is measured in seconds and d in centimeters. Discuss the motion of the weight.

Solution By definition, the motion is simple harmonic. The amplitude is $a = 10$ cm. Since $\omega = \pi/6$, the period is $2\pi/\omega = 2\pi/(\pi/6) = 12$. Thus, one complete oscillation takes 12 seconds. The frequency is $\omega/2\pi = (\pi/6)/2\pi = 1/12$; that is, 1/12 of an oscillation takes place each second. The following table indicates the position of Q at various times.

t	0	1	2	3	4	5	6
$(\pi/6)t$	0	$\pi/6$	$\pi/3$	$\pi/2$	$2\pi/3$	$5\pi/6$	π
d	10	$5\sqrt{3} \approx 8.7$	5	0	-5	$-5\sqrt{3} \approx -8.7$	-10

Note that the initial position of Q is 10 cm above the origin O. It then moves downward, gaining speed until it reaches O. In particular, Q travels approximately 1.3 cm during the first second, 3.7 cm during the next second, and 5 cm during the third second. It then slows down until it reaches a point 10 cm below O at the end of 6 seconds. It is left to the reader to verify that the direction of motion is then reversed, and the weight moves upward, gaining speed until it reaches O, after which it slows down until it returns to its original position at the end of 12 seconds. The direction of motion is then reversed again, and the same pattern is repeated indefinitely. ■

The preceding example is typical of simple harmonic motion. If the initial angle AOP in Figure 5.56 is ϕ at $t = 0$, then the position (x, y) of P on the circle is given by

$$x = a \cos(\omega t + \phi), \qquad y = a \sin(\omega t + \phi).$$

Points that vary on a coordinate line according to either of these formulas are also said to be in simple harmonic motion.

Simple harmonic motion takes place in many different types of wave motion, such as water waves, sound waves, radio waves, light waves, and distortional waves that are present in vibrating bodies. As a specific example, consider the waves made by holding one end of a long rope, as in Figure 5.58, and then causing it to vibrate by raising and lowering the hand in simple harmonic motion.

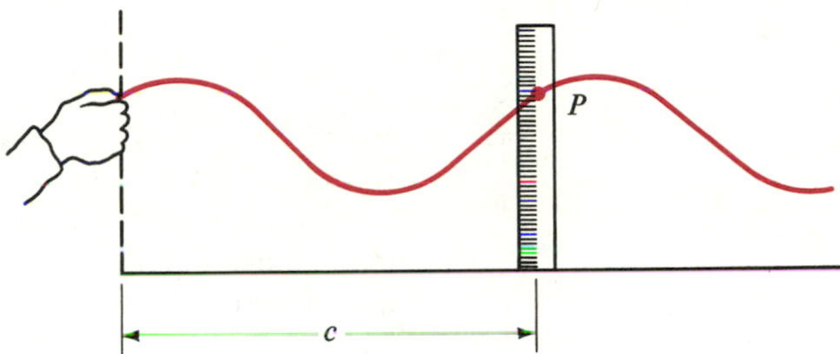

FIGURE 5.58

Waves appear to move along the rope, traveling away from the hand. A fixed, vertical measuring device placed a distance c from the end, as shown in the figure, may be used to study the upward and downward motion of a specific particle P in the rope. If a suitable horizontal axis is chosen, it can be shown that P is in simple harmonic motion, as if it were attached to a spring. A similar motion takes place in water ripples on a large pond.

The definition of simple harmonic motion is usually extended to include situations where d is *any* mathematical or physical quantity (not necessarily a distance). In the next illustration d is itself an angle.

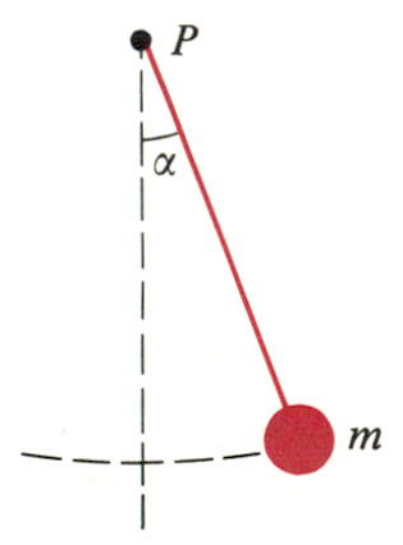

FIGURE 5.59

Historically, the first type of harmonic motion to be scientifically investigated involved pendulums. Figure 5.59 illustrates a *simple pendulum* consisting of a bob of mass m attached to one end of a string, with the other end of the string attached to a fixed point P. In the ideal case it is assumed that the string is weightless, that there is no air resistance, and that the only force acting on the bob is gravity. If the bob is displaced sideways and released, the pendulum oscillates back and forth in a vertical plane. Let α denote the *angular displacement* at time t (see Figure 5.59). If the bob moves through a small arc (say, $|\alpha| < 5°$), then it can be shown by using physical laws, that

$$\alpha = \beta \cos(\omega t + \alpha_0)$$

where α_0 is the initial displacement, ω is the frequency of oscillation of the angle α, and β is the (angular) amplitude of oscillation. This means that the *angle* α is in simple harmonic motion, and hence we refer to the motion as *angular* simple harmonic motion.

As a final illustration, in electric circuits an alternating electromotive force (emf) and the current may vary harmonically. For example, the emf e (measured in volts) at time t may be given by

$$e = E \sin \omega t$$

where E is the maximum value of e. If an emf of this type is impressed on a circuit containing only a resistance R, then by Ohm's Law, the current i at time t is

$$i = \frac{e}{R} = \frac{E}{R} \sin \omega t = I \sin \omega t$$

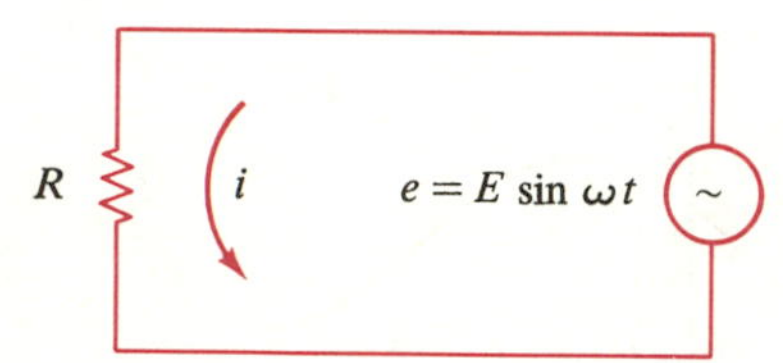

FIGURE 5.60

where $I = E/R$. A schematic drawing of an electric circuit of this type is illustrated in Figure 5.60.

In this case the maximum value I of i occurs at the same time as the maximum value E of e. In other situations these maximum values may occur at different times, in which case we say there is a **phase difference** between e and i. For example, if $e = E \sin \omega t$, we could have either

$$\text{(a)} \quad i = I \sin (\omega t - \phi), \quad \text{or} \quad \text{(b)} \quad i = I \sin (\omega t + \phi)$$

where $\phi > 0$. In case (a), the current is said to **lag** the emf by an amount ϕ/ω, and the graph of i can be obtained by shifting the graph of $i = I \sin \omega t$ and amount ϕ/ω to the *right*. In case (b) we shift the graph to the *left*, and the current is said to **lead** the emf by an amount ϕ/ω.

Numerous other physical illustrations involve simple harmonic motion. Interested students may find further information in books on physics and engineering.

EXERCISES 5.9

1 A wheel of diameter 40 cm is rotating about an axle at a rate of 100 revolutions per minute. If a coordinate system is introduced as in Figure 5.56, where P is a point on the rim of the wheel, find:

(a) the angular speed of OP.

(b) the position (x, y) of P after t minutes.

2 A wheel of radius 2 feet is rotating about an axle, and the angular speed of a ray from the center of the wheel to a point P on the rim is $5\pi/6$ radians per second.

(a) How many revolutions does the wheel make in 10 minutes?

(b) If a coordinate system is introduced as in Figure 5.56, find the position (x, y) of P after t seconds.

In Exercises 3–6 the given formula specifies the position of a point P that is moving harmonically on a vertical axis, where t is in seconds and d is in centimeters. Determine the amplitude, period, and frequency, and describe the motion of the point during one complete oscillation (starting at $t = 0$).

3 $d = 10 \sin 6\pi t$

4 $d = \frac{1}{3} \cos \frac{\pi}{4} t$

5 $d = 4 \cos \frac{3\pi}{2} t$

6 $d = 6 \sin \frac{2\pi}{3} t$

7 A point P in simple harmonic motion has a period of 3 seconds and an amplitude of 5 cm. Express the motion of P by means of an equation of the form $d = a \cos \omega t$.

8 A point P in simple harmonic motion has a frequency of $\frac{1}{2}$ oscillation per minute and amplitude of 4 feet. Express the motion of P by means of an equation of the form $d = a \sin \omega t$.

9 The electromotive force e and current i in a certain alternating current circuit are given by

$$e = 220 \sin 360\pi t,$$

$$i = 20 \sin (360\pi t - \tfrac{1}{4}\pi).$$

Sketch the graphs of e and i on the same coordinate axes and determine the lag or lead.

10 Rework Exercise 9 if

$$e = 110 \sin 120\pi t,$$

$$i = 15 \sin (120\pi t + \tfrac{1}{3}\pi).$$

SECTION 5.10
REVIEW

Define or discuss each of the following.

1 Angle
2 Initial and terminal sides of an angle
3 Coterminal sides of angles
4 Standard position of an angle
5 Positive and negative angles
6 Degree measure
7 Acute or obtuse angles
8 Radian measure
9 The relationship between radians and degrees
10 Trigonometric functions of acute angles
11 The Fundamental Identities
12 Trigonometric functions of any angle
13 Signs of the trigonometric functions
14 Trigonometric functions of real numbers
15 Trigonometric functions in terms of a unit circle
16 Domains and ranges of the trigonometric functions
17 Periodic function; period; periods of the trigonometric functions
18 Reference numbers and angles
19 Finding values of trigonometric functions using reference numbers and angles
20 Graphs of the trigonometric functions
21 The graphs of $f(x) = a \sin (bx + c)$ and $f(x) = a \cos (bx + c)$
22 Trigonometric solutions of right triangles
23 Harmonic motion

EXERCISES 5.10

1 Find the radian measures that correspond to the following degree measures: 330°, 405°, −150°, 240°, 36°.

2 Find the degree measures that correspond to the following radian measures: $9\pi/2$, $-2\pi/3$, $7\pi/4$, 5π, $\pi/5$.

3 (a) Find the reference number t' if t equals: $5\pi/4$, $-5\pi/6$, $-9\pi/8$.

(b) Find the reference angle for each of the following: 245°, 137°10′, 892°, $5\pi/4$, $-\pi/6$, $9\pi/8$.

4 A central angle θ is subtended by an arc 20 cm long on a circle of radius 2 meters. What is the radian measure of θ?

5 If θ is an acute angle of a right triangle and if the adjacent side and hypotenuse have lengths 4 and 7, respectively, find the values of the six trigonometric functions of θ.

6 Find the quadrant containing θ if:

(a) $\sec\theta < 0$ and $\sin\theta > 0$.

(b) $\cot\theta > 0$ and $\csc\theta < 0$.

(c) $\cos\theta > 0$, and $\tan\theta < 0$.

7 Find the values of the remaining trigonometric functions if:

(a) $\sin\theta = -\frac{4}{5}$ and $\cos\theta = \frac{3}{5}$.

(b) $\csc\theta = \sqrt{13}/2$ and $\cot\theta = -\frac{3}{2}$.

8 Find the values of the six trigonometric functions of θ if θ is in standard position and satisfies the stated condition.

(a) The point $(30, -40)$ is on the terminal side of θ.

(b) The terminal side of θ is in quadrant II and is parallel to the line $2x + 3y + 6 = 0$.

(c) $\theta = -90°$.

9 Without the use of tables or calculators, find the values of the trigonometric functions corresponding to the following.

(a) $\theta = 9\pi/2$ (b) $\theta = -5\pi/4$

(c) $\theta = 0$ (d) $\theta = 11\pi/6$

10 Find the following without the use of tables or calculators.

(a) $\cos 225°$ (b) $\tan 150°$

(c) $\sin(-\pi/6)$ (d) $\sec(4\pi/3)$

(e) $\cot(7\pi/4)$ (f) $\csc(300°)$

In Exercises 11–18 find the amplitude and period, and sketch the graph of f.

11 $f(x) = 5\cos x$ **12** $f(x) = \frac{2}{3}\sin x$

13 $f(x) = -\tan x$ **14** $f(x) = 4\sin 3x$

15 $f(x) = 3\cos 4x$ **16** $f(x) = -4\sin\frac{1}{3}x$

17 $f(x) = 2\sec\frac{1}{2}x$ **18** $f(x) = \frac{1}{2}\csc(2x)$

Sketch the graphs of the equations in Exercises 19–25.

19 $y = 2\sin\left(x - \dfrac{2\pi}{3}\right)$ **20** $y = -4\cos\left(x + \dfrac{\pi}{6}\right)$

21 $y = 5\cos\left(2x + \dfrac{\pi}{2}\right)$ **22** $y = \tan\left(x - \dfrac{\pi}{4}\right)$

23 $y = 2\sin x + \sin 2x$ **24** $y = 1 + x + \cos x$

25 $y = 2^{-x}\sin 2x$

Given the parts of triangle ABC in Exercises 26–28 with $\gamma = 90°$, approximate the remaining parts.

26 $\beta = 60°$, $b = 40$

27 $\alpha = 54°40'$, $b = 220$

28 $\alpha = 62$, $b = 25$

ANALYTIC TRIGONOMETRY

In this chapter we shall examine various aspects of trigonometry. In the first five sections the emphasis is on identities and equations. We then define the inverse trigonometric functions and study their properties. In Sections 6.7 and 6.8 we consider methods of solutions of oblique triangles. The trigonometric form for complex numbers and De Moivre's Theorem appear in Sections 6.9 and 6.10. The chapter concludes with a discussion of vectors. In the course of our work we shall derive many important trigonometric identities and formulas; for convenient reference, they are listed on the inside covers of the text.

SECTION 6.1
TRIGONOMETRIC IDENTITIES

A mathematical expression that contains symbols such as $\sin x$, $\cos \beta$, $\tan v$, etc., where the letters x, β, and v are variables, is referred to as a **trigonometric expression.** The following are examples of trigonometric expressions:

$$x + \sin x, \qquad \frac{\cos (3y + 1)}{x^2 + \tan^2 (z - y^2)}, \qquad \frac{\sqrt{\theta} + 2^{\sin \theta}}{\sec (\cot \theta)}.$$

As usual, we assume that the domain of each variable is the set of real numbers (or angles) for which the expressions are meaningful.

The Fundamental Identities that were introduced in Section 5.2 may sometimes be used to help simplify complicated trigonometric expressions. Let us begin by restating these important formulas and working an example. The variable t in each fundamental identity may represent either a real number or the measure of an angle.

THE FUNDAMENTAL IDENTITIES

$$\csc t = \frac{1}{\sin t} \qquad \tan t = \frac{\sin t}{\cos t} \qquad \sin^2 t + \cos^2 t = 1$$

$$\sec t = \frac{1}{\cos t} \qquad \cot t = \frac{\cos t}{\sin t} \qquad 1 + \tan^2 t = \sec^2 t$$

$$\cot t = \frac{1}{\tan t} \qquad 1 + \cot^2 t = \csc^2 t$$

EXAMPLE 1 Simplify the expression $(\sec \theta + \tan \theta)(1 - \sin \theta)$.

Solution Supply reasons for the following steps:

$$\begin{aligned}(\sec \theta + \tan \theta)(1 - \sin \theta) &= \left(\frac{1}{\cos \theta} + \frac{\sin \theta}{\cos \theta}\right)(1 - \sin \theta) \\ &= \left(\frac{1 + \sin \theta}{\cos \theta}\right)(1 - \sin \theta) \\ &= \frac{1 - \sin^2 \theta}{\cos \theta} \\ &= \frac{\cos^2 \theta}{\cos \theta} \\ &= \cos \theta\end{aligned}$$

■

There are other ways to treat the expression in Example 1. We could first multiply the two factors, and then simplify and combine terms. The method we employed—of changing all expressions to expressions that involve only sines and cosines—is often worthwhile. However, that technique does not always lead to the shortest possible simplification.

We have encountered several trigonometric identities in our previous work. Some of them, such as the Fundamental Identities, are basic and should be memorized. Other identities are introduced merely to supply practice in manipulating trigonometric expressions and therefore should not be memorized. We shall take the latter point of view in this section. Thus, the identities to be investigated are unimportant in their own right. The important thing is the manipulative practice that we will gain. The ability to carry out trigonometric manipulations is essential for solving problems that are encountered in advanced courses in mathematics and science.

We often use the phrase "verify an identity" instead of "prove that an equation is an identity." When verifying an identity, we shall use fundamental identities and algebraic manipulations to change the form of trigonometric expressions in a manner similar to the solution of Example 1. The preferred method of showing that an equation is an identity is to transform one side into the other, as illustrated in the next three examples. The reader should supply reasons for all steps in the solutions.

EXAMPLE 2 Verify the identity $\dfrac{\tan t + \cos t}{\sin t} = \sec t + \cot t$.

Solution We shall transform the left side into the right side. Thus,

$$\begin{aligned}\frac{\tan t + \cos t}{\sin t} &= \frac{\tan t}{\sin t} + \frac{\cos t}{\sin t}\\ &= \frac{\sin t/\cos t}{\sin t} + \cot t\\ &= \frac{1}{\cos t} + \cot t\\ &= \sec t + \cot t.\end{aligned}$$

■

EXAMPLE 3 Verify the identity $\sec \alpha - \cos \alpha = \sin \alpha \tan \alpha$.

Solution

$$\begin{aligned}\sec \alpha - \cos \alpha &= \frac{1}{\cos \alpha} - \cos \alpha\\ &= \frac{1 - \cos^2 \alpha}{\cos \alpha}\\ &= \frac{\sin^2\alpha}{\cos \alpha} = \sin \alpha \left(\frac{\sin \alpha}{\cos \alpha}\right)\\ &= \sin \alpha \tan \alpha.\end{aligned}$$

■

EXAMPLE 4 Verify the identity $\dfrac{\cos x}{1 - \sin x} = \dfrac{1 + \sin x}{\cos x}$.

Solution We begin by multiplying the numerator and denominator of the fraction on the left by $1 + \sin x$. Thus

$$\frac{\cos x}{1 - \sin x} = \frac{\cos x}{1 - \sin x} \cdot \frac{1 + \sin x}{1 + \sin x}$$
$$= \frac{\cos x\,(1 + \sin x)}{1 - \sin^2 x}$$
$$= \frac{\cos x\,(1 + \sin x)}{\cos^2 x}$$
$$= \frac{1 + \sin x}{\cos x}.$$ ■

Another technique for showing that an equation $p = q$ is an identity is to begin by transforming the left side p into another expression s, making sure that each step is *reversible* in the sense that it is possible to transform s back into p by reversing the procedure that has been used. In this case the equation $p = s$ is an identity. Next, as a *separate* exercise, we must show that the right side q can also be transformed to the expression s by means of reversible steps and hence that $q = s$ is an identity. It then follows that $p = q$ is an identity. This method is illustrated in the next example.

EXAMPLE 5 Verify the identity $(\tan\theta - \sec\theta)^2 = \dfrac{1 - \sin\theta}{1 + \sin\theta}$.

Solution We shall verify the identity by showing that each side of the equation can be transformed into the same expression. Starting with the left side, we may write

$$(\tan\theta - \sec\theta)^2 = \tan^2\theta - 2\tan\theta\sec\theta + \sec^2\theta$$
$$= \frac{\sin^2\theta}{\cos^2\theta} - \frac{2\sin\theta}{\cos^2\theta} + \frac{1}{\cos^2\theta}$$
$$= \frac{\sin^2\theta - 2\sin\theta + 1}{\cos^2\theta}.$$

The right side of the given equation may be changed by multiplying numerator and denominator by $1 - \sin\theta$. Thus,

$$\frac{1 - \sin\theta}{1 + \sin\theta} = \frac{1 - \sin\theta}{1 + \sin\theta} \cdot \frac{1 - \sin\theta}{1 - \sin\theta}$$
$$= \frac{1 - 2\sin\theta + \sin^2\theta}{1 - \sin^2\theta}$$
$$= \frac{1 - 2\sin\theta + \sin^2\theta}{\cos^2\theta}$$

The last expression is the same as that obtained for $(\tan\theta - \sec\theta)^2$. Since all steps are reversible, it follows that the given equation is an identity. ■

EXERCISES 6.1

Verify the identities in Exercises 1–88.

1 $\cos\theta\sec\theta = 1$

2 $\tan\alpha\cot\alpha = 1$

3 $\sin\theta\sec\theta = \tan\theta$

4 $\sin\alpha\cot\alpha = \cos\alpha$

5 $\dfrac{\csc x}{\sec x} = \cot x$

6 $\cot\beta\sec\beta = \csc\beta$

7 $(1 + \cos\alpha)(1 - \cos\alpha) = \sin^2\alpha$

8 $\cos^2 x(\sec^2 x - 1) = \sin^2 x$

9 $\cos^2 t - \sin^2 t = 2\cos^2 t - 1$

10 $(\tan\theta + \cot\theta)\tan\theta = \sec^2\theta$

11 $\dfrac{\sin t}{\csc t} + \dfrac{\cos t}{\sec t} = 1$

12 $1 - 2\sin^2 x = 2\cos^2 x - 1$

13 $(1 + \sin\alpha)(1 - \sin\alpha) = \dfrac{1}{\sec^2\alpha}$

14 $(1 - \sin^2 t)(1 + \tan^2 t) = 1$

15 $\sec\beta - \cos\beta = \tan\beta\sin\beta$

16 $\dfrac{\sin w + \cos w}{\cos w} = 1 + \tan w$

17 $\dfrac{\csc^2\theta}{1 + \tan^2\theta} = \cot^2\theta$

18 $\sin x + \cos x\cot x = \csc x$

19 $\sin t(\csc t - \sin t) = \cos^2 t$

20 $\cot t + \tan t = \csc t\sec t$

21 $\csc\theta - \sin\theta = \cot\theta\cos\theta$

22 $\cos\theta(\tan\theta + \cot\theta) = \csc\theta$

23 $\dfrac{\sec^2 u - 1}{\sec^2 u} = \sin^2 u$

24 $(\tan u + \cot u)(\cos u + \sin u) = \sec u + \csc u$

25 $(\cos^2 x - 1)(\tan^2 x + 1) = 1 - \sec^2 x$

26 $(\cot\alpha + \csc\alpha)(\tan\alpha - \sin\alpha) = \sec\alpha - \cos\alpha$

27 $\sec t\csc t + \cot t = \tan t + 2\cos t\csc t$

28 $\dfrac{1 + \cos^2 y}{\sin^2 y} = 2\csc^2 y - 1$

29 $\sec^2\theta\csc^2\theta = \sec^2\theta + \csc^2\theta$

30 $\dfrac{\sec x - \cos x}{\tan x} = \dfrac{\tan x}{\sec x}$

31 $\dfrac{1 + \cos t}{\sin t} + \dfrac{\sin t}{1 + \cos t} = 2\csc t$

32 $\tan^2\alpha - \sin^2\alpha = \tan^2\alpha\sin^2\alpha$

33 $\dfrac{1 + \tan^2 v}{\tan^2 v} = \csc^2 v$

34 $\dfrac{\sec\theta + \csc\theta}{\sec\theta - \csc\theta} = \dfrac{\sin\theta + \cos\theta}{\sin\theta - \cos\theta}$

35 $\dfrac{1 + \sin x}{1 - \sin x} - \dfrac{1 - \sin x}{1 + \sin x} = 4\tan x\sec x$

36 $\dfrac{1}{1 - \cos\gamma} + \dfrac{1}{1 + \cos\gamma} = 2\csc^2\gamma$

37 $\dfrac{1 + \csc\beta}{\sec\beta} - \cot\beta = \cos\beta$

38 $\dfrac{\cos x\cot x}{\cot x - \cos x} = \dfrac{\cot x + \cos x}{\cos x\cot x}$

39 $(\sec u - \tan u)(\csc u + 1) = \cot u$

40 $\dfrac{\cot\theta - \tan\theta}{\sin\theta + \cos\theta} = \csc\theta - \sec\theta$

41 $\dfrac{\cot\alpha - 1}{1 - \tan\alpha} = \cot\alpha$

42 $\dfrac{1 + \sec\beta}{\tan\beta + \sin\beta} = \csc\beta$

43 $\csc^4 t - \cot^4 t = \cot^2 t + \csc^2 t$

44 $\cos^4 \theta + \sin^2 \theta = \sin^4 \theta + \cos^2 \theta$

45 $\dfrac{\cos \beta}{1 - \sin \beta} = \sec \beta + \tan \beta$

46 $\dfrac{1}{\csc y - \cot y} = \csc y + \cot y$

47 $\dfrac{\tan^2 x}{\sec x + 1} = \dfrac{1 - \cos x}{\cos x}$

48 $\dfrac{\cot x}{\csc x + 1} = \dfrac{\csc x - 1}{\cot x}$

49 $\dfrac{\cot u - 1}{\cot u + 1} = \dfrac{1 - \tan u}{1 + \tan u}$

50 $\dfrac{1 + \sec x}{\sin x + \tan x} = \csc x$

51 $\sin^4 r - \cos^4 r = \sin^2 r - \cos^2 r$

52 $\sin^4 \theta + 2 \sin^2 \theta \cos^2 \theta + \cos^4 \theta = 1$

53 $\tan^4 k - \sec^4 k = 1 - 2 \sec^2 k$

54 $\sec^4 u - \sec^2 u = \tan^4 u + \tan^2 u$

55 $(\sec t + \tan t)^2 = \dfrac{1 + \sin t}{1 - \sin t}$

56 $\sec^2 \gamma + \tan^2 \gamma = (1 - \sin^4 \gamma)(\sec^4 \gamma)$

57 $(\sin^2 \theta + \cos^2 \theta)^3 = 1$

58 $\dfrac{\sin t}{1 - \cos t} = \csc t + \cot t$

59 $\dfrac{1 + \csc \beta}{\cot \beta + \cos \beta} = \sec \beta$

60 $\dfrac{\sin z \tan z}{\tan z - \sin z} = \dfrac{\tan z + \sin z}{\sin z \tan z}$

61 $\left(\dfrac{\sin^2 x}{\tan^4 x}\right)^3 \left(\dfrac{\csc^3 x}{\cot^6 x}\right)^2 = 1$

62 $\dfrac{\cos^3 x - \sin^3 x}{\cos x - \sin x} = 1 + \sin x \cos x$

63 $\dfrac{\sin \theta + \cos \theta}{\tan^2 \theta - 1} = \dfrac{\cos^2 \theta}{\sin \theta - \cos \theta}$

64 $(\csc t - \cot t)^4 (\csc t + \cot t)^4 = 1$

65 $(a \cos t - b \sin t)^2 + (a \sin t + b \cos t)^2 = a^2 + b^2$

66 $\sin^6 v + \cos^6 v = 1 - 3 \sin^2 v \cos^2 v$

67 $\dfrac{\sin \alpha \cos \beta + \cos \alpha \sin \beta}{\cos \alpha \cos \beta - \sin \alpha \sin \beta} = \dfrac{\tan \alpha + \tan \beta}{1 - \tan \alpha \tan \beta}$

68 $\dfrac{\tan u - \tan v}{1 + \tan u \tan v} = \dfrac{\cot v - \cot u}{1 + \cot u \cot v}$

69 $\sqrt{\dfrac{1 - \cos t}{1 + \cos t}} = \dfrac{1 - \cos t}{|\sin t|}$

70 $\sqrt{\dfrac{1 - \sin \theta}{1 + \sin \theta}} = \dfrac{|\cos \theta|}{1 + \sin \theta}$

71 $\dfrac{\sin \alpha}{1 + \cos \alpha} + \dfrac{1 + \cos \alpha}{\sin \alpha} = 2 \csc \alpha$

72 $\dfrac{\csc x}{1 + \csc x} - \dfrac{\csc x}{1 - \csc x} = 2 \sec^2 x$

73 $\dfrac{1}{\tan \beta + \cot \beta} = \sin \beta \cos \beta$

74 $\dfrac{\cot y - \tan y}{\sin y \cos y} = \csc^2 y - \sec^2 y$

75 $\sec \theta + \csc \theta - \cos \theta - \sin \theta = \sin \theta \tan \theta + \cos \theta \cot \theta$

76 $\sin^3 t + \cos^3 t = (1 - \sin t \cos t)(\sin t + \cos t)$

77 $(1 - \tan^2 \phi)^2 = \sec^4 \phi - 4 \tan^2 \phi$

78 $\cos^4 w + 1 - \sin^4 w = 2 \cos^2 w$

79 $\dfrac{\tan x}{1 - \cot x} + \dfrac{\cot x}{1 - \tan x} = 1 + \sec x \csc x$

80 $\dfrac{\cos \gamma}{1 - \tan \gamma} + \dfrac{\sin \gamma}{1 - \cot \gamma} = \cos \gamma + \sin \gamma$

81 $\sin(-t) \sec(-t) = -\tan t$

82 $\dfrac{\cot(-v)}{\csc(-v)} = \cos v$

83 $\log 10^{\tan t} = \tan t$

84 $10^{\log |\sin t|} = |\sin t|$

85 $\ln \cot x = -\ln \tan x$

86 $\ln \sec \theta = -\ln \cos \theta$

87 $-\ln |\sec \theta - \tan \theta| = \ln |\sec \theta + \tan \theta|$

88 $\ln |\csc x - \cot x| = -\ln |\csc x + \cot x|$

Show that the equations in Exercises 89–100 are not identities. (*Hint:* Find one number in the domain of t or θ for which the equation is false.)

89 $\cos t = \sqrt{1 - \sin^2 t}$

90 $\sqrt{\sin^2 t + \cos^2 t} = \sin t + \cos t$

91 $\sqrt{\sin^2 t} = \sin t$

92 $\sec t = \sqrt{\tan^2 t + 1}$

93 $(\sin \theta + \cos \theta)^2 = \sin^2 \theta + \cos^2 \theta$

94 $\log (1/\sin t) = 1/(\log \sin t)$

95 $\cos (-t) = -\cos t$

96 $\sin (t + \pi) = \sin t$

97 $\cos (\sec t) = 1$

98 $\cot (\tan \theta) = 1$

99 $\sin^2 t - 4 \sin t - 5 = 0$

100 $3 \cos^2 \theta + \cos \theta - 2 = 0$

SECTION 6.2
TRIGONOMETRIC EQUATIONS

A **trigonometric equation** is an equation that contains trigonometric expressions. Solutions of trigonometric equations may be expressed in terms of either real numbers or angles. If a trigonometric equation is not an identity, then techniques similar to those used for algebraic equations may be employed to find the solutions. The main difference is that we usually solve for $\sin x$, $\cos \theta$, and so on, and then find x and θ, as illustrated in the following examples.

EXAMPLE 1 Solve the equation $\sin \theta \tan \theta = \sin \theta$.

Solution Each of the following is equivalent to the given equation:

$$\sin \theta \tan \theta - \sin \theta = 0$$

$$\sin \theta (\tan \theta - 1) = 0$$

To find the solutions we set each factor on the left equal to zero, obtaining

$$\sin \theta = 0 \quad \text{and} \quad \tan \theta = 1.$$

The solutions of the equation $\sin \theta = 0$ consist of all multiples of π, that is, $\theta = n\pi$ where n is any integer.

Since the tangent function has period π, it is sufficient to find the solutions of the equation $\tan \theta = 1$ that are in the interval $[0, \pi)$, for once these are known, we may obtain all the others by adding multiples of π. The only solution of $\tan \theta = 1$ in $[0, \pi)$ is $\pi/4$, and hence every solution has the form

$$\theta = \frac{\pi}{4} + n\pi$$

where n is an integer.

It follows that the solutions of the given equation consist of all numbers of the forms

$$n\pi \quad \text{and} \quad \frac{\pi}{4} + n\pi$$

where n is any integer. Some particular solutions are 0, $\pm\pi$, $\pm 2\pi$, $\pm 3\pi$, $\pi/4$, $5\pi/4$, $-3\pi/4$, and $-7\pi/4$. ■

Note that in Example 1 it would have been incorrect to begin by dividing both sides by $\sin \theta$, for this manipulation would lose the solutions of $\sin \theta = 0$.

EXAMPLE 2 Solve the equation $2 \sin^2 t - \cos t - 1 = 0$.

Solution We first change the equation to an equation that involves only $\cos t$, and then factor as follows:

$$2(1 - \cos^2 t) - \cos t - 1 = 0$$
$$-2 \cos^2 t - \cos t + 1 = 0$$
$$2 \cos^2 t + \cos t - 1 = 0$$
$$(2 \cos t - 1)(\cos t + 1) = 0.$$

As in Example 1, the solutions of the last equation are the solutions of

$$2 \cos t - 1 = 0 \quad \text{and} \quad \cos t + 1 = 0$$

or equivalently, $\cos t = \frac{1}{2}$ and $\cos t = -1$.

It is sufficient to find the solutions that are in the interval $[0, 2\pi)$, for once these are known, all solutions may be found by adding multiples of 2π.

If $\cos t = \frac{1}{2}$, then the reference number (or reference angle) is $\pi/3$ (or 60°). Since $\cos t$ is positive, the angle of radian measure t is in quadrant I or quadrant IV. Hence, in the interval $[0, 2\pi)$,

$$t = \frac{\pi}{3} \quad \text{or} \quad t = 2\pi - \frac{\pi}{3} = \frac{5\pi}{3}.$$

If $\cos t = -1$, then $t = \pi$.

It follows that the solutions of the given equation are

$$\frac{\pi}{3} + 2\pi n, \qquad \frac{5\pi}{3} + 2\pi n, \qquad \pi + 2\pi n$$

where n is any integer. If we wish to express the solutions in terms of degrees, we may write

$$60^\circ + 360^\circ n, \qquad 300^\circ + 360^\circ n, \qquad 180^\circ + 360^\circ n.$$ ■

EXAMPLE 3 Find the solutions of $4 \sin^2 x \tan x - \tan x = 0$ that are in the interval $[0, 2\pi)$.

Solution Factoring the left side, we obtain

$$\tan x\,(4 \sin^2 x - 1) = 0.$$

Setting each factor equal to zero gives us

$$\tan x = 0 \quad \text{and} \quad \sin^2 x = \tfrac{1}{4}$$

or equivalently,

$$\tan x = 0, \quad \sin x = \tfrac{1}{2}, \quad \text{and} \quad \sin x = -\tfrac{1}{2}.$$

The equation $\tan x = 0$ has solutions 0 and π in the interval $[0, 2\pi)$. The equation $\sin x = \frac{1}{2}$ has solutions $\pi/6$ and $5\pi/6$. The equation $\sin x = -\frac{1}{2}$ leads to the numbers between π and 2π that have reference number $\pi/6$. These are

$$\pi + \frac{\pi}{6} = \frac{7\pi}{6} \quad \text{and} \quad 2\pi - \frac{\pi}{6} = \frac{11\pi}{6}.$$

Hence, the solutions of the given equation in the interval $[0, 2\pi)$ are

$$0, \quad \pi, \quad \frac{\pi}{6}, \quad \frac{5\pi}{6}, \quad \frac{7\pi}{6}, \quad \frac{11\pi}{6}.$$ ■

EXAMPLE 4 Find the solutions of the equation $\csc^4 2u - 4 = 0$.

Solution Factoring the left side, we obtain

$$(\csc^2 2u - 2)(\csc^2 2u + 2) = 0.$$

Setting each factor equal to 0 leads to

$$\csc^2 2u = 2 \quad \text{and} \quad \csc^2 2u = -2.$$

The equation $\csc^2 2u = -2$ has no real solutions. The solutions of $\csc^2 2u = 2$ consist of the solutions of the two equations

$$\csc 2u = \sqrt{2} \quad \text{and} \quad \csc 2u = -\sqrt{2}.$$

If $\csc 2u = \sqrt{2}$, then

$$2u = \frac{\pi}{4} + 2\pi n \quad \text{and} \quad 2u = \frac{3\pi}{4} + 2\pi n$$

where n is any integer. Dividing both sides of the last two equations by 2 gives us the solutions

$$u = \frac{\pi}{8} + \pi n \quad \text{and} \quad u = \frac{3\pi}{8} + \pi n.$$

Similarly, from $\csc 2u = -\sqrt{2}$ we obtain

$$2u = \frac{5\pi}{4} + 2\pi n \quad \text{and} \quad 2u = \frac{7\pi}{4} + 2\pi n$$

where n is any integer. Dividing both sides of the last two equations by 2 gives us

$$u = \frac{5\pi}{8} + \pi n \quad \text{and} \quad u = \frac{7\pi}{8} + \pi n.$$

Collecting the preceding information, we see that all solutions are

$$u = \frac{\pi}{8} + \frac{\pi}{4} n \quad \text{where } n \text{ is any integer.}$$

■

The next example illustrates the use of a calculator in solving a trigonometric equation.

EXAMPLE 5 Approximate, to the nearest minute, the solutions of the equation

$$5 \sin \theta \tan \theta - 10 \tan \theta + 3 \sin \theta - 6 = 0$$

in the degree interval $[0°, 360°)$.

Solution The equation may be factored by grouping terms as follows:

$$5 \tan \theta(\sin \theta - 2) + 3(\sin \theta - 2) = 0$$

$$(5 \tan \theta + 3)(\sin \theta - 2) = 0$$

Since the equation $\sin \theta = 2$ has no solutions (Why?), the solutions of the given equation are the same as those of

$$\tan \theta = -\tfrac{3}{5} = -0.6000.$$

Let us begin by approximating the reference angle, that is, the acute angle θ' such that $\tan \theta' = \frac{3}{5} = 0.6$. Using a calculator in degree mode, we obtain the following:

Enter:	0.6	(the value of $\tan\theta'$)
Press [INV] [tan]:	30.963757	(the degree measure of θ')

To change 0.963757° to minutes, we proceed as follows:

$$(0.963757)(60') \approx 57.825395 \approx 58'.$$

Hence $\theta' \approx 30°58'$. Since θ is in quadrant II or quadrant IV, this implies that

$$\theta \approx 180° - 30°58' = 149°2' \quad \text{or} \quad \theta \approx 360° - 30°58' = 329°2'.$$ ■

EXERCISES 6.2

In Exercises 1–16 find all solutions of the equations.

1 $2\cos t + 1 = 0$

2 $\cot\theta + 1 = 0$

3 $\tan^2 x = 1$

4 $4\cos\theta - 2 = 0$

5 $(\cos\theta - 1)(\sin\theta + 1) = 0$

6 $2\cos x = \sqrt{3}$

7 $\sec^2\alpha - 4 = 0$

8 $3 - \tan^2\beta = 0$

9 $\sqrt{3} + 2\sin\beta = 0$

10 $4\sin^2 x - 3 = 0$

11 $\cot^2 x - 3 = 0$

12 $(\sin t - 1)\cos t = 0$

13 $(2\sin\theta + 1)(2\cos\theta + 3) = 0$

14 $(2\sin u - 1)(\cos u - \sqrt{2}) = 0$

15 $\sin 2x(\csc 2x - 2) = 0$

16 $\tan\alpha + \tan^2\alpha = 0$

In Exercises 17–36 find the solutions of the equations in the interval $[0, 2\pi)$, and also find the degree measure of each solution.

17 $2 - 8\cos^2 t = 0$

18 $\cot^2\theta - \cot\theta = 0$

19 $2\sin^2 u = 1 - \sin u$

20 $2\cos^2 t + 3\cos t + 1 = 0$

21 $\tan^2 x \sin x = \sin x$

22 $\sec\beta\csc\beta = 2\csc\beta$

23 $2\cos^2\gamma + \cos\gamma = 0$

24 $\sin x - \cos x = 0$

25 $\sin^2\theta + \sin\theta - 6 = 0$

26 $2\sin^2 u + \sin u - 6 = 0$

27 $1 - \sin t = \sqrt{3}\cos t$

28 $\cos\theta - \sin\theta = 1$

29 $\cos\alpha + \sin\alpha = 1$

30 $2\tan t - \sec^2 t = 0$

31 $\tan\theta + \sec\theta = 1$

32 $\cot\alpha + \tan\alpha = \csc\alpha\sec\alpha$

33 $2\sin^3 x + \sin^2 x - 2\sin x - 1 = 0$

34 $\sec^5\theta = 4\sec\theta$

35 $2\tan t\csc t + 2\csc t + \tan t + 1 = 0$

36 $2\sin v\csc v - \csc v = 4\sin v - 2$

In Exercises 37–40 use a calculator to approximate, to the nearest multiple of ten minutes, the solutions of the equations in the interval $[0°, 360°)$.

37 $\sin^2 t - 4\sin t + 1 = 0$

38 $\tan^2\theta + 3\tan\theta + 2 = 0$

39 $12\sin^2 u - 5\sin u - 2 = 0$

40 $5\cos^2\alpha + 3\cos\alpha - 2 = 0$

SECTION 6.3

THE ADDITION AND SUBTRACTION FORMULAS

In this section we derive formulas involving trigonometric functions of $u + v$ or $u - v$, where u and v represent any real numbers or angles. These formulas are known as *addition or subtraction* formulas, respectively. We shall employ the unit circle approach to trigonometric functions, which was discussed in Section 5.3. Consider any two real numbers t_1, t_2, and let $P_1(t_1)$, $P_2(t_2)$ be the points on the unit circle U that correspond to t_1, t_2, respectively. Next consider the real number $t_1 - t_2$ and the corresponding point $P(t_1 - t_2)$ on U. Our goal is to obtain a formula for $\cos(t_1 - t_2)$ in terms of functional values of t_1 and t_2. For convenience, let us assume that t_1 and t_2 are between 0 and 2π and that $0 \leq t_1 - t_2 < t_2$. In this case, $t_2 \leq t_1$ and hence if A is the point (1, 0) on U, then the length t_1 of $\widehat{AP_1}$ is greater than or equal to the length t_2 of $\widehat{AP_2}$. Also the length $t_1 - t_2$ of $\widehat{AP}$ is less than the length t_2 of $\widehat{AP_2}$. Part (i) of Figure 6.1 illustrates one arrangement of points P_1, P_2, and P under these conditions; however, it is possible to extend our discussion to cover all values of t.

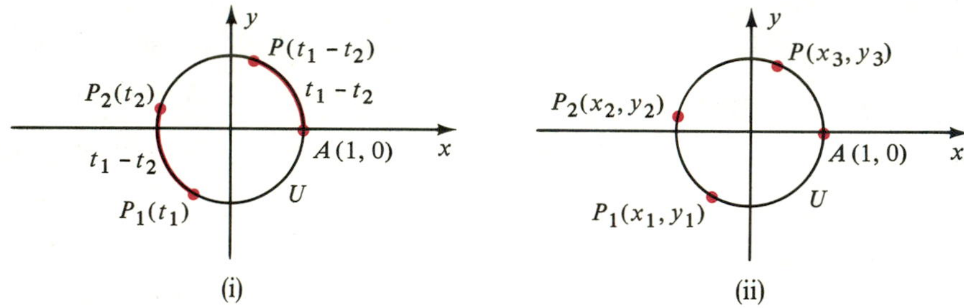

FIGURE 6.1

We shall denote the rectangular coordinates of the points $P_1(t_1)$, $P_2(t_2)$, and $P(t_1 - t_2)$ by $P_1(x_1, y_1)$, $P_2(x_2, y_2)$ and $P(x_3, y_3)$, respectively, as shown in (ii) of Figure 6.1. From our work with trigonometric functions in terms of a unit circle (see page 208),

$$(*) \quad \begin{aligned} &\cos t_1 = x_1, \quad &\cos t_2 = x_2, \quad &\cos(t_1 - t_2) = x_3, \\ &\sin t_1 = y_1, \quad &\sin t_2 = y_2, \quad &\sin(t_1 - t_2) = y_3, \end{aligned}$$

where the symbol (*) has been used for later reference to these formulas. Since the arc lengths of both $\widehat{P_2P_1}$ and $\widehat{AP}$ equal $t_1 - t_2$, the line segments P_2P_1 and AP also have the same length, that is.

$$d(A, P) = d(P_1, P_2).$$

Using the Distance Formula, the last equation may be written

$$\sqrt{(x_3 - 1)^2 + (y_3 - 0)^2} = \sqrt{(x_2 - x_1)^2 + (y_2 - y_1)^2}.$$

Squaring both sides and expanding the terms underneath the radical gives us

$$x_3^2 - 2x_3 + 1 + y_3^2 = x_2^2 - 2x_1x_2 + x_1^2 + y_2^2 - 2y_1y_2 + y_1^2.$$

Since P_1, P_2, and P lie on U, and since an equation for U is $x^2 + y^2 = 1$, we may substitute 1 for each of $x_1^2 + y_1^2$, $x_2^2 + y_2^2$, and $x_3^2 + y_3^2$. Doing this and simplifying, we obtain

$$2 - 2x_3 = 2 - 2x_1x_2 - 2y_1y_2$$

which reduces to $$x_3 = x_1x_2 + y_1y_2.$$

Substituting from the formulas stated in (*) at the beginning of this section gives us

$$\cos(t_1 - t_2) = \cos t_1 \cos t_2 + \sin t_1 \sin t_2.$$

To make the formula less cumbersome, we shall eliminate subscripts by changing the symbols for the variables from t_1 and t_2 to u and v, respectively. Our identity then takes on the following form.

SUBTRACTION FORMULA FOR COSINE

$$\cos(u - v) = \cos u \cos v + \sin u \sin v$$

EXAMPLE 1 Find the exact value of $\cos 15°$.

Solution If we write $15° = 60° - 45°$, we may use the subtraction formula with $u = 60°$ and $v = 45°$, as follows:

$$\begin{aligned}\cos 15° &= \cos(60° - 45°)\\ &= \cos 60° \cos 45° + \sin 60° \sin 45°\\ &= \left(\frac{1}{2}\right)\left(\frac{\sqrt{2}}{2}\right) + \left(\frac{\sqrt{3}}{2}\right)\left(\frac{\sqrt{2}}{2}\right)\\ &= \frac{\sqrt{2} + \sqrt{6}}{4}.\end{aligned}$$

■

It is easy to obtain a formula for $\cos(u + v)$. We begin by writing $u + v = u - (-v)$ and then employ the formula derived previously. Thus,

$$\begin{aligned}\cos(u + v) &= \cos[u - (-v)]\\ &= \cos u \cos(-v) + \sin u \sin(-v).\end{aligned}$$

By the Formulas for Negatives, given in Section 5.3, $\cos(-v) = \cos v$ and $\sin(-v) = -\sin v$ for all v, and hence we can state the following formula.

ADDITION FORMULA FOR COSINE

$$\cos(u+v) = \cos u \cos v - \sin u \sin v$$

EXAMPLE 2 Find $\cos(7\pi/12)$ by using $\pi/3$ and $\pi/4$.

Solution Writing $7\pi/12 = (\pi/3) + (\pi/4)$ and applying the formula for $\cos(u+v)$,

$$\begin{aligned}\cos\frac{7\pi}{12} &= \cos\left(\frac{\pi}{3}+\frac{\pi}{4}\right)\\ &= \cos\frac{\pi}{3}\cos\frac{\pi}{4} - \sin\frac{\pi}{3}\sin\frac{\pi}{4}\\ &= \frac{1}{2}\frac{\sqrt{2}}{2} - \frac{\sqrt{3}}{2}\frac{\sqrt{2}}{2}\\ &= \frac{\sqrt{2}-\sqrt{6}}{4}\end{aligned}$$

■

Identities that are analogous to those for cosine are true for the sine function. Let us first establish three identities that are of some interest in themselves:

$$\cos\left(\frac{\pi}{2}-u\right) = \sin u, \qquad \sin\left(\frac{\pi}{2}-u\right) = \cos u, \qquad \tan\left(\frac{\pi}{2}-u\right) = \cot u.$$

The first of these identities may be proved as follows:

$$\begin{aligned}\cos\left(\frac{\pi}{2}-u\right) &= \cos\frac{\pi}{2}\cos u + \sin\frac{\pi}{2}\sin u\\ &= 0\cdot\cos u + 1\cdot\sin u = \sin u.\end{aligned}$$

To obtain the second identity we substitute $(\pi/2) - v$ for u in the first identity, obtaining

$$\cos\left[\frac{\pi}{2} - \left(\frac{\pi}{2}-v\right)\right] = \sin\left(\frac{\pi}{2}-v\right)$$

or equivalently,

$$\cos v = \sin\left(\frac{\pi}{2}-v\right).$$

The third identity may be established as follows:

$$\tan\left(\frac{\pi}{2}-u\right) = \frac{\sin\left(\dfrac{\pi}{2}-u\right)}{\cos\left(\dfrac{\pi}{2}-u\right)} = \frac{\cos u}{\sin u} = \cot u.$$

If θ denotes the degree measure of an angle, we may write

$$\cos(90^\circ - \theta) = \sin\theta, \qquad \sin(90^\circ - \theta) = \cos\theta, \qquad \tan(90^\circ - \theta) = \cot\theta.$$

If θ is acute, then θ and $90^\circ - \theta$ are complementary, since their sum is 90°. It is customary to refer to the sine and cosine functions as **cofunctions** of one another. Similarly, the tangent and cotangent functions are cofunctions, as are the secant and cosecant. Consequently, the last three formulas constitute a partial description of the fact that *any functional value of θ equals the cofunction of the complementary angle* $90^\circ - \theta$.

The following identities may now be established.

ADDITION AND SUBTRACTION FORMULAS FOR SINE AND TANGENT

$$\sin(u + v) = \sin u \cos v + \cos u \sin v$$

$$\sin(u - v) = \sin u \cos v - \cos u \sin v$$

$$\tan(u + v) = \frac{\tan u + \tan v}{1 - \tan u \tan v}$$

$$\tan(u - v) = \frac{\tan u - \tan v}{1 + \tan u \tan v}$$

We shall prove the first and third and leave the proofs of the remaining two as exercises. The reader should supply reasons for each of the following steps.

$$\begin{aligned}\sin(u + v) &= \cos\left[\frac{\pi}{2} - (u + v)\right]\\ &= \cos\left[\left(\frac{\pi}{2} - u\right) - v\right]\\ &= \cos\left(\frac{\pi}{2} - u\right)\cos v + \sin\left(\frac{\pi}{2} - u\right)\sin v\\ &= \sin u \cos v + \cos u \sin v.\end{aligned}$$

To verify the formula for $\tan(u + v)$ we begin as follows:

$$\begin{aligned}\tan(u + v) &= \frac{\sin(u + v)}{\cos(u + v)}\\ &= \frac{\sin u \cos v + \cos u \sin v}{\cos u \cos v - \sin u \sin v}.\end{aligned}$$

Next, dividing numerator and denominator by $\cos u \cos v$ (assuming, of course, that $\cos u \cos v \neq 0$), we obtain

$$\tan(u+v) = \frac{\left(\dfrac{\sin u}{\cos u}\right)\left(\dfrac{\cos v}{\cos v}\right) + \left(\dfrac{\cos u}{\cos u}\right)\left(\dfrac{\sin v}{\cos v}\right)}{\left(\dfrac{\cos u}{\cos u}\right)\left(\dfrac{\cos v}{\cos v}\right) - \left(\dfrac{\sin u}{\cos u}\right)\left(\dfrac{\sin v}{\cos v}\right)}$$

$$= \frac{\tan u + \tan v}{1 - \tan u \tan v}.$$

EXAMPLE 3 Given $\sin\alpha = \frac{4}{5}$ where α is an angle in quadrant I, and $\cos\beta = -\frac{12}{13}$ where β is in quadrant II, find $\sin(\alpha+\beta)$, $\tan(\alpha+\beta)$, and the quadrant containing $\alpha+\beta$.

Solution It is convenient to represent α and β geometrically, as illustrated in Figure 6.2. There is no loss of generality in picturing α and β as positive angles between 0 and 2π as we have done in the figure. Since $\sin\alpha = \frac{4}{5}$, the point (3, 4) is on the terminal side of α. (Why?) Similarly, since $\cos\beta = -\frac{12}{13}$, we may choose the point $(-12, 5)$ on the terminal side of β. Referring to Figure 6.2 and using the definition of trigonometric functions of any angle (page 203),

$$\cos\alpha = \tfrac{3}{5}, \quad \tan\alpha = \tfrac{4}{3}, \quad \sin\beta = \tfrac{5}{13}, \quad \text{and} \quad \tan\beta = -\tfrac{5}{12}.$$

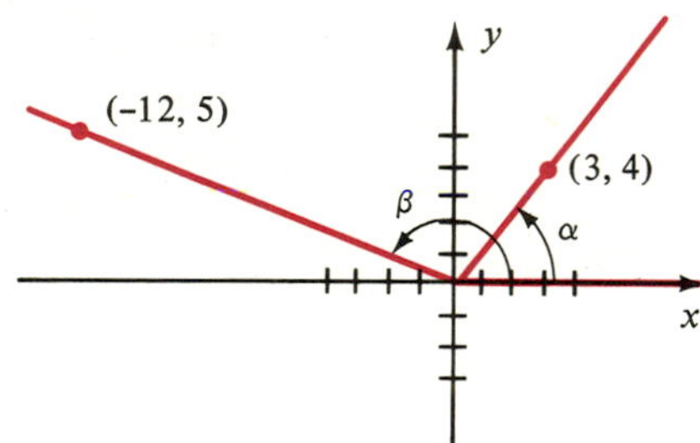

FIGURE 6.2

Using addition formulas:

$$\sin(\alpha+\beta) = \sin\alpha\cos\beta + \cos\alpha\sin\beta$$

$$= \left(\frac{4}{5}\right)\left(-\frac{12}{13}\right) + \left(\frac{3}{5}\right)\left(\frac{5}{13}\right) = -\frac{33}{65}.$$

$$\tan(\alpha+\beta) = \frac{\tan\alpha + \tan\beta}{1 - \tan\alpha\tan\beta}$$

$$= \frac{\frac{4}{3} + \left(-\frac{5}{12}\right)}{1 - \left(\frac{4}{3}\right)\left(-\frac{5}{12}\right)} = \frac{33}{56}.$$

Since $\sin(\alpha + \beta)$ is negative and $\tan(\alpha + \beta)$ is positive, it follows that $\alpha + \beta$ lies in quadrant III. ■

The addition formulas are very important in the study of calculus, which requires the type of simplification illustrated in the next example.

EXAMPLE 4 If $f(x) = \sin x$ and $h \neq 0$, prove that

$$\frac{f(x+h) - f(x)}{h} = \sin x\left(\frac{\cos h - 1}{h}\right) + \cos x\left(\frac{\sin h}{h}\right).$$

Solution Using the definition of f and the addition formula for the sine function, we obtain

$$\begin{aligned}\frac{f(x+h) - f(x)}{h} &= \frac{\sin(x+h) - \sin x}{h} \\ &= \frac{\sin x \cos h + \cos x \sin h - \sin x}{h} \\ &= \frac{\sin x(\cos h - 1) + \cos x \sin h}{h} \\ &= \sin x\left(\frac{\cos h - 1}{h}\right) + \cos x\left(\frac{\sin h}{h}\right).\end{aligned}$$ ■

EXAMPLE 5 Prove that for every x,

$$a \cos Bx + b \sin Bx = A \cos(Bx - C)$$

where $A = \sqrt{a^2 + b^2}$ and $\tan C = b/a$.

Solution Using the formula for $\cos(u - v)$ with $u = Bx$ and $v = C$,

$$A \cos(Bx - C) = A(\cos Bx \cos C + \sin Bx \sin C).$$

We shall complete the proof by determining conditions on a and b such that

$$a \cos Bx + b \sin Bx = A \cos Bx \cos C + A \sin Bx \sin C.$$

The last formula is true for every x if and only if

$$a = A \cos C \quad \text{and} \quad b = A \sin C$$

(to verify this let $x = 0$). Consequently,

$$\begin{aligned}a^2 + b^2 &= A^2 \cos^2 C + A^2 \sin^2 C \\ &= A^2(\cos^2 C + \sin^2 C) = A^2\end{aligned}$$

and we may choose $$A = \sqrt{a^2 + b^2}.$$

Finally, we note that

$$\frac{b}{a} = \frac{A \sin C}{A \cos C} = \frac{\sin C}{\cos C} = \tan C.$$

This establishes the identity. ■

EXAMPLE 6 The graph of $f(x) = \cos x + \sin x$ was obtained in Example 1 of Section 5.7 by adding y-coordinates (see Figure 5.47). Use the formula derived in Example 5 to find the amplitude, period, and phase shift of f.

Solution Letting $a = 1$, $b = 1$, and $B = 1$ in Example 5, we obtain

$$A = \sqrt{a^2 + b^2} = \sqrt{1 + 1} = \sqrt{2} \quad \text{and} \quad \tan C = b/a = 1/1 = 1.$$

Since $\tan C = 1$, we may choose $C = \pi/4$. Substitution in the formula gives us

$$\cos x + \sin x = \sqrt{2} \cos\left(x - \frac{\pi}{4}\right).$$

It follows from our work in Section 5.6 that the amplitude is $\sqrt{2}$, the period is 2π, and the phase shift is $\pi/4$. The reader may check these facts by referring to Figure 5.47. ■

EXERCISES 6.3

Write the expressions in Exercises 1–4 in terms of cofunctions of complementary angles.

1 (a) $\sin 46°37'$ (b) $\cos 73°12'$ (c) $\tan \pi/6$

2 (a) $\tan 24°12'$ (b) $\sin 89°41'$ (c) $\cos \pi/3$

3 (a) $\cos 7\pi/20$ (b) $\sin(\frac{1}{4})$ (c) $\tan 1$

4 (a) $\sin \pi/12$ (b) $\cos(0.64)$ (c) $\tan \sqrt{2}$

Find the exact functional values in Exercises 5-10.

5 (a) $\cos \pi/4 + \cos \pi/3$ (b) $\cos 7\pi/12$

6 (a) $\sin 2\pi/3 + \sin \pi/4$ (b) $\sin 11\pi/12$

7 (a) $\tan 60° + \tan 225°$ (b) $\tan 285°$

8 (a) $\cos 135° - \cos 60°$ (b) $\cos 75°$

9 (a) $\sin 3\pi/4 - \sin \pi/6$ (b) $\sin 7\pi/12$

10 (a) $\tan 3\pi/4 - \tan \pi/6$ (b) $\tan 7\pi/12$

Write the expressions in Exercises 11–16 in terms of one function of one angle.

11 $\cos 48° \cos 23° + \sin 48° \sin 23°$

12 $\cos 13° \cos 50° - \sin 13° \sin 50°$

13 $\cos 10° \sin 5° - \sin 10° \cos 5°$

14 $\sin 57° \cos 4° + \cos 57° \sin 4°$

15 $\cos 3 \sin(-2) - \cos 2 \sin 3$

16 $\sin(-5) \cos 2 + \cos 5 \sin(-2)$

17 If α and β are acute angles such that $\cos \alpha = \frac{4}{5}$ and $\tan \beta = \frac{8}{15}$, find $\cos(\alpha + \beta)$, $\sin(\alpha + \beta)$, and the quadrant containing $\alpha + \beta$.

18 If $\sin\alpha = -\frac{4}{5}$ and $\sec\beta = \frac{5}{3}$, where α is a third-quadrant angle and β is a first-quadrant angle, find $\sin(\alpha+\beta)$, $\tan(\alpha+\beta)$, and the quadrant containing $\alpha+\beta$.

19 If $\tan\alpha = -\frac{7}{24}$ and $\cot\beta = \frac{3}{4}$, where α is in the second quadrant and β is in the third quadrant, find $\sin(\alpha+\beta)$, $\cos(\alpha+\beta)$, $\tan(\alpha+\beta)$, $\sin(\alpha-\beta)$, $\cos(\alpha-\beta)$, and $\tan(\alpha-\beta)$.

20 If points $P(t_1)$ and $P(t_2)$ are in quadrant III, and if $\cos t_1 = -\frac{2}{5}$ and $\cos t_2 = -\frac{3}{5}$, find $\sin(t_1 - t_2)$, $\cos(t_1 - t_2)$, and the quadrant containing $P(t_1 - t_2)$.

Verify the identities in Exercises 21–40.

21 $\sin\left(x + \dfrac{\pi}{2}\right) = \cos x$

22 $\cos\left(x + \dfrac{\pi}{2}\right) = -\sin x$

23 $\cos\left(\theta + \dfrac{3\pi}{2}\right) = \sin\theta$

24 $\sin\left(\alpha - \dfrac{3\pi}{2}\right) = \cos\alpha$

25 $\sin\left(\theta + \dfrac{\pi}{4}\right) = \left(\dfrac{\sqrt{2}}{2}\right)(\sin\theta + \cos\theta)$

26 $\cos\left(\theta + \dfrac{\pi}{4}\right) = \left(\dfrac{\sqrt{2}}{2}\right)(\cos\theta - \sin\theta)$

27 $\tan\left(u + \dfrac{\pi}{4}\right) = \dfrac{1 + \tan u}{1 - \tan u}$

28 $\tan\left(x - \dfrac{\pi}{4}\right) = \dfrac{\tan x - 1}{\tan x + 1}$

29 $\tan\left(u + \dfrac{\pi}{2}\right) = -\cot u$

30 $\cot\left(t - \dfrac{\pi}{3}\right) = \dfrac{\sqrt{3}\tan t + 1}{\tan t - \sqrt{3}}$

31 $\sin(u+v)\cdot\sin(u-v) = \sin^2 u - \sin^2 v$

32 $\cos(u+v)\cdot\cos(u-v) = \cos^2 u - \sin^2 v$

33 $\cos(u+v) + \cos(u-v) = 2\cos u\cos v$

34 $\sin(u+v) + \sin(u-v) = 2\sin u\cos v$

35 $\sin 2u = 2\sin u\cos u$ (*Hint:* $2u = u + u$)

36 $\cos 2u = \cos^2 u - \sin^2 u$

37 $\dfrac{\sin(u+v)}{\cos(u-v)} = \dfrac{\tan u + \tan v}{1 + \tan u\tan v}$

38 $\dfrac{\cos(u+v)}{\cos(u-v)} = \dfrac{1 - \tan u\tan v}{1 + \tan u\tan v}$

39 $\dfrac{\sin(u+v)}{\sin(u-v)} = \dfrac{\tan u + \tan v}{\tan u - \tan v}$

40 $\tan u + \tan v = \dfrac{\sin(u+v)}{\cos u\cos v}$

41 Express $\sin(u+v+w)$ in terms of functions of u, v, and w. (*Hint:* Write $\sin(u+v+w) = \sin[(u+v)+w]$ and use an addition formula.)

42 Express $\tan(u+v+w)$ in terms of functions of u, v, and w.

43 Derive the formula $\cot(u+v) = \dfrac{\cot u\cot v - 1}{\cot u + \cot v}$

44 If α and β are complementary angles, prove that

$$\sin^2\alpha + \sin^2\beta = 1.$$

45 Derive the subtraction formulas for the sine and tangent functions.

46 Prove each of the following:

(a) $\sec\left(\dfrac{\pi}{2} - u\right) = \csc u$

(b) $\csc\left(\dfrac{\pi}{2} - u\right) = \sec u$

(c) $\cot\left(\dfrac{\pi}{2} - u\right) = \tan u$

47 If $f(x) = \cos x$, prove that

$$\frac{f(x+h) - f(x)}{h} = \cos x\left(\frac{\cos h - 1}{h}\right) - \sin x\left(\frac{\sin h}{h}\right).$$

48 If $f(x) = \tan x$, prove that

$$\frac{f(x+h) - f(x)}{h} = \sec^2 x\left(\frac{\sin h}{h}\right)\frac{1}{\cos h - \sin h\tan x}.$$

In Exercises 49 and 50 find the solutions of the equation that are in the interval $[0, 2\pi)$, and find the degree measure of each solution.

49 $\sin 4t\cos t = \sin t\cos 4t$

50 $\cos 5t\cos 3t = 2^{-1} + \sin(-5t)\sin 3t$

In Exercises 51–54 use the formula given in Example 5 to express f in terms of the cosine function, and determine the amplitude, period, and phase shift.

51 $f(x) = \sqrt{3}\cos 2x + \sin 2x$

52 $f(x) = \cos 4x + \sqrt{3}\sin 4x$

53 $f(x) = 2\cos 3x - 2\sin 3x$

54 $f(x) = -5\cos 10x + 5\sin 10x$

CALCULATOR EXERCISES 6.3

1 Demonstrate, by direct calculations, that the addition formulas for the sine, cosine, and tangent functions are true for the following values of u and v: $u = 1.46$, $v = 8.27$.

2 If u and v are in the interval $[0, \pi]$ such that $\cos u = -0.4163$ and $\sin v = 0.7216$, find $\cos(u + v)$ and $\sin(u + v)$.

SECTION 6.4

MULTIPLE-ANGLE FORMULAS

In the preceding section we were interested primarily in trigonometric identities that involve values of $u \pm v$. In this section we shall obtain formulas for finding values of nu, where n is an integer. The formulas are referred to as **multiple-angle formulas.** In particular, the following identities are called the **double-angle formulas,** because they contain the expression $2u$.

DOUBLE-ANGLE FORMULAS

$$\sin 2u = 2\sin u\cos u$$

$$\cos 2u = \cos^2 u - \sin^2 u = 1 - 2\sin^2 u = 2\cos^2 u - 1$$

$$\tan 2u = \frac{2\tan u}{1 - \tan^2 u}$$

These identities may be proved by letting $u = v$ in the appropriate addition formulas. If we use the formula for $\sin(u + v)$, then

$$\begin{aligned}\sin 2u &= \sin(u + u)\\ &= \sin u\cos u + \cos u\sin u\\ &= 2\sin u\cos u.\end{aligned}$$

Similarly, using the formula for $\cos(u + v)$,

$$\begin{aligned}\cos 2u &= \cos(u + u)\\ &= \cos u\cos u - \sin u\sin u\\ &= \cos^2 u - \sin^2 u.\end{aligned}$$

The other two forms for $\cos 2u$ may be obtained by using the Fundamental Identity $\sin^2 u + \cos^2 u = 1$. Thus,

$$\begin{aligned}\cos 2u &= \cos^2 u - \sin^2 u \\ &= (1 - \sin^2 u) - \sin^2 u \\ &= 1 - 2\sin^2 u.\end{aligned}$$

Similarly, if we substitute for $\sin^2 u$ instead of $\cos^2 u$, we obtain

$$\begin{aligned}\cos 2u &= \cos^2 u - (1 - \cos^2 u) \\ &= 2\cos^2 u - 1.\end{aligned}$$

The formula for $\tan 2u$ may be obtained by taking $u = v$ in the formula for $\tan (u + v)$.

EXAMPLE 1 Find $\sin 2\alpha$ and $\cos 2\alpha$ if $\sin \alpha = \frac{4}{5}$ and α is in quadrant I.

Solution As in Example 3 of the preceding section, $\cos \alpha = \frac{3}{5}$. Substitution in Double-Angle Formulas:

$$\begin{aligned}\sin 2\alpha &= 2\sin\alpha\cos\alpha = 2\left(\tfrac{4}{5}\right)\left(\tfrac{3}{5}\right) = \tfrac{24}{25} \\ \cos 2\alpha &= \cos^2\alpha - \sin^2\alpha = \tfrac{9}{25} - \tfrac{16}{25} = -\tfrac{7}{25}.\end{aligned}$$ ■

EXAMPLE 2 Express $\cos 3\theta$ in terms of $\cos \theta$.

Solution

$$\begin{aligned}\cos 3\theta &= \cos (2\theta + \theta) \\ &= \cos 2\theta \cos\theta - \sin 2\theta \sin\theta \\ &= (2\cos^2\theta - 1)\cos\theta - (2\sin\theta\cos\theta)\sin\theta \\ &= 2\cos^3\theta - \cos\theta - 2\cos\theta\sin^2\theta \\ &= 2\cos^3\theta - \cos\theta - 2\cos\theta\,(1 - \cos^2\theta) \\ &= 2\cos^3\theta - \cos\theta - 2\cos\theta + 2\cos^3\theta \\ &= 4\cos^3\theta - 3\cos\theta.\end{aligned}$$ ■

The next three identities are useful for simplifying certain expressions involving powers of trigonometric functions.

$$\sin^2 u = \frac{1-\cos 2u}{2}, \qquad \cos^2 u = \frac{1+\cos 2u}{2}, \qquad \tan^2 u = \frac{1-\cos 2u}{1+\cos 2u}.$$

The first and second of these identities may be verified by solving the equations

$$\cos 2u = 1 - 2\sin^2 u \quad \text{and} \quad \cos 2u = 2\cos^2 u - 1$$

for $\sin^2 u$ and $\cos^2 u$, respectively. The third identity may be obtained from the first two by using the fact that $\tan^2 u = \sin^2 u/\cos^2 u$.

EXAMPLE 3 Verify the identity $\sin^2 x \cos^2 x = \frac{1}{8}(1 - \cos 4x)$.

Solution Using the preceding identities for $\sin^2 u$ and $\cos^2 u$ with $u = x$, we obtain

$$\begin{aligned}\sin^2 x \cos^2 x &= \left(\frac{1 - \cos 2x}{2}\right)\left(\frac{1 + \cos 2x}{2}\right)\\ &= \tfrac{1}{4}(1 - \cos^2 2x)\\ &= \tfrac{1}{4}\sin^2 2x.\end{aligned}$$

Finally, using the formula $\sin^2 u = (1 - \cos 2u)/2$ with $u = 2x$,

$$\begin{aligned}\sin^2 x \cos^2 x &= \frac{1}{4}\left(\frac{1 - \cos 4x}{2}\right)\\ &= \tfrac{1}{8}(1 - \cos 4x).\end{aligned}$$

Another method of proof is to use the fact that $\sin 2x = 2 \sin x \cos x$ and, hence, that

$$\sin x \cos x = \tfrac{1}{2}\sin 2x.$$

Squaring both sides, we have

$$\sin^2 x \cos^2 x = \tfrac{1}{4}\sin^2 2x.$$

The remainder of the solution is the same as the first proof. ■

EXAMPLE 4 Express $\cos^4 t$ in terms of values of the cosine function with exponent 1.

Solution Supply reasons for the following steps:

$$\begin{aligned}\cos^4 t = (\cos^2 t)^2 &= \left(\frac{1 + \cos 2t}{2}\right)^2\\ &= \frac{1}{4}(1 + 2\cos 2t + \cos^2 2t)\\ &= \frac{1}{4}\left(1 + 2\cos 2t + \frac{1 + \cos 4t}{2}\right)\\ &= \frac{3}{8} + \frac{1}{2}\cos 2t + \frac{1}{8}\cos 4t.\end{aligned}$$

■

Substituting $v/2$ for u in the three formulas for $\sin^2 u$, $\cos^2 u$, and $\tan^2 u$ gives us

$$\sin^2 \frac{v}{2} = \frac{1 - \cos v}{2}, \qquad \cos^2 \frac{v}{2} = \frac{1 + \cos v}{2}, \qquad \tan^2 \frac{v}{2} = \frac{1 - \cos v}{1 + \cos v}.$$

If we take the square root of both sides of each of the last three equations and use the fact that $\sqrt{a^2} = |a|$ for every real number a, the following identities result. They are called *half-angle formulas* because of the expression $v/2$.

HALF-ANGLE FORMULAS

$$\left|\sin \frac{v}{2}\right| = \sqrt{\frac{1 - \cos v}{2}}, \qquad \left|\cos \frac{v}{2}\right| = \sqrt{\frac{1 + \cos v}{2}}, \qquad \left|\tan \frac{v}{2}\right| = \sqrt{\frac{1 - \cos v}{1 + \cos v}}.$$

The absolute value signs may be eliminated in the preceding formulas if more information is known about $v/2$. For example, if the angle determined by $v/2$ is in either quadrant I or II, then $\sin (v/2)$ is positive, and we may write

$$\sin \frac{v}{2} = \sqrt{\frac{1 - \cos v}{2}}.$$

However, if $v/2$ leads to an angle in either quadrant III or IV, then

$$\sin \frac{v}{2} = -\sqrt{\frac{1 - \cos v}{2}}.$$

Similar remarks are true for the other formulas.

An alternative form for $\tan (v/2)$ can be obtained. Multiplying numerator and denominator of the radicand in the third half-angle formula by $1 - \cos v$ gives us

$$\begin{aligned} \left|\tan \frac{v}{2}\right| &= \sqrt{\frac{1 - \cos v}{1 + \cos v} \cdot \frac{1 - \cos v}{1 - \cos v}} \\ &= \sqrt{\frac{(1 - \cos v)^2}{\sin^2 v}} \\ &= \frac{1 - \cos v}{|\sin v|}. \end{aligned}$$

The absolute value sign is unnecessary in the numerator since $1 - \cos v$ is never negative. It can be shown that $\tan (v/2)$ and $\sin v$ always have the same sign. For example, if $0 < v < \pi$, then $0 < v/2 < \pi/2$, and hence both $\sin v$ and $\tan (v/2)$ are positive. If $\pi < v < 2\pi$, then $\pi/2 < v/2 < \pi$, and hence both $\sin v$ and $\tan (v/2)$ are negative. It is possible to generalize these remarks to all values of v for which the expressions $\tan(v/2)$ and $(1 - \cos v)/|\sin v|$ have meaning. This gives us the first of the next two identities. The second identity for $\tan(v/2)$ may be obtained by multiplying numerator and denominator of the radicand in the third half-angle formula by $1 + \cos v$.

HALF-ANGLE FORMULAS FOR TANGENT

$$\tan\frac{v}{2} = \frac{1 - \cos v}{\sin v}, \qquad \tan\frac{v}{2} = \frac{\sin v}{1 + \cos v}.$$

EXAMPLE 5 Find the exact values of $\sin 22.5°$ and $\cos 22.5°$.

Solution Using the formula for $\sin(v/2)$ and the fact that 22.5° is in quadrant I,

$$\sin 22.5° = \sin\frac{45°}{2} = \sqrt{\frac{1 - \cos 45°}{2}}$$
$$= \sqrt{\frac{1 - \sqrt{2}/2}{2}} = \frac{\sqrt{2 - \sqrt{2}}}{2}$$
$$\cos 22.5° = \sqrt{\frac{1 + \cos 45°}{2}}$$
$$= \sqrt{\frac{1 + \sqrt{2}/2}{2}} = \frac{\sqrt{2 + \sqrt{2}}}{2}.$$

■

EXAMPLE 6 If $\tan\alpha = -\frac{4}{3}$ and α is in quadrant IV, find $\tan(\alpha/2)$.

Solution If we choose the point $(3, -4)$ on the terminal side of α as illustrated in Figure 6.3, then $\sin\alpha = -\frac{4}{5}$ and $\cos\alpha = \frac{3}{5}$. (Why?) Applying a half-angle formula,

$$\tan\frac{\alpha}{2} = \frac{1 - \cos\alpha}{\sin\alpha} = \frac{1 - \frac{3}{5}}{-\frac{4}{5}} = -\frac{1}{2}.$$

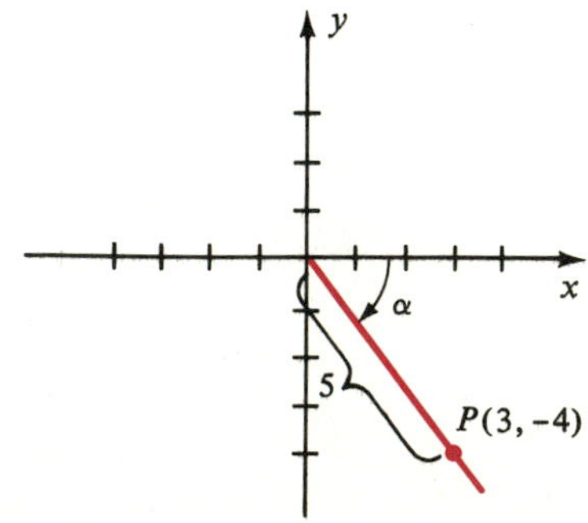

FIGURE 6.3

■

EXAMPLE 7 Find the solutions of the equation $\cos 2x + \cos x = 0$ that are in the interval $[0, 2\pi)$. Express the solutions both in radian measure and degree measure.

Solution We first use a double-angle formula to write the equation in terms of $\cos x$, and then slove by factoring as follows:

$$\cos 2x + \cos x = 0$$
$$(2\cos^2 x - 1) + \cos x = 0$$
$$2\cos^2 x + \cos x - 1 = 0$$
$$(2\cos x - 1)(\cos x + 1) = 0$$

Setting each factor equal to zero we obtain

$$\cos x = \tfrac{1}{2} \quad \text{and} \quad \cos x = -1.$$

The solutions of the last two equations (and hence of the given equation) in $[0, 2\pi)$ are

$$\pi/3, \quad 5\pi/3, \quad \text{and} \quad \pi.$$

The corresponding degree measures are

$$60°, \quad 300°, \quad \text{and} \quad 180°$$ ■

EXERCISES 6.4

In Exercises 1–4 find the exact values of $\sin 2\theta$, $\cos 2\theta$, and $\tan 2\theta$ for the given conditions.

1 $\cos \theta = \frac{3}{5}$ and θ acute

2 $\cot \theta = \frac{4}{3}$ and $180° < \theta < 270°$

3 $\sec \theta = -3$ and $90° < \theta < 180°$

4 $\sin \theta = -\frac{4}{5}$ and $270° < \theta < 360°$

In Exercises 5–8 find the exact values of $\sin(\theta/2)$, $\cos(\theta/2)$, and $\tan(\theta/2)$ for the given conditions.

5 $\sec \theta = \frac{5}{4}$ and θ acute

6 $\csc \theta = -\frac{5}{3}$ and $-90° < \theta < 0°$

7 $\tan \theta = 1$ and $-180° < \theta < -90°$

8 $\sec \theta = -4$ and $180° < \theta < 270°$

In Exercises 9 and 10, use half-angle formulas to find exact values.

9 (a) $\cos 67°30'$ (b) $\sin 15°$ (c) $\tan 3\pi/8$

10 (a) $\cos 165°$ (b) $\sin 157°30'$ (c) $\tan \pi/8$

Verify the identities stated in Exercises 11–28.

11 $\sin 10\theta = 2 \sin 5\theta \cos 5\theta$

12 $\cos^2 3x - \sin^2 3x = \cos 6x$

13 $4 \sin \dfrac{x}{2} \cos \dfrac{x}{2} = 2 \sin x$

14 $\dfrac{\sin^2 2\alpha}{\sin^2 \alpha} = 4 - 4\sin^2 \alpha$

15 $(\sin t + \cos t)^2 = 1 + \sin 2t$

16 $\csc 2u = \frac{1}{2} \sec u \csc u$

17 $\sin 3u = \sin u(3 - 4 \sin^2 u)$

18 $\sin 4t = 4 \cos t \sin t(1 - 2 \sin^2 t)$

19 $\cos 4\theta = 8 \cos^4 \theta - 8 \cos^2 \theta + 1$

20 $\cos 6t = 32 \cos^6 t - 48 \cos^4 t + 18 \cos^2 t - 1$

21 $\sin^4 t = \frac{3}{8} - \frac{1}{2} \cos 2t + \frac{1}{8} \cos 4t$

22 $\cos^4 x - \sin^4 x = \cos 2x$

23 $\sec 2\theta = \dfrac{\sec^2 \theta}{2 - \sec^2 \theta}$

24 $\cot 2u = \dfrac{\cot^2 u - 1}{2 \cot u}$

25 $2 \sin^2 2t + \cos 4t = 1$

26 $\tan \theta + \cot \theta = 2 \csc 2\theta$

27 $\tan 3u = \dfrac{(3 - \tan^2 u)\tan u}{1 - 3 \tan^2 u}$

28 $\dfrac{1 + \sin 2v + \cos 2v}{1 + \sin 2v - \cos 2v} = \cot v$

Write the expressions in Exercises 29 and 30 in terms of values of cosine with exponent 1.

29 $\cos^4 (\theta/2)$

30 $\sin^4 2x$

In Exercises 31–38 find all solutions of the equations in the interval $[0, 2\pi)$. Express the solutions both in radian measure and degree measure.

31 $\sin 2t + \sin t = 0$

32 $\cos t - \sin 2t = 0$

33 $\cos u + \cos 2u = 0$

34 $\cos 2\theta - \tan \theta = 1$

35 $\tan 2x = \tan x$

36 $\tan 2t - 2 \cos t = 0$

37 $\sin \frac{1}{2}u + \cos u = 1$

38 $2 - \cos^2 x = 4 \sin^2 \frac{1}{2}x$

39 If $a > 0$, $b > 0$, and $0 < u < \pi/2$, prove that

$$a \sin u + b \cos u = \sqrt{a^2 + b^2} \sin (u + v)$$

where $0 < v < \pi/2$, $\sin v = b/\sqrt{a^2 + b^2}$, and $\cos v = a/\sqrt{a^2 + b^2}$.

40 Use Exercise 39 to express $8 \sin u + 15 \cos u$ in the form $c \sin (u + v)$.

CALCULATOR EXERCISES 6.4

1 Approximate $\sin 2\theta$ if θ is a fourth-quadrant angle and $\sin \theta = -0.4637$.

2 Approximate $\cos (\theta/2)$ if θ is a third-quadrant angle and $\tan \theta = 3.8105$.

In Exercises 3–6 demonstrate, by direct calculations, that the formula is true for the stated value of x.

3 $\sin 2x = 2 \sin x \cos x$; $x = 3.624$

4 $\cos 2x = \cos^2 x - \sin^2 x$; $x = \sqrt{2\pi} + e^{\sqrt{7}}$

5 $\tan 2x = 2 \tan x/(1 - \tan^2 x)$; $x = \log (13.6)^5$

6 $|\sin (x/2)| = \sqrt{(1 - \cos x)/2}$; $x = \tan 0.1634$

SECTION 6.5
PRODUCT AND FACTORING FORMULAS

The following identities may be used to change the form of certain trigonometric expressions.

PRODUCT FORMULAS

$$\sin(u+v) + \sin(u-v) = 2\sin u \cos v$$
$$\sin(u+v) - \sin(u-v) = 2\cos u \sin v$$
$$\cos(u+v) + \cos(u-v) = 2\cos u \cos v$$
$$\cos(u-v) - \cos(u+v) = 2\sin u \sin v$$

Each of these identities is a consequence of our work in Section 6.3. For example, to verify the first product formula, we merely add the left- and right-hand sides of the identities we obtained for $\sin(u+v)$ and $\sin(u-v)$. The remaining product formulas are obtained in similar fashion.

By starting with the expression on the *right* side of each product formula, we can express certain products as sums, as illustrated in the next example.

EXAMPLE 1 Express each of the following as a sum:

(a) $\sin 4\theta \cos 3\theta$ (b) $\sin 3x \sin x$

Solution

(a) Using the first product formula with $u = 4\theta$ and $v = 3\theta$,

$$2\sin 4\theta \cos 3\theta = \sin(4\theta + 3\theta) + \sin(4\theta - 3\theta)$$

or

$$\sin 4\theta \cos 3\theta = \tfrac{1}{2}\sin 7\theta + \tfrac{1}{2}\sin \theta.$$

This relationship could also have been obtained by using the second product formula.

(b) Using the fourth product formula with $u = 3x$ and $v = x$,

$$2\sin 3x \sin x = \cos(3x - x) - \cos(3x + x)$$

or

$$\sin 3x \sin x = \tfrac{1}{2}\cos 2x - \tfrac{1}{2}\cos 4x.$$

■

Product formulas may also be employed to express a sum as a product. To obtain a form that can be applied more easily, we shall change the notation as follows. If we let

$$u + v = a \quad \text{and} \quad u - v = b$$

then $(u + v) + (u - v) = a + b$, which simplifies to

$$u = \frac{a+b}{2}.$$

Similarly, since $(u + v) - (u - v) = a - b$,

$$v = \frac{a-b}{2}.$$

If we now substitute for $u + v$ and $u - v$ on the left sides of the product formulas, and for u and v on the right sides, we obtain the following formulas.

FACTORING FORMULAS

$$\sin a + \sin b = 2 \sin \frac{a+b}{2} \cos \frac{a-b}{2}$$

$$\sin a - \sin b = 2 \cos \frac{a+b}{2} \sin \frac{a-b}{2}$$

$$\cos a + \cos b = 2 \cos \frac{a+b}{2} \cos \frac{a-b}{2}$$

$$\cos b - \cos a = 2 \sin \frac{a+b}{2} \sin \frac{a-b}{2}$$

EXAMPLE 2 Express $\sin 5x - \sin 3x$ as a product.

Solution Using the second factoring formula with $a = 5x$ and $b = 3x$,

$$\begin{aligned}\sin 5x - \sin 3x &= 2 \cos \frac{5x + 3x}{2} \sin \frac{5x - 3x}{2} \\ &= 2 \cos 4x \sin x.\end{aligned}$$

■

EXAMPLE 3 Verify the identity $\dfrac{\sin 3t + \sin 5t}{\cos 3t - \cos 5t} = \cot t.$

Solution Using the first and fourth factoring formulas,

$$\begin{aligned}\frac{\sin 3t + \sin 5t}{\cos 3t - \cos 5t} &= \frac{2 \sin \dfrac{3t + 5t}{2} \cos \dfrac{3t - 5t}{2}}{2 \sin \dfrac{5t + 3t}{2} \sin \dfrac{5t - 3t}{2}} \\ &= \frac{2 \sin 4t \cos (-t)}{2 \sin 4t \sin t} = \frac{\cos (-t)}{\sin t} \\ &= \frac{\cos t}{\sin t} = \cot t.\end{aligned}$$

■

EXAMPLE 4 Find the solutions of the equation $\cos t - \sin 2t - \cos 3t = 0.$

Solution Using the fourth factoring formula,

$$\begin{aligned}\cos t - \cos 3t &= 2 \sin \frac{3t + t}{2} \sin \frac{3t - t}{2} \\ &= 2 \sin 2t \sin t.\end{aligned}$$

Hence, the given equation is equivalent to

$$2 \sin 2t \sin t - \sin 2t = 0$$

or

$$\sin 2t(2 \sin t - 1) = 0.$$

It follows that the solutions of the given equation are the solutions of

$$\sin 2t = 0 \quad \text{and} \quad \sin t = \tfrac{1}{2}.$$

The first of these equations has solutions

$$2t = n\pi, \quad \text{or} \quad t = \frac{\pi}{2} n$$

where n is any integer. The second equation has solutions

$$t = \frac{\pi}{6} + 2n\pi \quad \text{and} \quad t = \frac{5\pi}{6} + 2n\pi$$

where n is any integer. ■

The addition formulas may also be employed to derive the so-called **reduction formulas.** The latter formulas can be used to write expressions such as

$$\sin\left(\theta + n \cdot \frac{\pi}{2}\right) \quad \text{and} \quad \cos\left(\theta + n \cdot \frac{\pi}{2}\right)$$

where n is any integer, in terms of only $\sin \theta$ or $\cos \theta$. Similar formulas are true for the other trigonometric functions. Instead of deriving general reduction formulas, we shall illustrate several special cases in the next example.

EXAMPLE 5 Express $\sin(\theta - 3\pi/2)$ and $\cos(\theta + \pi)$ in terms of a function of θ.

Solution We may proceed as follows:

$$\begin{aligned}\sin\left(\theta - \frac{3\pi}{2}\right) &= \sin \theta \cos \frac{3\pi}{2} - \cos \theta \sin \frac{3\pi}{2} \\ &= \sin \theta \cdot (0) - \cos \theta \cdot (-1) = \cos \theta \\ \cos(\theta + \pi) &= \cos \theta \cos \pi - \sin \theta \sin \pi \\ &= \cos \theta \cdot (-1) - \sin \theta \cdot (0) = -\cos \theta.\end{aligned}$$

■

We have derived many important identities in this chapter. For reference, these identities are listed on the inside of the covers of the text.

EXERCISES 6.5

In Exercises 1–8 express the product as a sum or difference.

1 $2 \sin 9\theta \cos 3\theta$ **2** $2 \cos 5\theta \sin 5\theta$

3 $\sin 7t \sin 3t$ **4** $\sin (-4x) \cos 8x$

5 $\cos 6u \cos (-4u)$ **6** $\sin 4t \sin 6t$

7 $3 \cos x \sin 2x$ **8** $5 \cos u \sin 5u$

In Exercises 9–16 write the expression as a product.

9 $\sin 6\theta + \sin 2\theta$ **10** $\sin 4\theta - \sin 8\theta$

11 $\cos 5x - \cos 3x$ **12** $\cos 5t + \cos 6t$

13 $\sin 3t - \sin 7t$ **14** $\cos \theta - \cos 5\theta$

15 $\cos x + \cos 2x$ **16** $\sin 8t + \sin 2t$

Verify the identities in Exercises 17–24.

17 $\dfrac{\sin 4t + \sin 6t}{\cos 4t - \cos 6t} = \cot t$

18 $\dfrac{\sin \theta + \sin 3\theta}{\cos \theta + \cos 3\theta} = \tan 2\theta$

19 $\dfrac{\sin u + \sin v}{\cos u + \cos v} = \tan \dfrac{u+v}{2}$

20 $\dfrac{\sin u - \sin v}{\cos u - \cos v} = -\cot \dfrac{u+v}{2}$

21 $\dfrac{\sin u - \sin v}{\sin u + \sin v} = \dfrac{\tan \frac{1}{2}(u - v)}{\tan \frac{1}{2}(u + v)}$

22 $\dfrac{\cos u - \cos v}{\cos u + \cos v} = \tan \frac{1}{2}(u + v) \tan \frac{1}{2}(u - v)$

23 $\sin 2x + \sin 4x + \sin 6x = 4 \cos x \cos 2x \sin 3x$

24 $\dfrac{\cos t + \cos 4t + \cos 7t}{\sin t + \sin 4t + \sin 7t} = \cot 4t$

25 Express $(\sin ax)(\cos bx)$ as a sum.

26 Express $(\cos mu)(\cos nu)$ as a sum.

In Exercises 27–30 find the solutions of the equations.

27 $\sin 5t + \sin 3t = 0$ **28** $\sin t + \sin 3t = \sin 2t$

29 $\cos x = \cos 3x$ **30** $\cos 4x - \cos 3x = 0$

In Exercises 31–36 verify the reduction formulas by using addition formulas.

31 $\sin (\theta + \pi) = -\sin \theta$

32 $\sin\left(\theta + \dfrac{3\pi}{2}\right) = -\cos \theta$

33 $\cos\left(\theta - \dfrac{5\pi}{2}\right) = \sin \theta$

34 $\cos (\theta - 3\pi) = -\cos \theta$

35 $\tan (\pi - \theta) = -\tan \theta$
$\left(\textit{Hint: } \tan (\pi - \theta) = \dfrac{\sin (\pi - \theta)}{\cos (\pi - \theta)}.\right)$

36 $\tan (\theta + \pi/2) = -\cot \theta$

CALCULATOR EXERCISES 6.5

1 Demonstrate, by direct calculations, that the first product formula is true if $u = 4.631$ and $v = -8.059$.

2 Demonstrate that the first factoring formula is true if $a = 10^{\sqrt{148}}$ and $b = \log 735$.

SECTION 6.6

THE INVERSE TRIGONOMETRIC FUNCTIONS

The concept of inverse function was defined in Section 2.6. Recall that if f is a one-to-one function with domain A and range B, then, as illustrated in (i) of Figure 6.4, for each number u in B there is exactly one number v

in A such that $f(v) = u$. In this case we define the function f^{-1} from B to A by letting $f^{-1}(u) = v$ (see (ii) of Figure 6.4).

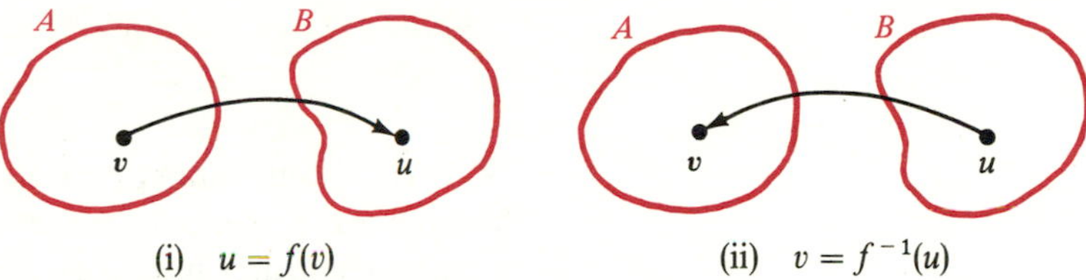

(i) $u = f(v)$ (ii) $v = f^{-1}(u)$

FIGURE 6.4

We call f^{-1} the **inverse function** of f. The function f^{-1} *reverses* the correspondence from A to B given by f in the sense that

$$v = f^{-1}(u) \quad \text{if and only if} \quad f(v) = u$$

for every u in B and v in A. Since the notation for the variables is immaterial, we shall use the traditional symbols x and y, and write

(I) $$y = f^{-1}(x) \quad \text{if and only if} \quad f(y) = x$$

for every x in the domain of f^{-1} and every y in the domain of f.

In the remainder of this section we shall use the relationship in (I) to obtain the *inverse trigonometric functions*. As pointed out in Section 2.6, the inverse function f^{-1} is characterized by the following identities:

(II) $$f^{-1}(f(y)) = y \quad \text{for every } y \text{ in the domain of } f$$
$$f(f^{-1}(x)) = x \quad \text{for every } x \text{ in the domain of } f^{-1}.$$

Since the trigonometric functions are not one-to-one, they do not have inverse functions. However, by restricting the domains, it is possible to obtain functions that have the same values as the trigonometric functions (over the smaller domains) and that do possess inverse functions. Given a trigonometric function, we shall choose a subset S of the domain throughout which the function is either increasing or decreasing, since this will guarantee the existence of an inverse. Of course, we must also choose S in such a way that the function takes on all of its values.

Let us first consider the sine function, whose domain is $\mathbb{R}$ and whose range is the set of real numbers in the interval $[-1, 1]$. The sine function is not one-to-one since, for example, numbers such as $\pi/6$, $5\pi/6$, and $-7\pi/6$ give rise to the same functional value, $\frac{1}{2}$. If we restrict y to the interval $[-\pi/2, \pi/2]$, then as y varies from $-\pi/2$ to $\pi/2$, $\sin y$ takes on each value between -1 and 1 once and only once. Hence the new function obtained by restricting the domain to $[-\pi/2, \pi/2]$ has an inverse function. Applying the relationship in (I) leads to the following definition.

DEFINITION

> The **inverse sine function,** denoted by $\sin^{-1}$, is defined by
>
> $$y = \sin^{-1} x \quad \text{if and only if} \quad \sin y = x$$
>
> where $-1 \le x \le 1$ and $-\pi/2 \le y \le \pi/2$.

It is also customary to refer to $\sin^{-1}$ as the **arcsine function** and to use the notation arcsin x in place of $\sin^{-1} x$. The expression arcsin x is used because if $y = \arcsin x$, then $\sin y = x$; that is, y is a number (or an *arc*-length) whose sine is x. Since both notations, $\sin^{-1}$ and arcsin, are commonly used in mathematics and its applications, we shall employ both of them in our work. Note that, by definition,

$$-\frac{\pi}{2} \le \sin^{-1} x \le \frac{\pi}{2}$$

or equivalently,

$$-\frac{\pi}{2} \le \arcsin x \le \frac{\pi}{2}.$$

EXAMPLE 1 Sketch the graph of $y = \sin^{-1} x$.

Solution By definition, the graph of $y = \sin^{-1} x$ is the same as the graph of $\sin y = x$, except that the variables are restricted as follows:

$$-\frac{\pi}{2} \le y \le \frac{\pi}{2} \quad \text{and} \quad -1 \le x \le 1.$$

Several coordinates of points on the graph are listed in the following table.

y	$-\frac{\pi}{2}$	$-\frac{\pi}{3}$	$-\frac{\pi}{6}$	0	$\frac{\pi}{6}$	$\frac{\pi}{6}$	$\frac{\pi}{2}$
x	-1	$-\frac{\sqrt{3}}{2}$	$-\frac{1}{2}$	0	$\frac{1}{2}$	$\frac{\sqrt{3}}{2}$	1

The graph is sketched in Figure 6.5. ■

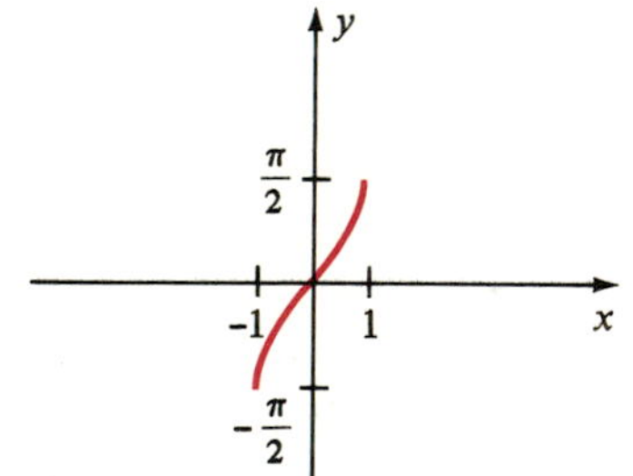

FIGURE 6.5 $y = \arcsin x = \sin^{-1} x$

Using the relationship stated in (II) gives us the following important identities:

$$\sin^{-1}(\sin y) = y \quad \text{if} \quad -\frac{\pi}{2} \le y \le \frac{\pi}{2}$$

$$\sin(\sin^{-1} x) = x \quad \text{if} \quad -1 \le x \le 1.$$

These may also be written in the form

$$\arcsin(\sin y) = y \quad \text{and} \quad \sin(\arcsin x) = x$$

provided that y and x are suitably restricted.

EXAMPLE 2 Use the definition of the inverse sine function to find $\sin^{-1}(\sqrt{2}/2)$, and arc sin $(-1/2)$.

Solution If $y = \sin^{-1}(\sqrt{2}/2)$, then, by definition, $\sin y = \sqrt{2}/2$, and consequently, $y = \pi/4$. Note that it is essential to choose y in the interval $[-\pi/2, \pi/2]$. A number such as $3\pi/4$ is incorrect, even though $\sin(3\pi/4) = \sqrt{2}/2$.

In like manner, if $y = \arcsin(-1/2)$, then $\sin y = -\frac{1}{2}$. Since, by definition, we must choose y in the interval $[-\pi/2, \pi/2]$, it follows that $y = -\pi/6$. ■

EXAMPLE 3 Find $\sin^{-1}(\tan 3\pi/4)$.

Solution If we let

$$y = \sin^{-1}\left(\tan\frac{3\pi}{4}\right) = \sin^{-1}(-1)$$

then, by definition, $\sin y = -1.$

Consequently, $y = -\pi/2$. ■

The other trigonometric functions may also be used to introduce inverse functions. The procedure is first to determine a convenient subset of the domain so that a one-to-one function is obtained, and then to apply the formulas in (I).

If the domain of the cosine function is restricted to the interval $[0, \pi]$, and if y increases from 0 to π, then $\cos y$ decreases and takes on each value between -1 and 1 once and only once. This leads to the following.

DEFINITION

The **inverse cosine function,** denoted by $\cos^{-1}$, is defined by

$$y = \cos^{-1} x \quad \text{if and only if} \quad \cos y = x$$

where $-1 \le x \le 1$ and $0 \le y \le \pi$.

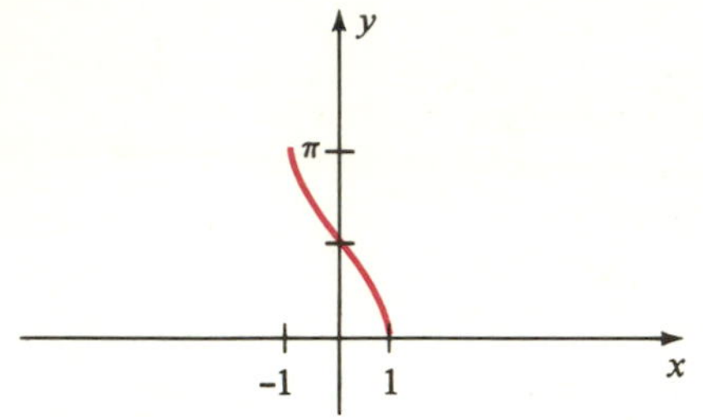

FIGURE 6.6 $y = \arccos x = \cos^{-1} x$

The inverse cosine function is also referred to as the **arccosine function,** and the notation arccos x is used interchangeably with $\cos^{-1} x$.

Applying the formulas in (II) we see that if $0 \le y \le \pi$ and $-1 \le x \le 1$, then

$$\cos(\cos^{-1} x) = \cos(\arccos x) = x$$

$$\cos^{-1}(\cos y) = \arccos(\cos y) = y$$

By definition, the graph of $y = \cos^{-1} x$ is the same as the graph of $\cos y = x$, where $0 \le y \le \pi$. It is left as an exercise to verify that the graph has the appearance indicated in Figure 6.6.

EXAMPLE 4 Use the definition of the inverse cosine function to find $\cos^{-1}(-\sqrt{3}/2)$.

Solution If $y = \cos^{-1}(-\sqrt{3}/2)$, then $\cos y = -\sqrt{3}/2$. The reference number for y is $\pi/6$. (Why?) By definition, y must be chosen in the interval $[0, \pi]$, and hence $y = \pi - (\pi/6) = 5\pi/6$. Thus,

$$\cos^{-1}(-\sqrt{3}/2) = 5\pi/6.$$ ■

If we restrict the domain of the tangent function to the open interval $(-\pi/2, \pi/2)$, then a one-to-one correspondence is obtained. We may, therefore, adopt the following definition.

DEFINITION

The **inverse tangent function** or **arctangent function,** denoted by $\tan^{-1}$ or arctan, is defined by

$$y = \tan^{-1} x = \arctan x \quad \text{if and only if} \quad \tan y = x$$

where x is any real number and $-\pi/2 < y < \pi/2$.

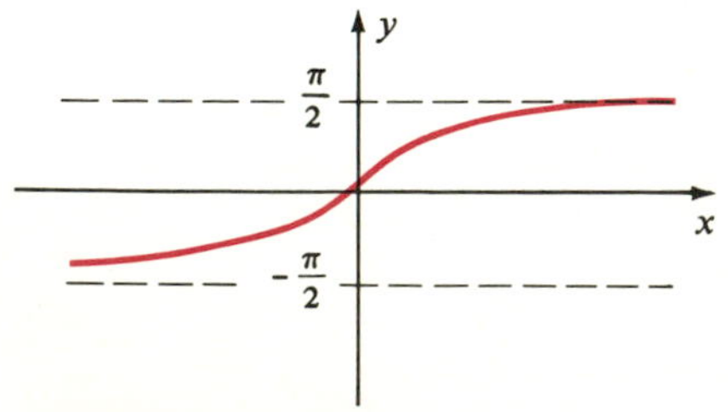

FIGURE 6.7 $y = \arctan x = \tan^{-1} x$

Note that the domain of the arctangent function is all of $\mathbb{R}$ and the range is the open interval $(-\pi/2, \pi/2)$. In Exercise 42 you are asked to verify the graph of the inverse tangent function, which is sketched in Figure 6.7.

An analogous procedure may be used for the remaining trigonometric functions. The functions we have defined are generally referred to as the **inverse trigonometric functions.**

EXAMPLE 5 Without using tables or calculators find sec (arctan $\frac{2}{3}$).

Solution If $y = \arctan \frac{2}{3}$, then $\tan y = \frac{2}{3}$. Since $\sec^2 y = 1 + \tan^2 y$

and $0 < y < \pi/2$, we may write

$$\sec y = \sqrt{1 + \tan^2 y} = \sqrt{1 + \left(\frac{2}{3}\right)^2}$$
$$= \sqrt{1 + \frac{4}{9}} = \sqrt{\frac{13}{9}} = \frac{\sqrt{13}}{3}.$$

Hence, $\sec(\arctan \frac{2}{3}) = \sec y = \sqrt{13}/3$. ∎

The next two examples illustrate some of the manipulations that can be carried out with the inverse trigonometric functions.

EXAMPLE 6 Evaluate $\sin(\arctan \frac{1}{2} - \arccos \frac{4}{5})$.

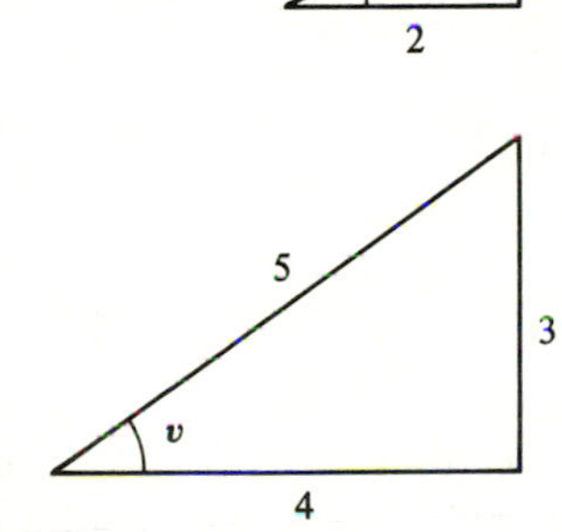

FIGURE 6.8

Solution If we let $u = \arctan \frac{1}{2}$ and $v = \arccos \frac{4}{5}$, then $\tan u = \frac{1}{2}$ and $\cos v = \frac{4}{5}$. We wish to find $\sin(u - v)$. Since u and v are in the interval $(0, \pi/2)$, they may be considered as the radian measure of positive acute angles, and other functional values of u and v may be found by referring to the right triangles in Figure 6.8. This gives us $\sin u = 1/\sqrt{5}$, $\cos u = 2/\sqrt{5}$, and $\sin v = 3/5$. Consequently,

$$\sin(u - v) = \sin u \cos v - \cos u \sin v$$
$$= \frac{1}{\sqrt{5}}\frac{4}{5} - \frac{2}{\sqrt{5}}\frac{3}{5}$$
$$= \frac{-2}{5\sqrt{5}} = \frac{-2\sqrt{5}}{25}.$$ ∎

EXAMPLE 7 Write $\cos(\sin^{-1} x)$ as an algebraic expression in x.

Solution Let $y = \sin^{-1} x$ so that $\sin y = x$. We wish to find an expression for $\cos y$. Since $-\pi/2 \le y \le \pi/2$, it follows that $\cos y \ge 0$, and hence

$$\cos y = \sqrt{1 - \sin^2 y} = \sqrt{1 - x^2}.$$

Consequently, $$\cos(\sin^{-1} x) = \sqrt{1 - x^2}.$$ ∎

EXAMPLE 8 Verify the identity

$$\frac{1}{2}\cos^{-1} x = \tan^{-1}\sqrt{\frac{1 - x}{1 + x}}, \qquad \text{where } |x| < 1.$$

Solution Let $y = \cos^{-1} x$. We wish to show that

$$\frac{1}{2} y = \tan^{-1}\sqrt{\frac{1 - x}{1 + x}}.$$

By a half-angle formula (see Section 6.4),

$$\left|\tan\frac{y}{2}\right| = \sqrt{\frac{1-\cos y}{1+\cos y}}.$$

Since $y = \cos^{-1} x$ and $|x| < 1$, it follows that $0 < y < \pi$, or $0 < y/2 < \pi/2$. Consequently, $\tan (y/2) > 0$ and we may drop the absolute value sign, obtaining

$$\tan\frac{y}{2} = \sqrt{\frac{1-\cos y}{1+\cos y}}.$$

Using the fact that $\cos y = x$ gives us

$$\tan\frac{y}{2} = \sqrt{\frac{1-x}{1+x}}$$

or equivalently

$$\frac{y}{2} = \tan^{-1}\sqrt{\frac{1-x}{1+x}},$$

which is what we wished to show. ■

Most of the trigonometric equations considered in Section 6.2 had solutions that were rational multiples of π, such as $\pi/3$, $3\pi/4$, π, etc. If solutions are not of that type we can sometimes use inverse functions to express them in exact form, as illustrated in the next example.

EXAMPLE 9 Find the solutions of the equation

$$5\sin^2 t + 3\sin t - 1 = 0$$

that are in the interval $[-\pi/2, \pi/2]$.

Solution The equation may be regarded as a quadratic equation in $\sin t$. Applying the Quadratic Formula,

$$\sin t = \frac{-3 \pm \sqrt{9+20}}{10} = \frac{-3 \pm \sqrt{29}}{10}.$$

Using the definition of the inverse sine function, we obtain the solutions

$$t = \sin^{-1}\tfrac{1}{10}(-3+\sqrt{29})$$

and

$$t = \sin^{-1}\tfrac{1}{10}(-3-\sqrt{29}).$$

If approximations are desired, we place a calculator in radian mode and proceed as follows:

Enter: $\frac{1}{10}(-3+\sqrt{29}) \approx 0.2385165$

Press INV sin: 0.240838

Hence $t \approx 0.2408.$

Similarly,

$$t = \sin^{-1}\tfrac{1}{10}(-3-\sqrt{29}) \approx \sin^{-1}(-0.8385165) \approx -0.9946. \quad ■$$

EXERCISES 6.6

Find the numbers in Exercises 1–10 without the use of tables or calculators.

1 (a) $\sin^{-1}\frac{1}{2}$ (b) $\cos^{-1}\frac{1}{2}$

2 (a) $\sin^{-1}\sqrt{3}/2$ (b) $\sin^{-1}(-\sqrt{3}/2)$

3 (a) $\arcsin 0$ (b) $\arccos 0$

4 (a) $\cos^{-1}\sqrt{2}/2$ (b) $\cos^{-1}(-\sqrt{2}/2)$

5 (a) $\arcsin 1$ (b) $\arccos 1$

6 (a) $\sin^{-1}(-1)$ (b) $\cos^{-1}(-1)$

7 (a) $\tan^{-1}\sqrt{3}$ (b) $\arctan(-\sqrt{3})$

8 (a) $\tan^{-1}(-1)$ (b) $\arccos(-\frac{1}{2})$

9 (a) $\arcsin(-\sqrt{2}/2)$ (b) $\tan^{-1}(-\sqrt{3}/3)$

10 (a) $\tan^{-1} 0$ (b) $\arctan 1$

Approximate the numbers in Exercises 11–14 by using (a) Table 4; (b) a calculator.

11 $\sin^{-1}(-0.6494)$

12 $\cos^{-1}(-0.9112)$

13 $\arctan(2.1775)$

14 $\arcsin(0.8004)$

Determine the numbers in Exercises 15–26 without the use of tables or calculators.

15 $\sin(\cos^{-1}\frac{1}{2})$

16 $\cos(\sin^{-1} 0)$

17 $\sin(\arccos\frac{4}{5})$

18 $\tan(\tan^{-1} 5)$

19 $\arcsin(\sin 5\pi/4)$

20 $\cos^{-1}(\cos 3\pi/4)$

21 $\cos(\sin^{-1}\frac{4}{5} + \tan^{-1}\frac{3}{4})$

22 $\sin(\arcsin\frac{1}{2} + \arccos 0)$

23 $\tan(\arctan\frac{4}{3} + \arccos\frac{8}{17})$

24 $\cos(2\sin^{-1}\frac{15}{17})$

25 $\sin[2\arccos(-\frac{3}{5})]$

26 $\cos(\frac{1}{2}\tan^{-1}\frac{8}{15})$

In Exercises 27–30 rewrite the expression as an algebraic expression in x.

27 $\sin(\tan^{-1} x)$

28 $\tan(\arccos x)$

29 $\cos(\frac{1}{2}\arccos x)$

30 $\cos(2\tan^{-1} x)$

Verify the identities in Exercises 31–38.

31 $\sin^{-1} x + \cos^{-1} x = \pi/2$
(*Hint:* let $\alpha = \sin^{-1} x$, $\beta = \cos^{-1} x$, and consider $\sin(\alpha+\beta)$.)

32 $\arctan x + \arctan(1/x) = \pi/2,\ x > 0$

33 $\arcsin\dfrac{2x}{1+x^2} = 2\arctan x,\ |x| < 1$

34 $2\cos^{-1} x = \cos^{-1}(2x^2 - 1),\ 0 \le x \le 1$

35 $\arcsin(-x) = -\arcsin x$

36 $\arccos(-x) = \pi - \arccos x$

37 $\sin^{-1} x = \tan^{-1}\dfrac{x}{\sqrt{1-x^2}}$

38 $\tan^{-1} u + \tan^{-1} v = \tan^{-1}\dfrac{u+v}{1-uv},\ |u| < 1$ and $|v| < 1$

39 Define $\cot^{-1}$ by restricting the domain of the cotangent function to the interval $(0, \pi)$.

40 Define $\sec^{-1}$ by restricting the domain of the secant function to $[0, \pi/2) \cup [\pi, 3\pi/2)$.

41 Verify the sketch in Figure 6.6.

42 Verify the sketch in Figure 6.7.

Sketch the graphs of the equations in Exercises 43–52.

43 $y = \sin^{-1} 2x$

44 $y = \cos^{-1} (x/2)$

45 $y = \frac{1}{2} \sin^{-1} x$

46 $y = 2 \cos^{-1} x$

47 $y = 2 \tan^{-1} x$

48 $y = \tan^{-1} 2x$

49 $y = 2 + \tan^{-1} x$

50 $y = \sin^{-1} (x + 1)$

51 $y = \sin (\arccos x)$

52 $y = \sin (\sin^{-1} x)$

Prove that the equations in Exercises 53 and 54 are not identities.

53 $\tan^{-1} x = \dfrac{1}{\tan x}$

54 $(\arcsin x)^2 + (\arccos x)^2 = 1$

In Exercises 55–60 use inverse trigonometric functions to find the solutions of the equations in the given intervals.

55 $2 \tan^2 t + 9 \tan t + 3 = 0;\quad (-\pi/2, \pi/2)$

56 $3 \sin^2 t + 7 \sin t + 3 = 0;\quad [-\pi/2, \pi/2]$

57 $15 \cos^4 x - 14 \cos^2 x + 3 = 0;\quad [0, \pi]$

58 $3 \tan^4 \theta - 19 \tan^2 \theta + 2 = 0;\quad (-\pi/2, \pi/2)$

59 $6 \sin^3 \theta + 18 \sin^2 \theta - 5 \sin \theta - 15 = 0;\quad (-\pi/2, \pi/2)$

60 $6 \sin 2x - 8 \cos x + 9 \sin x - 6 = 0;\quad (-\pi/2, \pi/2)$

CALCULATOR EXERCISES 6.6

1–6 Approximate the solutions of the equations in Exercises 55–60 to four decimal places.

7 Show, by calculations, that $\sin (\arcsin x) = x$ for the following values of x:

(a) 0.4631 (b) log 3.64 (c) $1/\sqrt{\pi}$

8 Show that $\tan^{-1} (\tan x) = x$ for the following values of x:

(a) 74.85 (b) $\log(\pi^2 + 94.7)$ (c) $10^{6.39}$

Exercises 9 and 10 employ the notation introduced in the Calculator Exercises of Section 5.5. Use a calculator to support the given statements by substituting the following values for t: 0.1, 0.01, 0.001, 0.0001.

9 $\dfrac{\arcsin 2x}{\arcsin x} \to 2 \quad \text{as} \quad x \to 0^+$

10 $\csc x \tan^{-1} x \to 1 \quad \text{as} \quad x \to 0^+$

SECTION 6.7
THE LAW OF SINES

A triangle that does not contain a right angle is referred to as an **oblique triangle.** Since it is always possible to divide such triangles into two right triangles, methods developed in Chapter 5 may be used to solve them; however, sometimes it is cumbersome to proceed in that manner. In this section and the next, we obtain formulas that aid in simplifying solutions of oblique triangles.

If two angles and a side of a triangle are known, or if two sides and an angle opposite one of them are known, then the remaining parts of the triangle may be found by means of the formula given in this section. We

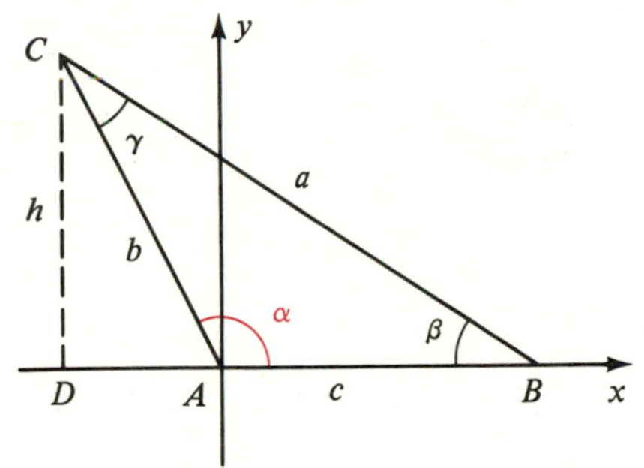

FIGURE 6.9

shall use the letters A, B, C, a, b, c, α, β, and γ for parts of triangles as they were used in Section 5.8. Given triangle ABC, let us place angle α in standard position on a rectangular coordinate system so that B is on the positive x-axis. The case where α is obtuse is illustrated in Figure 6.9. The type of argument we shall give may also be used if α is acute.

Consider the line through C parallel to the y-axis and intersecting the x-axis at point D. Suppose that $d(C, D) = h$, so that the y-coordinate of C is h. It follows that

$$\sin \alpha = \frac{h}{b}, \quad \text{or} \quad h = b \sin \alpha.$$

Referring to triangle BDC, we see that

$$\sin \beta = \frac{h}{a}, \quad \text{or} \quad h = a \sin \beta.$$

Consequently, $$a \sin \beta = b \sin \alpha,$$

which may be written $$\frac{a}{\sin \alpha} = \frac{b}{\sin \beta}.$$

If α is taken in standard position, so that C is on the positive x-axis, then by the same reasoning,

$$\frac{a}{\sin \alpha} = \frac{c}{\sin \gamma}.$$

The last two equalities give us the following result.

THE LAW OF SINES

If ABC is any oblique triangle labeled in the usual manner, then

$$\frac{a}{\sin \alpha} = \frac{b}{\sin \beta} = \frac{c}{\sin \gamma}.$$

The next example illustrates a method of applying the Law of Sines to the case in which two angles and a side of a triangle are known. We shall use the rules for rounding off answers discussed in Section 5.8.

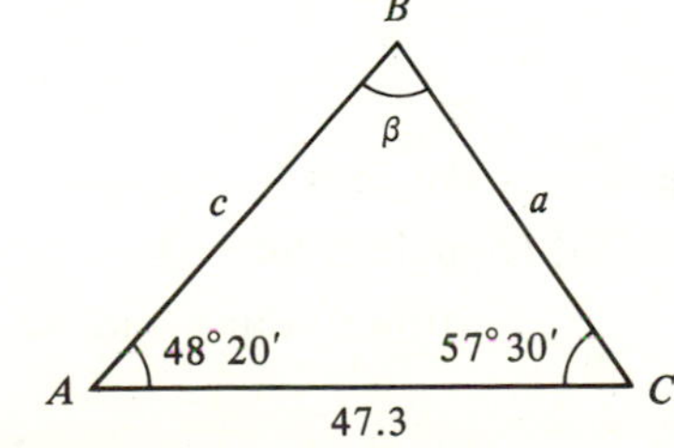

FIGURE 6.10

EXAMPLE 1 Given triangle ABC with $\alpha = 48°20'$, $\gamma = 57°30'$, and $b = 47.3$, approximate the remaining parts.

Solution The triangle is sketched in Figure 6.10. Since the sum of the angles is 180°,

$$\beta = 180° - (57°30' + 48°20') = 74°10'.$$

Applying the Law of Sines,

$$a = \frac{b \sin \alpha}{\sin \beta} = \frac{(47.3) \sin 48°20'}{\sin 74°10'}.$$

Using Table 4 or a calculator,

$$a \approx \frac{(47.3)(0.7470)}{(0.9621)} \approx 36.7.$$

Similarly, $$c = \frac{b \sin \gamma}{\sin \beta} = \frac{(47.3) \sin 57°30'}{\sin 74°10'}$$

$$\approx \frac{(47.3)(0.8434)}{(0.9621)} \approx 41.5.$$ ■

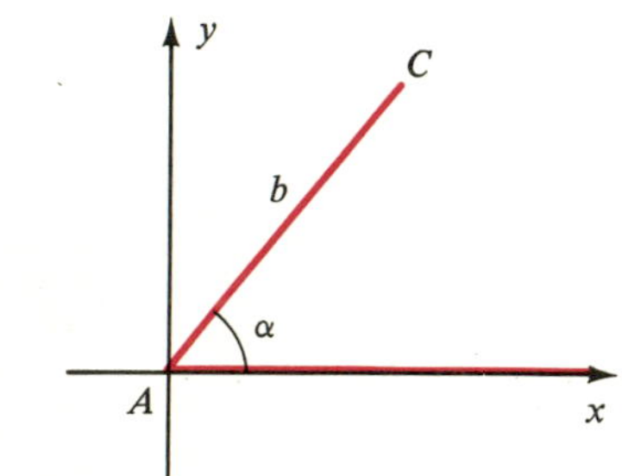

FIGURE 6.11

Data such as that in Example 1 always produces a unique triangle ABC. However, if two sides and an angle opposite one of them are given, a unique triangle is not always determined. To illustrate, suppose that a and b are to be lengths of sides of a triangle ABC. In addition, suppose that a specific angle α is to be opposite the side of length a. Let us consider the case in which α is acute. Place α in standard position on a rectangular coordinate system and consider the line segment AC of length b on the terminal side of α, as shown in Figure 6.11. The third vertex, B, should be somewhere on the x-axis. Since the length a of the side opposite α is given, B may be found by striking off a circular arc of length a with center at C. There are four possible outcomes for the construction, as illustrated in Figure 6.12, where the coordinate axes have been deleted.

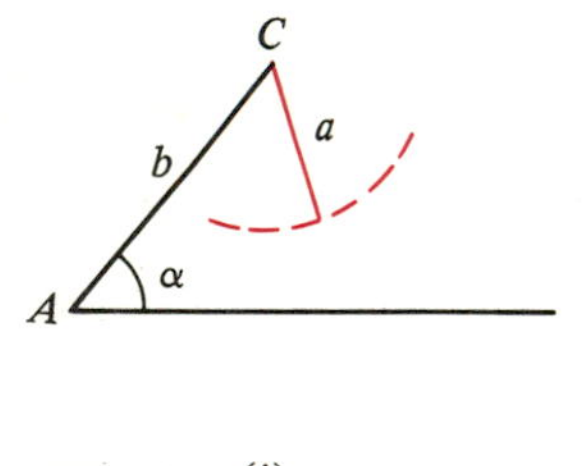

(i)

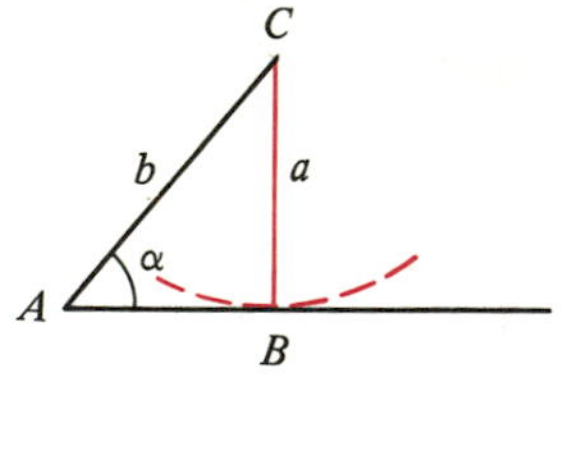

(ii)

(iii)

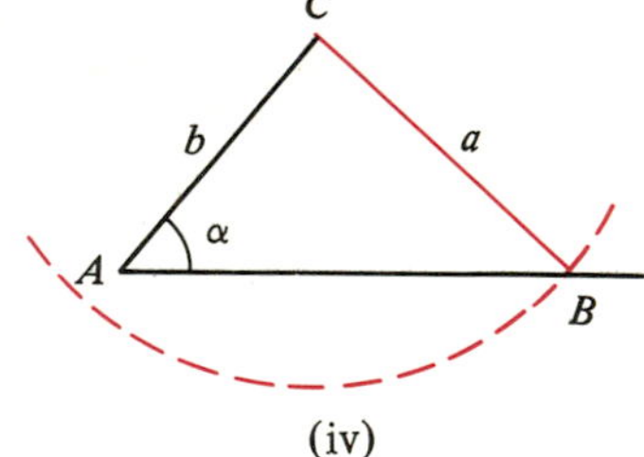

(iv)

FIGURE 6.12

The four possibilities may be listed as follows:

(i) The arc does not intersect the x-axis and no triangle is formed.

(ii) The arc is tangent to the x-axis and a right triangle is formed.

(iii) The arc intersects the positive x-axis in two distinct points and two triangles are formed.

(iv) The arc intersects both the positive and nonpositive parts of the x-axis and one triangle is formed.

Since the distance from C to the x-axis is $b \sin \alpha$ (Why?), we see that (i) occurs if $a < b \sin \alpha$, (ii) occurs if $a = b \sin \alpha$, (iii) occurs if $b \sin \alpha < a < b$, and (iv) occurs if $a \geq b$. It is unnecessary to memorize these facts, since in any specific problem the case that occurs will become evident when the solution is attempted. For example, when solving the equation

$$\frac{a}{\sin \alpha} = \frac{b}{\sin \beta}$$

suppose that we obtain $|\sin \beta| > 1$. This will indicate that no triangle exists. If the equation $\sin \beta = 1$ is obtained, then $\beta = 90°$ and hence case (ii) occurs. On the other hand, if $|\sin \beta| < 1$, then there are two possible choices for the angle β. By checking both possibilities, it will become apparent whether (iii) or (iv) occurs.

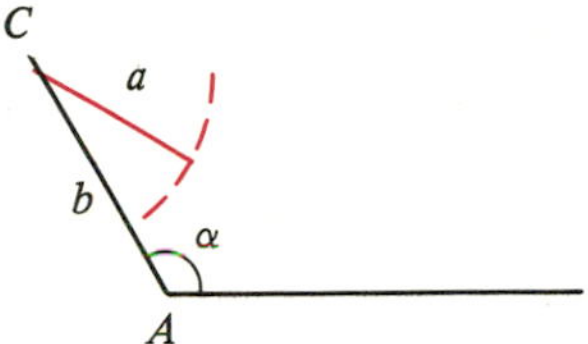

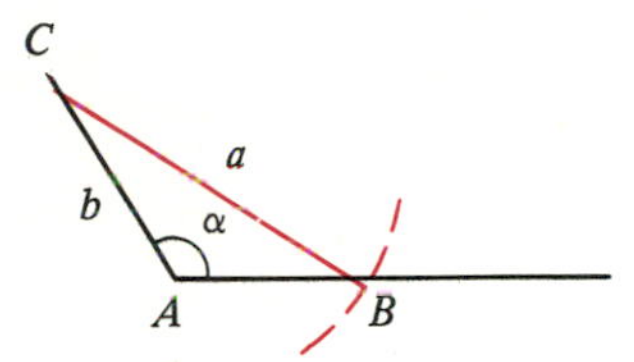

FIGURE 6.13

If the measure of α is greater than 90°, then, as shown in Figure 6.13, a triangle exists if and only if $a > b$. Of course, our discussion is independent of the symbols we have used; that is, we might be given b, c, β, or a, c, γ, etc.

Since different possibilities may arise, the case in which two sides and an angle opposite one of them are given is sometimes called the **ambiguous case.**

EXAMPLE 2 Solve triangle ABC if $\alpha = 67°$, $c = 125$, and $a = 100$.

Solution Using $\sin \gamma = \dfrac{c \sin \alpha}{a}$,

$$\sin \gamma = \frac{(125) \sin 67°}{100}$$

$$\approx \frac{(125)(0.9205)}{100} \approx 1.1506.$$

Since $\sin \gamma > 1$, no triangle can be formed with the given parts. ■

EXAMPLE 3 Approximate the remaining parts of triangle ABC if $a = 12.4$, $b = 8.7$, and $\beta = 36°40'$.

Solution Using the fact that $\sin \alpha = \dfrac{a \sin \beta}{b}$,

$$\sin \alpha = \frac{(12.4) \sin 36°40'}{8.7}$$

$$\approx \frac{(12.4)(0.5972)}{8.7} \approx 0.8512.$$

There are two possible angles α between 0° and 180° such that $\sin \alpha \approx 0.8512$. If we let α' denote the reference angle for α, then using Table 4 or a calculator, we obtain $\alpha' \approx 58°20'$. Consequently, the two possibilities for α are

$$\alpha_1 \approx 58°20' \quad \text{and} \quad \alpha_2 \approx 121°40'.$$

If we let γ_1 and γ_2 denote the third angles of the triangles corresponding to the angles α_1 and α_2, respectively, then

$$\gamma_1 \approx 180° - (36°40' + 58°20') = 85°$$

and

$$\gamma_2 \approx 180° - (36°40' + 121°40') = 21°40'.$$

Thus, there are two possible triangles that have the given parts. They are the triangles A_1BC and A_2BC shown in Figure 6.14.

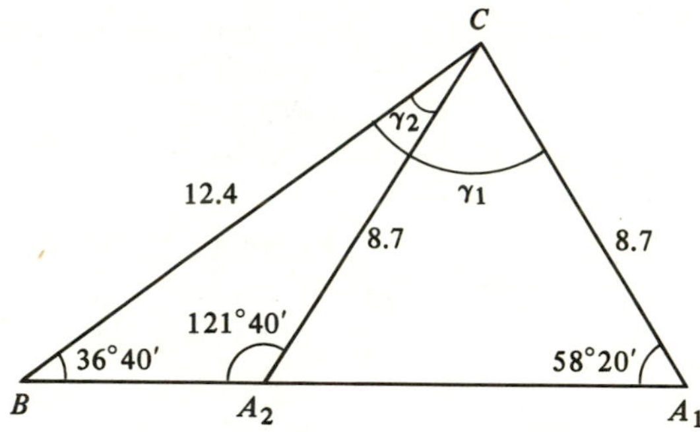

FIGURE 6.14

If c_1 is the side opposite γ_1 in triangle A_1BC, then

$$c_1 = \frac{a \sin \gamma_1}{\sin \alpha_1} \approx \frac{(12.4) \sin 85°}{\sin 58°20'}$$

$$\approx \frac{(12.4)(0.9962)}{0.8511} \approx 14.5.$$

Hence, the solution for triangle A_1BC is $\alpha_1 \approx 58°20'$, $\gamma_1 \approx 85°$, and $c_1 \approx 14.5$. Similarly, if c_2 is the side opposite angle γ_2 in triangle A_2BC, we have

$$c_2 = \frac{a \sin \gamma_2}{\sin \alpha_2} \approx \frac{(12.4) \sin 21°40'}{\sin 121°40'} \approx \frac{(12.4)(0.3692)}{0.8511} \approx 5.4.$$

Consequently, the solution for triangle A_2BC is $\alpha_2 \approx 121°40'$, $\gamma_2 \approx 21°40'$, and $c_2 \approx 5.4$. ■

EXERCISES 6.7

In Exercises 1–12 approximate the remaining parts of triangle ABC.

1 $\alpha = 41°$, $\gamma = 77°$, $a = 10.5$

2 $\beta = 20°$, $\gamma = 31°$, $b = 210$

3 $\alpha = 27°40'$, $\beta = 52°10'$, $a = 32.4$

4 $\alpha = 42°10'$, $\gamma = 61°20'$, $b = 19.7$

5 $\beta = 50°50'$, $\gamma = 70°30'$, $c = 537$

6 $\alpha = 7°10'$, $\beta = 11°40'$, $a = 2.19$

7 $\alpha = 65°10'$, $a = 21.3$, $b = 18.9$

8 $\beta = 30°$, $b = 17.9$, $a = 35.8$

9 $\gamma = 53°20'$, $a = 140$, $c = 115$

10 $\alpha = 27°30'$, $c = 52.8$, $a = 28.1$

11 $\beta = 113°10'$, $b = 248$, $c = 195$

12 $\gamma = 81°$, $c = 11$, $b = 12$

13 A person wishes to find the distance between two points A and B that lie on opposite banks of a river. A line segment AC of length 240 yards is laid off and the measures of angles BAC and ACB are found to be 63°20′ and 54°10′, respectively. Approximate the distance from A to B.

14 In order to determine the distance between two points A and B, a surveyor chooses a point C that is 375 yards from A and 530 yards from B. If Angle BAC has measure 49°30′, approximate the distance between A and B.

15 When the angle of elevation of the sun is 64°, a telephone pole that is tilted at an angle of 12° directly away from the sun casts a shadow 34 feet long on level ground. Approximate the length of the pole.

16 A straight road makes an angle of 15° with the horizontal. When the angle of elevation of the sun is 57°, a vertical pole at the side of the road casts a shadow 75 feet long directly down the road. Approximate the length of the pole.

17 The angles of elevation of a balloon from two points A and B on level ground are 24°10′ and 47°40′, respectively. If points A and B are 8.4 miles apart and the balloon is between the points, in the same vertical plane, approximate, to the nearest tenth of a mile, the height of the balloon above the ground.

18 An airport A is 480 miles due east of airport B. A pilot flew in the direction 235° from A to C and then in the direction 320° from C to B. Approximate, to the nearest mile, the total distance the pilot flew.

19 A forest ranger at an observation point A sights a fire in the direction N27°10′E. Another ranger at an observation point B, 6.0 miles due east of A, sights the same fire at N52°40′W. Approximate, to the nearest tenth of a mile, the distance from each of the observation points to the fire.

20 A surveyor notes that the direction from point A to point B is S63°W and the direction from A to C is S38°W. If the distance from A to B is 239 yards and the distance from B to C is 374 yards, approximate the distance from A to C.

21 A straight road makes an angle of 22° with the horizontal. From a certain point P on the road the angle of elevation of an airplane at point A is 57°. At the same instant, from another point Q, 100 meters farther up the road, the angle of elevation is 63°. If the points P, Q, and A lie in the same vertical plane, as indicated in the figure, approximate, to the nearest meter, the distance from P to the airplane.

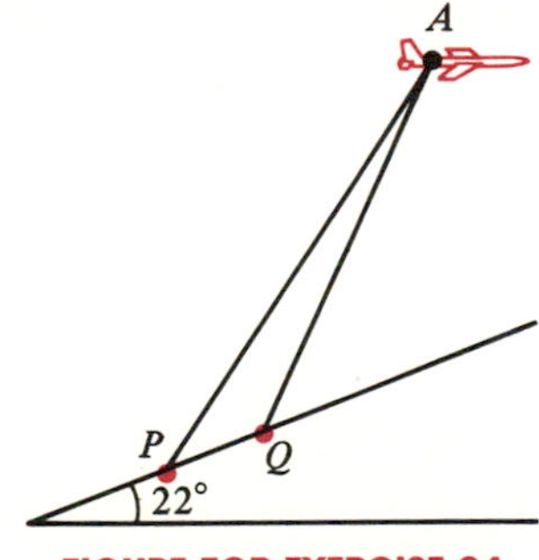

FIGURE FOR EXERCISE 21

22 A point P on level ground is 3.0 km due north of a point Q. If a person jogged in the direction N25°E from Q to a point R, and then from R to P in the direction S40°W, approximate, to the nearest km, the distance jogged.

CALCULATOR EXERCISES 6.7

In Exercises 1–6 approximate the remaining parts of triangle *ABC*.

1 $\beta = 25.6°$, $\gamma = 34.7°$, $b = 184.8$

2 $\alpha = 6.24°$, $\beta = 14.08°$, $a = 4.56$

3 $\alpha = 103.45°$, $\gamma = 27.19°$, $b = 38.84$

4 $\gamma = 47.74°$, $a = 131.08$, $c = 97.84$

5 $\beta = 121.624°$, $b = 0.283$, $c = 0.178$

6 $\alpha = 32.32°$, $c = 574.3$, $a = 263.6$

SECTION 6.8

THE LAW OF COSINES

If two sides and the included angle or the three sides of a triangle are given, the Law of Sines cannot be applied directly. We may, however, use the following result.

THE LAW OF COSINES

If *ABC* is any triangle labeled in the usual manner, then

$$a^2 = b^2 + c^2 - 2bc \cos \alpha$$
$$b^2 = a^2 + c^2 - 2ac \cos \beta$$
$$c^2 = a^2 + b^2 - 2ab \cos \gamma.$$

Instead of memorizing each of the three formulas, it is more convenient to remember the following statement, which takes all of them into account.

The square of the length of any side of a triangle equals the sum of the squares of the lengths of the other two sides minus twice the product of the lengths of the other two sides and the cosine of the angle between them.

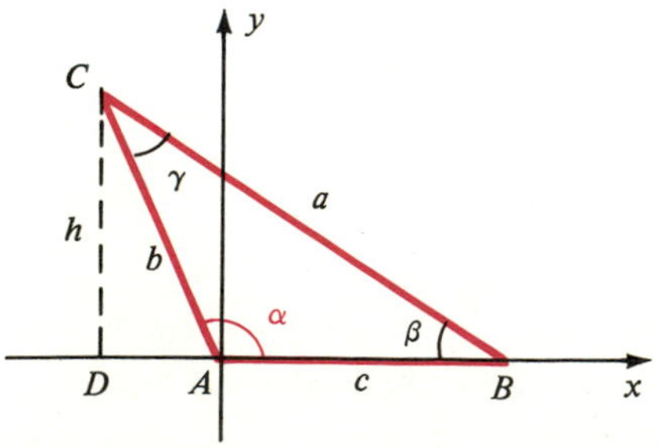

FIGURE 6.15

We shall use the Distance Formula to establish the Law of Cosines. We again place α in standard position on a rectangular coordinate system, as illustrated in Figure 6.15.

Although α is pictured as an obtuse angle, our development is also true if α is acute. Since the segment *AB* has length *c*, the coordinates of *B* are $(c, 0)$. It follows that the coordinates of *C* are $(b \cos \alpha, b \sin \alpha)$. (Why?) We also have $a = d(B, C)$ and hence $a^2 = [d(B, C)]^2$. Using the Distance Formula leads to the following equations:

$$\begin{aligned} a^2 &= (b \cos \alpha - c)^2 + (b \sin \alpha - 0)^2 \\ &= b^2 \cos^2 \alpha - 2bc \cos \alpha + c^2 + b^2 \sin^2 \alpha \\ &= b^2 (\cos^2 \alpha + \sin^2 \alpha) + c^2 - 2bc \cos \alpha \\ &= b^2 + c^2 - 2bc \cos \alpha. \end{aligned}$$

This gives us the first formula stated in the Law of Cosines. The second and third formulas may be obtained by placing β and γ, respectively, in standard position on a rectangular coordinate system and using a similar procedure.

EXAMPLE 1 Approximate all parts of triangle ABC if $a = 5.0$, $c = 8.0$, and $\beta = 77°10'$.

Solution Using the Law of Cosines,

$$\begin{aligned} b^2 &= (5.0)^2 + (8.0)^2 - 2(5.0)(8.0) \cos 77°10' \\ &\approx 25 + 64 - (80)(0.2221) \approx 71.2. \end{aligned}$$

Consequently, $b \approx \sqrt{71.2} \approx 8.44 \approx 8.4$.

We next use the Law of Sines to find α. Thus,

$$\sin \alpha = \frac{a \sin \beta}{b} \approx \frac{(5.0) \sin 77°10'}{8.44} \approx \frac{(5.0)(0.9750)}{8.44} \approx 0.5776.$$

Consulting Table 4 or using a calculator, we see that $\alpha \approx 35°20'$. Hence,

$$\gamma \approx 180° - (77°10' + 35°20') = 67°30'.$$

■

After the length of the third side was found in Example 1, the Law of Sines was used to find a second angle of the triangle. Whenever this procedure is followed, it is best to find the angle opposite the shortest side (as we did), since that angle is always acute. Of course, the Law of Cosines can also be used to find the remaining angles.

EXAMPLE 2 Given sides $a = 90$, $b = 70$, and $c = 40$ of triangle ABC, approximate angles α, β, and γ.

Solution According to the first formula stated in the Law of Cosines,

$$\begin{aligned} \cos \alpha &= \frac{b^2 + c^2 - a^2}{2bc} \\ &= \frac{4900 + 1600 - 8100}{5600} \approx -0.2857. \end{aligned}$$

Using Table 4 or a calculator, we see that the reference angle for α is approximately $73°20'$. Hence, $\alpha \approx 180° - 73°20' = 106°40'$.

Similarly, from the second formula stated in the Law of Cosines,

$$\cos \beta = \frac{a^2 + c^2 - b^2}{2ac}$$

$$= \frac{8100 + 1600 - 4900}{7200} \approx 0.6667$$

and hence $\beta \approx 48°10'$. Finally,

$$\gamma \approx 180° - (106°40' + 48°10') = 25°10'.$$

After finding the first angle in the preceding solution, we could have used the Law of Sines to find a second angle. Whenever this procedure is followed, it is best to use the Law of Cosines to find the angle opposite the longer side, that is, the largest angle of the triangle. This will guarantee that the remaining two angles are acute.

EXERCISES 6.8

In Exercises 1–10 approximate the remaining parts of triangle ABC.

1 $\alpha = 60°$, $b = 20$, $c = 30$

2 $\gamma = 45°$, $b = 10.0$, $a = 15.0$

3 $\beta = 150°$, $a = 150$, $c = 30.0$

4 $\beta = 73°50'$, $c = 14.0$, $a = 87.0$

5 $\gamma = 115°10'$, $a = 1.10$, $b = 2.10$

6 $\alpha = 23°40'$, $c = 4.30$, $b = 70.0$

7 $a = 2.0$, $b = 3.0$, $c = 4.0$

8 $a = 10$, $b = 15$, $c = 12$

9 $a = 25.0$, $b = 80.0$, $c = 60.0$

10 $a = 20.0$, $b = 20.0$, $c = 10.0$

11 A parallelogram has sides of length 30 cm. and 70 cm. If one of the angles has measure 65°, approximate the length of each diagonal.

12 The angle at one corner of a triangular plot of ground has measure $73°40'$. If the sides that meet at this corner are 175 feet and 150 feet long, approximate the length of the third side.

13 A vertical pole 40 feet tall stands on a hillside that makes an angle of 17° with the horizontal. Approximate the minimal length of rope that will reach from the top of the pole to a point directly down the hill 72 feet from the base of the pole.

14 To find the distance between two points A and B, a surveyor chooses a point C that is 420 yards from A and 540 yards from B. If angle ACB has measure $63°10'$, approximate the distance.

15 Two automobiles leave from the same point and travel along straight highways that differ in direction by 84°. If their speeds are 60 mph and 45 mph, respectively, approximately how far apart will they be at the end of 20 minutes?

16 A triangular plot of land has sides of length 420 feet, 350 feet, and 180 feet. Find the smallest angle between the sides.

17 A ship leaves P at 1:00 P.M. and travels S35°E at the rate of 24 mph. Another ship leaves P at 1:30 P.M. and

travels S20°W at 18 mph. Approximately how far apart are the ships at 3:00 P.M.?

18 An airplane flies 165 miles from point A in the direction 130° and then travels in the direction 245° for 80 miles. Approximately how far is the airplane from A?

19 A jogger, running at a constant speed of one mile every eight minutes, runs in the direction S40°E for 20 minutes and then in the direction N20°E for the next 16 minutes. Approximate, to the nearest tenth of a mile, the distance from the jogger to the starting point.

20 Two points P and Q on level ground are on opposite sides of a building. In order to find the distance d between the points, a surveyor chooses a point R that is 300 feet from P and 438 feet from Q, and then determines that angle PRQ has measure 37°40′. Approximate d.

21 A motor boat traveled along a triangular course of sides 2 km, 4 km, and 3 km, respectively. If the first side was traversed in the direction N20°W and the second in a direction SD°W, where D° is the degree measure of an acute angle, approximate, to the nearest minute, the direction that the third side was traversed.

22 If a rhombus has sides of length 100 cm and if the angle at one of the vertices is 70°, approximate the lengths of the diagonals to the nearest tenth of a centimeter.

23 Show that for every triangle ABC,

(a) $a^2 + b^2 + c^2 = 2(bc \cos \alpha + ac \cos \beta + ab \cos \gamma)$.

(b) $\dfrac{\cos \alpha}{a} + \dfrac{\cos \beta}{b} + \dfrac{\cos \gamma}{c} = \dfrac{a^2 + b^2 + c^2}{2abc}$.

24 Show that if ABC is a right triangle, then one part of the Law of Cosines reduces to the Pythagorean Theorem.

CALCULATOR EXERCISES 6.8

Approximate the remaining parts of triangle ABC in Exercises 1–6.

1 $\alpha = 48.3°$, $b = 24.7$, $c = 52.8$

2 $\beta = 137.8°$, $a = 178.2$, $c = 431.4$

3 $\gamma = 6.85°$, $a = 0.846$, $b = 0.364$

4 $a = 435$, $b = 482$, $c = 78$

5 $a = 5{,}150$, $b = 1{,}814$, $c = 3{,}429$

6 $a = 0.64$, $b = 0.27$, $c = 0.49$

SECTION 6.9

TRIGONOMETRIC FORM FOR COMPLEX NUMBERS

Real numbers may be represented geometrically by points on a coordinate line. We can also obtain geometric representations for complex numbers by using points in a coordinate plane. Specifically, each complex number $a + bi$ determines a unique ordered pair (a, b). The corresponding point $P(a, b)$ in a coordinate plane is called the **geometric representation of** $a + bi$. To emphasize that we are assigning complex numbers to points in a plane, the point $P(a, b)$ may be labeled $a + bi$. A coordinate plane with a complex number assigned to each point is referred to as the **complex plane** instead of the xy-plane. Also, according to this scheme, the x-axis is called the

real axis and the y-axis the **imaginary axis.** In Figure 6.16 we have indicated geometric representations of several complex numbers. Note that, to obtain the point corresponding to the conjugate $a - bi$ of any complex number $a + bi$, we simply reflect through the real axis.

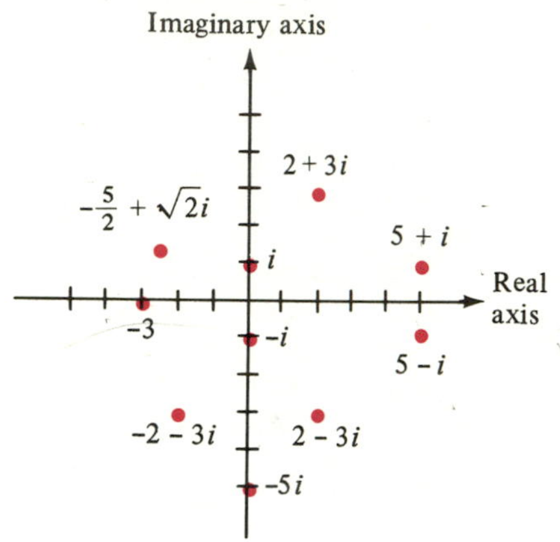

FIGURE 6.16

In Chapter 1 we defined the concept of the absolute value $|a|$ of a real number a and noted that, geometrically, a is the distance between the origin and the point on a coordinate line that corresponds to a. It is natural, therefore, to interpret the absolute value $|a + bi|$ of a complex number as the distance $\sqrt{a^2 + b^2}$ between the origin of a complex plane and the point (a, b) corresponding to $a + bi$, as in the following definition.

DEFINITION

> The **absolute value** of a complex number $a + bi$ is denoted by $|a + bi|$ and is defined to be the nonnegative real number $\sqrt{a^2 + b^2}$.

EXAMPLE 1 Find (a) $|2 - 6i|$; (b) $|3i|$.

Solution Using the definition of absolute value, we obtain

(a) $|2 - 6i| = \sqrt{4 + 36} = \sqrt{40} = 2\sqrt{10}$;

(b) $|3i| = \sqrt{0 + 9} = 3$. ■

It is worth noting that the points corresponding to all complex numbers that have a fixed absolute value lie on a circle with center at the origin in the complex plane. For example, the points corresponding to the complex numbers z with $|z| = 1$ lie on a unit circle.

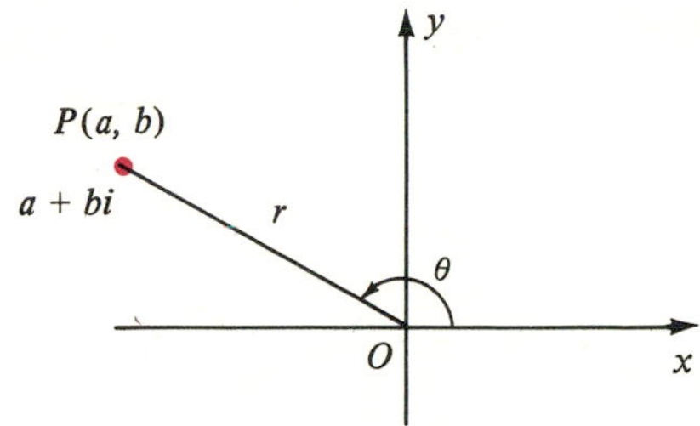

FIGURE 6.17
$z = a + bi = r(\cos\theta + i\sin\theta)$

The geometric representation of a complex number as a point in a coordinate plane leads to a method of representation that involves trigonometric functions. Let us consider a nonzero complex number $z = a + bi$ and its geometric representation $P(a, b)$. Let θ be any angle in standard position whose terminal side lies on the segment OP and let $r = |z|$, that is, $r = d(O, P) = \sqrt{a^2 + b^2}$. An illustration of this situation is shown in Figure 6.17.

Since $a = r\cos\theta$, $b = r\sin\theta$, and $z = a + bi$, we obtain

$$z = (r\cos\theta) + (r\sin\theta)i,$$

which may be written in the following *trigonometric form.*

TRIGONOMETRIC FORM FOR *a* + *bi*

$$z = a + bi = r(\cos\theta + i\sin\theta)$$

It is important to remember that the number r is the absolute value of z. If $z = 0$, then $r = 0$, and hence the trigonometric form may be used to represent the complex number 0 where θ is *any* angle. Sometimes we call $r(\cos\theta + i\sin\theta)$ the **polar form** for the complex number $z = a + bi$. The trigonometric form for z is not unique, since there are an unlimited number of different choices for the angle θ. When the trigonometric form is used, the absolute value r of z is often referred to as the **modulus** of z and the angle θ associated with z is called the **argument** (or **amplitude**) of z.

EXAMPLE 2 Express the following complex numbers in trigonometric form.

(a) $-4 + 4i$ (b) $2\sqrt{3} - 2i$ (c) $2 + 7i$

Solution The three complex numbers are represented geometrically in Figure 6.18.

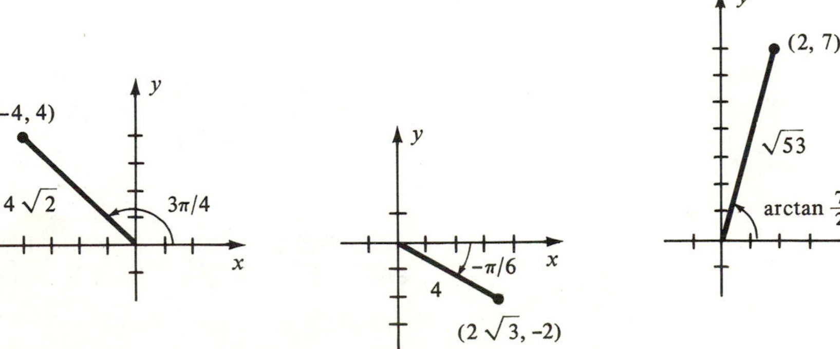

FIGURE 6.18

Using trigonometric form gives us the following:

(a) $$-4 + 4i = 4\sqrt{2}\left[\cos\frac{3\pi}{4} + i\sin\frac{3\pi}{4}\right]$$

(b) $$2\sqrt{3} - 2i = 4\left[\cos\left(-\frac{\pi}{6}\right) + i\sin\left(-\frac{\pi}{6}\right)\right]$$

(c) $$2 + 7i = \sqrt{53}\left[\cos\left(\arctan\frac{7}{2}\right) + i\sin\left(\arctan\frac{7}{2}\right)\right]$$ ■

If complex numbers are expressed in trigonometric form, then multiplications and divisions may be carried out in the simple manner indicated by the next theorem.

THEOREM If trigonometric forms for two complex numbers z_1 and z_2 are

$$z_1 = r_1(\cos\theta_1 + i\sin\theta_1) \quad \text{and} \quad z_2 = r_2(\cos\theta_2 + i\sin\theta_2)$$

then

(i) $$z_1z_2 = r_1r_2[\cos(\theta_1 + \theta_2) + i\sin(\theta_1 + \theta_2)]$$

(ii) $$\frac{z_1}{z_2} = \frac{r_1}{r_2}[\cos(\theta_1 - \theta_2) + i\sin(\theta_1 - \theta_2)], \quad z_2 \neq 0.$$

Proof We shall prove (i) and leave (ii) as an exercise. Thus,

$$\begin{aligned} z_1z_2 &= r_1(\cos\theta_1 + i\sin\theta_1)\cdot r_2(\cos\theta_2 + i\sin\theta_2) \\ &= r_1r_2\{[\cos\theta_1\cos\theta_2 - \sin\theta_1\sin\theta_2] + i[\sin\theta_1\cos\theta_2 + \cos\theta_1\sin\theta_2]\}. \end{aligned}$$

Applying the addition formulas for $\cos(\theta_1 + \theta_2)$ and $\sin(\theta_1 + \theta_2)$ gives us (i). □

Part (i) of the preceding theorem states that the modulus of a product of two complex numbers is the product of their moduli, and the argument is the sum of their arguments. An analogous statement can be made for (ii).

EXAMPLE 3 Use trigonometric forms to find z_1z_2 and z_1/z_2 if $z_1 = 2\sqrt{3} - 2i$ and $z_2 = -1 + \sqrt{3}i$. Check by using the methods of Section 3.4.

Solution From Example 2,

$$z_1 = 2\sqrt{3} - 2i = 4\left[\cos\left(-\frac{\pi}{6}\right) + i\sin\left(-\frac{\pi}{6}\right)\right].$$

Similarly, the reader may verify that

$$z_2 = -1 + \sqrt{3}i = 2\left[\cos\frac{2\pi}{3} + i\sin\frac{2\pi}{3}\right].$$

Applying (i) of the preceding theorem,

$$\begin{aligned} z_1z_2 &= 8\left[\cos\left(-\frac{\pi}{6} + \frac{2\pi}{3}\right) + i\sin\left(-\frac{\pi}{6} + \frac{2\pi}{3}\right)\right] \\ &= 8\left[\cos\frac{\pi}{2} + i\sin\frac{\pi}{2}\right] = 8i. \end{aligned}$$

As a check, using the methods of Section 3.4,

$$\begin{aligned} z_1z_2 &= (2\sqrt{3} - 2i)(-1 + \sqrt{3}i) \\ &= (-2\sqrt{3} + 2\sqrt{3}) + (2 + 6)i = 8i, \end{aligned}$$

which is in agreement with our previous answer.

Applying (ii) of the preceding theorem,

$$\begin{aligned} \frac{z_1}{z_2} &= 2\left[\cos\left(-\frac{\pi}{6} - \frac{2\pi}{3}\right) + i\sin\left(-\frac{\pi}{6} - \frac{2\pi}{3}\right)\right] \\ &= 2\left[\cos\left(-\frac{5\pi}{6}\right) + i\sin\left(-\frac{5\pi}{6}\right)\right] \\ &= 2\left[-\frac{\sqrt{3}}{2} + i\left(-\frac{1}{2}\right)\right] = -\sqrt{3} - i. \end{aligned}$$

Using the methods of Section 3.4,

$$\begin{aligned} \frac{z_1}{z_2} &= \frac{2\sqrt{3} - 2i}{-1 + \sqrt{3}i} \cdot \frac{-1 - \sqrt{3}i}{-1 - \sqrt{3}i} \\ &= \frac{(-2\sqrt{3} - 2\sqrt{3}) + (2 - 6)i}{4} = -\sqrt{3} - i. \end{aligned}$$

■

EXERCISES 6.9

Find the numbers in Exercises 1–10.

1 $|3 - 4i|$

2 $|5 + 8i|$

3 $|-6 - 7i|$

4 $|1 - i|$

5 $|8i|$

6 $|i^7|$

7 $|i^{500}|$

8 $|-15i|$

9 $|0|$

10 $|-15|$

Represent the complex numbers in Exercises 11–20 geometrically.

11 $4 + 2i$

12 $-5 + 3i$

13 $3 - 5i$

14 $-2 - 6i$

15 $-(3 - 6i)$

16 $(1 + 2i)^2$

17 $2i(2 + 3i)$

18 $(-3i)(2 - i)$

19 $(1 + i)^2$

20 $4(-1 + 2i)$

Change the complex numbers in Exercises 21–40 to trigonometric form.

21 $1 - i$

22 $\sqrt{3} + i$

23 $-4\sqrt{3} + 4i$

24 $-2 - 2i$

25 $-20i$

26 15

27 $-5(1 + \sqrt{3}i)$

28 $-6i$

29 -7

30 $2i(1 - \sqrt{3}i)$

31 $2 + i$

32 $4 - 3i$

33 $-4 - 4i$

34 $-10 + 10i$

35 $6i$

36 -5

37 $-4\sqrt{3} + 4i$

38 $-5 - 5\sqrt{3}i$

39 12

40 0

In Exercises 41–48 find z_1z_2 and z_1/z_2 by changing to trigonometric form and then using the last theorem of this section. Check by using the methods of Section 3.4.

41 $z_1 = -1 + i,\ z_2 = 1 + i$

42 $z_1 = \sqrt{3} - i,\ z_2 = -\sqrt{3} - i$

43 $z_1 = -2 - 2\sqrt{3}i,\ z_2 = 5i$

44 $z_1 = -5 + 5i,\ z_2 = -3i$

45 $z_1 = -10,\ z_2 = -4$

46 $z_1 = 2i,\ z_2 = -3i$

47 $z_1 = 4,\ z_2 = 2 - i$

48 $z_1 = -3,\ z_2 = 5 + 2i$

49 Extend (i) of the last theorem in this section to the case of three complex numbers. Generalize to n complex numbers.

50 Prove (ii) of the last theorem in this section.

SECTION 6.10

DE MOIVRE'S THEOREM AND *n*TH ROOTS OF COMPLEX NUMBERS

If z is a complex number and n is a positive integer, then a complex number w is called an ***n*th root** of z if $w^n = z$. In this section we shall show that every nonzero complex number has n distinct nth roots. Since $\mathbb{R}$ is in $\mathbb{C}$, it will also follow that every nonzero real number has n distinct nth roots. If a is a positive real number and $n = 2$, then we already know that the roots are $\sqrt{a}$ and $-\sqrt{a}$.

If, in the theorem on page 296, we let both z_1 and z_2 equal the complex number $z = r(\cos\theta + i\sin\theta)$, we obtain

$$z^2 = r^2(\cos 2\theta + i\sin 2\theta).$$

If we next apply the same theorem to z and z^2, then

$$z^2 \cdot z = (r^2 \cdot r)[\cos(2\theta + \theta) + i\sin(2\theta + \theta)]$$

or

$$z^3 = r^3(\cos 3\theta + i\sin 3\theta).$$

Applying the theorem to z^3 and z produces

$$z^4 = r^4(\cos 4\theta + i\sin 4\theta).$$

In general, we have the following result, named after the French mathematician Abraham De Moivre (1667–1754).

DE MOIVRE'S THEOREM

For every integer n,

$$[r(\cos\theta + i\sin\theta)]^n = r^n[\cos n\theta + i\sin n\theta].$$

A complete proof of the theorem may be given with the aid of the method of mathematical induction discussed in Chapter 9.

EXAMPLE 1 Find $(1+i)^{20}$.

Solution Introducing trigonometric form,

$$1 + i = \sqrt{2}\left(\cos\frac{\pi}{4} + i\sin\frac{\pi}{4}\right).$$

Next, by De Moivre's Theorem,

$$\begin{aligned}(1+i)^{20} &= (2^{1/2})^{20}\left[\cos 20\left(\frac{\pi}{4}\right) + i\sin 20\left(\frac{\pi}{4}\right)\right]\\ &= 2^{10}(\cos 5\pi + i\sin 5\pi) = -1024.\end{aligned}$$

■

If a nonzero complex number z has an nth root w, then $w^n = z$. If trigonometric forms for w and z are

$$w = s(\cos\alpha + i\sin\alpha) \quad\text{and}\quad z = r(\cos\theta + i\sin\theta),$$

then, applying De Moivre's Theorem to $w^n = z$,

$$s^n[\cos n\alpha + i\sin n\alpha] = r(\cos\theta + i\sin\theta).$$

If two complex numbers are equal, then so are their absolute values. Consequently, $s^n = r$, and since s and r are nonnegative, $s = \sqrt[n]{r}$. Substituting s^n for r in the last displayed equation and dividing both sides by s^n, we obtain

$$\cos n\alpha + i\sin n\alpha = \cos\theta + i\sin\theta.$$

It follows that $\quad \cos n\alpha = \cos\theta \quad\text{and}\quad \sin n\alpha = \sin\theta.$

Since both the sine and cosine functions have period 2π, the last two equations are true if and only if $n\alpha$ and θ differ by a multiple of 2π. Thus, for some integer k,

$$n\alpha = \theta + 2\pi k$$

and hence, $$\alpha = \frac{\theta + 2\pi k}{n}.$$

Substituting in the trigonometric form for w, we obtain the formula

$$w = \sqrt[n]{r}\left[\cos\left(\frac{\theta + 2\pi k}{n}\right) + i \sin\left(\frac{\theta + 2\pi k}{n}\right)\right].$$

If we substitute $k = 0, 1, \ldots, n - 1$ successively, we obtain n distinct values for w and hence n distinct nth roots of z. No other value of k will produce a new nth root. For example, if $k = n$, we obtain the angle $(\theta + 2\pi n)/n$, or $(\theta/n) + 2\pi$, which gives us the same nth root as $k = 0$. Similarly, $k = n + 1$ yields the same nth root as $k = 1$, and so on. The same is true for negative values of k. Our discussion also shows that the numbers given by the formula are the only possible nth roots of z. We have proved the following theorem.

THEOREM ON *n*th ROOTS

If $z = r(\cos\theta + i \sin\theta)$ is any nonzero complex number and if n is any positive integer, then z has precisely n distinct nth roots. Moreover, the roots are given by

$$\sqrt[n]{r}\left[\cos\left(\frac{\theta + 2\pi k}{n}\right) + i \sin\left(\frac{\theta + 2\pi k}{n}\right)\right]$$

where $k = 0, 1, \ldots, n - 1$.

Note that the nth roots of z all have modulus $\sqrt[n]{r}$ and hence their geometric representations lie on a circle of radius $\sqrt[n]{r}$ with center at O. Moreover, they are equispaced on this circle, since the difference in the arguments of successive nth roots is $2\pi/n$.

It is sometimes convenient to use degree measure for θ. In this event the formula in the preceding theorem becomes

$$\sqrt[n]{r}\left[\cos\left(\frac{\theta + k \cdot 360^\circ}{n}\right) + i \sin\left(\frac{\theta + k \cdot 360^\circ}{n}\right)\right]$$

where $k = 0, 1, \ldots, n - 1$.

EXAMPLE 2 Find the four fourth roots of $-8(1 + i\sqrt{3})$.

Solution The geometric representation of the complex number is shown in Figure 6.19. Introducing trigonometric form,

$$-8(1 + i\sqrt{3}) = 16(\cos 240^\circ + i \sin 240^\circ).$$

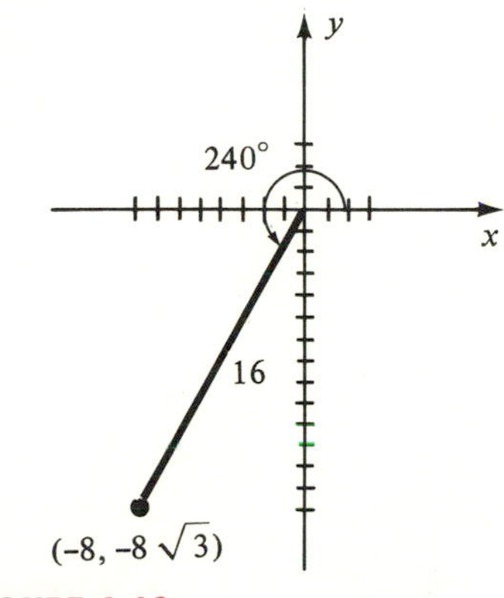

FIGURE 6.19

By the preceding theorem, with $n = 4$, the fourth roots are given by

$$2\left[\cos\left(\frac{240^\circ + k\cdot 360^\circ}{4}\right) + i\sin\left(\frac{240^\circ + k\cdot 360^\circ}{4}\right)\right]$$

where $k = 0$, 1, 2, and 3. The formula may be rewritten

$$2[\cos(60^\circ + k\cdot 90^\circ) + i\sin(60^\circ + k\cdot 90^\circ)].$$

Substituting 0, 1, 2, and 3 for k yields the following fourth roots:

$$2(\cos 60^\circ + i\sin 60^\circ) = 1 + \sqrt{3}i$$
$$2(\cos 150^\circ + i\sin 150^\circ) = -\sqrt{3} + i$$
$$2(\cos 240^\circ + i\sin 240^\circ) = -1 - \sqrt{3}i$$
$$2(\cos 330^\circ + i\sin 330^\circ) = \sqrt{3} - i.$$

■

EXAMPLE 3 Find the six sixth roots of -1.

Solution Writing $-1 = 1(\cos\pi + i\sin\pi)$ and using the Theorem on nth Roots with $n = 6$, we find that the sixth roots of -1 are given by

$$\cos\left(\frac{\pi + 2\pi k}{6}\right) + i\sin\left(\frac{\pi + 2\pi k}{6}\right)$$

or

$$\cos\left(\frac{\pi}{6} + \frac{\pi}{3}k\right) + i\sin\left(\frac{\pi}{6} + \frac{\pi}{3}k\right).$$

Substituting 0, 1, 2, 3, 4, and 5 for k gives us the sixth roots of -1, namely,

$$\cos\pi/6 + i\sin\pi/6 = \sqrt{3}/2 + (1/2)i$$
$$\cos\pi/2 + i\sin\pi/2 = i$$
$$\cos 5\pi/6 + i\sin 5\pi/6 = -\sqrt{3}/2 + (1/2)i$$
$$\cos 7\pi/6 + i\sin 7\pi/6 = -\sqrt{3}/2 - (1/2)i$$
$$\cos 3\pi/2 + i\sin 3\pi/2 = -i$$
$$\cos 11\pi/6 + i\sin 11\pi/6 = \sqrt{3}/2 - (1/2)i.$$

■

The special case in which $z = 1$ is of particular interest. The n distinct nth roots of 1 are called the ***n*th roots of unity.**

EXAMPLE 4 Find the three third roots of unity.

Solution Writing $1 = \cos 0 + i\sin 0$ and using the Theorem on nth

Roots with $n = 3$, we obtain the three roots

$$\cos\frac{2\pi k}{3} + i\sin\frac{2\pi k}{3}$$

where $k = 0$, 1, and 2. Substituting for k gives us

$$\cos 0 + i\sin 0 = 1$$

$$\cos\frac{2\pi}{3} + i\sin\frac{2\pi}{3} = -\frac{1}{2} + \left(\frac{\sqrt{3}}{2}\right)i$$

$$\cos\frac{4\pi}{3} + i\sin\frac{4\pi}{3} = -\frac{1}{2} - \left(\frac{\sqrt{3}}{2}\right)i.$$ ■

Compare the preceding solution with Example 6 of Section 3.4.

EXERCISES 6.10

In Exercises 1–12 use De Moivre's Theorem to express the given numbers in the form $a + bi$, where a and b are real numbers.

1 $(3 + 3i)^5$

2 $(1 + i)^{12}$

3 $(1 - i)^{10}$

4 $(-1 + i)^8$

5 $(1 - \sqrt{3}i)^3$

6 $(1 - \sqrt{3}i)^5$

7 $\left(-\frac{\sqrt{2}}{2} + \frac{\sqrt{2}}{2}i\right)^{15}$

8 $\left(\frac{\sqrt{2}}{2} + \frac{\sqrt{2}}{2}i\right)^{25}$

9 $\left(-\frac{\sqrt{3}}{2} - \frac{1}{2}i\right)^{20}$

10 $\left(-\frac{\sqrt{3}}{2} - \frac{1}{2}i\right)^{50}$

11 $(\sqrt{3} + i)^7$

12 $(-2 - 2i)^{10}$

13 Find the two square roots of $1 + \sqrt{3}i$.

14 Find the two square roots of $-9i$.

15 Find the four fourth roots of $-1 - \sqrt{3}i$.

16 Find the four fourth roots of $-8 + 8\sqrt{3}i$.

17 Find the three cube roots of $-27i$.

18 Find the three cube roots of $64i$.

19 Find the six sixth roots of unity.

20 Find the eight eighth roots of unity.

21 Find the five fifth roots of $1 + i$.

22 Find the five fifth roots of $-\sqrt{3} - i$.

Find the solutions of the equations in Exercises 23–30.

23 $x^4 - 16 = 0$

24 $x^6 - 64 = 0$

25 $x^6 + 64 = 0$

26 $x^5 + 1 = 0$

27 $x^3 + 8i = 0$

28 $x^3 - 64i = 0$

29 $x^5 - 243 = 0$

30 $x^4 + 81 = 0$

SECTION 6.11

VECTORS

In Chapter 1 we assigned (positive) directions to the x- and y-axes. In similar fashion, a **directed line segment** is a line segment to which a direction has been assigned. Another name for a directed line segment is a **vector.** If a vector extends from a point A (called the **initial point**) to a point B (called the **terminal point**), it is customary to place an arrowhead at B and use $\overrightarrow{AB}$ to represent the vector (see Figure 6.20). The length of the directed line segment is called the **magnitude** of the vector $\overrightarrow{AB}$ and is denoted by $|\overrightarrow{AB}|$. The vectors $\overrightarrow{AB}$ and $\overrightarrow{CD}$ are considered **equal,** and we write $\overrightarrow{AB} = \overrightarrow{CD}$ if and only if they have the same magnitude and direction, as illustrated in Figure 6.20. Consequently, vectors may be translated from one position to another, provided neither the magnitude nor direction is changed. Vectors of this type are often referred to as **free vectors.**

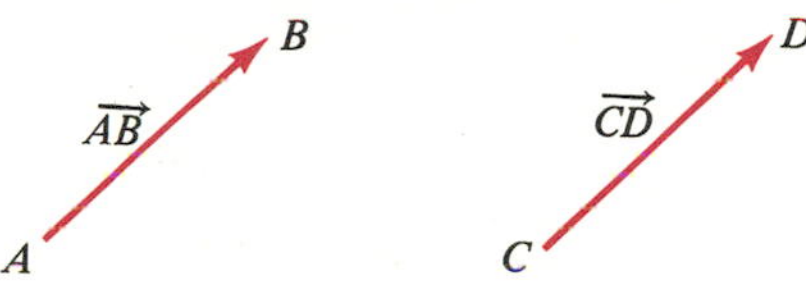

FIGURE 6.20

Many physical concepts may be represented by vectors. To illustrate, suppose an airplane is descending at a constant rate of 100 mph and its line of flight makes an angle of 20° with the horizontal. Both of these facts are represented by the vector in (i) of Figure 6.21, where units have been chosen so that the magnitude is 100. As shown in the figure, we will use a boldface letter such as **v** to denote a vector whose endpoints are not specified. The vector **v** in this illustration is called a **velocity vector.** The magnitude $|\mathbf{v}|$ of **v** is the speed of the airplane.

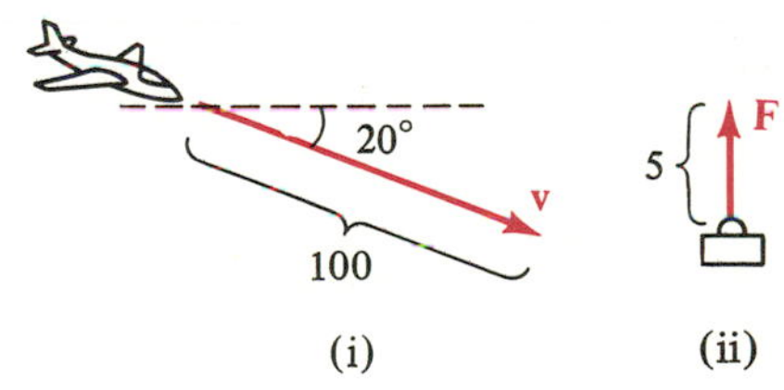

FIGURE 6.21

As a second illustration, suppose a person pulls directly upward on an object with a force of 5 kg, as would be the case in lifting a 5 kg weight. We may indicate this fact by the vector **F** in (ii) of Figure 6.21. A vector that represents a pull or push of some type is called a **force vector.**

Another use for vectors is to let $\overrightarrow{AB}$ represent the path of a point (or some physical particle) as it moves along the line segment from A to B. We then refer to $\overrightarrow{AB}$ as a **displacement** of the point (or particle). As illustrated in Figure 6.22, a displacement $\overrightarrow{AB}$ followed by a displacement $\overrightarrow{BC}$ will lead to the same point as the single displacement $\overrightarrow{AC}$. The vector $\overrightarrow{AC}$ is called the **sum** of the first two, written

$$\overrightarrow{AC} = \overrightarrow{AB} + \overrightarrow{BC}.$$

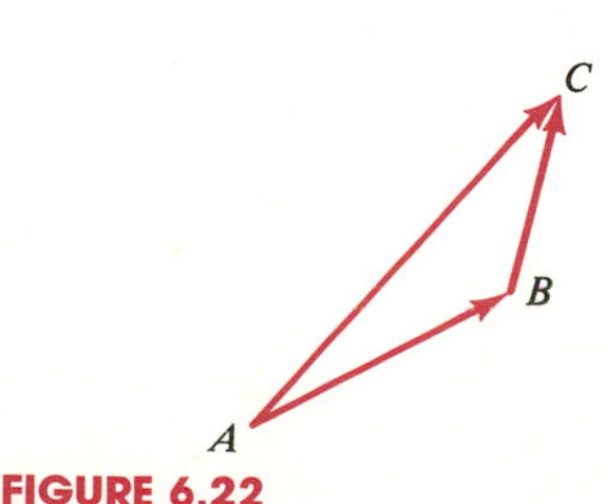

FIGURE 6.22

Since we are working with free vectors, any two vectors can be added by placing the initial point of one on the terminal point of the other and then finding $\overrightarrow{AC}$ as in Figure 6.22.

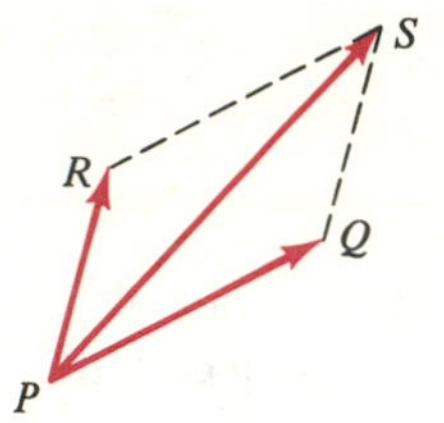

FIGURE 6.23

Another way to find the sum of two vectors is to consider vectors that are equal to the given ones and have the same initial point, as illustrated by $\overrightarrow{PQ}$ and $\overrightarrow{PR}$ in Figure 6.23. If we construct the parallelogram $RPQS$ with adjacent sides $\overrightarrow{PR}$ and $\overrightarrow{PQ}$, then since $\overrightarrow{PR} = \overrightarrow{QS}$, it follows that

$$\overrightarrow{PS} = \overrightarrow{PQ} + \overrightarrow{PR}.$$

If $\overrightarrow{PQ}$ and $\overrightarrow{PR}$ are two forces acting at P, then it can be shown experimentally that $\overrightarrow{PS}$ is the **resultant force**, that is, the single force that produces the same effect as the two combined forces.

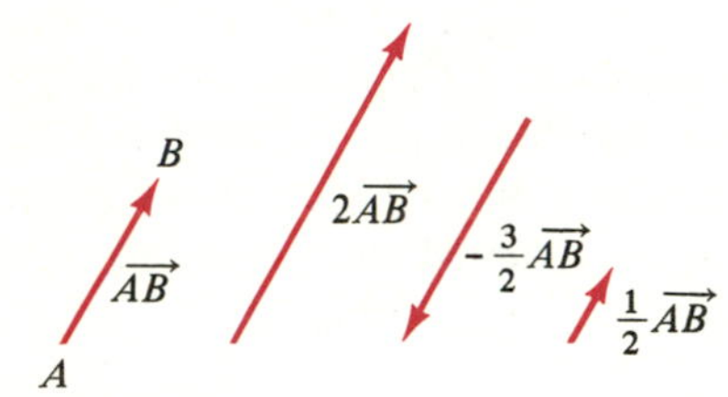

FIGURE 6.24

If c is a real number and $\overrightarrow{AB}$ is a vector, then the product $c\overrightarrow{AB}$ is defined as a vector whose magnitude is $|c|$ times the magnitude of $\overrightarrow{AB}$ and whose direction is the same as $\overrightarrow{AB}$ if $c > 0$, and opposite that of $\overrightarrow{AB}$ if $c < 0$. Geometric illustrations are given in Figure 6.24. We refer to c as a **scalar** and $c\overrightarrow{AB}$ as a **scalar multiple** of $\overrightarrow{AB}$.

Let us next introduce a coordinate plane and assume that all vectors under discussion are in that plane. Since the position of a vector may be changed, provided that the magnitude and direction are not altered, we may place the initial point of each vector at the origin. The terminal point of a typical vector $\overrightarrow{OP}$ may then be assigned coordinates (a, b), as shown in Figure 6.25. Conversely, every ordered pair (a, b) determines the vector $\overrightarrow{OP}$, where P has coordinates (a, b). We thus obtain a one-to-one correspondence between vectors and ordered pairs. This allows us to regard a vector in a plane as an ordered pair of real numbers instead of a directed line segment.

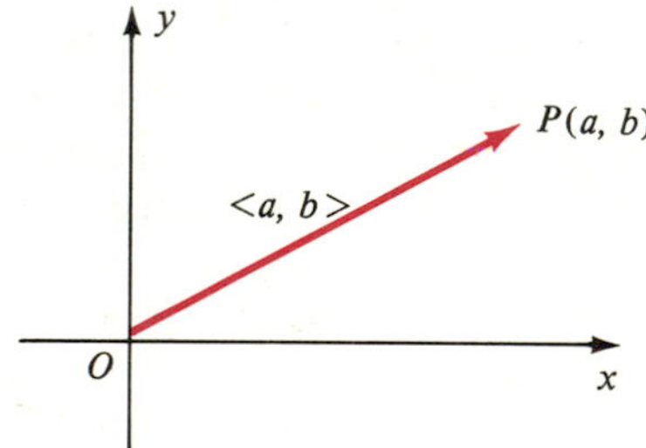

FIGURE 6.25

To avoid confusion with the notation for open intervals or points, we shall use the symbol $\langle a, b\rangle$ for an ordered pair that represents a vector. Moreover, we shall refer to $\langle a, b\rangle$ as a vector and denote it by a boldface letter. The numbers a and b are called the **components** of the vector $\langle a, b\rangle$. The magnitude of $\langle a, b\rangle$ is, by definition, the distance from the origin to the point $P(a, b)$. This may also be stated as follows.

DEFINITION

The **magnitude** $|\mathbf{v}|$ of the vector $\mathbf{v} = \langle a, b\rangle$ is $|\mathbf{v}| = \sqrt{a^2 + b^2}$.

EXAMPLE 1 Sketch the vectors corresponding to each of the following, and find the magnitude and the smallest positive angle θ from the positive x-axis to each vector.

(a) $\mathbf{a} = \langle -3, -3 \rangle$ (b) $\mathbf{b} = \langle 0, -2 \rangle$ (c) $\mathbf{c} = \langle \frac{4}{5}, \frac{3}{5} \rangle$

Solution The vectors are sketched in Figure 6.26.

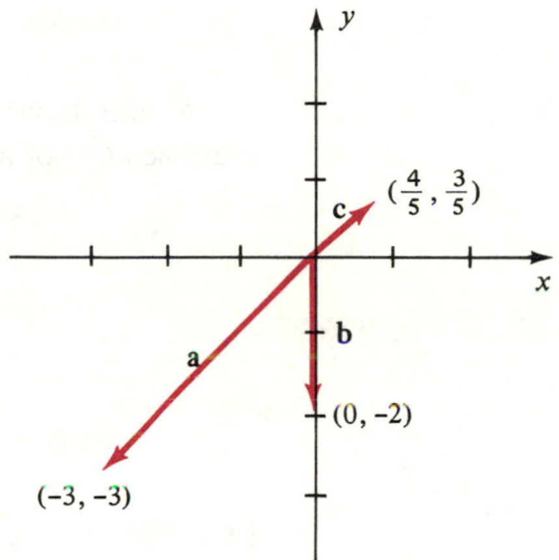

FIGURE 6.26

Applying the definition of magnitude, and our knowledge of trigonometry, we obtain:

(a) $|\mathbf{a}| = \sqrt{9 + 9} = 3\sqrt{2}; \quad \theta = 5\pi/4$

(b) $|\mathbf{b}| = \sqrt{0 + 4} = 2; \quad \theta = 3\pi/2$

(c) $|\mathbf{c}| = \sqrt{\frac{16}{25} + \frac{9}{25}} = 1; \quad \theta = \arctan \frac{3}{4}$. ■

If, as in Figure 6.27, we consider vectors $\overrightarrow{OQ}$ and $\overrightarrow{OR}$ corresponding to $\langle a, b \rangle$ and $\langle c, d \rangle$, respectively, and if we let $\overrightarrow{OS}$ be the vector corresponding to $\langle a + c, b + d \rangle$, it is easy to show, using slopes, that O, Q, S, and R are vertices of a parallelogram. It follows that

$$\overrightarrow{OQ} + \overrightarrow{OR} = \overrightarrow{OS}.$$

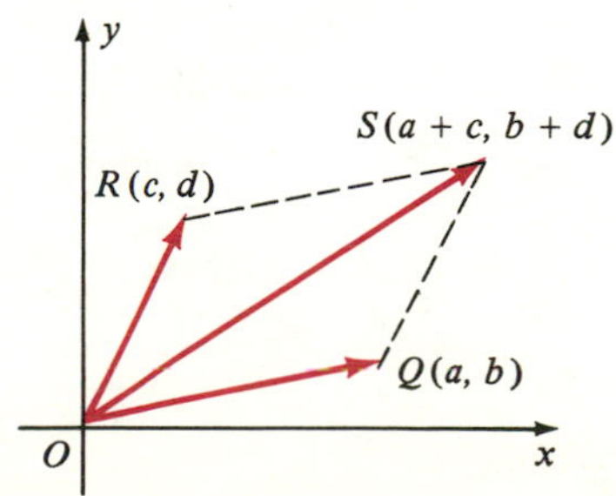

FIGURE 6.27 $\langle a, b \rangle + \langle c, d \rangle = \langle a + c, b + d \rangle$

Expressing this fact in terms of ordered pairs gives us the following rule for addition of vectors.

ADDITION OF VECTORS

$$\langle a, b\rangle + \langle c, d\rangle = \langle a + c, b + d\rangle$$

Although we shall not prove it, the rule corresponding to a scalar multiple $k\overrightarrow{OP}$ of a vector is as follows.

SCALAR MULTIPLES OF VECTORS

$$k\langle a, b\rangle = \langle ka, kb\rangle$$

EXAMPLE 2 If $\mathbf{a} = \langle 2, 1\rangle$, find $3\mathbf{a}$ and $-2\mathbf{a}$, and represent all three vectors geometrically.

Solution If $\mathbf{a} = \langle 2, 1\rangle$, then, by the preceding rule for scalar multiples, $3\mathbf{a} = \langle 6, 3\rangle$ and $-2\mathbf{a} = \langle -4, -2\rangle$. The geometric representations are shown in Figure 6.28.

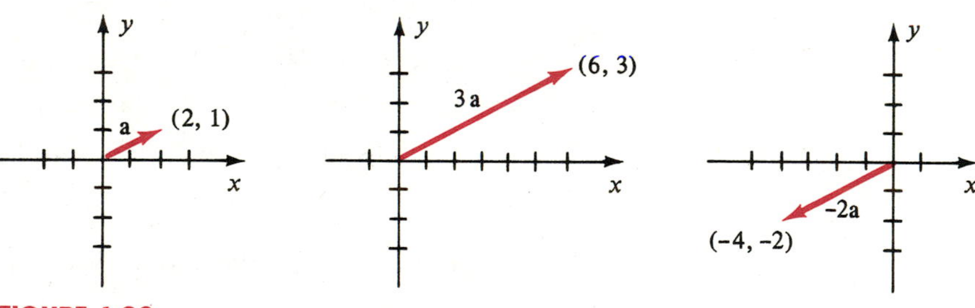

FIGURE 6.28 ■

EXAMPLE 3 If $\mathbf{a} = \langle 3, -2\rangle$ and $\mathbf{b} = \langle -6, 7\rangle$, find $\mathbf{a} + \mathbf{b}$, $4\mathbf{a}$, and $2\mathbf{a} + 3\mathbf{b}$.

Solution Using the rules for addition and scalar multiples of vectors,

$$\mathbf{a} + \mathbf{b} = \langle 3, -2\rangle + \langle -6, 7\rangle = \langle -3, 5\rangle$$

$$4\mathbf{a} = 4\langle 3, -2\rangle = \langle 12, -8\rangle$$

$$2\mathbf{a} + 3\mathbf{b} = \langle 6, -4\rangle + \langle -18, 21\rangle = \langle -12, 17\rangle.$$ ■

By definition, the **zero vector 0** corresponds to $\langle 0, 0\rangle$. If $\mathbf{a} = \langle a, b\rangle$ we define $-\mathbf{a} = \langle -a, -b\rangle$. Using these definitions, we may establish the following properties for arbitrary vectors **a**, **b**, and **c**.

PROPERTIES OF ADDITION OF VECTORS

$$\mathbf{a} + \mathbf{b} = \mathbf{b} + \mathbf{a}$$
$$\mathbf{a} + (\mathbf{b} + \mathbf{c}) = (\mathbf{a} + \mathbf{b}) + \mathbf{c}$$
$$\mathbf{a} + \mathbf{0} = \mathbf{a}$$
$$\mathbf{a} + (-\mathbf{a}) = \mathbf{0}$$

The proof of each property follows readily from the rule for addition of vectors and properties of real numbers. For example, if $\mathbf{a} = \langle a_1, a_2\rangle$ and if $\mathbf{b} = \langle b_1, b_2\rangle$, then since $a_1 + b_1 = b_1 + a_1$ and $a_2 + b_2 = b_2 + a_2$,

$$\begin{aligned}\mathbf{a} + \mathbf{b} &= \langle a_1 + b_1, a_2 + b_2\rangle \\ &= \langle b_1 + a_1, b_2 + a_2\rangle \\ &= \mathbf{b} + \mathbf{a}.\end{aligned}$$

The remainder of the proof is left as an exercise (see Exercise 15). The reader should also give geometric interpretations for each property.

The operation of **subtraction** of vectors, denoted by $-$, is defined as follows.

SUBTRACTION OF VECTORS

$$\mathbf{a} - \mathbf{b} = \mathbf{a} + (-\mathbf{b})$$

If we use the ordered pair notation for **a** and **b**, then $-\mathbf{b} = \langle -b_1, -b_2\rangle$, and

$$\mathbf{a} - \mathbf{b} = \langle a_1, a_2\rangle + \langle -b_1, -b_2\rangle,$$

or

$$\mathbf{a} - \mathbf{b} = \langle a_1 - b_1, a_2 - b_2\rangle.$$

Thus, to find $\mathbf{a} - \mathbf{b}$, we merely subtract the components of **b** from the corresponding components of **a**.

EXAMPLE 4 If $\mathbf{a} = \langle 2, -4 \rangle$ and $\mathbf{b} = \langle 6, 7 \rangle$, find $3\mathbf{a} - 5\mathbf{b}$.

Solution We may proceed as follows:

$$3\mathbf{a} - 5\mathbf{b} = \langle 6, -12 \rangle - \langle 30, 35 \rangle = \langle -24, -47 \rangle.$$ ■

The following properties can be proved for any vectors **a**, **b** and real numbers c, d.

PROPERTIES OF SCALAR MULTIPLES OF VECTORS

$$c(\mathbf{a} + \mathbf{b}) = c\mathbf{a} + c\mathbf{b}$$
$$(c + d)\mathbf{a} = c\mathbf{a} + d\mathbf{a}$$
$$(cd)\mathbf{a} = c(d\mathbf{a}) = d(c\mathbf{a})$$
$$1\mathbf{a} = \mathbf{a}$$
$$0\mathbf{a} = \mathbf{0} = c\mathbf{0}$$

We shall prove the first property and leave the remaining proofs as an exercise (see Exercise 16). Letting $\mathbf{a} = \langle a_1, a_2 \rangle$ and $\mathbf{b} = \langle b_1, b_2 \rangle$,

$$\begin{aligned} c(\mathbf{a} + \mathbf{b}) &= c\langle a_1 + b_1, a_2 + b_2 \rangle \\ &= \langle ca_1 + cb_1, ca_2 + cb_2 \rangle \\ &= \langle ca_1, ca_2 \rangle + \langle cb_1, cb_2 \rangle \\ &= c\mathbf{a} + c\mathbf{b}. \end{aligned}$$

The special vectors **i** and **j** are defined as follows:

DEFINITION OF i AND j

$$\mathbf{i} = \langle 1, 0 \rangle, \qquad \mathbf{j} = \langle 0, 1 \rangle.$$

The vectors **i** and **j** can be used to obtain an alternative way of denoting vectors. Specifically, if $\mathbf{a} = \langle a_1, a_2 \rangle$, then

$$\begin{aligned} \mathbf{a} &= \langle a_1, 0 \rangle + \langle 0, a_2 \rangle \\ &= a_1\langle 1, 0 \rangle + a_2\langle 0, 1 \rangle \end{aligned}$$

that is,

$$\mathbf{a} = \langle a_1, a_2 \rangle = a_1\mathbf{i} + a_2\mathbf{j}.$$

The vector sum on the right in the preceding formula is called a **linear combination** of **i** and **j**. If this notation is employed, then previous rules for addition, subtraction, and multiplication by a scalar may be written as follows, where $\mathbf{b} = \langle b_1, b_2 \rangle = b_1\mathbf{i} + b_2\mathbf{j}$:

$$(a_1\mathbf{i} + a_2\mathbf{j}) + (b_1\mathbf{i} + b_2\mathbf{j}) = (a_1 + b_1)\mathbf{i} + (a_2 + b_2)\mathbf{j}$$
$$(a_1\mathbf{i} + a_2\mathbf{j}) - (b_1\mathbf{i} + b_2\mathbf{j}) = (a_1 - b_1)\mathbf{i} + (a_2 - b_2)\mathbf{j}$$
$$c(a_1\mathbf{i} + a_2\mathbf{j}) = (ca_1)\mathbf{i} + (ca_2)\mathbf{j}.$$

These formulas show that linear combinations of **i** and **j** may be regarded as ordinary algebraic sums.

EXAMPLE 5 If $\mathbf{a} = 5\mathbf{i} + \mathbf{j}$ and $\mathbf{b} = 4\mathbf{i} - 7\mathbf{j}$, express $3\mathbf{a} - 2\mathbf{b}$ as a linear combination of **i** and **j**.

Solution

$$\begin{aligned} 3\mathbf{a} - 2\mathbf{b} &= 3(5\mathbf{i} + \mathbf{j}) - 2(4\mathbf{i} - 7\mathbf{j}) \\ &= (15\mathbf{i} + 3\mathbf{j}) - (8\mathbf{i} - 14\mathbf{j}) \\ &= 7\mathbf{i} + 17\mathbf{j} \end{aligned}$$

■

A **unit vector** is a vector of magnitude 1. The vectors **i** and **j** are unit vectors, as is the vector **c** in Example 1 of this section.

The formula $\mathbf{a} = a_1\mathbf{i} + a_2\mathbf{j}$ for the vector $\mathbf{a} = \langle a_1, a_2 \rangle$ has a useful geometric interpretation. Vectors corresponding to **i**, **j**, and **a** are illustrated in (i) of Figure 6.29. Since **i** and **j** are unit vectors, $a_1\mathbf{i}$ and $a_2\mathbf{j}$ may be represented by horizontal and vertical vectors of magnitudes $|a_1|$ and $|a_2|$, respectively, as illustrated in (ii) of Figure 6.29. The vector **a** may be regarded as the sum of these vectors. For this reason a_1 is called the **horizontal component** and a_2 the **vertical component** of the vector **a**.

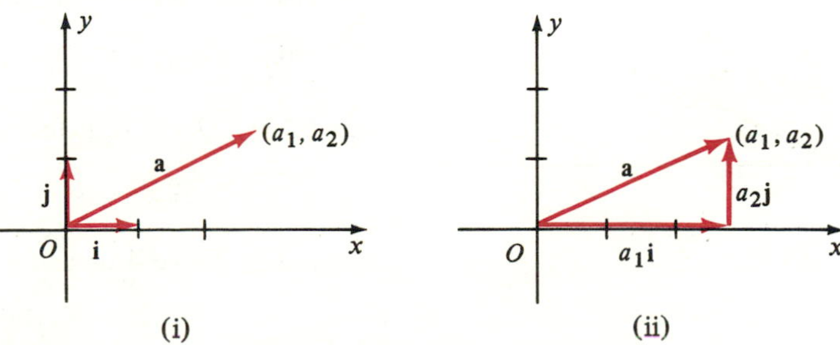

FIGURE 6.29

Let θ be an angle in standard position, measured from the positive x-axis to the terminal side of the vector corresponding to $\mathbf{a} = \langle a_1, a_2 \rangle$, as illustrated in Figure 6.30.

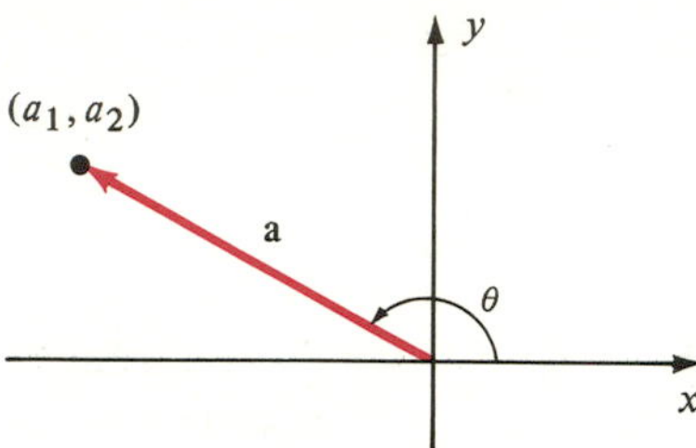

FIGURE 6.30

Since $\cos\theta = a_1/|\mathbf{a}|$ and $\sin\theta = a_2/|\mathbf{a}|$ we obtain the following useful formulas.

HORIZONTAL AND VERTICAL COMPONENTS OF $\mathbf{a} = \langle a_1, a_2 \rangle$

$$a_1 = |\mathbf{a}| \cos\theta, \qquad a_2 = |\mathbf{a}| \sin\theta.$$

Note, in addition, that $\tan\theta = a_2/a_1$. An application of these formulas is given in the next example.

EXAMPLE 6 Two forces $\overrightarrow{PQ}$ and $\overrightarrow{PR}$ of magnitudes 5.0 kg and 8.0 kg, respectively, act at a point P. If the direction of $\overrightarrow{PQ}$ is N20°E, and the direction of $\overrightarrow{PR}$ is N65°E, approximate the direction of the resultant vector $\overrightarrow{PS}$ to the nearest degree, and find $|\overrightarrow{PS}|$ to the nearest tenth.

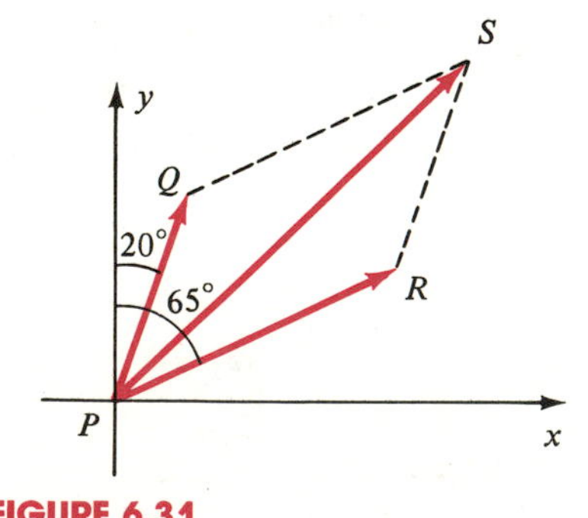

FIGURE 6.31

Solution To use components, let us introduce a coordinate system with origin at the point P, as in Figure 6.31. Note that the angles from the x-axis to $\overrightarrow{PQ}$ and $\overrightarrow{PR}$ have measures 70° and 25°, respectively. Using the formulas for horizontal and vertical components,

$$\overrightarrow{PQ} = \langle 5\cos 70°, 5\sin 70° \rangle, \qquad \overrightarrow{PR} = \langle 8\cos 25°, 8\sin 25° \rangle.$$

Since $\overrightarrow{PS} = \overrightarrow{PQ} + \overrightarrow{PR}$, it follows that

$$\overrightarrow{PS} = \langle 5\cos 70° + 8\cos 25°, 5\sin 70° + 8\sin 25° \rangle.$$

Referring to Table 4 or using a calculator gives us the approximation

$$\overrightarrow{PS} \approx \langle 8.9604, 8.0793 \rangle.$$

Consequently,

$$|\overrightarrow{PS}| \approx \sqrt{(8.9604)^2 + (8.0793)^2} \approx 12.1 \text{ kg}.$$

If θ is the angle from the positive x-axis to the resultant $\overrightarrow{PS} = \langle c_1, c_2 \rangle$, then

$$\tan \theta = \frac{c_2}{c_1} \approx \frac{8.0793}{8.9604} \approx 0.9017.$$

Using Table 4 or a calculator, $\theta \approx 42°$ and hence the direction of $\overrightarrow{PS}$ is approximately N48°E. ■

The technique used in Example 6 has many practical applications (see, for example, Exercises 43–48).

EXERCISES 6.11

In Exercises 1–10 find $\mathbf{a} + \mathbf{b}$, $\mathbf{a} - \mathbf{b}$, $4\mathbf{a} + 5\mathbf{b}$, and $4\mathbf{a} - 5\mathbf{b}$.

1 $\mathbf{a} = \langle 2, -3 \rangle$, $\mathbf{b} = \langle 1, 4 \rangle$

2 $\mathbf{a} = \langle -2, 6 \rangle$, $\mathbf{b} = \langle 2, 3 \rangle$

3 $\mathbf{a} = -\langle 7, -2 \rangle$, $\mathbf{b} = 4\langle -2, 1 \rangle$

4 $\mathbf{a} = 2\langle 5, -4 \rangle$, $\mathbf{b} = -\langle 6, 0 \rangle$

5 $\mathbf{a} = \mathbf{i} + 2\mathbf{j}$, $\mathbf{b} = 3\mathbf{i} - 5\mathbf{j}$

6 $\mathbf{a} = -3\mathbf{i} + \mathbf{j}$, $\mathbf{b} = -3\mathbf{i} + \mathbf{j}$

7 $\mathbf{a} = -(4\mathbf{i} - \mathbf{j})$, $\mathbf{b} = 2(\mathbf{i} - 3\mathbf{j})$

8 $\mathbf{a} = 8\mathbf{j}$, $\mathbf{b} = (-3)(-2\mathbf{i} + \mathbf{j})$

9 $\mathbf{a} = 2\mathbf{j}$, $\mathbf{b} = -3\mathbf{i}$

10 $\mathbf{a} = \mathbf{0}$, $\mathbf{b} = \mathbf{i} + \mathbf{j}$

In Exercises 11–14 sketch vectors corresponding to $\mathbf{a}$, $\mathbf{b}$, $\mathbf{a} + \mathbf{b}$, $2\mathbf{a}$, and $-3\mathbf{b}$.

11 $\mathbf{a} = 3\mathbf{i} + 2\mathbf{j}$, $\mathbf{b} = -\mathbf{i} + 5\mathbf{j}$

12 $\mathbf{a} = -5\mathbf{i} + 2\mathbf{j}$, $\mathbf{b} = \mathbf{i} - 3\mathbf{j}$

13 $\mathbf{a} = \langle -4, 6 \rangle$, $\mathbf{b} = \langle -2, 3 \rangle$

14 $\mathbf{a} = \langle 2, 0 \rangle$, $\mathbf{b} = \langle -2, 0 \rangle$

15 Prove the second, third, and fourth properties of addition of vectors listed on page 307.

16 Prove the second through the fifth properties of scalar multiples of vectors listed on page 308.

Prove the properties in Exercises 17–24, where $\mathbf{a} = \langle a_1, a_2 \rangle$, $\mathbf{b} = \langle b_1, b_2 \rangle$, and c is any real number.

17 $(-1)\mathbf{a} = -\mathbf{a}$

18 $(-c)\mathbf{a} = -c\mathbf{a}$

19 $-(\mathbf{a} + \mathbf{b}) = -\mathbf{a} - \mathbf{b}$

20 $c(\mathbf{a} - \mathbf{b}) = c\mathbf{a} - c\mathbf{b}$

21 If $\mathbf{a} + \mathbf{b} = \mathbf{0}$, then $\mathbf{b} = -\mathbf{a}$.

22 If $\mathbf{a} + \mathbf{b} = \mathbf{a}$, then $\mathbf{b} = \mathbf{0}$.

23 If $c\mathbf{a} = \mathbf{0}$ and $c \neq 0$, then $\mathbf{a} = \mathbf{0}$.

24 If $c\mathbf{a} = \mathbf{0}$ and $\mathbf{a} \neq \mathbf{0}$, then $c = 0$.

25 If $\mathbf{v} = \langle a, b \rangle$, prove each of the following.

(a) The magnitude of $2\mathbf{v}$ is twice the magnitude of $\mathbf{v}$.

(b) The magnitude of $\frac{1}{2}\mathbf{v}$ is one-half the magnitude of $\mathbf{v}$.

(c) The magnitude of $-2\mathbf{v}$ is twice the magnitude of $\mathbf{v}$.

(d) If k is any real number, then the magnitude of $k\mathbf{v}$ is $|k|$ times the magnitude of $\mathbf{v}$.

26 If $\mathbf{v} = \langle a, b \rangle$ and $\mathbf{w} = \langle c, d \rangle$, give a geometric interpretation for $\mathbf{v} - \mathbf{w}$.

If $\mathbf{v} = a_1\mathbf{i} + a_2\mathbf{j}$ and $\mathbf{u} = b_1\mathbf{i} + b_2\mathbf{j}$, the **dot product** $\mathbf{v} \cdot \mathbf{u}$ is defined by $\mathbf{v} \cdot \mathbf{u} = a_1b_1 + a_2b_2$. Prove the properties in Exercises 27–30.

27 $\mathbf{v} \cdot \mathbf{u} = \mathbf{u} \cdot \mathbf{v}$

28 $\mathbf{v} \cdot \mathbf{v} = |\mathbf{v}|^2$

29 $(c\mathbf{v}) \cdot \mathbf{u} = c(\mathbf{v} \cdot \mathbf{u})$, for every real number c

30 $\mathbf{v} \cdot (\mathbf{u} + \mathbf{w}) = \mathbf{v} \cdot \mathbf{u} + \mathbf{v} \cdot \mathbf{w}$, where $\mathbf{w} = c_1\mathbf{i} + c_2\mathbf{j}$

In Exercises 31–38 find the magnitude of $\mathbf{a}$ and the smallest positive angle θ from the positive x-axis to the vector $\overrightarrow{OP}$ corresponding to $\mathbf{a}$.

31 $\mathbf{a} = \langle 3, -3 \rangle$

32 $\mathbf{a} = \langle -2, -2\sqrt{3} \rangle$

33 $\mathbf{a} = \langle -5, 0 \rangle$

34 $\mathbf{a} = \langle 0, 10 \rangle$

35 $\mathbf{a} = -4\mathbf{i} + 5\mathbf{j}$

36 $\mathbf{a} = 10\mathbf{i} - 10\mathbf{j}$

37 $\mathbf{a} = -18\mathbf{j}$

38 $\mathbf{a} = 2\mathbf{i} - 3\mathbf{j}$

In Exercises 39–42, (a) and (b) represent the magnitudes and directions of two forces acting at a point P. Approximate the magnitude of the resultant (to two significant figures) and its direction (to the nearest degree).

39 (a) 90 kg, N75°W
(b) 60 kg, S5°E

40 (a) 20 kg, S17°W
(b) 50 kg, N82°W

41 (a) 6.0 lb, 110°
(b) 2.0 lb, 215°

42 (a) 70 lb, 320°
(b) 40 lb, 30°

43 An airplane with an airspeed of 200 mph is flying in the direction 50°, and a 40 mph wind is blowing directly from the west. As in the figure, these may be represented by vectors **p** and **w** of magnitudes 200 and 40, respectively. The direction of the resultant **p** + **w** gives the **true course** of the airplane relative to the ground, and the magnitude $|\mathbf{p} + \mathbf{w}|$ is called the **ground speed** of the airplane. Approximate the true course to the nearest degree and the ground speed to the nearest mph.

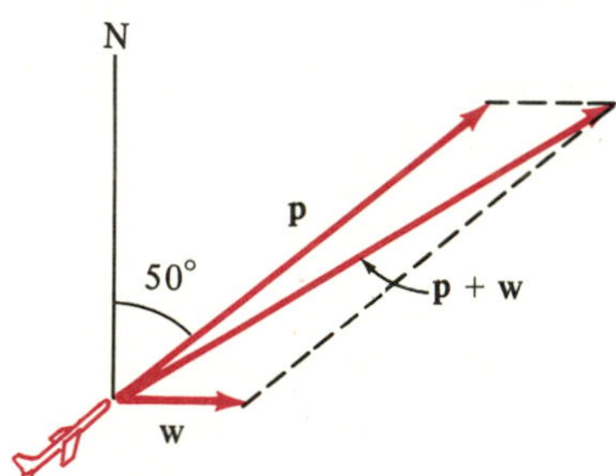

FIGURE FOR EXERCISE 43

44 An airplane is flying in the direction 140° with an airspeed of 500 mph, and a 30 mph wind is blowing in the direction 65°. Approximate the true course (to the nearest degree) and the ground speed (to the nearest mph).

45 An airplane pilot wishes to maintain a true course in the direction 250°, with a ground speed of 400 mph. If the wind is blowing directly north at 50 mph, approximate the required airspeed (to the nearest mph) and compass heading (to the nearest degree).

46 An airplane is flying in the direction 20° with an airspeed of 300 mph. Its ground speed and true course are 350 mph and 30°, respectively. Approximate the direction and speed of the wind.

47 The current in a river flows directly from the west at a rate of 1.5 ft/sec. A person who rows a boat at a rate of 4 ft/sec in still water wishes to row directly north across the river. Approximate, to the nearest 10′, the direction in which the person should row.

48 In order for a motor boat moving at a speed of 30 mph to travel directly north across a river, it must aim at a point that has a bearing of N15°E. If the current is flowing directly west, approximate the rate at which it flows, to the nearest mph.

SECTION 6.12
REVIEW

Define or discuss each of the following.

1 The Fundamental Identities
2 Verifying identities
3 Trigonometric equations
4 Addition and subtraction formulas
5 Double-angle formulas
6 Half-angle formulas
7 Product formulas
8 Factoring formulas
9 Reduction formulas
10 Inverse trigonometric functions
11 The Law of Sines
12 The Law of Cosines

13 Geometric representation of a complex number
14 The complex plane
15 Trigonometric form for a complex number
16 De Moivre's Theorem
17 Vector
18 Magnitude of a vector
19 Addition of vectors
20 Scalar multiple of a vector
21 Vectors as ordered pairs
22 Components of a vector
23 Subtraction of vectors
24 Unit vector

EXERCISES 6.12

Verify the identities in Exercise 1–16.

1 $(\cot^2 x + 1)(1 - \cos^2 x) = 1$

2 $\cos\theta + \sin\theta\tan\theta = \sec\theta$

3 $\dfrac{(\sec^2\theta - 1)\cot\theta}{\tan\theta\sin\theta + \cos\theta} = \sin\theta$

4 $(\tan x + \cot x)^2 = \sec^2 x \csc^2 x$

5 $\dfrac{1}{1 + \sin t} = (\sec t - \tan t)\sec t$

6 $\dfrac{\sin(\alpha - \beta)}{\cos(\alpha + \beta)} = \dfrac{\tan\alpha - \tan\beta}{1 - \tan\alpha\tan\beta}$

7 $\dfrac{2\cot u}{\csc^2 u - 2} = \tan 2u$

8 $\cos^2\dfrac{v}{2} = \dfrac{1 + \sec v}{2\sec v}$

9 $\dfrac{\tan^3\phi - \cot^3\phi}{\tan^2\phi + \csc^2\phi} = \tan\phi - \cot\phi$

10 $\dfrac{\sin u + \sin v}{\csc u + \csc v} = \dfrac{1 - \sin u \sin v}{-1 + \csc u \csc v}$

11 $\cos\left(x - \dfrac{5\pi}{2}\right) = \sin x$

12 $\tan\left(x + \dfrac{3\pi}{4}\right) = \dfrac{\tan x - 1}{\tan x + 1}$

13 $\frac{1}{4}\sin 4\beta = \sin\beta\cos^3\beta - \cos\beta\sin^3\beta$

14 $\tan\frac{1}{2}\theta = \csc\theta - \cot\theta$

15 $\sin 8\theta = 8\sin\theta\cos\theta(1 - 2\sin^2\theta)(1 - 8\sin^2\theta\cos^2\theta)$

16 $\arctan x = \frac{1}{2}\arctan\dfrac{2x}{1 - x^2}, \ |x| \le 1$

In Exercises 17–28 find the solutions of the equation that are in the interval $[0, 2\pi)$, and also find the degree measure of each solution.

17 $2\cos^3\theta - \cos\theta = 0$

18 $2\cos\alpha + \tan\alpha = \sec\alpha$

19 $\sin\theta = \tan\theta$

20 $\csc^5\theta - 4\csc\theta = 0$

21 $2\cos^3 t + \cos^2 t - 2\cos t - 1 = 0$

22 $\cos x \cot^2 x = \cos x$

23 $\sin\beta + 2\cos^2\beta = 1$

24 $\cos 2x + 3\cos x + 2 = 0$

25 $2\sec u \sin u + 2 = 4\sin u + \sec u$

26 $\sin 2u = \sin u$

27 $2\cos^2\frac{1}{2}\theta - 3\cos\theta = 0$

28 $\sec 2x \csc 2x = 2\csc 2x$

In Exercises 29–32 find the exact values without the use of tables or calculators.

29 $\cos 75°$ **30** $\tan 285°$

31 $\sin 195°$ **32** $\csc \pi/8$

If θ and ϕ are acute angles such that $\csc\theta = \frac{5}{3}$ and $\cos\phi = \frac{8}{17}$, find the numbers in Exercises 33–41 without the use of tables or calculators.

33 $\sin(\theta + \phi)$ **34** $\cos(\theta + \phi)$ **35** $\tan(\theta - \phi)$

36 $\sin(\phi - \theta)$ **37** $\sin 2\phi$ **38** $\cos 2\phi$

39 $\tan 2\theta$ **40** $\sin\theta/2$ **41** $\tan\theta/2$

42 Express $\cos(\alpha + \beta + \gamma)$ in terms of functions of α, β, and γ.

43 Express each product as a sum or difference.

(a) $\sin 7t \sin 4t$

(b) $\cos(u/4)\cos(-u/6)$

(c) $6\cos 5x \sin 3x$

44 Express each of the following as a product.

(a) $\sin 8u + \sin 2u$

(b) $\cos 3\theta - \cos 8\theta$

(c) $\sin(t/4) - \sin(t/5)$

Find the numbers in Exercises 45–53 without the use of tables or calculators.

45 $\cos^{-1}\left(\dfrac{-\sqrt{3}}{2}\right)$

46 $\sin^{-1}\left(\dfrac{\sqrt{2}}{2}\right)$

47 $\arccos(-1)$

48 $\arctan\left(\dfrac{-\sqrt{3}}{3}\right)$

49 $\sin \arccos\left(\dfrac{-\sqrt{3}}{2}\right)$

50 $\cos(\sin^{-1}\frac{15}{17} - \sin^{-1}\frac{8}{17})$

51 $\cos(2\sin^{-1}\frac{4}{5})$

52 $\sin(\sin^{-1}\frac{2}{3})$

53 $\cos^{-1}(\sin 0)$

Sketch the graphs of the equations in Exercises 54–56.

54 $y = \cos^{-1} 3x$

55 $y = 4\sin^{-1} x$

56 $y = 1 - \sin^{-1} x$

Without using tables or calculators, find the remaining parts of triangle ABC in Exercises 57–60.

57 $\alpha = 60°,\ \beta = 45°,\ b = 100$

58 $\gamma = 30°,\ a = 2\sqrt{3},\ c = 2$

59 $\alpha = 60°,\ b = 6,\ c = 7$

60 $a = 2,\ b = 3,\ c = 4$

Change the complex numbers in Exercises 61–66 to trigonometric form.

61 $-10 + 10i$

62 $2 - 2\sqrt{3}i$

63 -17

64 $-12i$

65 $-5\sqrt{3} - 5i$

66 $4 + 5i$

In each of Exercises 67–70 use De Moivre's Theorem to write the given number in the form $a + bi$, where a and b are real numbers.

67 $(-\sqrt{3} + i)^9$

68 $\left(\dfrac{\sqrt{2}}{2} - \dfrac{\sqrt{2}}{2}i\right)^{30}$

69 $(3 - 3i)^5$

70 $(2 + 2\sqrt{3}i)^{10}$

71 Find the three cube roots of -27.

72 Find the solutions of the equation $x^5 - 32 = 0$.

73 If $\mathbf{a} = \langle -4, 5\rangle$ and $\mathbf{b} = \langle 2, -8\rangle$, sketch vectors corresponding to $\mathbf{a} + \mathbf{b}$, $\mathbf{a} - \mathbf{b}$, $2\mathbf{a}$, and $-\frac{1}{2}\mathbf{b}$.

74 If $\mathbf{a} = 2\mathbf{i} + 5\mathbf{j}$ and $\mathbf{b} = 4\mathbf{i} - \mathbf{j}$, find the vectors or numbers corresponding to:

(a) $4\mathbf{a} + \mathbf{b}$

(b) $2\mathbf{a} - 3\mathbf{b}$

(c) $|\mathbf{a} - \mathbf{b}|$

(d) $|\mathbf{a}| - |\mathbf{b}|$.

7

SYSTEMS OF EQUATIONS AND INEQUALITIES

In certain types of mathematical problems it is necessary to work simultaneously with more than one equation in several variables. We then refer to the equations as a *system of equations.* It is usually desirable to find the solutions that are common to all equations in the system. In this chapter we shall develop methods for finding these common solutions. Of particular importance are the matrix techniques introduced for systems of linear equations. We shall also briefly discuss systems of inequalities and linear programming.

SECTION 7.1

SYSTEMS OF EQUATIONS

An ordered pair (a, b) is a **solution** of an equation in two variables x and y if a true statement is obtained when a and b are substituted for x and y, respectively. Two equations in x and y are **equivalent** if they have exactly the same solutions. For example, the following equations are equivalent:

$$x^2 - 4y = 3, \qquad x^2 - 3 = 4y$$

As we know, the **graph** of an equation in x and y consists of all points in a coordinate plane that correspond to solutions.

A **system** of two equations in x and y is any two equations in those variables. An ordered pair (a, b) is called a **solution of the system** if (a, b) is a solution of both equations. It follows that the points that correspond to the solutions are precisely the points at which the graphs of the two equations intersect.

As a concrete example, consider the system

$$\begin{cases} x^2 - y = 0 \\ y - 2x - 3 = 0. \end{cases}$$

The brace is used to indicate that the two equations are to be treated simultaneously. The following table exhibits some solutions of the equation $x^2 - y = 0$.

x	-2	-1	0	1	2	3	4
y	4	1	0	1	4	9	16

The graph is the parabola sketched in Figure 7.1. A table of solutions for $y - 2x - 3 = 0$ is:

x	-2	-1	0	1	2	3	4
y	-1	1	3	5	7	9	11

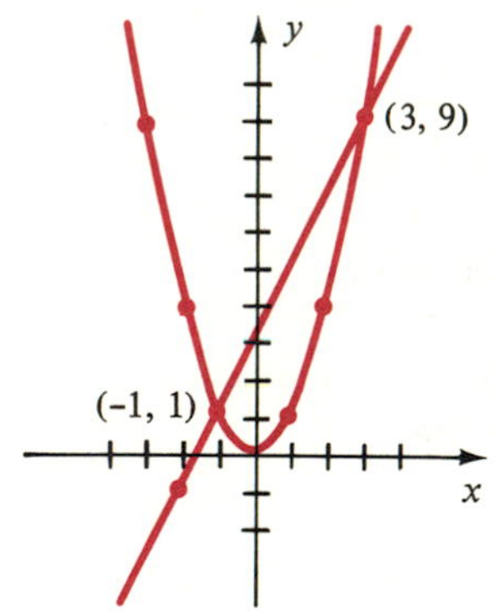

FIGURE 7.1 $\begin{cases} x^2 - y = 0 \\ y - 2x - 3 = 0 \end{cases}$

The graph is the line in Figure 7.1. Note that the pairs $(3, 9)$ and $(-1, 1)$ are solutions of both equations and hence are solutions of the system. We shall see later that they are the *only* solutions. As we pointed out before, the pairs that are solutions of the system represent points of intersection of the two graphs.

The graphical technique we have used is obviously of little value if the solutions of the system involve rational numbers, such as $\frac{37}{849}$, or irrational numbers, or if the equations are more complicated than those we have considered. Let us illustrate how the solutions of the preceding system can be found without reference to graphs. We shall begin by solving the first equation for y in terms of x, obtaining $y = x^2$. It follows that if an ordered pair (x, y) is a solution of the system, then it must be of the form (x, x^2). In particular, (x, x^2) must be a solution of the equation $y - 2x - 3 = 0$, that is,

$$x^2 - 2x - 3 = 0.$$

Factoring gives us $$(x - 3)(x + 1) = 0$$

and hence either $x = 3$ or $x = -1$. The corresponding values for y (obtained from $y = x^2$) are 9 and 1, respectively. Thus, the only possible solutions of the system are the ordered pairs (3, 9) and (−1, 1). That these are actually solutions can be seen by substitution.

To recapitulate, to find the solutions algebraically we began by solving one equation of the system for y in terms of x, and then we substituted for y in the other equation, obtaining an equation in one variable, x. The solutions of the latter equation were the only possible x-values for the solutions of the system. The corresponding y-values were found by means of the equation that expressed y in terms of x. This algebraic technique is called the *method of substitution*. Sometimes it is more convenient to begin by solving one equation for x in terms of y, and then substituting for x in the other equation. In general, the steps used in this process may be listed as follows.

METHOD OF SUBSTITUTION

(i) Solve one of the equations for one variable in terms of the other variable.

(ii) Substitute the expression obtained in step (i) into the other equation, obtaining an equation in one variable.

(iii) Find the solutions of the equation obtained in step (ii).

(iv) Use the solutions from step (iii), together with the expression obtained in step (i), to find the solutions of the system.

In the next example we find the solutions of the system considered at the beginning of this section by first solving one equation for x in terms of y.

EXAMPLE 1 Find the solutions of the system

$$\begin{cases} x^2 - y = 0, \\ y - 2x - 3 = 0. \end{cases}$$

Solution We may solve the second equation for x in terms of y as follows:

$$2x = y - 3, \quad \text{or} \quad x = \frac{y-3}{2}.$$

Substituting for x in the first equation gives us

$$\left(\frac{y-3}{2}\right)^2 - y = 0$$

$$\frac{y^2 - 6y + 9}{4} - y = 0$$

$$y^2 - 6y + 9 - 4y = 0$$

$$y^2 - 10y + 9 = 0$$

$$(y-9)(y-1) = 0.$$

The last equation has solutions $y = 9$ and $y = 1$. If we substitute 9 and 1 for y in the equation $x = (y - 3)/2$ we obtain the values $x = 3$ and $x = -1$, respectively. Hence, as before, the solutions of the system are (3, 9) and (−1, 1). ■

EXAMPLE 2 Find the solutions of the following system and then sketch the graph of each equation, showing the points of intersection:

$$\begin{cases} x^2 + y^2 = 25 \\ x^2 + y = 19. \end{cases}$$

Solution Solving the second equation for y, we obtain $y = 19 - x^2$. Substituting for y in the first equation leads to the following list of equivalent equations:

$$x^2 + (19 - x^2)^2 = 25$$

$$x^4 - 37x^2 + 336 = 0$$

$$(x^2 - 16)(x^2 - 21) = 0.$$

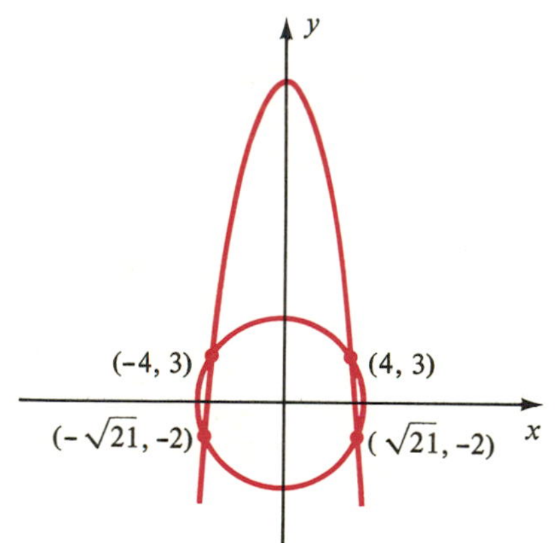

FIGURE 7.2 $\begin{cases} x^2 + y^2 = 25 \\ x^2 + y = 19 \end{cases}$

The solutions of the last equation are 4, −4, $\sqrt{21}$, and $-\sqrt{21}$. The corresponding values of y are found by substituting for x in $y = 19 - x^2$. Substitution of 4 or −4 for x gives us $y = 3$, whereas substitution of $\sqrt{21}$ or $-\sqrt{21}$ gives us $y = -2$. Hence, the only possible solutions of the system are

$$(4, 3), \quad (-4, 3), \quad (\sqrt{21}, -2), \quad \text{and} \quad (-\sqrt{21}, -2).$$

It can be seen by direct substitution in each of the given equations that all four pairs are solutions. The graph of $x^2 + y^2 = 25$ is a circle of radius 5 with center at the origin, and the graph of $y = 19 - x^2$ is a parabola with a vertical axis. The graphs and their points of intersection are illustrated in Figure 7.2. ■

EXAMPLE 3 Find the solutions of the system

$$\begin{cases} y - \log(x + 3) = 1 \\ \qquad 2 - y = \log x. \end{cases}$$

Solution From the second equation, we obtain $y = 2 - \log x$. Substituting in the first equation and simplifying, we obtain

$$\log x + \log(x + 3) = 1.$$

This leads to the following equivalent equations (supply the reasons):

$$\log x(x + 3) = 1$$

$$x(x + 3) = 10$$

$$x^2 + 3x - 10 = 0$$

$$(x + 5)(x - 2) = 0.$$

The only possible solutions of the last equation are $x = -5$ and $x = 2$. If we substitute -5 for x in the equation $y = 2 - \log x$, we arrive at the logarithm of a negative number. Since $\log x$ is defined only if $x > 0$, this value of x is extraneous. If we substitute 2 for x we get $y = 2 - \log 2$. Checking, we see that the system has one solution, $(2, 2 - \log 2)$. ■

We can also consider equations in three variables x, y, and z, such as

$$x^2y + x \log z + 3^y = 4z^3.$$

We say that the equation has a **solution** (a, b, c) if substitution of a, b, and c, for x, y, and z, respectively, produces a true statement. We refer to (a, b, c) as an **ordered triple** of real numbers. Equivalent equations are defined as before. A system of equations in three variables and the corresponding solutions are defined as in the two-variable case. In like manner, we can consider systems of *any* number of equations in *any* number of variables.

The method of substitution can be extended to these more complicated systems. For example, given three equations in three variables, suppose that it is possible to solve one of the equations for one variable in terms of the remaining two variables. By substituting that expression in each of the other equations, we obtain a system of two equations in two variables. The solutions of the two-variable system can then be used to find the solutions of the original system, as illustrated in the following example.

EXAMPLE 4 Find the solutions of the system

$$\begin{cases} x - y + z = 2 \\ xyz = 0 \\ 2y + z = 1. \end{cases}$$

Solution Solving the third equation for z, we see that

$$z = 1 - 2y.$$

Substituting for z in the first equations of the system we obtain the following system of two equations in two variables:

$$\begin{cases} x - y + (1 - 2y) = 2 \\ \qquad xy(1 - 2y) = 0. \end{cases}$$

This system is equivalent to

$$\begin{cases} x - 3y - 1 = 0 \\ xy(1 - 2y) = 0. \end{cases}$$

We now find the solutions of the last system. Solving the first equation for x in terms of y gives us

$$x = 3y + 1.$$

Substituting $3y + 1$ for x in the second equation $xy(1 - 2y) = 0$, we obtain

$$(3y + 1)y(1 - 2y) = 0,$$

which has as solutions the numbers $-\frac{1}{3}$, 0, and $\frac{1}{2}$. These are the only possible y-values for the solutions of the system. To obtain the corresponding x-values, we use the equation $x = 3y + 1$, obtaining $x = 0$, 1, and $\frac{5}{2}$, respectively. Finally, substituting in $z = 1 - 2y$ gives us the z-values $\frac{5}{3}$, 1, and 0. It follows that the solutions of the original system consist of the ordered triples

$$(0, -\tfrac{1}{3}, \tfrac{5}{3}), \quad (1, 0, 1), \quad \text{and} \quad (\tfrac{5}{2}, \tfrac{1}{2}, 0).$$ ■

EXERCISES 7.1

In Exercises 1–34 use the method of substitution to find the solutions of the system of equations.

1 $\begin{cases} y = x^2 - 4 \\ y = 2x - 1 \end{cases}$

2 $\begin{cases} y = x^2 + 1 \\ x + y = 3 \end{cases}$

3 $\begin{cases} y^2 = 1 - x \\ x + 2y = 1 \end{cases}$

4 $\begin{cases} y^2 = x \\ x + 2y + 3 = 0 \end{cases}$

5 $\begin{cases} 2y = x^2 \\ y = 4x^3 \end{cases}$

6 $\begin{cases} x - y^3 = 1 \\ 2x = 9y^2 + 2 \end{cases}$

7 $\begin{cases} x + 2y = -1 \\ 2x - 3y = 12 \end{cases}$

8 $\begin{cases} 3x - 4y + 20 = 0 \\ 3x + 2y + 8 = 0 \end{cases}$

9 $\begin{cases} 2x - 3y = 1 \\ -6x + 9y = 4 \end{cases}$

10 $\begin{cases} 4x - 5y = 2 \\ 8x - 10y = -5 \end{cases}$

11 $\begin{cases} x + 3y = 5 \\ x^2 + y^2 = 25 \end{cases}$

12 $\begin{cases} 3x - 4y = 25 \\ x^2 + y^2 = 25 \end{cases}$

13 $\begin{cases} x^2 + y^2 = 8 \\ y - x = 4 \end{cases}$

14 $\begin{cases} x^2 + y^2 = 25 \\ 3x + 4y = -25 \end{cases}$

15 $\begin{cases} x^2 + y^2 = 9 \\ y - 3x = 2 \end{cases}$

16 $\begin{cases} x^2 + y^2 = 16 \\ y + 2x = -1 \end{cases}$

17 $\begin{cases} x^2 + y^2 = 16 \\ 2y - x = 4 \end{cases}$

18 $\begin{cases} x^2 + y^2 = 1 \\ y + 2x = -3 \end{cases}$

19 $\begin{cases} (x - 1)^2 + (y + 2)^2 = 10 \\ x + y = 1 \end{cases}$

20 $\begin{cases} xy = 2 \\ 3x - y + 5 = 0 \end{cases}$

21 $\begin{cases} y = 20/x^2 \\ y = 9 - x^2 \end{cases}$

22 $\begin{cases} x = y^2 - 4y + 5 \\ x - y = 1 \end{cases}$

23 $\begin{cases} y^2 - 4x^2 = 4 \\ 9y^2 + 16x^2 = 140 \end{cases}$

24 $\begin{cases} 25y^2 - 16x^2 = 400 \\ 9y^2 - 4x^2 = 36 \end{cases}$

25 $\begin{cases} x^2 - y^2 = 4 \\ x^2 + y^2 = 12 \end{cases}$

26 $\begin{cases} 6x^3 - y^3 = 1 \\ 3x^3 + 4y^3 = 5 \end{cases}$

27 $\begin{cases} x + 2y - z = -1 \\ 2x - y + z = 9 \\ x + 3y + 3z = 6 \end{cases}$

28 $\begin{cases} 2x - 3y - z^2 = 0 \\ x - y - z^2 = -1 \\ x^2 - xy = 0 \end{cases}$

29 $\begin{cases} y = 5^x + 4 \\ y = 5^{2x} - 2 \end{cases}$

30 $\begin{cases} x + 3(2^y) = 2^{2y} \\ x - 2^{y+1} = 6 \end{cases}$

31 $\begin{cases} \log(3r + 1) = s + 5 \\ \log(r - 2) = s + 4 \end{cases}$

32 $\begin{cases} \log u = v - 3 \\ 5 = v + \log(u - 2) \end{cases}$

33 $\begin{cases} \log(3x + 5) - \log y = 1 \\ 5x - 2y = 6 \end{cases}$

34 $\begin{cases} y + 2 = \log(2x + 5) \\ y - 1 = \log x \end{cases}$

Solve Exercises 35–40 by introducing several variables and using a suitable system of equations.

35 Find two positive integers whose difference is 4 and whose squares differ by 88. (*Hint:* Let x and y denote the two integers.)

36 Find two real numbers whose difference and product both equal 4.

37 The perimeter of a rectangle is 40 inches and its area is 96 square inches. Find the length and width.

38 Generalize Exercise 37 to the case where the perimeter is P and area is A. Express the length and width in terms of P and A. What restrictions are necessary?

39 Find three numbers whose sum and product are 20 and 60, respectively, such that one of the numbers equals the sum of the other two.

40 Find the values of b such that the system

$$\begin{cases} x^2 + y^2 = 4 \\ y = x + b \end{cases}$$

has solutions consisting of (a) one real number, (b) two real numbers, (c) no real numbers. Interpret the three case geometrically.

SECTION 7.2

SYSTEMS OF LINEAR EQUATIONS IN TWO VARIABLES

An equation of the form $ax + by + c = 0$ (or, equivalently, $ax + by = -c$) is called a *linear equation* in x and y. Similarly, a linear equation in three variables x, y, and z is an equation of the form $ax + by + cz = d$, where the coefficients are real numbers. Linear equations in any number of variables may be defined in like manner. If a large number of variables occurs, a subscript notation is usually employed. If we let $x_1, x_2, \ldots, x_n$ denote variables, where n is any positive integer, then an expression of the form

$$a_1x_1 + a_2x_2 + \cdots + a_nx_n = a$$

where $a_1, a_2, \ldots, a_n$ and a are real numbers, is called a **linear equation in *n* variables** (with real coefficients).

In modern applications of mathematics, perhaps the most common systems of equations are those in which all the equations are linear. In this section we shall only consider systems of two linear equations in two variables. Systems involving more than two variables are discussed in Section 7.3.

One method for solving a system of equations in several variables is to replace the equations in the system by equivalent equations until equations are reached from which the solutions are easily obtained. Some general rules that allow us to transform a system of equations into an equivalent system are stated in the next theorem. Since we prefer not to specify the number or types of variables, we shall use the notation $p = 0$ and $q = 0$ for typical equations in a system. Bear in mind that the symbols p and q represent expressions in the variables under consideration. For example, if we let $p = 3x - 2y + 5z - 1$ and $q = 6x + y - 7z - 4$, then the equations $p = 0$ and $q = 0$ have the form

$$3x - 2y + 5z - 1 = 0, \qquad 6x + y - 7z - 4 = 0.$$

TRANSFORMATIONS THAT LEAD TO EQUIVALENT SYSTEMS OF EQUATIONS

The following transformations do not change the solutions of a system of equations:

(i) Interchanging the position of any two equations.

(ii) Multiplying both sides of an equation in the system by a nonzero real number.

(iii) Replacing an equation $q = 0$ of the system by $kp + q = 0$, where $p = 0$ is any other equation in the system and k is any real number.

Proof It is easy to show that (i) and (ii) do not change the solutions of the system, and therefore we shall omit the proofs.

To prove (iii), we first note that a solution of the original system is a solution of both equations $p = 0$ and $q = 0$. Accordingly, each of the expressions p and q equals zero if the variables are replaced by appropriate real numbers, and hence the expression $kp + q$ will also equal zero. Since none of the other equations has been changed, this shows that any solution of the original system is also a solution of the transformed system, which is obtained by replacing the equation $q = 0$ by $kp + q = 0$.

Conversely, given a solution of the transformed system, both of the expressions $kp + q$ and p equal zero if the variables are replaced by appropriate numbers. This implies, however, that $(kp + q) - kp$ equals 0. Since $(kp + q) - kp = q$, we see that q must also equal zero when the substitution is made. Thus, a solution of the transformed system is also a solution of the original system. This completes the proof. □

For convenience, we shall describe the process of using (iii) of the preceding theorem by the phrase "add to one equation of the system k times any other equation of the system." Of course, to *add* two equations means to add corresponding sides. To *multiply* an equation by k means to multiply both sides of the equation by k. The process may also be applied if 0 is not on one side of each equation, as illustrated in the next example.

EXAMPLE 1 Find the solutions of the system

$$\begin{cases} x + 3y = -1 \\ 2x - \ \ y = \ \ 5. \end{cases}$$

Solution By (ii) of the preceding theorem we may multiply the second equation by 3. This gives us the equivalent system

$$\begin{cases} x + 3y = -1 \\ 6x - 3y = \ \ 15. \end{cases}$$

Next, by (iii) of the theorem, we may add to the second equation 1 times the first. This gives us

$$\begin{cases} x + 3y = -1 \\ 7x \qquad = \ \ 14. \end{cases}$$

We see from the last equation that the only possible value for x is 2. The corresponding value for y may be found by substituting for x in the first equation. This gives us the following:

$$2 + 3y = -1, \qquad 3y = -3, \qquad y = -1.$$

Consequently, the given system has one solution, $(2, -1)$.

There are other methods of solution. For example, we could begin by multiplying the first equation by -2, obtaining

$$\begin{cases} -2x - 6y = 2 \\ \ \ \ 2x - \ \ y = 5. \end{cases}$$

If we next add the first equation to the second, we get

$$\begin{cases} -2x - 6y = 2 \\ \qquad - 7y = 7. \end{cases}$$

The last equation implies that $y = -1$. Substitution for y in the first equation gives us

$$-2x - 6(-1) = 2, \qquad -2x = -4, \qquad x = 2.$$

Again we see that the solution is $(2, -1)$.

The graphs of the two equations, showing the point of intersection $(2, -1)$, are sketched in Figure 7.3. ■

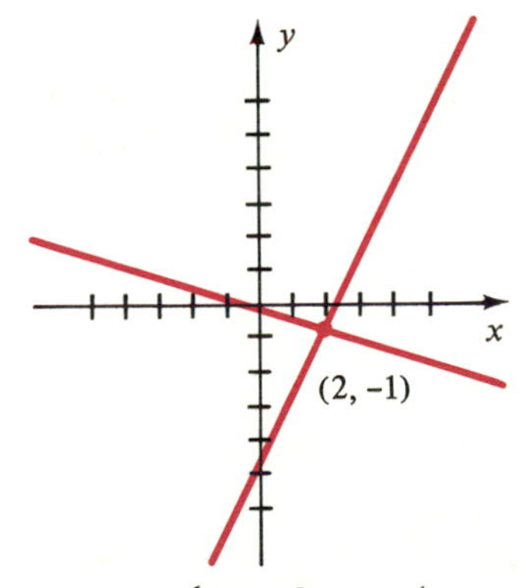

FIGURE 7.3 $\begin{cases} x + 3y = -1 \\ 2x - \ \ y = \ \ 5 \end{cases}$

The technique used in Example 1 is called the **method of elimination,** since it involves eliminating a variable from one of the equations. The method of elimination usually leads to solutions in fewer steps than the method of substitution discussed in Section 7.1.

EXAMPLE 2 Find the solutions of the system

$$\begin{cases} 3x + y = 6 \\ 6x + 2y = 12. \end{cases}$$

Solution Multiplying the second equation by $\frac{1}{2}$ gives us

$$\begin{cases} 3x + y = 6 \\ 3x + y = 6. \end{cases}$$

Consequently, (a, b) is a solution if and only if $3a + b = 6$, that is, $b = 6 - 3a$. It follows that the solutions consist of all ordered pairs of the form $(a, 6 - 3a)$ where a is a real number. If we wish to find particular solutions we may substitute various values for a. For example, a few solutions are $(0, 6)$, $(1, 3)$, $(3, -3)$, $(-2, 12)$, and $(\sqrt{2}, 6 - 3\sqrt{2})$. Note that the graph of each equation is the same line. ■

EXAMPLE 3 Find the solutions of the system

$$\begin{cases} 3x + y = 6 \\ 6x + 2y = 20. \end{cases}$$

Solution If we add to the second equation -2 times the first equation, we obtain the equivalent system

$$\begin{cases} 3x + y = 6 \\ \quad\ \ 0 = 8. \end{cases}$$

Since the last equation is never true, the system has no solutions. Of course, this means that the graphs of the given equations do not intersect (see Figure 7.4).

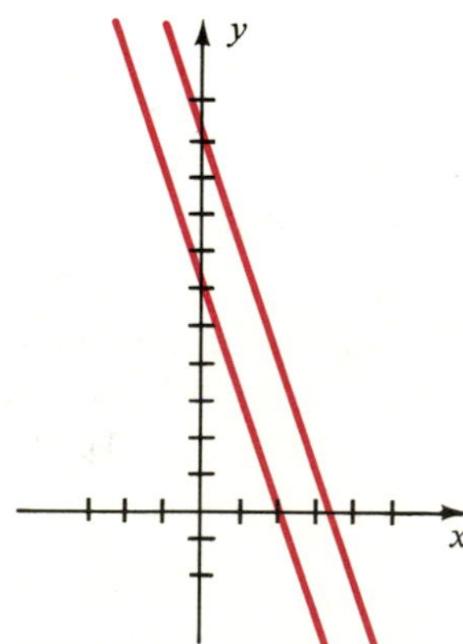

FIGURE 7.4 $\begin{cases} 3x + y = 6 \\ 6x + 2y = 20 \end{cases}$

■

Since the graph of every linear equation $ax + by = c$ is a line, it follows that for every system of two such equations, precisely one of the following three possibilities occurs:

(i) The lines intersect in exactly one point.
(ii) The lines are identical.
(iii) The lines are parallel.

Since the pairs that are solutions of the system exhibit the coordinates of the points of intersection of the graphs of the two equations, we may interpret statements (i)–(iii) in the following way.

THEOREM

For each system of two linear equations in two variables, one and only one of the following statements is true:

(i) The system has exactly one solution.
(ii) The system has an infinite number of solutions.
(iii) The system has no solution.

If (i) occurs the system is said to be **consistent.**
If (ii) occurs the equations are said to be **dependent.**
If (iii) occurs, the system is called **inconsistent.**

In practice, there should be little difficulty in determining which of the three cases occurs. The case of the unique solution (i) will become apparent when suitable transformations are applied to the system, as illustrated in Example 1. In case (ii) the solution is similar to that for Example 2, where one of the equations can be transformed into the other. In case (iii) the lack of a solution is indicated by an absurdity, such as the statement $0 = 8$, which appeared in Example 3.

Some applied problems can be solved by introducing systems of two linear equations, as illustrated in the next example.

EXAMPLE 4 A motor boat, operating at full throttle, made a trip 4 miles upstream (against a constant current) in 15 minutes. The return trip (with the same current and full throttle) took 12 minutes. Find the speed of the current and the equivalent speed of the boat in still water.

Solution We shall begin by introducing letters to denote the unknown quantities. Thus, let

$$x = \text{speed of boat (in mph)}$$

$$y = \text{speed of current (in mph)}.$$

We plan to use the formula $d = rt$, where d denotes the distance traveled, r the rate, and t the time. Since the current slows the boat as it travels upstream, but adds to its speed as it travels downstream, we obtain

$$\text{upstream rate} = x - y \text{ (in mph)}$$
$$\text{downstream rate} = x + y \text{ (in mph).}$$

The time (in hours) traveled in each direction is

$$\text{upstream time} = \tfrac{15}{60} = \tfrac{1}{4} \text{ hr}$$
$$\text{downstream time} = \tfrac{12}{60} = \tfrac{1}{5} \text{ hr.}$$

The distance is 4 miles for each trip. Substituting in $d = rt$ gives us the system

$$\begin{cases} 4 = (x - y)(\frac{1}{4}) \\ 4 = (x + y)(\frac{1}{5}) \end{cases}$$

or equivalently,
$$\begin{cases} x - y = 16 \\ x + y = 20. \end{cases}$$

Adding the last two equations, we see that $2x = 36$, or $x = 18$. Consequently, $y = 20 - x = 20 - 18 = 2$. Hence, the speed of the boat in still water is 18 mph and the speed of the current is 2 mph. ■

EXERCISES 7.2

Find the solutions of the systems in Exercises 1–20.

1 $\begin{cases} 2x + 3y = 2 \\ x - 2y = 8 \end{cases}$

2 $\begin{cases} 4x + 5y = 13 \\ 3x + y = -4 \end{cases}$

3 $\begin{cases} 2x + 5y = 16 \\ 3x - 7y = 24 \end{cases}$

4 $\begin{cases} 7x - 8y = 9 \\ 4x + 3y = -10 \end{cases}$

5 $\begin{cases} 3r + 4s = 3 \\ r - 2s = -4 \end{cases}$

6 $\begin{cases} 9u + 2v = 0 \\ 3u - 5v = 17 \end{cases}$

7 $\begin{cases} 5x - 6y = 4 \\ 3x + 7y = 8 \end{cases}$

8 $\begin{cases} 2x + 8y = 7 \\ 3x - 5y = 4 \end{cases}$

9 $\begin{cases} \frac{1}{3}c + \frac{1}{2}d = 5 \\ c - \frac{2}{3}d = -1 \end{cases}$

10 $\begin{cases} \frac{1}{2}t - \frac{1}{5}v = \frac{3}{2} \\ \frac{2}{3}t + \frac{1}{4}v = \frac{5}{12} \end{cases}$

11 $\begin{cases} \sqrt{3}x - \sqrt{2}y = 2\sqrt{3} \\ 2\sqrt{2}x + \sqrt{3}y = \sqrt{2} \end{cases}$

12 $\begin{cases} 0.11x - 0.03y = 0.25 \\ 0.12x + 0.05y = 0.70 \end{cases}$

13 $\begin{cases} 2x - 3y = 5 \\ -6x + 9y = 12 \end{cases}$

14 $\begin{cases} 3p - q = 7 \\ -12p + 4q = 3 \end{cases}$

15 $\begin{cases} 3m - 4n = 2 \\ -6m + 8n = -4 \end{cases}$

16 $\begin{cases} x - 5y = 2 \\ 3x - 15y = 6 \end{cases}$

17 $\begin{cases} 2y - 5x = 0 \\ 3y + 4x = 0 \end{cases}$

18 $\begin{cases} 3x + 7y = 9 \\ y = 5 \end{cases}$

19 $\begin{cases} \dfrac{2}{x} + \dfrac{3}{y} = -2 \\ \dfrac{4}{x} - \dfrac{5}{y} = 1 \end{cases}$ (*Hint:* Let $x' = 1/x$ and $y' = 1/y$.)

20 $\begin{cases} \dfrac{3}{x - 1} + \dfrac{4}{y + 2} = 2 \\ \dfrac{6}{x - 1} - \dfrac{7}{y + 2} = -3 \end{cases}$

21 Determine a and b so that the line $ax + by = 10$ passes through the points $P_1(-3, 4)$ and $P_2(3, 1)$.

22 Generalize Exercise 21 to the case where the graph of $ax + by = 10$ passes through the points $P_1(x_1, y_1)$ and $P_2(x_2, y_2)$. (Express a and b in terms of x_1, y_1, x_2, and y_2.) Is it necessary to restrict P_1 or P_2 in any way?

Solve Exercises 23–36 by employing systems of linear equations.

23 The price of admission for a certain event was \$2.25 for adults and \$1.50 for children. If 450 tickets were sold for a total of \$777.75, how many of each kind were purchased?

24 A chemist wishes to obtain 80 cc of a 30% acid solution by mixing a 50% solution and a 25% solution. How many cc of each solution should be used?

25 The sum of the digits of a certain two-digit number is 14. If the digits are reversed, the number is increased by 18. Find the number.

26 A man rows a boat 500 feet upstream against a constant current in 10 minutes. He then rows downstream (with the same current), covering 300 feet in 5 minutes. Find the speed of the current and the equivalent rate at which he can row in still water.

27 A collection of dimes and quarters amounts to \$7.00. If there are seven more dimes than quarters, find the number of quarters.

28 A woman has \$10,000 invested in two funds which pay simple interest rates of 8% and 7%, respectively. If she receives yearly interest of \$772, how much is invested in each fund?

29 A man receives income from two investments at simple interest rates of $6\frac{1}{2}\%$ and 8%, respectively. He has twice as much invested at $6\frac{1}{2}\%$ as at 8%. If his annual income from the two investments is \$698.25, find how much is invested at each rate.

30 Two hoses running at the same time can fill a swimming pool in 4 hours. If both hoses run for 2 hours and the first is then shut off, it takes 3 more hours for the second to fill the pool. How long does it take each hose to fill the pool by itself?

31 A silversmith has two alloys, the first containing 35% silver, and the second 60% silver. How much of each should be melted and combined to obtain 100 gm of an alloy containing 50% silver?

32 A merchant wishes to mix peanuts costing \$2.00 per pound with cashews costing \$3.50 per pound to obtain 60 pounds of a mixture costing \$2.65 per pound. How many pounds of each variety should be mixed?

33 An airplane, flying with a tail wind, travels 1200 miles in 2 hours. The return trip, against the wind, takes $2\frac{1}{2}$ hours. Find the speed of the plane and the speed of the wind (assume that both rates are constant).

34 A stationery company sells two types of notebooks to college bookstores, the first wholesaling for 50 ¢ and the second for 70 ¢. The company receives an order for 500 notebooks, together with a check for \$286. If the order fails to specify the number of each type, how should the company fill the order?

35 As a ball rolls down an inclined plane, its velocity $v(t)$ (in cm/sec) at time t (in sec) is given by $v(t) = v_0 + at$, where v_0 is the initial velocity and a is the acceleration (in cm/sec^2). If $v(2) = 16$ and $v(5) = 25$, find v_0 and a.

36 If an object is projected vertically upward from an altitude of s_0 feet with an initial velocity of v_0 ft/sec, then its distance $s(t)$ above the ground after t seconds is

$$s(t) = -16t^2 + v_0 t + s_0.$$

If $s(1) = 84$ and $s(2) = 116$, find v_0 and s_0.

SECTION 7.3

SYSTEMS OF LINEAR EQUATIONS IN MORE THAN TWO VARIABLES

For systems of linear equations containing more than two variables, we can use either the method of substitution explained in Section 7.1 or the method of elimination developed in Section 7.2. The method of elimination is the shorter and more straightforward technique for finding solutions. In addition, it leads to the matrix technique, which will be developed in this section.

EXAMPLE 1 Find the solutions of the system

$$\begin{cases} x - 2y + 3z = 4 \\ 2x + y - 4z = 3 \\ -3x + 4y - z = -2. \end{cases}$$

Solution We begin by eliminating x from the second and third equations. Adding to the second equation -2 times the first equation gives us the equivalent system

$$\begin{cases} x - 2y + 3z = 4 \\ 5y - 10z = -5 \\ -3x + 4y - z = -2. \end{cases}$$

Next, adding to the third equation 3 times the first equation, we obtain the equivalent system

$$\begin{cases} x - 2y + 3z = 4 \\ 5y - 10z = -5 \\ -2y + 8z = 10. \end{cases}$$

To simplify computations we multiply the second equation by $\frac{1}{5}$, obtaining

$$\begin{cases} x - 2y + 3z = 4 \\ y - 2z = -1 \\ -2y + 8z = 10. \end{cases}$$

We now eliminate y from the third equation by adding to it 2 times the second equation. This gives us the system

$$\begin{cases} x - 2y + 3z = 4 \\ y - 2z = -1 \\ 4z = 8. \end{cases}$$

The solutions of the last system are easy to obtain. From the third equation it is clear that the only possible z-value is 2. Substituting 2 for z in the second equation, $y - 2z = -1$, we get $y - 2(2) = -1$, or $y = 3$. Finally, the x-value is found by substituting for y and z in the first equation. This gives us

$$x - 2(3) + 3(2) = 4 \quad \text{and hence} \quad x = 4.$$

Thus, there is one solution, (4, 3, 2). ■

It can be shown that any system of three linear equations in three variables has either a *unique solution*, an *infinite number of solutions*, or *no solutions*. As with the case of two equations in two variables, the terminology used to describe these cases is *consistent*, *dependent*, or *inconsistent*, respectively.

If we analyze the solution of Example 1, we see that the symbols used for the variables really have little effect on the solution of the problem. The *coefficients* of the variables are the important things to consider. Thus, if different symbols such as r, s, and t are used for the variables, we obtain the system

$$\begin{cases} r - 2s + 3t = 4 \\ 2r + s - 4t = 3 \\ -3r + 4s - t = -2. \end{cases}$$

The method of elimination could then proceed exactly as before. Since this is true, it is possible to use a process that reduces the amount of labor involved. Specifically, we introduce a notation for keeping track of the coefficients in such a way that the variables do not have to be written down. Referring to the preceding system and checking that corresponding variables appear underneath one another, and terms not involving variables are to the right of the equal sign, we list the numbers that are involved in the equations in the following manner:

$$\begin{bmatrix} 1 & -2 & 3 & 4 \\ 2 & 1 & -4 & 3 \\ -3 & 4 & -1 & -2 \end{bmatrix}$$

If some variable had not appeared in one of the equations, we would have used a zero in the appropriate position. An array of numbers of this type is called a **matrix.** The **rows** of the matrix are the numbers that appear next to one another *horizontally*. Thus, the first row R_1 is 1 -2 3 4, the second row R_2 is 2 1 -4 3, and the third row R_3 is -3 4 -1 -2. The **columns** of the matrix are the numbers that form a *vertical* pattern. For example, the second column consists of the numbers -2, 1, 4 (in that order); the fourth column consists of 4, 3, -2; and so on.

Before discussing a method of solving a system of linear equation by using matrices, let us introduce a general definition of matrix. It will be convenient to use a **double subscript notation** for the symbols that represent the numbers in the matrix. Specifically, the symbol a_{ij} will denote the number that appears in row i and column j. We call i the **row subscript** and j the **column subscript** of a_{ij}.

DEFINITION

If m and n are positive integers, then an $\boldsymbol{m \times n}$ **matrix** is an array of the form

$$\begin{bmatrix} a_{11} & a_{12} & a_{13} & \cdots & a_{1n} \\ a_{21} & a_{22} & a_{23} & \cdots & a_{2n} \\ a_{31} & a_{32} & a_{33} & \cdots & a_{3n} \\ \vdots & \vdots & \vdots & & \vdots \\ a_{m1} & a_{m2} & a_{m3} & \cdots & a_{mn} \end{bmatrix}$$

where each a_{ij} is a real number.

The symbol $m \times n$ in the definition is read "m by n." It is also possible to consider matrices in which the symbols a_{ij} are complex numbers, polynomials, or other mathematical objects; however, we shall not do so in this text. The rows and columns of a matrix are defined as in the preceding discussion. Thus, the matrix in the definition has m rows and n columns. The reader should carefully examine the notation for a matrix, observing, for example, that the symbol a_{23} appears in row 2 and column 3, whereas a_{32} appears in row 3 and column 2. Each a_{ij} is called an **element of the matrix.** The elements $a_{11}, a_{22}, a_{33}, \ldots$ are called the **main diagonal elements.** If $m \neq n$, the matrix is sometimes referred to as a **rectangular matrix.** If $m = n$, we refer to the matrix as a **square matrix of order *n*.**

Let us return to the system of equations considered in Example 1. The 3×4 matrix on page 329 is called the **matrix of the system.** In certain cases we may wish to consider only the *coefficients* of the variables of a system of linear equations. The corresponding matrix is called the **coefficient matrix.** The coefficient matrix for our system is

$$\begin{bmatrix} 1 & -2 & 3 \\ 2 & 1 & -4 \\ -3 & 4 & -1 \end{bmatrix}.$$

The matrix of the *system* is sometimes called the **augmented matrix.** After forming the matrix of the system, we work with the rows of the matrix *just as though they were equations.* The only items missing are the symbols for the variables, the addition signs used between terms, and the equal signs. We simply keep in mind that the numbers in the first column are the coefficients of the first variable, the numbers in the second column are the coefficients of the second variable, and so on. Our rules for transforming a matrix are formulated in such a way that they always produce a matrix of an equivalent system of equations. As a matter of fact, the following theorem for matrices is a perfect analogue of the theorem on transformations of systems, stated on page 322.

MATRIX ROW TRANSFORMATION THEOREM

Given a matrix of a system of linear equations, each of the following transformations results in a matrix of an equivalent system of linear equations:

(i) Interchanging any two rows.

(ii) Multiplying all of the elements in a row by the same nonzero real number k.

(iii) Adding to the elements in a row k times the corresponding elements of any other row, where k is any real number.

We shall refer to (ii) as the process of "multiplying a row by the number k" and (iii) will be described by saying "add k times any other row to a row." Rules (i)–(iii) of the preceding theorem are called the **elementary row transformations** of a matrix.

It is convenient to use the following symbols to specify elementary row transformations.

Symbol	*Meaning*
R_{ij}	Interchange rows i and j
kR_i	Multiply the ith row by k
$kR_i + R_j$	Add to the jth row, k times the ith row

We shall next rework Example 1 using matrices. It may be enlightening to compare the two solutions, since analagous steps are employed in each case.

EXAMPLE 2 Find the solutions of the system

$$\begin{cases} x - 2y + 3z = 4 \\ 2x + y - 4z = 3 \\ -3x + 4y - z = -2. \end{cases}$$

Solution We shall begin by forming the matrix of the system and then changing its form by means of elementary row transformations. Arrows, together with the symbols we have defined, will specify each transformation.

Thus, in the first transformation used below, $-2R_1 + R_2$ indicates that we have added to the second row, -2 times the first row. In the second transformation, $3R_1 + R_3$ indicates that we have added to the third row, 3 times the first row. In the third transformation, $\frac{1}{5}R_2$ denotes the fact that the second row has been multiplied by $\frac{1}{5}$, etc.

$$\begin{bmatrix} 1 & -2 & 3 & 4 \\ 2 & 1 & -4 & 3 \\ -3 & 4 & -1 & -2 \end{bmatrix} \xrightarrow{-2R_1 + R_2} \begin{bmatrix} 1 & -2 & 3 & 4 \\ 0 & 5 & -10 & -5 \\ -3 & 4 & -1 & -2 \end{bmatrix}$$

$$\xrightarrow{3R_1 + R_3} \begin{bmatrix} 1 & -2 & 3 & 4 \\ 0 & 5 & -10 & -5 \\ 0 & -2 & 8 & 10 \end{bmatrix}$$

$$\xrightarrow{\frac{1}{5}R_2} \begin{bmatrix} 1 & -2 & 3 & 4 \\ 0 & 1 & -2 & -1 \\ 0 & -2 & 8 & 10 \end{bmatrix}$$

$$\xrightarrow{2R_2 + R_3} \begin{bmatrix} 1 & -2 & 3 & 4 \\ 0 & 1 & -2 & -1 \\ 0 & 0 & 4 & 8 \end{bmatrix}$$

We use the final matrix to return to the system of equations

$$\begin{cases} x - 2y + 3z = 4 \\ \quad\; y - 2z = -1 \\ \qquad\quad 4z = 8 \end{cases}$$

which is equivalent to the given system. The solutions of the last system may be found as in Example 1. ■

The last matrix in the solution of Example 2 has zeros everywhere below the main diagonal elements and is said to be in **echelon form.**

The method used in Example 2 is completely general. Beginning with *any* system of linear equations, we can use elementary row transformations of type (iii) many times, also interchanging rows if necessary, to introduce zeros everywhere in the first column after the first row. If the transformed matrix has a nonzero element in the second column and after the first row, then again using elementary row transformations we may transform the matrix into a form where only zeros appear below the element in the second row and second column. This process is continued until we obtain a matrix in echelon form; that is, all elements below the main diagonal are zero. The echelon matrix is then used to obtain a system of equations equivalent to the original system. The solutions are then found as in Example 1. Let us illustrate the technique by solving a system of four linear equations.

EXAMPLE 3 Find the solutions of the system

$$\begin{cases} x \qquad\quad - 2z + 2w = \ \ 1 \\ -2x + 3y + 4z \qquad\quad = -1 \\ \qquad\quad y + \ \ z - \ \ w = \ \ 0 \\ 3x + \ \ y - 2z - \ \ w = \ \ 3. \end{cases}$$

Solution Notice that we have arranged the equations so that the same variables appear in vertical columns. We begin with the matrix of the system and proceed as follows:

$$\begin{bmatrix} 1 & 0 & -2 & 2 & 1 \\ -2 & 3 & 4 & 0 & -1 \\ 0 & 1 & 1 & -1 & 0 \\ 3 & 1 & -2 & -1 & 3 \end{bmatrix} \xrightarrow{2R_1 + R_2} \begin{bmatrix} 1 & 0 & -2 & 2 & 1 \\ 0 & 3 & 0 & 4 & 1 \\ 0 & 1 & 1 & -1 & 0 \\ 3 & 1 & -2 & -1 & 3 \end{bmatrix}$$

$$\xrightarrow{-3R_1 + R_4} \begin{bmatrix} 1 & 0 & -2 & 2 & 1 \\ 0 & 3 & 0 & 4 & 1 \\ 0 & 1 & 1 & -1 & 0 \\ 0 & 1 & 4 & -7 & 0 \end{bmatrix}$$

$$\xrightarrow{R_{23}} \begin{bmatrix} 1 & 0 & -2 & 2 & 1 \\ 0 & 1 & 1 & -1 & 0 \\ 0 & 3 & 0 & 4 & 1 \\ 0 & 1 & 4 & -7 & 0 \end{bmatrix}$$

$$\xrightarrow{-3R_2 + R_3} \begin{bmatrix} 1 & 0 & -2 & 2 & 1 \\ 0 & 1 & 1 & -1 & 0 \\ 0 & 0 & -3 & 7 & 1 \\ 0 & 1 & 4 & -7 & 0 \end{bmatrix}$$

$$\xrightarrow{(-1)R_2 + R_4} \begin{bmatrix} 1 & 0 & -2 & 2 & 1 \\ 0 & 1 & 1 & -1 & 0 \\ 0 & 0 & -3 & 7 & 1 \\ 0 & 0 & 3 & -6 & 0 \end{bmatrix}$$

$$\xrightarrow{R_3 + R_4} \begin{bmatrix} 1 & 0 & -2 & 2 & 1 \\ 0 & 1 & 1 & -1 & 0 \\ 0 & 0 & -3 & 7 & 1 \\ 0 & 0 & 0 & 1 & 1 \end{bmatrix}$$

The final matrix is in echelon form and corresponds to the system of equations

$$\begin{cases} x \qquad - 2z + 2w = 1 \\ \quad y + \ \ z - \ \ w = 0 \\ \qquad\ \ - 3z + 7w = 1 \\ \qquad\qquad\qquad w = 1. \end{cases}$$

From the last equation we see that $w = 1$. Substituting in the third equation, we get $-3z + 7(1) = 1$, or $z = 2$. Using the second equation, $y + z - w = 0$, we obtain $y + 2 - 1 = 0$, or $y = -1$. Finally, from the first equation, we have $x - 2(2) + 2(1) = 1$, or $x = 3$. Hence, the system has only one solution, $x = 3$, $y = -1$, $z = 2$, and $w = 1$. ■

Sometimes it is necessary to consider systems in which the number of equations is not the same as the number of variables. The same matrix techniques are applicable, as illustrated in the next example.

EXAMPLE 4 Find the solutions of the system

$$\begin{cases} 2x + 3y + 4z = 1 \\ 3x + 4y + 5z = 3. \end{cases}$$

Solution In this example we shall go beyond the echelon form, and continue applying elementary row transformations until the number 1 is in each position on the main diagonal and there are zeros both below *and* above the main diagonal. Thus

$$\begin{bmatrix} 2 & 3 & 4 & 1 \\ 3 & 4 & 5 & 3 \end{bmatrix} \xrightarrow{-\frac{3}{2}R_1 + R_2} \begin{bmatrix} 2 & 3 & 4 & 1 \\ 0 & -\frac{1}{2} & -1 & \frac{3}{2} \end{bmatrix}$$

$$\xrightarrow{-2R_2} \begin{bmatrix} 2 & 3 & 4 & 1 \\ 0 & 1 & 2 & -3 \end{bmatrix}$$

$$\xrightarrow{-3R_2 + R_1} \begin{bmatrix} 2 & 0 & -2 & 10 \\ 0 & 1 & 2 & -3 \end{bmatrix}$$

$$\xrightarrow{\frac{1}{2}R_1} \begin{bmatrix} 1 & 0 & -1 & 5 \\ 0 & 1 & 2 & -3 \end{bmatrix}$$

The last matrix is the matrix of the system

$$\begin{cases} x \quad - z = 5 \\ \quad y + 2z = -3 \end{cases}$$

or equivalently,

$$\begin{cases} x = z + 5 \\ y = -2z - 3. \end{cases}$$

There are an infinite number of solutions to this system; they can be found by assigning z any value c and then using the last two equations to express x and y in terms of c. This gives us

$$x = c + 5, \qquad y = -2c - 3, \qquad z = c.$$

Thus, the solutions of the system consist of all ordered triples of the form

$$(c + 5, -2c - 3, c)$$

where c is any real number. These solutions may be checked by substituting $c + 5$ for x, $-2c - 3$ for y, and c for z in the two given equations. We can obtain any number of solutions for the system by substituting specific real numbers for c. For example, if $c = 0$, we obtain $(5, -3, 0)$; if $c = 2$, we have $(7, -7, 2)$; if $c = -\frac{1}{2}$, we get $(\frac{9}{2}, -2, -\frac{1}{2})$; etc. ■

A system of linear equations is said to be **homogeneous** if all the terms that do not contain variables are 0. A system of homogeneous equations always has the **trivial solution** obtained by substituting zero for all the variables. Nontrivial solutions sometimes exist. The procedure for finding solutions is the same as that used for nonhomogeneous systems.

EXAMPLE 5 Find the solutions of the system

$$\begin{cases} x - y + 4z = 0 \\ 2x + y - z = 0 \\ -x - y + 2z = 0. \end{cases}$$

Solution As in Example 3, we write

$$\begin{bmatrix} 1 & -1 & 4 & 0 \\ 2 & 1 & -1 & 0 \\ -1 & -1 & 2 & 0 \end{bmatrix} \xrightarrow{-2R_1 + R_2} \begin{bmatrix} 1 & -1 & 4 & 0 \\ 0 & 3 & -9 & 0 \\ -1 & -1 & 2 & 0 \end{bmatrix}$$

$$\xrightarrow{R_1 + R_3} \begin{bmatrix} 1 & -1 & 4 & 0 \\ 0 & 3 & -9 & 0 \\ 0 & -2 & 6 & 0 \end{bmatrix}$$

$$\xrightarrow{\frac{1}{3}R_2} \begin{bmatrix} 1 & -1 & 4 & 0 \\ 0 & 1 & -3 & 0 \\ 0 & -2 & 6 & 0 \end{bmatrix}$$

$$\xrightarrow{-\frac{1}{2}R_3} \begin{bmatrix} 1 & -1 & 4 & 0 \\ 0 & 1 & -3 & 0 \\ 0 & 1 & -3 & 0 \end{bmatrix}$$

$$\xrightarrow{-R_2 + R_3} \begin{bmatrix} 1 & -1 & 4 & 0 \\ 0 & 1 & -3 & 0 \\ 0 & 0 & 0 & 0 \end{bmatrix}$$

$$\xrightarrow{R_2 + R_1} \begin{bmatrix} 1 & 0 & 1 & 0 \\ 0 & 1 & -3 & 0 \\ 0 & 0 & 0 & 0 \end{bmatrix}$$

The last matrix corresponds to the system

$$\begin{cases} x \quad + \ z = 0 \\ \quad y - 3z = 0 \end{cases} \quad \text{or} \quad \begin{cases} x = -z \\ y = \ \ 3z \end{cases}$$

Assigning any value c to z, we obtain $x = -c$ and $y = 3c$. Thus, the solutions consist of all ordered triples of the form $(-c, 3c, c)$ where c is any real number. ■

EXAMPLE 6 Find the solutions of the system

$$\begin{cases} x + y + z = 0 \\ x - y + z = 0 \\ x - y - z = 0. \end{cases}$$

Solution Let us proceed as follows:

$$\begin{bmatrix} 1 & 1 & 1 & 0 \\ 1 & -1 & 1 & 0 \\ 1 & -1 & -1 & 0 \end{bmatrix} \xrightarrow{(-1)R_1 + R_2} \begin{bmatrix} 1 & 1 & 1 & 0 \\ 0 & -2 & 0 & 0 \\ 1 & -1 & -1 & 0 \end{bmatrix} \xrightarrow{(-1)R_1 + R_3} \begin{bmatrix} 1 & 1 & 1 & 0 \\ 0 & -2 & 0 & 0 \\ 0 & -2 & -2 & 0 \end{bmatrix}$$

$$\xrightarrow{-\frac{1}{2}R_2} \begin{bmatrix} 1 & 1 & 1 & 0 \\ 0 & 1 & 0 & 0 \\ 0 & -2 & -2 & 0 \end{bmatrix} \xrightarrow{-\frac{1}{2}R_3} \begin{bmatrix} 1 & 1 & 1 & 0 \\ 0 & 1 & 0 & 0 \\ 0 & 1 & 1 & 0 \end{bmatrix}$$

$$\xrightarrow{(-1)R_2 + R_1} \begin{bmatrix} 1 & 0 & 1 & 0 \\ 0 & 1 & 0 & 0 \\ 0 & 1 & 1 & 0 \end{bmatrix} \qquad \begin{bmatrix} 1 & 0 & 1 & 0 \\ 0 & 1 & 0 & 0 \\ 0 & 0 & 1 & 0 \end{bmatrix}$$

$$\xrightarrow{(-1)R_3 + R_1} \begin{bmatrix} 1 & 0 & 0 & 0 \\ 0 & 1 & 0 & 0 \\ 0 & 0 & 1 & 0 \end{bmatrix}$$

The last matrix corresponds to

$$x = 0, \qquad y = 0, \qquad z = 0.$$

Hence, the only solution for the given system is the trivial one, (0, 0, 0). ■

The next example is an illustration of an applied problem that can be solved by means of a system of three linear equations.

EXAMPLE 7 A merchant wishes to mix two grades of peanuts costing \$1.50 and \$2.50 per pound, respectively, with cashews costing \$4.00 per pound, to obtain 130 pounds of a mixture costing \$3.00 per pound. If the merchant also wants the amount of cheaper grade peanuts to be twice that of the better grade, how many pounds of each variety should be mixed?

Solution Let us introduce three variables as follows:

$x =$ pounds of peanuts at \$1.50 per pound

$y =$ pounds of peanuts at \$2.50 per pound

$z =$ pounds of cashews at \$4.00 per pound.

We refer to the statement of the problem and write the following system of equations

$$\begin{cases} x + y + z = 130 \\ 1.50x + 2.50y + 4.00z = 3.00(130) \\ x = 2y. \end{cases}$$

We leave it to the reader to verify that the solution to this system is $x = 40$, $y = 20$, $z = 70$. Thus, the merchant should use 40 pounds of the \$1.50 peanuts, 20 pounds of the \$2.50 peanuts, and 70 pounds of cashews. The check is left to the reader. ■

EXERCISES 7.3

Use matrices to solve the systems in Exercises 1–26.

1 $\begin{cases} x - 2y - 3z = -1 \\ 2x + y + z = 6 \\ x + 3y - 2z = 13 \end{cases}$

2 $\begin{cases} x + 3y - z = -3 \\ 3x - y + 2z = 1 \\ 2x - y + z = -1 \end{cases}$

3 $\begin{cases} 5x + 2y - z = -7 \\ x - 2y + 2z = 0 \\ 3y + z = 17 \end{cases}$

4 $\begin{cases} 4x - y + 3z = 6 \\ -8x + 3y - 5z = -6 \\ 5x - 4y = -9 \end{cases}$

5 $\begin{cases} 2x + 6y - 4z = 1 \\ x + 3y - 2z = 4 \\ 2x + y - 3z = -7 \end{cases}$

6 $\begin{cases} 2x - y = 5 \\ 5y + 3z = -2 \\ x - 7z = 3 \end{cases}$

7 $\begin{cases} 2x - 3y + 2z = -3 \\ -3x + 2y + z = 1 \\ 4x + y - 3z = 4 \end{cases}$

8 $\begin{cases} 2x - 3y + z = 2 \\ 3x + 2y - z = -5 \\ 5x - 2y + z = 0 \end{cases}$

9 $\begin{cases} x + 3y + z = 0 \\ x + y - z = 0 \\ x - 2y - 4z = 0 \end{cases}$

10 $\begin{cases} 2x - y + z = 0 \\ x - y - 2z = 0 \\ 2x - 3y + z = 0 \end{cases}$

11 $\begin{cases} 2x + y + z = 0 \\ x - 2y - 2z = 0 \\ x + y + z = 0 \end{cases}$

12 $\begin{cases} x + y - 2z = 0 \\ x - y - 4z = 0 \\ y + z = 0 \end{cases}$

13 $\begin{cases} 3x - 2y + 5z = 7 \\ x + 4y - z = -2 \end{cases}$

14 $\begin{cases} 2x - y + 4z = 8 \\ -3x + y - 2z = 5 \end{cases}$

15 $\begin{cases} 4x - 2y + z = 5 \\ 3x + y - 4z = 0 \end{cases}$

16 $\begin{cases} 5x + 2y - z = 10 \\ y + z = -3 \end{cases}$

17 $\begin{cases} x + 2y - z - 3w = 2 \\ 3x + y - 2z - w = 6 \\ x + y + 3z - 2w = -3 \\ 4x - 3y - z - 2w = -8 \end{cases}$

18 $\begin{cases} x - 2y - 5z + w = -1 \\ 2x - y + z + w = 1 \\ 3x - 2y - 4z - 2w = 1 \\ x + y + 3z - 2w = 2 \end{cases}$

19 $\begin{cases} 2x - y - 2z + 2s - 5t = 2 \\ x + 3y - 2z + s - 2t = -5 \\ -x + 4y + 2z - 3s + 8t = -4 \\ 3x - 2y - 4z + s - 3t = -3 \\ 4x - 6y + z - 2s + t = 10 \end{cases}$

20 $$\begin{cases} 3x + 2y + z + 3u + v + w = 1 \\ 2x + y - 2z + 3u - v + 4w = 6 \\ 6x + 3y + 4z - u + 2v + w = -6 \\ x + y + z + u - v - w = 8 \\ -2x - 2y + z - 3u + 2v - 3w = -10 \\ x - 3y + 2z + u + 3v + w = -1 \end{cases}$$

21 $$\begin{cases} 5x + 2z = 1 \\ y - 3z = 2 \\ 2x + y = 3 \end{cases}$$

22 $$\begin{cases} 2x - 5y = 4 \\ 3y + 2z = -3 \\ 7x - 3z = 1 \end{cases}$$

23 $$\begin{cases} 4x - 3y = 1 \\ 2x + y = -7 \\ -x + y = -1 \end{cases}$$

24 $$\begin{cases} 2x + 3y = -2 \\ x + y = 1 \\ x - 2y = 13 \end{cases}$$

25 $$\begin{cases} 2x + 3y = 5 \\ x - 3y = 4 \\ x + y = -2 \end{cases}$$

26 $$\begin{cases} 4x - y = 2 \\ 2x + 2y = 1 \\ 4x - 5y = 3 \end{cases}$$

27 A collection of nickels, dimes, and quarters amounts to \$8.00. If there are 72 coins in all and there are twice as many nickels as there are quarters, find the number of dimes.

28 A three-digit number is equal to 31 times the sum of its digits. If the three digits are reversed, the resulting number exceeds the given number by 99. The tens digit is three less than the sum of the hundreds and units digits. Find the number.

29 A chemist has three solutions containing a certain acid. The first contains 10% acid, the second 30%, and the third 50%. He wishes to use all three solutions to obtain a mixture of 50 liters containing 32% acid, using twice as much of the 50% solution as the 30% solution. How many liters of each solution should be used?

30 A swimming pool can be filled by three pipes *A*, *B*, and *C*. Pipe *A* alone can fill the pool in 8 hours. If pipes *A* and *C* are used together, the pool can be filled in 6 hours. If *B* and *C* are used together, it takes 10 hours. How long does it take to fill the pool if all three pipes are used?

31 A company has three machines *A*, *B*, and *C* that are each capable of producing a certain item. However, because of a lack of skilled operators, only two of the machines can be used simultaneously. The following table indicates production over a three-day period using various combinations of the machines.

Machines used	Hours used	Items produced
A and *B*	6	4500
A and *C*	8	3600
B and *C*	7	4900

How long would it take each machine, if used alone, to produce 1000 items?

32 In electrical circuits, the formula $1/R = (1/R_1) + (1/R_2)$ is used to find the total resistance R if two resistances R_1 and R_2 are connected in parallel. Given three resistances *A*, *B*, and *C*, suppose that the total resistance is 48 ohms if *A* and *B* are connected in parallel; 80 ohms if *B* and *C* are connected in parallel, and 60 ohms if *A* and *C* are connected in parallel. Find *A*, *B*, and *C*.

33 A supplier of lawn products has three types G_1, G_2, and G_3 of grass fertilizer having nitrogen contents of 30%, 20%, and 15%, respectively. The supplier plans to mix them, obtaining 600 pounds of fertilizer with a 25% nitrogen content. In addition, the mixture is to contain 100 pounds more of type G_3 than of type G_2. How much of each type should be used?

34 If a particle moves along a coordinate line with a constant acceleration a (cm/sec^2), then at time t (sec) its distance $s(t)$ (cm) from the origin is

$$s(t) = \frac{1}{2}at^2 + v_0t + s_0$$

where v_0 and s_0 are the velocity and distance from the origin, respectively, at $t = 0$. If the distances of the particle from the origin at $t = \frac{1}{2}$, $t = 1$, and $t = \frac{3}{2}$ are 7, 11, and 17, respectively, find a, v_0, and s_0.

35 Find a quadratic function f such that $f(1) = 3$, $f(2) = \frac{5}{2}$, and $f(-1) = 1$.

36 If $f(x) = ax^3 + bx + c$, determine a, b, and c such that that the graph of f passes through the points $(-3, -12)$, $(-1, 22)$, and $(2, 13)$.

37 Find an equation of the circle that passes through the three points $P_1(2, 1)$, $P_2(-1, -4)$, and $P_3(3, 0)$. (*Hint:* An equation of the circle has the form $x^2 + y^2 + ax + by + c = 0$.)

38 Determine a, b, and c such that the graph of the equation $y = ax^2 + bx + c$ passes through the points $P_1(3, -1)$, $P_2(1, -7)$, and $P_3(-2, 14)$.

SECTION 7.4

THE ALGEBRA OF MATRICES

It is possible to develop a comprehensive theory for matrices that has many mathematical and scientific applications. In this section we shall discuss several basic algebraic properties of matrices that serve as the starting point for such a theory.

To conserve space it is sometimes convenient to denote an $m \times n$ matrix A of the type displayed in the definition given in Section 7.3 by the symbol (a_{ij}). If (b_{ij}) denotes another $m \times n$ matrix B, then we say that A and B are **equal,** written

$$A = B \quad \text{if and only if} \quad a_{ij} = b_{ij}$$

for every i and j. For example,

$$\begin{bmatrix} 1 & 0 & 5 \\ \sqrt[3]{8} & 3^2 & -2 \end{bmatrix} = \begin{bmatrix} (-1)^2 & 0 & \sqrt{25} \\ 2 & 9 & -2 \end{bmatrix}.$$

If $A = (a_{ij})$ and $B = (b_{ij})$ are $m \times n$ matrices, then their **sum** $\boldsymbol{A + B}$ is defined as the $m \times n$ matrix $C = (c_{ij})$ where $c_{ij} = a_{ij} + b_{ij}$ for all i and j. Thus, to add two matrices we add the elements that appear in corresponding positions in each matrix. Note that two matrices can be added only if they have the same number of rows and the same number of columns. Using the parentheses notation, the definition of addition may be expressed as follows:

$$(a_{ij}) + (b_{ij}) = (a_{ij} + b_{ij}).$$

Although we have used the symbol $+$ in two different ways, there is little chance for confusion, since whenever $+$ appears between symbols for matrices it refers to matrix addition, and when $+$ is used between real numbers it denotes their sum. An example of the sum of two 3×2 matrices is

$$\begin{bmatrix} 4 & -5 \\ 0 & 4 \\ -6 & 1 \end{bmatrix} + \begin{bmatrix} 3 & 2 \\ 7 & -4 \\ -2 & 1 \end{bmatrix} = \begin{bmatrix} 7 & -3 \\ 7 & 0 \\ -8 & 2 \end{bmatrix}.$$

It is not difficult to prove that addition of matrices is both commutative and associative, that is,

$$A + B = B + A, \qquad A + (B + C) = (A + B) + C$$

for $m \times n$ matrices A, B, and C.

The $\boldsymbol{m \times n}$ **zero matrix,** denoted by O, is the matrix with m rows and n columns in which every element is 0. It is an identity element relative to addition, since

$$A + O = A$$

for every $m \times n$ matrix A. For example,

$$\begin{bmatrix} a_{11} & a_{12} \\ a_{21} & a_{22} \\ a_{31} & a_{32} \end{bmatrix} + \begin{bmatrix} 0 & 0 \\ 0 & 0 \\ 0 & 0 \end{bmatrix} = \begin{bmatrix} a_{11} & a_{12} \\ a_{21} & a_{22} \\ a_{31} & a_{32} \end{bmatrix}.$$

The **additive inverse** $-A$ of the matrix $A = (a_{ij})$ is, by definition, the matrix $(-a_{ij})$ obtained by changing the sign of each element of A. For example,

$$-\begin{bmatrix} 2 & -3 & 4 \\ -1 & 0 & 5 \end{bmatrix} = \begin{bmatrix} -2 & 3 & -4 \\ 1 & 0 & -5 \end{bmatrix}.$$

It follows that for every $m \times n$ matrix A,

$$A + (-A) = O.$$

Subtraction of two $m \times n$ matrices is defined by

$$A - B = A + (-B).$$

Using the parentheses notation for matrices, this implies that

$$(a_{ij}) - (b_{ij}) = (a_{ij}) + (-b_{ij}) = (a_{ij} - b_{ij}).$$

Thus, to subtract two matrices, we merely subtract the elements that are in corresponding positions.

The **product** of a real number c and an $m \times n$ matrix $A = (a_{ij})$ is defined by

$$cA = (ca_{ij}).$$

Thus, to find cA, we multiply each element of A by c. For example,

$$3\begin{bmatrix} 4 & -1 \\ 2 & 3 \end{bmatrix} = \begin{bmatrix} 12 & -3 \\ 6 & 9 \end{bmatrix}.$$

The following results may be established, where A and B are $m \times n$ matrices and c and d are any real numbers.

$$c(A + B) = cA + cB$$

$$(c + d)A = cA + dA$$

$$(cd)A = c(dA)$$

The following definition of the product of two matrices may appear unusual to the beginning student; however, there are many applications that justify the form of the definition. To define the product AB of two matrices A and B, *the number of columns of A must be the same as the number of rows* *of* B. Suppose that $A = (a_{ij})$ is $m \times n$ and $B = (b_{ij})$ is $n \times p$. To determine the element c_{ij} of the product, we single out row i of A and column j of B as follows:

$$\begin{bmatrix} a_{11} & a_{12} & \cdots & a_{1n} \\ \vdots & \vdots & & \vdots \\ a_{i1} & a_{i2} & \cdots & a_{in} \\ \vdots & \vdots & & \vdots \\ a_{m1} & a_{m2} & \cdots & a_{mn} \end{bmatrix} \begin{bmatrix} b_{11} & \cdots & b_{1j} & \cdots & b_{1p} \\ b_{21} & \cdots & b_{2j} & \cdots & b_{2p} \\ \vdots & & \vdots & & \vdots \\ b_{n1} & \cdots & b_{nj} & \cdots & b_{np} \end{bmatrix}$$

Next we multiply pairs of elements and then add them, using the formula

$$c_{ij} = a_{i1}b_{1j} + a_{i2}b_{2j} + \cdots + a_{in}b_{nj}.$$

For example, the element c_{11} in the first row and the first column of AB is

$$c_{11} = a_{11}b_{11} + a_{12}b_{21} + \cdots + a_{1n}b_{n1}.$$

The element c_{12} in the first row and second column of AB is

$$c_{12} = a_{11}b_{12} + a_{12}b_{22} + \cdots + a_{1n}b_{n2}.$$

By definition, the product AB has the same number of rows as A and the same number of columns as B. In particular, if A is $m \times n$ and B is $n \times p$, then AB is $m \times p$. This is illustrated by the following product of a 2×3 matrix and a 3×4 matrix.

$$\begin{bmatrix} 1 & 2 & -3 \\ 4 & 0 & -2 \end{bmatrix} \begin{bmatrix} 5 & -4 & 2 & 0 \\ -1 & 6 & 3 & 1 \\ 7 & 0 & 4 & 8 \end{bmatrix} = \begin{bmatrix} -18 & 8 & -4 & -22 \\ 6 & -16 & 0 & -16 \end{bmatrix}.$$

Here are some typical computations of the elements c_{ij} in the product:

$$c_{11} = (1)(5) + (2)(-1) + (-3)(7) = 5 - 2 - 21 = -18$$
$$c_{13} = (1)(2) + (2)(3) + (-3)(4) = 2 + 6 - 12 = -4$$
$$c_{23} = (4)(2) + (0)(3) + (-2)(4) = 8 + 0 - 8 = 0$$
$$c_{24} = (4)(0) + (0)(1) + (-2)(8) = 0 + 0 - 16 = -16.$$

The reader should check the remaining elements.

The product operation for matrices is not commutative. Indeed, if A is 2×3 and B is 3×4, then AB may be found, but BA is undefined, since the number of columns of B is different from the number of rows of A. Even if AB and BA are both defined, it is often true that these products are different. This is illustrated in the next example, along with the fact that the product of two nonzero matrices may equal a zero matrix.

EXAMPLE 1 If $A = \begin{bmatrix} 2 & 2 \\ -1 & -1 \end{bmatrix}$ and $B = \begin{bmatrix} 1 & 2 \\ 1 & 2 \end{bmatrix}$, show that $AB \neq BA$.

Solution Using the definition of product we obtain

$$AB = \begin{bmatrix} 4 & 8 \\ -2 & -4 \end{bmatrix} \quad \text{and} \quad BA = \begin{bmatrix} 0 & 0 \\ 0 & 0 \end{bmatrix}.$$

Hence, $AB \neq BA$. Note that BA is a zero matrix. ■

It is possible to show that matrix multiplication is associative in the sense that

$$A(BC) = (AB)C$$

provided that the indicated products are defined, which will be the case if A is $m \times n$, B is $n \times p$, and C is $p \times q$. The Distributive Properties also hold if the matrices involved have the proper number of rows and columns. Thus, if A_1 and A_2 are $m \times n$ matrices, and if B_1 and B_2 are $n \times p$ matrices, then

$$A_1(B_1 + B_2) = A_1B_1 + A_1B_2$$

and

$$(A_1 + A_2)B_1 = A_1B_1 + A_2B_1.$$

As a special case, if all matrices are square, of order n, then the Associative and Distributive Properties are true. We shall not give proofs for these theorems.

Throughout the remainder of this section we shall concentrate on square matrices. The symbol I_n will denote the square matrix of order n that has 1 in each position on the main diagonal and 0 elsewhere. For example,

$$I_2 = \begin{bmatrix} 1 & 0 \\ 0 & 1 \end{bmatrix}, \qquad I_3 = \begin{bmatrix} 1 & 0 & 0 \\ 0 & 1 & 0 \\ 0 & 0 & 1 \end{bmatrix},$$

etc. It can shown that if A is any square matrix of order n, then

$$AI_n = A = I_nA.$$

For that reason I_n is called an **identity matrix of order *n*.** To illustrate, if $A = (a_{ij})$ is of order 2, then a direct calculation shows that

$$\begin{bmatrix} a_{11} & a_{12} \\ a_{21} & a_{22} \end{bmatrix}\begin{bmatrix} 1 & 0 \\ 0 & 1 \end{bmatrix} = \begin{bmatrix} a_{11} & a_{12} \\ a_{21} & a_{22} \end{bmatrix} = \begin{bmatrix} 1 & 0 \\ 0 & 1 \end{bmatrix}\begin{bmatrix} a_{11} & a_{12} \\ a_{21} & a_{22} \end{bmatrix}.$$

Some, but not all, $n \times n$ matrices A have an **inverse** in the sense that there is a matrix B such that $AB = I_n = BA$. If A has an inverse we denote it by A^{-1} and write

$$AA^{-1} = I_n = A^{-1}A.$$

The symbol A^{-1} is read "A inverse." In matrix theory it is *not* acceptable to use the symbol $1/A$ in place of A^{-1}.

Let us describe a technique for finding the inverse of a square matrix A, whenever it exists. We shall not attempt to justify the following procedure, since that would require concepts beyond the scope of this text. Given the $n \times n$ matrix $A = (a_{ij})$, we begin by forming the $n \times (2n)$ matrix:

$$\begin{bmatrix} a_{11} & a_{12} & \cdots & a_{1n} & 1 & 0 & 0 & \cdots & 0 \\ a_{21} & a_{22} & \cdots & a_{2n} & 0 & 1 & 0 & \cdots & 0 \\ \cdot & \cdot & \cdots & \cdot & \cdot & \cdot & \cdot & \cdots & \cdot \\ \cdot & \cdot & \cdots & \cdot & \cdot & \cdot & \cdot & \cdots & \cdot \\ \cdot & \cdot & \cdots & \cdot & \cdot & \cdot & \cdot & \cdots & \cdot \\ a_{n1} & a_{n2} & \cdots & a_{nn} & 0 & 0 & 0 & \cdots & 1 \end{bmatrix}$$

where the $n \times n$ identity matrix I_n appears "to the right" of the matrix A, as indicated. We next apply a succession of elementary row transformations until we arrive at a matrix of the form

$$\begin{bmatrix} 1 & 0 & 0 & \cdots & 0 & b_{11} & b_{12} & \cdots & b_{1n} \\ 0 & 1 & 0 & \cdots & 0 & b_{21} & b_{22} & \cdots & b_{2n} \\ \cdot & \cdot & \cdot & \cdots & \cdot & \cdot & \cdot & \cdots & \cdot \\ \cdot & \cdot & \cdot & \cdots & \cdot & \cdot & \cdot & \cdots & \cdot \\ \cdot & \cdot & \cdot & \cdots & \cdot & \cdot & \cdot & \cdots & \cdot \\ 0 & 0 & 0 & \cdots & 1 & b_{n1} & b_{n2} & \cdots & b_{nn} \end{bmatrix}$$

where the identity matrix I_n appears "to the left" of the $n \times n$ matrix (b_{ij}). It can be shown that (b_{ij}) is the desired inverse A^{-1}.

EXAMPLE 2 Find A^{-1} if $A = \begin{bmatrix} 3 & 5 \\ 1 & 4 \end{bmatrix}$.

Solution We begin with the matrix $\begin{bmatrix} 3 & 5 & 1 & 0 \\ 1 & 4 & 0 & 1 \end{bmatrix}$ and then perform elementary row transformations until the identity matrix I_2 appears on the left, as follows:

$$\begin{bmatrix} 3 & 5 & 1 & 0 \\ 1 & 4 & 0 & 1 \end{bmatrix} \xrightarrow{R_{12}} \begin{bmatrix} 1 & 4 & 0 & 1 \\ 3 & 5 & 1 & 0 \end{bmatrix}$$

$$\xrightarrow{-3R_1 + R_2} \begin{bmatrix} 1 & 4 & 0 & 1 \\ 0 & -7 & 1 & -3 \end{bmatrix}$$

$$\xrightarrow{-\frac{1}{7}R_2} \begin{bmatrix} 1 & 4 & 0 & 1 \\ 0 & 1 & -\frac{1}{7} & \frac{3}{7} \end{bmatrix}$$

$$\xrightarrow{-4R_2 + R_1} \begin{bmatrix} 1 & 0 & \frac{4}{7} & -\frac{5}{7} \\ 0 & 1 & -\frac{1}{7} & \frac{3}{7} \end{bmatrix}$$

According to the previous discussion,

$$A^{-1} = \begin{bmatrix} \frac{4}{7} & -\frac{5}{7} \\ -\frac{1}{7} & \frac{3}{7} \end{bmatrix}.$$

Checking this fact we see that

$$\begin{bmatrix} 3 & 5 \\ 1 & 4 \end{bmatrix}\begin{bmatrix} \frac{4}{7} & -\frac{5}{7} \\ -\frac{1}{7} & \frac{3}{7} \end{bmatrix} = \begin{bmatrix} 1 & 0 \\ 0 & 1 \end{bmatrix} = \begin{bmatrix} \frac{4}{7} & -\frac{5}{7} \\ -\frac{1}{7} & \frac{3}{7} \end{bmatrix}\begin{bmatrix} 3 & 5 \\ 1 & 4 \end{bmatrix}.$$ ■

EXAMPLE 3 Find A^{-1} if $A = \begin{bmatrix} -1 & 3 & 1 \\ 2 & 5 & 0 \\ 3 & 1 & -2 \end{bmatrix}$.

Solution We proceed as follows:

$$\begin{bmatrix} -1 & 3 & 1 & 1 & 0 & 0 \\ 2 & 5 & 0 & 0 & 1 & 0 \\ 3 & 1 & -2 & 0 & 0 & 1 \end{bmatrix} \xrightarrow{(-1)R_1} \begin{bmatrix} 1 & -3 & -1 & -1 & 0 & 0 \\ 2 & 5 & 0 & 0 & 1 & 0 \\ 3 & 1 & -2 & 0 & 0 & 1 \end{bmatrix}$$

$$\xrightarrow{-2R_1+R_2} \begin{bmatrix} 1 & -3 & -1 & -1 & 0 & 0 \\ 0 & 11 & 2 & 2 & 1 & 0 \\ 3 & 1 & -2 & 0 & 0 & 1 \end{bmatrix}$$

$$\xrightarrow{-3R_1+R_3} \begin{bmatrix} 1 & -3 & -1 & -1 & 0 & 0 \\ 0 & 11 & 2 & 2 & 1 & 0 \\ 0 & 10 & 1 & 3 & 0 & 1 \end{bmatrix}$$

$$\xrightarrow{(-1)R_3+R_2} \begin{bmatrix} 1 & -3 & -1 & -1 & 0 & 0 \\ 0 & 1 & 1 & -1 & 1 & -1 \\ 0 & 10 & 1 & 3 & 0 & 1 \end{bmatrix}$$

$$\xrightarrow{3R_2+R_1} \begin{bmatrix} 1 & 0 & 2 & -4 & 3 & -3 \\ 0 & 1 & 1 & -1 & 1 & -1 \\ 0 & 10 & 1 & 3 & 0 & 1 \end{bmatrix}$$

$$\xrightarrow{-10R_2+R_3} \begin{bmatrix} 1 & 0 & 2 & -4 & 3 & -3 \\ 0 & 1 & 1 & -1 & 1 & -1 \\ 0 & 0 & -9 & 13 & -10 & 11 \end{bmatrix}$$

$$\xrightarrow{-\frac{1}{9}R_3} \begin{bmatrix} 1 & 0 & 2 & -4 & 3 & -3 \\ 0 & 1 & 1 & -1 & 1 & -1 \\ 0 & 0 & 1 & -\frac{13}{9} & \frac{10}{9} & -\frac{11}{9} \end{bmatrix}$$

$$\xrightarrow{-2R_3+R_1} \begin{bmatrix} 1 & 0 & 0 & -\frac{10}{9} & \frac{7}{9} & -\frac{5}{9} \\ 0 & 1 & 1 & -1 & 1 & -1 \\ 0 & 0 & 1 & -\frac{13}{9} & \frac{10}{9} & -\frac{11}{9} \end{bmatrix}$$

$$\xrightarrow{(-1)R_3+R_2} \begin{bmatrix} 1 & 0 & 0 & -\frac{10}{9} & \frac{7}{9} & -\frac{5}{9} \\ 0 & 1 & 0 & \frac{4}{9} & -\frac{1}{9} & \frac{2}{9} \\ 0 & 0 & 1 & -\frac{13}{9} & \frac{10}{9} & -\frac{11}{9} \end{bmatrix}$$

Consequently,

$$A^{-1} = \begin{bmatrix} -\frac{10}{9} & \frac{7}{9} & -\frac{5}{9} \\ \frac{4}{9} & -\frac{1}{9} & \frac{2}{9} \\ -\frac{13}{9} & \frac{10}{9} & -\frac{11}{9} \end{bmatrix} = \frac{1}{9}\begin{bmatrix} -10 & 7 & -5 \\ 4 & -1 & 2 \\ -13 & 10 & -11 \end{bmatrix}.$$

The reader should check the fact that

$$AA^{-1} = I_3 = A^{-1}A.$$

■

There are many uses for inverses of matrices. One application concerns solutions of systems of linear equations. To illustrate, let us consider the case of two linear equations in two unknowns:

$$\begin{cases} a_{11}x + a_{12}y = k_1 \\ a_{21}x + a_{22}y = k_2. \end{cases}$$

We may express the system by means of matrices as follows:

$$\begin{bmatrix} a_{11}x + a_{12}y \\ a_{21}x + a_{22}y \end{bmatrix} = \begin{bmatrix} k_1 \\ k_2 \end{bmatrix}.$$

If we let $A = \begin{bmatrix} a_{11} & a_{12} \\ a_{21} & a_{22} \end{bmatrix}$, $X = \begin{bmatrix} x \\ y \end{bmatrix}$, and $B = \begin{bmatrix} k_1 \\ k_2 \end{bmatrix}$,

this may be written in the form

$$AX = B.$$

If A^{-1} exists, then multiplying both sides of the last equation by A^{-1} gives us $A^{-1}AX = A^{-1}B$. Since $A^{-1}A = I_2$ and $I_2X = X$, this leads to

$$X = A^{-1}B$$

from which the solution (x, y) may be found. This technique may be extended to systems of n linear equations in n unknowns.

EXAMPLE 4 Solve the following system of equations:

$$\begin{cases} -x + 3y + z = 1 \\ 2x + 5y = 3 \\ 3x + y - 2z = -2. \end{cases}$$

Solution If we let

$$A = \begin{bmatrix} -1 & 3 & 1 \\ 2 & 5 & 0 \\ 3 & 1 & -2 \end{bmatrix}, \quad X = \begin{bmatrix} x \\ y \\ z \end{bmatrix}, \quad \text{and} \quad B = \begin{bmatrix} 1 \\ 3 \\ -2 \end{bmatrix},$$

then, as in the preceding discussion, the given system may be written in terms of matrices as $AX = B$. This implies that $X = A^{-1}B$. The matrix A^{-1} was found in Example 3. Substituting for X, A^{-1}, and B in the last equation gives us

$$\begin{bmatrix} x \\ y \\ z \end{bmatrix} = \frac{1}{9}\begin{bmatrix} -10 & 7 & -5 \\ 4 & -1 & 2 \\ -13 & 10 & -11 \end{bmatrix}\begin{bmatrix} 1 \\ 3 \\ -2 \end{bmatrix} = \frac{1}{9}\begin{bmatrix} 21 \\ -3 \\ 39 \end{bmatrix} = \begin{bmatrix} \frac{7}{3} \\ -\frac{1}{3} \\ \frac{13}{3} \end{bmatrix}.$$

It follows that $x = \frac{7}{3}$, $y = -\frac{1}{3}$, and $z = \frac{13}{3}$. Hence, the ordered triple $(\frac{7}{3}, -\frac{1}{3}, \frac{13}{3})$ is the solution of the given system. ■

The method of solution employed in Example 4 is beneficial only if A^{-1} is known, or if many systems with the same coefficient matrix are to be considered. The preferred technique for solving an arbitrary system of linear equations is the matrix method discussed in Section 7.3.

EXERCISES 7.4

In Exercises 1–8 find $A + B$, $A - B$, $2A$, and $-3B$.

1 $A = \begin{bmatrix} 5 & -2 \\ 1 & 3 \end{bmatrix}$, $B = \begin{bmatrix} 4 & 1 \\ -3 & 2 \end{bmatrix}$

2 $A = \begin{bmatrix} 3 & 0 \\ -1 & 2 \end{bmatrix}$, $B = \begin{bmatrix} 3 & -4 \\ 1 & 1 \end{bmatrix}$

3 $A = \begin{bmatrix} 6 & -1 \\ 2 & 0 \\ -3 & 4 \end{bmatrix}$, $B = \begin{bmatrix} 3 & 1 \\ -1 & 5 \\ 6 & 0 \end{bmatrix}$

4 $A = \begin{bmatrix} 0 & -2 & 7 \\ 5 & 4 & -3 \end{bmatrix}$, $B = \begin{bmatrix} 8 & 4 & 0 \\ 0 & 1 & 4 \end{bmatrix}$

5 $A = [4 \quad -3 \quad 2]$, $B = [7 \quad 0 \quad -5]$

6 $A = \begin{bmatrix} 7 \\ -16 \end{bmatrix}$, $B = \begin{bmatrix} -11 \\ 9 \end{bmatrix}$

7 $A = \begin{bmatrix} 0 & 4 & 0 & 3 \\ 1 & 2 & 0 & -5 \end{bmatrix}$,
$B = \begin{bmatrix} -3 & 0 & 1 & 3 \\ 2 & 0 & 7 & -2 \end{bmatrix}$

8 $A = [-7]$, $B = [9]$

In Exercises 9–18 find AB and BA.

9 $A = \begin{bmatrix} 2 & 6 \\ 3 & -4 \end{bmatrix}$, $B = \begin{bmatrix} 5 & -2 \\ 1 & 7 \end{bmatrix}$

10 $A = \begin{bmatrix} 4 & -2 \\ -2 & 1 \end{bmatrix}$, $B = \begin{bmatrix} 2 & 1 \\ 4 & 2 \end{bmatrix}$

11 $A = \begin{bmatrix} 3 & 0 & -1 \\ 0 & 4 & 2 \\ 5 & -3 & 1 \end{bmatrix}$, $B = \begin{bmatrix} 1 & -5 & 0 \\ 4 & 1 & -2 \\ 0 & -1 & 3 \end{bmatrix}$

12 $A = \begin{bmatrix} 5 & 0 & 0 \\ 0 & -3 & 0 \\ 0 & 0 & 2 \end{bmatrix}$, $B = \begin{bmatrix} 3 & 0 & 0 \\ 0 & 4 & 0 \\ 0 & 0 & -2 \end{bmatrix}$

13 $A = \begin{bmatrix} 4 & -3 & 1 \\ -5 & 2 & 2 \end{bmatrix}$, $B = \begin{bmatrix} 2 & 1 \\ 0 & 1 \\ -4 & 7 \end{bmatrix}$

14 $A = \begin{bmatrix} 2 & 1 & -1 & 0 \\ 3 & -2 & 0 & 5 \\ -2 & 1 & 4 & 2 \end{bmatrix}$, $B = \begin{bmatrix} 5 & -3 & 1 \\ 1 & 2 & 0 \\ -1 & 0 & 4 \\ 0 & -2 & 3 \end{bmatrix}$

15 $A = \begin{bmatrix} 1 & 2 & 3 \\ 4 & 5 & 6 \\ 7 & 8 & 9 \end{bmatrix}$, $B = \begin{bmatrix} 1 & 0 & 0 \\ 0 & 1 & 0 \\ 0 & 0 & 1 \end{bmatrix}$

16 $A = \begin{bmatrix} 1 & 2 & 3 \\ 2 & 3 & 1 \\ 3 & 1 & 2 \end{bmatrix}$, $B = \begin{bmatrix} 2 & 0 & 0 \\ 0 & 2 & 0 \\ 0 & 0 & 2 \end{bmatrix}$

17 $A = [-3 \quad 7 \quad 2]$, $B = \begin{bmatrix} 1 \\ 4 \\ -5 \end{bmatrix}$

18 $A = [4 \quad 8]$, $B = \begin{bmatrix} -3 \\ 2 \end{bmatrix}$

In Exercises 19–22 find AB.

19 $A = \begin{bmatrix} 4 & -2 \\ 0 & 3 \\ -7 & 5 \end{bmatrix}$, $B = \begin{bmatrix} 3 \\ 4 \end{bmatrix}$

20 $A = \begin{bmatrix} 4 \\ -3 \\ 2 \end{bmatrix}$, $B = [5 \quad 1]$

21 $A = \begin{bmatrix} 2 & 1 & 0 & -3 \\ -7 & 0 & -2 & 4 \end{bmatrix}$, $B = \begin{bmatrix} 4 & -2 & 0 \\ 1 & 1 & -2 \\ 0 & 0 & 5 \\ -3 & -1 & 0 \end{bmatrix}$

22 $A = \begin{bmatrix} 1 & 2 & -3 \\ 4 & -5 & 6 \end{bmatrix}$, $B = \begin{bmatrix} 1 & -1 & 0 & 2 \\ -2 & 3 & 1 & 0 \\ 0 & 4 & 0 & -3 \end{bmatrix}$

23 If $A = \begin{bmatrix} 1 & 2 \\ 0 & -3 \end{bmatrix}$, and $B = \begin{bmatrix} 2 & -1 \\ 3 & 1 \end{bmatrix}$, show that $(A + B)(A - B) \neq A^2 - B^2$, where $A^2 = AA$ and $B^2 = BB$.

24 If A and B are the matrices of Exercise 23, show that $(A + B)(A + B) \neq A^2 + 2AB + B^2$.

25 If A and B are the matrices of Exercise 23 and $C = \begin{bmatrix} 3 & 1 \\ -2 & 0 \end{bmatrix}$, prove that $A(B + C) = AB + AC$.

26 If A, B, and C are the matrices of Exercise 25, prove that $A(BC) = (AB)C$.

Prove the identities given in Exercises 27–30, where

$$A = \begin{bmatrix} a_{11} & a_{12} \\ a_{21} & a_{22} \end{bmatrix}, \quad B = \begin{bmatrix} b_{11} & b_{12} \\ b_{21} & b_{22} \end{bmatrix}, \quad C = \begin{bmatrix} c_{11} & c_{12} \\ c_{21} & c_{22} \end{bmatrix}$$

and c, d are real numbers.

27 $c(A + B) = cA + cB$

28 $(c + d)A = cA + dA$

29 $A(B + C) = AB + AC$

30 $A(BC) = (AB)C$

In Exercises 31–42 find the inverse of the matrix, if it exists.

31 $\begin{bmatrix} 2 & -4 \\ 1 & 3 \end{bmatrix}$

32 $\begin{bmatrix} 3 & 2 \\ 4 & 5 \end{bmatrix}$

33 $\begin{bmatrix} 2 & 4 \\ 4 & 8 \end{bmatrix}$

34 $\begin{bmatrix} 3 & -1 \\ 6 & -2 \end{bmatrix}$

35 $\begin{bmatrix} 3 & -1 & 0 \\ 2 & 2 & 0 \\ 0 & 0 & 4 \end{bmatrix}$

36 $\begin{bmatrix} 3 & 0 & 2 \\ 0 & 1 & 0 \\ -4 & 0 & 2 \end{bmatrix}$

37 $\begin{bmatrix} -2 & 2 & 3 \\ 1 & -1 & 0 \\ 0 & 1 & 4 \end{bmatrix}$

38 $\begin{bmatrix} 1 & 2 & 3 \\ -2 & 1 & 0 \\ 3 & -1 & 1 \end{bmatrix}$

39 $\begin{bmatrix} 2 & 0 & 0 \\ 0 & 4 & 0 \\ 0 & 0 & 6 \end{bmatrix}$

40 $\begin{bmatrix} 1 & 1 & 1 \\ 2 & 2 & 2 \\ 3 & 3 & 3 \end{bmatrix}$

41 $\begin{bmatrix} 1 & -1 & 0 & 1 \\ 0 & 1 & -2 & 0 \\ -1 & 2 & 1 & 2 \\ -2 & 1 & 2 & 0 \end{bmatrix}$

42 $\begin{bmatrix} 1 & 2 & 0 & 1 \\ 0 & -1 & 1 & -2 \\ 0 & 0 & 2 & 0 \\ 0 & 0 & 0 & 1 \end{bmatrix}$

43 State conditions on a and b which guarantee that the matrix $\begin{bmatrix} a & 0 \\ 0 & b \end{bmatrix}$ has an inverse, and find a formula for the inverse, if it exists.

44 If $abc \neq 0$, find the inverse of $\begin{bmatrix} a & 0 & 0 \\ 0 & b & 0 \\ 0 & 0 & c \end{bmatrix}$.

45 If $A = \begin{bmatrix} a_{11} & a_{12} & a_{13} \\ a_{21} & a_{22} & a_{23} \\ a_{31} & a_{32} & a_{33} \end{bmatrix}$, prove that $AI_3 = A = I_3A$.

46 Prove that $AI_4 = A = I_4A$ for every square matrix A of order 4.

Solve the systems in Exercises 47–50 by the method of Example 4. (Refer to inverses of matrices found in Exercises 31–38.)

47 $\begin{cases} 2x - 4y = 3 \\ x + 3y = 1 \end{cases}$

48 $\begin{cases} 3x + 2y = -1 \\ 4x + 5y = 1 \end{cases}$

49 $\begin{cases} -2x + 2y + 3z = 1 \\ x - y = 3 \\ y + 4z = -2 \end{cases}$

50 $\begin{cases} x + 2y + 3z = -1 \\ -2x + y = 4 \\ 3x - y + z = 2 \end{cases}$

SECTION 7.5

DETERMINANTS

Throughout this section and the next we will assume that all matrices under discussion are *square* matrices. Associated with each such matrix A is a number called the **determinant** of A. Determinants can be used to solve systems of linear equations if the number of equations is the same as the number of variables. In this section we shall state the definition and give some basic properties of determinants. It may be difficult to see exactly why the definitions to follow are used. One reason will be pointed out in Section 7.7, in which it is shown that our definitions arise naturally when solving systems of linear equations.

The determinant of a square matrix A will be denoted by $|A|$. This notation should not be confused with the symbol for the absolute value of a real number. To avoid any misunderstanding, the expression $\det A$ is used in some mathematics texts instead of $|A|$. We shall define $|A|$ by beginning with the case in which A has order 1 and then by increasing the order a step at a time.

If A is a square matrix of order 1, then A has only one element. Thus, $A = [a_{11}]$ and we define $|A| = a_{11}$. If A is a square matrix of order 2, then we may write

$$A = \begin{bmatrix} a_{11} & a_{12} \\ a_{21} & a_{22} \end{bmatrix}$$

and the determinant of A is defined by

$$|A| = a_{11}a_{22} - a_{21}a_{12}.$$

Another notation for $|A|$ is obtained by replacing the brackets in the symbol for A with vertical bars as follows.

DEFINITION

$$|A| = \begin{vmatrix} a_{11} & a_{12} \\ a_{21} & a_{22} \end{vmatrix} = a_{11}a_{22} - a_{21}a_{12}$$

EXAMPLE 1 Find $|A|$ if $A = \begin{bmatrix} 2 & -1 \\ 4 & -3 \end{bmatrix}$.

Solution By definition,

$$|A| = \begin{vmatrix} 2 & -1 \\ 4 & -3 \end{vmatrix} = (2)(-3) - (4)(-1) = -6 + 4 = -2.$$

■

For matrices of higher order it is convenient to introduce the following additional terminology.

DEFINITION

Let A be a matrix of order 3. The **minor** M_{ij} of the element a_{ij} of A is the determinant of the matrix of order 2 obtained by deleting row i and column j of A.

Thus, to determine the minor of an element, we discard the row and column in which the element appears and then find the determinant of the resulting matrix. To illustrate, given the general 3×3 matrix

$$A = \begin{bmatrix} a_{11} & a_{12} & a_{13} \\ a_{21} & a_{22} & a_{23} \\ a_{31} & a_{32} & a_{33} \end{bmatrix}$$

we obtain

$$M_{11} = \begin{vmatrix} a_{22} & a_{23} \\ a_{32} & a_{33} \end{vmatrix} = a_{22}a_{33} - a_{32}a_{23}$$

$$M_{12} = \begin{vmatrix} a_{21} & a_{23} \\ a_{31} & a_{33} \end{vmatrix} = a_{21}a_{33} - a_{31}a_{23}$$

$$M_{13} = \begin{vmatrix} a_{21} & a_{22} \\ a_{31} & a_{32} \end{vmatrix} = a_{21}a_{32} - a_{31}a_{22}$$

$$M_{23} = \begin{vmatrix} a_{11} & a_{12} \\ a_{31} & a_{32} \end{vmatrix} = a_{11}a_{32} - a_{31}a_{12}$$

and likewise for the other minors M_{21}, M_{22}, M_{31}, M_{32}, and M_{33}.

We shall also make use of the following concept.

DEFINITION

The **cofactor** A_{ij} of the element a_{ij} is

$$A_{ij} = (-1)^{i+j}M_{ij}.$$

By definition, to obtain the cofactor of a_{ij}, we find the minor and multiply it by 1 or -1, depending on whether the sum of i and j is even or odd, respectively. An easy way to remember the sign $(-1)^{i+j}$ associated with the cofactor A_{ij} is to consider the following "checkerboard" scheme:

$$\begin{bmatrix} + & - & + \\ - & + & - \\ + & - & + \end{bmatrix}$$

where we may regard the + signs as occurring on red squares and the − signs on black squares.

EXAMPLE 2 If

$$A = \begin{bmatrix} 1 & -3 & 3 \\ 4 & 2 & 0 \\ -2 & -7 & 5 \end{bmatrix}$$

find M_{11}, M_{21}, M_{22}, A_{11}, A_{21}, and A_{22}.

Solution By definition,

$$M_{11} = \begin{vmatrix} 2 & 0 \\ -7 & 5 \end{vmatrix} = (2)(5) - (-7)(0) = 10$$

$$M_{21} = \begin{vmatrix} -3 & 3 \\ -7 & 5 \end{vmatrix} = (-3)(5) - (-7)(3) = 6$$

$$M_{22} = \begin{vmatrix} 1 & 3 \\ -2 & 5 \end{vmatrix} = (1)(5) - (-2)(3) = 11.$$

All that is necessary to obtain the cofactors is to prefix the corresponding minors with the proper signs. Thus, using the definition of cofactor,

$$A_{11} = (-1)^{1+1}M_{11} = (1)(10) = 10$$
$$A_{21} = (-1)^{2+1}M_{21} = (-1)(6) = -6$$
$$A_{22} = (-1)^{2+2}M_{22} = (1)(11) = 11.$$

The checkerboard scheme can also be used to determine the proper signs. ■

The determinant $|A|$ of a square matrix of order 3 is defined as follows.

DEFINITION

$$|A| = \begin{vmatrix} a_{11} & a_{12} & a_{13} \\ a_{21} & a_{22} & a_{23} \\ a_{31} & a_{32} & a_{33} \end{vmatrix} = a_{11}A_{11} + a_{12}A_{12} + a_{13}A_{13}$$

Since $A_{11} = (-1)^{1+1}M_{11} = M_{11}$, $A_{12} = (-1)^{1+2}M_{12} = -M_{12}$, and $A_{13} = (-1)^{1+3}M_{13} = M_{13}$, the preceding definition may also be written

$$|A| = a_{11}M_{11} - a_{12}M_{12} + a_{13}M_{13}.$$

If we substitute for M_{11}, M_{12}, and M_{13}, we obtain the following formula for $|A|$ in terms of the elements of A:

$$|A| = a_{11}a_{22}a_{33} - a_{11}a_{32}a_{23} - a_{12}a_{21}a_{33} + a_{12}a_{31}a_{23} + a_{13}a_{21}a_{32} - a_{13}a_{31}a_{22}.$$

The definition of $|A|$ for a square matrix A of order 3 displays a pattern of multiplying each element in row 1 by its cofactor, and then adding to find $|A|$. This is referred to as *expanding* $|A|$ *by the first row.* By actually carrying out the computations, it is not difficult to show that $|A|$ *can be expanded in similar fashion by using any row or column.* As an illustration, the expansion by the second column is

$$\begin{aligned} |A| &= a_{12}A_{12} + a_{22}A_{22} + a_{32}A_{32} \\ &= a_{12}\left(-\begin{vmatrix} a_{21} & a_{23} \\ a_{31} & a_{32} \end{vmatrix}\right) + a_{22}\left(+\begin{vmatrix} a_{11} & a_{13} \\ a_{31} & a_{33} \end{vmatrix}\right) + a_{32}\left(-\begin{vmatrix} a_{11} & a_{13} \\ a_{21} & a_{23} \end{vmatrix}\right). \end{aligned}$$

Applying the definition to the determinants in parentheses, multiplying as indicated, and rearranging the terms in the sum, we could arrive at the formula for $|A|$ in terms of the elements of A. Similarly, the expansion by the third row is

$$|A| = a_{31}A_{31} + a_{32}A_{32} + a_{33}A_{33}.$$

Once again it can be shown that this agrees with previous expansions.

EXAMPLE 3 Find $|A|$ if

$$A = \begin{bmatrix} -1 & 3 & 1 \\ 2 & 5 & 0 \\ 3 & 1 & -2 \end{bmatrix}.$$

Solution Expanding $|A|$ by the second row,

$$|A| = (2)A_{21} + (5)A_{22} + (0)A_{23}.$$

Using the definition of cofactor,

$$A_{21} = (-1)^3M_{21} = -\begin{vmatrix} 3 & 1 \\ 1 & -2 \end{vmatrix} = -[(3)(-2) - (1)(1)] = 7$$

$$A_{22} = (-1)^4M_{22} = \begin{vmatrix} -1 & 1 \\ 3 & -2 \end{vmatrix} = [(-1)(-2) - (3)(1)] = -1.$$

Consequently,

$$|A| = (2)(7) + (5)(-1) + (0)A_{23} = 14 - 5 + 0 = 9.$$ ■

The method of defining determinants of matrices of arbitrary order n may be patterned after that used for order 3. Specifically, if a_{ij} is an element of a matrix of order $n > 1$, then the **minor** M_{ij} is defined as the determinant of the matrix of order $n - 1$ obtained by deleting row i and column j. The **cofactor** A_{ij} is defined as $(-1)^{i+j}M_{ij}$. The sign $(-1)^{i+j}$ associated with A_{ij} can be remembered by using a checkerboard similar to that used for order 3, extending the rows and columns as far as necessary. We then define the determinant $|A|$ of a matrix A of order n as the expansion by the first row, that is,

$$|A| = a_{11}A_{11} + a_{12}A_{12} + \cdots + a_{1n}A_{1n}$$

or, in terms of minors,

$$|A| = a_{11}M_{11} - a_{12}M_{12} + \cdots + a_{1n}(-1)^{1+n}M_{1n}.$$

It can be shown that the number $|A|$ may be found by using *any* row or column, as stated in the following theorem.

EXPANSION THEOREM FOR DETERMINANTS

If A is a square matrix of order $n > 1$, then the determinant $|A|$ may be found by multiplying the elements of any row (or column) by their respective cofactors, and adding the resulting products.

The proof of this theorem is difficult and may be found in texts on matrix theory. The theorem is quite useful if many zeros appear in a row or column, as illustrated in the following example.

EXAMPLE 4 Find $|A|$ if

$$A = \begin{bmatrix} 1 & 0 & 2 & 5 \\ -2 & 1 & 5 & 0 \\ 0 & 0 & -3 & 0 \\ 0 & -1 & 0 & 3 \end{bmatrix}.$$

Solution Note that all but one of the elements in the third row is zero. Hence if we expand $|A|$ by the third row, there will be at most one nonzero term. Specifically,

$$|A| = (0)A_{31} + (0)A_{32} + (-3)A_{33} + (0)A_{34} = -3A_{33}$$

where $$A_{33} = \begin{vmatrix} 1 & 0 & 5 \\ -2 & 1 & 0 \\ 0 & -1 & 3 \end{vmatrix}.$$

Expanding A_{33} by column 1, we obtain

$$A_{33} = (1)\begin{vmatrix} 1 & 0 \\ -1 & 3 \end{vmatrix} + (-2)\left(-\begin{vmatrix} 0 & 5 \\ -1 & 3 \end{vmatrix}\right) + 0\begin{vmatrix} 0 & 5 \\ 1 & 0 \end{vmatrix} = 3 + 10 + 0 = 13.$$

Therefore, $$|A| = -3A_{33} = (-3)(13) = -39.$$ ■

In general, if all but one element a in some row (or column) of A is zero, and if the determinant $|A|$ is expanded by that row (or column), then all terms drop out except the product of the element a with its cofactor. We will make important use of this fact in the next section.

If *every* element in a row (or column) of a matrix A is zero, then upon expanding $|A|$ by that row (or column), we obtain the number 0. This gives us the following result.

THEOREM If every element of a row (or column) of a square matrix A is zero, then $|A| = 0$.

EXERCISES 7.5

In Exercises 1–4 find all the minors and cofactors of the elements in the given matrix.

1 $\begin{bmatrix} 2 & 4 & -1 \\ 0 & 3 & 2 \\ -5 & 7 & 0 \end{bmatrix}$ **2** $\begin{bmatrix} 5 & -2 & 1 \\ 4 & 7 & 0 \\ -3 & 4 & -1 \end{bmatrix}$

3 $\begin{bmatrix} 7 & -1 \\ 5 & 0 \end{bmatrix}$ **4** $\begin{bmatrix} -6 & 4 \\ 3 & 2 \end{bmatrix}$

5–8 Find the determinants of the matrices in Exercises 1–4.

Find the determinants of the matrices in Exercises 9–20.

9 $\begin{bmatrix} -5 & 4 \\ -3 & 2 \end{bmatrix}$ **10** $\begin{bmatrix} 6 & 4 \\ -3 & 2 \end{bmatrix}$

11 $\begin{bmatrix} a & -a \\ b & -b \end{bmatrix}$ **12** $\begin{bmatrix} c & d \\ -d & c \end{bmatrix}$

13 $\begin{bmatrix} 3 & 1 & -2 \\ 4 & 2 & 5 \\ -6 & 3 & -1 \end{bmatrix}$ **14** $\begin{bmatrix} 2 & -5 & 1 \\ -3 & 1 & 6 \\ 4 & -2 & 3 \end{bmatrix}$

15 $\begin{bmatrix} -5 & 4 & 1 \\ 3 & -2 & 7 \\ 2 & 0 & 6 \end{bmatrix}$ **16** $\begin{bmatrix} 2 & 7 & -3 \\ 1 & 0 & 4 \\ 4 & -1 & -2 \end{bmatrix}$

17 $\begin{bmatrix} 3 & -1 & 2 & 0 \\ 4 & 0 & -3 & 5 \\ 0 & 6 & 0 & 0 \\ 1 & 3 & -4 & 2 \end{bmatrix}$ **18** $\begin{bmatrix} 2 & 5 & 1 & 0 \\ -4 & 0 & -3 & 0 \\ 3 & -2 & 1 & 6 \\ -1 & 4 & 2 & 0 \end{bmatrix}$

19 $\begin{bmatrix} 0 & b & 0 & 0 \\ 0 & 0 & c & 0 \\ a & 0 & 0 & 0 \\ 0 & 0 & 0 & d \end{bmatrix}$ **20** $\begin{bmatrix} a & u & v & w \\ 0 & b & x & y \\ 0 & 0 & c & z \\ 0 & 0 & 0 & d \end{bmatrix}$

Verify the identities in Exercises 21–28 by expanding each determinant.

21 $\begin{vmatrix} a & b \\ c & d \end{vmatrix} = -\begin{vmatrix} c & d \\ a & b \end{vmatrix}$ **22** $\begin{vmatrix} a & b \\ c & d \end{vmatrix} = -\begin{vmatrix} b & a \\ d & c \end{vmatrix}$

23 $\begin{vmatrix} a & kb \\ c & kd \end{vmatrix} = k\begin{vmatrix} a & b \\ c & d \end{vmatrix}$ **24** $\begin{vmatrix} a & b \\ kc & kd \end{vmatrix} = k\begin{vmatrix} a & b \\ c & d \end{vmatrix}$

25 $\begin{vmatrix} a & b \\ c & d \end{vmatrix} = \begin{vmatrix} a & b \\ ka + c & kb + d \end{vmatrix}$

26 $\begin{vmatrix} a & b \\ c & d \end{vmatrix} = \begin{vmatrix} a & ka + b \\ c & kc + d \end{vmatrix}$

27 $\begin{vmatrix} a & b \\ c & d \end{vmatrix} + \begin{vmatrix} a & e \\ c & f \end{vmatrix} = \begin{vmatrix} a & b + e \\ c & d + f \end{vmatrix}$

28 $\begin{vmatrix} a & b \\ c & d \end{vmatrix} + \begin{vmatrix} a & b \\ e & f \end{vmatrix} = \begin{vmatrix} a & b \\ c + e & d + f \end{vmatrix}$

29 Prove that if a square matrix A of order 2 has two identical rows or columns, then $|A| = 0$.

30 Repeat Exercise 29 for a matrix of order 3.

In Exercises 31–34 let $I = I_2$ be the identity matrix of order 2, and let $f(x) = |A - xI|$. Find (a) the polynomial $f(x)$, and (b) the zeros of $f(x)$. (In the study of matrices, $f(x)$ is called the **characteristic polynomial** of A, and the zeros of $f(x)$ are called the **characteristic values,** or **eigenvalues,** of A.

31 $A = \begin{bmatrix} 1 & 2 \\ 3 & 2 \end{bmatrix}$

32 $A = \begin{bmatrix} 3 & 1 \\ 2 & 2 \end{bmatrix}$

33 $A = \begin{bmatrix} -3 & -2 \\ 2 & 2 \end{bmatrix}$

34 $A = \begin{bmatrix} 2 & -4 \\ -3 & 5 \end{bmatrix}$

In Exercises 35–38 let $I = I_3$ and let $f(x) = |A - xI|$. Find (a) the polynomial $f(x)$, and (b) the zeros of $f(x)$.

35 $A = \begin{bmatrix} 1 & 0 & 0 \\ 1 & 0 & -2 \\ -1 & 1 & -3 \end{bmatrix}$

36 $A = \begin{bmatrix} 2 & 1 & 0 \\ -1 & 0 & 0 \\ 1 & 3 & 2 \end{bmatrix}$

37 $A = \begin{bmatrix} 0 & 2 & -2 \\ -1 & 3 & 1 \\ -3 & 3 & 1 \end{bmatrix}$

38 $A = \begin{bmatrix} 3 & 2 & 2 \\ 1 & 0 & 2 \\ -1 & -1 & 0 \end{bmatrix}$

In Exercises 39–42 express the determinants in the form $ai + bj + ck$, where a, b, and c are real numbers.

39 $\begin{vmatrix} i & j & k \\ 2 & -1 & 6 \\ -3 & 5 & 1 \end{vmatrix}$

40 $\begin{vmatrix} i & j & k \\ 1 & -2 & 3 \\ 2 & 1 & -4 \end{vmatrix}$

41 $\begin{vmatrix} i & j & k \\ 5 & -6 & -1 \\ 3 & 0 & 1 \end{vmatrix}$

42 $\begin{vmatrix} i & j & k \\ 4 & -6 & 2 \\ -2 & 3 & -1 \end{vmatrix}$

SECTION 7.6

PROPERTIES OF DETERMINANTS

The method of evaluating a determinant by means of the Expansion Theorem, stated in Section 7.5, is inefficient for matrices of high order. For example, if a determinant of a matrix of order 10 is expanded by any row, a sum of 10 terms is obtained, and each term contains the determinant of a matrix of order 9, which is a cofactor of the original matrix. If any of the latter determinants is expanded by a row (or column), a sum of 9 terms is obtained, each containing the determinant of a matrix of order 8. Hence, at this stage there are 90 determinants of matrices of order 8 to evaluate! The process could be continued until only determinants of matrices of order 2 remain. Unless many elements of the original matrix are zero, it is an enormous task to carry out all of the computations.

In this section we shall consider rules that simplify the process of evaluating determinants. The main use for these rules is to introduce zeros into the determinant. They may also be used to change the determinant to echelon form, that is, a form in which the elements below the main diagonal elements are all zero. The transformations on rows stated in the next theorem are the same as the elementary row transformations of a matrix introduced in Section 7.3. However, for determinants we may also employ similar transformations on columns.

THEOREM ON ROW AND COLUMN TRANSFORMATIONS OF A DETERMINANT

Let A be a matrix of order n.

(i) If a matrix B is obtained from A by interchanging two rows (or columns), then $|B| = -|A|$.

(ii) If B is obtained from A by multiplying every element of one row (or column) of A by a real number k, then $|B| = k|A|$.

(iii) If B is obtained from A by adding to any row (or column) of A, k times another row (or column), where k is any real number, then $|B| = |A|$.

We shall not prove this theorem. When using the theorem to justify manipulations with determinants, we refer to the rows (or columns) of the *determinant* in the obvious way. For example, property (iii) may be phrased: "Adding the product of k times another row (or column) to any row (or column) of a determinant does not affect the value of the determinant."

Row transformations of determinants will be specified by means of the symbols R_{ij}, $k\text{R}_i$, and $k\text{R}_i + \text{R}_j$, which were introduced in Section 7.3. Analogous symbols are used for column transformations. For example, $k\text{C}_i + \text{C}_j$ means "add to the jth column, k times the ith column." The following are illustrations of the preceding theorem, with the reason for each equality stated at the right.

$$\begin{vmatrix} 2 & 0 & 1 \\ 6 & 4 & 3 \\ 0 & 3 & 5 \end{vmatrix} = -\begin{vmatrix} 6 & 4 & 3 \\ 2 & 0 & 1 \\ 0 & 3 & 5 \end{vmatrix} \qquad \text{R}_{12}$$

$$\begin{vmatrix} 2 & 0 & 1 \\ 6 & 4 & 3 \\ 0 & 3 & 5 \end{vmatrix} = -\begin{vmatrix} 1 & 0 & 2 \\ 3 & 4 & 6 \\ 5 & 3 & 0 \end{vmatrix} \qquad \text{C}_{13}$$

$$\begin{vmatrix} 1 & -3 & 4 \\ 2 & -1 & 0 \\ 3 & 1 & 6 \end{vmatrix} = \begin{vmatrix} 1 & -3 & 4 \\ 0 & 5 & -8 \\ 3 & 1 & 6 \end{vmatrix} \qquad -2\text{R}_1 + \text{R}_2$$

$$\begin{vmatrix} 1 & -3 & 4 \\ 2 & -1 & 0 \\ 3 & 1 & 6 \end{vmatrix} = \begin{vmatrix} -5 & -3 & 4 \\ 0 & -1 & 0 \\ 5 & 1 & 6 \end{vmatrix} \qquad 2\text{C}_2 + \text{C}_1$$

THEOREM

If two rows (or columns) of a square matrix A are identical, then $|A| = 0$.

Proof If B is the matrix obtained from A by interchanging the two identical rows (or columns), then B and A are the same and, consequently,

$|B| = |A|$. However, by (i) of the Theorem on Row and Column Transformations of a Determinant, $|B| = -|A|$ and hence $-|A| = |A|$, which implies that $|A| = 0$. □

EXAMPLE 1 Find $|A|$ if

$$A = \begin{bmatrix} 2 & 3 & 0 & 4 \\ 0 & 5 & -1 & 6 \\ 1 & 0 & -2 & 3 \\ -3 & 2 & 0 & -5 \end{bmatrix}.$$

Solution We plan to use (iii) of the Theorem on Row and Column Transformations to introduce many zeros in some row or column. To do this, it is convenient to work with an element of the matrix that equals 1 or -1, since this enables us to avoid the use of fractions. If no such element appears in the original matrix, it is always possible to introduce the number 1 by using (iii) or (ii) of the theorem. In this example this is unnecessary, since 1 appears in row 3, and -1 in row 2. Let us proceed as follows, where the reason for each equality is stated at the right.

$$\begin{vmatrix} 2 & 3 & 0 & 4 \\ 0 & 5 & -1 & 6 \\ 1 & 0 & -2 & 3 \\ -3 & 2 & 0 & -5 \end{vmatrix} = \begin{vmatrix} 0 & 3 & 4 & -2 \\ 0 & 5 & -1 & 6 \\ 1 & 0 & -2 & 3 \\ -3 & 2 & 0 & -5 \end{vmatrix} \qquad -2R_3 + R_1$$

$$= \begin{vmatrix} 0 & 3 & 4 & -2 \\ 0 & 5 & -1 & 6 \\ 1 & 0 & -2 & 3 \\ 0 & 2 & -6 & 4 \end{vmatrix} \qquad 3R_3 + R_4$$

$$= (1)\begin{vmatrix} 3 & 4 & -2 \\ 5 & -1 & 6 \\ 2 & -6 & 4 \end{vmatrix} \qquad \text{Expand by the first column}$$

$$= \begin{vmatrix} 23 & 4 & -2 \\ 0 & -1 & 6 \\ -28 & -6 & 4 \end{vmatrix} \qquad 5C_2 + C_1$$

$$= \begin{vmatrix} 23 & 4 & 22 \\ 0 & -1 & 0 \\ -28 & -6 & -32 \end{vmatrix} \qquad 6C_2 + C_3$$

$$= (-1)\begin{vmatrix} 23 & 22 \\ -28 & -32 \end{vmatrix} \qquad \text{Expand by the second row}$$

$$= (-1)[(23)(-32) - (-28)(22)] \qquad \text{Definition of determinant}$$

$$= 120$$

■

Part (ii) of the Theorem on Row and Column Transformations is useful for finding factors of determinants. To illustrate, for a determinant of a matrix of order 3, we have the following:

$$\begin{vmatrix} a_{11} & a_{12} & a_{13} \\ ka_{21} & ka_{22} & ka_{23} \\ a_{31} & a_{32} & a_{33} \end{vmatrix} = k \begin{vmatrix} a_{11} & a_{12} & a_{13} \\ a_{21} & a_{22} & a_{23} \\ a_{31} & a_{32} & a_{33} \end{vmatrix}.$$

Similar formulas hold if k is a common factor of the elements of some other row or column. When the theorem is used in this way, we often use the phrase "k is a common factor in the row (or column)."

EXAMPLE 2 Find $|A|$ if $A = \begin{bmatrix} 14 & -6 & 4 \\ 4 & -5 & 12 \\ -21 & 9 & -6 \end{bmatrix}$.

Solution

$$|A| = 2\begin{vmatrix} 7 & -3 & 2 \\ 4 & -5 & 12 \\ -21 & 9 & -6 \end{vmatrix}$$ 2 is a common factor in row 1.

$$= (2)(-3)\begin{vmatrix} 7 & -3 & 2 \\ 4 & -5 & 12 \\ 7 & -3 & 2 \end{vmatrix}$$ -3 is a common factor in row 3.

$$= 0$$ Two rows are identical. ■

EXAMPLE 3 Without expanding, show that $a - b$ is a factor of

$$\begin{vmatrix} 1 & 1 & 1 \\ a & b & c \\ a^2 & b^2 & c^2 \end{vmatrix}.$$

Solution

$$\begin{vmatrix} 1 & 1 & 1 \\ a & b & c \\ a^2 & b^2 & c^2 \end{vmatrix} = \begin{vmatrix} 0 & 1 & 1 \\ a-b & b & c \\ a^2-b^2 & b^2 & c^2 \end{vmatrix}$$ $(-1)C_2 + C_1$

$$= (a-b)\begin{vmatrix} 0 & 1 & 1 \\ 1 & b & c \\ a+b & b^2 & c^2 \end{vmatrix}$$ $a - b$ is a common factor of column 1 ■

EXERCISES 7.6

Without expanding, explain why the statements in Exercises 1–14 are true.

1 $\begin{vmatrix} 1 & 0 & 1 \\ 0 & 1 & 1 \\ 1 & 1 & 0 \end{vmatrix} = -\begin{vmatrix} 1 & 0 & 1 \\ 1 & 1 & 0 \\ 0 & 1 & 1 \end{vmatrix}$

2 $\begin{vmatrix} 1 & 0 & 1 \\ 0 & 1 & 1 \\ 1 & 1 & 0 \end{vmatrix} = -\begin{vmatrix} 1 & 1 & 0 \\ 0 & 1 & 1 \\ 1 & 0 & 1 \end{vmatrix}$

3 $\begin{vmatrix} 1 & 0 & 1 \\ 2 & 1 & 0 \\ 1 & 1 & 2 \end{vmatrix} = \begin{vmatrix} 1 & 0 & 1 \\ 2 & 1 & 0 \\ 0 & 1 & 1 \end{vmatrix}$

4 $\begin{vmatrix} 1 & 1 & 2 \\ 1 & 0 & 1 \\ 2 & 1 & 1 \end{vmatrix} = \begin{vmatrix} 0 & 1 & 1 \\ 1 & 0 & 1 \\ 2 & 1 & 1 \end{vmatrix}$

5 $\begin{vmatrix} 2 & 4 & 2 \\ 1 & 2 & 4 \\ 2 & 6 & 4 \end{vmatrix} = 4\begin{vmatrix} 1 & 2 & 1 \\ 1 & 2 & 4 \\ 1 & 3 & 2 \end{vmatrix}$

6 $\begin{vmatrix} 2 & 1 & 6 \\ 4 & 3 & 3 \\ 2 & 1 & 3 \end{vmatrix} = 6\begin{vmatrix} 1 & 1 & 2 \\ 2 & 3 & 1 \\ 1 & 1 & 1 \end{vmatrix}$

7 $\begin{vmatrix} 1 & -1 & 2 \\ 1 & 2 & -1 \\ 1 & -1 & 2 \end{vmatrix} = 0$

8 $\begin{vmatrix} 1 & -1 & 1 \\ 0 & 1 & 0 \\ -1 & 0 & -1 \end{vmatrix} = 0$

9 $\begin{vmatrix} 1 & 5 \\ -3 & 2 \end{vmatrix} = -\begin{vmatrix} 1 & 5 \\ 3 & -2 \end{vmatrix}$

10 $\begin{vmatrix} 2 & -2 \\ 1 & 1 \end{vmatrix} = -\begin{vmatrix} -2 & 2 \\ 1 & 1 \end{vmatrix}$

11 $\begin{vmatrix} 0 & 0 & 1 \\ 1 & 0 & 0 \\ 0 & 0 & 2 \end{vmatrix} = 0$

12 $\begin{vmatrix} 1 & 0 & 1 \\ 0 & 0 & 0 \\ 1 & 1 & 0 \end{vmatrix} = 0$

13 $\begin{vmatrix} 1 & -1 & -2 \\ -1 & 2 & 1 \\ 0 & 1 & 1 \end{vmatrix} = \begin{vmatrix} 1 & -1 & 0 \\ -1 & 2 & -1 \\ 0 & 1 & 1 \end{vmatrix}$

14 $\begin{vmatrix} a & 0 & 0 \\ 0 & b & 0 \\ 0 & 0 & c \end{vmatrix} = -\begin{vmatrix} 0 & 0 & a \\ 0 & b & 0 \\ c & 0 & 0 \end{vmatrix}$

In each of Exercises 15–24 find the determinant of the matrix after introducing zeros, as in Example 1.

15 $\begin{bmatrix} 3 & 1 & 0 \\ -2 & 0 & 1 \\ 1 & 3 & -1 \end{bmatrix}$

16 $\begin{bmatrix} -3 & 0 & 4 \\ 1 & 2 & 0 \\ 4 & 1 & -1 \end{bmatrix}$

17 $\begin{bmatrix} 5 & 4 & 3 \\ -3 & 2 & 1 \\ 0 & 7 & -2 \end{bmatrix}$

18 $\begin{bmatrix} 0 & 2 & -6 \\ 5 & 1 & -3 \\ 6 & -2 & 5 \end{bmatrix}$

19 $\begin{bmatrix} 2 & 2 & -3 \\ 3 & 6 & 9 \\ -2 & 5 & 4 \end{bmatrix}$

20 $\begin{bmatrix} 3 & 8 & 5 \\ 5 & 3 & -6 \\ 2 & 4 & -2 \end{bmatrix}$

21 $\begin{bmatrix} 3 & 1 & -2 & 2 \\ 2 & 0 & 1 & 4 \\ 0 & 1 & 3 & 5 \\ -1 & 2 & 0 & -3 \end{bmatrix}$

22 $\begin{bmatrix} 3 & 2 & 0 & 4 \\ -2 & 0 & 5 & 0 \\ 4 & -3 & 1 & 6 \\ 2 & -1 & 2 & 0 \end{bmatrix}$

23 $\begin{bmatrix} 2 & -2 & 0 & 0 & -3 \\ 3 & 0 & 3 & 2 & -1 \\ 0 & 1 & -2 & 0 & 2 \\ -1 & 2 & 0 & 3 & 0 \\ 0 & 4 & 1 & 0 & 0 \end{bmatrix}$

24 $\begin{bmatrix} 2 & 0 & -1 & 0 & 2 \\ 1 & 3 & 0 & 0 & 1 \\ 0 & 4 & 3 & 0 & -1 \\ -1 & 2 & 0 & -2 & 0 \\ 0 & 1 & 5 & 0 & -4 \end{bmatrix}$

25 Prove that

$$\begin{vmatrix} 1 & 1 & 1 \\ a & b & c \\ a^2 & b^2 & c^2 \end{vmatrix} = (a-b)(b-c)(c-a).$$

(*Hint:* See Example 3.)

26 Prove that

$$\begin{vmatrix} 1 & 1 & 1 \\ a & b & c \\ a^3 & b^3 & c^3 \end{vmatrix} = (a-b)(b-c)(c-a)(a+b+c).$$

27 If A is a matrix of order 4 of the form

$$A = \begin{bmatrix} a_{11} & a_{12} & a_{13} & a_{14} \\ 0 & a_{22} & a_{23} & a_{24} \\ 0 & 0 & a_{33} & a_{34} \\ 0 & 0 & 0 & a_{44} \end{bmatrix}$$

show that $|A| = a_{11}a_{22}a_{33}a_{44}$.

28 If

$$A = \begin{bmatrix} a & b & 0 & 0 \\ c & d & 0 & 0 \\ 0 & 0 & e & f \\ 0 & 0 & g & h \end{bmatrix}$$

prove that

$$|A| = \begin{vmatrix} a & b \\ c & d \end{vmatrix} \begin{vmatrix} e & f \\ g & h \end{vmatrix}.$$

29 If $A = (a_{ij})$ and $B = (b_{ij})$ are arbitrary square matrices of order 2, prove that $|AB| = |A|\,|B|$.

30 If $A = (a_{ij})$ is a square matrix of order n and k is any real number, prove that $|kA| = k^n|A|$. (*Hint:* Use (ii) of the Theorem on Row and Column Transformations of a Determinant.)

SECTION 7.7
CRAMER'S RULE

Determinants arise naturally in the study of solutions of systems of linear equations. To illustrate, let us consider the following case of two linear equations in two unknowns:

$$\begin{cases} a_{11}x + a_{12}y = k_1 \\ a_{21}x + a_{22}y = k_2 \end{cases}$$

where at least one nonzero coefficient appears in each equation. We may as well assume that $a_{11} \neq 0$, for otherwise $a_{12} \neq 0$ and we could regard y as the "first" variable instead of x. We shall use elementary row transformations to obtain the matrix of an equivalent system as follows:

$$\begin{bmatrix} a_{11} & a_{12} & k_1 \\ a_{21} & a_{22} & k_2 \end{bmatrix} \xrightarrow{-\frac{a_{21}}{a_{11}}\mathrm{R}_1 + \mathrm{R}_2} \begin{bmatrix} a_{11} & a_{12} & k_1 \\ 0 & a_{22} - \left(\dfrac{a_{12}a_{21}}{a_{11}}\right) & k_2 - \left(\dfrac{a_{21}k_1}{a_{11}}\right) \end{bmatrix}$$

$$\xrightarrow{a_{11}\mathrm{R}_2} \begin{bmatrix} a_{11} & a_{12} & k_1 \\ 0 & (a_{11}a_{22} - a_{12}a_{21}) & (a_{11}k_2 - a_{21}k_1) \end{bmatrix}$$

Thus, the given system is equivalent to

$$\begin{cases} a_{11}x + a_{12}y = k_1 \\ (a_{11}a_{22} - a_{12}a_{21})y = a_{11}k_2 - a_{21}k_1, \end{cases}$$

which may also be written

$$\begin{cases} a_{11}x + a_{12}y = k_1 \\ \begin{vmatrix} a_{11} & a_{12} \\ a_{21} & a_{22} \end{vmatrix} y = \begin{vmatrix} a_{11} & k_1 \\ a_{21} & k_2 \end{vmatrix}. \end{cases}$$

If $\begin{vmatrix} a_{11} & a_{12} \\ a_{21} & a_{22} \end{vmatrix} \neq 0$, we can solve the second equation for y, obtaining

$$y = \frac{\begin{vmatrix} a_{11} & k_1 \\ a_{21} & k_2 \end{vmatrix}}{\begin{vmatrix} a_{11} & a_{12} \\ a_{21} & a_{22} \end{vmatrix}}$$

The corresponding value for x may be found by substituting for y in the first equation. It can be shown that this leads to

$$x = \frac{\begin{vmatrix} k_1 & a_{12} \\ k_2 & a_{22} \end{vmatrix}}{\begin{vmatrix} a_{11} & a_{12} \\ a_{21} & a_{22} \end{vmatrix}}$$

This proves that *if the determinant of the coefficient matrix of a system of two linear equations in two variables is not zero, then the system has a unique solution.* The last two formulas for x and y as quotients of certain determinants constitute what is known as **Cramer's Rule.**

There is an easy way to remember Cramer's Rule. Let

$$D = \begin{bmatrix} a_{11} & a_{12} \\ a_{21} & a_{22} \end{bmatrix}$$

be the coefficient matrix of the system and let D_x denote the matrix obtained from D by replacing the coefficients a_{11}, a_{21} of x by the numbers k_1, k_2, respectively. Similarly, let D_y denote the matrix obtained from D by replacing the coefficients a_{12}, a_{22} of y by the numbers k_1, k_2, respectively. Thus,

$$D_x = \begin{bmatrix} k_1 & a_{12} \\ k_2 & a_{22} \end{bmatrix}, \qquad D_y = \begin{bmatrix} a_{11} & k_1 \\ a_{21} & k_2 \end{bmatrix}.$$

According to the previous discussion, if $|D| \neq 0$, the solution (x, y) is given by

CRAMER'S RULE

$$x = \frac{|D_x|}{|D|}, \qquad y = \frac{|D_y|}{|D|}.$$

EXAMPLE 1 Use Cramer's Rule to solve the system

$$\begin{cases} 2x - 3y = -4 \\ 5x + 7y = 1. \end{cases}$$

Solution The determinant of the coefficient matrix is

$$|D| = \begin{vmatrix} 2 & -3 \\ 5 & 7 \end{vmatrix} = 29.$$

Using the notation introduced previously,

$$|D_x| = \begin{vmatrix} -4 & -3 \\ 1 & 7 \end{vmatrix} = -25, \qquad |D_y| = \begin{vmatrix} 2 & -4 \\ 5 & 1 \end{vmatrix} = 22.$$

Hence,
$$x = \frac{|D_x|}{|D|} = \frac{-25}{29}, \qquad y = \frac{|D_y|}{|D|} = \frac{22}{29}.$$

Thus, the system has the unique solution $(-\frac{25}{29}, \frac{22}{29})$. ■

Cramer's Rule can be extended to systems of n linear equations in n variables $x_1, x_2, \ldots x_n$, where each equation is written in the form

$$a_1x_1 + a_2x_2 + \cdots + a_nx_n = a.$$

To solve such a system, let D denote the coefficient matrix and let D_{x_i} denote the matrix obtained by replacing the coefficients of x_i in D by the column of numbers $k_1, \ldots, k_n$ that appears to the right of the equal signs in the system. It can be shown that if $|D| \neq 0$, then the system has a unique solution given by the following:

CRAMER'S RULE (GENERAL FORM)

$$x_1 = \frac{|D_{x_1}|}{|D|}, \quad x_2 = \frac{|D_{x_2}|}{|D|}, \quad \ldots, \quad x_n = \frac{|D_{x_n}|}{|D|}.$$

EXAMPLE 2 Use Cramer's Rule to solve the system

$$\begin{cases} x \qquad - 2z = 3 \\ \quad - y + 3z = 1 \\ 2x \qquad + 5z = 0. \end{cases}$$

Solution We shall merely list the various determinants, leaving the reader to check the answers:

$$|D| = \begin{vmatrix} 1 & 0 & -2 \\ 0 & -1 & 3 \\ 2 & 0 & 5 \end{vmatrix} = -9, \qquad |D_x| = \begin{vmatrix} 3 & 0 & -2 \\ 1 & -1 & 3 \\ 0 & 0 & 5 \end{vmatrix} = -15,$$

$$|D_y| = \begin{vmatrix} 1 & 3 & -2 \\ 0 & 1 & 3 \\ 2 & 0 & 5 \end{vmatrix} = 27, \qquad |D_z| = \begin{vmatrix} 1 & 0 & 3 \\ 0 & -1 & 1 \\ 2 & 0 & 0 \end{vmatrix} = 6.$$

By Cramer's Rule, the solution is

$$x = \frac{|D_x|}{|D|} = \frac{-15}{-9} = \frac{5}{3}, \quad y = \frac{|D_y|}{|D|} = \frac{27}{-9} = -3, \quad z = \frac{|D_z|}{|D|} = \frac{6}{-9} = -\frac{2}{3}.$$

■

Cramer's Rule is an inefficient method to apply if there are a large number of equations, since many determinants of matrices of high order must be evaluated. Note also that Cramer's Rule cannot be used directly if $|D| = 0$ or if the number of equations is not the same as the number of variables. In general, the matrix method is far superior to Cramer's Rule.

EXERCISES 7.7

1–18 Use Cramer's Rule to solve the systems in Exercises 1–18 of Section 7.2.

19–26 Use Cramer's Rule to solve the systems in Exercises 1–8 of Section 7.3.

27–30 Use Cramer's Rule to solve the systems in Exercises 17, 18, 21, and 22 of Section 7.3.

SECTION 7.8
SYSTEMS OF INEQUALITIES

The discussion of inequalities in Chapter 1 was restricted to inequalities in one variable. Inequalities in several variables can be considered in a manner similar to our work with equations in several variables. For example, expressions of the form

$$3x + y < 5y^2 + 1$$

$$2x^2 \geq 4 - 3y$$

are called **inequalities in x and y.** As with equations, a **solution** of an inequality in x and y is defined as an ordered pair (a, b) that produces a true statement if a and b are substituted for x and y, respectively. The **graph of an inequality** is the graph of all the solutions. Two inequalities are **equivalent** if they have exactly the same solutions. An inequality in x and y can often be simplified by adding an expression in x and y to both sides or by multiplying both sides by some expression (provided care is taken with regard to signs). Similar remarks apply to inequalities in more than two variables. We shall restrict our discussion, however, to the case of inequalities in two variables.

EXAMPLE 1 Find the solutions and sketch the graph of the inequality

$$3x - 3 < 5x - y.$$

Solution The inequalities in the following list are equivalent:

$$\begin{aligned} 3x - 3 &< 5x - y \\ y + 3x - 3 &< 5x \\ y &< 5x - (3x - 3) \\ y &< 2x + 3 \end{aligned}$$

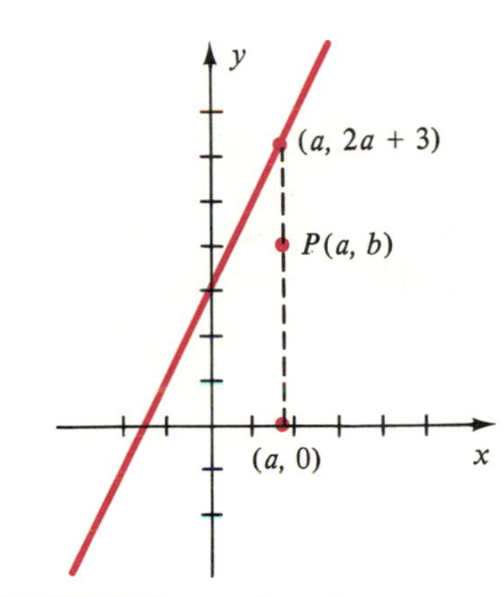

FIGURE 7.5 $y = 2x + 3$

Hence, the solutions consist of all ordered pairs (x, y) such that $y < 2x + 3$. It is convenient to denote the solutions as follows:

$$\{(x, y): y < 2x + 3\}.$$

There is a close relationship between the graph of the inequality $y < 2x + 3$ and the graph of the equation $y = 2x + 3$. The graph of the equation is the line sketched in Figure 7.5. For each real number a, the point on the line with x-coordinate a has coordinates $(a, 2a + 3)$. A point $P(a, b)$ belongs to the graph of the *inequality* if and only if $b < 2a + 3$; that is, if and only if the point $P(a, b)$ lies directly below the point with coordinates $(a, 2a + 3)$ as shown in Figure 7.5. It follows that the graph of the inequality $y < 2x + 3$ consists of all points that lie below the line $y = 2x + 3$. In Figure 7.6. we have shaded a portion of the graph of the inequality. Dashes used for the line indicate that it is not part of the graph. ■

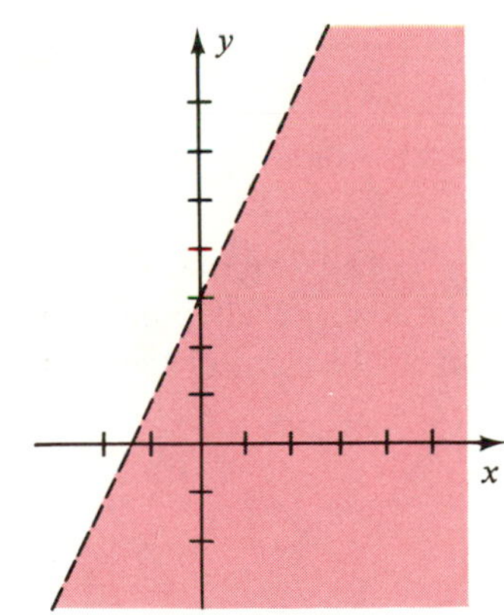

FIGURE 7.6 $y < 2x + 3$

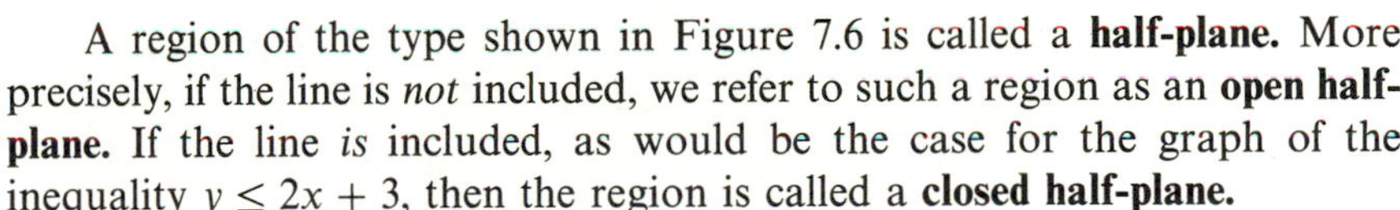
A region of the type shown in Figure 7.6 is called a **half-plane.** More precisely, if the line is *not* included, we refer to such a region as an **open half-plane.** If the line *is* included, as would be the case for the graph of the inequality $y \leq 2x + 3$, then the region is called a **closed half-plane.**

By an argument similar to that used in Example 1, it can be shown that the graph of the inequality $y > 2x + 3$ is the open half-plane that lies *above* the line $y = 2x + 3$.

If an inequality involves only polynomials of the first degree in x and y, as was the case in Example 1, it is called a **linear inequality.**

The procedure used in Example 1 can be generalized to inequalities of the form $y < f(x)$ for any function f. Specifically, the following theorem can be established.

THEOREM

> If f is a function, then the graph of the inequality $y < f(x)$ is the set of points that lie *below* the graph of the equation $y = f(x)$. Similarly, the graph of $y > f(x)$ is the set of points that lie *above* the graph of $y = f(x)$.

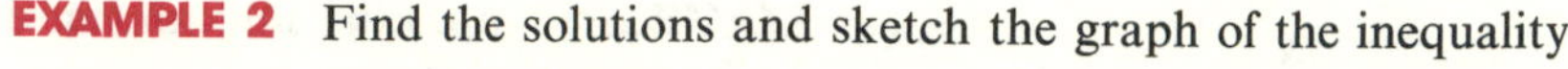

EXAMPLE 2 Find the solutions and sketch the graph of the inequality

$$x(x+1) - 2y > 3(x - y).$$

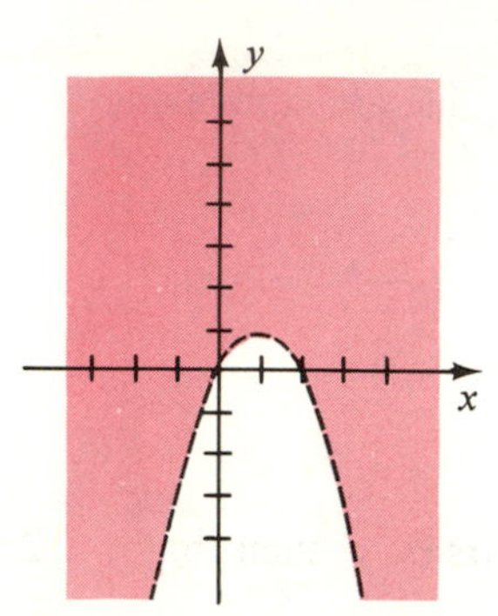

FIGURE 7.7 $y > 2x - x^2$

Solution The inequality is equivalent to

$$x^2 + x - 2y > 3x - 3y.$$

Adding $3y - x^2 - x$ to both sides, we obtain

$$y > 2x - x^2.$$

Hence, the solutions are $\{(x, y): y > 2x - x^2\}$. To find the graph, we begin by sketching the graph of $y = 2x - x^2$ (a parabola) with dashes as illustrated in Figure 7.7. By the preceding theorem, the graph is the region above the parabola, as indicated by the shaded portion of the figure. ■

The following result can also be proved.

THEOREM

> If g is a function, then the graph of the inequality $x < g(y)$ is the set of points to the *left* of the graph of the equation $x = g(y)$. Similarly, the graph of $x > g(y)$ is the set of points to the *right* of the graph of $x = g(y)$.

EXAMPLE 3 Sketch the graph of $x \geq y^2$.

Solution The graph of the equation $x = y^2$ is a parabola. By the preceding theorem, the graph of the inequality consists of all points on the parabola, together with the points in the region to the right of the parabola (see Figure 7.8).

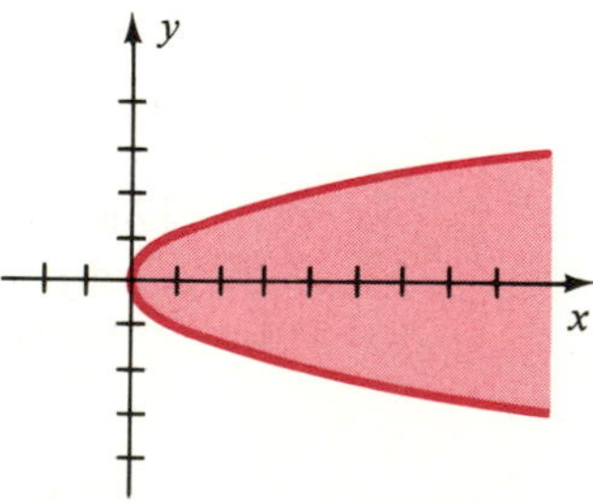

FIGURE 7.8 $x \geq y^2$ ■

It is sometimes necessary to work simultaneously with several inequalities in two variables. In this case we refer to the given inequalities as a **system of inequalities.** The **solutions of a system** of inequalities are, by definition, the common solutions of all the inequalities in the system. It should be clear how to define **equivalent systems** and the **graph of a system** of inequalities. The following examples illustrate a method for solving systems of inequalities.

EXAMPLE 4 Find the solutions and sketch the graph of the system

$$\begin{cases} x + y \le 4 \\ 2x - y \le 4. \end{cases}$$

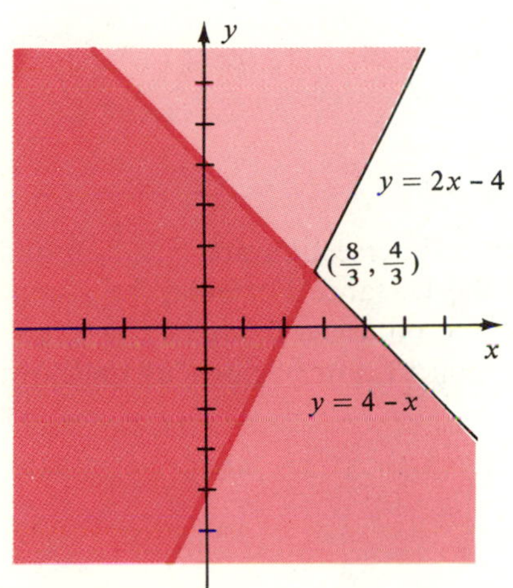

FIGURE 7.9

Solution The system is equivalent to

$$\begin{cases} y \le 4 - x \\ y \ge 2x - 4. \end{cases}$$

We begin by sketching the graphs of the lines $y = 4 - x$ and $y = 2x - 4$. The lines intersect at the point $(\frac{8}{3}, \frac{4}{3})$ shown in Figure 7.9. The graph of $y \le 4 - x$ includes the points on the graph of $y = 4 - x$, together with the points that lie below this line. The graph of $y \ge 2x - 4$ includes the points on the graph of $y = 2x - 4$, together with the points that lie above this line. A portion of each of these regions is shown in Figure 7.9. The graph of the system consists of the points that are in *both* regions as indicated by the double-shaded portion of the figure. ■

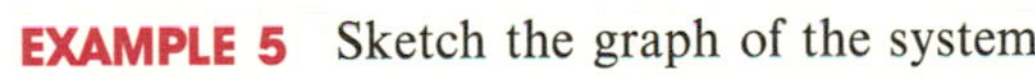

EXAMPLE 5 Sketch the graph of the system

$$\begin{cases} x + y \le 4 \\ 2x - y \le 4 \\ x \ge 0 \\ y \ge 0. \end{cases}$$

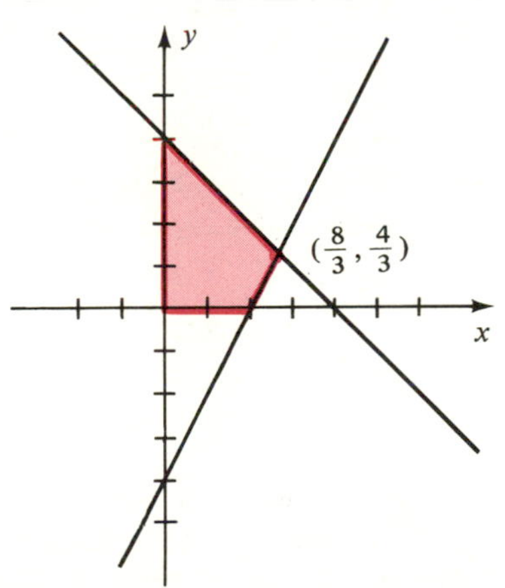

FIGURE 7.10

Solution The first two inequalities are the same as those considered in Example 4, and hence the points on the graph of the system must lie within the double-shaded region shown in Figure 7.9. In addition, the third and fourth inequalities in the system tell us that the points must lie in the first quadrant. This gives us the region shown in Figure 7.10. ■

EXAMPLE 6 Sketch the graph of the system

$$\begin{cases} x^2 + y^2 \le 1 \\ (x - 1)^2 + y^2 \le 1. \end{cases}$$

Solution The graph of the equation $x^2 + y^2 = 1$ is a unit circle with center at the origin, and the graph of $(x - 1)^2 + y^2 = 1$ is a unit circle with center at the point $C(1, 0)$. To find the points of intersection of the two circles,

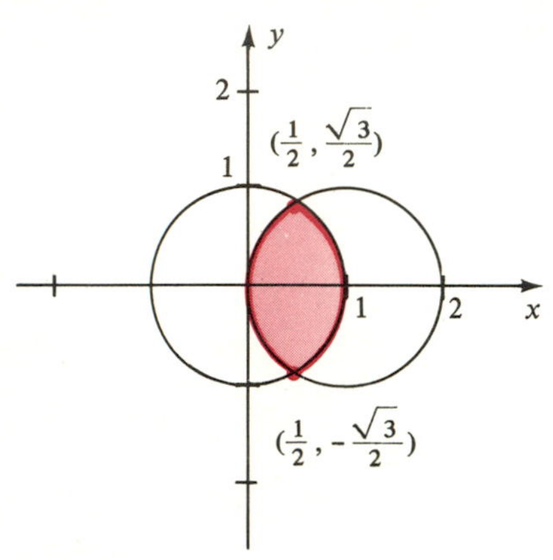

FIGURE 7.11

let us solve the equation $x^2 + y^2 = 1$ for y^2, obtaining $y^2 = 1 - x^2$. Substituting for y^2 in $(x - 1)^2 + y^2 = 1$ leads to the following equations:

$$\begin{aligned}(x - 1)^2 + (1 - x^2) &= 1\\ x^2 - 2x + 1 + 1 - x^2 &= 1\\ -2x &= -1\\ x &= \tfrac{1}{2}.\end{aligned}$$

The corresponding values for y are given by

$$y^2 = 1 - x^2 = 1 - (\tfrac{1}{2})^2 = \tfrac{3}{4}$$

and hence $y = \pm\sqrt{3}/2$. Thus, the points of intersection are $(\frac{1}{2}, \sqrt{3}/2)$ and $(\frac{1}{2}, -\sqrt{3}/2)$ as shown in Figure 7.11. By the Distance Formula, the graphs of the inequalities are the regions within and on the two circles. The graph of the system consists of the points common to both regions, as indicated by the shaded portion of the figure. ■

EXERCISES 7.8

In Exercises 1–10 find the solutions and sketch the graph of the given inequality.

1 $3x - 2y < 6$

2 $4x + 3y < 12$

3 $2x + 3y \geq 2y + 1$

4 $2x - y > 3$

5 $y + 2 < x^2$

6 $y^2 - x \leq 0$

7 $x^2 + 1 \leq y$

8 $y - x^3 < 1$

9 $yx^2 \geq 1$

10 $x^2 + 4 \geq y$

In Exercises 11–24 sketch the graph of the system of inequalities.

11 $\begin{cases} 3x + y < 3 \\ 4 - y < 2x \end{cases}$

12 $\begin{cases} y + 2 < 2x \\ y - x > 4 \end{cases}$

13 $\begin{cases} y - x < 0 \\ 2x + 5y < 10 \end{cases}$

14 $\begin{cases} 2y - x \leq 4 \\ 3y + 2x < 6 \end{cases}$

15 $\begin{cases} 3x + y \leq 6 \\ y - 2x \geq 1 \\ x \geq -2 \\ y \leq 4 \end{cases}$

16 $\begin{cases} 3x - 4y \geq 12 \\ x - 2y \leq 2 \\ x \geq 9 \\ y \leq 5 \end{cases}$

17 $\begin{cases} x^2 + y^2 \leq 4 \\ x + y \geq 1 \end{cases}$

18 $\begin{cases} x^2 + y^2 > 1 \\ x^2 + y^2 < 4 \end{cases}$

19 $\begin{cases} x^2 \leq 1 - y \\ x \geq 1 + y \end{cases}$

20 $\begin{cases} x - y^2 < 0 \\ x + y^2 > 0 \end{cases}$

21 $\begin{cases} y < 3^x \\ y > 2^x \\ x \geq 0 \end{cases}$

22 $\begin{cases} y \geq \log x \\ y - x \leq 1 \\ x \geq 1 \end{cases}$

23 $\begin{cases} y \leq \log x \\ y + x \geq 1 \\ x \leq 10 \end{cases}$

24 $\begin{cases} y \leq 3^{-x} \\ y \geq 2^{-x} \\ y < 9 \end{cases}$

25 The manager of a baseball team wishes to buy bats and balls, costing \$13 and \$3, respectively. If the maximum amount available is \$160 and if at least twelve balls and five bats are required, find a system of inequalities that describes all the possibilities, and sketch the graph.

26 An office worker wishes to purchase some 40-cent postage stamps and some 50-cent stamps, totaling not more than \$35. Moreover, there should be at least twice as many 40-cent stamps as 50-cent stamps and more than ten 40-cent stamps. Find a system of inequalities that describes all the possibilities, and sketch the graph.

27 A store sells two brands of television sets. Customer demand indicates that it is necessary to stock at least twice as many sets of brand *A* as of brand *B*. It is also necessary to have on hand at least 20 of brand *A* and 10 of brand *B*. If there is room for not more than 100 sets in the store, find a system of inequalities that describes all possibilities, and sketch the graph.

28 An auditorium contains 600 seats. For a certain event it is planned to charge \$8.00 for some seats and \$5.00 for others. At least 225 tickets are to be sold for \$5.00, and total sales of more than \$3,000 is desired. Find a system of inequalities that describes all possibilities, and sketch the graph.

29 A woman wishes to invest \$15,000 in two different savings accounts. She also wants to have at least \$2,000 in each account, with the amount in one account being at least three times that in the other. Find a system of inequalities that describes all possibilities, and sketch the graph.

30 The manager of a college bookstore stocks two types of notebooks, the first wholesaling for 55 cents and the second for 85 cents. If the maximum amount to be spent is \$600 and if an inventory of at least 300 of the 85-cent variety and 400 of the 55-cent variety is desired, find a system of inequalities that describes all possibilities, and sketch the graph.

SECTION 7.9
LINEAR PROGRAMMING

In applications, problems sometimes arise that require finding solutions of systems of inequalities. A typical problem consists of finding maximum and minimum values of expressions involving variables that are subject to various constraints. If all the expressions and inequalities are linear in the variables, then a technique called **linear programming** may be used to help solve such problems. This technique has become very important in businesses where decisions must be made concerning the best use of stock, parts, or manufacturing processes. Usually the objective of management is to maximize profit or to minimize cost. Since there are often many choices, it may be extremely difficult to arrive at a correct decision. A mathematical theory such as that afforded by linear programming can simplify the task considerably. The logical development of the theorems and techniques that are needed would take us beyond the objectives of this text. We shall, therefore, limit ourselves to several examples.

EXAMPLE 1 A manufacturer of a certain product has two warehouses, W_1 and W_2. There are 80 units of the product stored at W_1 and 70 units at W_2. Two customers, *A* and *B*, order 35 units and 60 units, respectively. The shipping cost from each warehouse to *A* and *B* is determined according to the following table. How should the order be filled to minimize the total shipping cost?

Warehouse	Customer	Shipping cost per unit
W_1	A	\$ 8
W_1	B	12
W_2	A	10
W_2	B	13

Solution If we let x denote the number of units to be sent to A from W_1, then $35 - x$ units must be sent from W_2 to A. Similarly, if y denotes the number of units to be sent from W_1 to B, then $60 - y$ units must be sent from W_2 to B. We wish to determine values for x and y that make the total shipping costs minimal. We first note that, since x and y are between 35 and 60, respectively, the pair (x, y) must be a solution of the following system of inequalities:

$$0 \le x \le 35, \qquad 0 \le y \le 60.$$

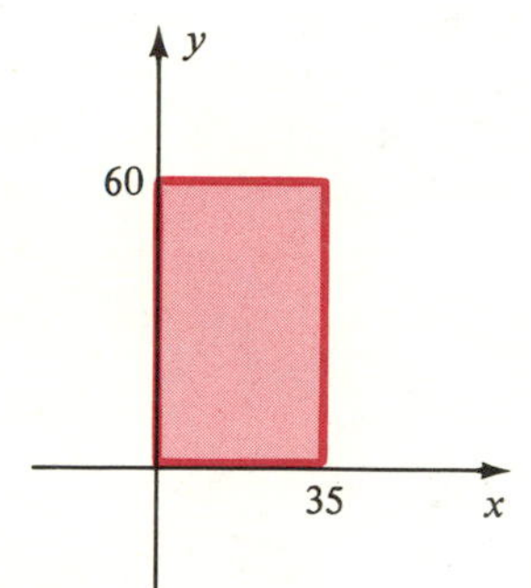

FIGURE 7.12

The graph of this system is the rectangular region shown in Figure 7.12.

There are further constraints on x and y that make it possible to reduce the size of the region in Figure 7.12. Since the total number of units shipped from W_1 cannot exceed 80 and the total shipped from W_2 cannot exceed 70, the pair (x, y) must also be a solution of the system

$$\begin{cases} x + y \le 80 \\ (35 - x) + (60 - y) \le 70. \end{cases}$$

This system is equivalent to

$$\begin{cases} x + y \le 80 \\ x + y \ge 25. \end{cases}$$

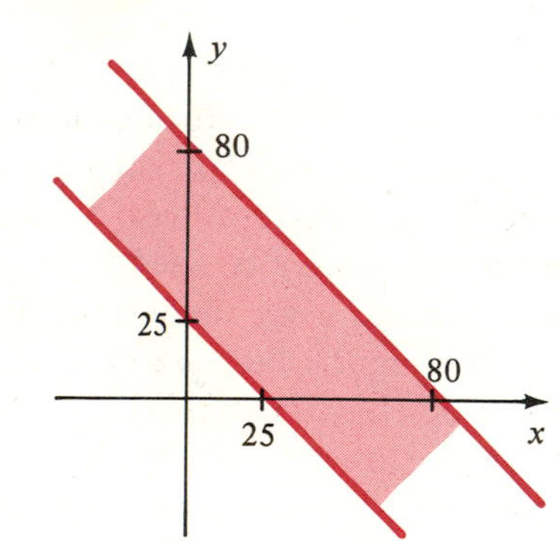

FIGURE 7.13

The graph of the last system is the region between the parallel lines $x + y = 80$ and $x + y = 25$ (see Figure 7.13). Since the pair (x, y) that we seek must be a solution of this system, and also of the system $0 \le x \le 35$, $0 \le y \le 60$, the corresponding point must lie in the region shown in Figure 7.14.

Let C denote the total cost (in dollars) of shipping the merchandise to A and B. We see from the table at the top of the page that the cost of shipping the 35 units to A is $8x + 10(35 - x)$, whereas the cost of shipping the 60 units to B is $12y + 13(60 - y)$. Hence, the total cost is

$$C = (8x + 350 - 10x) + (12y + 780 - 13y)$$

or

$$C = 1130 - 2x - y.$$

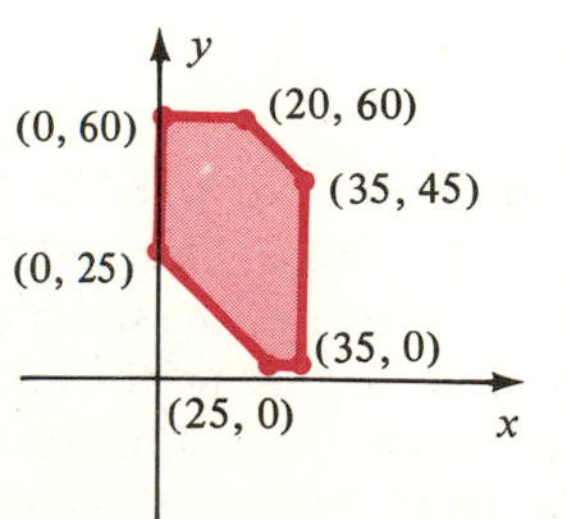

FIGURE 7.14

For each point (x, y) of the region shown in Figure 7.14 there corresponds a value for C. For example, at (20, 40)

$$C = 1130 - 40 - 40 = 1050;$$

and at (10, 50), $C = 1130 - 20 - 50 = 1060.$

Since x and y must be integers, there is only a finite number of possible values for C. By checking each possibility, we could find the pair (x, y) that produces the smallest cost. However, since there is a very large number of pairs, the task of checking each one would be very tedious. This is where the theory developed in linear programming is helpful. It can be shown that if we are interested in the value C of a linear expression $ax + by + c$, where each pair (x, y) is a solution of a system of linear inequalities, and hence corresponds to a point that is in the intersection of half-planes, then C takes on its maximum and minimum values at a point of intersection of the lines that determine the half-planes. This means that to determine the minimum (or maximum) value of C we need only check the points (0, 25), (0, 60), (20, 60), (35, 45), (35, 0), and (25, 0) shown in Figure 7.14. The values are displayed in the following table.

Point	$1130 - 2x - y = C$
(0, 25)	$1130 - 2(0) - 25 = 1105$
(0, 60)	$1130 - 2(0) - 60 = 1070$
(20, 60)	$1130 - 2(20) - 60 = 1030$
(35, 45)	$1130 - 2(35) - 45 = 1015$
(35, 0)	$1130 - 2(35) - 0 = 1060$
(25, 0)	$1130 - 2(25) - 0 = 1080$

According to our remarks, the minimal shipping cost \$1,015 occurs if $x = 35$ and $y = 45$. This means that the manufacturer should ship all of the units to A from W_1. In addition, the manufacturer should ship 45 units to B from W_1 and 15 units to B from W_2. Note that the *maximum* shipping cost will occur if $x = 0$ and $y = 25$, that is, if all 35 units are shipped to A from W_2 and if B receives 25 units from W_1 and 35 units from W_2. ■

The preceding example demonstrates how linear programming can be used to minimize the cost in a certain situation. The next example illustrates maximization of profit.

EXAMPLE 2 A firm manufactures two products X and Y. For each product it is necessary to use three different machines A, B, and C. To manufacture one unit of product X, machine A must be used for 3 hours, machine B for 1 hour, and machine C for 1 hour. To manufacture one unit of product Y requires 2 hours on A, 2 hours on B, and 1 hour on C. The profit on product X is \$500 per unit and the profit on product Y is \$350 per unit. Machine A is available for a total of 24 hours per day; however, B can only be used for 16 hours and C for 9 hours. If the machines are available when needed (subject to the noted total hour restrictions), determine the number of units of each product that should be manufactured each day in order to maximize the profit.

Solution The following table summarizes the data given in the statement of the problem.

Machine	Hours required for 1 unit of X	Hours required for one unit of Y	Hours available
A	3	2	24
B	1	2	16
C	1	1	9

Let x and y denote the number of units of products X and Y, respectively, to be produced per day. If each unit of product X requires 3 hours on machine A, then x units require $3x$ hours. Similarly, if each unit of product Y requires 2 hours on A, then y units require $2y$ hours. Hence, the total number of hours per day that machine A must be used is $3x + 2y$. Since A can be used for at most 24 hours per day,

$$3x + 2y \le 24.$$

Using the same type of reasoning on rows two and three of the table, we see that

$$x + 2y \le 16$$

$$x + y \le 9.$$

This system of three linear inequalities, together with the obvious inequalities

$$x \ge 0, \qquad y \ge 0$$

states, in mathematical form, the restraints that occur in the manufacturing process. The graph of the preceding system of five linear inequalities is sketched in Figure 7.15. The points shown in the figure are found by solving

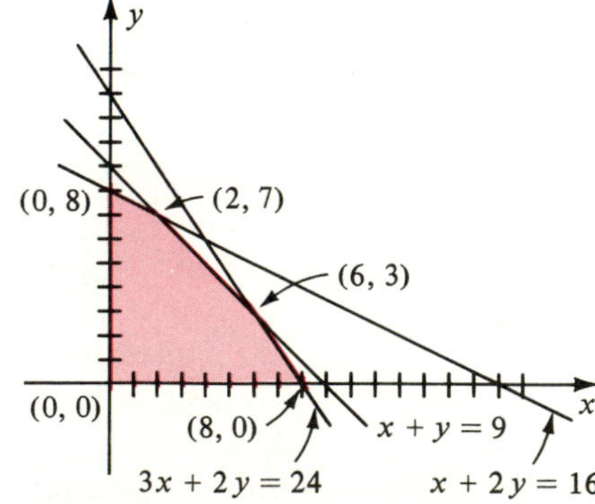

FIGURE 7.15

systems of linear equations. Specifically, (6, 3) is a solution of the system $3x + 2y = 24$, $x + y = 9$, and (2, 7) is a solution of the system $x + 2y = 16$, $x + y = 9$.

Since the production of each unit of product X yields a profit of $500, and each unit of product Y yields a profit of $350, the profit P obtained by producing x units of X together with y units of Y is

$$P = 500x + 350y.$$

The maximum value of P must occur at one of the points shown in Figure 7.15. The values of P at all the points are shown in the following table.

(x, y)	$500x + 350y = P$
(0, 8)	$500(0) + 350(8) = 2800$
(2, 7)	$500(2) + 350(7) = 3450$
(6, 3)	$500(6) + 350(3) = 4050$
(8, 0)	$500(8) + 350(0) = 4000$
(0, 0)	$500(0) + 350(0) = 0$

We see from the table that a maximum profit of $4,050 occurs for a daily production of 6 units of product X and 3 units of product Y. ■

The two illustrations given in this section are elementary problems in linear programming that can be solved by rather crude methods. The much more complicated problems that occur in practice are usually solved by employing matrix techniques that are adapted for solutions by computers.

EXERCISES 7.9

1 A manufacturer of tennis rackets makes a profit of $15 on each Set Point racket and $8 on each Double Fault racket. To meet dealer demand, daily production of Double Faults should be between 30 and 80, whereas the number of Set Points produced should be between 10 and 30. To maintain high quality, the total number of rackets produced should not exceed 80 per day. How many of each type should be manufactured daily to maximize the profit?

2 A manufacturer of CB radios makes a profit of $25 on a deluxe model and $30 on a standard model. The company wishes to produce at least 80 deluxe models and at least 100 standard models per day. To maintain high quality, the daily production should not exceed 200 radios. How many of each type should be produced daily in order to maximize the profit?

3 Two substances S and T each contain two types of ingredients I and G. One pound of S contains 2 ounces of I and 4 ounces of G. One pound of T contains 2 ounces of I and 6 ounces of G. A manufacturer plans to combine quantities of the two substances to obtain a mixture that contains at least 9 ounces of I and 20 ounces of G. If the cost of S is $3.00 per pound and the cost of T is $4.00 per pound, how much of each substance should be used to keep the cost to a minimum?

4 A stationery company makes two types of notebooks: Type M sells for $1.25 and type N sells for $0.90. It costs the company $1.00 to produce one notebook of type M and $0.75 to produce one of type N. The company has the facilities to manufacture between 2,000 and 3,000 of type M and between 3,000 and 6,000 of type N, but not more than 7,000 altogether. How many notebooks of

each type should be manufactured to maximize the difference between the selling prices and the costs of production?

5 Refer to Example 1 of this section. If the shipping costs are \$12 per unit from W_1 to A, \$10 per unit from W_2 to A, \$16 per unit from W_1 to B, and \$12 per unit from W_2 to B, determine how the order should be filled to minimize shipping costs.

6 A coffee company purchases mixed lots of coffee beans and then grades them into premium, regular, and unusable beans. The company needs at least 280 tons of premium-grade and 200 tons of regular-grade coffee beans. The company can purchase ungraded coffee from two suppliers in any amount desired. Samples from the two suppliers contain the following percentages of premium, regular, and unusable beans:

Supplier	Premium	Regular	Unusable
A	20%	50%	30%
B	40%	20%	40%

If supplier A charges \$125 per ton and B charges \$200 per ton, how much should the company purchase from each supplier to fulfill its needs at minimum cost?

7 A farmer has 100 acres available for planting two crops A and B. The seed for crop A costs \$4 per acre and the seed for crop B costs \$6 per acre. The total cost of labor will amount to \$20 per acre for crop A and \$10 per acre for crop B. The expected income from crop A is \$110 per acre, and from B, \$150 per acre. If the farmer does not wish to spend more than \$480 for seed and \$1,400 for labor, how many acres of each crop should be planted to obtain the maximum profit?

8 A firm manufactures two products A and B. For each product it is necessary to employ two different machines X and Y. To manufacture product A, machine X must be used for $\frac{1}{2}$ hour and machine Y for 1 hour. To manufacture product B, machine X must be used for 2 hours and machine Y for 2 hours. The profit on product A is \$20 per unit and the profit on B is \$50 per unit. If machine X can be used for 8 hours per day and machine Y for 12 hours per day, determine how many units of each product should be manufactured each day to maximize profit.

9 Three substances X, Y, and Z each contain four ingredients A, B, C, and D. The percentage of each ingredient and the cost in cents per ounce of substances are given in the following table.

Substance	*A*	*B*	*C*	*D*	Cost/ Ounce
X	20%	10%	25%	45%	25¢
Y	20%	40%	15%	25%	35¢
Z	10%	20%	25%	45%	50¢

If the cost is to be minimum, how many ounces of each substance should be combined to obtain a mixture of 20 ounces containing at least 14% A, 16% B, and 20% C? What combination would make the cost greatest?

10 A man plans to operate a stand at a one-day fair at which he will sell bags of peanuts and bags of candy. He has \$100 available to purchase his stock, which will cost 10¢ per bag of peanuts and 20¢ per bag of candy. He intends to sell the peanuts at 25¢ and the candy at 40¢ per bag. His stand can accomodate up to 500 bags of peanuts and 400 bags of candy. From past experience he knows that he will sell no more than a total of 700 bags. Find the number of bags of each that he should have available in order to maximize his profit. What is the maximum profit?

SECTION 7.10
REVIEW

Define or discuss each of the following.

1 System of equations
2 Solution of a system of equations
3 Equivalent systems of equations
4 System of linear equations
5 An $m \times n$ matrix
6 A square matrix of order n

7 The coefficient matrix of a system of linear equations; the augmented matrix
8 Elementary row transformations
9 Homogeneous system of linear equations
10 The sum and product of two matrices
11 Zero matrix
12 Identity matrix
13 Inverse of a matrix
14 Determinant
15 Minor
16 Cofactor
17 Properties of determinants
18 Cramer's Rule
19 System of inequalities
20 Linear programming

EXERCISES 7.10

Find the solutions of the systems of equations in Exercises 1–16.

1 $\begin{cases} 2x - 3y = 4 \\ 5x + 4y = 1 \end{cases}$

2 $\begin{cases} x - 3y = 4 \\ -2x + 6y = 2 \end{cases}$

3 $\begin{cases} y + 4 = x^2 \\ 2x + y = -1 \end{cases}$

4 $\begin{cases} x^2 + y^2 = 25 \\ x - y = 7 \end{cases}$

5 $\begin{cases} 9x^2 + 16y^2 = 140 \\ x^2 - 4y^2 = 4 \end{cases}$

6 $\begin{cases} 2x = y^2 + 3z \\ x = y^2 + z - 1 \\ x^2 = xz \end{cases}$

7 $\begin{cases} \dfrac{1}{x} + \dfrac{3}{y} = 7 \\ \dfrac{4}{x} - \dfrac{2}{y} = 1 \end{cases}$

8 $\begin{cases} 2^x + 3^{y+1} = 10 \\ 2^{x+1} - 3^y = 5 \end{cases}$

9 $\begin{cases} 3x + y - 2z = -1 \\ 2x - 3y + z = 4 \\ 4x + 5y - z = -2 \end{cases}$

10 $\begin{cases} x + 3y = 0 \\ y - 5z = 3 \\ 2x + z = -1 \end{cases}$

11 $\begin{cases} 4x - 3y - z = 0 \\ x - y - z = 0 \\ 3x - y + 3z = 0 \end{cases}$

12 $\begin{cases} 2x + y - z = 0 \\ x - 2y + z = 0 \\ 3x + 3y + 2z = 0 \end{cases}$

13 $\begin{cases} 4x + 2y - z = 1 \\ 3x + 2y + 4z = 2 \end{cases}$

14 $\begin{cases} 2x + y = 6 \\ x - 3y = 17 \\ 3x + 2y = 7 \end{cases}$

15 $\begin{cases} \dfrac{4}{x} + \dfrac{1}{y} + \dfrac{2}{z} = 4 \\ \dfrac{2}{x} + \dfrac{3}{y} - \dfrac{1}{z} = 1 \\ \dfrac{1}{x} + \dfrac{1}{y} + \dfrac{1}{z} = 4 \end{cases}$

16 $\begin{cases} 2x - y + 3z - w = -3 \\ 3x + 2y - z + w = 13 \\ x - 3y + z - 2w = -4 \\ -x + y + 4z + 3w = 0 \end{cases}$

Find the solutions and sketch the graphs of the systems in Exercises 17–20.

17 $\begin{cases} x^2 + y^2 < 16 \\ y - x^2 > 0 \end{cases}$

18 $\begin{cases} y - x \leq 0 \\ y + x \geq 2 \\ x \leq 5 \end{cases}$

19 $\begin{cases} x - 2y \leq 2 \\ y - 3x \leq 4 \\ 2x + y \leq 4 \end{cases}$

20 $\begin{cases} x^2 - y < 0 \\ y - 2x < 5 \\ xy < 0 \end{cases}$

Find the determinants of the matrices in Exercises 21–30.

21 $[-6]$

22 $\begin{bmatrix} 3 & 4 \\ -6 & -5 \end{bmatrix}$

23 $\begin{bmatrix} 3 & -4 \\ 6 & 8 \end{bmatrix}$

24 $\begin{bmatrix} 0 & 4 & -3 \\ 2 & 0 & 4 \\ -5 & 1 & 0 \end{bmatrix}$

25 $\begin{bmatrix} 2 & -3 & 5 \\ -4 & 1 & 3 \\ 3 & 2 & -1 \end{bmatrix}$

26 $\begin{bmatrix} 3 & 1 & -2 \\ -5 & 2 & -4 \\ 7 & 3 & -6 \end{bmatrix}$

27 $\begin{bmatrix} 5 & 0 & 0 & 0 \\ 6 & -3 & 0 & 0 \\ 1 & 4 & -4 & 0 \\ 7 & 2 & 3 & 2 \end{bmatrix}$

28 $\begin{bmatrix} 1 & 2 & 0 & 3 & 1 \\ -2 & -1 & 4 & 1 & 2 \\ 3 & 0 & -1 & 0 & -1 \\ 2 & -3 & 2 & -4 & 2 \\ -1 & 1 & 0 & 1 & 3 \end{bmatrix}$

29 $\begin{bmatrix} 2 & 0 & 1 & 0 & -1 \\ 0 & 1 & 0 & 1 & 2 \\ 2 & -2 & 1 & -2 & 0 \\ 0 & 0 & -2 & 0 & 1 \\ 1 & -1 & 0 & -1 & 0 \end{bmatrix}$

30 $\begin{bmatrix} 1 & 2 & 0 & 0 & 0 \\ 3 & 4 & 0 & 0 & 0 \\ 0 & 0 & 1 & 2 & 3 \\ 0 & 0 & 2 & -1 & 1 \\ 0 & 0 & 1 & 3 & -1 \end{bmatrix}$

31 Find the determinant of the $n \times n$ matrix (a_{ij}) where $a_{ij} = 0$ if $i \neq j$.

32 Without expanding, show that

$$\begin{vmatrix} 1 & a & b+c \\ 1 & b & a+c \\ 1 & c & a+b \end{vmatrix} = 0.$$

Find the inverses of the matrices in Exercises 33–36.

33 $\begin{bmatrix} 5 & -4 \\ -3 & 2 \end{bmatrix}$

34 $\begin{bmatrix} 2 & -1 & 0 \\ 1 & 4 & 2 \\ 3 & -2 & 1 \end{bmatrix}$

35 $\begin{bmatrix} 3 & -1 & 0 & 0 \\ 1 & 2 & 0 & 0 \\ 0 & 0 & -1 & -2 \\ 0 & 0 & 5 & 3 \end{bmatrix}$

36 $\begin{bmatrix} 2 & 0 & 0 & 0 \\ 0 & 3 & 0 & 0 \\ 0 & 0 & 4 & 0 \\ 0 & 0 & 0 & 5 \end{bmatrix}$

In Exercises 37–46 express as a single matrix.

37 $\begin{bmatrix} 2 & -1 & 0 \\ 3 & 0 & -2 \end{bmatrix} \begin{bmatrix} 2 & -1 & 3 \\ 0 & 3 & 0 \\ 1 & 4 & 2 \end{bmatrix}$

38 $\begin{bmatrix} 4 & 2 \\ 5 & -3 \end{bmatrix} \begin{bmatrix} 3 \\ 7 \end{bmatrix}$

39 $\begin{bmatrix} 2 & 0 \\ 1 & 4 \\ -2 & 3 \end{bmatrix} \begin{bmatrix} 0 & 2 & -3 \\ 4 & 5 & 1 \end{bmatrix}$

40 $\begin{bmatrix} 0 & -2 & 3 \\ 4 & 1 & 2 \end{bmatrix} \begin{bmatrix} 2 & 0 \\ 3 & 8 \\ 2 & -7 \end{bmatrix}$

41 $2\begin{bmatrix} 0 & -1 & -4 \\ 3 & 2 & 1 \end{bmatrix} - 3\begin{bmatrix} 4 & -2 & 1 \\ 0 & 5 & -1 \end{bmatrix}$

42 $\begin{bmatrix} 1 & 3 \\ 2 & 4 \end{bmatrix} \begin{bmatrix} a & 0 \\ 0 & a \end{bmatrix}$

43 $\begin{bmatrix} a & 0 \\ 0 & b \end{bmatrix} \begin{bmatrix} 1 & 3 \\ 2 & 4 \end{bmatrix}$

44 $\begin{bmatrix} 3 & 2 \\ 0 & 0 \end{bmatrix} \begin{bmatrix} -2 & 0 \\ 3 & 0 \end{bmatrix}$

45 $\begin{bmatrix} 1 & 2 \\ 3 & 4 \end{bmatrix} \left\{ \begin{bmatrix} 2 & -4 \\ 3 & 7 \end{bmatrix} + \begin{bmatrix} 1 & 5 \\ -2 & -3 \end{bmatrix} \right\}$

46 $\begin{bmatrix} 3 & 2 & 5 \\ -3 & 4 & 7 \\ 6 & 5 & 1 \end{bmatrix} \begin{bmatrix} 3 & 2 & 5 \\ -3 & 4 & 7 \\ 6 & 5 & 1 \end{bmatrix}^{-1}$

Verify Exercises 47 and 48 without expanding the determinants.

47 $\begin{vmatrix} 2 & 4 & -6 \\ 1 & 4 & 3 \\ 2 & 2 & 0 \end{vmatrix} = 12 \begin{vmatrix} 1 & 1 & -1 \\ 1 & 2 & 1 \\ 2 & 1 & 0 \end{vmatrix}$

48 $\begin{vmatrix} a & b & c \\ d & e & f \\ g & h & k \end{vmatrix} = \begin{vmatrix} d & e & f \\ g & h & k \\ a & b & c \end{vmatrix}$

49 Suppose that $A = (a_{ij})$ is a square matrix of order n such that $a_{ij} = 0$ if $i < j$. Prove that

$$|A| = a_{11}a_{22} \cdots a_{nn}.$$

50 If $A = (a_{ij})$ is any 2×2 matrix such that $|A| \neq 0$, prove that A has an inverse and find a general formula for A^{-1}.

SEQUENCES AND SERIES

The method of *mathematical induction*, which is considered in the first section of this chapter, is needed to prove that certain statements are true for every positive integer. In particular, in Section 8.2 it is used to prove the *Binomial Theorem*. Section 8.3 contains a discussion of *sequences* and *summation notation*. Of special interest are the *arithmetic* and *geometric sequences* considered in Sections 8.4 and 8.5.

SECTION 8.1

MATHEMATICAL INDUCTION

If n is a positive integer, let P_n denote the statement

$$(xy)^n = x^n y^n$$

where x and y are real numbers. Thus, P_1 represents $(xy)^1 = x^1 y^1$, P_2 denotes $(xy)^2 = x^2 y^2$, P_3 is $(xy)^3 = x^3 y^3$, etc. It is easy to show that P_1, P_2, and P_3 are *true* statements. However, since the set of positive integers is infinite, it is impossible to check the validity of P_n for every positive integer n. To give a proof, the method of mathematical induction is required. This method is based on the following fundamental axiom.

AXIOM OF MATHEMATICAL INDUCTION

Suppose a set S of positive integers has the following two properties:

(i) S contains the integer 1.

(ii) Whenever S contains a positive integer k, S also contains $k + 1$.

Then S contains every positive integer.

If S is a set of positive integers that satisfies property (ii), then whenever S contains an arbitrary positive integer k, it must also contain the next positive integer, $k + 1$. If S also satisfies property (i), then S contains 1 and hence by (ii), S contains $1 + 1$, or 2. Applying (ii) again, we see that S contains $2 + 1$, or 3. Once again, S must contain $3 + 1$, or 4. If we continue in this manner, it can be argued that if n is any *specific* positive integer, then n is in S, since we can proceed a step at a time, eventually reaching n. Although this argument does not *prove* the axiom, it certainly makes it plausible.

We shall use the preceding axiom to establish the following fundamental principle.

PRINCIPLE OF MATHEMATICAL INDUCTION

If with each positive integer n there is associated a statement P_n, then all the statements P_n are true, provided the following two conditions are satisfied:

(i) P_1 is true.

(ii) Whenever k is a positive integer such that P_k is true, then P_{k+1} is also true.

Proof Assume that conditions (i) and (ii) of the Principle hold, and let S denote the set of all positive integers n such that P_n is true. By assumption, P_1 is true, and consequently, 1 is in S. Thus, S satisfies property (i) of the Axiom of Mathematical Induction. Whenever S contains a positive integer k, then by the definition of S, P_k is true and hence from condition (ii) of the Principle, P_{k+1} is also true. This means that S contains $k + 1$. We have shown that whenever S contains a positive integer k, then S also contains $k + 1$. Consequently, property (ii) of the Axiom of Mathematical Induction is true, and hence S contains every positive integer; that is, P_n is true for every positive integer n. □

There are other variations of the Principle of Mathematical Induction. One is stated later in this section.

When applying the Principle of Mathematical Induction always follow these two steps:

Step (i) Prove that P_1 is true.

Step (ii) Assume that P_k is true and prove that P_{k+1} is true.

Step (ii) is usually the most confusing for the beginning student. We do not *prove* that P_k is true (except for $k = 1$). Instead, we show that *if* P_k is true, then the statement P_{k+1} is true. That is all that is necessary according to the Principle of Mathematical Induction. The assumption that P_k is true is referred to as the **induction hypothesis.**

Many interesting formulas about positive integers can be established by using mathematical induction; two such formulas are illustrated in Examples 1 and 2. Others appear in the Exercises.

EXAMPLE 1 Prove that for every positive integer n, the sum of the first n positive integers is $n(n + 1)/2$.

Solution If n is any positive integer, let P_n denote the statement

$$1 + 2 + 3 + \cdots + n = \frac{n(n + 1)}{2}$$

where, by convention, if $n \leq 4$, then the left side is adjusted so that there are precisely n terms in the sum. The following are some special cases of P_n:

If $n = 2$, then P_2 is

$$1 + 2 = \frac{2(2 + 1)}{2}, \quad \text{or} \quad 3 = 3.$$

If $n = 3$, then P_3 is

$$1 + 2 + 3 = \frac{3(3 + 1)}{2}, \quad \text{or} \quad 6 = 6.$$

If $n = 5$, then P_5 is

$$1 + 2 + 3 + 4 + 5 = \frac{5(5 + 1)}{2}, \quad \text{or} \quad 15 = 15.$$

Although it is instructive to check the validity of P_n for several values of n as we have done, it unnecessary to do so. We need only apply the two step process outlined prior to this example. Thus, we proceed as follows:

Step (i): If we substitute $n = 1$ in P_n, then, by convention, the left side collapses to 1 and the right side is $\frac{1(1 + 1)}{2}$, which also equals 1. This proves that P_1 is true.

Step (*ii*): Assume that P_k is true. Thus, the induction hypothesis is

$$1 + 2 + 3 + \cdots + k = \frac{k(k+1)}{2}.$$

Our goal is to prove that P_{k+1} is true, that is,

$$1 + 2 + 3 + \cdots + (k+1) = \frac{(k+1)[(k+1)+1]}{2}.$$

By the induction hypothesis we already have a formula for the sum of the first k positive integers. Hence, a formula for the sum of the first $k + 1$ positive integers may be found simply by adding $(k + 1)$ to both sides. Doing so and simplifying, we obtain

$$\begin{aligned} 1 + 2 + 3 + \cdots + k + (k+1) &= \frac{k(k+1)}{2} + (k+1) \\ &= \frac{k(k+1) + 2(k+1)}{2} \\ &= \frac{k^2 + 3k + 2}{2} \\ &= \frac{(k+1)(k+2)}{2} \\ &= \frac{(k+1)[(k+1)+1]}{2}. \end{aligned}$$

We have shown that P_{k+1} is true and, therefore, the proof by mathematical induction is complete. ■

EXAMPLE 2 Prove that for each positive integer n,

$$1^2 + 3^2 + \cdots + (2n-1)^2 = \frac{n(2n-1)(2n+1)}{3}.$$

Solution For each positive integer n, let P_n denote the given statement. Note that this is a formula for the sum of the squares of the first n odd positive integers. We again follow the two-step procedure used in Example 1.

Step (*i*): Substituting 1 for n in P_n, we obtain

$$1^2 = \frac{(1)(2-1)(2+1)}{3} = \frac{3}{3} = 1.$$

This shows that P_1 is true.

Step (ii): Assume that P_k is true. Thus, the induction hypothesis is

$$1^2 + 3^2 + \cdots + (2k-1)^2 = \frac{k(2k-1)(2k+1)}{3}.$$

We wish to prove that P_{k+1} is true, that is,

$$1^2 + 3^2 + \cdots + [2(k+1)-1]^2 = \frac{(k+1)[2(k+1)-1][2(k+1)+1]}{3}.$$

This equation for P_{k+1} simplifies to

$$1^2 + 3^2 + \cdots + (2k+1)^2 = \frac{(k+1)(2k+1)(2k+3)}{3}.$$

Observe that the second from the last term on the left-hand side is $(2k-1)^2$. (Why)? In a manner similar to the solution of Example 1, we may obtain the left side of P_{k+1} by adding $(2k+1)^2$ to both sides of the induction hypothesis P_k. This gives us

$$1^2 + 3^2 + \cdots + (2k-1)^2 + (2k+1)^2 = \frac{k(2k-1)(2k+1)}{3} + (2k+1)^2.$$

It is left to the reader to show that the right side of the preceding equation may be written in the form of the right side of P_{k+1}. This proves that P_{k+1} is true, and hence P_n is true for every n. ■

Consider a positive integer j, and suppose that with each integer $n \geq j$ there is associated a statement P_n. For example, if $j = 6$, then the statements are numbered $P_6, P_7, P_8, \ldots$. The principle of mathematical induction may be extended to cover this situation. Just as before, two steps are used. Specifically, to prove that the statements S_n are true for $n \geq j$, we use the following steps.

EXTENDED PRINCIPLE OF MATHEMATICAL INDUCTION

(i′) Prove that S_j is true.

(ii′) Assume that S_k is true for $k \geq j$ and prove that S_{k+1} is true.

EXAMPLE 3 Let a be a nonzero real number such that $a > -1$. Prove that $(1 + a)^n > 1 + na$ for every integer $n \geq 2$.

Solution For each positive integer n, let P_n denote the inequality $(1 + a)^n > 1 + na$. Note that P_1 is *false*, since $(1 + a)^1 = 1 + (1)(a)$. However, we can show that P_n is true for $n \geq 2$ by using the Extended Principle with $j = 2$.

Step (i′): We first note that $(1 + a)^2 = 1 + 2a + a^2$. Since $a \neq 0$, we have $a^2 > 0$ and therefore $1 + 2a + a^2 > 1 + 2a$. This gives us $(1 + a)^2 > 1 + 2a$, and hence P_2 is true.

Step (ii′): Assume that P_k is true. Thus, the induction hypothesis is

$$(1 + a)^k > 1 + ka.$$

We wish to show that P_{k+1} is true, that is,

$$(1 + a)^{k+1} > 1 + (k + 1)a.$$

Since $a > -1$, we have $1 + a > 0$, and hence multiplying both sides of the induction hypothesis by $1 + a$ will not change the inequality sign. Consequently,

$$(1 + a)^k(1 + a) > (1 + ka)(1 + a),$$

which may be rewritten as

$$(1 + a)^{k+1} > 1 + ka + a + ka^2$$

or as

$$(1 + a)^{k+1} > 1 + (k + 1)a + ka^2.$$

Since $ka^2 > 0$,

$$1 + (k + 1)a + ka^2 > 1 + (k + 1)a$$

and therefore,

$$(1 + a)^{k+1} > 1 + (k + 1)a.$$

Thus, P_{k+1} is true and the proof is complete. ■

EXERCISES 8.1

In Exercises 1–18 prove that the formula is true for every positive integer n.

1 $2 + 4 + 6 + \cdots + 2n = n(n + 1)$

2 $1 + 4 + 7 + \cdots + (3n - 2) = \dfrac{n(3n - 1)}{2}$

3 $1 + 3 + 5 + \cdots + (2n - 1) = n^2$

4 $3 + 9 + 15 + \cdots + (6n - 3) = 3n^2$

5 $2 + 7 + 12 + \cdots + (5n - 3) = \frac{1}{2}n(5n - 1)$

6 $2 + 6 + 18 + \cdots + 2 \cdot 3^{n-1} = 3^n - 1$

7 $1 + 2 \cdot 2 + 3 \cdot 2^2 + 4 \cdot 2^3 + \cdots + n \cdot 2^{n-1} = 1 + (n - 1) \cdot 2^n$

8 $(-1)^1 + (-1)^2 + (-1)^3 + \cdots + (-1)^n = \dfrac{(-1)^n - 1}{2}$

9 $1^2 + 2^2 + 3^2 + \cdots + n^2 = \dfrac{n(n + 1)(2n + 1)}{6}$

10 $1^3 + 2^3 + 3^3 + \cdots + n^3 = \left[\dfrac{n(n + 1)}{2}\right]^2$

11 $\dfrac{1}{1 \cdot 2} + \dfrac{1}{2 \cdot 3} + \dfrac{1}{3 \cdot 4} + \cdots + \dfrac{1}{n(n + 1)} = \dfrac{n}{n + 1}$

12 $\dfrac{1}{1 \cdot 2 \cdot 3} + \dfrac{1}{2 \cdot 3 \cdot 4} + \dfrac{1}{3 \cdot 4 \cdot 5} + \cdots + \dfrac{1}{n(n + 1)(n + 2)} = \dfrac{n(n + 3)}{4(n + 1)(n + 2)}$

13 $3 + 3^2 + 3^3 + \cdots + 3^n = \frac{3}{2}(3^n - 1)$

14 $1^3 + 3^3 + 5^3 + \cdots + (2n - 1)^3 = n^2(2n^2 - 1)$

15 $n < 2^n$

16 $1 + 2n \leq 3^n$

17 $1 + 2 + 3 + \cdots + n < \frac{1}{8}(2n + 1)^2$

18 If $0 < a < b$, then $\left(\dfrac{a}{b}\right)^{n+1} < \left(\dfrac{a}{b}\right)^n$.

Prove that the statements in Exercises 19–22 are true for every positive integer n.

19 3 is a factor of $n^3 - n + 3$.

20 2 is a factor of $n^2 + n$.

21 4 is a factor of $5^n - 1$.

22 9 is a factor of $10^{n+1} + 3 \cdot 10^n + 5$.

23 Use mathematical induction to prove that if a is any real number greater than 1, then $a^n > 1$ for every positive integer n.

24 Prove that

$$a + ar + ar^2 + \cdots + ar^{n-1} = \frac{a(1 - r^n)}{1 - r}$$

where n is any positive integer and a and r are real numbers with $r \neq 1$.

25 Use mathematical induction to prove that $a - b$ is a factor of $a^n - b^n$ for every positive integer n. (*Hint:* $a^{k+1} - b^{k+1} = a^k(a - b) + (a^k - b^k)b$.)

26 Prove that $a + b$ is a factor of $a^{2n-1} + b^{2n-1}$ for every positive integer n.

27 Use mathematical induction to prove De Moivre's Theorem:

$$[r(\cos\theta + i\sin\theta)]^n = r^n[\cos n\theta + i\sin n\theta]$$

for every positive integer n.

28 Prove that for every positive integer $n \geq 3$, the sum of the interior angles of a polygon of n sides is $(n - 2) \cdot 180°$.

Prove that the formulas in Exercises 29 and 30 are true for every positive integer n.

29 $\sin(\theta + n\pi) = (-1)^n \sin\theta$

30 $\cos(\theta + n\pi) = (-1)^n \cos\theta$

SECTION 8.2
THE BINOMIAL THEOREM

A sum $a + b$ is sometimes called a **binomial.** It is often necessary to work with powers of the form $(a + b)^n$, where a and b are mathematical expressions of some type and n is a large positive integer. A general formula for *expanding* $(a + b)^n$, that is, for expressing it as a sum, is given by the **Binomial Theorem.** To obtain the formula let us first consider cases where n is a small positive integer. By examining the patterns that emerge, we shall make an educated guess about the nature of the general formula. Finally, we shall prove by mathematical induction that our guess is correct.

If we actually perform the multiplications, the following expansions of $(a + b)^n$ are obtained for the cases $n = 2, 3, 4$, and 5:

$$(a + b)^2 = a^2 + 2ab + b^2$$
$$(a + b)^3 = a^3 + 3a^2b + 3ab^2 + b^3$$
$$(a + b)^4 = a^4 + 4a^3b + 6a^2b^2 + 4ab^3 + b^4$$
$$(a + b)^5 = a^5 + 5a^4b + 10a^3b^2 + 10a^2b^3 + 5ab^4 + b^5.$$

Let us now make some observations regarding these expansions of $(a + b)^n$. We see that there are always $n + 1$ terms, the first being a^n and the last b^n. Each intermediate term contains a product of the form $a^i b^j$ where $i + j = n$. Moreover, as we move from one term to the next, the exponent associated with a decreases by 1, whereas the exponent associated with b increases by 1. The pattern for the coefficients is interesting: If we start at one end of the expansion and consider successive terms, the coefficients match those obtained by starting at the other end of the expansion and proceeding in the reverse direction.

It requires some ingenuity to find a formula for the general term in the expansion. The second term of the preceding special cases for $(a + b)^n$ is always $na^{n-1}b$. The third term is

$$\frac{n(n-1)}{2}a^{n-2}b^2.$$

If there are more than three terms, the fourth term is

$$\frac{n(n-1)(n-2)}{3 \cdot 2}a^{n-3}b^3.$$

It appears that, for each term, *if we take the product of the coefficient and the exponent of a and then divide by the number of the term, we obtain the coefficient of the next term* in the expansion. For example, applying this rule to the general fourth term gives us the following fifth term (if it exists):

$$\frac{n(n-1)(n-2)(n-3)}{4 \cdot 3 \cdot 2}a^{n-4}b^4.$$

The denominators in these terms may be abbreviated by employing the **factorial notation.** If n is any positive integer, then the symbol $n!$ (read "n factorial") is defined as follows.

DEFINITION OF $n!$

$$n! = n(n-1)(n-2) \cdots 1$$

Observe that there are n factors on the right side of the preceding formula. As special cases of $n!$ we have

$$1! = 1, \quad 2! = 2 \cdot 1, \quad 3! = 3 \cdot 2 \cdot 1 = 6,$$
$$4! = 4 \cdot 3 \cdot 2 \cdot 1 = 24, \quad 5! = 5 \cdot 4 \cdot 3 \cdot 2 \cdot 1 = 120.$$

To ensure that certain formulas will be true for all *nonnegative* integers, we define $0! = 1$. Note that for any positive integer k, $(k + 1)! = (k + 1)k!$.

If we compare the coefficients with the exponents in the general fourth and fifth terms of $(a + b)^n$, we are led to believe that, for a positive integer

r, the $(r+1)$st term in the expansion of $(a+b)^n$ is

$$\frac{n(n-1)(n-2)\cdots(n-r+1)}{r!}a^{n-r}b^r.$$

If $r = n$, the coefficient reduces to $n!/n!$ and we obtain b^n, the last term in the expansion. If $r = n - 1$, we obtain nab^{n-1}, which is the second from the last term. Hence, we *conjecture* that the following formula is true for every positive integer n and for all real or complex numbers a and b.

THE BINOMIAL THEOREM

$$(a+b)^n = a^n + na^{n-1}b + \frac{n(n-1)}{2!}a^{n-2}b^2 + \cdots + \frac{n(n-1)(n-2)\cdots(n-r+1)}{r!}a^{n-r}b^r + \cdots + nab^{n-1} + b^n$$

Proof We shall use mathematical induction as follows. For each positive integer n, let P_n denote the statement given in the Binomial Theorem.

Step (i): If $n = 1$, the statement reduces to $(a+b)^1 = a^1 + b^1$. Consequently, P_1 is true.

Step (ii): Assume that P_k is true. Thus, the induction hypothesis is

$$(a+b)^k = a^k + ka^{k-1}b + \frac{k(k-1)}{2!}a^{k-2}b^2 + \cdots + \frac{k(k-1)(k-2)\cdots(k-r+2)}{(r-1)!}a^{k-r+1}b^{r-1} + \frac{k(k-1)(k-2)\cdots(k-r+1)}{r!}a^{k-r}b^r + \cdots + kab^{k-1} + b^k.$$

We have shown both the rth and the $(r+1)$st terms in the expansion.

Multiplying both sides of the last equation by $(a+b)$, we obtain

$$(a+b)^{k+1} = \left[a^{k+1} + ka^k b + \frac{k(k-1)}{2!}a^{k-1}b^2 + \cdots + \frac{k(k-1)\cdots(k-r+1)}{r!}a^{k-r+1}b^r + \cdots + ab^k\right] + \left[a^k b + ka^{k-1}b^2 + \cdots + \frac{k(k-1)\cdots(k-r+2)}{(r-1)!}a^{k-r+1}b^r + \cdots + kab^k + b^{k+1}\right]$$

where the terms in the first pair of brackets result from multiplying the right side of the induction hypothesis by a and the terms in the second pair of brackets result from multiplying by b. Rearranging and combining terms,

$$(a+b)^{k+1} = a^{k+1} + (k+1)a^k b + \left[\frac{k(k-1)}{2!} + k\right]a^{k-1}b^2 + \cdots$$
$$+ \left[\frac{k(k-1)\cdots(k-r+1)}{r!} + \frac{k(k-1)\cdots(k-r+2)}{(r-1)!}\right]a^{k-r+1}b^r$$
$$+ \cdots + (1+k)ab^k + b^{k+1}.$$

It is left to the reader to show that if the coefficients are simplified we obtain statement P_n with $k+1$ substituted for n. Thus, P_{k+1} is true and therefore P_n holds for every positive integer n. □

EXAMPLE 1 Find the binomial expansion of $(2x+3y^2)^4$.

Solution Using the Binomial Theorem with $a = 2x$, $b = 3y^2$, and $n = 4$,

$$(2x+3y^2)^4 = (2x)^4 + 4(2x)^3(3y^2) + \frac{4\cdot 3}{2!}(2x)^2(3y^2)^2$$
$$+ \frac{4\cdot 3\cdot 2}{3!}(2x)(3y^2)^3 + \frac{4\cdot 3\cdot 2\cdot 1}{4!}(3y^2)^4.$$

This simplifies to

$$(2x+3y^2)^4 = 16x^4 + 96x^3y^2 + 216x^2y^4 + 216xy^6 + 81y^8.$$ ■

The symbol $\binom{n}{r}$ is often used to denote the coefficient of $a^{n-r}b^r$ in the Binomial Theorem, as indicated in the following definition.

DEFINITION

Let n be a positive integer and let r be any integer between 0 and n. The **binomial coefficient** $\binom{n}{r}$ is

$$\binom{n}{r} = \frac{n(n-1)\cdots(n-r+1)}{r!} = \frac{n!}{r!(n-r)!}.$$

If we substitute n for r in the definition, we obtain

$$\binom{n}{n} = \frac{n!}{n!} = 1.$$

It is also convenient to define

$$\binom{n}{0} = 1.$$

Thus, the formulas for $\binom{n}{r}$ may be used if r is any integer such that $0 \le r \le n$. Employing this notation gives us the following alternative form of the Binomial Theorem.

THE BINOMIAL THEOREM (ALTERNATIVE FORM)

$$(a+b)^n = \binom{n}{0}a^n b^0 + \binom{n}{1}a^{n-1}b + \binom{n}{2}a^{n-2}b^2 + \cdots + \binom{n}{r}a^{n-r}b^r + \cdots + \binom{n}{n-1}a^{n-(n-1)}b^{n-1} + \binom{n}{n}a^0 b^n$$

As a special case, if $n = 4$,

$$(a+b)^4 = \binom{4}{0}a^4b^0 + \binom{4}{1}a^3b + \binom{4}{2}a^2b^2 + \binom{4}{3}ab^3 + \binom{4}{4}a^0b^4.$$

It is easy to show that this reduces to the expansion of $(a + b)^4$ given at the beginning of this section.

The next example illustrates the fact that if one of a or b is negative, then the terms of the expansion are alternately positive and negative.

EXAMPLE 2 Expand $\left(\frac{1}{x} - 2\sqrt{x}\right)^5$.

Solution Letting $a = 1/x$, $b = -2\sqrt{x}$, and $n = 5$ in the Binomial Theorem,

$$\left(\frac{1}{x} - 2\sqrt{x}\right)^5 = \left(\frac{1}{x}\right)^5 + 5\left(\frac{1}{x}\right)^4(-2\sqrt{x}) + \frac{5\cdot 4}{2!}\left(\frac{1}{x}\right)^3(-2\sqrt{x})^2 + \frac{5\cdot 4\cdot 3}{3!}\left(\frac{1}{x}\right)^2(-2\sqrt{x})^3 + \frac{5\cdot 4\cdot 3\cdot 2}{4!}\left(\frac{1}{x}\right)(-2\sqrt{x})^4 + \frac{5\cdot 4\cdot 3\cdot 2\cdot 1}{5!}(-2\sqrt{x})^5.$$

This simplifies to

$$\left(\frac{1}{x} - 2\sqrt{x}\right)^5 = \frac{1}{x^5} - \frac{10}{x^{7/2}} + \frac{40}{x^2} - \frac{80}{x^{1/2}} + 80x - 32x^{5/2}.$$ ■

Solving certain problems only requires finding a specific term in the expansion of $(a + b)^n$. To work such problems we first find the exponent r that is to be assigned to b. Notice that, by the Binomial Theorem, *the exponent of b is always one less than the number of the term.* Once r is found, the exponent of a is $n - r$. Referring to the form of the $(r + 1)$st term, we see that the coefficient of the term involving $a^{n-r}b^r$ is of the form $p/r!$, *where p is the product of r factors and where the first factor is n and each factor is one less than the preceding factor.*

EXAMPLE 3 Find the fifth term in the expansion of $(x^3 + \sqrt{y})^{13}$.

Solution We let $a = x^3$ and $b = \sqrt{y}$. The exponent of b in the fifth term is 4 and hence the exponent of a is 9. From the discussion of the preceding paragraph we obtain

$$\frac{13 \cdot 12 \cdot 11 \cdot 10}{4!}(x^3)^9(\sqrt{y})^4, \quad \text{or} \quad 715x^{27}y^2. \qquad \blacksquare$$

Finally, there is an interesting triangular array of numbers, called **Pascal's Triangle,** that can be used to obtain binomial coefficients. The numbers are arranged as follows.

$$\begin{array}{ccccccccccccc}
 & & & & & & 1 & & & & & & \\
 & & & & & 1 & & 1 & & & & & \\
 & & & & 1 & & 2 & & 1 & & & & \\
 & & & 1 & & 3 & & 3 & & 1 & & & \\
 & & 1 & & 4 & & 6 & & 4 & & 1 & & \\
 & 1 & & 5 & & 10 & & 10 & & 5 & & 1 & \\
1 & & 6 & & 15 & & 20 & & 15 & & 6 & & 1 \\
 & \cdot & & \cdot & & \cdot & & \cdot & & \cdot & & \cdot & \\
\cdot & & \cdot & & \cdot & & \cdot & & \cdot & & \cdot & & \cdot
\end{array}$$

The numbers in the second row are the coefficients in the expansion of $(a + b)^1$; those in the third row are the coefficients determined by $(a + b)^2$; those in the fourth row are obtained from $(a + b)^3$, and so on. Each number in the array that is different from 1 can be found by adding the two numbers in the previous row that appear above and immediately to the left and right of the number.

EXERCISES 8.2

Expand and simplify the expressions in Exercises 1–12.

1 $(a + b)^6$

2 $(a + b)^7$

3 $(a - b)^8$

4 $(a - b)^9$

5 $(3x - 5y)^4$

6 $(2t - s)^5$

7 $(u^2 + 4v)^5$

8 $(\frac{1}{2}c + d^3)^4$

9 $(r^{-2} - 2r)^6$

10 $(x^{1/2} - y^{-1/2})^6$

11 $(1+x)^{10}$ **12** $(1-x)^{10}$

13 Find the first four terms in the binomial expansion of $(3c^{2/5}+c^{4/5})^{25}$.

14 Find the first three terms and the last three terms in the binomial expansion of $(x^3+5x^{-2})^{20}$.

15 Find the last two terms in the expansion of $(4b^{-1}-3b)^{15}$.

16 Find the last three terms in the expansion of $(s-2t^3)^{12}$.

Solve Exercises 17–28 without expanding completely.

17 Find the fifth term in the expansion of $(3a^2+\sqrt{b})^9$.

18 Find the sixth term in the expansion of $\left(\frac{2}{c}+\frac{c^2}{3}\right)^7$.

19 Find the seventh term in the expansion of $(\frac{1}{2}u-2v)^{10}$.

20 Find the fourth term in the expansion of $(2x^3-y^2)^6$.

21 Find the middle term in the expansion of $(x^{1/3}+y^{1/3})^{12}$.

22 Find the two middle terms in the expansion of $(rs+t)^7$.

23 Find the term that does not contain x in the expansion of $\left(6x-\frac{1}{2x}\right)^{10}$.

24 Find the term involving x^8 in the expansion of $(y+3x^2)^6$.

25 Find the term containing y^6 in the expansion of $(x-2y^3)^4$.

26 Find the term containing b^9 in the expansion of $(5a+2b^3)^4$.

27 Find the term containing c^3 in the expansion of $(\sqrt{c}+\sqrt{d})^{10}$.

28 In the expansion of $(xy-2y^{-3})^8$ find the term that does not contain y.

29 Use the first four terms in the binomial expansion of $(1+0.02)^{10}$ to approximate $(1.02)^{10}$. Compare with the answer obtained using a calculator.

30 Use the first four terms in the binomial expansion of $(1-0.01)^4$ to approximate $(0.99)^4$. Compare with the answer obtained using a calculator.

SECTION 8.3

INFINITE SEQUENCES AND SUMMATION NOTATION

A function f from a set X to a set Y is a correspondence that associates with each element x of X a unique element $f(x)$ of Y. Up to now the domain X has usually been a set of real numbers. In this section we shall consider a different class of functions.

DEFINITION

> An **infinite sequence** is a function whose domain is the set of positive integers.

For convenience we sometimes refer to infinite sequences merely as *sequences*. In this book the range of an infinite sequence will be a set of real numbers.

If f is an infinite sequence, then to each positive integer n there corresponds a real number $f(n)$. These numbers in the range of f may be represented by writing

$$f(1), f(2), f(3), \ldots, f(n), \ldots$$

where the dots at the end indicate that the sequence does not terminate. The number $f(1)$ is called the **first term** of the sequence, $f(2)$ the **second term** and, in general, $f(n)$ the ***n*th term** of the sequence. It is customary to use a subscript notation instead of the functional notational and write these numbers as

$$a_1, a_2, a_3, \ldots, a_n, \ldots$$

where it is understood that for each positive integer n, the symbol a_n denotes the real number $f(n)$. In this way we obtain an infinite collection of real numbers that is *ordered* in the sense that there is a first number, a second number, a forty-fifth number, etc. Although sequences are functions, an ordered collection of the type displayed above will also be referred to as an infinite sequence. If we wish to convert the collection to a function f, we let $f(n) = a_n$ for all positive integers n.

From the definition of equality of functions we see that a sequence

$$a_1, a_2, a_3, \ldots, a_n, \ldots$$

is **equal** to a sequence $\quad b_1, b_2, b_3, \ldots, b_n, \ldots$

if and only if $a_i = b_i$ for every positive integer i. Infinite sequences are often defined by stating a formula for the nth term, as in the following example.

EXAMPLE 1 List the first four terms and the tenth term of the sequence whose nth term is as follows:

(a) $a_n = \dfrac{n}{n+1}$ (b) $a_n = 2 + (0.1)^n$

(c) $a_n = (-1)^{n+1}\dfrac{n^2}{3n-1}$ (d) $a_n = 4$

Solution To find the first four terms we substitute, successively, $n = 1$, 2, 3, and 4 in the formula for a_n. The tenth term is found by substituting 10 for n. Doing this and simplifying gives us the following:

	nth term	*First four terms*	*Tenth term*
(a)	$\dfrac{n}{n+1}$	$\dfrac{1}{2}, \dfrac{2}{3}, \dfrac{3}{4}, \dfrac{4}{5}$	$\dfrac{10}{11}$
(b)	$2 + (0.1)^n$	2.1, 2.01, 2.001, 2.0001	2.0000000001
(c)	$(-1)^{n+1}\dfrac{n^2}{3n-1}$	$\dfrac{1}{2}, -\dfrac{4}{5}, \dfrac{9}{8}, -\dfrac{16}{11}$	$-\dfrac{100}{29}$
(d)	4	4, 4, 4, 4	4

■

Some infinite sequences are described by stating the first term a_1, together with a rule that shows how to obtain any term a_{k+1} from the preceding term a_k whenever $k \geq 1$. A description of this type is called a **recursive definition,** and the sequence is said to be defined **recursively.**

EXAMPLE 2 Find the first four terms and the nth term of the infinite sequence defined as follows:

$$a_1 = 3 \quad \text{and} \quad a_{k+1} = 2a_k, \qquad \text{for} \quad k \geq 1.$$

Solution The sequence is defined recursively since the first term is given and, moreover, whenever a term a_k is known, then the next term a_{k+1} can be found. Thus,

$$\begin{aligned} a_1 &= 3 \\ a_2 &= 2a_1 = 2 \cdot 3 = 6 \\ a_3 &= 2a_2 = 2 \cdot 2 \cdot 3 = 2^2 \cdot 3 = 12 \\ a_4 &= 2a_3 = 2 \cdot 2 \cdot 2 \cdot 3 = 2^3 \cdot 3 = 24. \end{aligned}$$

We have written the terms as products in order to gain some insight into the nature of the nth term. Continuing, we obtain $a_5 = 2^4 \cdot 3$ and $a_6 = 2^5 \cdot 3$; and it appears that

$$a_n = 2^{n-1} \cdot 3$$

for every positive integer n. We shall prove that this guess is correct by mathematical induction. If we let P_n denote the statement $a_n = 2^{n-1} \cdot 3$, then P_1 is true since $a_1 = 2^0 \cdot 3 = 3$. Next, *assume* that P_k is true, that is, $a_k = 2^{k-1} \cdot 3$. We then have

$$\begin{aligned} a_{k+1} &= 2a_k && \text{(definition of } a_{k+1}\text{)} \\ &= 2 \cdot 2^{k-1} \cdot 3 && \text{(induction hypothesis)} \\ &= 2^k \cdot 3 && \text{(a law of exponents)} \\ &= 2^{(k+1)-1} \cdot 3 && \text{(Why?)} \end{aligned}$$

which shows that P_{k+1} is true. Hence, $a_n = 2^{n-1} \cdot 3$ for every positive integer n. ■

It is important to observe that if only the first few terms of an infinite sequence are known, then it is impossible to predict additional terms. For example, if we were given 3, 6, 9, . . . and asked to find the fourth term, we could not proceed without further information. The infinite sequence with nth term

$$a_n = 3n + (1 - n)^3(2 - n)^2(3 - n)$$

has for its first four terms 3, 6, 9, and 120. It is possible to describe sequences in which the first three terms are 3, 6, and 9 and the fourth term is *any* given number. This shows that when we work with infinite sequence it is essential to have specific information about the nth term or to know a general scheme for obtaining each term from the preceding one.

It is often desirable to find the sum of many terms of an infinite sequence. For ease in expressing such sums we use the following **summation notation.** Given an infinite sequence

$$a_1, a_2, a_3, \ldots, a_n, \ldots$$

the symbol $\sum_{i=1}^{m} a_i$ represents the sum of the first m terms, that is,

SUMMATION NOTATION

$$\sum_{i=1}^{m} a_i = a_1 + a_2 + a_3 + \cdots + a_m.$$

The Greek capital letter $\sum$ (sigma) indicates a sum and the symbol a_i represents the ith term. The letter i is called the **index of summation** or the **summation variable,** and the numbers 1 and m indicate the extreme values of the summation variable.

EXAMPLE 3 Find $\sum_{i=1}^{4} i^2(i-3)$.

Solution In this case, $a_i = i^2(i-3)$. To find the sum we merely substitute, in succession, the integers 1, 2, 3, and 4 for i and add the resulting terms.

$$\begin{aligned}\sum_{i=1}^{4} i^2(i-3) &= 1^2(1-3) + 2^2(2-3) + 3^2(3-3) + 4^2(4-3)\\ &= (-2) + (-4) + 0 + 16 = 10.\end{aligned}$$

■

The letter used for the summation variable is arbitrary. To illustrate, if j is the summation variable, then

$$\sum_{j=1}^{m} a_j = a_1 + a_2 + a_3 + \cdots + a_m$$

which is the same as $\sum_{i=1}^{m} a_i$. Other symbols can also be used. As a numerical example, the sum in Example 3 can be written

$$\sum_{k=1}^{4} k^2(k-3).$$

If n is a positive integer, then the sum of the first n terms of an infinite sequence will be denoted by S_n. For example, given $a_1, a_2, a_3, \ldots, a_n, \ldots$,

$$S_1 = a_1$$
$$S_2 = a_1 + a_2$$
$$S_3 = a_1 + a_2 + a_3$$
$$S_4 = a_1 + a_2 + a_3 + a_4$$

and, in general, $$S_n = \sum_{i=1}^{n} a_i = a_1 + a_2 + \cdots + a_n.$$

The real number S_n is called the **nth partial sum** of the infinite sequence $a_1, a_2, a_3, \ldots, a_n, \ldots$, and the sequence

$$S_1, S_2, S_3, \ldots, S_n, \ldots$$

is called a **sequence of partial sums.** Sequences of partial sums are very important in calculus, where the concept of *infinite series* is studied. We shall discuss some special types of infinite series in Section 8.5.

EXAMPLE 4 Find the first four terms and the nth term of the sequence of partial sums associated with the sequence $1, 2, 3, \ldots, n, \ldots$ of positive integers.

Solution The first four terms of the sequence of partial sums are

$$S_1 = 1$$
$$S_2 = 1 + 2 = 3$$
$$S_3 = 1 + 2 + 3 = 6$$
$$S_4 = 1 + 2 + 3 + 4 = 10.$$

From Example 1 of Section 8.1 we see that

$$S_n = 1 + 2 + 3 + \cdots + n = \frac{n(n+1)}{2}.$$ ■

If a_i is the same for every positive integer i, say $a_i = c$, where c is a real number, then

$$\sum_{i=1}^{n} a_i = a_1 + a_2 + a_3 + \cdots + a_n$$
$$= c + c + c + \cdots + c = nc.$$

This gives us the following result.

THEOREM

$$\sum_{i=1}^{n} c = nc$$

The preceding formula can also be established by mathematical induction (see Exercise 56).

The domain of the summation variable does not have to begin at 1. For example, the following is self-explanatory:

$$\sum_{i=4}^{8} a_i = a_4 + a_5 + a_6 + a_7 + a_8.$$

As another variation, if the first term of an infinite sequence is a_0, as in

$$a_0, a_1, a_2, \ldots, a_n, \ldots,$$

then sums of the form

$$\sum_{i=0}^{n} a_i = a_0 + a_1 + a_2 + \cdots + a_n$$

may be considered. Note that this is the sum of the first $n + 1$ terms of the sequence.

EXAMPLE 5 Find $\sum_{i=0}^{3} \frac{2^i}{(i+1)}$.

Solution

$$\sum_{i=0}^{3} \frac{2^i}{(i+1)} = \frac{2^0}{(0+1)} + \frac{2^1}{(1+1)} + \frac{2^2}{(2+1)} + \frac{2^3}{(3+1)}$$

$$= 1 + 1 + \frac{4}{3} + 2 = \frac{16}{3}.$$ ■

Summation notation can be used to denote polynomials compactly. For example, in place of

$$f(x) = a_0 + a_1x + a_2x^2 + \cdots + a_nx^n$$

we may write

$$f(x) = \sum_{i=0}^{n} a_ix^i.$$

As another illustration, the rather cumbersome formula for the Binomial Theorem can be written

$$(a+b)^n = \sum_{r=0}^{n} \binom{n}{r} a^{n-r}b^r.$$

The following theorem concerning sums has many uses in advanced courses in mathematics such as calculus.

THEOREM ON SUMS

If $a_1, a_2, \ldots, a_n, \ldots$ and $b_1, b_2, \ldots, b_n, \ldots$ are infinite sequences, then for every positive integer n,

(i) $\displaystyle\sum_{i=1}^{n} (a_i + b_i) = \sum_{i=1}^{n} a_i + \sum_{i=1}^{n} b_i$

(ii) $\displaystyle\sum_{i=1}^{n} (a_i - b_i) = \sum_{i=1}^{n} a_i - \sum_{i=1}^{n} b_i$

(iii) $\displaystyle\sum_{i=1}^{n} ca_i = c\left(\sum_{i=1}^{n} a_i\right)$, for every number c.

Proof Although the theorem can be proved by mathematical induction, we shall use an argument that makes the truth of the formulas transparent. We begin as follows:

$$\sum_{i=1}^{n} (a_i + b_i) = (a_1 + b_1) + (a_2 + b_2) + (a_3 + b_3) + \cdots + (a_n + b_n).$$

Using commutative and associative properties many times, we may rearrange the terms on the right to produce

$$\sum_{i=1}^{n} (a_i + b_i) = (a_1 + a_2 + a_3 + \cdots + a_n) + (b_1 + b_2 + b_3 + \cdots + b_n).$$

Expressing the right side in summation notation gives us formula (i).

For formula (iii) we have

$$\begin{aligned}\sum_{i=1}^{n} (ca_i) &= ca_1 + ca_2 + ca_3 + \cdots + ca_n \\ &= c(a_1 + a_2 + a_3 + \cdots + a_n) \\ &= c\left(\sum_{i=1}^{n} a_i\right).\end{aligned}$$

The proof of (ii) is left as an exercise. □

EXERCISES 8.3

In Exercises 1–16 find the first five terms and the eighth term of the sequence that has the given nth term.

1 $a_n = 12 - 3n$

2 $a_n = \dfrac{3}{5n - 2}$

3 $a_n = \dfrac{3n - 2}{n^2 + 1}$

4 $a_n = 10 + \dfrac{1}{n}$

5 $a_n = 9$

6 $a_n = (n-1)(n-2)(n-3)$

7 $a_n = 2 + (-0.1)^n$

8 $a_n = 4 + (0.1)^n$

9 $a_n = (-1)^{n-1}\dfrac{n + 7}{2n}$

10 $a_n = (-1)^n\dfrac{6 - 2n}{\sqrt{n + 1}}$

11 $a_n = 1 + (-1)^{n+1}$

12 $a_n = (-1)^{n+1} + (0.1)^{n-1}$

13 $a_n = \dfrac{2^n}{n^2 + 2}$

14 $a_n = \sqrt{2}$

15 a_n is the number of decimal places in $(0.1)^n$.

16 a_n is the number of positive integers less than n^3.

Find the first five terms of the infinite sequences defined recursively in Exercises 17–24.

17 $a_1 = 2,\ a_{k+1} = 3a_k - 5$

18 $a_1 = 5,\ a_{k+1} = 7 - 2a_k$

19 $a_1 = -3,\ a_{k+1} = a_k^2$

20 $a_1 = 128,\ a_{k+1} = a_k/4$

21 $a_1 = 5,\ a_{k+1} = ka_k$

22 $a_1 = 3,\ a_{k+1} = 1/a_k$

23 $a_1 = 2,\ a_{k+1} = (a_k)^k$

24 $a_1 = 2,\ a_{k+1} = (a_k)^{1/k}$

25 A test question lists the first four terms of a sequence as 2, 4, 6, and 8 and asks for the fifth term. Show that the fifth term can be any real number a by finding the nth term of a sequence that has for its first five terms, 2, 4, 6, 8, and a.

26 The number of bacteria in a certain culture doubles every day. If the initial number of bacteria is 500, how many are present after one day? Two days? Three days? Find a formula for the number of bacteria present after n days.

Find the number given in each of Exercises 27–42.

27 $\sum_{k=1}^{5} (2k - 7)$

28 $\sum_{k=1}^{6} (10 - 3k)$

29 $\sum_{k=1}^{4} (k^2 - 5)$

30 $\sum_{k=1}^{10} [1 + (-1)^k]$

31 $\sum_{k=0}^{5} k(k - 2)$

32 $\sum_{k=0}^{4} (k - 1)(k - 3)$

33 $\sum_{k=3}^{6} \frac{k - 5}{k - 1}$

34 $\sum_{k=1}^{6} \frac{3}{k + 1}$

35 $\sum_{k=1}^{5} (-3)^{k-1}$

36 $\sum_{k=0}^{4} 3(2^k)$

37 $\sum_{k=1}^{100} 100$

38 $\sum_{k=1}^{1000} 5$

39 $\sum_{k=1}^{n} (k^2 + 3k + 5)$ (*Hint*: Use the Theorem on Sums to write the sum as $\sum_{k=1}^{n} k^2 + 3\sum_{k=1}^{n} k + \sum_{k=1}^{n} 5$. Next employ Exercise 9 and Example 1 of Section 8.1, together with the formula for $\sum_{k=1}^{n} c$.)

40 $\sum_{k=1}^{n} (3k^2 - 2k + 1)$

41 $\sum_{k=1}^{n} (2k - 3)^2$

42 $\sum_{k=1}^{n} (k^3 + 2k^2 - k + 4)$ (*Hint*: See Exercise 10 of Section 8.1).

Express the sums in Exercises 43–52 in terms of summation notation.

43 $1 + 5 + 9 + 13 + 17$

44 $2 + 5 + 8 + 11 + 14$

45 $\frac{1}{2} + \frac{2}{5} + \frac{3}{8} + \frac{4}{11}$

46 $\frac{1}{4} + \frac{2}{9} + \frac{3}{14} + \frac{4}{19}$

47 $1 - \frac{x^2}{2} + \frac{x^4}{4} - \frac{x^6}{6} + \cdots + (-1)^n \frac{x^{2n}}{2n}$

48 $2 - 4 + 8 - 16 + 32 - 64$

49 $1 - \frac{1}{2} + \frac{1}{3} - \frac{1}{4} + \frac{1}{5} - \frac{1}{6} + \frac{1}{7}$

50 $1 + x + \frac{x^2}{2} + \frac{x^3}{3} + \cdots + \frac{x^n}{n}$

51 $\frac{1}{1 \cdot 2} + \frac{1}{2 \cdot 3} + \frac{1}{3 \cdot 4} + \cdots + \frac{1}{99 \cdot 100}$

52 $\frac{1}{1 \cdot 2 \cdot 3} + \frac{1}{2 \cdot 3 \cdot 4} + \frac{1}{3 \cdot 4 \cdot 5} + \cdots + \frac{1}{98 \cdot 99 \cdot 100}$

53 Prove (ii) of the Theorem on Sums.

54 Extend (i) of the Theorem on Sums to $\sum_{i=1}^{n} (a_i + b_i + c_i)$.

55 Prove the Theorem on Sums by mathematical induction.

56 Prove $\sum_{i=1}^{n} c = nc$ by mathematical induction.

SECTION 8.4
ARITHMETIC SEQUENCES

In this section and the next we shall concentrate on two special types of sequences. The first may be defined as follows.

DEFINITION

An **arithmetic sequence** is a sequence such that successive terms differ by the same real number.

Arithmetic sequences are also called **arithmetic progressions.** By definition, a sequence

$$a_1, a_2, a_3, \ldots, a_n, \ldots$$

is arithmetic if and only if there is a real number d such that

$$a_{k+1} - a_k = d$$

for every positive integer k. The number d is called the **common difference** associated with the arithmetic sequence.

EXAMPLE 1 Show that the sequence

$$1, 4, 7, 10, \ldots, 3n - 2, \ldots$$

is arithmetic, and find the common difference.

Solution Since $a_n = 3n - 2$, it follows that for every positive integer k,

$$\begin{aligned} a_{k+1} - a_k &= [3(k + 1) - 2] - (3k - 2) \\ &= 3k + 3 - 2 - 3k + 2 = 3. \end{aligned}$$

Hence, by definition, the given sequence is arithmetic with common difference 3. ■

Given an arithmetic sequence, we know that

$$a_{k+1} = a_k + d$$

for every positive integer k. This provides a recursive formula for obtaining successive terms. Beginning with any real number a_1, we can obtain an arithmetic sequence with common difference d simply by adding d to a_1, then to $a_1 + d$, and so on, obtaining

$$a_1, \; a_1 + d, \; a_1 + 2d, \; a_1 + 3d, \; a_1 + 4d, \ldots$$

It is evident that the nth term a_n of this sequence is given by the next formula.

***n*TH TERM OF AN ARITHMETIC SEQUENCE**

$$a_n = a_1 + (n - 1)d$$

EXAMPLE 2 Find the fifteenth term of the arithmetic sequence whose first three terms are 20, 16.5, and 13.

Solution The common difference is -3.5. (Why?) Substituting $a_1 = 20$, $d = -3.5$, and $n = 15$ in the formula $a_n = a_1 + (n-1)d$,

$$a_{15} = 20 + (15 - 1)(-3.5) = 20 - 49 = -29.$$

■

EXAMPLE 3 If the fourth term of an arithmetic sequence is 5 and the ninth term is 20, find the sixth term.

Solution Substituting $n = 4$ and $n = 9$ in the formula $a_n = a_1 + (n-1)d$, and using the fact that $a_4 = 5$ and $a_9 = 20$, we obtain the following system of linear equations in the variables a_1 and d:

$$\begin{cases} 5 = a_1 + (4 - 1)d \\ 20 = a_1 + (9 - 1)d. \end{cases}$$

This system has the solution $d = 3$ and $a_1 = -4$. (Verify this fact.) Substitution in the formula $a_n = a_1 + (n-1)d$ gives us

$$a_6 = (-4) + (6 - 1)(3) = 11.$$

■

THEOREM

If $a_1, a_2, \ldots, a_n, \ldots$ is an arithmetic sequence with common difference d, then the nth partial sum S_n is given by both

$$S_n = \frac{n}{2}[2a_1 + (n-1)d] \quad \text{and} \quad S_n = \frac{n}{2}(a_1 + a_n).$$

Proof We may write

$$\begin{aligned} S_n &= a_1 + a_2 + a_3 + \cdots + a_n \\ &= a_1 + (a_1 + d) + (a_1 + 2d) + \cdots + [a_1 + (n-1)d]. \end{aligned}$$

Employing commutative and associative properties many times, we obtain

$$S_n = (a_1 + a_1 + a_1 + \cdots + a_1) + [d + 2d + \cdots + (n-1)d]$$

where a_1 appears n times within the first parentheses. It follows that

$$S_n = na_1 + d[1 + 2 + \cdots + (n-1)].$$

The expression within brackets is the sum of the first $n - 1$ positive integers. From Example 1 of Section 8.1 (with $n - 1$ in place of n),

$$1 + 2 + \cdots + (n-1) = \frac{(n-1)n}{2}.$$

Substituting in the last equation for S_n and factoring,

$$S_n = na_1 + d\frac{n(n-1)}{2} = \frac{n}{2}[2a_1 + (n-1)d].$$

Since $a_n = a_1 + (n-1)d$, this is the same as

$$S_n = \frac{n}{2}(a_1 + a_n).$$

□

EXAMPLE 4 Find the sum of all the even integers from 2 through 100.

Solution This problem is equivalent to finding the sum of the first fifty terms of the arithmetic sequence 2, 4, 6, . . . , $2n$, ... Substituting $n = 50$, $a_1 = 2$, and $a_{50} = 100$ in the second formula of the preceding theorem,

$$S_{50} = \frac{50}{2}(2 + 100) = 2550.$$

As a check on our work we may use the first formula of the theorem. Thus,

$$S_{50} = \frac{50}{2}[2 \cdot 2 + (50 - 1)2] = 25[4 + 98] = 2550.$$

■

The **arithmetic mean** of two numbers a and b is defined as $(a + b)/2$. This is also called the **average** of a and b. Note that

$$a, \frac{a+b}{2}, b$$

is an arithmetic sequence. This concept may be generalized as follows. If $c_1, c_2, \ldots, c_k$ are real numbers such that

$$a, c_1, c_2, \ldots, c_k, b$$

is an arithmetic sequence, then $c_1, c_2, \ldots, c_k$ are called the ***k* arithmetic means** of the numbers a and b. The process of determining these numbers is referred to as *inserting* k arithmetic means between a and b.

EXAMPLE 5 Insert three arithmetic means between 2 and 9.

Solution We wish to find three real numbers c_1, c_2, and c_3 such that 2, c_1, c_2, c_3, 9 is an arithmetic sequence. The common difference d may be found by using the formula $a_n = a_1 + (n-1)d$ with $n = 5$, $a_5 = 9$, and $a_1 = 2$. This gives us

$$9 = 2 + (5 - 1)d, \quad \text{or} \quad d = \tfrac{7}{4}.$$

The three arithmetic means are

$$c_1 = a_1 + d = 2 + \tfrac{7}{4} = \tfrac{15}{4}$$
$$c_2 = c_1 + d = \tfrac{15}{4} + \tfrac{7}{4} = \tfrac{22}{4} = \tfrac{11}{2}$$
$$c_3 = c_2 + d = \tfrac{11}{2} + \tfrac{7}{4} = \tfrac{29}{4}.$$

■

EXERCISES 8.4

In Exercises 1–8 find the fifth term, the tenth term, and the nth term of the arithmetic sequence.

1 2, 6, 10, 14, . . .

2 16, 13, 10, 7, . . .

3 3, 2.7, 2.4, 2.1, . . .

4 $-6, -4.5, -3, -1.5, \ldots$

5 $-7, -3.9, -0.8, 2.3, \ldots$

6 $x - 8, x - 3, x + 2, x + 7, \ldots$

7 ln 3, ln 9, ln 27, ln 81, . . .

8 log 1000, log 100, log 10, 0, . . .

9 Find the twelfth term of the arithmetic sequence whose first two terms are 9.1 and 7.5.

10 Find the eleventh term of the arithmetic sequence whose first two terms are $2 + \sqrt{2}$ and 3.

11 The sixth and seventh terms of an arithmetic sequence are 2.7 and 5.2. Find the first term.

12 Given an arithmetic sequence with $a_3 = 7$ and $a_{20} = 43$, find a_{15}.

In Exercises 13–16 find the sum S_n of the arithmetic sequence that satisfies the stated conditions.

13 $a_1 = 40,\ d = -3,\ n = 30$

14 $a_1 = 5,\ d = 0.1,\ n = 40$

15 $a_1 = -9,\ a_{10} = 15,\ n = 10$

16 $a_7 = \tfrac{7}{3},\ d = -\tfrac{2}{3},\ n = 15$

Find the sums in Exercises 17–20.

17 $\sum_{k=1}^{20} (3k - 5)$

18 $\sum_{k=1}^{12} (7 - 4k)$

19 $\sum_{k=1}^{18} \left(\frac{1}{2}k + 7\right)$

20 $\sum_{k=1}^{10} \left(\frac{1}{4}k + 3\right)$

21 How many integers between 32 and 395 are divisible by 6? Find their sum.

22 How many negative integers greater than -500 are divisible by 33? Find their sum.

23 How many terms are in an arithmetic sequence with first term -2, common difference $\tfrac{1}{4}$, and sum 21?

24 How many terms are in an arithmetic sequence with sixth term -3, common difference 0.2, and sum -33?

25 Insert five arithmetic means between 2 and 10.

26 Insert three arithmetic means between 3 and -5.

27 A pile of logs has 24 logs in the first layer, 23 in the second, 22 in the third, and so on. The top layer contains 10 logs. Find the total number of logs in the pile.

28 A seating section in a certain athletic stadium has 30 seats in the first row, 32 seats in the second, 34 in the third, and so on, until the tenth row is reached, after which there are 10 more rows, each containing 50 seats. Find the total number of seats in the section.

29 A man wishes to construct a ladder with nine rungs that diminish uniformly from 24 inches at the base to 18 inches at the top. Determine the lengths of the seven intermediate rungs.

30 A boy on a bicycle coasts down a hill, covering 4 feet the first second and in each succeeding second 5 feet more than in the preceding second. If he reaches the bottom of the hill in 11 seconds, find the total distance traveled.

SECTION 8.5

GEOMETRIC SEQUENCES

Another important type of infinite sequence is defined as follows.

DEFINITION

A sequence $a_1, a_2, \ldots, a_n, \ldots$ is a **geometric sequence** if and only if there is a real number $r \neq 0$ such that

$$\frac{a_{k+1}}{a_k} = r$$

for every positive integer k.

Geometric sequences are also called **geometric progressions.** The number r in the definition is called the **common ratio** associated with the geometric sequence. Thus $a_1, a_2, \ldots, a_n, \ldots$ is geometric (with common ratio r) if and only if

$$a_{k+1} = a_k r$$

for every positive integer k. As with arithmetic sequences, this provides a recursive method for obtaining terms. Beginning with any nonzero real number a_1, we *multiply* by the number r successively, obtaining

$$a_1,\ a_1 r,\ a_1 r^2,\ a_1 r^3,\ \ldots$$

The nth term a_n of this sequence is given by the next formula.

***n*th TERM OF A GEOMETRIC SEQUENCE**

$$a_n = a_1 r^{n-1}$$

EXAMPLE 1 Find the first five terms and the tenth term of the geometric sequence having first term 3 and common ratio $-\frac{1}{2}$.

Solution If we let $a_1 = 3$ and $r = -\frac{1}{2}$, then the first five terms are

$$3,\ -\frac{3}{2},\ \frac{3}{4},\ -\frac{3}{8},\ \frac{3}{16}.$$

Using the formula $a_n = a_1 r^{n-1}$ with $n = 10$,

$$a_{10} = 3\left(-\frac{1}{2}\right)^9 = -\frac{3}{512}.$$

■

EXAMPLE 2 If the third term of a geometric sequence is 5 and the sixth term is -40, find the eighth term.

Solution We are given $a_3 = 5$ and $a_6 = -40$. Substituting $n = 3$ and $n = 6$ in the formula $a_n = a_1 r^{n-1}$ leads to the following system of equations:

$$\begin{cases} 5 = a_1 r^2 \\ -40 = a_1 r^5. \end{cases}$$

Since $r \neq 0$, the first equation is equivalent to $a_1 = 5/r^2$. Substituting for a_1 in the second equation,

$$-40 = \left(\frac{5}{r^2}\right) \cdot r^5 = 5r^3.$$

Hence, $r^3 = -8$ and $r = -2$. If we now substitute -2 for r in the equation $5 = a_1 r^2$, we obtain $a_1 = \frac{5}{4}$. Finally, using $a_n = a_1 r^{n-1}$ with $n = 8$,

$$a_8 = \left(\frac{5}{4}\right)(-2)^7 = -160.$$ ■

Let us find a formula for S_n, the nth partial sum of a geometric sequence. If the first term is a_1 and the common ratio is r,

$$S_n = a_1 + a_1 r + a_1 r^2 + \cdots + a_1 r^{n-2} + a_1 r^{n-1}.$$

If $r = 1$, then $S_n = n a_1$. Next suppose that $r \neq 1$. Multiplying both sides of the general formula for S_n by r,

$$rS_n = a_1 r + a_1 r^2 + a_1 r^3 + \cdots + a_1 r^{n-1} + a_1 r^n.$$

If we subtract the preceding equation from that for S_n, then many terms on the right side drop out, leaving

$$S_n - rS_n = a_1 - a_1 r^n \quad \text{or} \quad (1 - r)S_n = a_1(1 - r^n).$$

Since $r \neq 1$, we have $1 - r \neq 0$, and dividing both sides by $1 - r$ gives us

$$S_n = \frac{a_1(1 - r^n)}{1 - r}.$$

We have proved the following theorem.

THEOREM

The nth partial sum of a geometric sequence with first term a_1 and common ratio $r \neq 1$ is

$$S_n = a_1 \frac{(1 - r^n)}{1 - r}.$$

EXAMPLE 3 Find the sum of the first five terms of the geometric sequence that begins as follows: 1, 0.3, 0.09, 0.027, . . .

Solution If we let $a_1 = 1$, $r = 0.3$, and $n = 5$ in the formula for S_n, then

$$S_5 = 1\left(\frac{1 - (0.3)^5}{1 - 0.3}\right).$$

This reduces to $S_5 = 1.4251$. ■

EXAMPLE 4 A man wishes to save money by setting aside 1 cent the first day, 2 cents the second day, 4 cents the third day and so on, doubling the amount each day. If this is continued, how much must be set aside on the fifteenth day? Assuming he does not run out of money, what is the total amount saved at the end of 30 days?

Solution The amount (in cents) set aside on successive days forms a geometric sequence

$$1, 2, 4, 8, \ldots$$

with first term 1 and common ratio 2. The amount needed for the fifteenth day is found by using $a_n = a_1 r^{n-1}$ with $a_1 = 1$ and $n = 15$. This gives us $1 \cdot 2^{14}$, or \$163.84. To find the total amount set aside after 30 days, we use the formula for S_n with $n = 30$. Thus,

$$S_{30} = 1\,\frac{(1 - 2^{30})}{1 - 2},$$

which simplifies to \$10,737,418.23. ■

Given the geometric series with first term a_1 and common ratio $r \neq 1$, we may write the formula for S_n of the last theorem in the form

$$S_n = \frac{a_1}{1 - r} - \frac{a_1}{1 - r}\,r^n.$$

Although we shall not do it here, it can be shown that if $|r| < 1$, then *r^n approaches 0 as n increases* without bound; that is, we can make r^n as close as we wish to 0 by taking n sufficiently large. It follows that S_n approaches $a_1/(1 - r)$ as n increases without bound. Using the notation introduced in our work with rational functions in Chapter 3, this may be expressed symbolically as

$$S_n \to \frac{a_1}{1 - r} \quad \text{as} \quad n \to \infty.$$

The number $a_1/(1 - r)$ is called the *sum* of the **infinite geometric series**

$$a_1 + a_1 r + a_1 r^2 + \cdots + a_1 r^{n-1} + \cdots$$

This gives us the next result.

THEOREM

If $|r| < 1$, then the infinite geometric series

$$a_1 + a_1r + a_1r^2 + \cdots + a_1r^{n-1} + \cdots$$

has the sum $a_1/(1 - r)$.

Intuitively, the preceding theorem implies that if we add more and more terms of the indicated infinite geometric series, the sums get closer and closer to $a_1/(1 - r)$. The next example illustrates how the theorem can be used to show that every real number represented by a repeating decimal is rational.

EXAMPLE 5 Find the rational number that corresponds to the infinite repeating decimal $5.4\overline{27}$, where the bar means that the block of digits underneath is repeated indefinitely.

Solution From the decimal expression 5.4272727 . . . , we obtain the infinite series

$$5.4 + 0.027 + 0.00027 + 0.0000027 + \cdots$$

The part of the expression after the first term is

$$0.027 + 0.00027 + 0.0000027 + \cdots,$$

which has the form given in the last theorem with $a_1 = 0.027$ and $r = 0.01$. Hence, the sum of this infinite geometric series is

$$\frac{0.027}{1 - 0.01} = \frac{0.027}{0.99} = \frac{27}{990} = \frac{3}{110}.$$

Thus, it appears that the desired number is $5.4 + \frac{3}{110}$, or $\frac{597}{110}$. A check by division shows that $\frac{597}{110}$ does equal the given repeating decimal. ■

In general, given any infinite sequence $a_1, a_2, \ldots, a_n, \ldots$, the expression

$$a_1 + a_2 + \cdots + a_n + \cdots$$

is called an **infinite series,** or simply a **series.** In summation notation, this series is denoted by

$$\sum_{n=1}^{\infty} a_n.$$

Each number a_i is called a **term** of the series and a_n is called the ***n*th term.** Since only finite sums may be added algebraically, it is necessary to *define* what is meant by an infinite sum. A natural way to do so is to use the sequence of partial sums

$$S_1, S_2, \ldots, S_n, \ldots$$

If there is a number S such that $S_n \to S$ as $n \to \infty$, then, as in our discussion of infinite geometric series, we call S the **sum** of the series and write

$$S = a_1 + a_2 + \cdots + a_n + \cdots$$

Thus, in Example 5 we may write

$$\frac{597}{110} = 5.4 + 0.027 + 0.00027 + 0.0000027 + \cdots$$

The next example provides an illustration of a nongeometric infinite series. Since a complete discussion is beyond the scope of this text, the solution is presented in an intuitive manner.

EXAMPLE 6 Show that the following infinite series has a sum:

$$\frac{1}{1 \cdot 2} + \frac{1}{2 \cdot 3} + \frac{1}{3 \cdot 4} + \cdots + \frac{1}{n(n+1)} + \cdots$$

Solution The nth term $a_n = 1/n(n+1)$ has the partial fraction decomposition

$$a_n = \frac{1}{n(n+1)} = \frac{1}{n} - \frac{1}{n+1}.$$

Consequently, the nth partial sum of the series may be written

$$\begin{aligned} S_n &= a_1 + a_2 + a_3 + \cdots + a_n \\ &= \left(1 - \frac{1}{2}\right) + \left(\frac{1}{2} - \frac{1}{3}\right) + \left(\frac{1}{3} - \frac{1}{4}\right) + \cdots + \left(\frac{1}{n} - \frac{1}{n+1}\right) \\ &= 1 - \frac{1}{n+1}. \end{aligned}$$

Since $1/(n+1) \to 0$ as $n \to \infty$, it follows that $S_n \to 1$, and we may write

$$1 = \frac{1}{1 \cdot 2} + \frac{1}{2 \cdot 3} + \frac{1}{3 \cdot 4} + \cdots + \frac{1}{n(n+1)} + \cdots$$

This means that, as we add more and more terms of the series, the sums get closer and closer to 1. ■

If the terms of an infinite sequence are alternately positive and negative, and if we consider the expression

$$a_1 + (-a_2) + a_3 + (-a_4) + \cdots + [(-1)^{n+1}a_n] + \cdots$$

where all the a_i are positive real numbers, then this expression is referred to as an **alternating infinite series,** and we write it in the form

$$a_1 - a_2 + a_3 - a_4 + \cdots + (-1)^{n+1}a_n + \cdots$$

Illustrations of alternating infinite series can be obtained by using infinite geometric series with negative common ratio.

Infinite series have many applications in mathematics and the sciences. The reader is referred to texts on calculus for a rigorous treatment of this important concept.

EXERCISES 8.5

In Exercises 1–12 find the fifth term, the eighth term, and the nth term of the geometric sequence.

1 $8, 4, 2, 1, \ldots$

2 $4, 1.2, 0.36, 0.108, \ldots$

3 $300, -30, 3, -0.3, \ldots$

4 $1, -\sqrt{3}, 3, -\sqrt{27}, \ldots$

5 $5, 25, 125, 625, \ldots$

6 $2, 6, 18, 54, \ldots$

7 $4, -6, 9, -13.5, \ldots$

8 $162, -54, 18, -6, \ldots$

9 $1, -x^2, x^4, -x^6, \ldots$

10 $1, -\frac{x}{3}, \frac{x^2}{9}, -\frac{x^3}{27}, \ldots$

11 $2, 2^{x+1}, 2^{2x+1}, 2^{3x+1}, \ldots$

12 $10, 10^{2x-1}, 10^{4x-3}, 10^{6x-5}, \ldots$

13 Find the sixth term of the geometric sequence whose first two terms are 4 and 6.

14 Find the seventh term of the geometric sequence that has 2 and $-\sqrt{2}$ for its second and third terms, respectively.

15 In a certain geometric sequence $a_5 = \frac{1}{16}$ and $r = \frac{3}{2}$. Find a_1 and S_5.

16 Given a geometric sequence such that $a_4 = 4$ and $a_7 = 12$, find r and a_{10}.

Find the sums in Exercises 17–20.

17 $\sum_{k=1}^{10} 3^k$

18 $\sum_{k=1}^{9} (-\sqrt{5})^k$

19 $\sum_{k=0}^{9} (-\frac{1}{2})^{k+1}$

20 $\sum_{k=1}^{7} (3^{-k})$

21 A vacuum pump removes one-half of the air in a container at each stroke. After 10 strokes, what percentage of the original amount of air remains in the container?

22 The yearly depreciation of a certain machine is 25% of its value at the beginning of the year. If the original cost of the machine is \$20,000, find its value after 6 years.

23 A culture of bacteria increases 20% every hour. If the original culture contains 10,000 bacteria, find a formula for the number of bacteria present after t hours. How many bacteria are in the culture at the end of 10 hours?

24 If an amount P of money is deposited in a savings account that pays interest at a rate of r percent per year compounded quarterly, and if the principal and accumulated interest are left in the account, find a formula for the total amount in the account after n years.

Find the sums of the infinite geometric series in Exercises 25–30, whenever they exist.

25 $1 - \frac{1}{2} + \frac{1}{4} - \frac{1}{8} + \cdots$

26 $2 + \frac{2}{3} + \frac{2}{9} + \frac{2}{27} + \cdots$

27 $1.5 + 0.015 + 0.00015 + \cdots$

28 $1 - 0.1 + 0.01 - 0.001 + \cdots$

29 $\sqrt{2} - 2 + \sqrt{8} - 4 + \cdots$

30 $250 - 100 + 40 - 16 + \cdots$

In Exercises 31–38 find the rational number represented by the repeating decimal.

31 $0.\overline{23}$

32 $0.0\overline{71}$

33 $2.4\overline{17}$

34 $10.\overline{55}$

35 $5.\overline{146}$

36 $3.2\overline{394}$

37 $1.\overline{6124}$

38 $123.61\overline{83}$

39 A rubber ball is dropped from a height of 10 meters. If it rebounds approximately one-half the distance after each fall, use an infinite geometric series to approximate the total distance the ball travels before coming to rest.

40 The bob of a pendulum swings through an arc 24 cm long on its first swing. If each successive swing is approximately five-sixths the length of the preceding swing, use an infinite geometric series to approximate the total distance it travels before coming to rest.

SECTION 8.6
REVIEW

Define or discuss each of the following.

1 Axiom of Mathematical Induction
2 Principle of Mathematical Induction
3 The Binomial Theorem
4 Binomial coefficients
5 Infinite sequence
6 Summation notation
7 The nth partial sum of an infinite sequence
8 Arithmetic sequence
9 Arithmetic mean of two numbers
10 Geometric sequence
11 Infinite geometric series
12 Infinite series

EXERCISES 8.6

Prove that the statements in Exercises 1–5 are true for every positive integer n.

1 $2 + 5 + 8 + \cdots + (3n - 1) = \dfrac{n(3n + 1)}{2}$

2 $2^2 + 4^2 + 6^2 + \cdots + (2n)^2 = \dfrac{2n(2n + 1)(n + 1)}{3}$

3 $\dfrac{1}{1 \cdot 3} + \dfrac{1}{3 \cdot 5} + \dfrac{1}{5 \cdot 7} + \cdots + \dfrac{1}{(2n - 1)(2n + 1)} = \dfrac{n}{2n + 1}$

4 $1 \cdot 2 + 2 \cdot 3 + 3 \cdot 4 + \cdots + n(n + 1) = \dfrac{n(n + 1)(n + 2)}{3}$

5 3 is a factor of $n^3 + 2n$.

6 Prove that $2^n > n^2$ for every positive integer $n \geq 5$.

7 Expand and simplify $(x^2 - 3y)^6$.

8 Find the first four terms in the binomial expansion of $(a^{2/5} + 2a^{-3/5})^{20}$.

9 Find the sixth term in the expansion of $(b^3 - \frac{1}{2}c^2)^9$.

10 In the expansion of $(2c^3 + 5c^{-2})^{10}$ find the term that does not contain c.

In Exercises 11–14 find the first four terms and the seventh term of the sequence that has the given nth term.

11 $a_n = \dfrac{5n}{3 - 2n^2}$

12 $a_n = (-1)^{n+1} - (0.1)^n$

13 $a_n = 1 + (-\frac{1}{2})^{n-1}$

14 $a_n = \dfrac{2^n}{(n+1)(n+2)(n+3)}$

Find the first five terms of the infinite sequences defined recursively in Exercises 15–18.

15 $a_1 = 10,\ a_{k+1} = 1 + (1/a_k)$

16 $a_1 = 2,\ a_{k+1} = a_k!$

17 $a_1 = 9,\ a_{k+1} = \sqrt{a_k}$

18 $a_1 = 1,\ a_{k+1} = (1 + a_k)^{-1}$

Find the numbers represented by the sums in Exercises 19–22.

19 $\sum_{k=1}^{5} (k^2 + 4)$

20 $\sum_{k=2}^{6} \dfrac{2k - 8}{k - 1}$

21 $\sum_{k=1}^{100} 10$

22 $\sum_{i=1}^{4} (2^i - 10)$

In Exercises 23–26 use summation notation to represent the sums.

23 $3 + 6 + 9 + 12 + 15$

24 $2 + 4 + 8 + 16 + 32 + 64 + 128$

25 $100 - 95 + 90 - 85 + 80$

26 $a_0 + a_4x^4 + a_8x^8 + \cdots + a_{100}x^{100}$

27 Find the tenth term and the sum of the first ten terms of the arithmetic sequence whose first two terms are $4 + \sqrt{3}$ and 3.

28 Find the sum of the first eight terms of an arithmetic sequence in which the fourth term is 9 and the common difference is -5.

29 The fifth and thirteenth terms of an arithmetic sequence are 5 and 77, respectively. Find the first term and the tenth term.

30 Insert four arithmetic means between 20 and -10.

31 Find the tenth term of the geometric sequence whose first two terms are $\frac{1}{8}$ and $\frac{1}{4}$.

32 If a geometric sequence has 3 and -0.3 as its third and fourth terms, find the eighth term.

33 Find a positive number c such that 4, c, 8 are successive terms of a geometric sequence.

34 In a certain geometric sequence the eighth term is 100 and the common ratio is $-\frac{3}{2}$. Find the first term.

Find the sums in Exercises 35–38.

35 $\sum_{k=1}^{15} (5k - 2)$

36 $\sum_{k=1}^{10} (6 - \frac{1}{2}k)$

37 $\sum_{k=1}^{10} (2^k - \frac{1}{2})$

38 $\sum_{k=1}^{8} (\frac{1}{2} - 2^k)$

39 Find the sum of the infinite geometric series

$$1 - \frac{2}{5} + \frac{4}{25} - \frac{8}{125} + \cdots$$

40 Find the rational number whose decimal representation is $6.\overline{274}$.

TOPICS IN ANALYTIC GEOMETRY

Plane geometry includes the study of figures, such as lines, circles, and triangles, that lie in a plane. Theorems are proved by reasoning deductively from certain postulates. In *analytic* geometry, plane geometric figures are investigated by introducing a coordinate system and then using equations and formulas of various types. If the study of analytic geometry were to be summarized by means of one statement, perhaps the following would be appropriate: "Given an equation, find its graph and, conversely, given a graph, find its equation." In this chapter we shall apply coordinate methods to several basic plane figures.

SECTION 9.1

CONIC SECTIONS

Each of the geometric figures discussed in Sections 9.1–9.5 can be obtained by intersecting a double-napped right circular cone with a plane. For this reason they are called **conic sections** or simply **conics.** If, as in (i) of Figure 9.1, the plane cuts entirely across one nappe of the cone and is not perpendicular to the axis, then the curve of intersection is called an **ellipse.** If the plane is perpendicular to the axis of the cone, a **circle** results. If the plane does not cut across one entire nappe and does not intersect both nappes, as illustrated in (ii) of Figure 9.1, then the curve of intersection is a **parabola.** If the plane cuts through both nappes of the cone, as in (iii) of Figure 9.1, then the resulting figure is called a **hyperbola.**

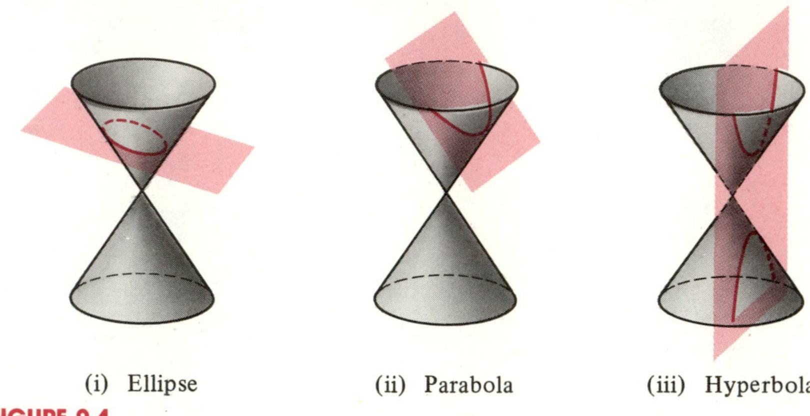

(i) Ellipse (ii) Parabola (iii) Hyperbola

FIGURE 9.1

By changing the position of the plane and the shape of the cone, conics can be made to vary considerably. Certain positions of the plane result in **degenerate conics.** For example, if the plane intersects the cone only at the vertex, then the conic consists of one point. If the axis of the cone lies on the plane, then a pair of intersecting lines is obtained. Finally, if we begin with the parabolic case, as in (ii) of Figure 9.1, and move the plane parallel to its initial position until it coincides with one of the generators of the cone, we obtain a line.

The conic sections were studied extensively by the early Greek mathematicians, who discovered the properties that enable us to define conics in terms of points (foci) and lines (directrices) in the plane of the conic. Reconciliation of these definitions with the previous discussion requires proofs that we shall not go into here.

A remarkable fact about conic sections is that although they were studied thousands of years ago, they are far from obsolete. Indeed, they are important tools for present-day investigations in outer space and for the study of the behavior of atomic particles. It is shown in physics that if a particle moves under the influence of what is called an *inverse square force field*, then its path may be described by means of a conic section. Examples of inverse square fields are gravitational and electromagnetic fields. Planetary orbits are elliptical. If the ellipse is very "flat," the curve resembles the path of a comet. The hyperbola is useful for describing the path of an alpha particle in the electric field of the nucleus of an atom. Parabolic mirrors are sometimes used to collect solar energy. The interested reader can find many other applications of conic sections.

SECTION 9.2

PARABOLAS

Parabolas were discussed in Section 2.5; however, the definition was not stated at that time. Moreover, we concentrated on parabolas with vertical axes. We shall now define parabola and derive equations for parabolas that have either vertical or horizontal axes.

DEFINITION

A **parabola** is the set of all points in a plane equidistant from a fixed point F (the **focus**) and a fixed line l (the **directrix**) in the plane.

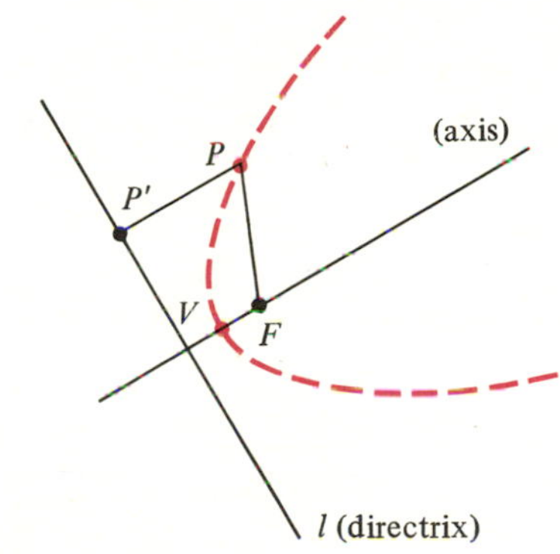

FIGURE 9.2

We shall assume that F is not on l, for otherwise, a line is obtained. If P is any point in the plane and P' is the point on l determined by a line through P that is perpendicular to l (see Figure 9.2), then by definition, P is on the parabola if and only if $d(P, F) = d(P, P')$. The dashes in Figure 9.2 indicate possible positions of P. The line through F, perpendicular to the directrix, is called the **axis** of the parabola. The point V on the axis, half-way from F to l, is called the **vertex** of the parabola.

A simple equation for a parabola can be obtained by choosing the y-axis along the axis of the parabola, with the origin at the vertex V, as illustrated in Figure 9.3. In this case, the focus F has coordinates $(0, p)$ for some real number $p \neq 0$, and the equation of the directrix is $y = -p$. By the Distance Formula, a point $P(x, y)$ is on the parabola if and only if

$$\sqrt{(x-0)^2 + (y-p)^2} = \sqrt{(x-x)^2 + (y+p)^2}.$$

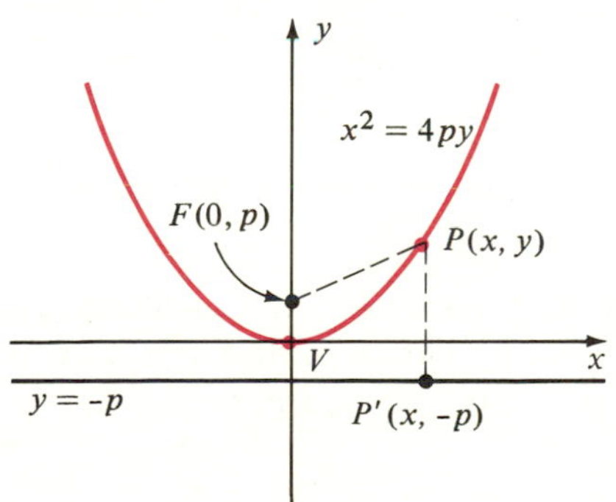

FIGURE 9.3

Squaring both sides gives us

$$(x-0)^2 + (y-p)^2 = (y+p)^2,$$

or

$$x^2 + y^2 - 2py + p^2 = y^2 + 2py + p^2.$$

This simplifies to

$$x^2 = 4py.$$

We have shown that the coordinates of every point (x, y) on the parabola satisfy $x^2 = 4py$. Conversely, if (x, y) is a solution of this equation, then by reversing the previous steps we see that the point (x, y) is on the parabola. If $p > 0$, the parabola opens upward, as in Figure 9.3, whereas if $p < 0$, the parabola opens downward. Note that the graph is symmetric with respect to the y-axis, since the solutions of the equation $x^2 = 4py$ are unchanged if $-x$ is substituted for x.

An analogous situation exists if the axis of the parabola is taken along the x-axis. If the vertex is $V(0, 0)$, the focus is $F(p, 0)$, and the directrix has equation $x = -p$ (see Figure 9.4), then using the same type of argument we obtain the equation $y^2 = 4px$. If $p > 0$, the parabola opens to the right, whereas if $p < 0$, it opens to the left. In this case the graph is symmetric with respect to the x-axis.

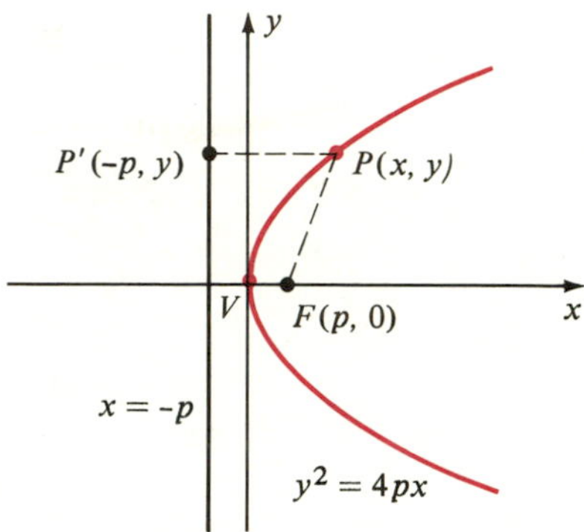

FIGURE 9.4

The following theorem summarizes our discussion.

THEOREM

> The graph of each of the following equations is a parabola that has its vertex at the origin and has the indicated focus and directrix.
>
> (i) $x^2 = 4py$: focus $F(0, p)$, directrix $y = -p$.
>
> (ii) $y^2 = 4px$: focus $F(p, 0)$, directrix $x = -p$.

EXAMPLE 1 Find the focus and directrix of the parabola that has the equation $y^2 = -6x$, and sketch the graph.

Solution The equation $y^2 = -6x$ has form (ii) of the preceding theorem with $4p = -6$, and hence $p = -\frac{3}{2}$. Thus the focus is $F(p, 0)$, that is, $F(-\frac{3}{2}, 0)$. The equation of the directrix is $x = -p$, or $x = \frac{3}{2}$. The graph is sketched in Figure 9.5. ■

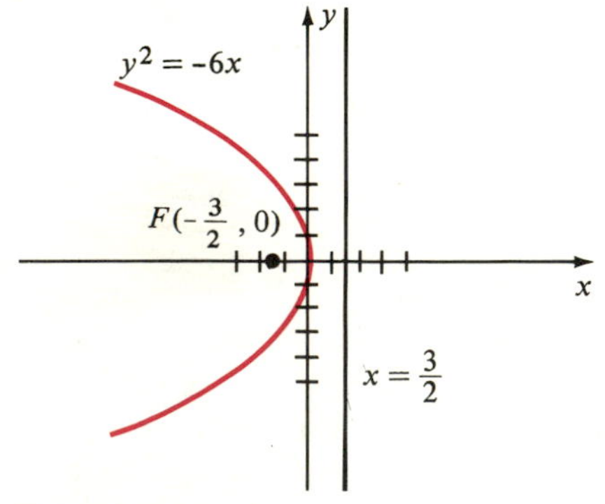

FIGURE 9.5

EXAMPLE 2 Find an equation of the parabola that has its vertex at the origin, opens upward, and passes through the point $P(-3, 7)$.

Solution The general form of the equation is $x^2 = 4py$ (see (i) of the theorem). If P is on the parabola, then $(-3, 7)$ is a solution of the equation. Hence we must have $(-3)^2 = 4p(7)$, or $p = \frac{9}{28}$. Substituting for p in $x^2 = 4py$ leads to the desired equation $x^2 = \frac{9}{7}y$, or $7x^2 = 9y$. ■

We may use a technique called **translation of axes** to extend our discussion to the case in which the vertex of the parabola is not at the origin. Recall that if a and b are the coordinates of two points A and B, respectively, on a coordinate line l, then the distance between A and B is $d(A, B) = |b - a|$. If we wish to take into account the direction of l, then we use the **directed distance** $\overline{AB}$ from A to B, which is, by definition,

$$\overline{AB} = b - a.$$

Since $\overline{BA} = a - b$, we have $\overline{AB} = -\overline{BA}$. If the positive direction on l is to the right, then B is to the right of A if and only if $\overline{AB} > 0$, and B is to the left of A if and only if $\overline{AB} < 0$. If C is any other point on l with coordinate c, it follows that

$$\overline{AC} = \overline{AB} + \overline{BC}$$

since $c - a = (b - a) + (c - b)$. We shall use this formula in the following discussion to develop formulas for translation of axes.

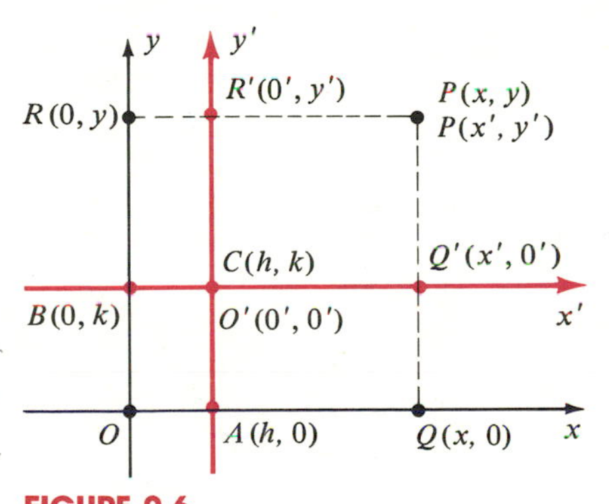

FIGURE 9.6

Suppose that $C(h, k)$ is an arbitrary point in an xy-coordinate plane. Let us introduce a new $x'y'$-coordinate system with origin O' at C such that the x'- and y'-axes are parallel to, and have the same unit lengths and positive directions as, the x- and y-axes, respectively. A typical situation of this type is illustrated in Figure 9.6 where, for simplicity, we have placed C in the first quadrant. We shall use primes on letters to denote coordinates of points in the $x'y'$-coordinate system, to distinguish them from coordinates with respect to the xy-coordinate system. Thus the point $P(x, y)$ in the xy-system will be denoted by $P(x', y')$ in the $x'y'$-system. If we label projections of P on the various axes as indicated in Figure 9.6, and let A and B denote projections of C on the x- and y-axes, respectively, then using directed distances we obtain

$$x = \overline{OQ} = \overline{OA} + \overline{AQ} = \overline{OA} + \overline{O'Q'} = h + x'$$

$$y = \overline{OR} = \overline{OB} + \overline{BR} = \overline{OB} + \overline{O'R'} = k + y'.$$

This proves the following result.

TRANSLATION OF AXES FORMULAS

If (x, y) are the coordinates of a point P relative to an xy-coordinate system, and if (x', y') are the coordinates of P relative to an $x'y'$-coordinate system with origin at the point $C(h, k)$ of the xy-system, then

(i) $x = x' + h, \quad y = y' + k$

(ii) $x' = x - h, \quad y' = y - k.$

The Translation of Axes Formulas enable us to go from either coordinate system to the other. Their major use is to change the form of equations of graphs. To be specific, if, in the xy-plane, a certain collection of points is the graph of an equation in x and y, then to find an equation in x' and y' that has the same graph in the $x'y'$-plane, we may substitute $x' + h$ for x and $y' + k$ for y in the given equation. Conversely, if a set of points in the $x'y'$-plane is the graph of an equation in x' and y', then to find the corresponding equation in x and y we substitute $x - h$ for x' and $y - k$ for y'.

As a simple illustration of the preceding remarks, the equation

$$(x')^2 + (y')^2 = r^2$$

has, for its graph in the $x'y'$-plane, a circle of radius r with center at the origin O'. Using the Translation of Axes Formulas, an equation for this circle in the xy-plane is

$$(x - h)^2 + (y - k)^2 = r^2,$$

which is in agreement with the formula for a circle of radius r with center at $C(h, k)$ in the xy-plane.

As another illustration, we know that

$$(x')^2 = 4py'$$

is an equation of a parabola with a vertex at the origin O' of the $x'y'$-plane. Using the Translation of Axes Formulas, we see that

$$(x - h)^2 = 4p(y - k)$$

is an equation of the same parabola in the xy-plane with vertex $V(h, k)$. This is in agreement with the equation we obtained in Section 2.5 for a parabola with a vertical axis. The focus is $F(h, k + p)$ and the directrix is $y = k - p$. Similarly, starting with the equation $(y')^2 = 4px'$ gives us $(y - k)^2 = 4p(x - h)$. The next theorem summarizes this discussion.

THEOREM

The graph of each of the following equations is a parabola that has vertex $V(h, k)$ and has the indicated focus and directrix.

(i) $(x - h)^2 = 4p(y - k)$: focus $F(h, k + p)$, directrix $y = k - p$.

(ii) $(y - k)^2 = 4p(x - h)$: focus $F(h + p, k)$, directrix $x = h - p$.

In each case the axis of the parabola is parallel to a coordinate axis. The parabola having equation (i) opens upward or downward, whereas the parabola in (ii) opens to the right or left. Typical graphs are sketched in Figure 9.7.

Squaring the left side of the equation in (i) of the preceding theorem and simplifying leads to an equation of the form

$$y = ax^2 + bx + c$$

for real numbers a, b, and c. Conversely, the graph of such an equation is a parabola with a vertical axis. As in Section 2.5, we may complete the square in x to find the vertex (see Examples 3 and 4 of Section 2.5).

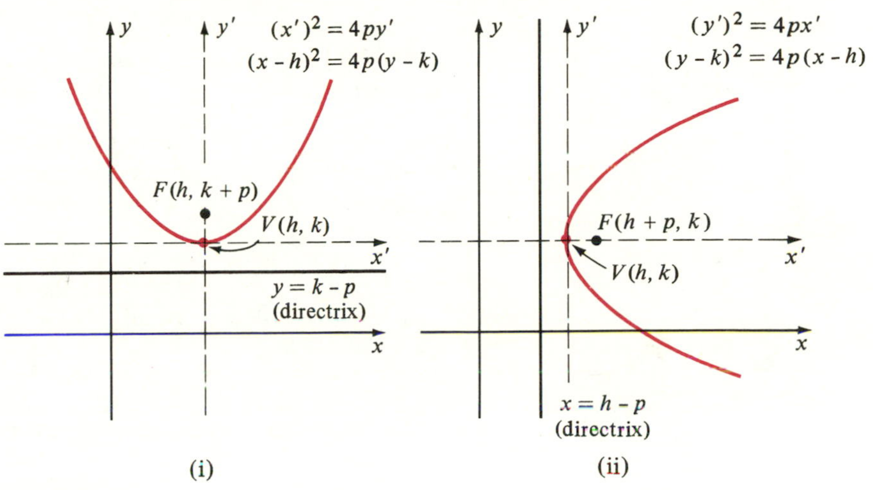

FIGURE 9.7

The equation in (ii) of the theorem may be written as

$$x = ay^2 + by + c$$

for real numbers a, b, and c. Conversely, the last equation can be expressed in form (ii) by completing the square in y as illustrated next, in Example 3. Hence, if $a \neq 0$, the graph of $x = ay^2 + by + c$ is a parabola with a horizontal axis.

EXAMPLE 3 Discuss and sketch the graph of the equation

$$2x = y^2 + 8y + 22.$$

Solution By our previous remarks, the graph is a parabola with a horizontal axis. Writing

$$y^2 + 8y = 2x - 22,$$

we complete the square on the left by adding 16 to both sides. This gives us

$$y^2 + 8y + 16 = 2x - 6.$$

The last equation may be written

$$(y + 4)^2 = 2(x - 3),$$

which is in form (ii) of the theorem, with $h = 3$, $k = -4$, and $4p = 2$, or $p = \frac{1}{2}$. Hence the vertex is $V(3, -4)$. Since $p = \frac{1}{2} > 0$, the parabola opens

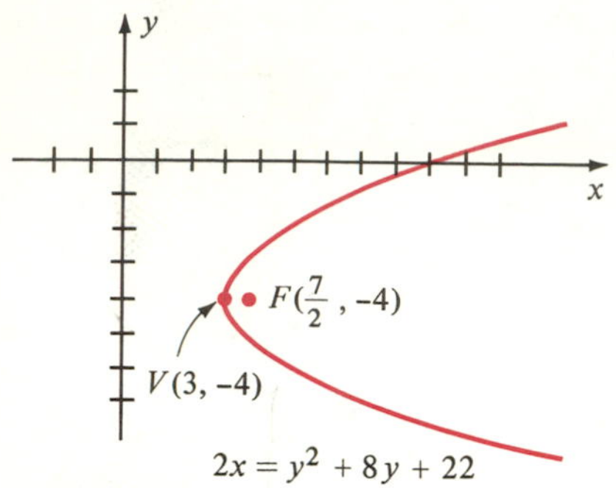

FIGURE 9.8

to the right with focus at $F(h+p, k)$, that is, $F(\frac{7}{2}, -4)$. The equation of the directrix is $x = h - p$, or $x = \frac{5}{2}$. The parabola is sketched in Figure 9.8. ■

EXAMPLE 4 Find an equation of the parabola with vertex $V(-4, 2)$ and directrix $y = 5$.

Solution The vertex and directrix are shown in Figure 9.9. The dashes indicate a possible position for the parabola. It follows that an equation of the parabola is

$$(x-h)^2 = 4p(y-k)$$

where $h = -4$, $k = 2$, and $p = -3$. (Why?) This gives us

$$(x+4)^2 = -12(y-2)$$

The last equation can be expressed in the form $y = ax^2 + bx + c$ as follows:

$$x^2 + 8x + 16 = -12y + 24$$

$$12y = -x^2 - 8x + 8$$

$$y = -\tfrac{1}{12}x^2 - \tfrac{2}{3}x + \tfrac{2}{3}.$$ ■

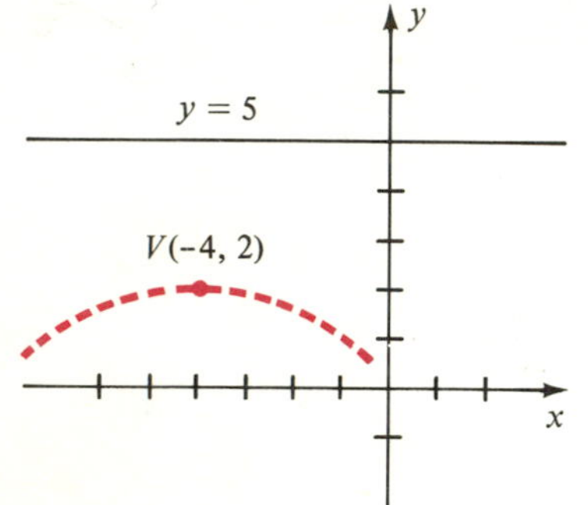

FIGURE 9.9

Parabolas are very useful in applications of mathematics to the physical world. For example, it can be shown that if a projectile is fired, and we assume that it is acted upon only by the force of gravity (that is, air resistance and other outside factors are ignored), then the path of the projectile is parabolic. Properties of parabolas are used in the design of mirrors for telescopes and searchlights and in the construction of radar antenna. These are only a few of many physical applications.

EXERCISES 9.2

In Exercises 1–16 find the vertex, focus, and directrix of the parabola with the given equation, and sketch the graph.

1 $x^2 = -12y$

2 $y^2 = \frac{1}{2}x$

3 $2y^2 = -3x$

4 $x^2 = -3y$

5 $8x^2 = y$

6 $y^2 = -100x$

7 $y^2 - 12 = 12x$

8 $y = 40x - 97 - 4x^2$

9 $y = x^2 - 4x + 2$

10 $y = 8x^2 + 16x + 10$

11 $y^2 - 4y - 2x - 4 = 0$

12 $y^2 + 14y + 4x + 45 = 0$

13 $4x^2 + 40x + y + 106 = 0$

14 $y^2 - 20y + 100 = 6x$

15 $x^2 + 20y = 10$

16 $4x^2 + 4x + 4y + 1 = 0$

In Exercises 17–22 find an equation of the parabola that satisfies the given conditions.

17 Focus $(2, 0)$, directrix $x = -2$

18 Focus $(0, -4)$, directrix $y = 4$

19 Focus $(6, 4)$, directrix $y = -2$

20 Focus $(-3, -2)$, directrix $y = 1$

21 Vertex at the origin, symmetric to the y-axis, and passing through the point $A(2, -3)$

22 Vertex $V(-3, 5)$, axis parallel to the x-axis, and passing through $A(5, 9)$

23 A searchlight reflector is designed so that a cross section through its axis is a parabola and the light source is at the focus. Find the focus if the reflector is 3 feet across at the opening and 1 foot deep.

24 One section of a suspension bridge has its weight uniformly distributed between twin towers that are 400 feet apart and rise 90 feet above the horizontal roadway (see figure). A cable strung between the tops of the towers has the shape of a parabola and its center point is 10 feet above the roadway. Suppose coordinate axes are introduced as shown in the figure.

(a) Find an equation for the parabola.

(b) If nine equispaced vertical cables are used to support the bridge (see figure), find the total length of these supports.

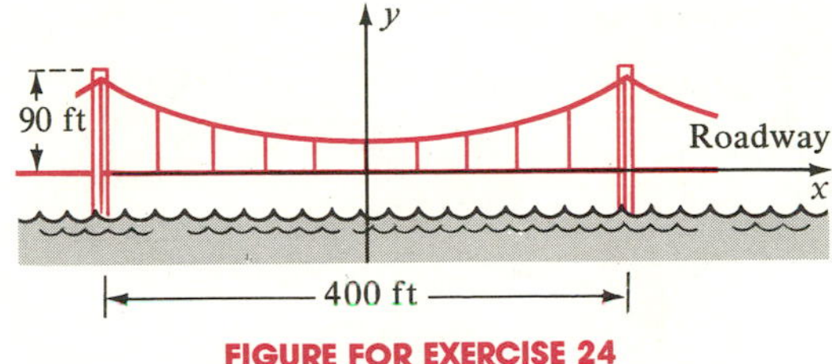

FIGURE FOR EXERCISE 24

25 Find an equation of the parabola that has a vertical axis and passes through the points $A(2, 3)$, $B(-1, 6)$, and $C(1, 0)$.

26 Prove that there is exactly one line of a given slope m that intersects the parabola $x^2 = 4py$ in exactly one point, and that its equation is $y = mx - pm^2$.

SECTION 9.3

ELLIPSES

An ellipse may be defined as follows.

DEFINITION An **ellipse** is the set of all points in a plane, the sum of whose distances from two fixed points in the plane (the **foci**) is constant.

It is known that the orbits of planets in the solar system are elliptical, with the sun at one of the foci. This is only one of many important applications of ellipses.

There is an easy way to construct an ellipse on paper. We begin by inserting two thumbtacks in the paper at points labeled F and F' and fastening the ends of a piece of string to the thumbtacks. If the string is now looped around a pencil and drawn taut at point P, as in Figure 9.10, then moving the pencil and at the same time keeping the string taut, the sum of the distances $d(F, P)$ and $d(F', P)$ is the length of the string, and hence is constant. The pencil will, therefore, trace out an ellipse with foci at F and F'. By varying the positions of F and F', but keeping the length of string fixed, the shape of the ellipse can be made to change considerably. If F and F' are far

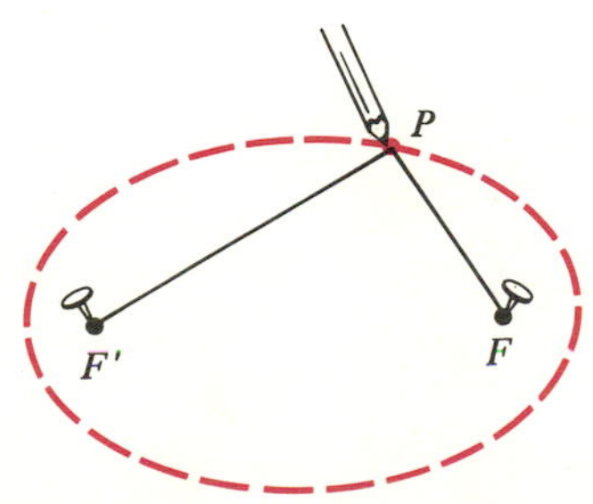

FIGURE 9.10

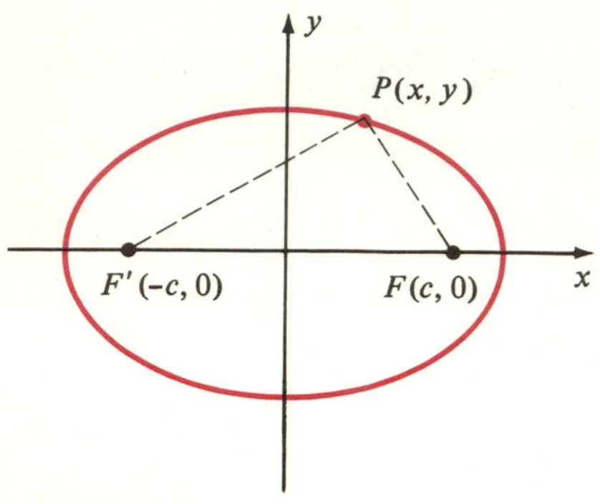

FIGURE 9.11

apart, in the sense that $d(F, F')$ is almost the same as the length of the string, then the ellipse is quite flat. If $d(F, F')$ is close to zero, the ellipse is almost circular. Indeed, if $F = F'$, a circle is obtained.

By introducing suitable coordinate systems, we may derive simple equations for ellipses. Let us choose the x-axis as the line through the two foci F and F', with the origin at the midpoint of the segment $F'F$. This point is called the **center** of the ellipse. If F has coordinates $(c, 0)$, where $c > 0$, then, as shown in Figure 9.11, F' has coordinates $(-c, 0)$ and hence the distance between F and F' is $2c$. Let the constant sum of the distances of P from F and F' be denoted by $2a$, where in order to get points that are not on the x-axis we must have $2a > 2c$, that is, $a > c$. (Why?) By definition, $P(x, y)$ is on the ellipse if and only if

$$d(P, F) + d(P, F') = 2a$$

or, by the Distance Formula,

$$\sqrt{(x-c)^2 + (y-0)^2} + \sqrt{(x+c)^2 + (y-0)^2} = 2a.$$

Writing the preceding equation as

$$\sqrt{(x-c)^2 + y^2} = 2a - \sqrt{(x+c)^2 + y^2}$$

and squaring both sides, we obtain

$$x^2 - 2cx + c^2 + y^2 = 4a^2 - 4a\sqrt{(x+c)^2 + y^2} + x^2 + 2cx + c^2 + y^2,$$

which simplifies to

$$a\sqrt{(x+c)^2 + y^2} = a^2 + cx.$$

Squaring both sides gives us

$$a^2(x^2 + 2cx + c^2 + y^2) = a^4 + 2a^2cx + c^2x^2,$$

which may be written in the form

$$x^2(a^2 - c^2) + a^2y^2 = a^2(a^2 - c^2).$$

Dividing both sides by $a^2(a^2 - c^2)$ leads to

$$\frac{x^2}{a^2} + \frac{y^2}{a^2 - c^2} = 1.$$

For convenience, we let

$$b^2 = a^2 - c^2 \qquad \text{where } b > 0$$

in the preceding equation, obtaining

$$\frac{x^2}{a^2} + \frac{y^2}{b^2} = 1.$$

Since $c > 0$ and $b^2 = a^2 - c^2$, it follows that $a^2 > b^2$ and hence $a > b$.

We have shown that the coordinates of every point (x, y) on the ellipse in Figure 9.11 satisfy the equation $x^2/a^2 + y^2/b^2 = 1$. Conversely, if (x, y) is a solution of this equation, then by reversing the preceding steps we see that the point (x, y) is on the ellipse.

The x-intercepts may be found by setting $y = 0$. Doing so gives us $x^2/a^2 = 1$, or $x^2 = a^2$, and consequently the x-intercepts are a and $-a$. The corresponding points $V(a, 0)$ and $V'(-a, 0)$ on the graph are called the **vertices** of the ellipse, and the line segment $V'V$ is referred to as the **major axis.** Similarly, letting $x = 0$ in the equation of the ellipse, we obtain $y^2/b^2 = 1$, or $y^2 = b^2$. Hence the y-intercepts are b and $-b$. The segment from $M'(0, -b)$ to $M(0, b)$ is called the **minor axis** of the ellipse. Note that the major axis is longer than the minor axis, since $a > b$.

Applying tests for symmetry we see that the ellipse is symmetric to both the x-axis and the y-axis. It is also a symmetric with respect to the origin, since substitution of $-x$ for x and $-y$ for y does not change the equation.

The preceding discussion may be summarized as follows.

THEOREM

The graph of the equation

$$\frac{x^2}{a^2} + \frac{y^2}{b^2} = 1,$$

for $a^2 > b^2$, is an ellipse with vertices $(\pm a, 0)$. The endpoints of the minor axis are $(0, \pm b)$. The foci are $(\pm c, 0)$, where $c^2 = a^2 - b^2$.

EXAMPLE 1 Discuss and sketch the graph of the equation

$$4x^2 + 18y^2 = 36.$$

Solution To obtain the form in the theorem we divide both sides of the equation by 36 and simplify. This leads to

$$\frac{x^2}{9} + \frac{y^2}{2} = 1,$$

which is in the proper form with $a^2 = 9$ and $b^2 = 2$. Thus $a = 3$, $b = \sqrt{2}$, and hence the endpoints of the major axis are $(\pm 3, 0)$ and the endpoints of the minor axis are $(0, \pm\sqrt{2})$. Since

$$c^2 = a^2 - b^2 = 9 - 2 = 7, \quad \text{or} \quad c = \sqrt{7},$$

the foci are $(\pm\sqrt{7}, 0)$. The graph is sketched in Figure 9.12. ■

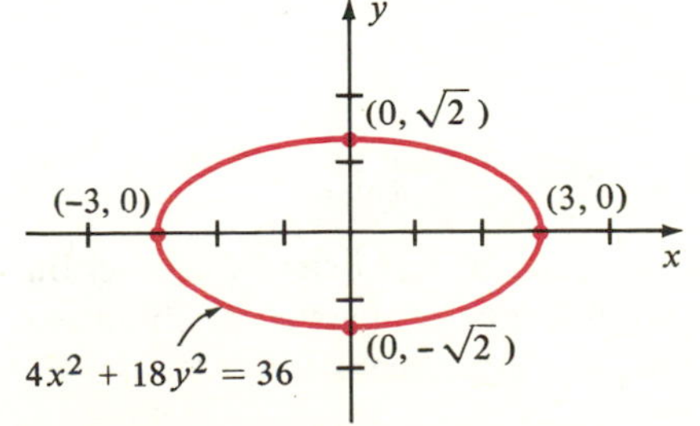

FIGURE 9.12

EXAMPLE 2 Find an equation of the ellipse with vertices $(\pm 4, 0)$ and foci $(\pm 2, 0)$.

Solution Using the notation of the theorem, $a = 4$ and $c = 2$. Since $c^2 = a^2 - b^2$, we see that $b^2 = a^2 - c^2 = 16 - 4 = 12$. This gives us

$$\frac{x^2}{16} + \frac{y^2}{12} = 1.$$

Multiplying both sides by 48 leads to $3x^2 + 4y^2 = 48$. ■

It is sometimes convenient to choose the major axis of the ellipse along the y-axis. If the foci are $(0, \pm c)$, then by the same type of argument used previously, we obtain the following.

THEOREM

The graph of the equation

$$\frac{x^2}{b^2} + \frac{y^2}{a^2} = 1,$$

for $a^2 > b^2$, is an ellipse with vertices $(0, \pm a)$. The endpoints of the minor axis are $(\pm b, 0)$. There foci are $(0, \pm c)$, where $c^2 = a^2 - b^2$.

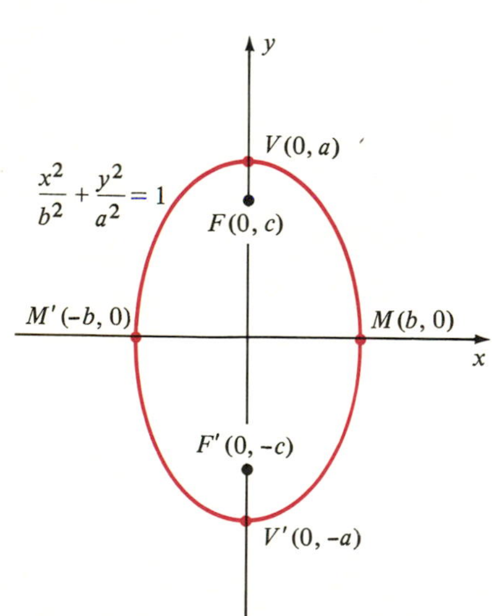

FIGURE 9.13

A typical graph is sketched in Figure 9.13.

The preceding discussion shows that an equation of an ellipse with center at the origin and foci on a coordinate axis can always be written in the form

$$\frac{x^2}{p} + \frac{y^2}{q} = 1, \quad \text{or} \quad qx^2 + py^2 = pq$$

where p and q are positive. If $p > q$, the major axis lies on the x-axis, whereas if $q > p$, the major axis is on the y-axis. It is unnecessary to memorize these facts, since in any given problem the major axis can be determined by examining the x- and y-intercepts.

EXAMPLE 3 Sketch the graph of the equation $9x^2 + 4y^2 = 25$.

Solution The graph is an ellipse with center at the origin and foci on one of the coordinate axes. To find the x-intercepts, we let $y = 0$, obtaining $9x^2 = 25$, or $x = \pm\frac{5}{3}$. Similarly, to find the y-intercepts, we let $x = 0$,

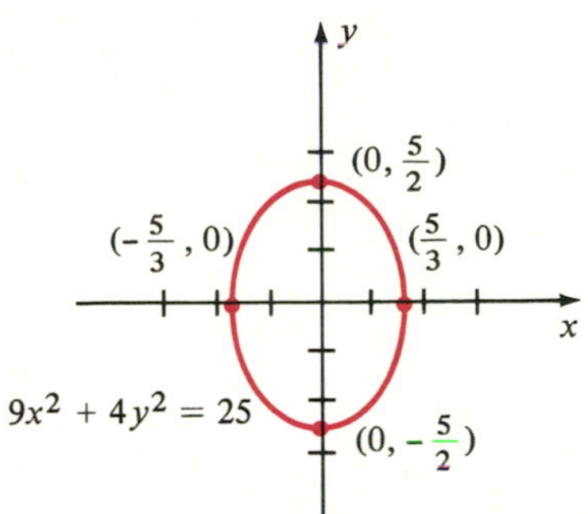

FIGURE 9.14

obtaining $4y^2 = 25$, or $y = \pm\frac{5}{2}$. This enables us to sketch the ellipse (see Figure 9.14). Since $\frac{5}{3} < \frac{5}{2}$, the major axis is on the y-axis. ■

By using the Translation of Axes Formulas we can extend our work to an ellipse with center at any point $C(h, k)$ in the xy-plane. For example, since the graph of

$$\frac{(x')^2}{a^2} + \frac{(y')^2}{b^2} = 1$$

is an ellipse with center at O' in an $x'y'$-plane (see Figure 9.15), then its equation relative to the xy-coordinate system is

$$\frac{(x-h)^2}{a^2} + \frac{(y-k)^2}{b^2} = 1.$$

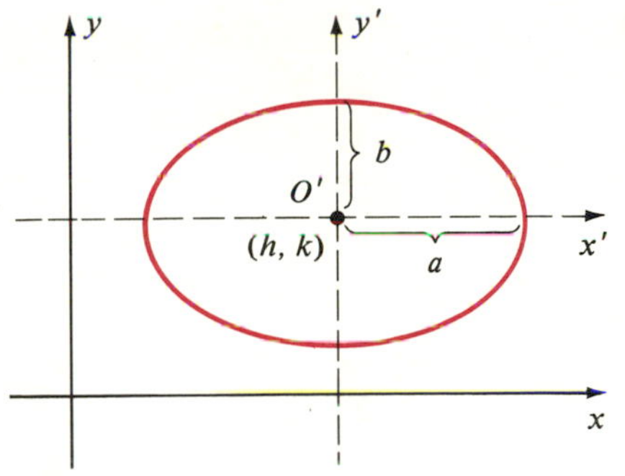

FIGURE 9.15

Squaring the indicated terms in the last equation and simplifying gives us an equation of the form

$$Ax^2 + Cy^2 + Dx + Ey + F = 0$$

where the coefficients are real numbers and A and C are both positive. Conversely, if we start with such an equation, then by completing squares we can obtain a form that displays the center of the ellipse and the lengths of the major and minor axes. This technique is illustrated in the next example.

EXAMPLE 4 Discuss and sketch the graph of the equation

$$16x^2 + 9y^2 + 64x - 18y - 71 = 0$$

Solution We begin by writing the equation in the form

$$16(x^2 + 4x) + 9(y^2 - 2y) = 71.$$

Next, we complete the squares for the expressions within parentheses, obtaining

$$16(x^2 + 4x + 4) + 9(y^2 - 2y + 1) = 71 + 64 + 9.$$

Note that by adding 4 to the expression within the first parentheses we have added 64 to the left side of the equation, and hence must compensate by adding 64 to the right side. Similarly, by adding 1 to the expression within the second parentheses, 9 is added to the left side and consequently 9 must also be added to the right side. The last equation may be written

$$16(x + 2)^2 + 9(y - 1)^2 = 144.$$

Dividing by 144, we obtain

$$\frac{(x+2)^2}{9} + \frac{(y-1)^2}{16} = 1,$$

which is of the form $$\frac{(x')^2}{9} + \frac{(y')^2}{16} = 1$$

with $x' = x + 2$ and $y' = y - 1$. This corresponds to letting $h = -2$ and $k = 1$ in the Translation of Axes Formulas. Since the graph of the equation $(x')^2/9 + (y')^2/16 = 1$ is an ellipse with center at the origin O' in the $x'y'$-plane, it follows that the graph of the given equation is an ellipse with center $C(-2, 1)$ in the xy-plane and with axes parallel to the coordinate axes. The graph is sketched in Figure 9.16.

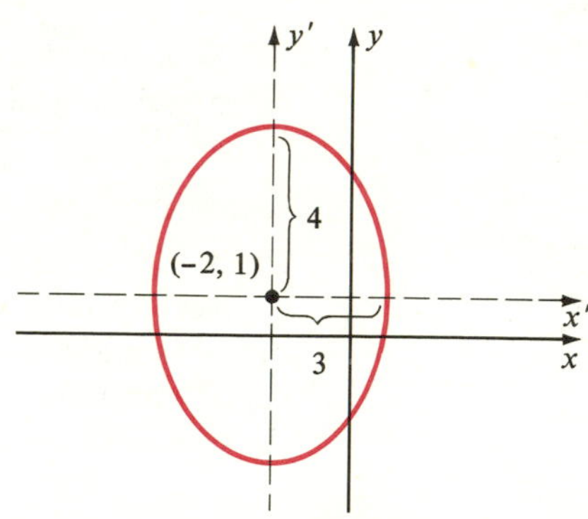

FIGURE 9.16 $\frac{(x+2)^2}{9} + \frac{(y-1)^2}{16} = 1$ ■

EXERCISES 9.3

In Exercises 1–14 sketch the graph of the equation and find the coordinates of the vertices and foci.

1 $\frac{x^2}{9} + \frac{y^2}{4} = 1$

2 $\frac{x^2}{25} + \frac{y^2}{16} = 1$

3 $4x^2 + y^2 = 16$

4 $y^2 + 9x^2 = 9$

5 $5x^2 + 2y^2 = 10$

6 $\frac{1}{2}x^2 + 2y^2 = 8$

7 $4x^2 + 25y^2 = 1$

8 $10y^2 + x^2 = 5$

9 $4x^2 + 9y^2 - 32x - 36y + 64 = 0$

10 $x^2 + 2y^2 + 2x - 20y + 43 = 0$

11 $9x^2 + 16y^2 + 54x - 32y - 47 = 0$

12 $4x^2 + 9y^2 + 24x + 18y + 9 = 0$

13 $25x^2 + 4y^2 - 250x - 16y + 541 = 0$

14 $4x^2 + y^2 = 2y$

In Exercises 15–20 find an equation for the ellipse satisfying the given conditions.

15 Vertices $V(\pm 8, 0)$, foci $F(\pm 5, 0)$

16 Vertices $V(0, \pm 7)$, foci $F(0, \pm 2)$

17 Vertices $V(0, \pm 5)$, length of minor axis 3

18 Foci $F(\pm 3, 0)$, length of minor axis 2

19 Vertices $V(0, \pm 6)$, passing through (3, 2)

20 Center at the origin, symmetric with respect to both axes, passing through $A(2, 3)$ and $B(6, 1)$

In Exercises 21 and 22 find the points of intersection of the graphs of the given equations. Sketch both graphs on the same coordinate axes, showing points of intersection.

21 $\begin{cases} x^2 + 4y^2 = 20 \\ x + 2y = 6 \end{cases}$ **22** $\begin{cases} x^2 + 4y^2 = 36 \\ x^2 + y^2 = 12 \end{cases}$

23 An arch of a bridge is semi-elliptical with major axis horizontal. The base of the arch is 30 feet across and the highest part of the arch is 10 feet above the horizontal roadway. Find the height of the arch 6 feet from the center of the base.

24 The **eccentricity** of an ellipse is defined as the ratio $(\sqrt{a^2 - b^2})/a$. If a is fixed and b varies, describe the general shape of the ellipse when the eccentricity is close to 1 and when it is close to zero.

25 A line segment of length $a + b$ moves with its endpoints A and B attached to the coordinate axes, as illustrated in the figure. Prove that if $a \neq b$, then the point P traces an ellipse.

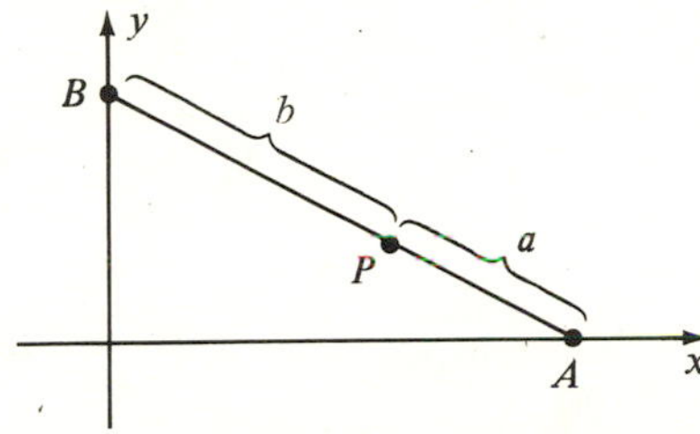

FIGURE FOR EXERCISE 25

26 Consider the ellipse $px^2 + qy^2 = pq$ where $p > 0$ and $q > 0$. Prove that if m is any real number, there are exactly two lines of slope m that intersect the ellipse in precisely one point, and that equations of the lines are $y = mx \pm \sqrt{p + qm^2}$.

SECTION 9.4
HYPERBOLAS

The definition of a hyperbola is similar to that of an ellipse. The only change is that instead of using the *sum* of distances from two fixed points, we use the *difference*.

DEFINITION

A **hyperbola** is the set of all points in a plane, the difference of whose distances from two fixed points in the plane (the **foci**) is a positive constant.

To find a simple equation for a hyperbola, we choose a coordinate system with foci at $F(c, 0)$ and $F'(-c, 0)$, and denote the (constant) distance by $2a$. Referring to Figure 9.17, we see that a point $P(x, y)$ is on the hyperbola if and only if either one of the following is true:

$$d(P, F') - d(P, F) = 2a$$

$$d(P, F) - d(P, F') = 2a.$$

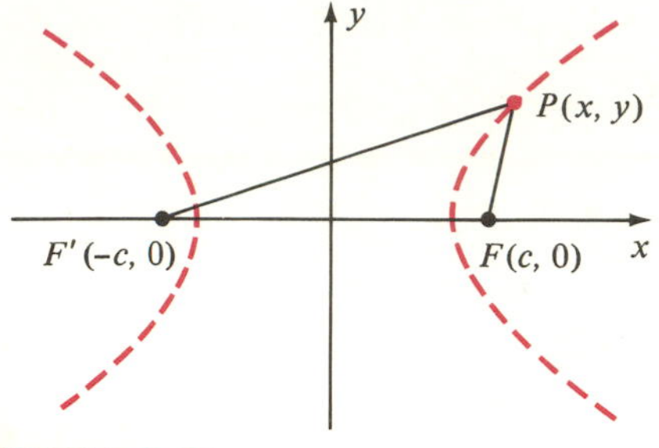

FIGURE 9.17

For hyperbolas (unlike ellipses) we need $a < c$ to obtain points on the hyperbola that are not on the x-axis, for if P is such a point, then from Figure 9.17 we see that

$$d(P, F) < d(F', F) + d(P, F')$$

because the length of one side of a triangle is always less than the sum of the lengths of the other two sides. Similarly,

$$d(P, F') < d(F', F) + d(P, F).$$

Equivalent forms for the previous two inequalities are

$$d(P, F) - d(P, F') < d(F', F)$$
$$d(P, F') - d(P, F) < d(F', F).$$

Since the differences on the left both equal $2a$, and since $d(F', F) = 2c$, the last two inequalities imply that $2a < 2c$, or $a < c$.

The equations $d(P, F') - d(P, F) = 2a$ and $d(P, F) - d(P, F') = 2a$ may be replaced by the single equation

$$|d(P, F) - d(P, F')| = 2a.$$

It follows from the Distance Formula that an equation of the hyperbola is

$$\left|\sqrt{(x-c)^2 + (y-0)^2} - \sqrt{(x+c)^2 + (y-0)^2}\right| = 2a.$$

Employing the type of simplification procedure used to derive an equation for an ellipse, we arrive at the equivalent equation

$$\frac{x^2}{a^2} - \frac{y^2}{c^2 - a^2} = 1.$$

For convenience, let

$$b^2 = c^2 - a^2 \qquad \text{where } b > 0$$

in the preceding equation, obtaining

$$\frac{x^2}{a^2} - \frac{y^2}{b^2} = 1.$$

We have shown that the coordinates of every point (x, y) on the hyperbola in Figure 9.17 satisfy the last equation. Conversely, if (x, y) is a solution of that equation, then by reversing the preceding steps we see that the point (x, y) is on the hyperbola.

By the Tests for Symmetry, the hyperbola is symmetric with respect to both axes and the origin. The x-intercepts are $\pm a$. The corresponding points $V(a, 0)$ and $V'(-a, 0)$ are called the **vertices,** and the line segment $V'V$ is known as the **transverse axis** of the hyperbola. The origin is called the **center** of the hyperbola. There are no y-intercepts, since the equation $-y^2/b^2 = 1$ has no solutions.

The preceding discussion may be summarized as follows.

THEOREM

The graph of the equation

$$\frac{x^2}{a^2} - \frac{y^2}{b^2} = 1$$

is a hyperbola with vertices $(\pm a, 0)$. The foci are $(\pm c, 0)$, where $c^2 = a^2 + b^2$.

If the equation $x^2/a^2 - y^2/b^2 = 1$ is solved for y, we obtain

$$y = \pm\frac{b}{a}\sqrt{x^2 - a^2}.$$

Hence, there are no points (x, y) on the graph if $x^2 - a^2 < 0$, that is, if $-a < x < a$. However, there *are* points $P(x, y)$ on the graph if $x \geq a$ or $x \leq -a$. If $x \geq a$, we may write the last equation in the form

$$y = \pm\frac{b}{a}x\sqrt{1 - \frac{a^2}{x^2}}.$$

It follows that if x is large (in comparison to a), the radicand is close to 1, and hence the y-coordinate y of the point $P(x, y)$ on the hyperbola is close to either $(b/a)x$ or $-(b/a)x$. This means that the point $P(x, y)$ is close to the line with equation $y = (b/a)x$ when y is positive, or the line with equation $y = -(b/a)x$ when y is negative. As x increases, we say that the point $P(x, y)$ *approaches* one of these lines. A corresponding situation exists when $x \leq -a$. The lines with equations

$$y = \pm\frac{b}{a}x$$

are called the **asymptotes** of the hyperbola.

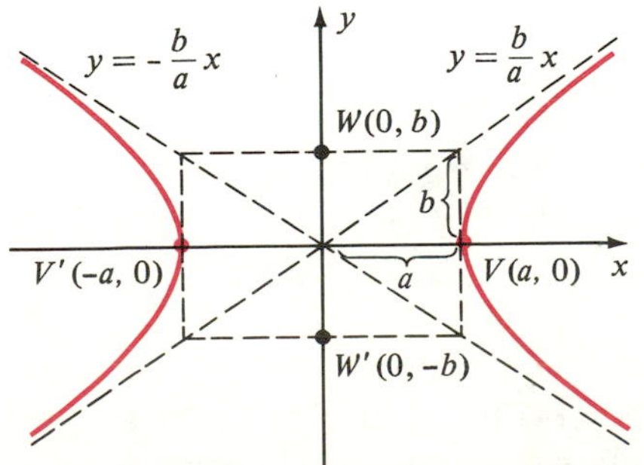

FIGURE 9.18

The asymptotes serve as excellent guides for sketching the graph. A convenient way to sketch the asymptotes is to first plot the vertices $V(a, 0)$, $V'(-a, 0)$ and the points $W(0, b)$, $W'(0, -b)$ (see Figure 9.18). The line segment $W'W$ of length $2b$ is called the **conjugate axis** of the hyperbola. If horizontal and vertical lines are drawn through the endpoints of the conjugate and transverse axes, respectively, then the diagonals of the resulting rectangle have slopes b/a and $-b/a$. Hence, by extending these diagonals we obtain lines with equations $y = (\pm b/a)x$. The hyperbola is then sketched as in Figure 9.18, using the asymptotes as guides. The two curves that make up the hyperbola are called the **branches** of the hyperbola.

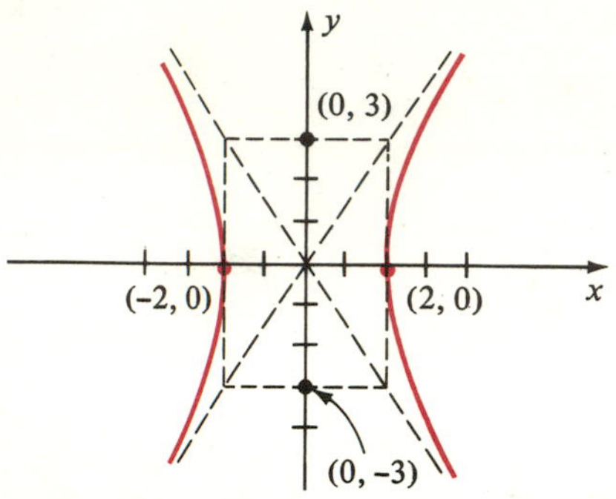

FIGURE 9.19

EXAMPLE 1 Discuss and sketch the graph of the equation $9x^2 - 4y^2 = 36$.

Solution Dividing both sides by 36, we have

$$\frac{x^2}{4} - \frac{y^2}{9} = 1,$$

which is of the form stated in the theorem with $a^2 = 4$ and $b^2 = 9$. Hence, $a = 2$ and $b = 3$. The vertices $(\pm 2, 0)$ and the endpoints $(0, \pm 3)$ of the conjugate axis determine a rectangle whose diagonals (extended) give us the asymptotes. The graph of the equation is sketched in Figure 9.19. The equations of the asymptotes, $y = \pm\frac{3}{2}x$, can be found by referring to the graph or to the equations $y = (\pm b/a)x$. Since $c^2 = a^2 + b^2 = 4 + 9 = 13$, the foci are $(\pm\sqrt{13}, 0)$. ■

The preceding example indicates that for hyperbolas it is not always true that $a < b$, as is the case for ellipses. Indeed, we may have $a < b$, $a > b$, or $a = b$.

EXAMPLE 2 Find an equation, the foci, and the asymptotes of a hyperbola that has vertices $(\pm 3, 0)$ and passes through the point $P(5, 2)$.

Solution Substituting $a = 3$ in $x^2/a^2 - y^2/b^2 = 1$, we obtain the equation

$$\frac{x^2}{9} - \frac{y^2}{b^2} = 1.$$

If (5, 2) is a solution of this equation, then

$$\frac{25}{9} - \frac{4}{b^2} = 1.$$

This gives us $b^2 = \frac{9}{4}$, and hence the desired equation is

$$\frac{x^2}{9} - \frac{4y^2}{9} = 1,$$

or equivalently, $x^2 - 4y^2 = 9$.

Since $c^2 = a^2 + b^2 = 9 + \frac{9}{4} = \frac{45}{4}$, the foci are $(\pm\frac{3}{2}\sqrt{5}, 0)$. Substituting for b and a in $y = \pm(b/a)x$ and simplifying, we obtain equations $y = \pm\frac{1}{2}x$ for the asymptotes. ■

If the foci of a hyperbola are the points $(0, \pm c)$ on the y-axis, then by the same type of argument used previously, we obtain the following theorem.

THEOREM

> The graph of the equation
>
> $$\frac{y^2}{a^2} - \frac{x^2}{b^2} = 1$$
>
> is a hyperbola with vertices $(0, \pm a)$. The foci are $(0, \pm c)$, where $c^2 = a^2 + b^2$.

In this case the endpoints of the conjugate axis are $W(b, 0)$ and $W'(-b, 0)$. The asymptotes are found, as before, by using the diagonals of the rectangle determined by these points, the vertices, and lines parallel to the coordinate axes. The graph is sketched in Figure 9.20. The equations of the asymptotes are $y = (\pm a/b)x$. Note the difference between these equations and the equations $y = (\pm b/a)x$ for the asymptotes of the hyperbola considered first in this section.

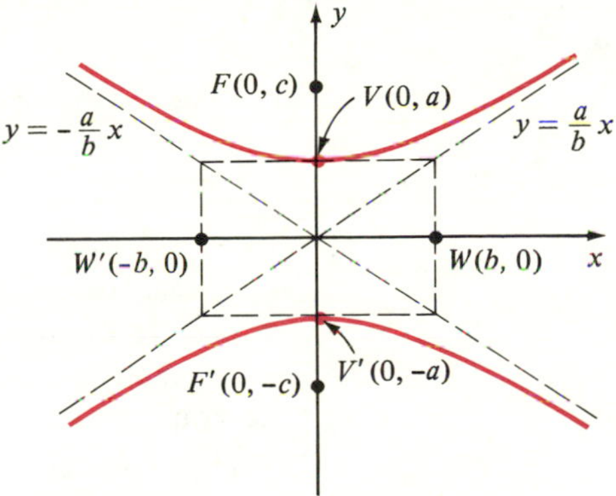

FIGURE 9.20

EXAMPLE 3 Discuss and sketch the graph of the equation $4y^2 - 2x^2 = 1$.

Solution The form in the last theorem may be obtained by writing the equation as

$$\frac{y^2}{\frac{1}{4}} - \frac{x^2}{\frac{1}{2}} = 1.$$

Thus $a^2 = \frac{1}{4}$, $b^2 = \frac{1}{2}$, and $c^2 = \frac{1}{4} + \frac{1}{2} = \frac{3}{4}$. Consequently $a = \frac{1}{2}$, $b = \sqrt{2}/2$, and $c = \sqrt{3}/2$. The vertices are $(0, \pm\frac{1}{2})$ and the foci are $(0, \pm\sqrt{3}/2)$. The graph has the general appearance of the graph shown in Figure 9.20.

As was the case for ellipses, we may use translations of axes to generalize our work. The following example illustrates this technique.

EXAMPLE 4 Discuss and sketch the graph of the equation

$$9x^2 - 4y^2 - 54x - 16y + 29 = 0.$$

Solution We arrange our work as follows:

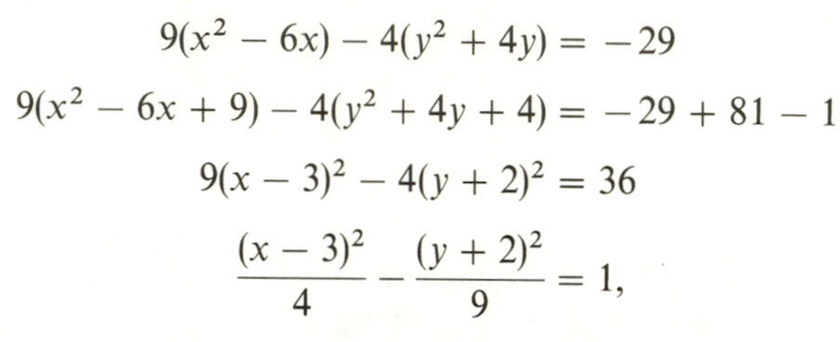

$$9(x^2 - 6x) - 4(y^2 + 4y) = -29$$

$$9(x^2 - 6x + 9) - 4(y^2 + 4y + 4) = -29 + 81 - 16$$

$$9(x - 3)^2 - 4(y + 2)^2 = 36$$

$$\frac{(x-3)^2}{4} - \frac{(y+2)^2}{9} = 1,$$

which is of the form
$$\frac{(x')^2}{4} - \frac{(y')^2}{9} = 1$$

with $x' = x - 3$ and $y' = y + 2$. By translating the x- and y-axes to the new origin $C(3, -2)$, we obtain the sketch shown in Figure 9.21. ■

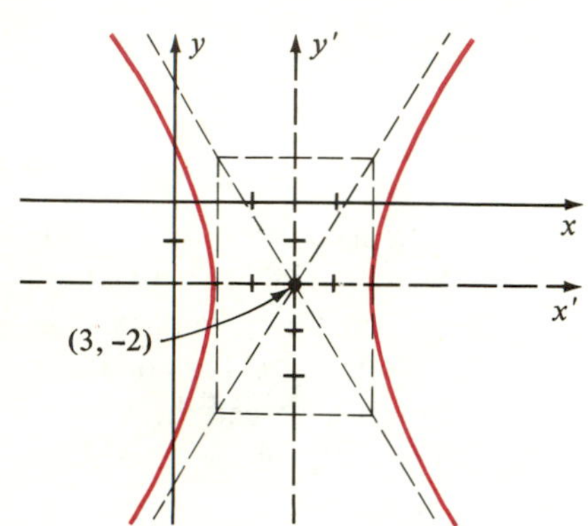

FIGURE 9.21

The results of the last three sections indicate that the graph of every equation of the form

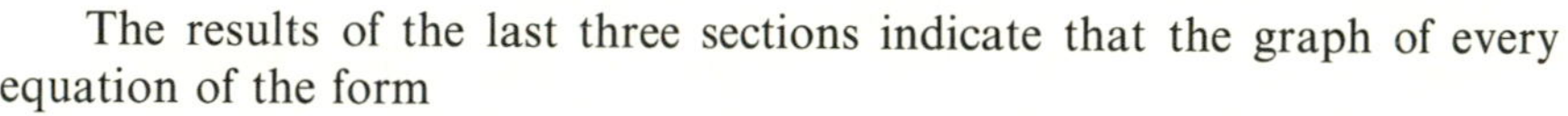

$$Ax^2 + Cy^2 + Dx + Ey + F = 0$$

is a conic, except for certain degenerate cases in which points, lines, or no graphs are obtained. Although we have only considered special examples, our methods are perfectly general. If A and C are equal and not zero, then the graph, when it exists, is a circle or, in exceptional cases, a point. If A and C are unequal but have the same sign, then by completing squares and properly translating axes, we obtain an equation whose graph, when it exists, is an ellipse (or a point). If A and C have opposite signs, an equation of a hyperbola is obtained, or possibly, in the degenerate case, two intersecting straight lines. Finally, if either A or C (but not both) is zero, the graph is a parabola or, in certain cases, a pair of parallel straight lines.

EXERCISES 9.4

In Exercises 1–18 sketch the graph of the equation, find the coordinates of the vertices and foci, and write equations for the asymptotes.

1 $\frac{x^2}{9} - \frac{y^2}{4} = 1$

2 $\frac{y^2}{49} - \frac{x^2}{16} = 1$

3 $\frac{y^2}{9} - \frac{x^2}{4} = 1$

4 $\frac{x^2}{49} - \frac{y^2}{16} = 1$

5 $y^2 - 4x^2 = 16$

6 $x^2 - 2y^2 = 8$

7 $x^2 - y^2 = 1$

8 $y^2 - 16x^2 = 1$

9 $x^2 - 5y^2 = 25$

10 $4y^2 - 4x^2 = 1$

11 $3x^2 - y^2 = -3$

12 $16x^2 - 36y^2 = 1$

13 $25x^2 - 16y^2 + 250x + 32y + 109 = 0$

14 $y^2 - 4x^2 - 12y - 16x + 16 = 0$

15 $4y^2 - x^2 + 40y - 4x + 60 = 0$

16 $25x^2 - 9y^2 - 100x - 54y + 10 = 0$

17 $9y^2 - x^2 - 36y + 12x - 36 = 0$

18 $4x^2 - y^2 + 32x - 8y + 49 = 0$

In Exercises 19–26 find an equation for the hyperbola satisfying the given conditions.

19 Foci $F(0, \pm 4)$, vertices $V(0, \pm 1)$

20 Foci $F(\pm 8, 0)$, vertices $V(\pm 5, 0)$

21 Foci $F(\pm 5, 0)$ vertices $V(\pm 3, 0)$

22 Foci $F(0, \pm 3)$, vertices $V(0, \pm 2)$

23 Foci $F(0, \pm 5)$, length of conjugate axis 4

24 Vertices $V(\pm 4, 0)$, passing through $P(8, 2)$

25 Vertices $V(\pm 3, 0)$, equations of asymptotes $y = \pm 2x$

26 Foci $F(0, \pm 10)$, equations of asymptotes $y = \pm \frac{1}{3}x$

In Exercises 27 and 28 find the points of intersection of the graphs of the given equations and sketch both graphs on the same coordinate axes, showing points of intersection.

27 $\begin{cases} y^2 - 4x^2 = 16 \\ y - x = 4 \end{cases}$ **28** $\begin{cases} x^2 - y^2 = 4 \\ y^2 - 3x = 0 \end{cases}$

29 The graphs of the equations

$$\frac{x^2}{a^2} - \frac{y^2}{b^2} = 1 \quad \text{and} \quad \frac{x^2}{a^2} - \frac{y^2}{b^2} = -1$$

are called **conjugate hyperbolas.** Sketch the graphs of both equations on the same coordinate system with $a = 2$ and $b = 5$. Describe the relationship between the two graphs.

30 Find an equation of the hyperbola with foci $(h \pm c, k)$ and vertices $(h \pm a, k)$, where $0 < a < c$ and $c^2 = a^2 + b^2$.

SECTION 9.5

ROTATION OF AXES

The $x'y'$-coordinate system used in a translation of axes may be thought of as having been obtained by moving the origin O of the xy-system to a new position $C(h, k)$ while, at the same time, not changing the positive directions of the axes or the units of length. We shall next introduce a new coordinate system by keeping the origin O fixed and rotating the x- and y-axes about O to another position denoted by x' and y'. A transformation of this type will be referred to as a **rotation of axes.**

Consider a rotation of axes as shown in Figure 9.22, and let φ denote the angle through which the positive x-axis must be rotated in order to coincide with the positive x'-axis. If (x, y) are the coordinates of a point P relative to the xy-plane, then (x', y') will denote its coordinates relative to the new $x'y'$-coordinate system.

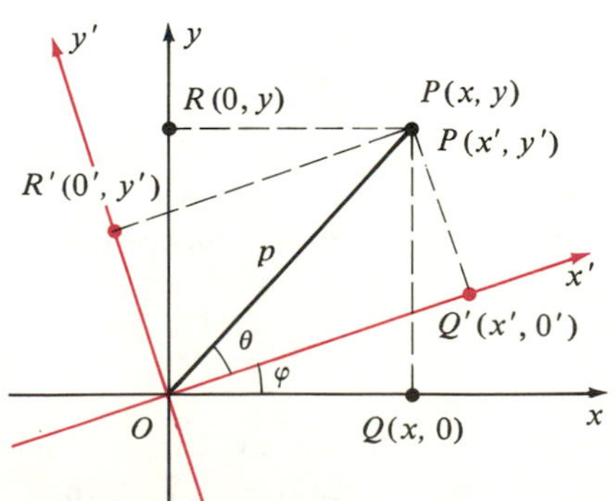

FIGURE 9.22

Let the projections of P on the various axes be denoted as in Figure 9.22, and let θ denote angle POQ'. If $p = d(O, P)$, then

$$x' = p \cos \theta, \qquad y' = p \sin \theta,$$

$$x = p \cos (\theta + \varphi), \qquad y = p \sin (\theta + \varphi).$$

Applying the addition formulas for the sine and cosine, we see that

$$x = p \cos \theta \cos \varphi - p \sin \theta \sin \varphi$$

$$y = p \sin \theta \cos \varphi + p \cos \theta \sin \varphi.$$

Using the fact that $x' = p\cos\theta$ and $y' = p\sin\theta$ gives us (i) of the next theorem. The formulas in (ii) may be obtained from (i) by solving for x' and y'.

ROTATION OF AXES FORMULAS

If the x- and y-axes are rotated about the origin O, through an angle φ, then the coordinates (x, y) and (x', y') of a point P in the two systems are related as follows:

(i) $x = x'\cos\varphi - y'\sin\varphi, \quad y = x'\sin\varphi + y'\cos\varphi$

(ii) $x' = x\cos\varphi + y\sin\varphi, \quad y' = -x\sin\varphi + y\cos\varphi.$

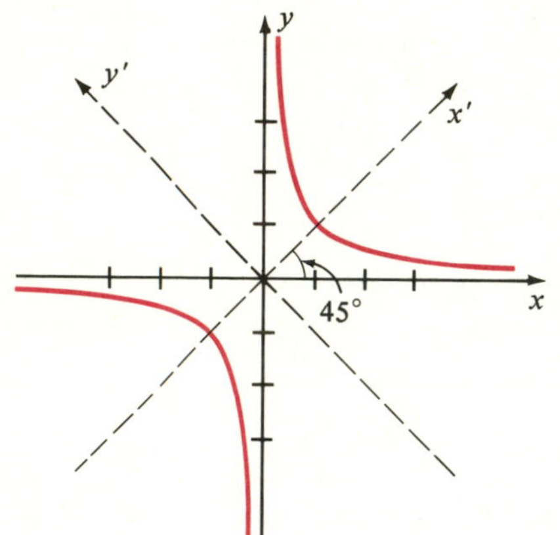

FIGURE 9.23

EXAMPLE 1 The graph of the equation $xy = 1$, or equivalently $y = 1/x$, is sketched in Figure 9.23. If the coordinate axes are rotated through an angle of 45°, find an equation of the graph relative to the new $x'y'$-coordinate system.

Solution Letting $\varphi = 45°$ in (i) of the Rotation of Axes Formulas,

$$x = x'\left(\frac{\sqrt{2}}{2}\right) - y'\left(\frac{\sqrt{2}}{2}\right) = \left(\frac{\sqrt{2}}{2}\right)(x' - y')$$

$$y = x'\left(\frac{\sqrt{2}}{2}\right) + y'\left(\frac{\sqrt{2}}{2}\right) = \left(\frac{\sqrt{2}}{2}\right)(x' + y').$$

Substituting for x and y in the equation $xy = 1$ gives us

$$\left(\frac{\sqrt{2}}{2}\right)(x' - y')\left(\frac{\sqrt{2}}{2}\right)(x' + y') = 1.$$

This reduces to
$$\frac{(x')^2}{2} - \frac{(y')^2}{2} = 1,$$

which is an equation of a hyperbola with vertices $(\pm\sqrt{2}, 0)$ on the x'-axis. Note that the asymptotes for the hyperbola have equations $y' = \pm x'$ in the new system. These correspond to the original x- and y-axes. ■

Example 1 illustrates a method for eliminating a term of an equation that contains the product xy. This method can be used to transform any equation of the form

$$Ax^2 + Bxy + Cy^2 + Dx + Ey + F = 0,$$

where $B \neq 0$, into an equation in x' and y' that contains no $x'y'$ term. Let us prove that this may always be done. If we rotate the axes through an angle φ, then using (i) of the Rotation of Axes Formulas to substitute for x and y

gives us

$$A(x' \cos \varphi - y' \sin \varphi)^2 + B(x' \cos \varphi - y' \sin \varphi)(x' \sin \varphi + y' \cos \varphi) + C(x' \sin \varphi + y' \cos \varphi)^2 + D(x' \cos \varphi - y' \sin \varphi) + E(x' \sin \varphi + y' \cos \varphi) + F = 0.$$

This equation may be written in the form

$$A'(x')^2 + B'x'y' + C'(y')^2 + D'x' + E'y' + F = 0$$

where the coefficient B' of $x'y'$ is

$$B' = 2(C - A) \sin \varphi \cos \varphi + B(\cos^2 \varphi - \sin^2 \varphi).$$

To eliminate the $x'y'$ term, we must select φ such that

$$2(C - A) \sin \varphi \cos \varphi + B(\cos^2 \varphi - \sin^2 \varphi) = 0.$$

Using double-angle formulas, the last equation may be written

$$(C - A) \sin 2\varphi + B \cos 2\varphi = 0,$$

which is equivalent to $\quad \cot 2\varphi = \dfrac{A - C}{B}.$

This proves the next result.

THEOREM

To eliminate the xy-term from the equation

$$Ax^2 + Bxy + Cy^2 + Dx + Ey + F = 0$$

where $B \neq 0$, choose an angle φ such that

$$\cot 2\varphi = \frac{A - C}{B}$$

and use the Rotation of Axes Formulas.

It follows that the graph of any equation in x and y of the type displayed in the preceding theorem is a conic, except for certain degenerate cases. (Why?)

EXAMPLE 2 Discuss and sketch the graph of the equation

$$41x^2 - 24xy + 34y^2 - 25 = 0.$$

Solution Using the notation of the preceding theorem, we have

$$A = 41, \quad B = -24, \quad C = 34,$$

and

$$\cot 2\varphi = \frac{41 - 34}{-24} = -\frac{7}{24}.$$

Since $\cot 2\varphi$ is negative, we may choose 2φ such that $90° < 2\varphi < 180°$, and consequently $\cos 2\varphi = -\frac{7}{25}$. (Why?) We now use the half-angle formulas to obtain

$$\sin \varphi = \sqrt{\frac{1 - \cos 2\varphi}{2}} = \sqrt{\frac{1 - (-\frac{7}{25})}{2}} = \frac{4}{5}$$

$$\cos \varphi = \sqrt{\frac{1 + \cos 2\varphi}{2}} = \sqrt{\frac{1 + (-\frac{7}{25})}{2}} = \frac{3}{5}.$$

Thus, the desired rotation formulas are

$$x = \tfrac{3}{5}x' - \tfrac{4}{5}y', \qquad y = \tfrac{4}{5}x' + \tfrac{3}{5}y'.$$

We leave it to the reader to show that after substituting for x and y in the given equation and simplifying, we obtain the equation

$$(x')^2 + 2(y')^2 = 1.$$

Thus the graph is an ellipse with vertices at $(\pm 1, 0)$ on the x'-axis. Since $\tan \varphi = \sin \varphi/\cos \varphi = (\frac{4}{5})/(\frac{3}{5}) = \frac{4}{3}$, we obtain $\varphi = \tan^{-1} (\frac{4}{3})$. To the nearest minute, $\varphi \approx 53°8'$. The graph is sketched in Figure 9.24. ■

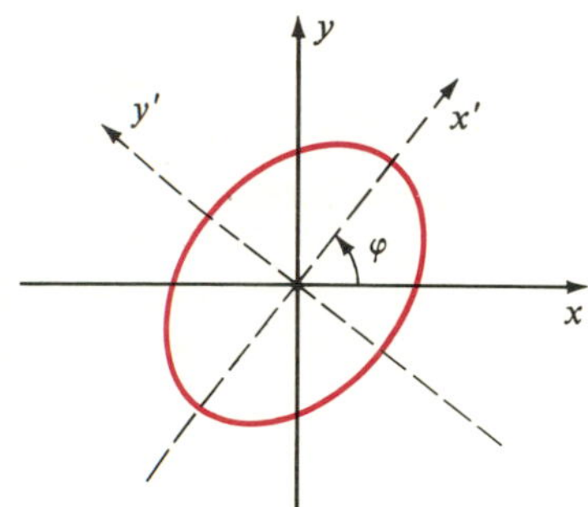

FIGURE 9.24

In some cases, after eliminating the xy-term, the resulting equation will contain an x' or y' term. It is then necessary to translate the axes of the $x'y'$-coordinate system to obtain the graph. Several problems of this type are contained in the following exercises.

EXERCISES 9.5

After a suitable rotation of axes, describe and sketch the graph of each of the equations in Exercises 1–14.

1 $32x^2 - 72xy + 53y^2 = 80$

2 $7x^2 - 48xy - 7y^2 = 225$

3 $11x^2 + 10\sqrt{3}xy + y^2 = 4$

4 $x^2 - xy + y^2 = 3$

5 $5x^2 - 8xy + 5y^2 = 9$

6 $11x^2 - 10\sqrt{3}xy + y^2 = 20$

7 $16x^2 - 24xy + 9y^2 - 60x - 80y + 100 = 0$

8 $x^2 + 2\sqrt{3}xy + 3y^2 + 8\sqrt{3}x - 8y + 32 = 0$

9 $5x^2 + 6\sqrt{3}xy - y^2 + 8x - 8\sqrt{3}y - 12 = 0$

10 $18x^2 - 48xy + 82y^2 + 6\sqrt{10}x + 2\sqrt{10}y - 80 = 0$

11 $x^2 + 4xy + 4y^2 + 6\sqrt{5}x - 18\sqrt{5}y + 45 = 0$

12 $15x^2 + 20xy - 4\sqrt{5}x + 8\sqrt{5}y - 100 = 0$

13 $40x^2 - 36xy + 25y^2 - 8\sqrt{13}x - 12\sqrt{13}y = 0$

14 $64x^2 - 240xy + 225y^2 + 1020x - 544y = 0$

15 Prove that, except for degenerate cases, the graph of $Ax^2 + Bxy + Cy^2 + Dx + Ey + F = 0$ is

(a) a parabola if $B^2 - 4AC = 0$.

(b) an ellipse if $B^2 - 4AC < 0$.

(c) a hyperbola if $B^2 - 4AC > 0$.

16 Use the results of Exercise 15 to determine the nature of the graphs in Exercises 1–14.

17 Let the matrices X, X', and A be defined by

$$X = \begin{bmatrix} x \\ y \end{bmatrix}, \quad X' = \begin{bmatrix} x' \\ y' \end{bmatrix}, \quad A = \begin{bmatrix} \cos\varphi & -\sin\varphi \\ \sin\varphi & \cos\varphi \end{bmatrix}.$$

Show that the Rotation of Axes Formulas may be written

(i) $X = AX'$ (ii) $X' = A^{-1}X$.

SECTION 9.6

POLAR COORDINATE SYSTEMS

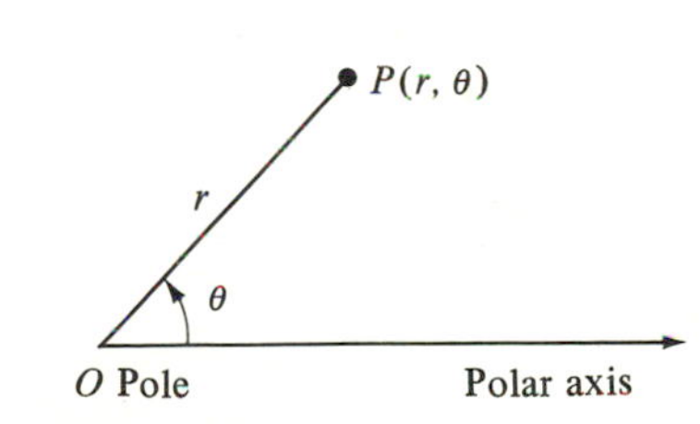

FIGURE 9.25

We have previously specified points in a plane in terms of rectangular coordinates, using the ordered pair (a, b) to denote the point whose directed distances from the x- and y-axes are b and a, respectively. Another important method for representing points is by means of *polar coordinates.* To introduce a system of polar coordinates in a plane, we begin with a fixed point O (called the **origin,** or **pole**) and a directed half-line (called the **polar axis**) with endpoint O. Next we consider any point P in the plane different from O. If, as illustrated in Figure 9.25, $r = d(O, P)$ and θ denotes the measure of any angle determined by the polar axis and OP, then r and θ are called **polar coordinates** of P and the symbols (r, θ) or $P(r, \theta)$ are used to denote P. As usual, θ is considered positive if the angle is generated by a counterclockwise rotation of the polar axis and negative if the rotation is clockwise. Either radians or degrees may be used for the measure of θ.

If the polar axis is the initial side, then there are many angles with the same terminal side. Hence the polar coordinates of a point are not unique. For example, $(3, \pi/4)$, $(3, 9\pi/4)$, and $(3, -7\pi/4)$ all represent the same point (see Figure 9.26). We shall also allow r to be negative. In this event, instead of measuring $|r|$ units along the terminal side of the angle θ, we measure along the half-line with endpoint O that has direction opposite to that of the terminal side. The points corresponding to the pairs $(-3, 5\pi/4)$ and $(-3, -3\pi/4)$ are also plotted in Figure 9.26.

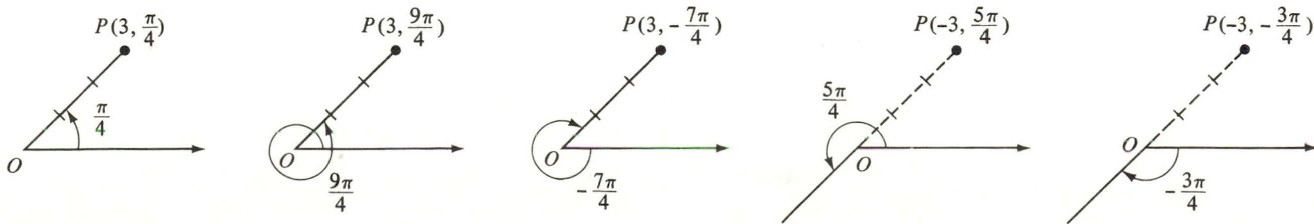

FIGURE 9.26

Finally, we agree that the pole O has polar coordinates $(0, \theta)$ for *any* θ. An assignment of ordered pairs of the form (r, θ) to points in a plane will be referred to as a **polar coordinate system** and the plane will be called an **$r\theta$-plane.**

A **polar equation** is an equation in r and θ. A **solution** of a polar equation is an ordered pair (a, b) that leads to equality if a is substituted for r and b for θ. The **graph** of a polar equation is the set of all points (in an $r\theta$-plane) that correspond to the solutions.

EXAMPLE 1 Sketch the graph of the polar equation $r = 4 \sin \theta$.

Solution The following table contains some solutions of the equation.

θ	0	$\pi/6$	$\pi/4$	$\pi/3$	$\pi/2$	$2\pi/3$	$3\pi/4$	$5\pi/6$	π
r	0	2	$2\sqrt{2}$	$2\sqrt{3}$	4	$2\sqrt{3}$	$2\sqrt{2}$	2	0

In rectangular coordinates, the graph of the equation consists of sine waves of amplitude 4 and period 2π. However, if polar coordinates are used, then the points that correspond to the pairs in the table lie on a circle of radius 2 and we draw the graph accordingly (see Figure 9.27). As an aid to plotting points, we have extended the polar axis in the negative direction and introduced a vertical line through the pole. The proof that the graph is a circle will be given in Example 5. Additional points obtained by letting θ vary from π to 2π lie on the same circle. For example, the solution $(-2, 7\pi/6)$ gives us the same point as $(2, \pi/6)$; the point corresponding to $(-2\sqrt{2}, 5\pi/4)$ is the same as that obtained from $(2\sqrt{2}, \pi/4)$, etc. If we let θ increase through all real numbers, we obtain the same points again and again because of the periodicity of the sine function.

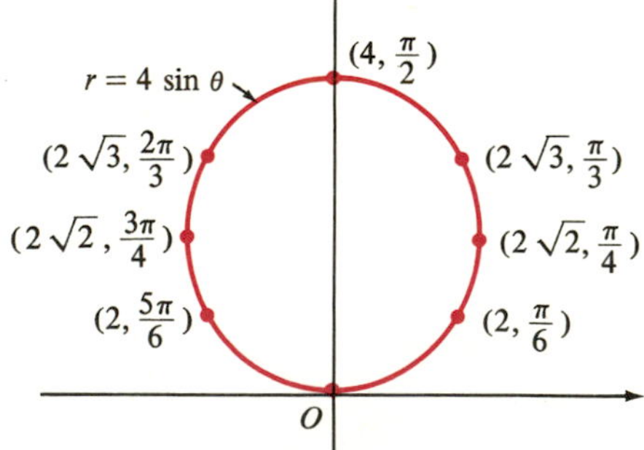

FIGURE 9.27

■

EXAMPLE 2 Sketch the graph of the equation $r = 2 + 2 \cos \theta$.

Solution Since the cosine function decreases from 1 to -1 as θ varies from 0 to π, it follows that r decreases from 4 to 0 in this θ-interval. The following table exhibits some solutions of the equation.

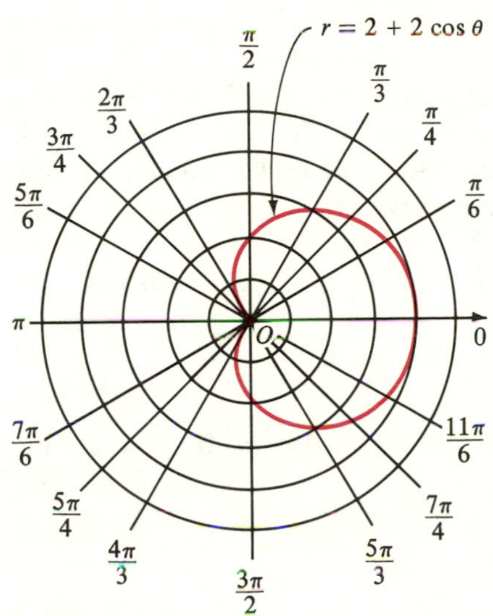

FIGURE 9.28

θ	0	$\pi/6$	$\pi/4$	$\pi/3$	$\pi/2$	$2\pi/3$	$3\pi/4$	$5\pi/6$	π
r	4	$2+\sqrt{3}$	$2+\sqrt{2}$	3	2	1	$2-\sqrt{2}$	$2-\sqrt{3}$	0

If θ increases from π to 2π, then $\cos\theta$ increases from -1 to 1, and consequently r increases from 0 to 4. Plotting points leads to the sketch shown in Figure 9.28, where we have used polar coordinate graph paper, which displays lines through O at various angles and circles with centers at the pole. The same graph may be obtained by taking other intervals for θ. ■

The heart-shaped graph in Example 2 is called a **cardioid.** In general, the graph of any polar equation of the form

$$r = a(1 + \cos\theta), \qquad r = a(1 + \sin\theta)$$
$$r = a(1 - \cos\theta), \qquad r = a(1 - \sin\theta)$$

where a is a real number, is a cardioid.

The graph of an equation of the form

$$r = a + b\cos\theta \quad \text{or} \quad r = a + b\sin\theta,$$

where $a \neq b$, is called a **limaçon.** The graph is similar in shape to a cardioid; however, there may be an added "loop" as shown in the next example.

EXAMPLE 3 Sketch the graph of $r = 2 + 4\cos\theta$.

Solution Some points corresponding to $0 \leq \theta \leq \pi$ are exhibited in the following table.

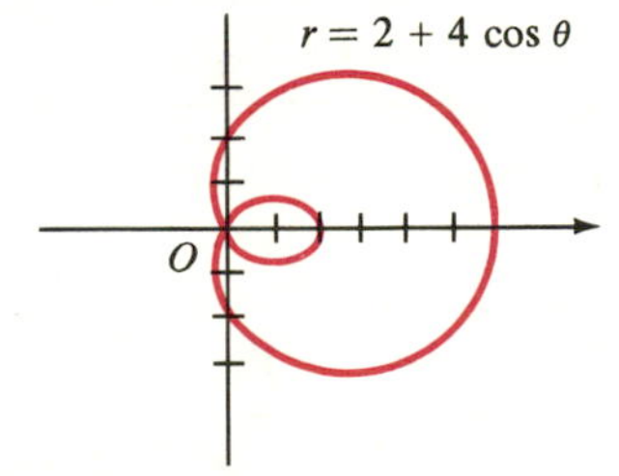

FIGURE 9.29

θ	0	$\pi/6$	$\pi/4$	$\pi/3$	$\pi/2$	$2\pi/3$	$3\pi/4$	$5\pi/6$	π
r	6	$2+2\sqrt{3}$	$2+2\sqrt{2}$	4	2	0	$2-2\sqrt{2}$	$2-2\sqrt{3}$	-2

Note that $r = 0$ at $\theta = 2\pi/3$. The values of r are negative if $2\pi/3 < \theta \leq \pi$, and this leads to the lower half of the small loop in Figure 9.29. (Verify this fact.) Letting θ range from π to 2π gives us the upper half of the small loop and the lower half of the large loop. ■

EXAMPLE 4 Sketch the graph of the equation $r = a\sin 2\theta$ where $a > 0$.

Solution Instead of tabulating solutions, let us reason as follows: If θ increases from 0 to $\pi/4$, then 2θ varies from 0 to $\pi/2$ and hence $\sin 2\theta$ increases from 0 to 1. It follows that r increases from 0 to a in the θ-interval $[0, \pi/4]$. If we next let θ increase from $\pi/4$ to $\pi/2$, then 2θ changes from $\pi/2$ to π. Consequently, r decreases from a to 0 in the θ-interval $[\pi/4, \pi/2]$. (Why?) The corresponding points on the graph constitute the first-quadrant

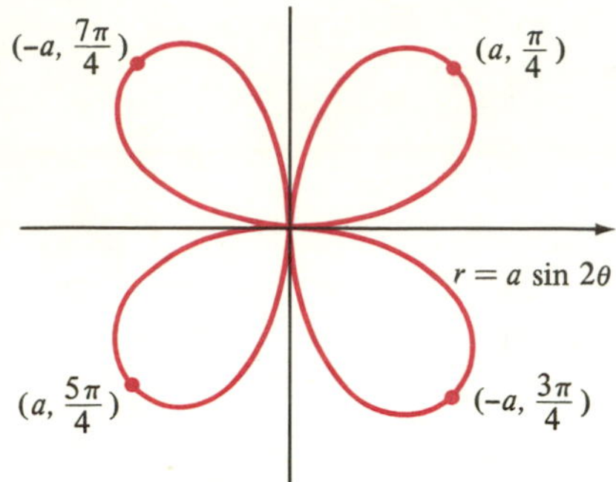

FIGURE 9.30

loop illustrated in Figure 9.30. Note that the point $P(r, \theta)$ traces the loop in the *counterclockwise* direction as θ increases from 0 to $\pi/2$.

If we let θ vary from $\pi/2$ to π, then 2θ varies from π to 2π and, therefore, $r = a \sin 2\theta \leq 0$. Thus, the points $P(r, \theta)$ are in the fourth quadrant if $\pi/2 < \theta < \pi$. We shall leave it to the reader to verify that if θ increases from $\pi/2$ to π, then $P(r, \theta)$ traces (in a counterclockwise direction) the loop shown in the fourth quadrant.

Similar loops are obtained for the θ-intervals $[\pi, 3\pi/2]$ and $[3\pi/2, 2\pi]$. In Figure 9.30 we have plotted only those points on the graph that correspond to the largest numerical values of r. ■

The graph in Example 4 is called a **four-leafed rose.** In general, any equation of the form

$$r = a \sin n\theta \quad \text{or} \quad r = a \cos n\theta$$

where n is a positive integer greater than 1 and a is a real number, has a graph that consists of a number of loops attached to the origin. If n is even there are $2n$ loops, whereas if n is odd there are n loops (see, for example, Exercises 9, 10, and 22).

Many other interesting graphs result from polar equations. Some are included in the exercises at the end of this section. Polar coordinates are very useful in applications involving circles with centers at the origin or lines that pass through the origin, since equations that have these graphs may be written in the simple forms $r = k$ or $\theta = k$ for some fixed number k. (Verify this fact.)

Let us next superimpose an xy-plane on an $r\theta$-plane such that the positive x-axis coincides with the polar axis. Any point P in the plane may then be assigned rectangular coordinates (x, y) or polar coordinates (r, θ). It is not difficult to obtain formulas that specify the relationship between the two coordinate systems. If $r > 0$ we have a situation similar to that illustrated in (i) of Figure 9.31. If $r < 0$ we have that shown in (ii) of the figure where,

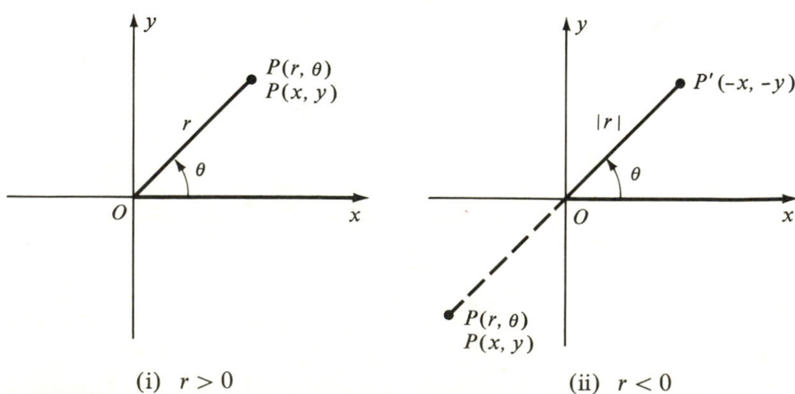

FIGURE 9.31

for later purposes, we have also plotted the point P' having polar coordinates $(|r|, \theta)$ and rectangular coordinates $(-x, -y)$.

The following result specifies relationships between (x, y) and (r, θ). In the statement of the theorem it is assumed that the positive x-axis coincides with the polar axis.

THEOREM

The rectangular coordinates (x, y) and polar coordinates (r, θ) of a point P are related as follows:

$$\text{(i)} \quad x = r \cos \theta, \quad y = r \sin \theta$$

$$\text{(ii)} \quad \tan \theta = \frac{y}{x}, \quad r^2 = x^2 + y^2.$$

Proof Although we have pictured θ as an acute angle in Figure 9.31, the discussion that follows is valid for all angles. On the one hand, if $r > 0$ as in (i) of Figure 9.31, then

$$x = r \cos \theta, \quad y = r \sin \theta.$$

On the other hand, if $r < 0$ then $|r| = -r$, and from (ii) of Figure 9.31 we see that

$$\cos \theta = \frac{-x}{|r|} = \frac{-x}{-r} = \frac{x}{r},$$

$$\sin \theta = \frac{-y}{|r|} = \frac{-y}{-r} = \frac{y}{r}.$$

Multiplication by r produces (i) of the theorem and, therefore, these formulas hold whether r is positive or negative. If $r = 0$, then the point is the pole and we again see that the formulas in (i) are true.

The formulas in (ii) follow readily from Figure 9.31. □

We may use the preceding theorem to change from one system of coordinates to the other. A more important use is for transforming a polar equation to an equation in x and y, and vice versa. This is illustrated in the next two examples.

EXAMPLE 5 Find an equation in x and y that has the same graph as the polar equation $r = 4 \sin \theta$.

Solution The equation $r = 4 \sin \theta$ was considered in Example 1. It is convenient to multiply both sides by r, obtaining $r^2 = 4r \sin \theta$. Applying the theorem gives us $x^2 + y^2 = 4y$, which is equivalent to $x^2 + (y - 2)^2 = 4$. Thus, the graph is a circle of radius 2 with center at (0, 2) in the xy-plane. ■

EXAMPLE 6 Find a polar equation of an arbitrary line.

Solution We know that every line in an xy-coordinate system is the graph of a linear equation $ax + by + c = 0$. Using the formula $x = r \cos \theta$ and $y = r \sin \theta$ leads to the polar equation

$$r(a \cos \theta + b \sin \theta) + c = 0.$$ ■

If we superimpose an xy-plane on an $r\theta$-plane, then the graph of a polar equation may be symmetric with respect to the x-axis, the y-axis, or the origin. It is left to the reader to show that if a substitution listed in the following table does not change the solutions of a polar equation, then the graph has the indicated symmetry.

TESTS FOR SYMMETRY

Substitution	*Symmetry*
$-\theta$ for θ	the line $\theta = 0$ (x-axis)
$-r$ for r	the pole (origin)
$\pi + \theta$ for θ	the pole (origin)
$\pi - \theta$ for θ	the line $\theta = \pi/2$ (y-axis)

To illustrate, since $\cos(-\theta) = \cos \theta$, the graph of the equation in Example 2 (see Figure 9.28) is symmetric with respect to the x-axis. Since $\sin(\pi - \theta) = \sin \theta$, the graph in Example 1 is symmetric with respect to the y-axis. The graph in Example 4 is symmetric to both axes and the origin. Other tests for symmetry may be stated; however, those listed above are among the easiest to apply.

EXERCISES 9.6

Sketch the graphs of the polar equations in Exercises 1–24.

1 $r = 5$

2 $\theta = \pi/4$

3 $\theta = -\pi/6$

4 $r = -2$

5 $r = 4 \cos \theta$

6 $r = -2 \sin \theta$

7 $r = 4(1 - \sin \theta)$

8 $r = 1 + 2 \cos \theta$

9 $r = 8 \cos 3\theta$

10 $r = 2 \sin 4\theta$

11 $r^2 = 4 \cos 2\theta$ (lemniscate)

12 $r = 6 \sin^2 (\frac{1}{2}\theta)$

13 $r = 4 \csc \theta$

14 $r = -3 \sec \theta$

15 $r = 2 - \cos \theta$

16 $r = 2 + 2 \sec \theta$ (conchoid)

17 $r = 2^\theta,\ \theta \geq 0$ (spiral)

18 $r\theta = 1,\ \theta > 0$ (spiral)

19 $r = -6(1 + \cos\theta)$

20 $r = e^{2\theta}$ (logarithmic spiral)

21 $r = 2 + 4\sin\theta$

22 $r = 8\cos 5\theta$

23 $r^2 = -16\sin 2\theta$

24 $4r = \theta$

In Exercises 25–32 find a polar equation that has the same graph as the given equation.

25 $x = -3$ **26** $y = 2$

27 $x^2 + y^2 = 16$ **28** $x^2 = 8y$

29 $y = 6$ **30** $y = 6x$

31 $x^2 - y^2 = 16$ **32** $9x^2 + 4y^2 = 36$

In Exercises 33–48 find an equation in x and y that has the same graph as the given polar equation and use it as an aid in sketching the graph in a polar coordinate system.

33 $r\cos\theta = 5$ **34** $r\sin\theta = -2$

35 $r - 6\sin\theta = 0$ **36** $r = 6\cot\theta$

37 $r = 2$ **38** $\theta = \pi/4$

39 $r = \tan\theta$ **40** $r = 4\sec\theta$

41 $r^2(4\sin^2\theta - 9\cos^2\theta) = 36$

42 $r^2(\cos^2\theta + 4\sin^2\theta) = 16$

43 $r^2\cos 2\theta = 1$

44 $r^2\sin 2\theta = 4$

45 $r(\sin\theta - 2\cos\theta) = 6$

46 $r(\sin\theta + r\cos^2\theta) = 1$

47 $r = 1/(1 + \cos\theta)$

48 $r = 4/(2 + \sin\theta)$

49 If $P_1(r_1, \theta_1)$ and $P_2(r_2, \theta_2)$ are points in an $r\theta$-plane, use the Law of Cosines to prove that

$$[d(P_1, P_2)]^2 = r_1^2 + r_2^2 - 2r_1r_2\cos(\theta_2 - \theta_1).$$

50 Prove that the graphs of the following polar equations are circles and find the center and radius in each case.

(a) $r = a\sin\theta,\ a \neq 0$

(b) $r = b\cos\theta,\ b \neq 0$

(c) $r = a\sin\theta + b\cos\theta,\ ab \neq 0$

SECTION 9.7
POLAR EQUATIONS OF CONICS

The following theorem provides another method for describing conic sections.

THEOREM

Let F be a fixed point and l a fixed line in a plane. The set of all points P in the plane such that the ratio $d(P, F)/d(P, Q)$ is a positive constant e, where $d(P, Q)$ is the distance from P to l, is a conic section. Moreover, the conic is a parabola if $e = 1$, an ellipse if $0 < e < 1$, and a hyperbola if $e > 1$.

The constant e is called the **eccentricity** of the conic. It will be seen that the point F is a focus of the conic. The line l is called a **directrix.** A typical

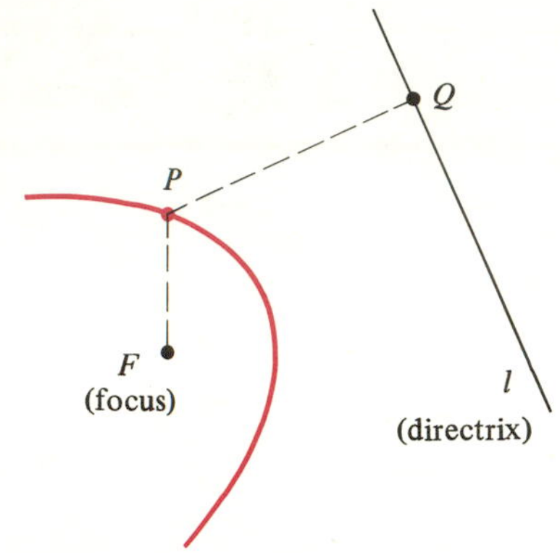

FIGURE 9.32

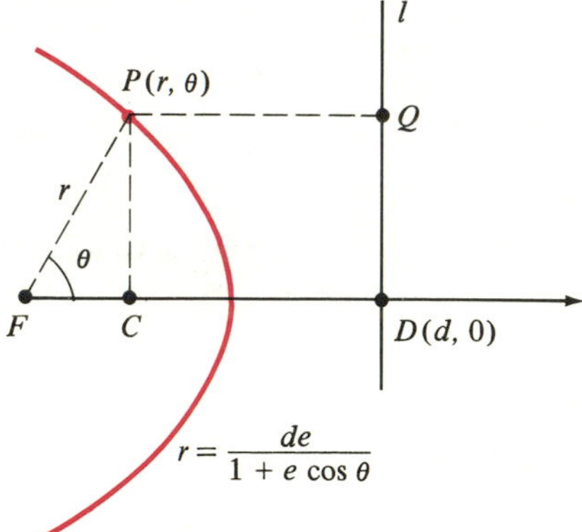

FIGURE 9.33

situation is illustrated in Figure 9.32, where the curve indicates possible positions of P.

If $e = 1$, then $d(P, F) = d(P, Q)$ and, by definition, we obtain a parabola with focus F and directrix l.

Suppose next that $0 < e < 1$. It is convenient to introduce a polar coordinate system in the plane with F as the pole and with l perpendicular to the polar axis at the point $D(d, 0)$ where $d > 0$. If $P(r, \theta)$ is a point in the plane such that $d(P, F)/d(P, Q) = e < 1$, then from Figure 9.33 we see that P lies to the left of l. Let C be the projection of P on the polar axis. Since

$$d(P, F) = r \quad \text{and} \quad d(P, Q) = \overline{FD} - \overline{FC} = d - r\cos\theta,$$

it follows that P satisfies the condition in the theorem if and only if

$$\frac{r}{d - r\cos\theta} = e$$

or

$$r = de - er\cos\theta.$$

Solving for r gives us

$$r = \frac{de}{1 + e\cos\theta},$$

which is a polar equation of the graph. Actually, the same equation is obtained if $e = 1$; however, there is no point (r, θ) on the graph if $1 + \cos\theta = 0$.

The rectangular equation corresponding to $r = de - er\cos\theta$ is

$$\pm\sqrt{x^2 + y^2} = de - ex.$$

Squaring both sides and rearranging terms leads to

$$(1 - e^2)x^2 + 2de^2x + y^2 = d^2e^2.$$

Completing the square in the previous equation and simplifying, we obtain

$$\left(x + \frac{de^2}{1 - e^2}\right)^2 + \frac{y^2}{1 - e^2} = \frac{d^2e^2}{(1 - e^2)^2}.$$

Finally, dividing both sides by $d^2e^2/(1 - e^2)^2$ leads to an equation of the form

$$\frac{(x - h)^2}{a^2} + \frac{y^2}{b^2} = 1$$

where $h = -de^2/(1 - e^2)$. Consequently, the graph is an ellipse with center at the point $(-de^2/(1 - e^2), 0)$ on the x-axis and where

$$a^2 = \frac{d^2e^2}{(1 - e^2)^2}, \qquad b^2 = \frac{d^2e^2}{1 - e^2}.$$

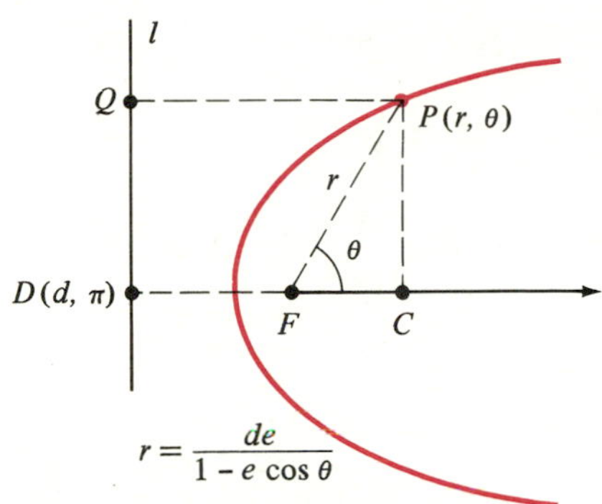

FIGURE 9.34

Since

$$c^2 = a^2 - b^2 = \frac{d^2e^4}{(1-e^2)^2},$$

we have $c = de^2/(1 - e^2)$. This proves that F is a focus of the ellipse. It also follows that $e = c/a$. A similar proof may be given for the case $e > 1$.

It can be shown, conversely, that every conic that is not a circle may be described by means of the statement in the theorem. This gives us a formulation of conic sections that is equivalent to the approach used previously. Since the theorem includes all three types of conics, it is sometimes regarded as a definition for the conic sections.

If we had chosen the focus F to the *right* of the directrix, as illustrated in Figure 9.34 (where $d > 0$), then the equation $r = de/(1 - e\cos\theta)$ would have resulted. (Note the minus sign in place of the plus sign.) Other sign changes occur if d is allowed to be negative.

If l is taken *parallel* to the polar axis through one of the points $(d, \pi/2)$ or $(d, 3\pi/2)$, as illustrated in Figure 9.35, then the corresponding equations would contain $\sin\theta$ instead of $\cos\theta$. The proofs of these facts are left to the reader.

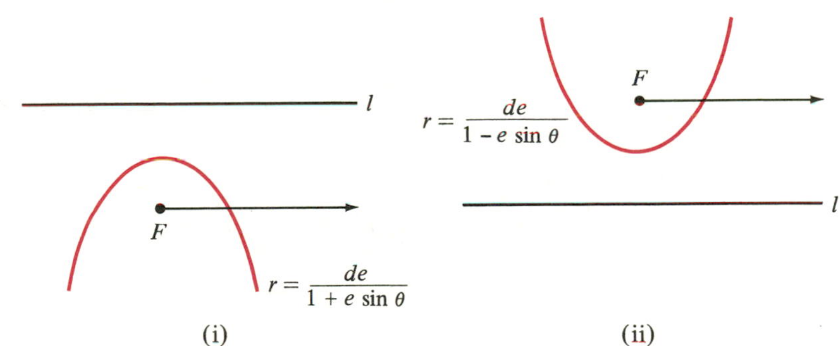

FIGURE 9.35

The following theorem summarizes our discussion.

THEOREM

A polar equation that has one of the four forms

$$r = \frac{de}{1 \pm e\cos\theta}, \qquad r = \frac{de}{1 \pm e\sin\theta}$$

is a conic section. Moreover, the conic is a parabola if $e = 1$, an ellipse if $0 < e < 1$, or a hyperbola if $e > 1$.

EXAMPLE 1 Describe and sketch the graph of the equation

$$r = \frac{10}{3 + 2\cos\theta}.$$

Solution Dividing numerator and denominator of the fraction by 3 gives us

$$r = \frac{\frac{10}{3}}{1 + \frac{2}{3}\cos\theta},$$

which has one of the forms in the preceding theorem with $e = \frac{2}{3}$. Thus the graph is an ellipse with focus F at the pole and major axis along the polar axis. The endpoints of the major axis may be found by setting θ equal to 0 and π. This gives us $V(2, 0)$ and $V'(10, \pi)$. Hence

$$2a = d(V', V) = 12, \quad \text{or} \quad a = 6.$$

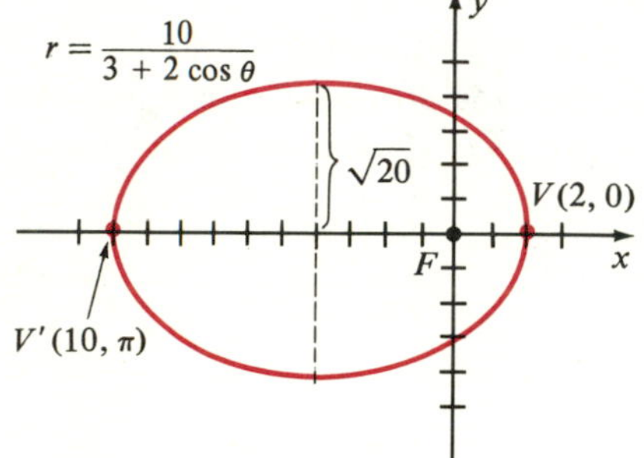

FIGURE 9.36

The center of the ellipse is the midpoint of the segment $V'V$, namely $(4, \pi)$. Using the fact that $e = c/a$, we obtain

$$c = ae = 6(\tfrac{2}{3}) = 4.$$

Hence,
$$b^2 = a^2 - c^2 = 36 - 16 = 20,$$

that is, the semiminor axis has length $\sqrt{20}$. The graph is sketched in Figure 9.36 where, for reference, we have superimposed a rectangular coordinate system on the polar system. ■

EXAMPLE 2 Describe and sketch the graph of the equation

$$r = \frac{10}{2 + 3\sin\theta}.$$

Solution To express the equation in the proper form we divide numerator and denominator of the fraction by 2, obtaining

$$r = \frac{5}{1 + \frac{3}{2}\sin\theta}.$$

Thus $e = \frac{3}{2}$ and the graph is a hyperbola with a focus at the pole. The expression $\sin\theta$ tells us that the transverse axis of the hyperbola is perpendicular to the polar axis. To find the vertices we let θ equal $\pi/2$ and $3\pi/2$ in the given equation. This gives us the points $V(2, \pi/2)$, $V'(-10, 3\pi/2)$, and hence

$$2a = d(V, V') = 8, \quad \text{or} \quad a = 4.$$

The points $(5, 0)$ and $(5, \pi)$ on the graph can be used to get a rough estimate of the lower branch of the hyperbola. The upper branch is obtained by sym-

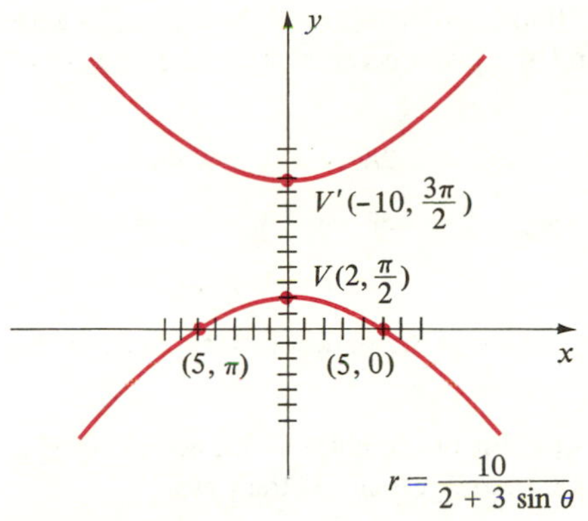

FIGURE 9.37

metry, as illustrated in Figure 9.37. If more accuracy or additional information is desired, we may calculate

$$c = ae = 4(\tfrac{3}{2}) = 6$$

and
$$b^2 = c^2 - a^2 = 36 - 16 = 20.$$

Asymptotes may then be constructed in the usual way. ■

EXAMPLE 3 Sketch the graph of the equation

$$r = \frac{15}{4 - 4\cos\theta}.$$

Solution To obtain the proper form we divide numerator and denominator by 4, obtaining

$$r = \frac{\frac{15}{4}}{1 - \cos\theta}.$$

Consequently, $e = 1$ and the graph is a parabola with focus at the pole. A rough sketch can be found by plotting the points that correspond to the x- and y- intercepts. These are indicated in the following table.

θ	0	$\pi/2$	π	$3\pi/2$
r	—	$15/4$	$15/8$	$15/4$

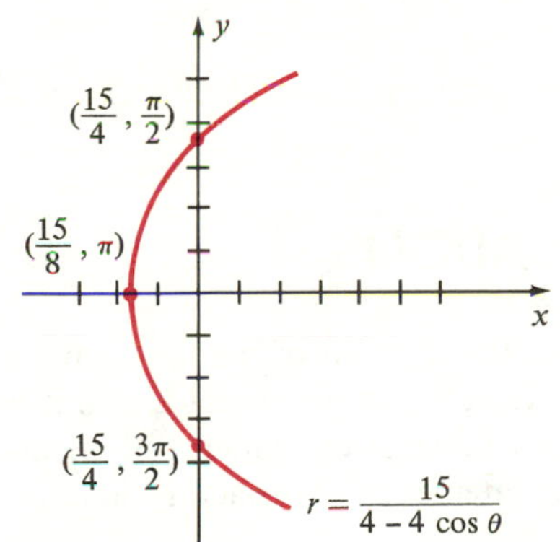

FIGURE 9.38

Note that there is no point on the graph corresponding to $\theta = 0$. (Why?) Plotting the three points and using the fact that the graph is a parabola with focus at the pole gives us the sketch in Figure 9.38. ■

If only a rough sketch of a conic is desired, then the technique employed in Example 3 is recommended. To use this method we plot (if possible) points corresponding to $\theta = 0$, $\pi/2$, π, and $3\pi/2$. These points, together with the type of conic (obtained from the value of e), readily lead to the sketch.

EXERCISES 9.7

In Exercises 1–10 identify and sketch the graph of the equation.

1 $r = \dfrac{12}{6 + 2\sin\theta}$

2 $r = \dfrac{12}{6 - 2\sin\theta}$

3 $r = \dfrac{12}{2 - 6\cos\theta}$

4 $r = \dfrac{12}{2 + 6\cos\theta}$

5 $r = \dfrac{3}{2 + 2\cos\theta}$

6 $r = \dfrac{3}{2 - 2\sin\theta}$

7 $r = \dfrac{4}{\cos\theta - 2}$

8 $r = \dfrac{4\sec\theta}{2\sec\theta - 1}$

9 $r = \dfrac{6 \csc \theta}{2 \csc \theta + 3}$

10 $r = \csc \theta(\csc \theta - \cot \theta)$

11–20 Find rectangular equations for the graphs in Exercises 1–10.

In Exercises 21–26 express the equation in polar form and then find the eccentricity and an equation for the directrix.

21 $y^2 = 4 - 4x$

22 $x^2 = 1 - 2y$

23 $3y^2 - 16y - x^2 + 16 = 0$

24 $5x^2 + 9y^2 = 32x + 64$

25 $8x^2 + 9y^2 + 4x = 4$

26 $4x^2 - 5y^2 + 36y - 36 = 0$

In Exercises 27–32 find a polar equation of the conic with focus at the pole and the given eccentricity and equation of directrix.

27 $e = \frac{1}{3},\ r = 2 \sec \theta$

28 $e = \frac{2}{5},\ r = 4 \csc \theta$

29 $e = 4,\ r = -3 \csc \theta$

30 $e = 3,\ r = -4 \sec \theta$

31 $e = 1,\ r \cos \theta = 5$

32 $e = 1,\ r \sin \theta = -2$

33 Find a polar equation of the parabola with focus at the pole and vertex $(4, \pi/2)$.

34 Find a polar equation of the ellipse with eccentricity $\frac{2}{3}$, a vertex at $(1, 3\pi/2)$, and a focus at the pole.

35 Prove the theorem of this section for the case $e > 1$.

36 Derive the formulas $r = de/(1 \pm e \sin \theta)$ discussed in this section.

SECTION 9.8

PLANE CURVES AND PARAMETRIC EQUATIONS

The graph of an equation $y = f(x)$, where the domain of the function f is an interval I, is often called a *plane curve*. However, to use this as a definition is unnecessarily restrictive since it rules out most of the conic sections and many other useful graphs. The following statement is satisfactory for most applications.

DEFINITION

A **plane curve** is a set C of ordered pairs of the form

$$(f(t), g(t))$$

where f and g are functions defined on an interval I.

For simplicity, we shall often refer to a plane curve as a **curve.** The **graph** of the curve C in the definition is the set of all points $P(t) = (f(t), g(t))$ in a rectangular coordinate system that corresponds to the ordered pairs. Each $P(t)$ is referred to as a *point* on the curve. We shall use the term *curve* interchangeably with *graph of a curve*. Sometimes it is convenient to imagine that the point $P(t)$ traces the curve C as t varies through the interval I. This is especially true in applications where t represents time and $P(t)$ is the position of a moving particle at time t.

The graphs of several curves are sketched in Figure 9.39 for the case where I is a closed interval $[a, b]$. If, as in (i) of the figure, $P(a) \neq P(b)$, then $P(a)$ and $P(b)$ are called the **endpoints** of C. Note that the curve illustrated in (i) intersects itself in the sense that two different values of t give rise to the same point. If $P(a) = P(b)$, as illustrated in (ii) of Figure 9.39, then C is called a **closed curve.** If $P(a) = P(b)$ and C does not intersect itself at any other point, as illustrated in (iii) of the figure, then C is called a **simple closed curve.**

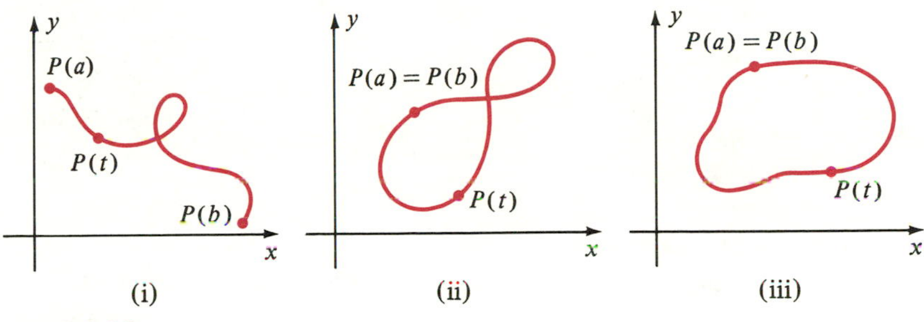

FIGURE 9.39

It is convenient to represent curves as indicated in the next definition.

DEFINITION

Let C be the curve consisting of all ordered pairs $(f(t), g(t))$, where f and g are defined on an interval I. The equations

$$x = f(t), \qquad y = g(t)$$

where t is in I, are called **parametric equations** for C, and t is called a **parameter.**

If we are given parametric equations, then as t varies through I, the point $P(x, y)$ traces the curve. As shown in the next example, it is sometimes possible to eliminate the parameter and obtain an equation for C that involves the variables x and y.

EXAMPLE 1 Sketch and identify the graph of the curve C given parametrically by

$$x = 2t, \quad y = t^2 - 1, \quad \text{where } -1 \leq t \leq 2.$$

Solution The parametric equations can be used to tabulate coordinates for points $P(x, y)$ on C as in the following table.

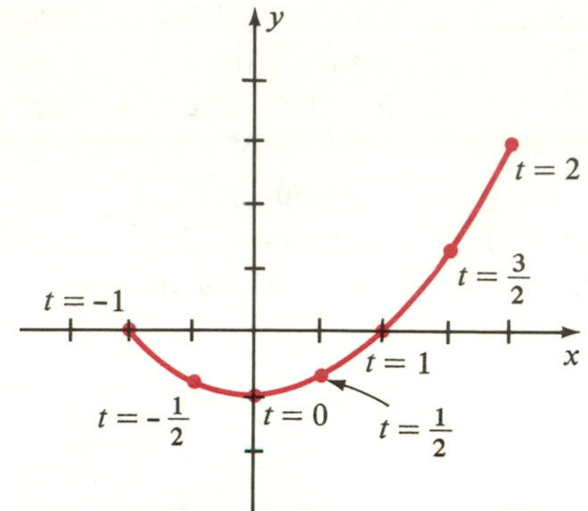

FIGURE 9.40

t	-1	$-\frac{1}{2}$	0	$\frac{1}{2}$	1	$\frac{3}{2}$	2
x	-2	-1	0	1	2	3	4
y	0	$-\frac{3}{4}$	-1	$-\frac{3}{4}$	0	$\frac{5}{4}$	3

Plotting points leads to the sketch in Figure 9.40.

A precise description of the graph may be obtained by eliminating the parameter. To illustrate, solving the first parametric equation for t, we obtain $t = x/2$. If we next substitute for t in the second equation the result is

$$y = \left(\frac{x}{2}\right)^2 - 1, \quad \text{or} \quad y + 1 = \frac{1}{4}x^2.$$

The graph of this equation is a parabola with vertical axis and vertex at the point $(0, -1)$. The curve C is that part of the parabola shown in Figure 9.40. ■

Parametric equations of curves are never unique. The reader should verify that the curve C in Example 1 is given by any of the following parametric representations:

$$x = 2t, \quad y = t^2 - 1; \qquad -1 \le t \le 2$$

$$x = t, \quad y = \tfrac{1}{4}t^2 - 1; \qquad -2 \le t \le 4$$

$$x = t^3, \quad y = \tfrac{1}{4}t^6 - 1; \qquad \sqrt[3]{-2} \le t \le \sqrt[3]{4}.$$

The next example illustrates the fact that it is often useful to eliminate the parameter *before* plotting points.

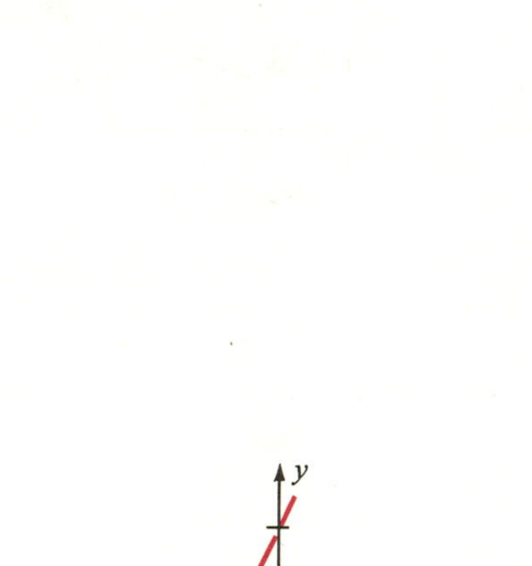

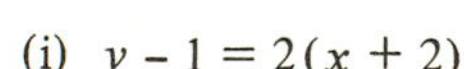
(i) $y - 1 = 2(x + 2)$

EXAMPLE 2 Sketch the graph of the curve C that is given parametrically by

$$x = -2 + t^2, \quad y = 1 + 2t^2$$

where t is in $\mathbb{R}$.

Solution To eliminate the parameter we note, from the first equation, that $t^2 = x + 2$. Substituting for t^2 in the second equation gives us

$$y = 1 + 2(x + 2), \quad \text{or} \quad y - 1 = 2(x + 2).$$

This is an equation of the line of slope 2 through the point $(-2, 1)$, as indicated by the dashes in (i) of Figure 9.41. However, since $t^2 \ge 0$,

$$x = -2 + t^2 \ge -2 \quad \text{and} \quad y = 1 + 2t^2 \ge 1.$$

It follows that the graph of C is that part of the line to the right of $(-2, 1)$, as shown in (ii) of the figure. This fact may also be verified by plotting several points. ■

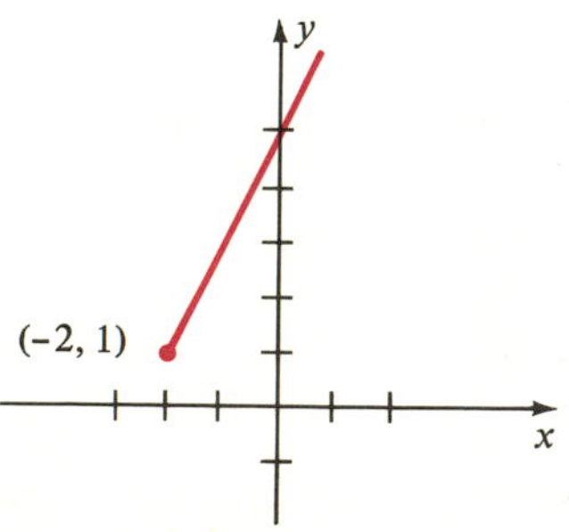

(ii) $x = -2 + t^2$, $y = 1 + 2t^2$

FIGURE 9.41

EXAMPLE 3 Describe the graph of the curve C having parametric equations

$$x = \cos t, \quad y = \sin t, \qquad 0 \le t \le 2\pi.$$

Solution We may use the identity $\cos^2 t + \sin^2 t = 1$ to eliminate the parameter. This gives us $x^2 + y^2 = 1$, and hence points on C are on the unit circle with center at the origin. As t increases from 0 to 2π, $P(t)$ starts at the point $A(1, 0)$ and traverses the circle once in the counterclockwise direction. In this example the parameter may be interpreted geometrically as the length of arc from A to P, as illustrated in Figure 5.19 (page 208). ■

If a curve C is described by means of an equation $y = f(x)$, where f is a function, then an easy way to obtain parametric equations is to let

$$x = t, \quad y = f(t)$$

where t is in the domain of f. For example, if $y = x^3$, then parametric equations are

$$x = t, \quad y = t^3 \qquad \text{where } t \text{ is in } \mathbb{R}.$$

We can use many different substitutions for x, provided that as t varies through some interval, x takes on all values in the domain of f. Thus the graph of $y = x^3$ is also given by

$$x = t^{1/3}, \quad y = t \qquad \text{where } t \text{ is in } \mathbb{R}.$$

Note, however, that the parametric equations

$$x = \sin t, \quad y = \sin^3 t, \qquad \text{where } t \text{ is in } \mathbb{R},$$

give only that part of the graph of $y = x^3$ that lies between the points $(-1, -1)$ and $(1, 1)$.

EXAMPLE 4 Find three different parametric representations for the line of slope m through the point (x_1, y_1).

Solution By the Point-Slope Form, an equation for the line is

$$y - y_1 = m(x - x_1).$$

If we let $x = t$, then $y - y_1 = m(t - x_1)$ and we obtain the parametric equations

$$x = t, \quad y = y_1 + m(t - x_1) \qquad \text{where } t \text{ is in } \mathbb{R}.$$

Another pair of parametric equations results if we let $x - x_1 = t$. In this case $y - y_1 = mt$ and hence the line is given parametrically by

$$x = x_1 + t, \quad y = y_1 + mt \qquad \text{where } t \text{ is in } \mathbb{R}.$$

As a third illustration, if we let $x - x_1 = \tan t$ we obtain

$$x = x_1 + \tan t, \quad y = y_1 + m \tan t \qquad \text{where } -\frac{\pi}{2} < t < \frac{\pi}{2}.$$

There are many other ways to represent the line parametrically. ■

EXAMPLE 5 The curve traced by a fixed point P on the circumference of a circle as the circle rolls along a line in a plane is called a **cycloid.** Find parametric equations for a cycloid.

Solution Suppose that the circle has radius a and that it rolls along (and above) the x-axis in the positive direction. If one position of P is the origin, then Figure 9.42 displays part of the curve and a possible position of the circle.

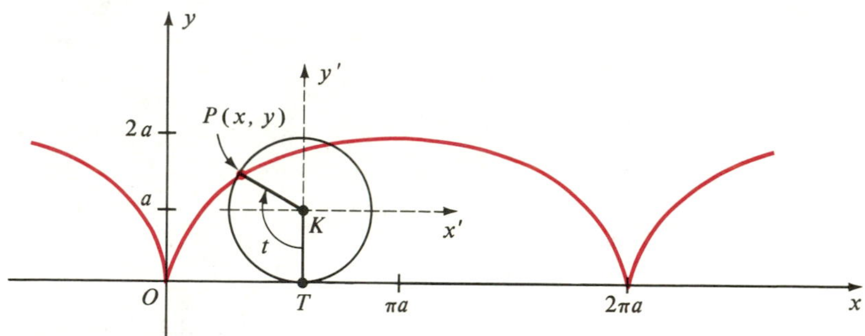

FIGURE 9.42

Let K denote the center of the circle and T the point of tangency with the x-axis. We introduce a parameter t as the radian measure of angle TKP. Since $\overline{OT}$ is the distance the circle has rolled, $\overline{OT} = at$. Consequently, the coordinates of K are (at, a). If we consider an $x'y'$-coordinate system with origin at $K(at, a)$, and if $P(x', y')$ denotes the point P relative to this system, then by the Translation of Axes Formulas, with $h = at$ and $k = a$,

$$x = at + x', \quad y = a + y'.$$

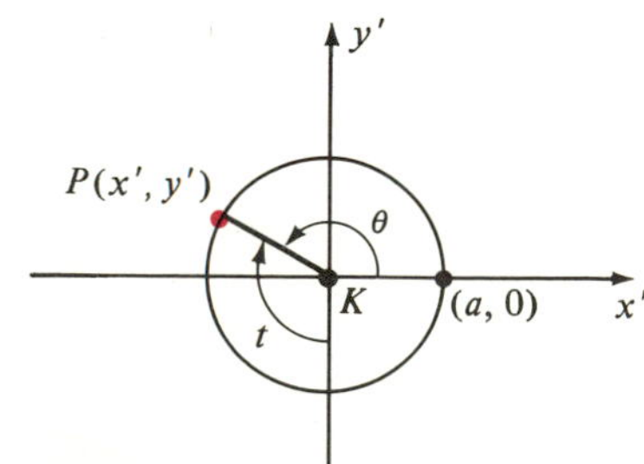

FIGURE 9.43

If, as in Figure 9.43, θ denotes an angle in standard position on the $x'y'$-system, then $\theta = 3\pi/2 - t$. Hence

$$x' = a\cos\theta = a\cos(3\pi/2 - t) = -a\sin t$$
$$y' = a\sin\theta = a\sin(3\pi/2 - t) = -a\cos t,$$

and substitution in $x = at + x'$, $y = a + y'$ gives us parametric equations for the cycloid, namely

$$x = a(t - \sin t), \quad y = a(1 - \cos t)$$

where t is in $\mathbb{R}$. ■

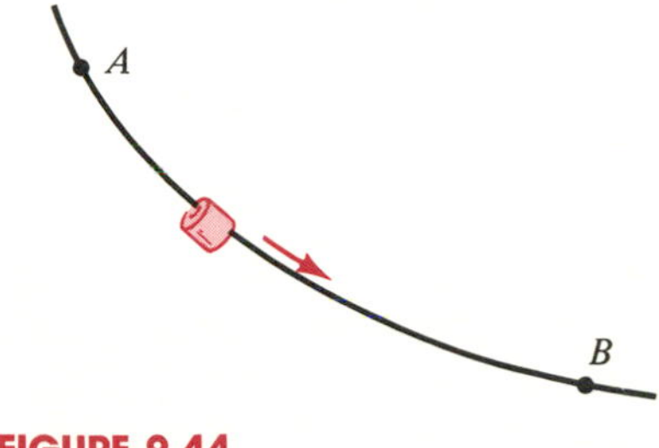

FIGURE 9.44

If $a < 0$, then the graph of $x = a(t - \sin t)$, $y = a(1 - \cos t)$ is the inverted cycloid that results if the circle of Example 5 rolls *below* the x-axis. This curve has a number of important physical properties. In particular, suppose a thin wire passes through two fixed points A and B as illustrated in Figure 9.44, and that the shape of the wire can be changed by bending it in any manner. Suppose further, that a bead is allowed to slide along the wire and the only force acting on the bead is gravity. We now ask which of all the possible paths will allow the bead to slide from A to B in the least amount of time. It is natural to conjecture that the desired path is the straight line segment from A to B; however, this is not the correct answer. The path that requires the least time coincides with the graph of the inverted cycloid with A at the origin and B the lowest point on the curve. The proof of this result will not be given in this text.

To cite another interesting property of this curve, suppose that A is the origin and B is the point with x-coordinate $\pi|a|$, that is, the lowest point on the cycloid occurring in the first arc to the right of A. It can be shown that if the bead is released at *any* point between A and B, the time required for it to reach B is always the same!

Variations of the cycloid occur in practical problems. For example, if a motorcycle wheel rolls along a straight road, then the curve traced by a fixed point on one of the spokes is a cycloid-like curve. In this case, the curve does not have sharp corners, nor does it intersect the road (the x-axis) as does the graph of a cycloid. In like manner, if the wheel of a train rolls along a railroad track, then the curve traced by a fixed point on the circumference of the wheel (which extends below the track) contains loops at regular intervals (see Exercise 38). Several other cycloids are defined in Exercises 33 and 34.

EXERCISES 9.8

In Exercises 1–18, (a) sketch the graph of the curve C having the indicated parametric equations, and (b) find a rectangular equation of a graph that contains the points on C.

1 $x = t - 2$, $y = 2t + 3$; $0 \le t \le 5$

2 $x = 1 - 2t$, $y = 1 + t$; $-1 \le t \le 4$

3 $x = t^2 + 1$, $y = t^2 - 1$; $-2 \le t \le 2$

4 $x = t^3 + 1$, $y = t^3 - 1$; $-2 \le t \le 2$

5 $x = 4t^2 - 5$, $y = 2t + 3$; t in $\mathbb{R}$

6 $x = t^3$, $y = t^2$; t in $\mathbb{R}$

7 $x = e^t$, $y = e^{-2t}$; t in $\mathbb{R}$

8 $x = \sqrt{t}$, $y = 3t + 4$; $t \ge 0$

9 $x = 2 \sin t,\ y = 3 \cos t;\quad 0 \le t \le 2\pi$

10 $x = \cos t - 2,\ y = \sin t + 3;\quad 0 \le t \le 2\pi$

11 $x = \sec t,\ y = \tan t;\quad -\pi/2 < t < \pi/2$

12 $x = \cos 2t,\ y = \sin t;\quad -\pi \le t \le \pi$

13 $x = t^2,\ y = 2 \ln t;\quad t > 0$

14 $x = \cos^3 t,\ y = \sin^3 t;\quad 0 \le t \le 2\pi$

15 $x = \sin t,\quad y = \csc t;\quad 0 < t \le \pi/2$

16 $x = e^t,\ y = e^{-t};\quad t$ in $\mathbb{R}$

17 $x = t,\ y = \sqrt{t^2 - 1};\quad |t| \ge 1$

18 $x = -2\sqrt{1 - t^2},\ y = t;\quad |t| \le 1$

In Exercises 19–22 sketch the graph of the curve C having the indicated parametric equations.

19 $x = t,\ y = \sqrt{t^2 - 2t + 1};\quad 0 \le t \le 4$

20 $x = 2t,\ y = 8t^3;\quad -1 \le t \le 1$

21 $x = (t + 1)^3,\ y = (t + 2)^2;\quad 0 \le t \le 2$

22 $x = \tan t,\ y = 1;\quad -\pi/2 < t < \pi/2$

In Exercises 23 and 24 curves C_1, C_2, C_3, and C_4 are given parametrically, where t is in $\mathbb{R}$. Sketch the graphs of C_1, C_2, C_3, C_4, and discuss their similarities and differences.

23 C_1: $x = t^2,\ y = t$
C_2: $x = t^4,\ y = t^2$
C_3: $x = \sin^2 t,\ y = \sin t$
C_4: $x = e^{2t},\ y = -e^t$

24 C_1: $x = t,\ y = 1 - t$
C_2: $x = 1 - t^2,\ y = t^2$
C_3: $x = \cos^2 t,\ y = \sin^2 t$
C_4: $x = \ln t - t,\ y = 1 + t - \ln t$

25 If $P_1(x_1, y_1)$ and $P_2(x_2, y_2)$ are distinct points, show that

$$x = (x_2 - x_1)t + x_1,\quad y = (y_2 - y_1)t + y_1,$$

where t is in $\mathbb{R}$, are parametric equations of the line l through P_1 and P_2. Find three other pairs of parametric equations for the line l. Show that there is an infinite number of different pairs of parametric equations for l.

26 Show that

$$x = a \cos t + h,\quad y = b \sin t + k,$$

where $0 \le t \le 2\pi$, are parametric equations of an ellipse with center at the point (h, k) and semi-axes of lengths a and b.

27 A circle C of radius b rolls on the inside of a second circle having equation $x^2 + y^2 = a^2$, where $b < a$. Let P be a fixed point on C and let the initial position of P be $A(a, 0)$ as shown in the figure. If the parameter t is the angle from the positive x-axis to the line segment from O to the center of C, show that parametric equations for the curve traced by P (called a **hypocycloid**) are

$$x = (a - b) \cos t + b \cos \frac{a - b}{b} t$$

$$y = (a - b) \sin t - b \sin \frac{a - b}{b} t$$

where $0 \le t \le 2\pi$. If $b = a/4$, show that

$$x = a \cos^3 t,\quad y = a \sin^3 t$$

and sketch the graph of the curve.

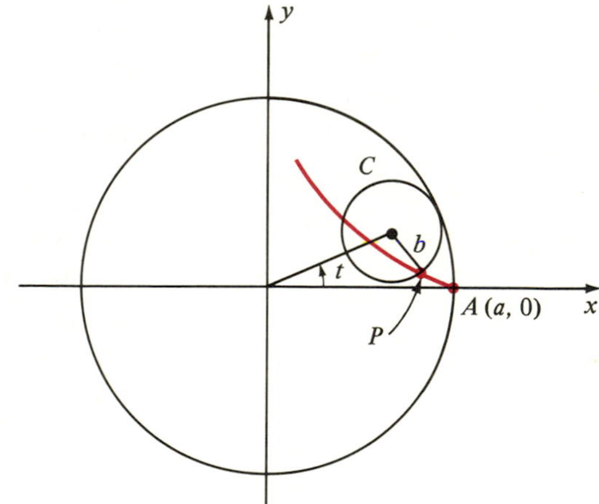

FIGURE FOR EXERCISE 27

28 If the circle C of Exercise 27 rolls on the outside of the second circle (see figure) then the curve traced by P is called an **epicycloid.** Show that parametric equations for this curve are

$$x = (a + b) \cos t - b \cos \frac{a + b}{b} t$$

$$y = (a + b) \sin t - b \sin \frac{a + b}{b} t.$$

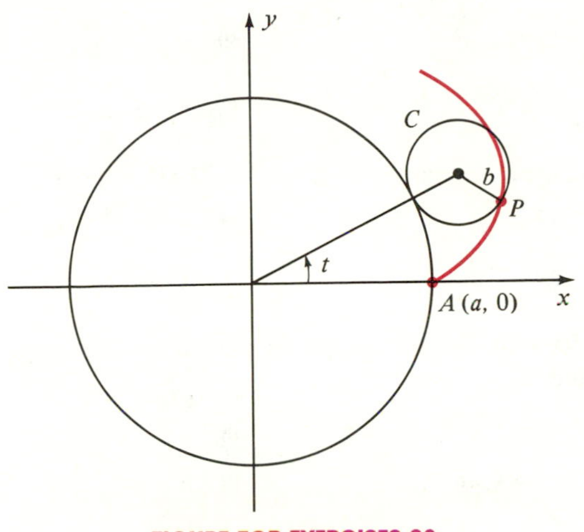

FIGURE FOR EXERCISES 28

29 If $b = a/3$ in Exercise 28, find parametric equations for the epicycloid and sketch the graph.

30 Problem 17 of the 1982 SAT examination was as follows: The radius of circle A is one-third that of circle B. How many revolutions will circle A make as it rolls around circle B until it reaches its starting point? Use Exercise 29 to show that the answer is 4. (To the embarassment of the Educational Testing Service of Princeton, NJ, this was not one of the choices given as an answer.)

31 If a string is unwound from around a circle and is kept taut in the plane of the circle, then a fixed point P on the string will trace a curve called the **involute of the circle.** If the circle is chosen as in Exercise 27 and the parameter t is the angle from the positive x-axis to the point of tangency of the string, show that parametric equations for the involute are

$$x = a(\cos t + t \sin t), \quad y = a(\sin t - t \cos t).$$

32 Generalize the cycloid of Example 5 to the case where P is any point on a fixed line through the center C of the circle. If $b = d(C, P)$, derive the parametric equations

$$x = at - b \sin t, \quad y = a - b \cos t.$$

Sketch a typical graph if $b < a$ (a **curtate cycloid**) and if $b > a$ (a **prolate cycloid**). The term **trochoid** is sometimes used for either of these curves.

SECTION 9.9
REVIEW

Define or discuss each of the following.

1 Conic sections
2 Parabola
3 Focus, directrix, vertex, and axis of a parabola
4 Ellipse
5 Major and minor axes of an ellipse
6 Foci and vertices of an ellipse
7 Hyperbola
8 Transverse and conjugate axes of a hyperbola
9 Foci and vertices of a hyperbola
10 Asymptotes of a hyperbola
11 Translation of axes
12 Rotation of axes
13 Polar coordinates of a point
14 The relationship between polar and rectangular coordinates
15 Graphs of polar equations
16 Polar equations of conics
17 Plane curve
18 Parametric equations of a curve

EXERCISES 9.9

In Exercises 1–10 find the foci and vertices and sketch the graph of the conic that has the given equation.

1 $y^2 = 64x$

2 $y - 1 = 8(x + 2)^2$

3 $9y^2 = 144 - 16x^2$

4 $9y^2 = 144 + 16x^2$

5 $x^2 - y^2 - 4 = 0$

6 $25x^2 + 36y^2 = 1$

7 $25y = 100 - x^2$

8 $3x^2 + 4y^2 - 18x + 8y + 19 = 0$

9 $x^2 - 9y^2 + 8x + 90y - 210 = 0$

10 $x = 2y^2 + 8y + 3$

Find equations for the conics in Exercises 11–18.

11 The hyperbola with vertices $V(0, \pm 7)$ and endpoints of conjugate axes $(\pm 3, 0)$

12 The parabola with focus $F(-4, 0)$ and directrix $x = 4$

13 The parabola with focus $(0, -10)$ and directrix $y = 10$

14 The parabola with vertex at the origin, symmetric to the x-axis, and passing through the point $(5, -1)$

15 The ellipse with vertices $V(0, \pm 10)$ and foci $F(0, \pm 5)$

16 The hyperbola with foci $F(\pm 10, 0)$ and vertices $V(\pm 5, 0)$

17 The hyperbola with vertices $V(0, \pm 6)$ and asymptotes that have equations $y = \pm 9x$

18 The ellipse with foci $F(\pm 2, 0)$ and passing through the point $(2, \sqrt{2})$

Discuss and sketch the graph of each of the equations in Exercises 19–24 after making a suitable translation of axes.

19 $4x^2 + 9y^2 + 24x - 36y + 36 = 0$

20 $4x^2 - y^2 - 40x - 8y + 88 = 0$

21 $y^2 - 8x + 8y + 32 = 0$

22 $4x^2 + y^2 - 24x + 4y + 36 = 0$

23 $y^2 - 2x^2 + 6y + 8x - 3 = 0$

24 $x^2 - 9y^2 + 8x + 7 = 0$

Sketch the graphs of the equations in Exercises 25–32.

25 $r = -4 \sin \theta$ **26** $r = 3 \sin 5\theta$

27 $r = 6 - 3 \cos \theta$ **28** $r^2 = 9 \sin 2\theta$

29 $2r = \theta$ **30** $r = \dfrac{8}{1 - 3 \sin \theta}$

31 $r = 6 - r \cos \theta$ **32** $r = 8 \sec \theta$

Change the equations in Exercises 33–36 to polar equations.

33 $y^2 = 4x$ **34** $x^2 + y^2 - 3x + 4y = 0$

35 $2x - 3y = 8$ **36** $x^2 + y^2 = 2xy$

In Exercises 37–40 change each equation to an equation in x and y.

37 $r^2 = \tan \theta$ **38** $r = 2 \cos \theta + 3 \sin \theta$

39 $r^2 = 4 \sin 2\theta$ **40** $\theta = \sqrt{3}$

In Exercises 41–43 sketch the graph of the curve and find a rectangular equation of a graph that contains the points on the curve.

41 $x = (1/t) + 1, \; y = (2/t) - t; \quad 0 < t \leq 4$

42 $x = \cos^2 t - 2, \; y = \sin t + 1; \quad 0 \leq t \leq 2\pi$

43 $x = \sqrt{t}, \; y = 2^{-t}; \quad t \geq 0$

44 Let the curves C_1, C_2, C_3, and C_4 be given parametrically by

$$C_1\colon x = t, \; y = \sqrt{t}$$
$$C_2\colon x = t^2, \; y = t$$
$$C_3\colon x = 1 - \sin^2 t, \; y = \cos t$$
$$C_4\colon x = e^{2t}, \; y = -e^t$$

where t varies through $\mathbb{R}$. Sketch the graphs of C_1, C_2, C_3, and C_4 and discuss their similarities and differences.

After making a suitable rotation of axes, describe and sketch the graphs of the equations in Exercises 45 and 46.

45 $x^2 - 8xy + 16y^2 - 12\sqrt{17}x - 3\sqrt{17}y = 0$

46 $8x^2 + 12xy + 17y^2 - 16\sqrt{5}x - 12\sqrt{5}y = 0$

APPENDICES

APPENDIX I

USING LOGARITHMIC AND TRIGONOMETRIC TABLES

If x is any positive real number and we write

$$x = c \cdot 10^k$$

where $1 \leq c < 10$ and k is an integer, then applying (iii) of the Laws of Logarithms,

$$\log x = \log c + \log 10^k.$$

Since $\log 10^k = k$, we see that

$$\log x = \log c + k.$$

The last equation tells us that to find $\log x$ for any positive real number x it is sufficient to know the logarithms of numbers between 1 and 10. The number $\log c$, where $1 \leq c < 10$, is called the **mantissa,** and the integer k is called the **characteristic** of $\log x$.

If $1 \leq c < 10$, then, since $\log x$ increases as x increases,

$$\log 1 \leq \log c < \log 10,$$

or equivalently,

$$0 \leq \log c < 1.$$

Hence, the mantissa of a logarithm is a number between 0 and 1. In numerical problems, it is usually necessary to approximate logarithms. For example, it can be shown that

$$\log 2 = 0.3010299957\ldots$$

where the decimal is nonrepeating and nonterminating. We often round off such logarithms to four decimal places and write

$$\log 2 \approx 0.3010.$$

If a number between 0 and 1 is written as a finite decimal, it is sometimes referred to as a **decimal fraction.** Thus, the equation $\log x = \log c + k$ implies that if x is any positive real number, then $\log x$ *may be approximated by the sum of a positive decimal fraction (the mantissa) and an integer k (the characteristic).* We shall refer to this representation as the **standard form** for $\log x$.

Common logarithms of many of the numbers between 1 and 10 have been calculated. Table 1 of Appendix II contains four-decimal-place approximations for logarithms of numbers between 1.00 and 9.99 at intervals of 0.01. This table can be used to find the common logarithm of any three-digit number to four-decimal-place accuracy. The use of Table 1 is illustrated in the following examples.

EXAMPLE 1 Approximate each of the following:

(a) $\log 43.6$ (b) $\log 43{,}600$ (c) $\log 0.0436$

Solution

(a) Since $43.6 = (4.36)10^1$, the characteristic of $\log 43.6$ is 1. Referring to Table 1, we find that the mantissa of $\log 4.36$ may be approximated by 0.6395. Hence, as in the preceding discussion,

$$\log 43.6 \approx 0.6395 + 1 = 1.6395.$$

(b) Since $43{,}600 = (4.36)10^4$, the mantissa is the same as in part (a); however, the characteristic is 4. Consequently,

$$\log 43{,}600 \approx 0.6395 + 4 = 4.6395.$$

(c) If we write $0.0436 = (4.36)10^{-2}$, then

$$\log 0.0436 = \log 4.36 + (-2).$$

Hence, $$\log 0.0436 \approx 0.6395 + (-2).$$

We could subtract 2 from 0.6395 and obtain

$$\log 0.0436 \approx -1.3605$$

but this is not standard form, since $-1.3605 = -0.3605 + (-1)$, a number in which the decimal part is *negative*. A common error is to write $0.6395 + (-2)$ as -2.6395. This is incorrect since $-2.6395 = -0.6395 + (-2)$, which is not the same as $0.6395 + (-2)$. ■

If a logarithm has a negative characteristic, we usually either leave it in standard form or rewrite the logarithm, keeping the decimal part positive. To illustrate the latter technique, let us add and subtract 8 on the right side of the equation

$$\log 0.0436 \approx 0.6395 + (-2).$$

This give us

$$\log 0.0436 \approx 0.6395 + (8 - 8) + (-2)$$

or $$\log 0.0436 \approx 8.6395 - 10.$$

We could also write

$$\log 0.0436 \approx 18.6395 - 20 = 43.6395 - 45$$

and so on, as long as the *integral part* of the logarithm is -2.

EXAMPLE 2 Approximate each of the following:

(a) $\log (0.00652)^2$ (b) $\log (0.00652)^{-2}$ (c) $\log (0.00652)^{1/2}$

Solution

(a) By (iii) of the Laws of Logarithms,

$$\log (0.00652)^2 = 2 \log 0.00652.$$

Since $0.00652 = (6.52)10^{-3}$,

$$\log 0.00652 = \log 6.52 + (-3).$$

Referring to Table 1, we see that log 6.52 is approximately 0.8142 and, therefore,

$$\log 0.00652 \approx 0.8142 + (-3).$$

Hence,
$$\begin{aligned}\log (0.00652)^2 &= 2 \log 0.00652\\ &\approx 2[0.8142 + (-3)]\\ &= 1.6284 + (-6).\end{aligned}$$

The standard form is $0.6284 + (-5)$.

(b) Again using Law (iii) and the value for log 0.00652 found in part (a),

$$\begin{aligned}\log (0.00652)^{-2} &= -2 \log 0.00652\\ &\approx -2[0.8142 + (-3)]\\ &= -1.6284 + 6.\end{aligned}$$

It is important to note that -1.6284 means $-0.6284 + (-1)$ and, consequently, the decimal part is negative. To obtain the standard form, we may write

$$\begin{aligned}-1.6284 + 6 &= 6.0000 - 1.6284\\ &= 4.3716.\end{aligned}$$

This shows that the mantissa is 0.3716 and the characteristic is 4.

(c) By Law (iii),

$$\begin{aligned}\log (0.00652)^{1/2} &= \tfrac{1}{2} \log 0.00652\\ &\approx \tfrac{1}{2}[0.8142 + (-3)].\end{aligned}$$

If we multiply by $\frac{1}{2}$, the standard form is not obtained, since neither number in the resulting sum is the characteristic. In order to avoid this, we may adjust the expression within brackets by adding and subtracting a suitable number. If we use 1 in this way, we obtain

$$\begin{aligned}\log (0.00652)^{1/2} &\approx \tfrac{1}{2}[1.8142 + (-4)]\\ &= 0.9071 + (-2),\end{aligned}$$

which is in standard form. We could also have added and subtracted a number other than 1. For example,

$$\begin{aligned}\tfrac{1}{2}[0.8142 + (-3)] &= \tfrac{1}{2}[17.8142 + (-20)]\\ &= 8.9071 + (-10).\end{aligned}$$ ■

Table 1 can be used to find an approximation to x if $\log x$ is given, as illustrated in the following example.

EXAMPLE 3 Find a decimal approximation to x for each $\log x$:

(a) $\log x = 1.7959$ (b) $\log x = -3.5918$

Solution

(a) The mantissa 0.7959 determines the sequence of digits in x and the characteristic determines the position of the decimal point. Referring to the *body* of Table 1, we see that the mantissa 0.7959 is the logarithm of 6.25. Since the characteristic is 1, x lies between 10 and 100. Consequently, $x \approx 62.5$.

(b) To find x from Table 1, $\log x$ must be written in standard form. To change $\log x = -3.5918$ to standard form, we may add and subtract 4, obtaining

$$\begin{aligned} \log x &= (4 - 3.5918) - 4 \\ &= 0.4082 - 4. \end{aligned}$$

Referring to Table 1, we see that the mantissa 0.4082 is the logarithm of 2.56. Since the characteristic of $\log x$ is -4, it follows that $x \approx 0.000256$. ■

If a calculator with a log key is used to determine common logarithms, then the standard form for $\log x$ is obtained only if $x \geq 1$. For example, to find log 43.6 on a typical calculator, we enter 43.6 and press log, obtaining the standard form

1.6394865

If we find log 0.0436 in similar fashion, then the following number appears on the display panel:

-1.3605135

This is not the standard form for the logarithm, but is similar to that which occurred in the solution to Example 1(c). To find the standard form we could add 2 to the logarithm (using a calculator) and then subtract 2 as follows:

$$\begin{aligned} \log 0.0436 &\approx -1.3605135 \\ &= (-1.3605135 + 2) - 2 \\ &= 0.6394865 - 2 \\ &= 0.6394865 + (-2) \end{aligned}$$

The only common logarithms that can be found *directly* from Table 1 are logarithms of numbers that contain at most three nonzero digits. If *four* nonzero digits are involved, then it is possible to obtain an approximation by using the method of linear interpolation described next. The terminology **linear interpolation** is used because, as we shall see, the method is based upon approximating portions of the graph of $y = \log x$ by line segments.

To illustrate the process of linear interpolation, and at the same time give some justification for it, let us consider the specific example log 12.64. Since the logarithmic function with base 10 is increasing, this number lies between $\log 12.60 \approx 1.1004$ and $\log 12.70 \approx 1.1038$. Examining the graph of

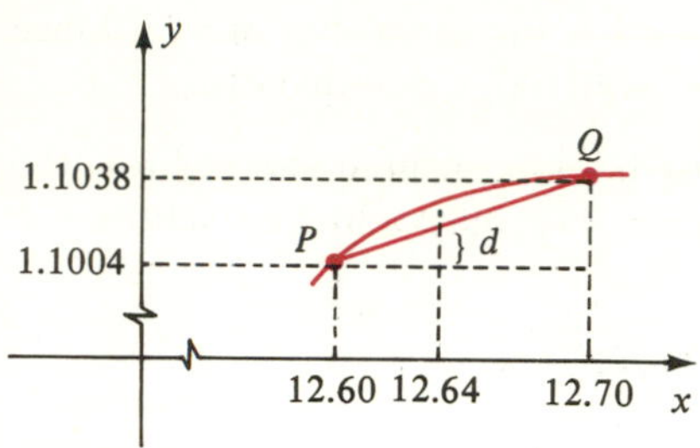

FIGURE A1.1

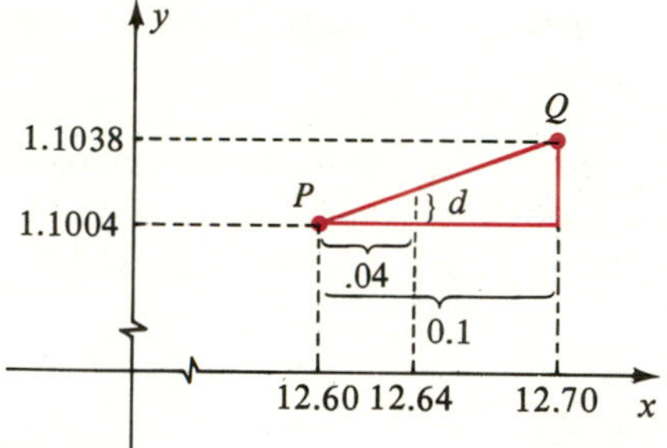

FIGURE A1.2

$y = \log x$, we have the situation shown in Figure A1.1, where we have distorted the units on the x- and y-axes and also the portion of the graph shown. A more accurate drawing would indicate that the graph of $y = \log x$ is much closer to the line segment joining $P(12.60, 1.1004)$ to $Q(12.70, 1.1038)$ than is shown in the figure. Since log 12.64 is the y-coordinate of the point on the graph having x-coordinate 12.64, it can be approximated by the y-coordinate of the point with x-coordinate 12.64 on the *line segment PQ*. Referring to Figure A1.1, we see that the latter y-coordinate is $1.1004 + d$. The number d can be approximated by using similar triangles. Referring to Figure A1.2, where the graph of $y = \log x$ has been deleted, we may form the following proportion:

$$\frac{d}{0.0034} = \frac{0.04}{0.1}.$$

Hence
$$d = \frac{(0.04)(0.0034)}{0.1} = 0.00136.$$

When using this technique, we always round off decimals to the same number of places as appear in the body of the table. Consequently, $d \approx 0.0014$ and

$$\log 12.64 \approx 1.1004 + 0.0014 = 1.1018.$$

Hereafter we shall not sketch a graph when interpolating. Instead we shall use the scheme illustrated in the next example.

EXAMPLE 4 Approximate log 572.6.

Solution It is convenient to arrange our work as follows:

$$1.0\left\{\begin{array}{l} 0.6\left\{\begin{array}{l}\log 572.0 \approx 2.7574 \\ \log 572.6 = ?\end{array}\right\}d \\ \qquad \log 573.0 \approx 2.7582\end{array}\right\}0.0008$$

where differences are indicated by appropriate symbols alongside of the braces. This leads to the proportion

$$\frac{d}{0.0008} = \frac{0.6}{1.0} = \frac{6}{10}, \quad \text{or} \quad d = \left(\frac{6}{10}\right)(0.0008) = 0.00048 \approx 0.0005.$$

Hence,
$$\log 572.6 \approx 2.7574 + 0.0005 = 2.7579.$$

Another way of working this type of problem is to reason that since 572.6 is $\frac{6}{10}$ of the way from 572.0 to 573.0, then log 572.6 is (appproximately) $\frac{6}{10}$ of the way from 2.7574 to 2.7582. Hence,

$$\log 572.6 \approx 2.7574 + (\tfrac{6}{10})(0.0008) \approx 2.7574 + 0.0005 = 2.7579. \quad ■$$

EXAMPLE 5 Approximate log 0.003678.

Solution We begin by arranging our work as in the solution of Example 1. Thus,

$$10\left\{\begin{array}{l} 8\left\{\begin{array}{l}\log 0.003670 \approx 0.5647 + (-3) \\ \log 0.003678 = ?\end{array}\right\} d \\ \log 0.003680 \approx 0.5658 + (-3)\end{array}\right\} 0.0011$$

Since we are only interested in ratios, we have used the numbers 8 and 10 on the left side because their ratio is the same as the ratio of 0.000008 to 0.000010. This leads to the proportion

$$\frac{d}{0.0011} = \frac{8}{10} = 0.8, \quad \text{or} \quad d = (0.0011)(0.8) = 0.00088 \approx 0.0009.$$

Hence,
$$\begin{aligned}\log 0.003678 &\approx [0.5647 + (-3)] + 0.0009 \\ &= 0.5656 + (-3).\end{aligned}$$

If a number x is written in the form $x = c \cdot 10^k$, where $1 \leq c < 10$, then before using Table 1 to find $\log x$ by interpolation, c should be rounded off to three decimal places. Another way of saying this is that x should be rounded off to four **significant figures.** Some examples will help to clarify the procedure. If $x = 36.4635$, we round off to 36.46 before approximating $\log x$. The number 684,279 should be rounded off to 684,300. For a decimal such as 0.096202 we use 0.09620. The reason for doing this is that Table 1 does not guarantee more than four-digit accuracy, since the mantissas that appear in it are approximations. This means that if *more* than four-digit accuracy is required in a problem, then Table 1 cannot be used. If, in more extensive tables, the logarithm of a number containing n digits can be found directly, then interpolation is allowed for numbers involving $n + 1$ digits, and numbers should be rounded off accordingly.

The method of interpolation can also be used to find x when we are given $\log x$. If we use Table 1, then x may be found to four significant figures. In this case we are given the *y-coordinate* of a point on the graph of $y = \log x$ and are asked to find the *x-coordinate.* A geometric argument similar to the one given earlier can be used to justify the procedure illustrated in the next example.

EXAMPLE 6 Find x to four significant figures if $\log x = 1.7949$.

Solution The mantissa 0.7949 does not appear in Table 1, but it can be isolated between adjacent entries, namely the mantissas corresponding to 6.230 and 6.240. We shall arrange our work as follows:

$$0.1\left\{\begin{array}{l} r\left\{\begin{array}{l}\log 62.30 \approx 1.7945 \\ \log x \quad\;\; = 1.7949\end{array}\right\} 0.0004 \\ \log 62.40 \approx 1.7952\end{array}\right\} 0.0007.$$

This leads to the proportion

$$\frac{r}{0.1} = \frac{0.0004}{0.0007} = \frac{4}{7} \quad \text{or} \quad r = (0.1)\left(\frac{4}{7}\right) \approx 0.06.$$

Hence,
$$x \approx 62.30 + 0.06 = 62.36. \qquad \blacksquare$$

The following examples illustrate the use of interpolation in trigonometric tables.

EXAMPLE 7 Approximate $\tan 24°16'$.

Solution We consult Table 4 and interpolate as follows:

$$10'\left\{\begin{array}{l} 6'\left\{\begin{array}{l}\tan 24°10' \approx 0.4487\\ \tan 24°16' = ?\end{array}\right\}d \\ \quad\ \tan 24°20' \approx 0.4522\end{array}\right\}0.0035$$

$$\frac{d}{0.0035} = \frac{6}{10}, \quad \text{or} \quad d = \frac{6}{10}(0.0035) \approx 0.0021$$

$$\tan 24°16' \approx 0.4487 + 0.0021 = 0.4508 \qquad \blacksquare$$

EXAMPLE 8 Approximate $\cos(-117°47')$.

Solution The angle is in quadrant III. The reader should check that the reference angle is $62°13'$. Consequently, $\cos(-117°47') = -\cos 62°13'$. Interpolating in Table 4 we have

$$10'\left\{\begin{array}{l} 3'\left\{\begin{array}{l}\cos 62°10' \approx 0.4669\\ \cos 62°13' = ?\end{array}\right\}d \\ \quad\ \cos 62°20' \approx 0.4643\end{array}\right\}0.0026$$

$$\frac{d}{0.0026} = \frac{3}{10}, \quad \text{or} \quad d \approx 0.0008$$

Since the cosine function is decreasing,

$$\cos 62°13' \approx 0.4669 - 0.0008 = 0.4661.$$

Hence, $\cos(-117°47') \approx -0.4661$. $\blacksquare$

EXAMPLE 9 Approximate the smallest positive real number t such that $\sin t = 0.6635$.

Solution We locate 0.6635 between successive entries in the sine column of Table 4 and interpolate as follows:

$$0.0029\left\{\begin{array}{l} d\left\{\begin{array}{ll}\sin(0.7243) \approx 0.6626 \\ \sin t = 0.6635\end{array}\right\} 0.0009 \\ \sin(0.7272) \approx 0.6648\end{array}\right\} 0.0022$$

$$\frac{d}{0.0029} = \frac{0.0009}{0.0022}, \quad \text{or} \quad d = \frac{9}{22}(0.0029) \approx 0.0012.$$

Hence, $$t \approx 0.7243 + 0.0012 = 0.7255.$$

EXAMPLE 10 If $\sin\theta = -0.7963$, approximate the degree measure of all angles θ that are in the interval $[0°, 360°]$.

Solution Let θ' be the reference angle, so that $\sin\theta' = 0.7963$. Interpolating in Table 4,

$$10'\left\{\begin{array}{l} d\left\{\begin{array}{ll}\sin 52°40' \approx 0.7951 \\ \sin\theta' = 0.7963\end{array}\right\} 0.0012 \\ \sin 52°50' \approx 0.7969\end{array}\right\} 0.0018$$

$$\frac{d}{10} = \frac{0.0012}{0.0018}, \quad \text{or} \quad d \approx 7'$$

$$\theta' \approx 52°47'.$$

Since $\sin\theta$ is negative, θ lies in quadrant III or IV. Using the reference angle $52°47'$, we have

$$\theta \approx 180° + 52°47' = 232°47'$$

$$\theta \approx 360° - 52°47' = 307°13'.$$

EXERCISES A.1

In Exercises 1–16 use Table 1 and the Laws of Logarithms to approximate the common logarithms of the numbers.

1 347; 0.00347; 3.47

2 86.2; 8,620; 0.862

3 0.54; 540; 540,000

4 208; 2.08; 20,800

5 60.2; 0.0000602; 602

6 5; 0.5; 0.0005

7 $(44.9)^2$; $(44.9)^{1/2}$; $(44.9)^{-2}$

8 $(1810)^4$; $(1810)^{40}$; $(1810)^{1/4}$

9 $(0.943)^3$; $(0.943)^{-3}$; $(0.943)^{1/3}$

10 $(0.017)^{10}$; $10^{0.017}$; $10^{1.43}$

11 $(638)(17.3)$

12 $\dfrac{(2.73)(78.5)}{621}$

13 $\dfrac{(47.4)^3}{(29.5)^2}$

14 $\dfrac{(897)^4}{\sqrt{17.8}}$

15 $\sqrt[3]{20.6(371)^3}$ **16** $\dfrac{(0.0048)^{10}}{\sqrt{0.29}}$

In Exercises 17–30 use Table 1 to find a decimal approximation to x.

17 $\log x = 3.6274$ **18** $\log x = 1.8965$

19 $\log x = 0.9469$ **20** $\log x = 0.5729$

21 $\log x = 5.2095$ **22** $\log x = 6.7300 - 10$

23 $\log x = 9.7348 - 10$ **24** $\log x = 7.6739 - 10$

25 $\log x = 8.8306 - 10$ **26** $\log x = 4.9680$

27 $\log x = 2.2765$ **28** $\log x = 3.0043$

29 $\log x = -1.6253$ **30** $\log x = -2.2118$

Use interpolation in Table 1 to approximate the common logarithms of the numbers in Exercises 31–50.

31 25.48 **32** 421.6

33 5363 **34** 0.3817

35 0.001259 **36** 69,450

37 123,400 **38** 0.0212

39 0.7786 **40** 1.203

41 384.7 **42** 54.44

43 0.9462 **44** 7259

45 66,590 **46** 0.001428

47 0.04321 **48** 300,100

49 3.003 **50** 1.236

In Exercises 51–70 use interpolation in Table 1 to approximate x.

51 $\log x = 1.4437$ **52** $\log x = 3.7455$

53 $\log x = 4.6931$ **54** $\log x = 0.5883$

55 $\log x = 9.1664 - 10$ **56** $\log x = 8.3902 - 10$

57 $\log x = 3.8153 - 6$ **58** $\log x = 5.9306 - 9$

59 $\log x = 2.3705$ **60** $\log x = 4.2867$

61 $\log x = 0.1358$ **62** $\log x = 0.0194$

63 $\log x = 8.9752 - 10$ **64** $\log x = 2.4979 - 5$

65 $\log x = 5.0409$ **66** $\log x = 1.3796$

67 $\log x = -2.8712$ **68** $\log x = -1.8164$

69 $\log x = -0.6123$ **70** $\log x = -3.1426$

In Exercises 71–82 use interpolation in Table 4 to approximate the numbers.

71 $\sin (0.46)$ **72** $\cos (0.82)$

73 $\tan 3$ **74** $\cot 6$

75 $\sec \frac{1}{4}$ **76** $\csc (1.54)$

77 $\cos 37°43'$ **78** $\sin 22°34'$

79 $\cot 62°27'$ **80** $\tan 57°16'$

81 $\csc 16°55'$ **82** $\sec 9°12'$

In Exercises 83–88 use interpolation in Table 4 to approximate the smallest positive number t for which the equality is true.

83 $\cos t = 0.8620$ **84** $\sin t = 0.6612$

85 $\tan t = 4.501$ **86** $\sec t = 3.641$

87 $\csc t = 1.436$ **88** $\cot t = 1.165$

In Exercises 89–96 use interpolation in Table 4 to approximate, to the nearest minute, the degree measure of all angles θ that lie in the interval $[0°, 360°]$.

89 $\sin \theta = 0.3672$ **90** $\cos \theta = 0.8426$

91 $\tan \theta = 0.5042$ **92** $\cot \theta = 1.348$

93 $\cos \theta = 0.3465$ **94** $\csc \theta = 1.219$

95 $\sec \theta = 1.385$ **96** $\sin \theta = 0.7534$

TABLES

Table 1 COMMON LOGARITHMS

N	0	1	2	3	4	5	6	7	8	9
1.0	.0000	.0043	.0086	.0128	.0170	.0212	.0253	.0294	.0334	.0374
1.1	.0414	.0453	.0492	.0531	.0569	.0607	.0645	.0682	.0719	.0755
1.2	.0792	.0828	.0864	.0899	.0934	.0969	.1004	.1038	.1072	.1106
1.3	.1139	.1173	.1206	.1239	.1271	.1303	.1335	.1367	.1399	.1430
1.4	.1461	.1492	.1523	.1553	.1584	.1614	.1644	.1673	.1703	.1732
1.5	.1761	.1790	.1818	.1847	.1875	.1903	.1931	.1959	.1987	.2014
1.6	.2041	.2068	.2095	.2122	.2148	.2175	.2201	.2227	.2253	.2279
1.7	.2304	.2330	.2355	.2380	.2405	.2430	.2455	.2480	.2504	.2529
1.8	.2553	.2557	.2601	.2625	.2648	.2672	.2695	.2718	.2742	.2765
1.9	.2788	.2810	.2833	.2856	.2878	.2900	.2923	.2945	.2967	.2989
2.0	.3010	.3032	.3054	.3075	.3096	.3118	.3139	.3160	.3181	.3201
2.1	.3222	.3243	.3263	.3284	.3304	.3324	.3345	.3365	.3385	.3404
2.2	.3424	.3444	.3464	.3483	.3502	.3522	.3541	.3560	.3579	.3598
2.3	.3617	.3636	.3655	.3674	.3692	.3711	.3729	.3747	.3766	.3784
2.4	.3802	.3820	.3838	.3856	.3874	.3892	.3909	.3927	.3945	.3962
2.5	.3979	.3997	.4014	.4031	.4048	.4065	.4082	.4099	.4116	.4113
2.6	.4150	.4166	.4183	.4200	.4216	.4232	.4249	.4265	.4281	.4298
2.7	.4314	.4330	.4346	.4362	.4378	.4393	.4409	.4425	.4440	.4456
2.8	.4472	.4487	.4502	.4518	.4533	.4548	.4564	.4579	.4594	.4609
2.9	.4624	.4639	.4654	.4669	.4683	.4698	.4713	.4728	.4742	.4757
3.0	.4771	.4786	.4800	.4814	.4829	.4843	.4857	.4871	.4886	.4900
3.1	.4914	.4928	.4942	.4955	.4969	.4983	.4997	.5011	.5024	.5038
3.2	.5051	.5065	.5079	.5092	.5105	.5119	.5132	.5145	.5159	.5172
3.3	.5185	.5198	.5211	.5224	.5237	.5250	.5263	.5276	.5289	.5302
3.4	.5315	.5328	.5340	.5353	.5366	.5378	.5391	.5403	.5416	.5428
3.5	.5441	.5453	.5465	.5478	.5490	.5502	.5514	.5527	.5539	.5551
3.6	.5563	.5575	.5587	.5599	.5611	.5623	.5635	.5647	.5658	.5670
3.7	.5682	.5694	.5705	.5717	.5729	.5740	.5752	.5763	.5775	.5786
3.8	.5798	.5809	.5821	.5832	.5843	.5855	.5866	.5877	.5888	.5899
3.9	.5911	.5922	.5933	.5944	.5955	.5966	.5977	.5988	.5999	.6010
4.0	.6021	.6031	.6042	.6053	.6064	.6075	.6085	.6096	.6107	.6117
4.1	.6128	.6138	.6149	.6160	.6170	.6180	.6191	.6201	.6212	.6222
4.2	.6232	.6243	.6253	.6263	.6274	.6284	.6294	.6304	.6314	.6325
4.3	.6335	.6345	.6355	.6365	.6375	.6385	.6395	.6405	.6415	.6425
4.4	.6435	.6444	.6454	.6464	.6474	.6484	.6493	.6503	.6513	.6522
4.5	.6532	.6542	.6551	.6561	.6571	.6580	.6590	.6599	.6609	.6118
4.6	.6628	.6637	.6646	.6656	.6665	.6675	.6684	.6693	.6702	.6712
4.7	.6721	.6730	.6739	.6749	.6758	.6767	.6776	.6785	.6794	.6803
4.8	.6812	.6821	.6830	.6839	.6848	.6857	.6866	.6875	.6884	.6893
4.9	.6902	.6911	.6920	.6928	.6937	.6946	.6955	.6964	.6972	.6981
5.0	.6990	.6998	.7007	.7016	.7024	.7033	.7042	.7050	.7059	.7067
5.1	.7076	.7084	.7093	.7101	.7110	.7118	.7126	.7135	.7143	.7152
5.2	.7160	.7168	.7177	.7185	.7193	.7202	.7210	.7218	.7226	.7235
5.3	.7243	.7251	.7259	.7267	.7275	.7284	.7292	.7300	.7308	.7316
5.4	.7324	.7332	.7340	.7348	.7356	.7364	.7372	.7380	.7388	.7396
5.5	.7404	.7412	.7419	.7427	.7435	.7443	.7451	.7459	.7466	.7474
5.6	.7482	.7490	.7497	.7505	.7513	.7520	.7528	.7536	.7543	.7551
5.7	.7559	.7566	.7574	.7582	.7589	.7597	.7604	.7612	.7619	.7627
5.8	.7634	.7642	.7649	.7657	.7664	.7672	.7679	.7686	.7694	.7701
5.9	.7709	.7716	.7723	.7731	.7738	.7745	.7752	.7760	.7767	.7774
6.0	.7782	.7789	.7796	.7803	.7810	.7818	.7825	.7832	.7839	.7846
6.1	.7853	.7860	.7868	.7875	.7882	.7889	.7896	.7903	.7910	.7917
6.2	.7924	.7931	.7938	.7945	.7952	.7959	.7966	.7973	.7980	.7987
6.3	.7993	.8000	.8007	.8014	.8021	.8028	.8035	.8041	.8048	.8055
6.4	.8062	.8069	.8075	.8082	.8089	.8096	.8102	.8109	.8116	.8122
6.5	.8129	.8136	.8142	.8149	.8156	.8162	.8169	.8176	.8182	.8189
6.6	.8195	.8202	.8209	.8215	.8222	.8228	.8235	.8241	.8248	.8254
6.7	.8261	.8267	.8274	.8280	.8287	.8293	.8299	.8306	.8312	.8319
6.8	.8325	.8331	.8338	.8344	.8351	.8357	.8363	.8370	.8376	.8382
6.9	.8388	.8395	.8401	.8407	.8414	.8420	.8426	.8432	.8439	.8445
7.0	.8451	.8457	.8463	.8470	.8476	.8482	.8488	.8494	.8500	.8506
7.1	.8513	.8519	.8525	.8531	.8537	.8543	.8549	.8555	.8561	.8567
7.2	.8573	.8579	.8585	.8591	.8597	.8603	.8609	.8615	.8621	.8627
7.3	.8633	.8639	.8645	.8651	.8657	.8663	.8669	.8675	.8681	.8686
7.4	.8692	.8698	.8704	.8710	.8716	.8722	.8727	.8733	.8739	.8745
7.5	.8751	.8756	.8762	.8768	.8774	.8779	.8785	.8791	.8797	.8802
7.6	.8808	.8814	.8820	.8825	.8831	.8837	.8842	.8848	.8854	.8859
7.7	.8865	.8871	.8876	.8882	.8887	.8893	.8899	.8904	.8910	.8915
7.8	.8921	.8927	.8932	.8938	.8943	.8949	.8954	.8960	.8965	.8971
7.9	.8976	.8982	.8987	.8993	.8998	.9004	.9009	.9015	.9020	.9025
8.0	.9031	.9036	.9042	.9047	.9053	.9058	.9063	.9069	.9074	.9079
8.1	.9085	.9090	.9096	.9101	.9106	.9112	.9117	.9122	.9128	.9133
8.2	.9138	.9143	.9149	.9154	.9159	.9165	.9170	.9175	.9180	.9186
8.3	.9191	.9196	.9201	.9206	.9212	.9217	.9222	.9227	.9232	.9238
8.4	.9243	.9248	.9253	.9258	.9263	.9269	.9274	.9279	.9284	.9289
8.5	.9294	.9299	.9304	.9309	.9315	.9320	.9325	.9330	.9335	.9340
8.6	.9345	.9350	.9355	.9360	.9365	.9370	.9375	.9380	.9385	.9390
8.7	.9395	.9400	.9405	.9410	.9415	.9420	.9425	.9430	.9435	.9440
8.8	.9445	.9450	.9455	.9460	.9465	.9469	.9474	.9479	.9484	.9489
8.9	.9494	.9499	.9504	.9509	.9513	.9518	.9523	.9528	.9533	.9538
9.0	.9542	.9547	.9552	.9557	.9562	.9566	.9571	.9576	.9581	.9586
9.1	.9590	.9595	.9600	.9605	.9609	.9614	.9619	.9624	.9628	.9633
9.2	.9638	.9643	.9647	.9652	.9657	.9661	.9666	.9671	.9675	.9680
9.3	.9685	.9689	.9694	.9699	.9703	.9708	.9713	.9717	.9722	.9727
9.4	.9731	.9736	.9741	.9745	.9750	.9754	.9759	.9763	.9768	.9773
9.5	.9777	.9782	.9786	.9791	.9795	.9800	.9805	.9809	.9814	.9818
9.6	.9823	.9827	.9832	.9836	.9841	.9845	.9850	.9854	.9859	.9863
9.7	.9868	.9872	.9877	.9881	.9886	.9890	.9894	.9899	.9903	.9908
9.8	.9912	.9917	.9921	.9926	.9930	.9934	.9939	.9943	.9948	.9952
9.9	.9956	.9961	.9965	.9969	.9974	.9978	.9983	.9987	.9991	.9996

Table 2 NATURAL EXPONENTIAL FUNCTION

x	e^x	e^{-x}	x	e^x	e^{-x}
0.00	1.0000	1.0000	2.50	12.182	0.0821
0.05	1.0513	0.9512	2.60	13.464	0.0743
0.10	1.1052	0.9048	2.70	14.880	0.0672
0.15	1.1618	0.8607	2.80	16.445	0.0608
0.20	1.2214	0.8187	2.90	18.174	0.0550
0.25	1.2840	0.7788	3.00	20.086	0.0498
0.30	1.3499	0.7408	3.10	22.198	0.0450
0.35	1.4191	0.7047	3.20	24.533	0.0408
0.40	1.4918	0.6703	3.30	27.113	0.0369
0.45	1.5683	0.6376	3.40	29.964	0.0334
0.50	1.6487	0.6065	3.50	33.115	0.0302
0.55	1.7333	0.5769	3.60	36.598	0.0273
0.60	1.8221	0.5488	3.70	40.447	0.0247
0.65	1.9155	0.5220	3.80	44.701	0.0224
0.70	2.0138	0.4966	3.90	49.402	0.0202
0.75	2.1170	0.4724	4.00	54.598	0.0183
0.80	2.2255	0.4493	4.10	60.340	0.0166
0.85	2.3396	0.4274	4.20	66.686	0.0150
0.90	2.4596	0.4066	4.30	73.700	0.0136
0.95	2.5857	0.3867	4.40	81.451	0.0123
1.00	2.7183	0.3679	4.50	99.017	0.0111
1.10	3.0042	0.3329	4.60	99.484	0.0101
1.20	3.3201	0.3012	4.70	109.95	0.0091
1.30	3.6693	0.2725	4.80	121.51	0.0082
1.40	4.0552	0.2466	4.90	134.29	0.0074
1.50	4.4817	0.2231	5.00	148.41	0.0067
1.60	4.9530	0.2019	6.00	403.43	0.0025
1.70	5.4739	0.1827	7.00	1096.6	0.0009
1.80	6.0496	0.1653	8.00	2981.0	0.0003
1.90	6.6859	0.1496	9.00	8103.1	0.0001
2.00	7.3891	0.1353	10.00	22026.0	0.00005
2.10	8.1662	0.1225			
2.20	9.0250	0.1108			
2.30	9.9742	0.1003			
2.40	11.0232	0.0907			

Table 3 NATURAL LOGARITHMS

n	0.0	0.1	0.2	0.3	0.4	0.5	0.6	0.7	0.8	0.9
0*		7.697	8.391	8.796	9.084	9.307	9.489	9.643	9.777	9.895
1	0.000	0.095	0.182	0.262	0.336	0.405	0.470	0.531	0.588	0.642
2	0.693	0.742	0.788	0.833	0.875	0.916	0.956	0.993	1.030	1.065
3	1.099	1.131	1.163	1.194	1.224	1.253	1.281	1.308	1.335	1.361
4	1.386	1.411	1.435	1.459	1.482	1.504	1.526	1.548	1.569	1.589
5	1.609	1.629	1.649	1.668	1.686	1.705	1.723	1.740	1.758	1.775
6	1.792	1.808	1.825	1.841	1.856	1.872	1.887	1.902	1.917	1.932
7	1.946	1.960	1.974	1.988	2.001	2.015	2.028	2.041	2.054	2.067
8	2.079	2.092	2.104	2.116	2.128	2.140	2.152	2.163	2.175	2.186
9	2.197	2.208	2.219	2.230	2.241	2.251	2.262	2.272	2.282	2.293
10	2.303	2.313	2.322	2.332	2.342	2.351	2.361	2.370	2.380	2.389

* Subtract 10 if $n < 1$; for example, $\ln 0.3 \approx 8.796 - 10 = -1.204$.

Table 4 VALUES OF THE TRIGONOMETRIC FUNCTIONS

t	t degrees	sin t	cos t	tan t	cot t	sec t	csc t		
.0000	**0°00′**	.0000	1.0000	.0000	—	1.000	—	**90°00′**	1.5708
.0029	10	.0029	1.0000	.0029	343.8	1.000	343.8	50	1.5679
.0058	20	.0058	1.0000	.0058	171.9	1.000	171.9	40	1.5650
.0087	30	.0087	1.0000	.0087	114.6	1.000	114.6	30	1.5621
.0116	40	.0116	.9999	.0116	85.94	1.000	85.95	20	1.5592
.0145	50	.0145	.9999	.0145	68.75	1.000	68.76	10	1.5563
.0175	**1°00′**	.0175	.9998	.0175	57.29	1.000	57.30	**89°00′**	1.5533
.0204	10	.0204	.9998	.0204	49.10	1.000	49.11	50	1.5504
.0233	20	.0233	.9997	.0233	42.96	1.000	42.98	40	1.5475
.0262	30	.0262	.9997	.0262	38.19	1.000	38.20	30	1.5446
.0291	40	.0291	.9996	.0291	34.37	1.000	34.38	20	1.5417
.0320	50	.0320	.9995	.0320	31.24	1.001	31.26	10	1.5388
.0349	**2°00′**	.0349	.9994	.0349	28.64	1.001	28.65	**88°00′**	1.5359
.0378	10	.0378	.9993	.0378	26.43	1.001	26.45	50	1.5330
.0407	20	.0407	.9992	.0407	24.54	1.001	24.56	40	1.5301
.0436	30	.0436	.9990	.0437	22.90	1.001	22.93	30	1.5272
.0465	40	.0465	.9989	.0466	21.47	1.001	21.49	20	1.5243
.0495	50	.0494	.9988	.0495	20.21	1.001	20.23	10	1.5213
.0524	**3°00′**	.0523	.9986	.0524	19.08	1.001	19.11	**87°00′**	1.5184
.0553	10	.0552	.9985	.0553	18.07	1.002	18.10	50	1.5155
.0582	20	.0581	.9983	.0582	17.17	1.002	17.20	40	1.5126
.0611	30	.0610	.9981	.0612	16.35	1.002	16.38	30	1.5097
.0640	40	.0640	.9980	.0641	15.60	1.002	15.64	20	1.5068
.0669	50	.0669	.9978	.0670	14.92	1.002	14.96	10	1.5039
.0698	**4°00′**	.0698	.9976	.0699	14.30	1.002	14.34	**86°00′**	1.5010
.0727	10	.0727	.9974	.0729	13.73	1.003	13.76	50	1.4981
.0756	20	.0756	.9971	.0758	13.20	1.003	13.23	40	1.4952
.0785	30	.0785	.9969	.0787	12.71	1.003	12.75	30	1.4923
.0814	40	.0814	.9967	.0816	12.25	1.003	12.29	20	1.4893
.0844	50	.0843	.9964	.0846	11.83	1.004	11.87	10	1.4864
.0873	**5°00′**	.0872	.9962	.0875	11.43	1.004	11.47	**85°00′**	1.4835
.0902	10	.0901	.9959	.0904	11.06	1.004	11.10	50	1.4806
.0931	20	.0929	.9957	.0934	10.71	1.004	10.76	40	1.4777
.0960	30	.0958	.9954	.0963	10.39	1.005	10.43	30	1.4748
.0989	40	.0987	.9951	.0992	10.08	1.005	10.13	20	1.4719
.1018	50	.1016	.9948	.1022	9.788	1.005	9.839	10	1.4690
.1047	**6°00′**	.1045	.9945	.1051	9.514	1.006	9.567	**84°00′**	1.4661
.1076	10	.1074	.9942	.1080	9.255	1.006	9.309	50	1.4632
.1105	20	.1103	.9939	.1110	9.010	1.006	9.065	40	1.4603
.1134	30	.1132	.9936	.1139	8.777	1.006	8.834	30	1.4573
.1164	40	.1161	.9932	.1169	8.556	1.007	8.614	20	1.4544
.1193	50	.1190	.9929	.1198	8.345	1.007	8.405	10	1.4515
.1222	**7°00′**	.1219	.9925	.1228	8.144	1.008	8.206	**83°00′**	1.4486
		cos t	sin t	cot t	tan t	csc t	sec t	t degrees	t

t	t degrees	sin t	cos t	tan t	cot t	sec t	csc t		
.1222	**7°00′**	.1219	.9925	.1228	8.144	1.008	8.206	**83°00′**	1.4486
.1251	10	.1248	.9922	.1257	7.953	1.008	8.016	50	1.4457
.1280	20	.1276	.9918	.1287	7.770	1.008	7.834	40	1.4428
.1309	30	.1305	.9914	.1317	7.596	1.009	7.661	30	1.4399
.1338	40	.1334	.9911	.1346	7.429	1.009	7.496	20	1.4370
.1367	50	.1363	.9907	.1376	7.269	1.009	7.337	10	1.4341
.1396	**8°00′**	.1392	.9903	.1405	7.115	1.010	7.185	**82°00′**	1.4312
.1425	10	.1421	.9899	.1435	6.968	1.010	7.040	50	1.4283
.1454	20	.1449	.9894	.1465	6.827	1.011	6.900	40	1.4254
.1484	30	.1478	.9890	.1495	6.691	1.011	6.765	30	1.4224
.1513	40	.1507	.9886	.1524	6.561	1.012	6.636	20	1.4195
.1542	50	.1536	.9881	.1554	6.435	1.012	6.512	10	1.4166
.1571	**9°00′**	.1564	.9877	.1584	6.314	1.012	6.392	**81°00′**	1.4137
.1600	10	.1593	.9872	.1614	6.197	1.013	6.277	50	1.4108
.1629	20	.1622	.9868	.1644	6.084	1.013	6.166	40	1.4079
.1658	30	.1650	.9863	.1673	5.976	1.014	6.059	30	1.4050
.1687	40	.1679	.9858	.1703	5.871	1.014	5.955	20	1.4021
.1716	50	.1708	.9853	.1733	5.769	1.015	5.855	10	1.3992
.1745	**10°00′**	.1736	.9848	.1763	5.671	1.015	5.759	**80°00′**	1.3963
.1774	10	.1765	.9843	.1793	5.576	1.016	5.665	50	1.3934
.1804	20	.1794	.9838	.1823	5.485	1.016	5.575	40	1.3904
.1833	30	.1822	.9833	.1853	5.396	1.017	5.487	30	1.3875
.1862	40	.1851	.9827	.1883	5.309	1.018	5.403	20	1.3846
.1891	50	.1880	.9822	.1914	5.226	1.018	5.320	10	1.3817
.1920	**11°00′**	.1908	.9816	.1944	5.145	1.019	5.241	**79°00′**	1.3788
.1949	10	.1937	.9811	.1974	5.066	1.019	5.164	50	1.3759
.1978	20	.1965	.9805	.2004	4.989	1.020	5.089	40	1.3730
.2007	30	.1994	.9799	.2035	4.915	1.020	5.016	30	1.3701
.2036	40	.2022	.9793	.2065	4.843	1.021	4.945	20	1.3672
.2065	50	.2051	.9787	.2095	4.773	1.022	4.876	10	1.3643
.2094	**12°00′**	.2079	.9781	.2126	4.705	1.022	4.810	**78°00′**	1.3614
.2123	10	.2108	.9775	.2156	4.638	1.023	4.745	50	1.3584
.2153	20	.2136	.9769	.2186	4.574	1.024	4.682	40	1.3555
.2182	30	.2164	.9763	.2217	4.511	1.024	4.620	30	1.3526
.2211	40	.2193	.9757	.2247	4.449	1.025	4.560	20	1.3497
.2240	50	.2221	.9750	.2278	4.390	1.026	4.502	10	1.3468
.2269	**13°00′**	.2250	.9744	.2309	4.331	1.026	4.445	**77°00′**	1.3439
.2298	10	.2278	.9737	.2339	4.275	1.027	4.390	50	1.3410
.2327	20	.2306	.9730	.2370	4.219	1.028	4.336	40	1.3381
.2356	30	.2334	.9724	.2401	4.165	1.028	4.284	30	1.3352
.2385	40	.2363	.9717	.2432	4.113	1.029	4.232	20	1.3323
.2414	50	.2391	.9710	.2462	4.061	1.030	4.182	10	1.3294
.2443	**14°00′**	.2419	.9703	.2493	4.011	1.031	4.134	**76°00′**	1.3265
		cos t	sin t	cot t	tan t	csc t	sec t	t degrees	t

Table 4 VALUES OF THE TRIGONOMETRIC FUNCTIONS *(cont'd.)*

t	t degrees	$\sin t$	$\cos t$	$\tan t$	$\cot t$	$\sec t$	$\csc t$		
.2443	**14°00′**	.2419	.9703	.2493	4.011	1.031	4.134	**76°00′**	1.3265
.2473	10	.2447	.9696	.2524	3.962	1.031	4.086	50	1.3235
.2502	20	.2476	.9689	.2555	3.914	1.032	4.039	40	1.3206
.2531	30	.2504	.9681	.2586	3.867	1.033	3.994	30	1.3177
.2560	40	.2532	.9674	.2617	3.821	1.034	3.950	20	1.3148
.2589	50	.2560	.9667	.2648	3.776	1.034	3.906	10	1.3119
.2618	**15°00**	.2588	.9659	.2679	3.732	1.035	3.864	**75°00′**	1.3090
.2647	10	.2616	.9652	.2711	3.689	1.036	3.822	50	1.3061
.2676	20	.2644	.9644	.2742	3.647	1.037	3.782	40	1.3032
.2705	30	.2672	.9636	.2773	3.606	1.038	3.742	30	1.3003
.2734	40	.2700	.9628	.2805	3.566	1.039	3.703	20	1.2974
.2763	50	.2728	.9621	.2836	3.526	1.039	3.665	10	1.2945
.2793	**16°00′**	.2756	.9613	.2867	3.487	1.040	3.628	**74°00**	1.2915
.2822	10	.2784	.9605	.2899	3.450	1.041	3.592	50	1.2886
.2851	20	.2812	.9596	.2931	3.412	1.042	3.556	40	1.2857
.2880	30	.2840	.9588	.2962	3.376	1.043	3.521	30	1.2828
.2909	40	.2868	.9580	.2994	3.340	1.044	3.487	20	1.2799
.2938	50	.2896	.9572	.3026	3.305	1.045	3.453	10	1.2770
.2967	**17°00′**	.2924	.9563	.3057	3.271	1.046	3.420	**73°00′**	1.2741
.2996	10	.2952	.9555	.3089	3.237	1.047	3.388	50	1.2712
.3025	20	.2979	.9546	.3121	3.204	1.048	3.356	40	1.2683
.3054	30	.3007	.9537	.3153	3.172	1.049	3.326	30	1.2654
.3083	40	.3035	.9528	.3185	3.140	1.049	3.295	20	1.2625
.3113	50	.3062	.9520	.3217	3.108	1.050	3.265	10	1.2595
.3142	**18°00**	.3090	.9511	.3249	3.078	1.051	3.236	**72°00**	1.2566
.3171	10	.3118	.9502	.3281	3.047	1.052	3.207	50	1.2537
.3200	20	.3145	.9492	.3314	3.018	1.053	3.179	40	1.2508
.3229	30	.3173	.9483	.3346	2.989	1.054	3.152	30	1.2479
.3258	40	.3201	.9474	.3378	2.960	1.056	3.124	20	1.2450
.3287	50	.3228	.9465	.3411	2.932	1.057	3.098	10	1.2421
.3316	**19°00′**	.3256	.9455	.3443	2.904	1.058	3.072	**71°00′**	1.2392
.3345	10	.3283	.9446	.3476	2.877	1.059	3.046	50	1.2363
.3374	20	.3311	.9436	.3508	2.850	1.060	3.021	40	1.2334
.3403	30	.3338	.9426	.3541	2.824	1.061	2.996	30	1.2305
.3432	40	.3365	.9417	.3574	2.798	1.062	2.971	20	1.2275
.3462	50	.3393	.9407	.3607	2.773	1.063	2.947	10	1.2246
.3491	**20°00′**	.3420	.9397	.3640	2.747	1.064	2.924	**70°00′**	1.2217
.3520	10	.3448	.9387	.3673	2.723	1.065	2.901	50	1.2188
.3549	20	.3475	.9377	.3706	2.699	1.066	2.878	40	1.2159
.3578	30	.3502	.9367	.3739	2.675	1.068	2.855	30	1.2130
.3607	40	.3529	.9356	.3772	2.651	1.069	2.833	20	1.2101
.3636	50	.3557	.9346	.3805	2.628	1.070	2.812	10	1.2072
.3665	**21°00′**	.3584	.9336	.3839	2.605	1.071	2.790	**69°00′**	1.2043
		$\cos t$	$\sin t$	$\cot t$	$\tan t$	$\csc t$	$\sec t$	t degrees	t

t	t degrees	$\sin t$	$\cos t$	$\tan t$	$\cot t$	$\sec t$	$\csc t$		
.3665	**21°00′**	.3584	.9336	.3839	2.605	1.071	2.790	**69°00′**	1.2043
.3694	10	.3611	.9325	.3872	2.583	1.072	2.769	50	1.2014
.3723	20	.3638	.9315	.3906	2.560	1.074	2.749	40	1.1985
.3752	30	.3665	.9304	.3939	2.539	1.075	2.729	30	1.1956
.3782	40	.3692	.9293	.3973	2.517	1.076	2.709	20	1.1926
.3811	50	.3719	.9283	.4006	2.496	1.077	2.689	10	1.1897
.3840	**22°00′**	.3746	.9272	.4040	2.475	1.079	2.669	**68°00′**	1.1868
.3869	10	.3773	.9261	.4074	2.455	1.080	2.650	50	1.1839
.3898	20	.3800	.9250	.4108	2.434	1.081	2.632	40	1.1810
.3927	30	.3827	.9239	.4142	2.414	1.082	2.613	30	1.1781
.3956	40	.3854	.9228	.4176	2.394	1.084	2.595	20	1.1752
.3985	50	.3881	.9216	.4210	2.375	1.085	2.577	10	1.1723
.4014	**23°00′**	.3907	.9205	.4245	2.356	1.086	2.559	**67°00′**	1.1694
.4043	10	.3934	.9194	.4279	2.337	1.088	2.542	50	1.1665
.4072	20	.3961	.9182	.4314	2.318	1.089	2.525	40	1.1636
.4102	30	.3987	.9171	.4348	2.300	1.090	2.508	30	1.1606
.4131	40	.4014	.9159	.4383	2.282	1.092	2.491	20	1.1577
.4160	50	.4041	.9147	.4417	2.264	1.093	2.475	10	1.1548
.4189	**24°00′**	.4067	.9135	.4452	2.246	1.095	2.459	**66°00′**	1.1519
.4218	10	.4094	.9124	.4487	2.229	1.096	2.443	50	1.1490
.4247	20	.4120	.9112	.4522	2.211	1.097	2.427	40	1.1461
.4276	30	.4147	.9100	.4557	2.194	1.099	2.411	30	1.1432
.4305	40	.4173	.9088	.4592	2.177	1.100	2.396	20	1.1403
.4334	50	.4200	.9075	.4628	2.161	1.102	2.381	10	1.1374
.4363	**25°00′**	.4226	.9063	.4663	2.145	1.103	2.366	**65°00′**	1.1345
.4392	10	.4253	.9051	.4699	2.128	1.105	2.352	50	1.1316
.4422	20	.4279	.9038	.4734	2.112	1.106	2.337	40	1.1286
.4451	30	.4305	.9026	.4770	2.097	1.108	2.323	30	1.1257
.4480	40	.4331	.9013	.4806	2.081	1.109	2.309	20	1.1228
.4509	50	.4358	.9001	.4841	2.066	1.111	2.295	10	1.1199
.4538	**26°00′**	.4384	.8988	.4877	2.050	1.113	2.281	**64°00′**	1.1170
.4567	10	.4410	.8975	.4913	2.035	1.114	2.268	50	1.1141
.4596	20	.4436	.8962	.4950	2.020	1.116	2.254	40	1.1112
.4625	30	.4462	.8949	.4986	2.006	1.117	2.241	30	1.1083
.4654	40	.4488	.8936	.5022	1.991	1.119	2.228	20	1.1054
.4683	50	.4514	.8923	.5059	1.977	1.121	2.215	10	1.1025
.4712	**27°00′**	.4540	.8910	.5095	1.963	1.122	2.203	**63°00′**	1.0996
.4741	10	.4566	.8897	.5132	1.949	1.124	2.190	50	1.0966
.4771	20	.4592	.8884	.5169	1.935	1.126	2.178	40	1.0937
.4800	30	.4617	.8870	.5206	1.921	1.127	2.166	30	1.0908
.4829	40	.4643	.8857	.5243	1.907	1.129	2.154	20	1.0879
.4858	50	.4669	.8843	.5280	1.894	1.131	2.142	10	1.0850
.4887	**28°00′**	.4695	.8829	.5317	1.881	1.133	2.130	**62°00′**	1.0821
		$\cos t$	$\sin t$	$\cot t$	$\tan t$	$\csc t$	$\sec t$	t degrees	t

Table 4 VALUES OF THE TRIGONOMETRIC FUNCTIONS *(cont'd.)*

t	t degrees	$\sin t$	$\cos t$	$\tan t$	$\cot t$	$\sec t$	$\csc t$		
.4887	**28°00′**	.4695	.8829	.5317	1.881	1.133	2.130	**62°00′**	1.0821
.4916	10	.4720	.8816	.5354	1.868	1.134	2.118	50	1.0792
.4945	20	.4746	.8802	.5392	1.855	1.136	2.107	40	1.0763
.4974	30	.4772	.8788	.5430	1.842	1.138	2.096	30	1.0734
.5003	40	.4797	.8774	.5467	1.829	1.140	2.085	20	1.0705
.5032	50	.4823	.8760	.5505	1.816	1.142	2.074	10	1.0676
.5061	**29°00′**	.4848	.8746	.5543	1.804	1.143	2.063	**61°00′**	1.0647
.5091	10	.4874	.8732	.5581	1.792	1.145	2.052	50	1.0617
.5120	20	.4899	.8718	.5619	1.780	1.147	2.041	40	1.0588
.5149	30	.4924	.8704	.5658	1.767	1.149	2.031	30	1.0559
.5178	40	.4950	.8689	.5696	1.756	1.151	2.202	20	1.0530
.5207	50	.4975	.8675	.5735	1.744	1.153	2.010	10	1.0501
.5236	**30°00′**	.5000	.8660	.5774	1.732	1.155	2.000	**60°00′**	1.0472
.5265	10	.5025	.8646	.5812	1.720	1.157	1.990	50	1.0443
.5294	20	.5050	.8631	.5851	1.709	1.159	1.980	40	1.0414
.5323	30	.5075	.8616	.5890	1.698	1.161	1.970	30	1.0385
.5352	40	.5100	.8601	.5930	1.686	1.163	1.961	20	1.0356
.5381	50	.5125	.8587	.5969	1.675	1.165	1.951	10	1.0327
.5411	**31°00′**	.5150	.8572	.6009	1.664	1.167	1.942	**59°00′**	1.0297
.5440	10	.5175	.8557	.6048	1.653	1.169	1.932	50	1.0268
.5469	20	.5200	.8542	.6088	1.643	1.171	1.923	40	1.0239
.5498	30	.5225	.8526	.6128	1.632	1.173	1.914	30	1.0210
.5527	40	.5250	.8511	.6168	1.621	1.175	1.905	20	1.0181
.5556	50	.5275	.8496	.6208	1.611	1.177	1.896	10	1.0152
.5585	**32°00′**	.5299	.8480	.6249	1.600	1.179	1.887	**58°00′**	1.0123
.5614	10	.5324	.8465	.6289	1.590	1.181	1.878	50	1.0094
.5643	20	.5348	.8450	.6330	1.580	1.184	1.870	40	1.0065
.5672	30	.5373	.8434	.6371	1.570	1.186	1.861	30	1.0036
.5701	40	.5398	.8418	.6412	1.560	1.188	1.853	20	1.0007
.5730	50	.5422	.8403	.6453	1.550	1.190	1.844	10	.9977
.5760	**33°00′**	.5446	.8387	.6494	1.540	1.192	1.836	**57°00′**	.9948
.5789	10	.5471	.8371	.6536	1.530	1.195	1.828	50	.9919
.5818	20	.5495	.8355	.6577	1.520	1.197	1.820	40	.9890
.5847	30	.5519	.8339	.6619	1.511	1.199	1.812	30	.9861
.5876	40	.5544	.8323	.6661	1.501	1.202	1.804	20	.9832
.5905	50	.5568	.8307	.6703	1.492	1.204	1.796	10	.9803
.5934	**34°00′**	.5592	.8290	.6745	1.483	1.206	1.788	**56°00′**	.9774
.5963	10	.5616	.8274	.6787	1.473	1.209	1.781	50	.9745
.5992	20	.5640	.8258	.6830	1.464	1.211	1.773	40	.9716
.6021	20	.5664	.8241	.6873	1.455	1.213	1.766	30	.9687
.6050	40	.5688	.8225	.6916	1.446	1.216	1.758	20	.9657
.6080	50	.5712	.8208	.6959	1.437	1.218	1.751	10	.9628
.6109	**35°00′**	.5736	.8192	.7002	1.428	1.221	1.743	**55°00′**	.9599
		$\cos t$	$\sin t$	$\cot t$	$\tan t$	$\csc t$	$\sec t$	t degrees	t

t	t degrees	$\sin t$	$\cos t$	$\tan t$	$\cot t$	$\sec t$	$\csc t$		
.6109	**35°00′**	.5736	.8192	.7002	1.428	1.221	1.743	**55°00′**	.9599
.6138	10	.5760	.8175	.7046	1.419	1.223	1.736	50	.9570
.6167	20	.5783	.8158	.7089	1.411	1.226	1.729	40	.9541
.6196	30	.5807	.8141	.7133	1.402	1.228	1.722	30	.9512
.6225	40	.5831	.8124	.7177	1.393	1.231	1.715	20	.9483
.6254	50	.5854	.8107	.7221	1.385	1.233	1.708	10	.9454
.6283	**36°00′**	.5878	.8090	.7265	1.376	1.236	1.701	**54°00′**	.9425
.6312	10	.5901	.8073	.7310	1.368	1.239	1.695	50	.9396
.6341	20	.5925	.8056	.7355	1.360	1.241	1.688	40	.9367
.6370	30	.5948	.8039	.7400	1.351	1.244	1.681	30	.9338
.6400	40	.5972	.8021	.7445	1.343	1.247	1.675	20	.9308
.6429	50	.5995	.8004	.7490	1.335	1.249	1.668	10	.9279
.6458	**37°00′**	.6018	.7986	.7536	1.327	1.252	1.662	**53°00′**	.9250
.6487	10	.6041	.7969	.7581	1.319	1.255	1.655	50	.9221
.6516	20	.6065	.7951	.7627	1.311	1.258	1.649	40	.9192
.6545	30	.6088	.7934	.7673	1.303	1.260	1.643	30	.9163
.6574	40	.6111	.7916	.7720	1.295	1.263	1.636	20	.9134
.6603	50	.6134	.7898	.7766	1.288	1.266	1.630	10	.9105
.6632	**38°00′**	.6157	.7880	.7813	1.280	1.269	1.624	**52°00′**	.9076
.6661	10	.6180	.7862	.7860	1.272	1.272	1.618	50	.9047
.6690	20	.6202	.7844	.7907	1.265	1.275	1.612	40	.9018
.6720	30	.6225	.7826	.7954	1.257	1.278	1.606	30	.8988
.6749	40	.6248	.7808	.8002	1.250	1.281	1.601	20	.8959
.6778	50	.6271	.7790	.8050	1.242	1.284	1.595	10	.8930
.6807	**39°00′**	.6293	.7771	.8098	1.235	1.287	1.589	**51°00′**	.8901
.6836	10	.6316	.7753	.8146	1.228	1.290	1.583	50	.8872
.6865	20	.6338	.7735	.8195	1.220	1.293	1.578	40	.8843
.6894	30	.6361	.7716	.8243	1.213	1.296	1.572	30	.8814
.6923	40	.6383	.7698	.8292	1.206	1.299	1.567	20	.8785
.6952	50	.6406	.7679	.8342	1.199	1.302	1.561	10	.8756
.6981	**40°00′**	.6428	.7660	.8391	1.192	1.305	1.556	**50°00′**	.8727
.7010	10	.6450	.7642	.8441	1.185	1.309	1.550	50	.8698
.7039	20	.6472	.7623	.8491	1.178	1.312	1.545	40	.8668
.7069	30	.6494	.7604	.8541	1.171	1.315	1.540	30	.8639
.7098	40	.6517	.7585	.8591	1.164	1.318	1.535	20	.8610
.7127	50	.6539	.7566	.8642	1.157	1.322	1.529	10	.8581
.7156	**41°00′**	.6561	.7547	.8693	1.150	1.325	1.524	**49°00′**	.8552
.7185	10	.6583	.7528	.8744	1.144	1.328	1.519	50	.8523
.7214	20	.6604	.7509	.8796	1.137	1.332	1.514	40	.8494
.7243	30	.6626	.7490	.8847	1.130	1.335	1.509	30	.8465
.7272	40	.6648	.7470	.8899	1.124	1.339	1.504	20	.8436
.7301	50	.6670	.7451	.8952	1.117	1.342	1.499	10	.8407
.7330	**42°00′**	.6691	.7431	.9004	1.111	1.346	1.494	**48°00′**	.8378
		$\cos t$	$\sin t$	$\cot t$	$\tan t$	$\csc t$	$\sec t$	t degrees	t

Table 4 VALUES OF THE TRIGONOMETRIC FUNCTIONS *(cont'd.)*

t	t degrees	sin t	cos t	tan t	cot t	sec t	csc t		
.7330	**42°00′**	.6691	.7431	.9004	1.111	1.346	1.494	**48°00′**	.8378
.7359	10	.6713	.7412	.9057	1.104	1.349	1.490	50	.8348
.7389	20	.6734	.7392	.9110	1.098	1.353	1.485	40	.8319
.7418	30	.6756	.7373	.9163	1.091	1.356	1.480	30	.8290
.7447	40	.6777	.7353	.9217	1.085	1.360	1.476	20	.8261
.7476	50	.6799	.7333	.9271	1.079	1.364	1.471	10	.8232
.7505	**43°00′**	.6820	.7314	.9325	1.072	1.367	1.466	**47°00′**	.8203
.7534	10	.6841	.7294	.9380	1.066	1.371	1.462	50	.8174
.7563	20	.6862	.7274	.9435	1.060	1.375	1.457	40	.8145
.7592	30	.6884	.7254	.9490	1.054	1.379	1.453	30	.8116
.7621	40	.6905	.7234	.9545	1.048	1.382	1.448	20	.8087
.7650	50	.6926	.7214	.9601	1.042	1.386	1.444	10	.8058
.7679	**44°00′**	.6947	.7193	.9657	1.036	1.390	1.440	**46°00′**	.8029
.7709	10	.6967	.7173	.9713	1.030	1.394	1.435	50	.7999
.7738	20	.6988	.7153	.9770	1.024	1.398	1.431	40	.7970
.7767	30	.7009	.7133	.9827	1.018	1.402	1.427	30	.7941
.7796	40	.7030	.7112	.9884	1.012	1.406	1.423	20	.7912
.7825	50	.7050	.7092	.9942	1.006	1.410	1.418	10	.7883
.7854	**45°00′**	.7071	.7071	1.0000	1.0000	1.414	1.414	**45°00′**	.7854
		cos t	sin t	cot t	tan t	csc t	sec t	t degrees	t

Table 5 TRIGONOMETRIC FUNCTIONS OF RADIANS AND REAL NUMBERS

t	$\sin t$	$\cos t$	$\tan t$	$\cot t$	$\sec t$	$\csc t$
.00	.0000	1.0000	.0000	—	1.000	—
.01	.0100	1.0000	.0100	99.997	1.000	100.00
.02	.0200	.9998	.0200	49.993	1.000	50.00
.03	.0300	.9996	.0300	33.323	1.000	33.34
.04	.0400	.9992	.0400	24.987	1.001	25.01
.05	.0500	.9988	.0500	19.983	1.001	20.01
.06	.0600	.9982	.0601	16.647	1.002	16.68
.07	.0699	.9976	.0701	14.262	1.002	14.30
.08	.0799	.9968	.0802	12.473	1.003	12.51
.09	.0899	.9960	.0902	11.081	1.004	11.13
.10	.9998	.9950	.1003	9.967	1.005	10.02
.11	.1098	.9940	.1104	9.054	1.006	9.109
.12	.1197	.9928	.1206	8.293	1.007	8.353
.13	.1296	.9916	.1307	7.649	1.009	7.714
.14	.1395	.9902	.1409	7.096	1.010	7.166
.15	.1494	.9888	.1511	6.617	1.011	6.692
.16	.1593	.9872	.1614	6.197	1.013	6.277
.17	.1692	.9856	.1717	5.862	1.015	5.911
.18	.1790	.9838	.1820	5.495	1.016	5.586
.19	.1889	.9820	.1923	5.200	1.018	5.295
.20	.1987	.9801	.2027	4.933	1.020	5.033
.21	.2085	.9780	.2131	4.692	1.022	4.797
.22	.2182	.9759	.2236	4.472	1.025	4.582
.23	.2280	.9737	.2341	4.271	1.027	4.386
.24	.2377	.9713	.2447	4.086	1.030	4.207
.25	.2474	.9689	.2553	3.916	1.032	4.042
.26	.2571	.9664	.2660	3.759	1.035	3.890
.27	.2667	.9638	.2768	3.613	1.038	3.749
.28	.2764	.9611	.2876	3.478	1.041	3.619
.29	.2860	.9582	.2984	3.351	1.044	3.497
.30	.2955	.9553	.3093	3.233	1.047	3.384
.31	.3051	.9523	.3203	3.122	1.050	3.278
.32	.3146	.9492	.3314	3.018	1.053	3.179
.33	.3240	.9460	.3425	2.920	1.057	3.086
.34	.3335	.9428	.3537	2.827	1.061	2.999
.35	.3429	.9394	.3650	2.740	1.065	2.916
.36	.3523	.9359	.3764	2.657	1.068	2.839
.37	.3616	.9323	.3879	2.578	1.073	2.765
.38	.3709	.9287	.3994	2.504	1.077	2.696
.39	.3802	.9249	.4111	2.433	1.081	2.630

t	$\sin t$	$\cos t$	$\tan t$	$\cot t$	$\sec t$	$\csc t$
.40	.3894	.9211	.4228	2.365	1.086	2.568
.41	.3986	.9171	.4346	2.301	1.090	2.509
.42	.4078	.9131	.4466	2.239	1.095	2.452
.43	.4169	.9090	.4586	2.180	1.100	2.399
.44	.4259	.9048	.4708	2.124	1.105	2.348
.45	.4350	.9004	.4831	2.070	1.111	2.299
.46	.4439	.8961	.4954	2.018	1.116	2.253
.47	.4529	.8916	.5080	1.969	1.122	2.208
.48	.4618	.8870	.5206	1.921	1.127	2.166
.49	.4706	.8823	.5334	1.875	1.133	2.125
.50	.4794	.8776	.5463	1.830	1.139	2.086
.51	.4882	.8727	.5594	1.788	1.146	2.048
.52	.4969	.8678	.5726	1.747	1.152	2.013
.53	.5055	.8628	.5859	1.707	1.159	1.978
.54	.5141	.8577	.5994	1.668	1.166	1.945
.55	.5227	.8525	.6131	1.631	1.173	1.913
.56	.5312	.8473	.6269	1.595	1.180	1.883
.57	.5396	.8419	.6410	1.560	1.188	1.853
.58	.5480	.8365	.6552	1.526	1.196	1.825
.59	.5564	.8309	.6696	1.494	1.203	1.797
.60	.5646	.8253	.6841	1.462	1.212	1.771
.61	.5729	.8196	.6989	1.431	1.220	1.746
.62	.5810	.8139	.7139	1.401	1.229	1.721
.63	.5891	.8080	.7291	1.372	1.238	1.697
.64	.5972	.8021	.7445	1.343	1.247	1.674
.65	.6052	.7961	.7602	1.315	1.256	1.652
.66	.6131	.7900	.7761	1.288	1.266	1.631
.67	.6210	.7838	.7923	1.262	1.276	1.610
.68	.6288	.7776	.8087	1.237	1.286	1.590
.69	.6365	.7712	.8253	1.212	1.297	1.571
.70	.6442	.7648	.8423	1.187	1.307	1.552
.71	.6518	.7584	.8595	1.163	1.319	1.534
.72	.6594	.7518	.8771	1.140	1.330	1.517
.73	.6669	.7452	.8949	1.117	1.342	1.500
.74	.6743	.7385	.9131	1.095	1.354	1.483
.75	.6816	.7317	.9316	1.073	1.367	1.467
.76	.6889	.7248	.9505	1.052	1.380	1.452
.77	.6961	.7179	.9697	1.031	1.393	1.437
.78	.7033	.7109	.9893	1.011	1.407	1.422
.79	.7104	.7038	1.009	.9908	1.421	1.408

Table 5 TRIGONOMETRIC FUNCTIONS OF RADIANS AND REAL NUMBERS *(cont'd.)*

t	sin *t*	cos *t*	tan t	cot *t*	sec *t*	csc *t*
.80	.7174	.6967	1.030	.9712	1.435	1.394
.81	.7243	.6895	1.050	.9520	1.450	1.381
.82	.7311	.6822	1.072	.9331	1.466	1.368
.83	.7379	.6749	1.093	.9146	1.482	1.355
.84	.7446	.6675	1.116	.8964	1.498	1.343
.85	.7513	.6600	1.138	.8785	1.515	1.331
.86	.7578	.6524	1.162	.8609	1.533	1.320
.87	.7643	.6448	1.185	.8437	1.551	1.308
.88	.7707	.6372	1.210	.8267	1.569	1.297
.89	.7771	.6294	1.235	.8100	1.589	1.287
.90	.7833	.6216	1.260	.7936	1.609	1.277
.91	.7895	.6137	1.286	.7774	1.629	1.267
.92	.7956	.6058	1.313	.7615	1.651	1.257
.93	.8016	.5978	1.341	.7458	1.673	1.247
.94	.8076	.5898	1.369	.7303	1.696	1.238
.95	.8134	.5817	1.398	.7151	1.719	1.229
.96	.8192	.5735	1.428	.7001	1.744	1.221
.97	.8249	.5653	1.459	.6853	1.769	1.212
.98	.8305	.5570	1.491	.6707	1.795	1.204
.99	.8360	.5487	1.524	.6563	1.823	1.196
1.00	.8415	.5403	1.557	.6421	1.851	1.188
1.01	.8468	.5319	1.592	.6281	1.880	1.181
1.02	.8521	.5234	1.628	.6142	1.911	1.174
1.03	.8573	.5148	1.665	.6005	1.942	1.166
1.04	.8624	.5062	1.704	.5870	1.975	1.160
1.05	.8674	.4976	1.743	.5736	2.010	1.153
1.06	.8724	.4889	1.784	.5604	2.046	1.146
1.07	.8772	.4801	1.827	.5473	2.083	1.140
1.08	.8820	.4713	1.871	.5344	2.122	1.134
1.09	.8866	.4625	1.917	.5216	2.162	1.128
1.10	.8912	.4536	1.965	.5090	2.205	1.122
1.11	.8957	.4447	2.014	.4964	2.249	1.116
1.12	.9001	.4357	2.066	.4840	2.295	1.111
1.13	.9044	.4267	2.120	.4718	2.344	1.106
1.14	.9086	.4176	2.176	.4596	2.395	1.101
1.15	.9128	.4085	2.234	.4475	2.448	1.096
1.16	.9168	.3993	2.296	.4356	2.504	1.091
1.17	.9208	.3902	2.360	.4237	2.563	1.086
1.18	.9246	.3809	2.427	.4120	2.625	1.082
1.19	.9284	.3717	2.498	.4003	2.691	1.077

t	sin *t*	cos *t*	tan *t*	cot *t*	sec *t*	csc *t*
1.20	.9320	.3624	2.572	.3888	2.760	1.073
1.21	.9356	.3530	2.650	.3773	2.833	1.069
1.22	.9391	.3436	2.733	.3659	2.910	1.065
1.23	.9425	.3342	2.820	.3546	2.992	1.061
1.24	.9458	.3248	2.912	.3434	3.079	1.057
1.25	.9490	.3153	3.010	.3323	3.171	1.054
1.26	.9521	.3058	3.113	.3212	3.270	1.050
1.27	.9551	.2963	3.224	.3102	3.375	1.047
1.28	.9580	.2867	3.341	.2993	3.488	1.044
1.29	.9608	.2771	3.467	.2884	3.609	1.041
1.30	.9636	.2675	3.602	.2776	3.738	1.038
1.31	.9662	.2579	3.747	.2669	3.878	1.035
1.32	.9687	.2482	3.903	.2562	4.029	1.032
1.33	.9711	.2385	4.072	.2456	4.193	1.030
1.34	.9735	.2288	4.256	.2350	4.372	1.027
1.35	.9757	.2190	4.455	.2245	4.566	1.025
1.36	.9779	.2092	4.673	.2140	4.779	1.023
1.37	.9799	.1994	4.913	.2035	5.014	1.021
1.38	.9819	.1896	5.177	.1931	5.273	1.018
1.39	.9837	.1798	5.471	.1828	5.561	1.017
1.40	.9854	.1700	5.798	.1725	5.883	1.015
1.41	.9871	.1601	6.165	.1622	6.246	1.013
1.42	.9887	.1502	6.581	.1519	6.657	1.011
1.43	.9901	.1403	7.055	.1417	7.126	1.010
1.44	.9915	.1304	7.602	.1315	7.667	1.009
1.45	.9927	.1205	8.238	.1214	8.299	1.007
1.46	.9939	.1106	8.989	.1113	9.044	1.006
1.47	.9949	.1006	9.887	.1011	9.938	1.005
1.48	.9959	.0907	10.983	.0910	11.029	1.004
1.49	.9967	.0807	12.350	.0810	12.390	1.003
1.50	.9975	.0707	14.101	.0709	14.137	1.003
1.51	.9982	.0608	16.428	.0609	16.458	1.002
1.52	.9987	.0508	19.670	.0508	19.695	1.001
1.53	.9992	.0408	24.498	.0408	24.519	1.001
1.54	.9995	.0308	32.461	.0308	32.476	1.000
1.55	.9998	.0208	48.078	.0208	48.089	1.000
1.56	.9999	.0108	92.620	.0108	92.626	1.000
1.57	1.0000	.0008	1255.8	.0008	1255.8	1.000

ANSWERS TO ODD-NUMBERED EXERCISES

CHAPTER 1

Exercises 1.1, page 9

1 Commutative **3** Associative **5** Identity

7 Identity **9** Inverse **11** $\frac{29}{12}$ **13** $\frac{5}{9}$

15 $\frac{11}{12}$ **17** $\frac{13}{10}$ **19** $\frac{10}{21}$

21 Two examples are $2-3 \neq 3-2$ and $2-(3-4) \neq (2-3)-4$.

23 (a) $<$ (b) $>$ (c) $=$ **25** (a) $>$ (b) $<$ (c) $>$

27 $-8 < -5$ **29** $0 > -1$ **31** $x < 0$

33 $3 < a < 5$ **35** $b \geq 2$ **37** $c \leq 1$

39 (a) 5 (b) -5 (c) 13 **41** (a) 0 (b) $4-\pi$ (c) -1

43 (a) 4 (b) 6 (c) 6 (d) 10

45 (a) 12 (b) 3 (c) 3 (d) 9 **47** $x-5$

49 $4+x^2$ **51** Not true; for example, $\frac{1}{2}+\frac{1}{3} \neq 1/(2+3)$

53 If $a=-a$, then $a+a=0$, or $2a=0$. Multiplying both sides by $\frac{1}{2}$ produces $a=0$. Conversely, if $a=0$, then also $(-1)a=(-1)0=0$ and hence $-a=0$. Thus $a=0=-a$.

55 If $ab=0$ and $a \neq 0$, then $(1/a)ab=(1/a)0=0$, and hence $b=0$.

57 $a(b-c)=a[b+(-c)]=ab+a(-c)$
$=ab+[-(ac)]=ab-ac$

59 $\dfrac{-a}{-b}=-\left(\dfrac{a}{-b}\right)=-\left(-\dfrac{a}{b}\right)=\dfrac{a}{b}$

Exercises 1.2, page 19

1 $\frac{16}{81}$ **3** $\frac{9}{8}$ **5** $-\frac{71}{9}$ **7** $\frac{1}{8}$ **9** 0.04

11 $8x^9$ **13** $6/x$ **15** $-2a^{14}$ **17** $\frac{9}{2}$

19 $12u^{11}/v^2$ **21** $4/xy$ **23** $9y^6/x^8$ **25** $81y^6/64$

27 $s^6/(4r^8)$ **29** $20y/x^3$ **31** $9x^{10}y^{14}$ **33** $8a^2$

35 $24x^{3/2}$ **37** $1/(9a^4)$ **39** $8/x^{1/2}$ **41** $4x^2y^4$

43 $3/x^3y^2$ **45** 1 **47** 9 **49** $-2\sqrt[3]{2}$

51 $\sqrt[3]{4}/2$ **53** $3y^3/x^2$ **55** $2a^2/b$

57 $(3a^2/b)\sqrt[3]{2ab}$ **59** $\sqrt{3uv}/3u^2v$ **61** $3x^5/y^2$

63 $(2x/y^2)\sqrt[5]{x^2y^4}$ **65** $-3tv^2$ **67** $(a+\sqrt{ab})/(a-b)$

69 $(t-8\sqrt{t}+16)/(t-16)$

71 $2/(\sqrt{2(x+h)+1}+\sqrt{2x+1})$ **73** 1.7×10^{-24}

75 (a) 4.27×10^5 (b) 9.8×10^{-8} (c) 8.1×10^8

77 (a) 830,000 (b) 0.0000000000029 (c) 563,000,000

79 6.93 **81** 1.125×10^{24} **83** 3.6×10^9

85 5.87×10^{12} miles

Exercises 1.3, page 32

1 $x^4-3x^3+4x^2+4x+5$; 4 **3** $-8x$; 1

5 $6x^3-25x^2+46x-35$; 3

7 $2x^5+2x^4+3x^3+4x^2+3x+10$; 5

9 $15x^4-29x^2y-14y^2$ **11** $36t^2-25v^2$

13 $9r^2+60rs+100s^2$ **15** $16x^4-40x^2y^2+25y^4$

17 $x^3-6x^2y+12xy^2-8y^3$

19 $27r^3+108r^2s+144rs^2+64s^3$

21 $x^6+3x^4y^2+3x^2y^4+y^6$ **23** $(3x-2)(x+4)$

25 $3x(4x-5)(x+3)$ **27** $(5x-3)(5x+3)$

29 $(3x-1)(9x^2+3x+1)$

31 $(x+3)(x-3)(x+2)(x-2)$ **33** $(x+1)^2$

35 x^2+x+1 **37** $(x^2+1)(x+1)$

39 $(x^2+3)(x^2-3)(x^2+1)(x+1)(x-1)$

41 $(x-1)(x^2+x+1)(x+1)(x^2-x+1)$

43 $(2x+3)(2x-3)(x-2)$ **45** $(x+2)/(x+3)$

47 $3y/(5y+1)$ **49** $-(a+2)/(2a+1)$

51 $(2x-3)/x(5x-2)$ **53** $34r/(3r-1)(2r+5)$

55 $6x/(2x-1)(x+1)$ **57** $3/2(4t^2+6t+9)$

59 $a/(a^2+4)(5a+2)$ **61** $(6x-7)/(3x+1)^2$

63 $(3-2c)/c^3$ **65** $(5x^2-2x-6)/(x-1)^3$

67 $(18x^3-10x^2-66x+63)/x^2(2x-3)^2$

69 $(p^2+2p+4)/(p-3)$ **71** $11/(7x-3)(2x+1)^2$

73 $a-b$ **75** $(x^2+xy+y^2)/(x+y)$

77 $(2x^2+7x+15)/(x^2+10x+7)$ **79** $2x+h-7$

81 $-(3x^2+3xh+h^2)/x^3(x+h)^3$
83 $-2/(2x+2h+3)(2x+3)$
85 $(3x+2)^3(36x^2-33x+6)$
87 $(2x+1)^2(8x^2+x-24)/(x^2-4)^{1/2}$
89 $(3x+1)^5(39x-89)/(2x-5)^{1/2}$
91 $(27x^2-24x+2)/(6x+1)^4$
93 $(x^2+12)/(x^2+4)^{4/3}$ **95** $6(3-2x)/(4x^2+9)^{3/2}$

Exercises 1.4, page 43

1 $\frac{5}{2}$ **3** $\frac{9}{38}$ **5** $\frac{2}{3}, -\frac{5}{2}$ **7** $-\frac{5}{4}, -6$ **9** $0, \frac{5}{3}$
11 $-3, -\frac{3}{2}, \frac{3}{2}$ **13** No solution **15** $\frac{5}{9}$ **17** $-\frac{57}{5}$
19 $(2, 5)$ **21** $(-1, 3]$ **23** $[1, 4]$
25 $(-1, \infty)$ **27** $(-\infty, 2]$ **29** $-1 < x < 7$
31 $8 < x \le 9$ **33** $5 < x$ **35** $(\frac{17}{5}, \infty)$
37 $[-2, \infty)$ **39** $(\frac{5}{2}, \infty)$ **41** $(-\infty, \frac{44}{3}]$
43 $(-3, 1)$ **45** $[1, 5]$ **47** $[2, \frac{8}{3})$ **49** $(\frac{3}{19}, \infty)$
51 $(-\frac{2}{3}, \infty)$ **53** $(-\frac{7}{3}, \infty)$ **55** $\frac{20}{9} \le x \le 4$
57 88 **59** 64 sec; 96 m and 128 m, respectively
61 After an additional 50 games **63** $\frac{20}{3}$ liters
65 400 mi **67** (a) $\frac{5}{9}$ mph (b) $2\frac{2}{9}$ m
69 (a) $d = \sqrt{(400t)^2 + (200t + 100)^2} = 100\sqrt{20t^2 + 4t + 1}$
(b) 3:30 P.M.

Exercises 1.5, page 51

1 $(-2, 2)$ **3** $(6, \infty) \cup (-\infty, -6)$ **5** $[-10, 10]$
7 $(9.95, 10.05)$ **9** $[-10, 2]$ **11** $(-\frac{2}{3}, 4)$
13 $(-\infty, -6) \cup (2, \infty)$ **15** $(-\infty, 1] \cup [4, \infty)$
17 $(-2, 3)$ **19** $(-\infty, -\frac{5}{2}) \cup (1, \infty)$ **21** $[0, 10]$
23 $(-\infty, 4)$ **25** $(-\frac{7}{2}, \infty)$ **27** $(-\infty, -\frac{1}{2}) \cup (\frac{10}{3}, \infty)$
29 $(-3, 3)$ **31** $(4, \infty) \cup (-4, 0]$
33 $[-5, -\frac{3}{2}) \cup (\frac{3}{2}, 5]$ **35** $(-\infty, \frac{1}{2}) \cup (\frac{7}{5}, \infty)$
37 $(-\infty, -1) \cup (0, 1)$ **39** $[-2, 1] \cup [2, \infty)$
41 $(-\infty, -1] \cup (1, 2] \cup (3, \infty)$
43 $(a - \delta, a) \cup (a, a + \delta)$ **45** $[\frac{1}{2}, 4]$

Exercises 1.6, page 53

1 (a) $-\frac{5}{12}$ (b) $\frac{39}{20}$ (c) $-\frac{13}{56}$ (d) $\frac{5}{8}$
3 (a) $x < 0$ (b) $\frac{1}{3} < a < \frac{1}{2}$ (c) $|x| \le 4$
5 (a) 5 (b) 5 (c) 7 **7** $18a^5b^5$ **9** $xy^5/9$
11 $-p^8/2q$ **13** x^3z/y^{10} **15** b^6/a^2 **17** $s + r$
19 x^8/y^2 **21** $\sqrt[3]{2}/2$ **23** $2x^2y\sqrt[3]{x}$ **25** $(1 - \sqrt{t})/t$
27 $2x/y^2$ **29** $(1 - 2\sqrt{x} + x)/(1 - x)$
31 (a) 9.37×10^{10} (b) 4.02×10^{-6}
33 $-x^3 - 3x^2 + 6x - 9$; 3 **35** $x^3 - 8x^2 + 19x - 20$; 3
37 $4x^4 - 12x^2y + 9y^2$ **39** $(4x - 1)(2x + 3)$
41 $x(3x - 2)(x^2 + 4)$ **43** $2(x^2 + 4)(x + 2)(x - 2)$

45 $(3x-5)/(2x+1)$ **47** $(3x+2)/x(x-2)$
49 $(5x^2-6x-20)/x(x+2)^2$
51 $-(2x^2+x+3)/x(x+1)(x+3)$
53 $(x^2+1)^{1/2}(x+5)^3(7x^2+15x+4)$ **55** $-\frac{5}{6}$
57 $-\frac{1}{4}, \frac{2}{3}$ **59** $-\frac{5}{2}, 4$ **61** $\frac{13}{4}$
63 The interval $(-\frac{11}{4}, \frac{9}{4})$ **65** $(-\infty, -\frac{3}{16})$
67 $(-\infty, \frac{11}{3}] \cup [7, \infty)$ **69** $(-\infty, -\frac{3}{2}) \cup (\frac{2}{5}, \infty)$
71 $(-\infty, -\frac{3}{2}) \cup (2, 9)$ **73** $(1, \infty)$ **75** 64 mph
77 16 ml **79** 9 in., 11 in.

CHAPTER 2

Exercises 2.1, page 64

1 $5, (4, -\frac{1}{2})$ **3** $\sqrt{26}, (-\frac{1}{2}, -\frac{9}{2})$ **5** $5, (-\frac{11}{2}, -2)$
7 35 **9** $d(A, D) = d(B, C)$ and $d(A, B) = d(C, D)$
11 $a > 4$ or $a < \frac{2}{5}$
15 **17** **19**

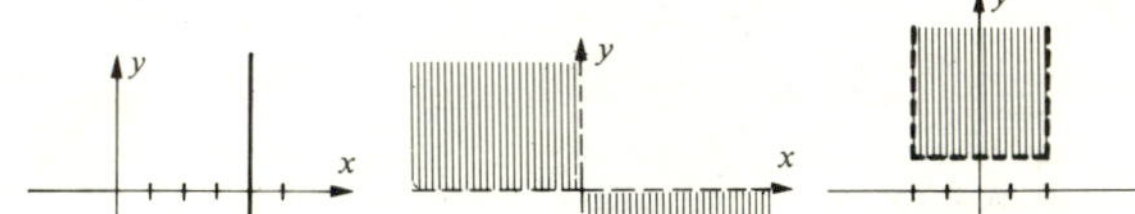

21 **23** **25** Symmetry: y-axis

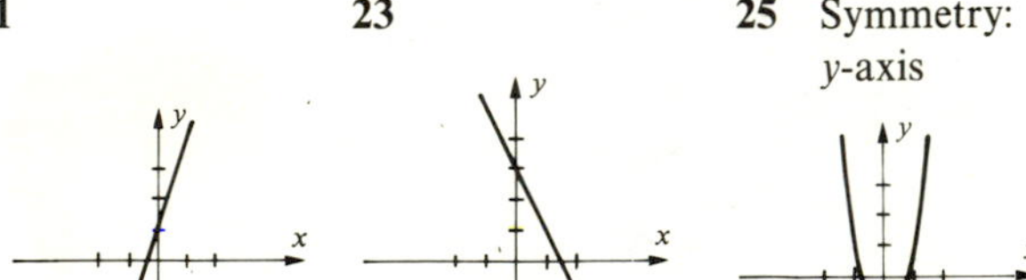

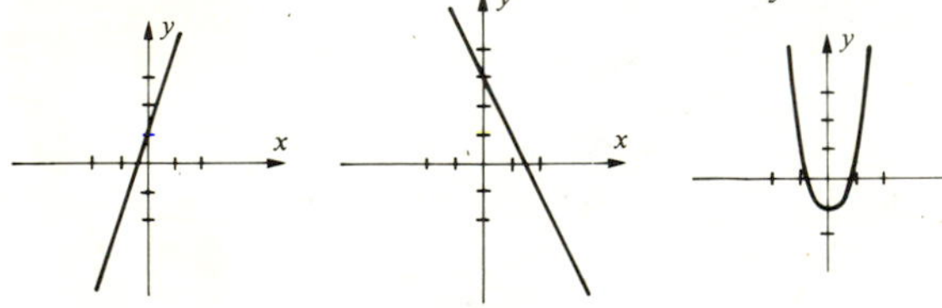

27 Symmetry: y-axis **29** Symmetry: origin **31**

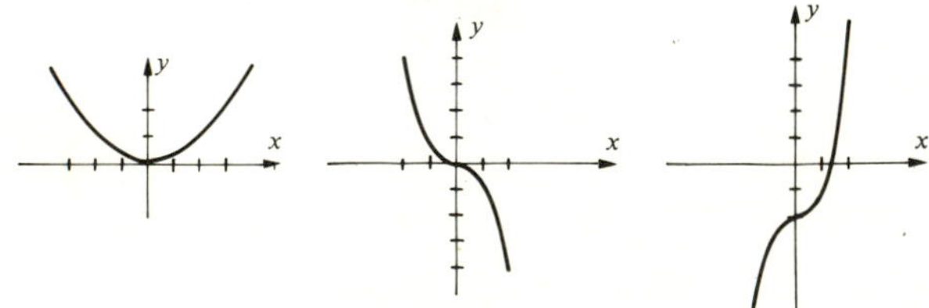

33 **35**

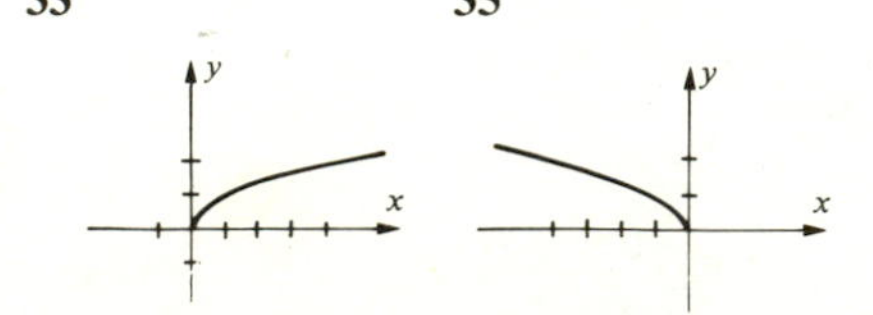

37 Symmetry: x-axis, y-axis, origin **39** Symmetry: y-axis **41** Symmetry: x-axis

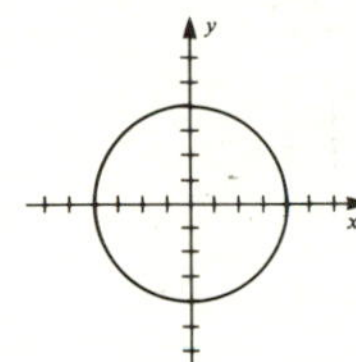

43 $(x-3)^2+(y+2)^2=16$ **45** $x^2+y^2=34$
47 $(x+4)^2+(y-2)^2=4$ **49** $(x-1)^2+(y-2)^2=34$
51 $(-2, 3), 3$ **53** $(-3, 0), 3$ **55** $(\frac{1}{4}, -\frac{1}{4}), \sqrt{26}/4$

Calculator Exercises 2.1, page 65

1

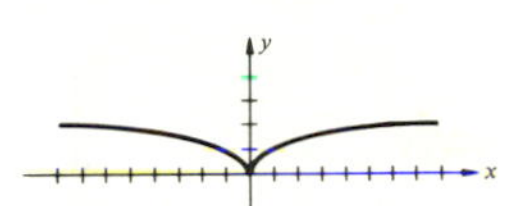

3

5

7

Exercises 2.2 page 72

1 4 **3** The slope does not exist.
5 The slopes of opposite sides are equal.
7 *Hint:* Show that opposite sides are parallel and that two adjacent sides are perpendicular.
9 $(-12, 0)$ **11** $x-2y-14=0$
13 $3x-8y-41=0$ **15** $x-8y-24=0$
17 (a) $x=10$ (b) $y=-6$ **19** $5x+2y-29=0$
21 $5x-7y+15=0$
23 $x+6y-9=0$; $3x-5y+5=0$; $4x+y-4=0$; $(\frac{15}{23}, \frac{32}{23})$
25 $m=\frac{3}{4}, b=2$ **27** $m=-\frac{1}{2}, b=0$

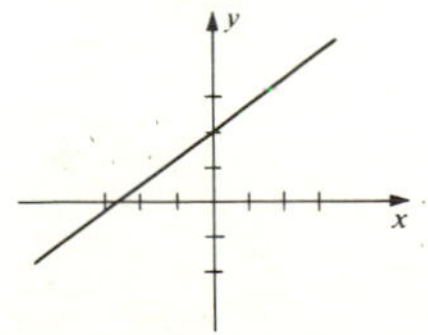

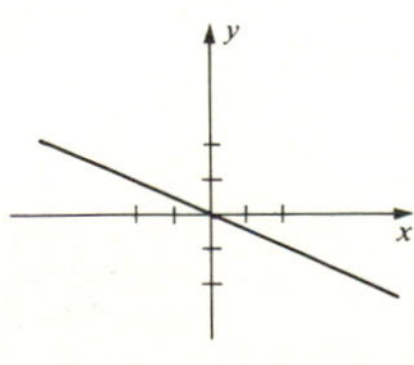

29 $m=0, b=4$ **31** $m=-\frac{5}{4}, b=5$

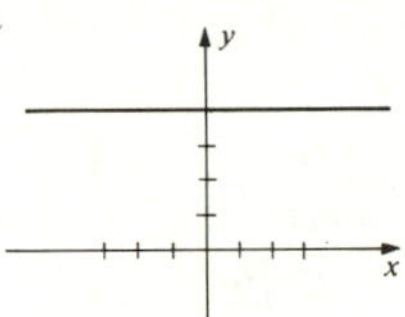

33 $m=\frac{1}{3}, b=-\frac{7}{3}$ **35** $k=-3$

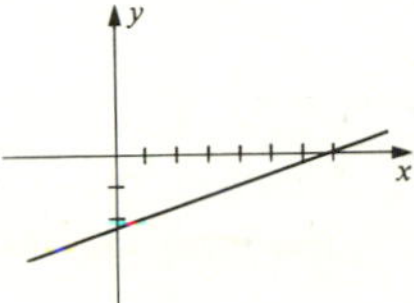

37 $x/\frac{3}{2}+y/(-3)=1$ **39** $r<1$ or $r>2$
41 $y=59{,}000+6000x$, where y is the value of the house and x is the number of years after the purchase date; 2 years and 4 months after the purchase date.

Exercises 2.3, page 79

1 3, 9, 4, 6 **3** $2, \sqrt{2}+6, 12, 23$
5 (a) $5a-2$ (b) $-5a-2$ (c) $-5a+2$ (d) $5a+5h-2$ (e) $5a+5h-4$ (f) 5
7 (a) $2a^2-a+3$ (b) $2a^2+a+3$ (c) $-2a^2+a-3$ (d) $2a^2+4ah+2h^2-a-h+3$ (e) $2a^2-a+2h^2-h+6$ (f) $4a+2h-1$
9 (a) $3/a^2$ (b) $1/3a^2$ (c) $3a^4$ (d) $9a^4$ (e) $3a$ (f) $\sqrt{3a^2}$
11 (a) $2a/(a^2+1)$ (b) $(a^2+1)/2a$ (c) $2a^2/(a^4+1)$ (d) $4a^2/(a^4+2a^2+1)$ (e) $2\sqrt{a}/(a+1)$ (f) $\sqrt{2a^3+2a}/(a^2+1)$
13 $[\frac{5}{3}, \infty)$ **15** $[-2, 2]$
17 All real numbers except 0, 3, and -3
19 All nonnegative real numbers except 4 and $\frac{3}{2}$
21 $\frac{9}{7}$; $(a+5)/7$; $\mathbb{R}$
23 19; a^2+3; all nonnegative real numbers
25 $\sqrt[3]{4}$; $\sqrt[3]{a}$; $\mathbb{R}$ **27** Yes **29** No **31** Yes
33 No **35** Odd **37** Even **39** Even
41 Neither **43** Neither
45 The equation $y=f(x)$ is not changed if $-x$ is substituted for x.
47 $r=C/2\pi$; $6/\pi \approx 1.9$ in. **49** $P=4\sqrt{A}$
51 $V=4x^3-100x^2+600x$ **53** $d=2\sqrt{t^2+2500}$
55 $A=20x$ if $x<50$ and $A=20x-0.02x^2$ if $50 \le x \le 600$
57 (a) $V=\frac{2}{3}\pi x^2(15+2x)$ (b) $S=4\pi x(5+x)$
59 (a) $h=3(4-r)$ (b) $V=3\pi r^2(4-r)$

Exercises 2.4, page 89

In Exercises 1–19, X denotes the domain of f and Y denotes the range.

1 $X = \mathbb{R}$, $Y = \mathbb{R}$; increasing on $\mathbb{R}$

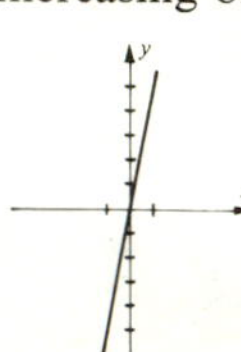

3 $X = \mathbb{R}$, $Y = \mathbb{R}$; decreasing on $\mathbb{R}$

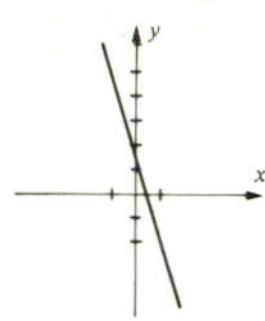

5 $X = \mathbb{R}$, $Y = (-\infty, 3]$; increasing on $(-\infty, 0]$, decreasing on $[0, \infty)$

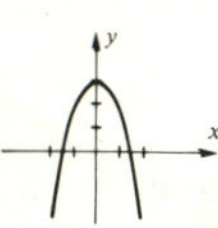

7 $X = \mathbb{R}$, $Y = [-4, \infty)$; decreasing on $(-\infty, 0]$, increasing on $[0, \infty)$

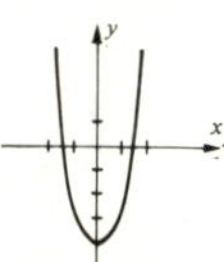

9 $X = [-4, \infty), Y = [0, \infty)$; increasing on $[-4, \infty)$

11 $X = [0, \infty)$, $Y = [2, \infty)$; increasing on $[0, \infty)$

13 $X = (-\infty, 0) \cup (0, \infty) = Y$; decreasing on $(-\infty, 0)$ and on $(0, \infty)$

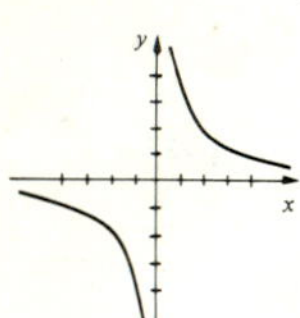

15 $X = \mathbb{R}$, $Y = [0, \infty)$; decreasing on $(-\infty, 2]$, increasing on $[2, \infty)$

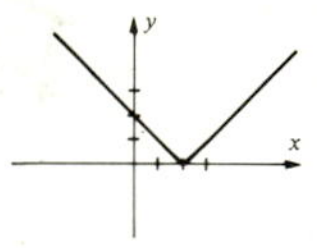

17 $X = \mathbb{R}$, $Y = [-2, \infty)$; decreasing on $(-\infty, 0]$, increasing on $[0, \infty)$

19 $X = (-\infty, 0) \cup (0, \infty)$, $Y = \{-1, 1\}$; Neither increasing nor decreasing

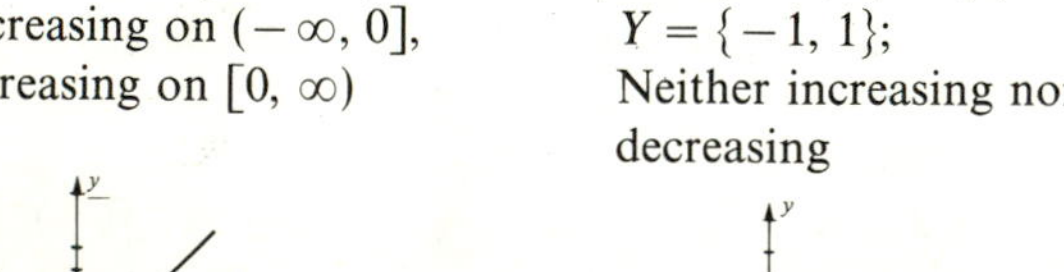

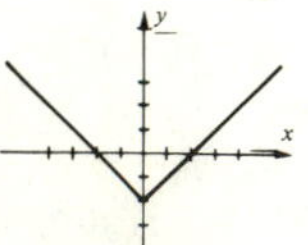

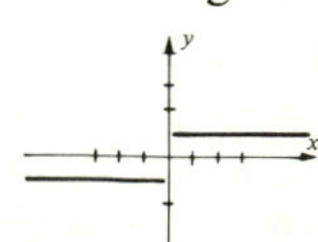

21

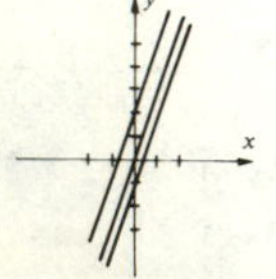

23

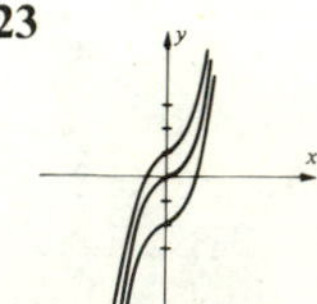

25

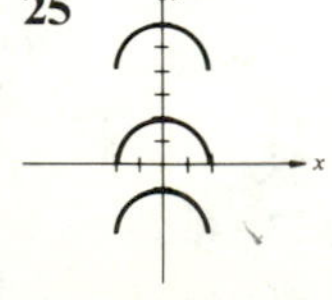

27

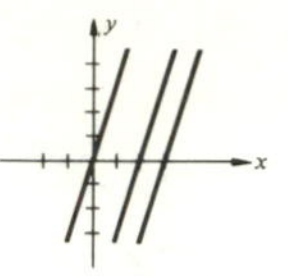

29

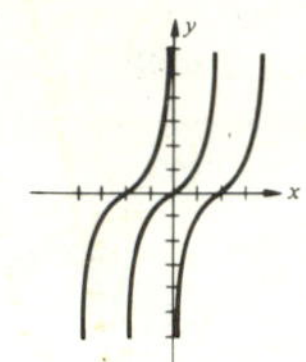

31 (a) (b)

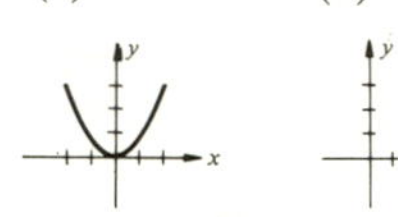

(c) (d)

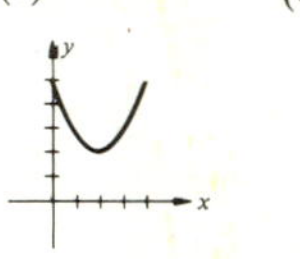

(e)

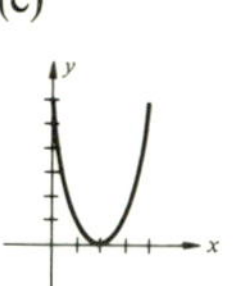

(f)

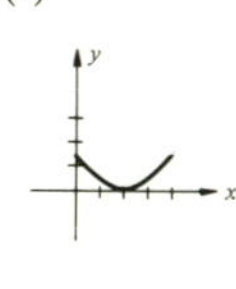

(g) (h)

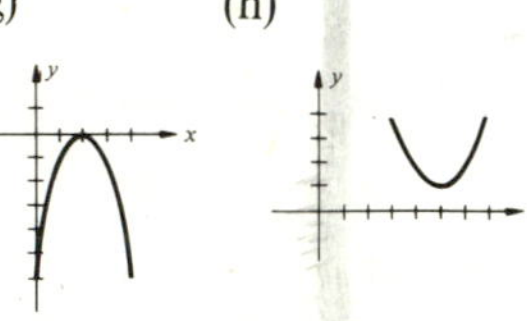

33

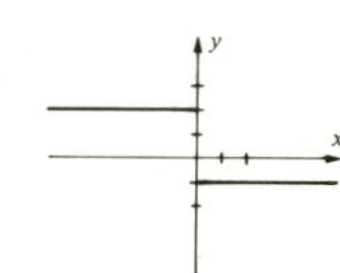

35

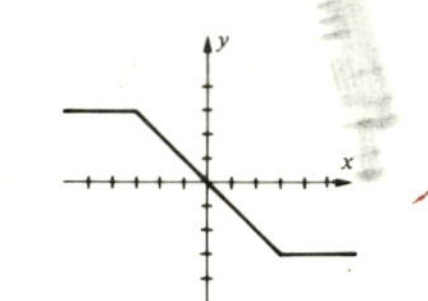

37

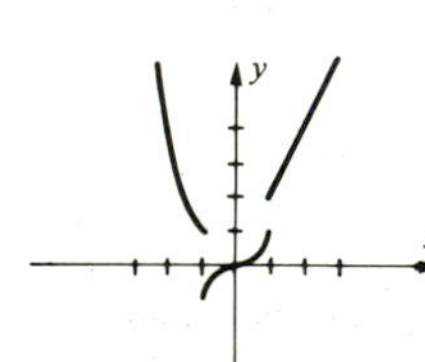

39

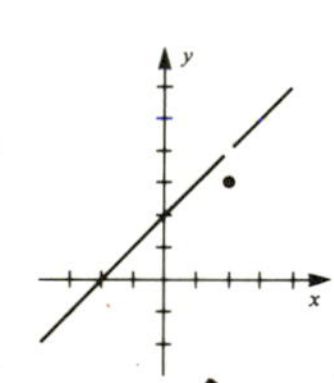

41 (a)

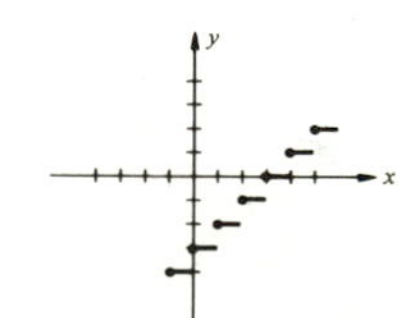

(b) (c)

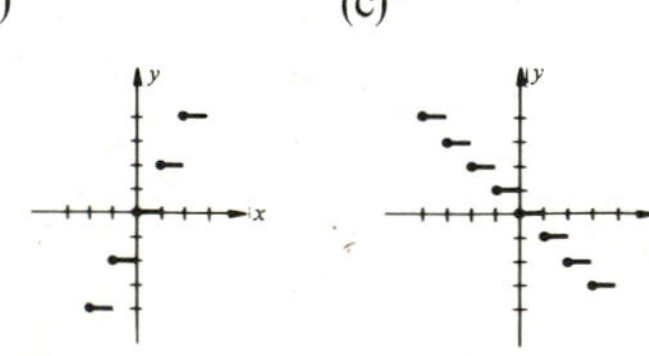

43

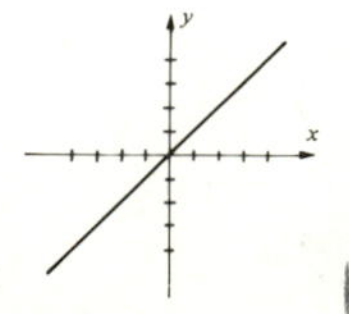

45 $f(x) = -\sqrt{1 - x^2}$

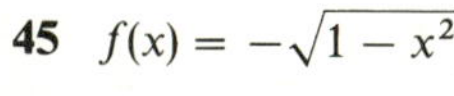

47 If $a < b$, then $f(a) > f(b)$; and if $b < a$, then $f(b) > f(a)$. Hence, if $a \neq b$, then $f(a) \neq f(b)$.

Exercises 2.5, page 95

1 (a)

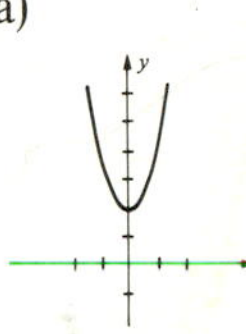

(b)

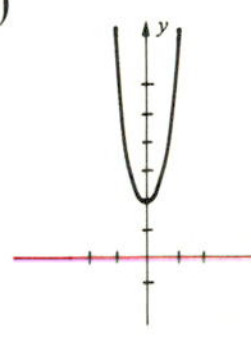

(c)

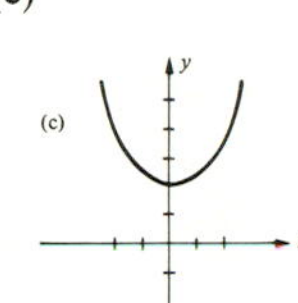

(d)

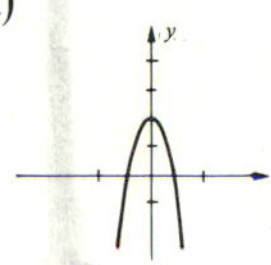

3 Vertex: $(0, -9)$ **5** Vertex: $(0, 9)$

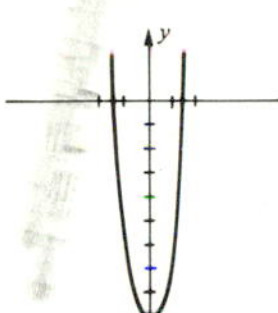

7 $3, -\frac{1}{4}$ **9** $\frac{2}{3}$ **11** $(-1 \pm \sqrt{41})/4$

13 $(-9 \pm \sqrt{85})/2$

15 $2(x-4)^2 - 9$; minimum: $f(4) = -9$

17 $-5(x+1)^2 + 8$; maximum: $f(-1) = 8$

19 $4(x - \frac{1}{2})^2 + 1$; minimum: $f(\frac{1}{2}) = 1$

21 min: $f(-\frac{5}{2}) = -\frac{9}{4}$ **23** max: $f(4) = 4$

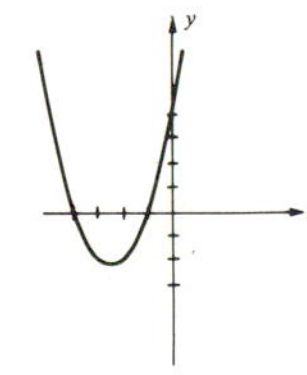

25 min: $f(-\frac{1}{2}) = \frac{11}{4}$ **27** min: $f(2) = 4$

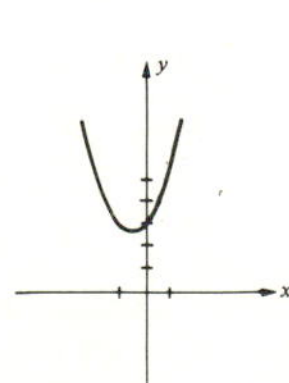

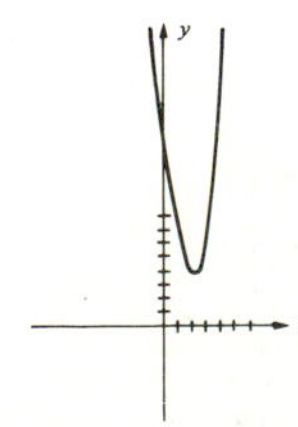

29 max: $f(-1) = 1$ **31** 10, 2 **33** $-\frac{1}{2}$

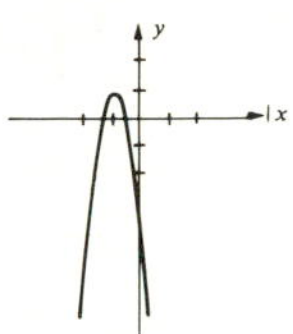

35 20, -20

39 125 yards by 250 yards, with intermediate fences parallel to the short side

Exercises 2.6, page 102

1 $6x - 1$, $6x + 3$ **3** $36x^2 - 5$, $12x^2 - 15$

5 $12x^2 - 8x + 1$, $6x^2 + 4x - 1$

7 $x^3 - 1$, $x^3 - 3x^2 + 3x - 1$

9 $x + 9 + 9\sqrt{x+9}$, $\sqrt{x^2 + 9x + 9}$

11 $x^2/(2 - 5x^2)$, $4x^2 - 20x + 25$ **13** 5, -5

15 $1/x^4$, $1/x^4$ **17** x, x

19 and **21** Show that $f(g(x)) = x = g(f(x))$.

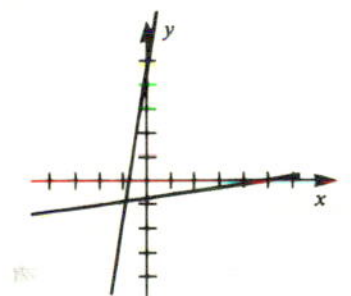

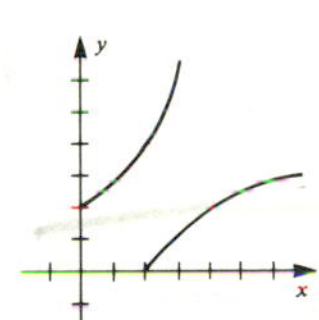

23 $f^{-1}(x) = (x + 3)/4$ **25** $f^{-1}(x) = (1 - 5x)/2x$, $x > 0$

27 $f^{-1}(x) = \sqrt{9 - x}$, $x \le 9$ **29** $f^{-1}(x) = \sqrt[3]{(x + 2)/5}$

31 $f^{-1}(x) = (x^2 + 5)/3$, $x \ge 0$ **33** $f^{-1}(x) = (x - 8)^3$

35 $f^{-1}(x) = x$

37 (a) $f^{-1}(x) = (x - b)/a$ (b) No (not one-to-one)

39 If g and h are both inverse functions of f, then $f(g(x)) = x = f(h(x))$ for all x. Since f is one-to-one this implies that $g(x) = h(x)$ for all x, that is, $g = h$.

Exercises 2.7, page 104

1 Area = 10 **3** The points in quadrants II and IV

7 $(x + 5)^2 + (y + 1)^2 = 81$

9 (a) $18x + 6y - 7 = 0$ (b) $2x - 6y - 3 = 0$

11 **13** **15**

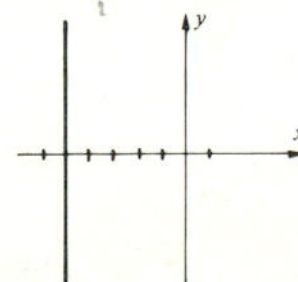

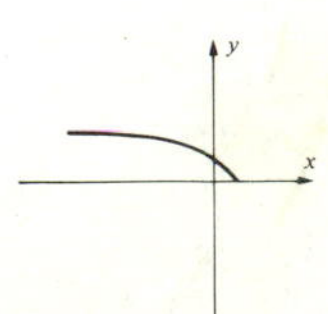

17 **19** **21**

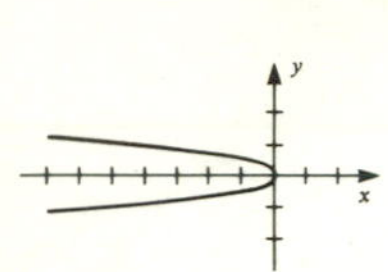

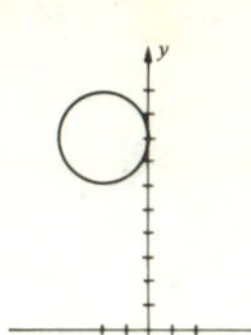

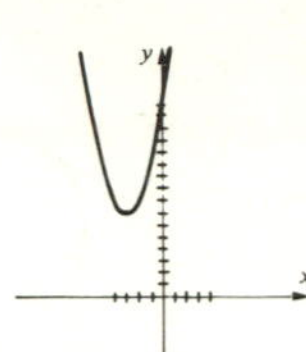

23 (a) $\frac{1}{2}$ (b) $-\sqrt{2}/2$ (c) 0 (d) $-x/\sqrt{3-x}$
(e) $-x/\sqrt{x+3}$ (f) $x^2/\sqrt{x^2+3}$ (g) $x^2/(x+3)$

25 $X = \mathbb{R}$, $Y = \mathbb{R}$; decreasing on $\mathbb{R}$

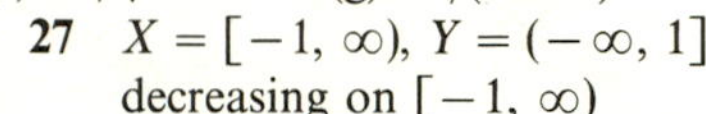

27 $X = [-1, \infty)$, $Y = (-\infty, 1]$ decreasing on $[-1, \infty)$

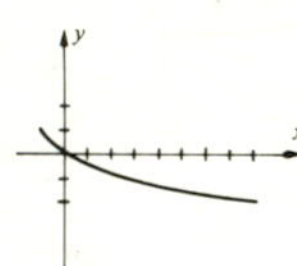

29 $X = \mathbb{R}$, $Y = \{1{,}000\}$

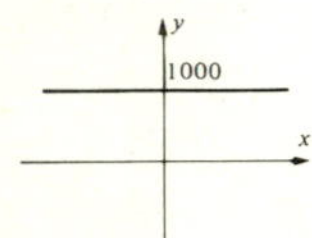

31 (a)

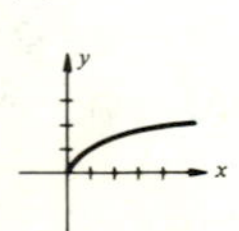

(b)

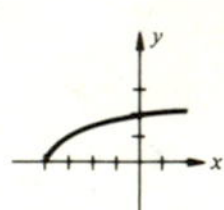

(c)

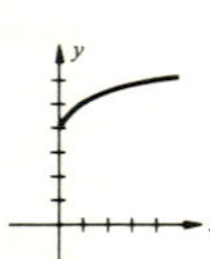

(d)

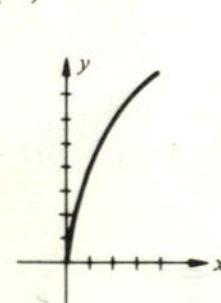

(e)

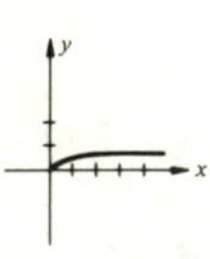

(f)

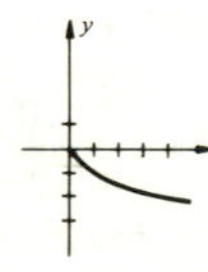

33 $(2 \pm \sqrt{14})/2$

35 $f(x) = 5(x+3)^2 + 4$; min: $f(-3) = 4$

37 Radius of semicircle is $1/8\pi$ mi; length of rectangle is $\frac{1}{8}$ mi.

39 $18x^2 + 9x - 1$; $6x^2 - 15x + 5$

41 $f^{-1}(x) = (10 - x)/15$ **43** $V = C^3/8\pi^2$

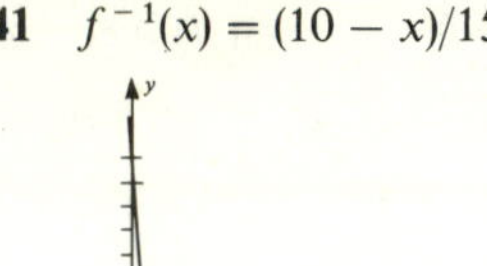

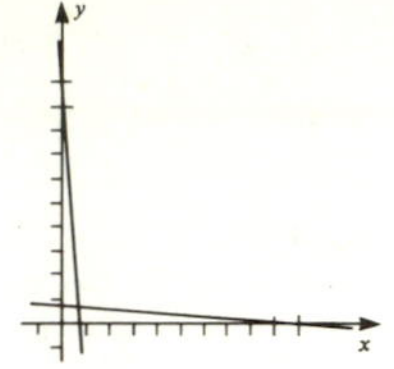

CHAPTER 3

Exercises 3.1, page 111

1 (a)

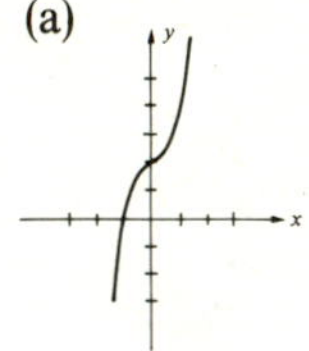

(b)

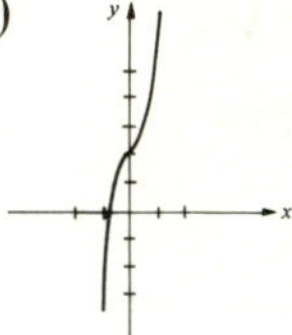

(c)

(d)

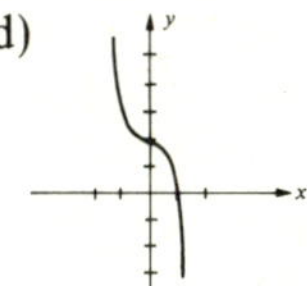

3 $f(x) > 0$ if $x > 2$
$f(x) < 0$ if $x < 2$

5 $f(x) > 0$ for all x

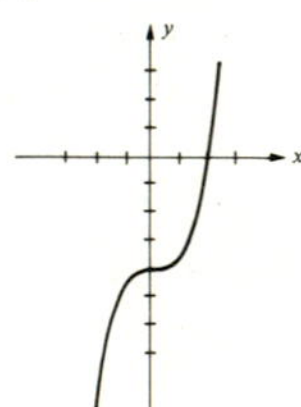

7 $f(x) > 0$ if $-3 < x < 0$ or $x > 3$
$f(x) < 0$ if $x < -3$ or $0 < x < 3$

9 $f(x) > 0$ if $x < -2$ or $0 < x < 1$
$f(x) < 0$ if $-2 < x < 0$ or $x > 1$

7

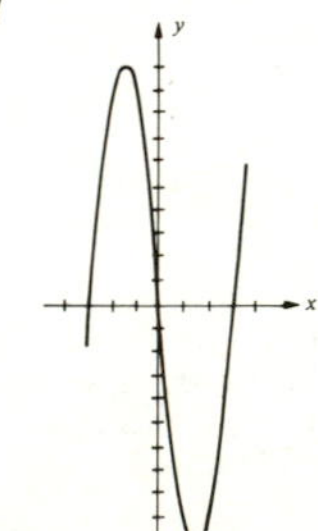

9

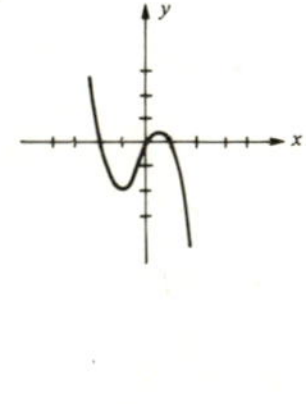

11 $f(x) > 0$ if $-4 < x < 1$ or $x > 5$
$f(x) < 0$ if $x < -4$ or $1 < x < 5$

13 $f(x) > 0$ if $x < -2$ or $x > 2$
$f(x) < 0$ if $-2 < x < 2$

11

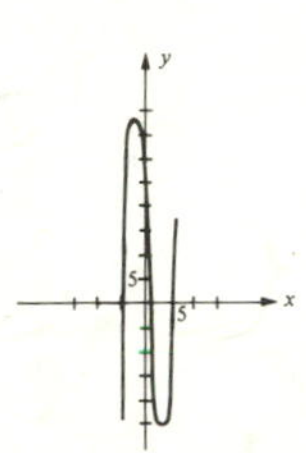

13

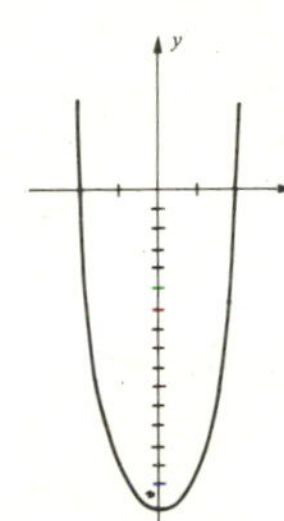

15 $f(x) > 0$ if $-1 < x < 1$
$f(x) < 0$ if $x < -1$ or $x > 1$

17 $f(x) > 0$ if $x < -3$, $-1 < x < 0$, or $x > 2$
$f(x) < 0$ if $-3 < x < -1$ or $0 < x < 2$

15

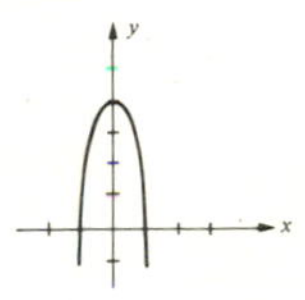

17

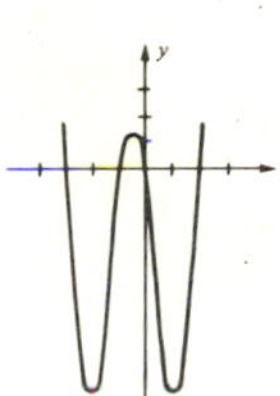

19 $k = -\frac{4}{3}$

Calculator Exercises 3.1, page 112

In Exercises 3–9 there is a zero between the indicated numbers.

3 2.51 and 2.52 **5** -1.10 and -1.09
7 1.331 and 1.332 **9** -1.733 and -1.732

Exercises 3.2, page 116

1 $x^2 + x + 2,\ -2x + 13$ **3** $\frac{5}{2}x,\ -\frac{9}{2}x$
5 $0,\ 7x^3 - 5x + 2$ **7** 95 **9** -23 **11** 0
13 $k = \frac{17}{12}$ **15** $f(2) = 0$ **17** $f(c) > 0$
19 If $f(x) = x^n - y^n$, then $f(y) = 0$. If n is even, then $f(-y) = 0$.

Exercises 3.3, page 120

1 $2x^2 + x + 6,\ 7$ **3** $x^2 - 3x + 1,\ -8$
5 $3x^4 - 6x^3 + 12x^2 - 18x + 36,\ -65$
7 $4x^3 + 2x^2 - 4x - 2,\ 0$
9 $x^{n-1} + x^{n-2} + \cdots + x + 1,\ 0$ **11** $-23,\ 53$
13 3277 **15** $8 + 7\sqrt{3}$

Exercises 3.4, page 127

1 $-2 + 6i$ **3** $-2 + 4i$ **5** $7 - 5i$ **7** $4 + 7i$
9 $-6 + 3i$ **11** $-10 + 5i$ **13** $20 - 10i$
15 25 **17** $-72 - 36i$ **19** $32 - 44i$ **21** 15
23 $\frac{3}{13} - \frac{2}{13}i$ **25** $\frac{35}{61} + \frac{42}{61}i$ **27** $-\frac{1}{5} - \frac{11}{10}i$
29 $-\frac{7}{13} + \frac{17}{13}i$ **31** $-7 - 21i$ **33** $\frac{12}{17} - \frac{54}{17}i$
35 $(3 \pm \sqrt{31}i)/2$ **37** $-1 \pm 2i$ **39** $(-1 \pm \sqrt{47}i)/8$
41 $5, (-5 \pm 5\sqrt{3}i)/2$ **43** $2,\ -2,\ -1 \pm \sqrt{3}i,\ 1 \pm \sqrt{3}i$
45 $\pm 2i,\ \pm\frac{3}{2}i$ **47** $0, (-3 \pm \sqrt{7}i)/2$

Exercises 3.5, page 134

1 $-\frac{1}{15}x^3 + \frac{19}{15}x + 2$ **3** $-2x^3 + 12x^2 - 22x + 12$
5 $\frac{3}{10}x^3 - \frac{33}{10}x + 6$ **7** $x^4 - 2x^3 - 11x^2 + 12x + 36$
9 $\frac{2}{9}x^8 - \frac{4}{3}x^7 + \frac{8}{3}x^6 - \frac{16}{9}x^5$
11 -4 (multiplicity 3); $\frac{4}{3}$ (multiplicity 1)
13 0 (multiplicity 3); 5, -1 (each of multiplicity 1)
15 $\pm\frac{5}{3}$ (each of multiplicity 4); $\pm 4i$ (each of multiplicity 1)
17 -2 (multiplicity 3); 1 (multiplicity 2); 2 (multiplicity 1)
19 $(x + 3)^2(x + 2)(x - 1)$ **21** $(x - 1)^5(x + 1)$

In Exercises 23–29 the types of possible solutions are listed in the following order: positive, negative, nonreal complex.

23 Either 3, 0, 0 or 1, 0, 2 **25** 0, 1, 2
27 Either 2, 0, 2; 2, 2, 0; 0, 2, 2; or 0, 0, 4
29 Either 2, 3, 0; 2, 1, 2; 0, 3, 2; or 0, 1, 4
31 Upper 5, lower -2 **33** Upper 2, lower -2
35 Upper 3, lower -3

Exercises 3.6, page 141

1 $x^2 - 8x + 17$ **3** $x^3 - 10x^2 + 37x - 52$
5 $x^4 - 6x^3 + 30x^2 - 78x + 85$
7 $x^5 - 2x^4 + 27x^3 - 50x^2 + 50x$
9 No. If i is a root, then $-i$ is also a root. Hence, the polynomial would have factors $x - 1$, $x + 1$, $x - i$, $x + i$ and, therefore, would be of degree greater than 3.
11 $-1, -2, 4$ **13** $2, -3, \frac{5}{2}$ **15** $4, -7, \pm\sqrt{2}$
17 $4, -2, \frac{3}{2}$ **19** $\frac{1}{2}, -\frac{2}{3}, -3$ **21** $-\frac{3}{4}, (-3 \pm 3\sqrt{7}i)/4$
23 $4, -3, \pm\sqrt{3}i$ **25** $1, -2, -\frac{4}{3}, \frac{2}{3}$
27 $3, -4, -2, \frac{1}{2}, -\frac{1}{2}$
37 Complex zeros occur in conjugate pairs.

Exercises 3.7, page 151

1

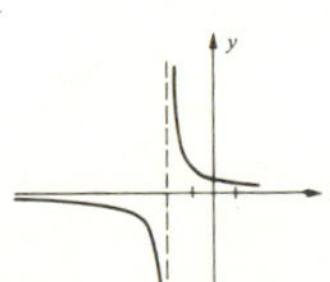

3

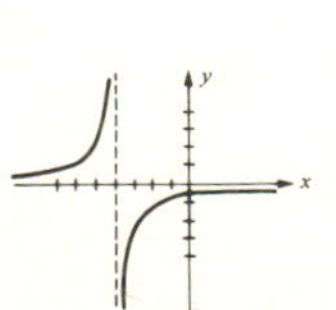

5

7 9 11

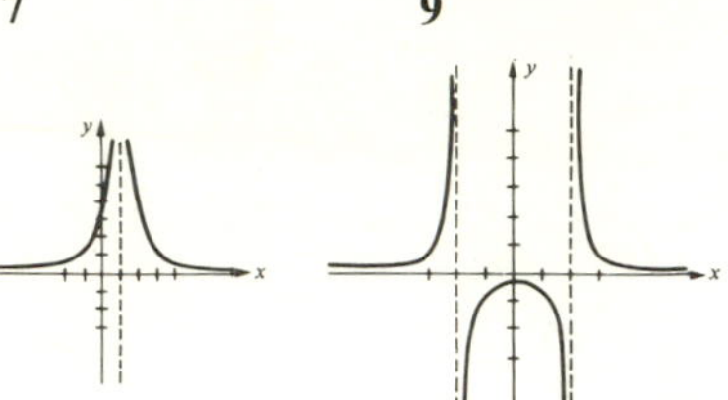

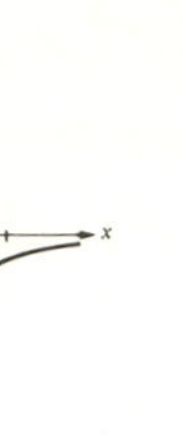

13 15 17

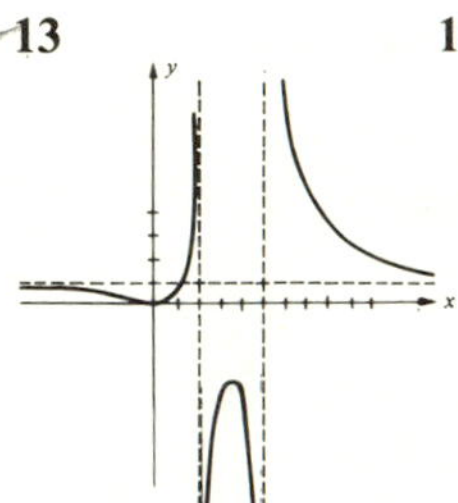

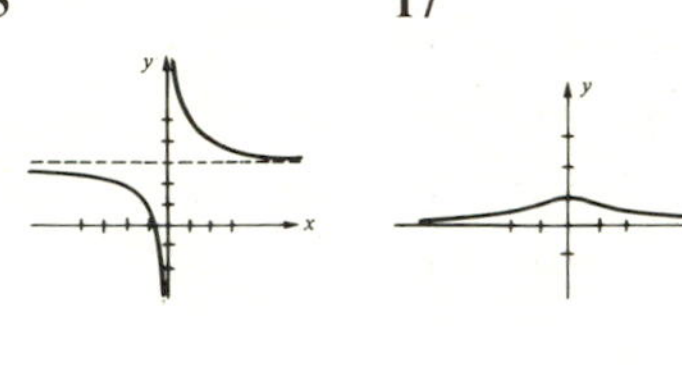

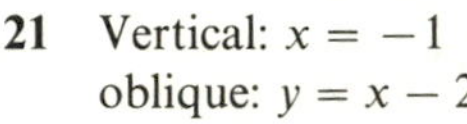

19

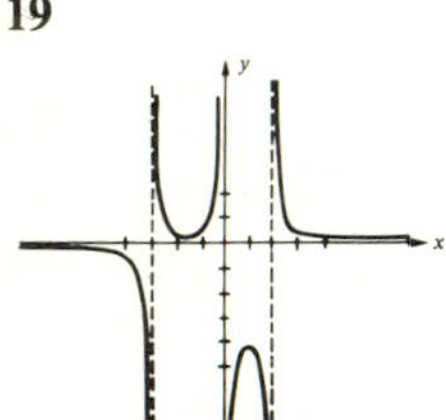

21 Vertical: $x = -1$
oblique: $y = x - 2$

23 Vertical: $x = 0$
oblique: $y = -\frac{1}{2}x$

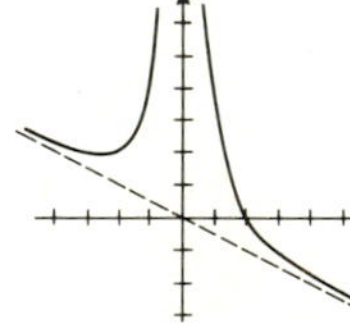

Exercises 3.8, page 157

1 $3/(x-2) + 5/(x+3)$ 3 $5/(x-6) - 4/(x+2)$
5 $2/(x-1) + 3/(x+2) - 1/(x-3)$
7 $(3/x) + 2/(x-5) - 1/(x+1)$
9 $2/(x-1) + 5/(x-1)^2$
11 $(-7/x) + (5/x^2) + 40/(3x-5)$
13 $-\frac{23}{25}/(2x-1) + \frac{24}{25}/(x+2) + \frac{2}{5}/(x+2)^2$
15 $(5/x) - 2/(x+1) + 3/(x+1)^3$
17 $-2/(x-1) + (3x+4)/(x^2+1)$
19 $(4/x) + (5x-3)/(x^2+2)$
21 $(4x-1)/(x^2+1) + 3/(x^2+1)^2$
23 $2x + 3x/(x^2+1) + 1/(x-1)$
25 $(2x+3) + 2/(x-1) - 3/(2x+1)$

Exercises 3.9, page 158

1 3 5

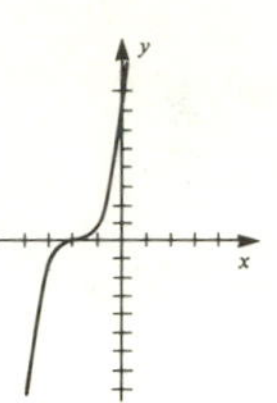

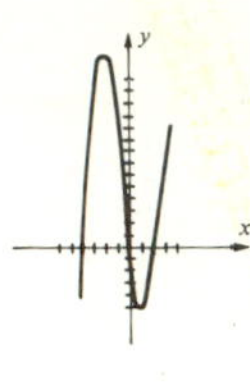

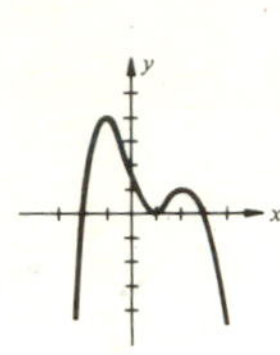

7 Vertical: $x = -1$
horizontal: $y = 0$

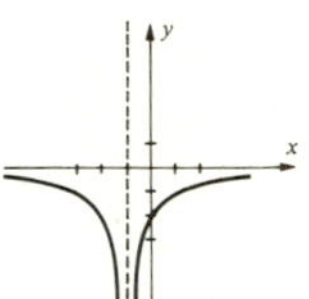

9 Vertical: $x = 4$, $x = -4$
horizontal: $y = -3$

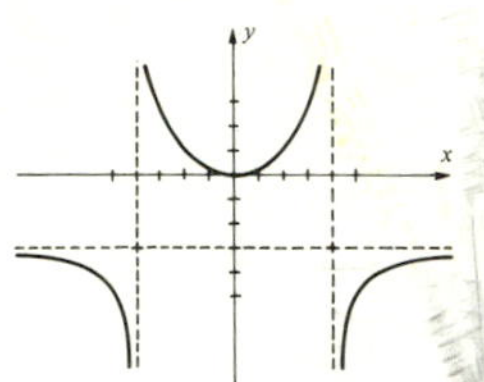

11 Vertical: $x = -3$
oblique: $y = x - 1$

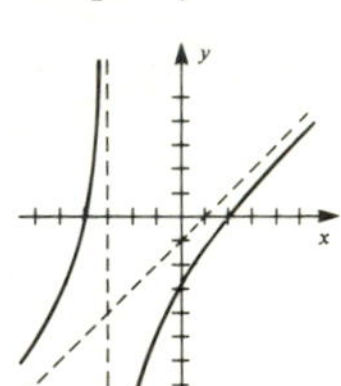

13 $3x^2 + 2$, $-21x^2 + 5x - 9$

15 $\frac{9}{2}, \frac{53}{2}$ 17 -132
19 $6x^4 - 12x^3 + 24x^2 - 52x + 104$, -200
21 $-1 + 8i$ 23 $-55 + 48i$ 25 $-\frac{9}{53} - \frac{48}{53}i$
27 $\frac{2}{41}x^3 + \frac{14}{41}x^2 + \frac{80}{41}x + \frac{68}{41}$ 29 $x^7 + 6x^6 + 9x^5$
31 1 (multiplicity 5); -3 (multiplicity 1)
33 (a) Either 3 positive and 1 negative or 1 positive, 1 negative, and 2 nonreal complex
(b) Upper bound 3; lower bound -1
37 $-\frac{1}{2}, \frac{1}{4}, \frac{3}{2}$ 39 $8/(x-1) - 3/(x+5) - 1/(x+3)$

CHAPTER 4

Exercises 4.1, page 167

1 3 5 7

9 **11** 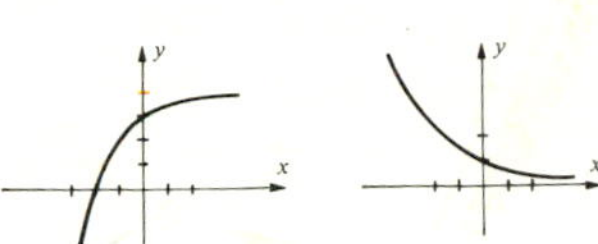 **13** 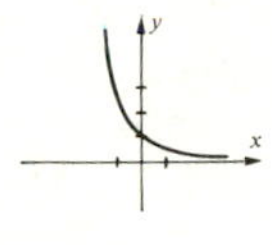 **15** 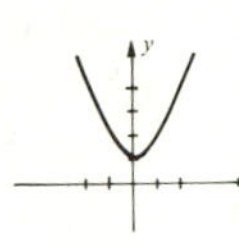

17 **19**  **21** **23** 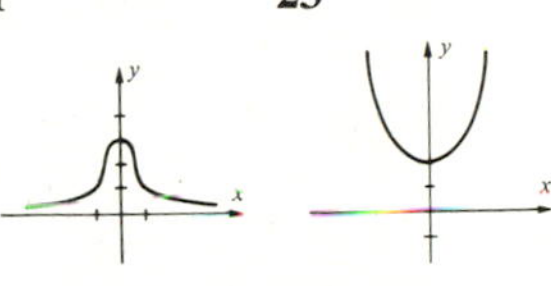

25 -1 **27** $0, -\frac{3}{4}$  **29** $4/(e^x + e^{-x})^2$

31 

33 (a) 1,039; 3,118; 5,400 (b)

35 (a) 50 gm, 25 gm, $(100)2^{-5/2} \approx 17.7$ gm 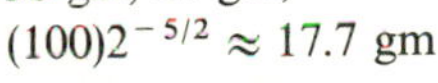 (b)

37 $-1/1600$

39 \$1,010.00; \$1,020.10; \$1,061.52; \$1,126.83

41 (a) \$7,800 (b) \$4,790 (c) \$2,942

43 a^x is not always real if $a < 0$.

45 Reflection through the x-axis.

Calculator Exercises 4.1, page 168

1 4.7288, 4.6555, 4.7070, 4.7277, 4.7287

3 $<$ **7** 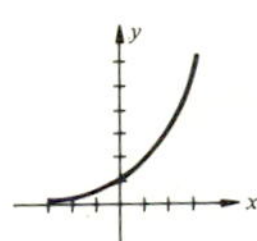

9 (a) \$1,061.36 (b) \$1,126.49 (c) \$1,346.86 (d) \$1,814.02

11 (a) 5 years (b) 8 years (c) 12.5 years

13 (a) 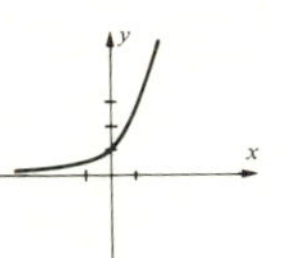 (b) 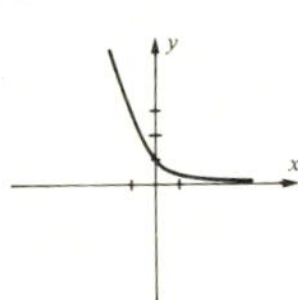 (c) 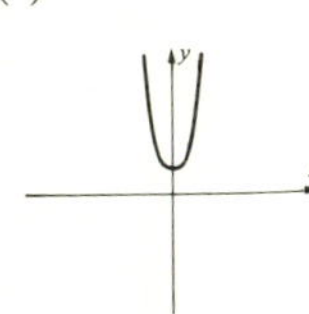

(d) 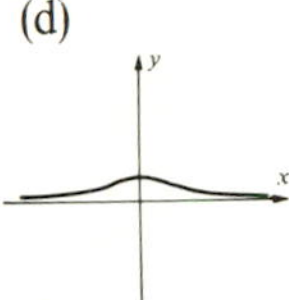 (e) 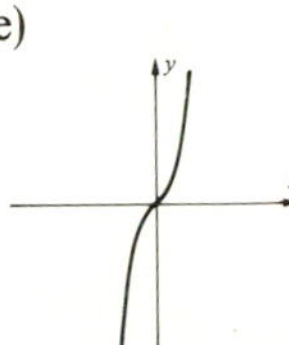 (f) 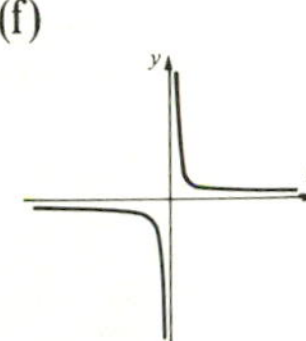

15 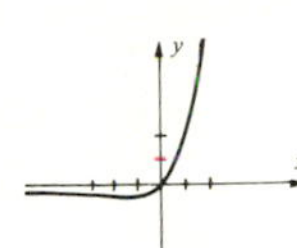 **17** 7.44 **19** 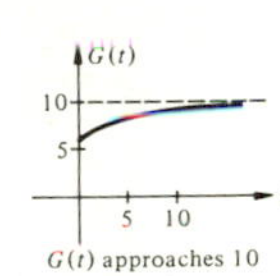

$G(t)$ approaches 10

Exercises 4.2, page 175

1 $\log_4 64 = 3$ **3** $\log_2 128 = 7$

5 $\log_{10}(0.001) = -3$ **7** $\log_t s = r$ **9** $10^3 = 1000$

11 $3^{-5} = \frac{1}{243}$ **13** $7^0 = 1$ **15** $t^p = r$ **17** -2

19 2 **21** 5 **23** $\frac{1}{3}$ **25** -1 **27** -3

29 16 **31** 0.3 **33** 2.4 **25** -0.25 **37** 1.8

39 1.25 **41** 13 **43** 27 **45** $\frac{1}{5}, -\frac{1}{5}$

47 2, 3 **49** $\sqrt[3]{7}$ **51** $\frac{7}{2}$ **53** No solution

55 1 **57** $2 + \sqrt{8}$ **59** $2 \log_a x + \log_a y - 3 \log_a z$

61 $\frac{1}{2}\log_a x + 2\log_a z - 4\log_a y$

63 $\frac{2}{3}\log_a x - \frac{1}{3}\log_a y - \frac{5}{3}\log_a z$

65 $\frac{1}{2}\log_a x + \frac{1}{4}\log_a y + \frac{3}{4}\log_a z$

67 $\log_a (x^2 \sqrt[3]{x-2}/(2x+3)^5)$

69 $\log_a (x^4/y^{5/3})$

Calculator Exercises 4.2, page 176

7 1.8928 **9** 2.1252

Exercises 4.3, page 181

1 (a) (b) 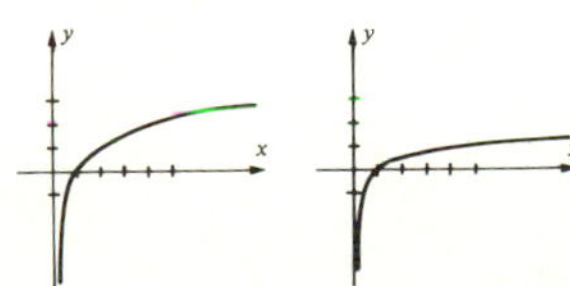 **3** (a) (b) 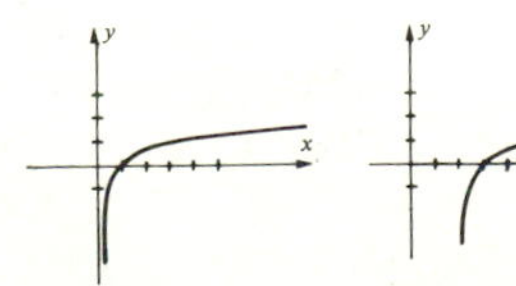

5 (a) (b)

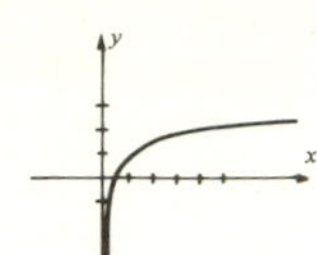
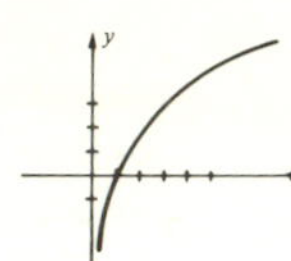

7 (a) (b)

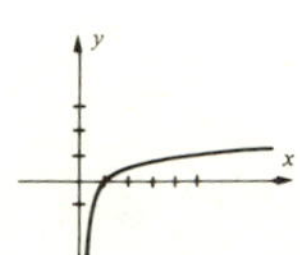
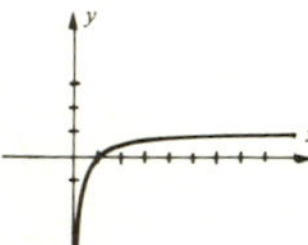

9 (a) (b)

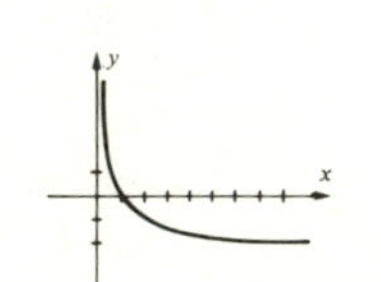
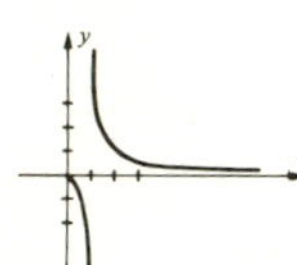

11 They are reflections of one another through the line $y = x$. They are inverse functions of one another.
13 $t = -1600 \log_2 (q/q_0)$
15 $t = (1/k)(\ln Q - \ln Q_0) = (1/k) \ln (Q/Q_0)$
17 (a) 10 (b) 30 (c) 40 **19** (a) 2 (b) 4 (c) 5

Calculator Exercises 4.3, page 182

1 (a) (b) (c)

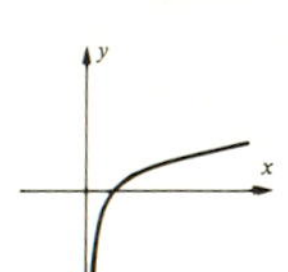
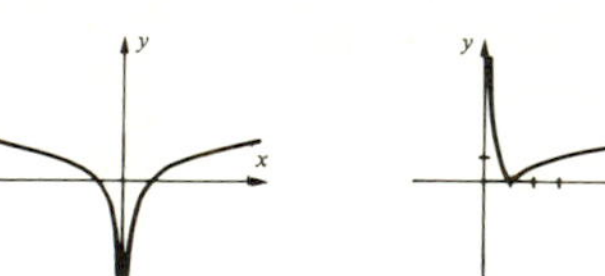

5 (a) 1.2619 (b) 0.3074 (c) 0.1781 (d) 2.5821 (e) −0.1203

Exercises 4.4, page 187

1 (a) 2.5403, −2.4597, 0.5403
(b) 5.8493, −5.6636, 1.2442
3 (a) −0.2676, 2.7324, 5.7324
(b) −0.6162, 6.2916, 13.1993
5 (a) 1.7796, −4.2204, 2.7796
(b) 4.0977, −9.7178, 6.4003
7 4,240 **9** 8.85 **11** 0.0237 **13** 9.97
15 1.05 **17** 0.202 **19** (a) 2.2 (b) 5 (c) 8.3
21 Acidic if pH < 7, basic if pH > 7 **23** $\alpha \approx 54.5$
25 $10^{4.5} I_0$ (approximately 31,623 times that of I_0)
27 log 7/log 10 **29** log 3/log 4
31 4 − (log 5/log 3) **33** (log 2 − log 81)/log 24
35 −3 **37** 5 **39** $\frac{2}{3}\sqrt{\frac{1001}{111}}$
41 100, 1 **43** 10^{100} **45** 10,000
47 $x = \log (y \pm \sqrt{y^2 - 1})$
49 $x = \frac{1}{2} \log[(1 + y)/(1 - y)]$
51 $x = \ln (y + \sqrt{y^2 + 1})$
53 $x = \frac{1}{2} \ln [(1 + y)/(1 - y)]$
55 $t = -(L/R) \ln (1 - RI/E)$ **57** (a) $I = I_0 10^{\alpha/10}$
59 (a) $(\ln 2)/1.4 \approx 0.495$ m (b) $(\ln 10)/1.4 \approx 1.645$ m

Calculator Exercises 4.4, page 188

1 $x \approx 1.2091$ **3** $x \approx \pm 1.347$ **5** 139 months
7 110 days **9** 0.807

Exercises 4.5, page 189

1 −4 **3** 4 **5** 6 **7** $\frac{1}{2}$ **9** 5
11 **13** **15**

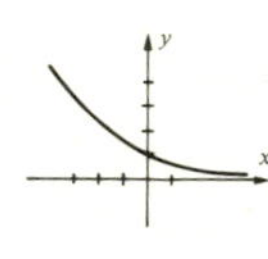
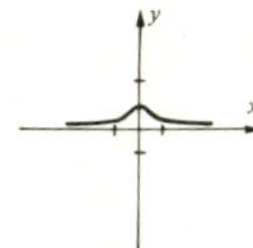

17 **19**

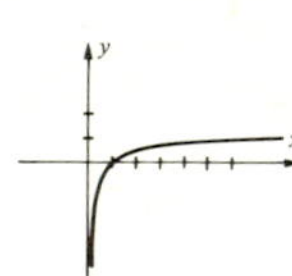
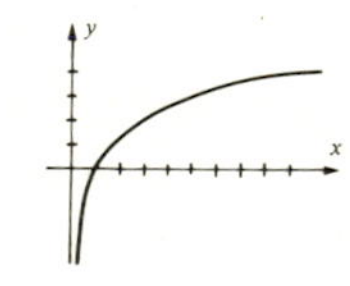

21 9 **23** 1 **25** $\log \frac{16}{3}/\log 2$ **27** $\log \frac{3}{8}/\log \frac{32}{9}$
29 0, 1, −1 **31** $4 \log x + \frac{2}{3} \log y - \frac{1}{3} \log z$
33 $x = \frac{1}{2}\log[(y + 1)/(y - 1)]$ **35** 78.3
37 6.0496 **39** 1.887

CHAPTER 5

Exercises 5.1, page 196

1 480°, 840°, −240°, −600°
3 495°, 855°, −225°, −585°
5 330°, 690°, −390°, −750°

7 $260°, 980°, -100°, -460°$
9 $17\pi/6, 29\pi/6, -7\pi/6, -19\pi/6$
11 $7\pi/4, 15\pi/4, -9\pi/4, -17\pi/4$ **13** $5\pi/6$ **15** $-\pi/3$
17 $5\pi/4$ **19** $5\pi/2$ **21** $2\pi/5$ **23** $5\pi/9$
25 $120°$ **27** $330°$ **29** $135°$ **31** $-630°$
33 $1260°$ **35** $20°$ **37** $114°35'30''$
39 $286°28'44''$ **41** (a) 1.75 (b) 100.27°
43 6.98 meters **45** 8.59 km
47 Approximations (in miles): (a) 4189 (b) 3142 (c) 2094 (d) 698 (e) 70
49 $\frac{1}{8}$ radian $= (45/2\pi)° \approx 7°10'$

Calculator Exercises 5.1, page 197

1 1.2811 **3** 8.4173 **5** 149.7651°
7 Approximations (in miles): (a) 4150.26 (b) 3112.69 (c) 2075.13 (d) 691.71 (e) 69.17
9 Approximately 0.126 radians, or 7.219°

Exercises 5.2, page 206

Answers to Exercises 1–17 are in the order sin, cos, tan, csc, sec, cot.

1 $\frac{4}{5}, \frac{3}{5}, \frac{4}{3}, \frac{5}{4}, \frac{5}{3}, \frac{3}{4}$
3 $2\sqrt{13}/13, 3\sqrt{13}/13, \frac{2}{3}, \sqrt{13}/2, \sqrt{13}/3, \frac{3}{2}$
5 $\frac{2}{5}, \sqrt{21}/5, 2\sqrt{21}/21, \frac{5}{2}, 5\sqrt{21}/21, \sqrt{21}/2$
7 $a/\sqrt{a^2+b^2}, b/\sqrt{a^2+b^2}, a/b, \sqrt{a^2+b^2}/a, \sqrt{a^2+b^2}/b, b/a$
9 $b/c, \sqrt{c^2-b^2}/c, b/\sqrt{c^2-b^2}, c/b, c/\sqrt{c^2-b^2}, \sqrt{c^2-b^2}/b$
11 $\frac{3}{5}, \frac{4}{5}, \frac{3}{4}, \frac{5}{3}, \frac{5}{4}, \frac{4}{3}$
13 $\sqrt{11}/6, \frac{5}{6}, \sqrt{11}/5, 6\sqrt{11}/11, \frac{6}{5}, 5\sqrt{11}/11$
15 $\sqrt{2}/2, \sqrt{2}/2, 1, \sqrt{2}, \sqrt{2}, 1$
17 $b/\sqrt{a^2+b^2}, a/\sqrt{a^2+b^2}, b/a, \sqrt{a^2+b^2}/b, \sqrt{a^2+b^2}/a, a/b$
19 $\cot\theta = \sqrt{1-\sin^2\theta}/\sin\theta$ **21** $\sec\theta = 1/\sqrt{1-\sin^2\theta}$
23 $\tan\theta = \sqrt{\sec^2\theta - 1}$

Answers to Exercises 27–43 are in the order sin, cos, tan, csc, sec, cot.

27 $-\frac{3}{5}, \frac{4}{5}, -\frac{3}{4}, -\frac{5}{3}, \frac{5}{4}, -\frac{4}{3}$
29 $-5\sqrt{29}/29, -2\sqrt{29}/29, \frac{5}{2}, -\sqrt{29}/5, -\sqrt{29}/2, \frac{2}{5}$
31 $4\sqrt{17}/17, -\sqrt{17}/17, -4, \sqrt{17}/4, -\sqrt{17}, -\frac{1}{4}$
33 $-7\sqrt{53}/53, -2\sqrt{53}/53, \frac{7}{2}, -\sqrt{53}/7, -\sqrt{53}/2, \frac{2}{7}$
35 $\frac{4}{5}, \frac{3}{5}, \frac{4}{3}, \frac{5}{4}, \frac{5}{3}, \frac{3}{4}$ **37** 1, 0, —, 1, —, 0
39 0, −1, 0, —, −1, — **41** 0, 1, 0, —, 1, —
43 −1, 0, —, −1, —, 0 **45** IV **47** III
49 II **51** III **53** I **55** No; $|\sin\theta| \le 1$

Exercises 5.3, page 213

1 (a) $(-1, 0)$ (b) 0, −1, 0, —, −1, —
3 (a) $(0, -1)$ (b) −1, 0, —, −1, —, 0
5 (a) $(-\sqrt{2}/2, \sqrt{2}/2)$
(b) $\sqrt{2}/2, -\sqrt{2}/2, -1, \sqrt{2}, -\sqrt{2}, -1$
7 (a) $(\sqrt{3}/2, \frac{1}{2})$ (b) $\frac{1}{2}, \sqrt{3}/2, \sqrt{3}/3, 2, 2\sqrt{3}/3, \sqrt{3}$
9 (a) $(-\sqrt{3}/2, \frac{1}{2})$
(b) $\frac{1}{2}, -\sqrt{3}/2, -\sqrt{3}/3, 2, -2\sqrt{3}/3, -\sqrt{3}$
11 (a) $-\frac{1}{2}$ (b) $\sqrt{2}/2$ (c) $-\sqrt{3}/3$
13 (a) $\csc(-t) = 1/\sin(-t) = 1/(-\sin t) = -(1/\sin t) = -\csc t$
15 Using the hint with $t = 0$ gives us $\sin k = \sin 0 = 0$. Since $0 < k < 2\pi$, it follows that $k = \pi$, and hence $\sin(t+\pi) = \sin t$ for all t. In particular, if $t = \pi/2$, then $\sin(3\pi/2) = \sin(\pi/2)$, or $-1 = 1$, an absurdity.
17 Use $\csc t = 1/\sin t$ and the result of Exercise 15.
19 If $y/x = a$, where $x^2 + y^2 = 1$, then $\pm\sqrt{1-x^2}/x = a$. Hence $(1-x^2)/x^2 = a^2$, and $1 - x^2 = a^2x^2$, or $1 = a^2x^2 + x^2 = (a^2+1)x^2$. Consequently, $x^2 = 1/(a^2+1)$, or $x = \pm 1/\sqrt{a^2+1}$. If $a > 0$, choose the point $P(x, y)$ on U where $x = 1/\sqrt{a^2+1}$. If $a < 0$, choose $P(x, y)$ such that $x = -1/\sqrt{a^2+1}$. Thus, there is always a point $P(x, y)$ on U such that $y/x = a$.

Exercises 5.4, page 221

1 (a) $\pi/4$ (b) $\pi/3$ (c) $\pi/6$
3 (a) $\pi/4$ (b) $\pi/6$ (c) $\pi/3$
5 (a) 1.5 (b) $2\pi - 5$
7 (a) 60° (b) 20° (c) 70°
9 (a) 49°20′ (b) 45° (c) 80°35′
11 (a) $\sqrt{3}/2$ (b) $-\sqrt{3}/2$ **13** (a) −1 (b) −1
15 (a) $-2\sqrt{3}/3$ (b) −2 **17** 0.2644 **19** 1.540
21 1.431 **23** 0.7826 **25** 6.197 **27** −1.019
29 −0.3035 **31** 0.4618 **33** −2.1850
35 1.032 **37** −0.1343 **39** 0.5217 **41** 3.4499
43 (a) 30.46° (b) 0.53 **45** (a) 74.88° (b) 1.31
47 (a) 24.94° (b) 0.44 **49** (a) 76.38° (b) 1.33

Exercises 5.5, page 227

1 (c) $[-2\pi, -3\pi/2), (-3\pi/2, -\pi], [0, \pi/2), (\pi/2, \pi]$
(d) $[-\pi, -\pi/2), (-\pi/2, 0], [\pi, 3\pi/2), (3\pi/2, 2\pi]$
(e) $\sec x \to \infty$ as $x \to (\pi/2)^-$; $\sec x \to -\infty$ as $x \to (\pi/2)^+$
(f) Symmetric with respect to the y-axis

5

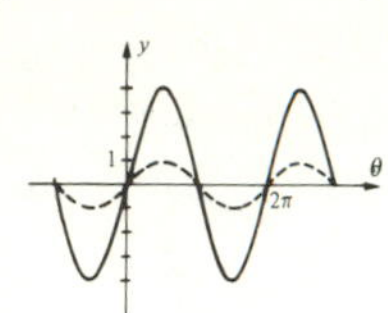

7

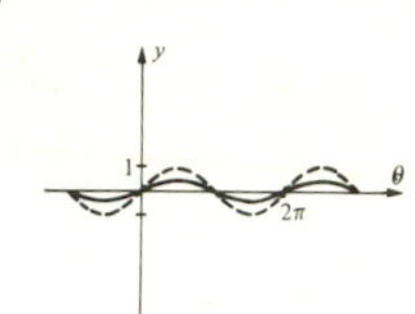

9

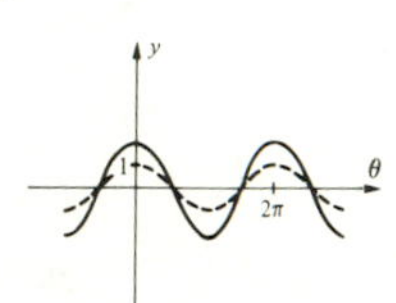

11

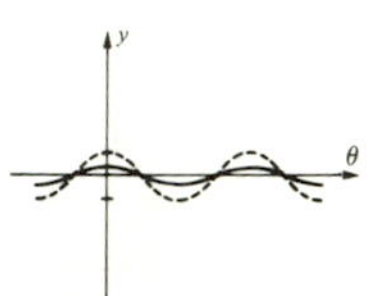

13

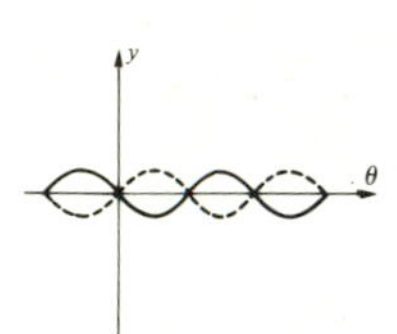

21 **23**

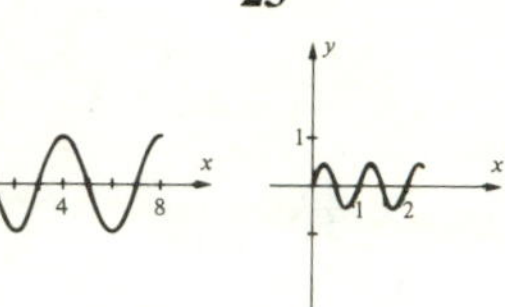

25

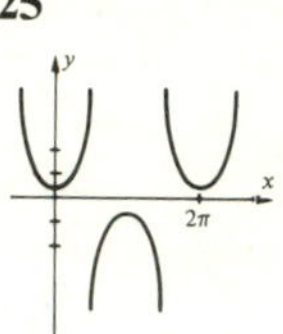

27

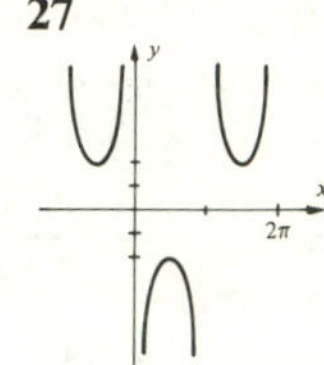

29

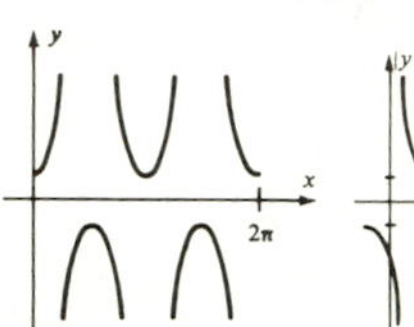

31

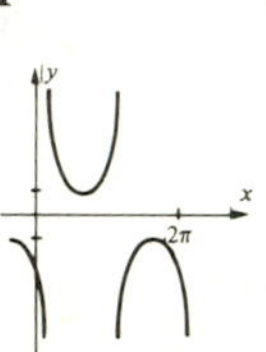

33 **35**

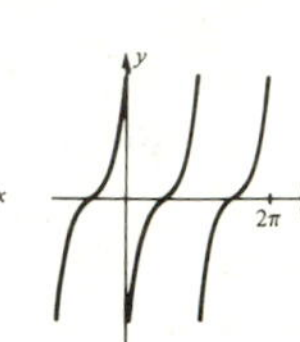

37

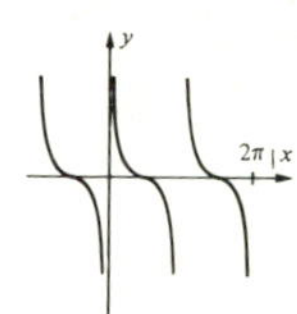

39 **41**

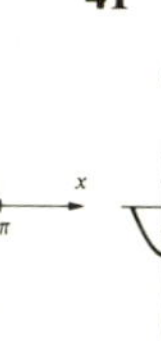

43

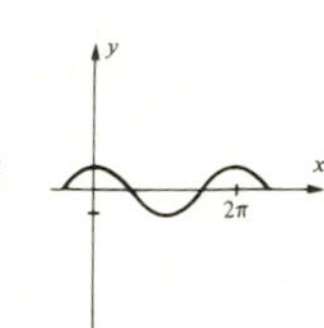

Exercises 5.6, page 231

1 The amplitudes and periods are:
(a) 4, 2π (b) 1, $\pi/2$ (c) $\frac{1}{4}$, 2π (d) 1, 8π
(e) 2, 8π (f) $\frac{1}{2}$, $\pi/2$ (g) 4, 2π (h) 1, $\pi/2$

3 The amplitudes and periods are:
(a) 3, 2π (b) 1, $2\pi/3$ (c) $\frac{1}{3}$, 2π (d) 1, 6π
(e) 2, 6π (f) $\frac{1}{3}$, π (g) 3, 2π (h) 1, $2\pi/3$

5 **7**

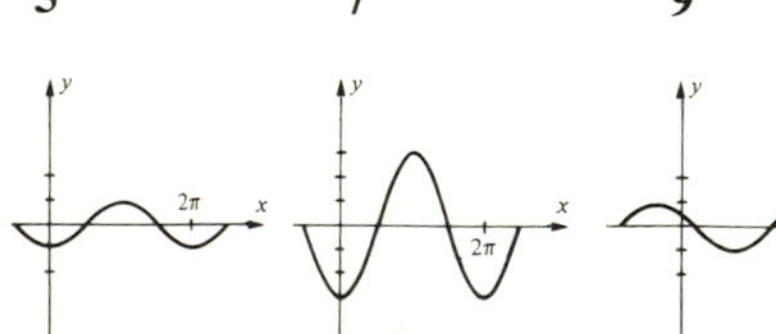

9 **11**

13 **15**

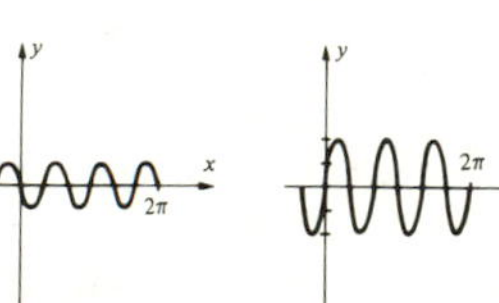

17

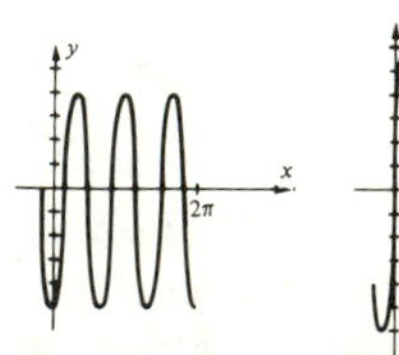

19

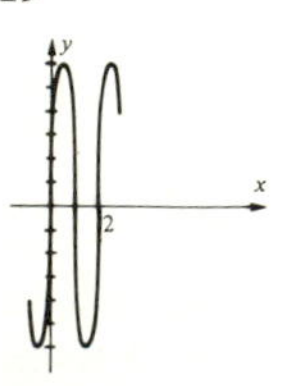

Exercises 5.7, page 234

1 **3**

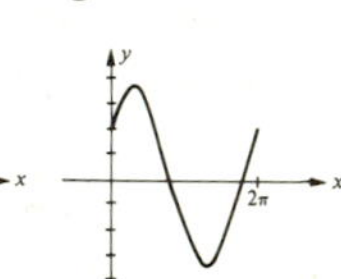

5

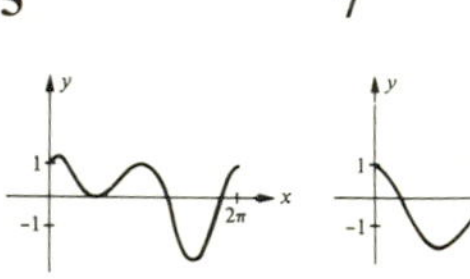

7

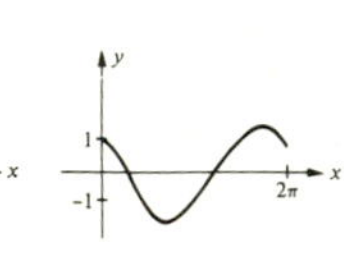

9

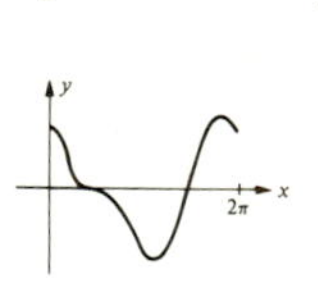

11 **13**

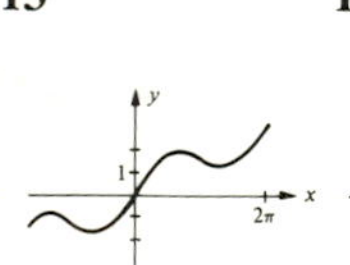

15

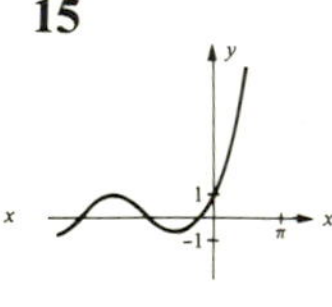

17 **19**

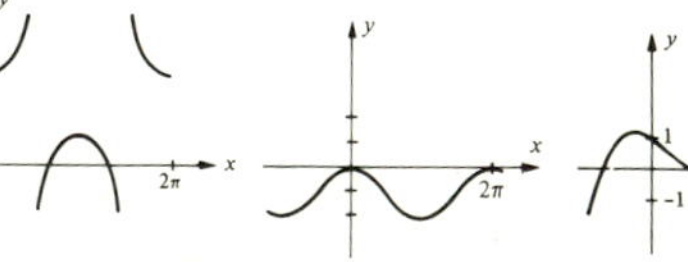

21 **23**

25 **27**

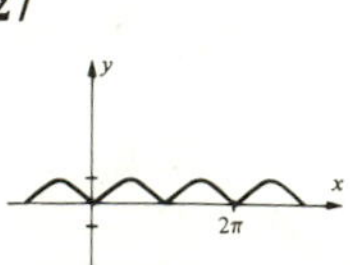

29

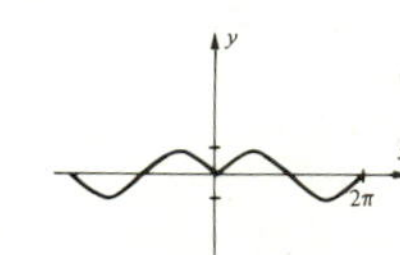

Calculator Exercises 5.7, page 234

1

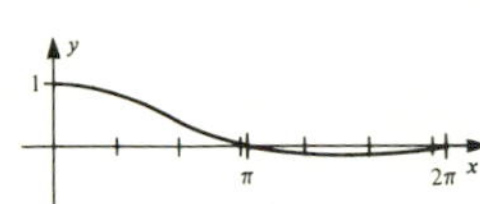

Exercises 5.8, page 239

1 $\beta = 60°, a = 20\sqrt{3}/3 \approx 12, c = 40\sqrt{3}/3 \approx 23$
3 $\alpha = 38°, b \approx 19, c \approx 24$
5 $\beta = 72°20', b \approx 14.1, c \approx 14.8$
7 $\alpha = 18°9', a \approx 78.67, c \approx 252.6$
9 $\alpha \approx 29°, \beta \approx 61°, c \approx 51$
11 $\alpha \approx 69°, \beta \approx 21°, a \approx 5.4$
13 $\beta = 52°14', a \approx 396.7, c \approx 647.7$
15 $\alpha \approx 49°40', \beta \approx 40°20', b \approx 522$ 17 51°
19 70.6 meters 21 20.2 meters 23 29.7 km
25 $d \approx 160$ meters 27 $192(\sin 22°30') \approx 73.5$ cm
29 9,659 ft 31 126 mph 33 28,400 ft
35 55 miles 37 325 miles 39 $h = d(\tan\beta - \tan\alpha)$

Calculator Exercises 5.8, page 240

1 $\beta = 48.73°, b \approx 358.5, c \approx 476.9$
3 $\alpha \approx 52.94°, b \approx 0.3484, c \approx 0.5781$
5 $\alpha \approx 87.29°, a \approx 151{,}000, c \approx 151{,}200$
7 $c \approx 48.67, \alpha \approx 74.36°, \beta \approx 15.64°$
9 $\alpha \approx 60.97°, \beta \approx 29.03°, a \approx 4437$

Exercises 5.9, page 244

1 (a) $\omega = 200\pi$ radians/min
(b) In cm: $x = 40\cos 200\pi t$, $y = 40 \sin 200\pi t$

3 Amplitude 10 cm, period $\frac{1}{3}$ sec, frequency 3 osc/sec. The point is at the origin at $t = 0$. It then moves upward with decreasing speed, reaching the point with coordinate 10 at $t = \frac{1}{12}$. It then reverses direction and moves downward, gaining speed until it reaches the origin at $t = \frac{1}{6}$. It continues downward, with decreasing speed, reaching the point with coordinate -10 at $t = \frac{1}{4}$. It then reverses direction and moves upward with increasing speed, returning to the origin at $t = \frac{1}{3}$.

5 Amplitude 4 cm, period $\frac{4}{3}$ sec, frequency $\frac{3}{4}$ osc/sec. The motion is similar to that in Exercise 3; however, the point starts 4 units above the origin and moves downward, reaching O at $t = \frac{1}{3}$, and the point with coordinate -4 at $t = \frac{2}{3}$. It then reverses direction and moves upward, reaching O at $t = 1$ and its initial point at $t = \frac{4}{3}$.

7 $d = 5\cos(2\pi/3)t$

9 Current lags emf by $\frac{1}{1440}$ sec.

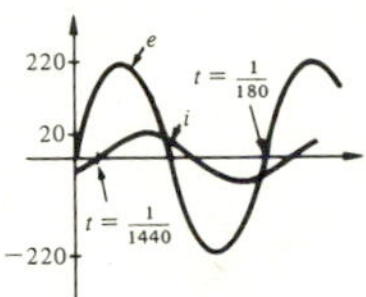

Exercises 5.10, page 245

1 $11\pi/6, 9\pi/4, -5\pi/6, 4\pi/3, \pi/5$
3 (a) $\pi/4, 5\pi/6, \pi/8$ (b) $65°, 42°50', 8°, \pi/4, \pi/6, \pi/8$

Answers for Exercises 5–9 are in the order sin, cos, tan, csc, sec, cot.

5 $\sqrt{33}/7, \frac{4}{7}, \sqrt{33}/4, 7\sqrt{33}/33, \frac{7}{4}, 4\sqrt{33}/33$
7 (a) $-\frac{4}{5}, \frac{3}{5}, -\frac{4}{3}, -\frac{5}{4}, \frac{5}{3}, -\frac{3}{4}$
(b) $2\sqrt{13}/13, -3\sqrt{13}/13, -\frac{2}{3}, \sqrt{13}/2, -\sqrt{13}/3, -\frac{3}{2}$
9 (a) 1, 0, —, 1, —, 0
(b) $\sqrt{2}/2, -\sqrt{2}/2, -1, \sqrt{2}, -\sqrt{2}, -1$
(c) 0, 1, 0, —, 1, —
(d) $-\frac{1}{2}, \sqrt{3}/2, -\sqrt{3}/3, -2, 2\sqrt{3}/3, -\sqrt{3}$

11

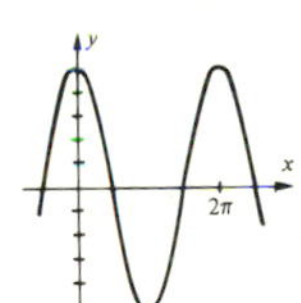

13

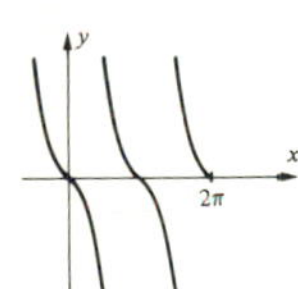

15

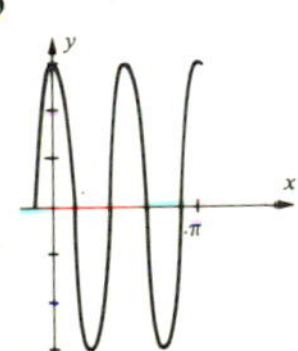

17

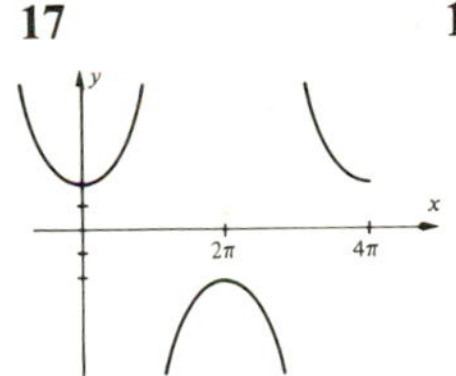

19

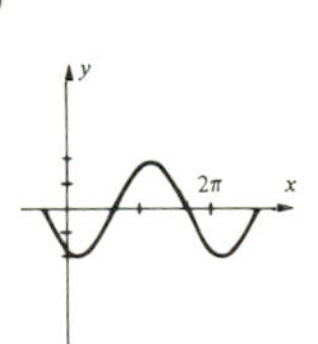

21

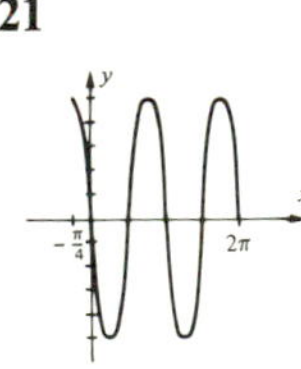

23

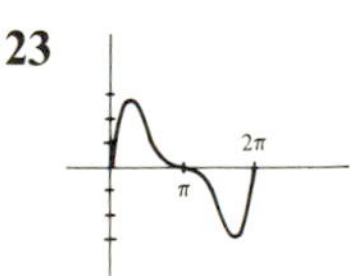

25

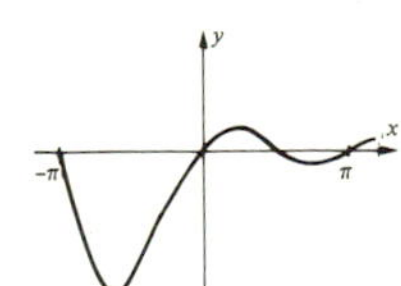

27 $\beta = 35°20', a \approx 310, c \approx 380$

CHAPTER 6

Exercises 6.2, page 257

In the following, n denotes any integer.

1 $(2\pi/3) + 2\pi n, (4\pi/3) + 2\pi n$ 3 $(\pi/4) + (\pi/2)n$

5 $2\pi n$, $(3\pi/2) + 2\pi n$ **7** $(\pi/3) + \pi n$, $(2\pi/3) + \pi n$
9 $(4\pi/3) + 2\pi n$, $(5\pi/3) + 2\pi n$
11 $(\pi/6) + \pi n$, $(5\pi/6) + \pi n$
13 $(7\pi/6) + 2\pi n$, $(11\pi/6) + 2\pi n$
15 $(\pi/12) + n\pi$, $(5\pi/12) + n\pi$
17 $\pi/3, 2\pi/3, 4\pi/3, 5\pi/3$; $60°, 120°, 240°, 300°$
19 $\pi/6, 5\pi/6, 3\pi/2$; $30°, 150°, 270°$
21 $0, \pi, \pi/4, 3\pi/4, 5\pi/4, 7\pi/4$; $0°, 180°, 45°, 135°, 225°, 315°$
23 $\pi/2, 3\pi/2, 2\pi/3, 4\pi/3$; $90°, 270°, 120°, 240°$
25 No solutions **27** $11\pi/6, \pi/2$; $330°, 90°$
29 $0, \pi/2$; $0°, 90°$ **31** 0; $0°$
33 $7\pi/6, 11\pi/6, \pi/2, 3\pi/2$; $210°, 330°, 90°, 270°$
35 $3\pi/4, 7\pi/4$; $135°, 315°$ **37** $15°30'$, $164°30'$
39 $41°50'$, $138°10'$, $194°30'$, $345°30'$

Exercises 6.3, page 264

1 (a) $\cos 43°23'$ (b) $\sin 16°48'$ (c) $\cot \pi/3$
3 (a) $\sin 3\pi/20$ (b) $\cos (2\pi - 1)/4$ (c) $\cot (\pi - 2)/2$
5 (a) $(\sqrt{2} + 1)/2$ (b) $(\sqrt{2} - \sqrt{6})/4$
7 (a) $\sqrt{3} + 1$ (b) $-2 - \sqrt{3}$
9 (a) $(\sqrt{2} - 1)/2$ (b) $(\sqrt{6} + \sqrt{2})/4$
11 $\cos 25°$ **13** $\sin (-5°)$ **15** $-\sin 5$
17 $\frac{36}{85}, \frac{77}{85}$; I **19** $\frac{3}{5}, \frac{4}{5}, \frac{3}{4}, -\frac{117}{125}, \frac{44}{125}, -\frac{117}{44}$
41 $\sin u \cos v \cos w + \cos u \sin v \cos w + \cos u \cos v \sin w - \sin u \sin v \sin w$
49 $0, \pi/3, 2\pi/3, \pi, 4\pi/3, 5\pi/3$; $0°, 60°, 120°, 180°, 240°, 300°$
51 $f(x) = 2 \cos (2x - \pi/6)$; $2, \pi, \pi/12$
53 $f(x) = 2\sqrt{2} \cos (3x + \pi/4)$; $2\sqrt{2}, 2\pi/3, -\pi/12$

Exercises 6.4, page 271

1 $\frac{24}{25}, -\frac{7}{25}, -\frac{24}{7}$ **3** $-4\sqrt{2}/9, -\frac{7}{9}, 4\sqrt{2}/7$
5 $\sqrt{10}/10, 3\sqrt{10}/10, \frac{1}{3}$
7 $-\sqrt{2 + \sqrt{2}}/2, \sqrt{2 - \sqrt{2}}/2, -\sqrt{2} - 1$
9 (a) $\sqrt{2 - \sqrt{2}}/2$ (b) $\sqrt{2 - \sqrt{3}}/2$ (c) $\sqrt{2} + 1$
29 $\frac{3}{8} + \frac{1}{2} \cos \theta + \frac{1}{8} \cos 2\theta$
31 $0, 2\pi/3, \pi, 4\pi/3$; $0°, 120°, 180°, 240°$
33 $\pi/3, \pi, 5\pi/3$; $60°, 180°, 300°$ **35** $0, \pi$; $0°, 180°$
37 $0, \pi/3, 5\pi/3$; $0°, 60°, 300°$

Calculator Exercises 6.4, page 272

1 -0.8217

Exercises 6.5, page 276

1 $\sin 12\theta + \sin 6\theta$ **3** $\frac{1}{2} \cos 4t - \frac{1}{2} \cos 10t$
5 $\frac{1}{2} \cos 10u + \frac{1}{2} \cos 2u$ **7** $\frac{3}{2} \sin 3x + \frac{3}{2} \sin x$
9 $2 \cos 2\theta \sin 4\theta$ **11** $-2 \sin 4x \sin x$
13 $-2 \cos 5t \sin 2t$ **15** $2 \cos \frac{3}{2}x \cos \frac{1}{2}x$
25 $\frac{1}{2} \sin (a + b)x + \frac{1}{2} \sin (a - b)x$
27 $n\pi/4$ where n is any integer.
29 $n\pi/2$ where n is any integer.

Exercises 6.6, page 283

1 (a) $\pi/6$ (b) $\pi/3$ **3** (a) 0 (b) $\pi/2$
5 (a) $\pi/2$ (b) 0 **7** (a) $\pi/3$ (b) $-\pi/3$
9 (a) $-\pi/4$ (b) $-\pi/6$
11 (a) -0.7069 (b) -0.7067952
13 (a) 1.1403 (b) 1.1402832 **15** $\sqrt{3}/2$ **17** $\frac{3}{5}$
19 $-\pi/4$ **21** 0 **23** $-77/36$
25 $-24/25$ **27** $x\sqrt{1 + x^2}/(1 + x^2)$ **29** $\sqrt{2 + 2x}/2$
39 $\cot^{-1} u = v$ if and only if $\cot v = u$.

43

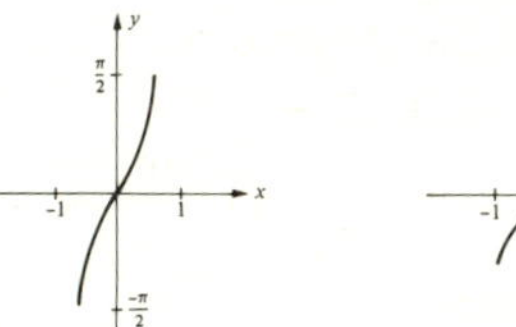

45

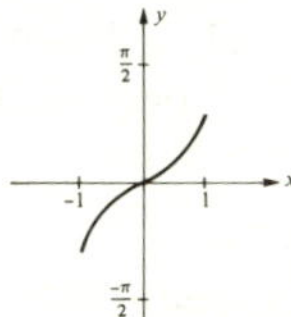

47

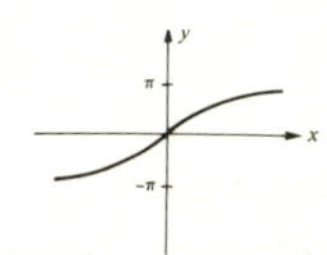

49

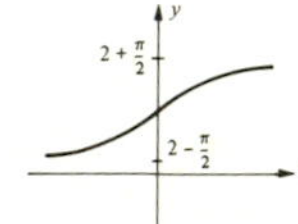

51

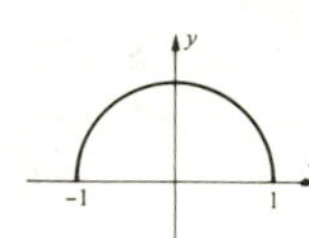

55 $\arctan (-9 \pm \sqrt{57})/4$
57 $\arccos (\pm\sqrt{15}/5)$, $\arccos (\pm\sqrt{3}/3)$
59 $\arcsin (\pm\sqrt{30}/6)$

Calculator Exercises 6.6, page 284

1 $-0.3478, -1.3336$ **3** $0.6847, 2.4569, 0.9553, 2.1863$
5 ± 1.1503

Exercises 6.7, page 289

1 $\beta \approx 62°$, $b \approx 14.1$, $c \approx 15.6$
3 $\gamma \approx 100°10'$, $b \approx 55.1$, $c \approx 68.7$
5 $\alpha \approx 58°40'$, $a \approx 487$, $b \approx 442$
7 $\beta \approx 53°40'$, $\gamma \approx 61°10'$, $c \approx 20.6$
9 $\alpha \approx 77°30'$, $\beta \approx 49°10'$, $b \approx 108$; $\alpha \approx 102°30'$, $\beta = 24°10'$, $b \approx 58.7$ **11** $\alpha \approx 20°30'$, $\gamma \approx 46°20'$, $a \approx 94.5$
13 219 yd **15** 50 ft **17** 2.7 mi
19 3.7 mi from A and 5.4 mi from B **21** 628 meters

Calculator Exercises 6.7, page 290

1 $\alpha = 119.7°$, $a \approx 371.5$, $c \approx 243.5$
3 $\beta = 49.36°$, $a \approx 49.78$, $c \approx 23.39$
5 $\alpha \approx 25.993°$, $\gamma \approx 32.383°$, $a \approx 0.146$

Exercises 6.8, page 292

1 $a \approx 26$, $\beta \approx 41°$, $\gamma \approx 79°$
3 $b \approx 177$, $\alpha \approx 25°10'$, $\gamma \approx 4°50'$
5 $c \approx 2.75$, $\alpha \approx 21°10'$, $\beta \approx 43°40'$
7 $\alpha \approx 29°$, $\beta \approx 47°$, $\gamma \approx 104°$
9 $\alpha \approx 12°30'$, $\beta \approx 136°30'$, $\gamma \approx 31°$ **11** 63, 87 cm
13 92 ft **15** 24 mi **17** 39 mi
19 Approximately 2.3 mi **21** Approximately N55°31′E

Calculator Exercises 6.8, page 293

1 $a \approx 40.8$, $\beta \approx 26.9°$, $\gamma \approx 104.8°$
3 $c \approx 0.487$, $\beta \approx 5.11°$, $\alpha \approx 168.04°$
5 $\alpha \approx 157°16'$, $\beta \approx 7°49'$, $\gamma \approx 14°55'$

Exercises 6.9, page 297

1 5 **3** $\sqrt{85}$ **5** 8 **7** 1 **9** 0

The geometric representations in Exercises 11–19 are the following points:

11 $P(4, 2)$ **13** $P(3, -5)$ **15** $P(-3, 6)$
17 $P(-6, 4)$ **19** $P(0, 2)$
21 $\sqrt{2}(\cos 7\pi/4 + i \sin 7\pi/4$
23 $8(\cos 5\pi/6 + i \sin 5\pi/6)$
25 $20(\cos 3\pi/2 + i \sin 3\pi/2)$
27 $10(\cos 4\pi/3 + i \sin 4\pi/3)$
29 $7(\cos \pi + i \sin \pi)$
31 $\sqrt{5}(\cos \theta + i \sin \theta)$, where $\theta = \arctan \frac{1}{2}$.
33 $4\sqrt{2}(\cos 5\pi/4 + i \sin 5\pi/4)$
35 $6(\cos \pi/2 + i \sin \pi/2)$ **37** $8(\cos 5\pi/6 + i \sin 5\pi/6)$
39 $12(\cos 0 + i \sin 0)$ **41** $-2, i$
43 $10\sqrt{3} - 10i$, $(-2\sqrt{3}/5) + \frac{2}{5}i$ **45** $40, \frac{5}{2}$
47 $8 - 4i$, $\frac{8}{5} + \frac{4}{5}i$
49 $z_1z_2z_3 = r_1r_2r_3[\cos(\theta_1 + \theta_2 + \theta_3) + i \sin(\theta_1 + \theta_2 + \theta_3)]$,
$z_1z_2 \cdots z_n = (r_1r_2 \cdots r_n)[\cos(\theta_1 + \theta_2 + \cdots + \theta_n) + i \sin(\theta_1 + \theta_2 + \cdots + \theta_n)]$

Exercises 6.10, page 302

1 $-972 - 972i$ **3** $-32i$ **5** -8
7 $-(\sqrt{2}/2) - (\sqrt{2}/2)i$ **9** $-\frac{1}{2} - (\sqrt{3}/2)i$
11 $-64\sqrt{3} - 64i$
13 $(\sqrt{6}/2) + (\sqrt{2}/2)i$, $-(\sqrt{6}/2) - (\sqrt{2}/2)i$
15 $(\sqrt[4]{2}/2) + (\sqrt[4]{18}/2)i, -(\sqrt[4]{18}/2) + (\sqrt[4]{4}/2)i$,
$(\sqrt[4]{18}/2) - (\sqrt[4]{2}/2)i, -(\sqrt[4]{2}/2) - (\sqrt[4]{18}/2)i$
17 $3i$, $-(3\sqrt{3}/2) - \frac{3}{2}i$, $(3\sqrt{3}/2) - \frac{3}{2}i$
19 ± 1, $\frac{1}{2} \pm (\sqrt{3}/2)i$, $-\frac{1}{2} \pm (\sqrt{3}/2)i$
21 $\sqrt[5]{2}(\cos \theta + i \sin \theta)$, where $\theta = 9°, 81°, 153°, 225°, 279°$
23 $\pm 2, \pm 2i$ **25** $\pm 2i, \pm\sqrt{3} + i, \pm\sqrt{3} - i$
27 $2i, -\sqrt{3} - i, \sqrt{3} - i$
29 $3 \cos \theta + 3 \sin \theta i$ where $\theta = 0, 2\pi/5, 4\pi/5, 6\pi/5, 8\pi/5$

Exercises 6.11, page 311

1 $\langle 3, 1\rangle, \langle 1, -7\rangle, \langle 13, 8\rangle, \langle 3, -32\rangle$
3 $\langle -15, 6\rangle, \langle 1, -2\rangle, \langle -68, 28\rangle, \langle 12, -12\rangle$
5 $4\mathbf{i} - 3\mathbf{j}$, $-2\mathbf{i} + 7\mathbf{j}$, $19\mathbf{i} - 17\mathbf{j}$, $-11\mathbf{i} + 33\mathbf{j}$
7 $-2\mathbf{i} - 5\mathbf{j}$, $-6\mathbf{i} + 7\mathbf{j}$, $-6\mathbf{i} - 26\mathbf{j}$, $-26\mathbf{i} + 34\mathbf{j}$
9 $-3\mathbf{i} + 2\mathbf{j}$, $3\mathbf{i} + 2\mathbf{j}$, $-15\mathbf{i} + 8\mathbf{j}$, $15\mathbf{i} + 8\mathbf{j}$

Answers for Exercises 11 and 13 are terminal points of the vectors.

11 $(3, 2), (-1, 5), (2, 7), (6, 4), (3, -15)$
13 $(-4, 6), (-2, 3), (-6, 9), (-8, 12), (6, -9)$
31 $3\sqrt{2}, 7\pi/4$ **33** $5, \pi$
35 $\sqrt{41}$, $\arccos(-4\sqrt{41}/41)$ **37** $18, 3\pi/2$
39 89 kg, S66°W **41** 5.8 lb, 129° **43** 56°, 232 mph
45 420 mph, 244° **47** N22°W

Exercises 6.12, page 313

17 $\pi/2, 3\pi/2, \pi/4, 3\pi/4, 5\pi/4, 7\pi/4$;
90°, 270°, 45°, 135°, 225°, 315°
19 $0, \pi$; 0°, 180°
21 $0, \pi, 2\pi/3, 4\pi/3$; 0°, 180°, 120°, 240°
23 $\pi/2, 7\pi/6, 11\pi/6$; 90°, 210°, 330°
25 $\pi/6, 5\pi/6, \pi/3, 5\pi/3$; 30°, 150°, 60°, 300°
27 $\pi/3, 5\pi/3$; 60°, 300° **29** $\sqrt{2 - \sqrt{3}}/2$ or $(\sqrt{2} - \sqrt{6})/4$
31 $\sqrt{2 - \sqrt{3}}/2$ or $(\sqrt{2} - \sqrt{6})/4$ **33** $\frac{84}{85}$ **35** $-\frac{36}{77}$
37 $\frac{240}{289}$ **39** $\frac{24}{7}$ **41** $\frac{1}{3}$
43 (a) $\frac{1}{2}\cos 3t - \frac{1}{2}\cos 11t$ (b) $\frac{1}{2}\cos\frac{1}{12}u + \frac{1}{2}\cos\frac{5}{12}u$
(c) $3 \sin 8x - 3 \sin 2x$
45 $5\pi/6$ **47** π **49** $\frac{1}{2}$ **51** $-\frac{7}{25}$ **53** $\pi/2$
55

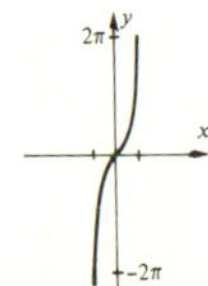

57 $a = 50\sqrt{6}, c = 50(1 + \sqrt{3}), \gamma = 75^\circ$
59 $a = \sqrt{43}, \beta = \arccos \frac{11}{16}, \gamma = \arccos(-\frac{1}{4})$
61 $10\sqrt{2}\,(\cos 3\pi/4 + i \sin 3\pi/4)$ **63** $17\,(\cos \pi + i \sin \pi)$
65 $10(\cos 7\pi/6 + i \sin 7\pi/6)$ **67** $-512i$
69 $-972 + 972i$ **71** $-3, \frac{3}{2} \pm (3\sqrt{3}/2)i$
73 Terminal points are $(-2, -3), (-6, 13), (-8, 10), (-1, 4)$.

CHAPTER 7

Exercises 7.1, page 320

1 $(3, 5), (-1, -3)$ **3** $(1, 0), (-3, 2)$
5 $(0, 0), (\frac{1}{8}, \frac{1}{128})$ **7** $(3, -2)$ **9** No solutions
11 $(-4, 3), (5, 0)$ **13** $(-2, 2)$
15 $((-6 - \sqrt{86})/10, (2 - 3\sqrt{86})/10)$, $((-6 + \sqrt{86})/10, (2 + 3\sqrt{86})/10)$
17 $(-4, 0), (\frac{12}{5}, \frac{16}{5})$ **19** $(0, 1), (4, -3)$
21 $(\pm 2, 5), (\pm\sqrt{5}, 4)$ **23** $(\sqrt{2}, \pm 2\sqrt{3}), (-\sqrt{2}, \pm 2\sqrt{3})$
25 $(2\sqrt{2}, \pm 2), (-2\sqrt{2}, \pm 2)$ **27** $x = 3, y = -1, z = 2$
29 $(\log 3/\log 5, 7)$ **31** $r = 3, s = -4$
33 $x = \frac{35}{22}, y = \frac{43}{44}$ **35** 13, 9 **37** 8 in., 12 in.
39 $5 - \sqrt{19}, 5 + \sqrt{19}, 10$

Exercises 7.2, page 326

1 $(4, -2)$ **3** $(8, 0)$ **5** $(-1, \frac{3}{2})$ **7** $(\frac{76}{53}, \frac{28}{53})$
9 $(\frac{51}{13}, \frac{96}{13})$ **11** $(\frac{8}{7}, -3\sqrt{6}/7)$ **13** No solution
15 All ordered pairs (m, n) such that $3m - 4n = 2$
17 $(0, 0)$ **19** $(-\frac{22}{7}, -\frac{11}{5})$ **21** $a = 2, b = 4$
23 137 adults, 313 children **25** 68
27 25 dimes, 18 quarters
29 \$6,650 at 6.5% and \$3,325 at 8%
31 40 gm of 35% alloy, 60 gm of 60% alloy
33 540 mph, 60 mph **35** $v_0 = 10, a = 3$

Exercises 7.3, page 337

1 $(2, 3, -1)$ **3** $(-2, 4, 5)$ **5** No solution
7 $(\frac{2}{3}, \frac{31}{21}, \frac{1}{21})$

There are other forms for the answers in Exercises 9–15.

9 $(2c, -c, c)$ where c is any real number
11 $(0, -c, c)$ where c is any real number
13 $((12 - 9c)/7, (8c - 13)/14, c)$, where c is any real number
15 $((7c + 5)/10, (19c - 15)/10, c)$ where c is any real number
17 $(1, 3, -1, 2)$ **19** $x = 2, y = -1, z = 3, s = 4, t = 1$
21 $(\frac{1}{11}, \frac{31}{11}, \frac{3}{11})$ **23** $(-2, -3)$ **25** No solution
27 24 dimes **29** 17 L of 10%, 11 L of 30%, 22 L of 50%
31 4 hr for A, 2 hr for B, 5 hr for C
33 380 lb of G_1, 60 lb of G_2, 160 lb of G_3
35 $f(x) = -\frac{1}{2}x^2 + x + \frac{5}{2}$ **37** $x^2 + y^2 - x + 3y - 6 = 0$

Exercises 7.4, page 346

1 $\begin{bmatrix} 9 & -1 \\ -2 & 5 \end{bmatrix}, \begin{bmatrix} 1 & -3 \\ 4 & 1 \end{bmatrix}, \begin{bmatrix} 10 & -4 \\ 2 & 6 \end{bmatrix}, \begin{bmatrix} -12 & -3 \\ 9 & -6 \end{bmatrix}$

3 $\begin{bmatrix} 9 & 0 \\ 1 & 5 \\ 3 & 4 \end{bmatrix}, \begin{bmatrix} 3 & -2 \\ 3 & -5 \\ -9 & 4 \end{bmatrix}, \begin{bmatrix} 12 & -2 \\ 4 & 0 \\ -6 & 8 \end{bmatrix}, \begin{bmatrix} -9 & -3 \\ 3 & -15 \\ -18 & 0 \end{bmatrix}$

5 $[11 \;\; -3 \;\; -3], [-3 \;\; -3 \;\; 7], [8 \;\; -6 \;\; 4]$, $[-21 \;\; 0 \;\; 15]$

7 $\begin{bmatrix} -3 & 4 & 1 & 6 \\ 3 & 2 & 7 & -7 \end{bmatrix}, \begin{bmatrix} 3 & 4 & -1 & 0 \\ -1 & 2 & -7 & -3 \end{bmatrix}$, $\begin{bmatrix} 0 & 8 & 0 & 6 \\ 2 & 4 & 0 & -10 \end{bmatrix}, \begin{bmatrix} 9 & 0 & -3 & -9 \\ -6 & 0 & -21 & 6 \end{bmatrix}$

9 $\begin{bmatrix} 16 & 38 \\ 11 & -34 \end{bmatrix}, \begin{bmatrix} 4 & 38 \\ 23 & -22 \end{bmatrix}$

11 $\begin{bmatrix} 3 & -14 & -3 \\ 16 & 2 & -2 \\ -7 & -29 & 9 \end{bmatrix}, \begin{bmatrix} 3 & -20 & -11 \\ 2 & 10 & -4 \\ 15 & -13 & 1 \end{bmatrix}$

13 $\begin{bmatrix} 4 & 8 \\ -18 & 11 \end{bmatrix}, \begin{bmatrix} 3 & -4 & 4 \\ -5 & 2 & 2 \\ -51 & 26 & 10 \end{bmatrix}$

15 $\begin{bmatrix} 1 & 2 & 3 \\ 4 & 5 & 6 \\ 7 & 8 & 9 \end{bmatrix}, \begin{bmatrix} 1 & 2 & 3 \\ 4 & 5 & 6 \\ 7 & 8 & 9 \end{bmatrix}$

17 $[15], \begin{bmatrix} -3 & 7 & 2 \\ -12 & 28 & 8 \\ 15 & -35 & -10 \end{bmatrix}$ **19** $\begin{bmatrix} 4 \\ 12 \\ -1 \end{bmatrix}$

21 $\begin{bmatrix} 18 & 0 & -2 \\ -40 & 10 & -10 \end{bmatrix}$ **31** $\dfrac{1}{10}\begin{bmatrix} 3 & 4 \\ -1 & 2 \end{bmatrix}$

33 Does not exist. **35** $\dfrac{1}{8}\begin{bmatrix} 2 & 1 & 0 \\ -2 & 3 & 0 \\ 0 & 0 & 2 \end{bmatrix}$

37 $\dfrac{1}{3}\begin{bmatrix} -4 & -5 & 3 \\ -4 & -8 & 3 \\ 1 & 2 & 0 \end{bmatrix}$ **39** $\begin{bmatrix} \frac{1}{2} & 0 & 0 \\ 0 & \frac{1}{4} & 0 \\ 0 & 0 & \frac{1}{6} \end{bmatrix}$

41 $\dfrac{1}{6}\begin{bmatrix} -8 & -7 & 4 & -9 \\ -8 & -4 & 4 & -6 \\ -4 & -5 & 2 & -3 \\ 6 & 3 & 0 & 3 \end{bmatrix}$ **43** $ab \neq 0; \begin{bmatrix} 1/a & 0 \\ 0 & 1/b \end{bmatrix}$

47 $(\frac{13}{10}, -\frac{1}{10})$ **49** $(-\frac{25}{3}, -\frac{34}{3}, \frac{7}{3})$

Exercises 7.5, page 353

1 $M_{11} = -14$, $M_{21} = 7$, $M_{31} = 11$, $M_{12} = 10$, $M_{22} = -5$, $M_{32} = 4$, $M_{13} = 15$, $M_{23} = 34$, $M_{33} = 6$; $A_{11} = -14$, $A_{21} = -7$, $A_{31} = 11$, $A_{12} = -10$, $A_{22} = -5$, $A_{32} = -4$, $A_{13} = 15$, $A_{23} = -34$, $A_{33} = 6$
3 $M_{11} = 0$, $M_{12} = 5$, $M_{21} = -1$, $M_{22} = 7$; $A_{11} = 0$, $A_{12} = -5$, $A_{21} = 1$, $A_{22} = 7$
5 -83 **7** 5 **9** 2 **11** 0 **13** -125
15 48 **17** -216 **19** $abcd$
31 (a) $x^2 - 3x - 4$ (b) $-1, 4$
33 (a) $x^2 + x - 2$ (b) $1, -2$
35 (a) $-x^3 - 2x^2 + x + 2$ (b) $1, -1, -2$
37 (a) $-x^3 + 4x^2 + 4x - 16$ (b) $4, 2, -2$
39 $-31i - 20j + 7k$ **41** $-6i - 8j + 18k$

Exercises 7.6, page 358

1 R_{23} **3** $(-1)R_1 + R_3$
5 2 is a common factor of rows 1 and 3.
7 Two rows are identical. **9** $(-1)R_2$
11 Every number in column 2 is 0. **13** $2C_1 + C_3$
15 -10 **17** -142 **19** -183 **21** 44
23 359

Exercises 7.8, page 366

1

3

5

7

9

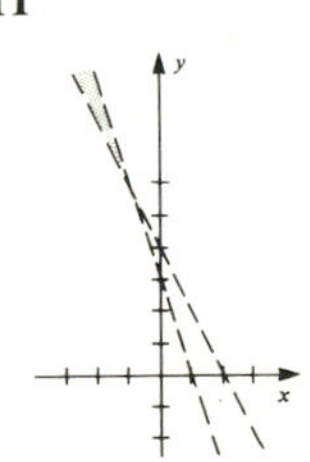

11

13

15

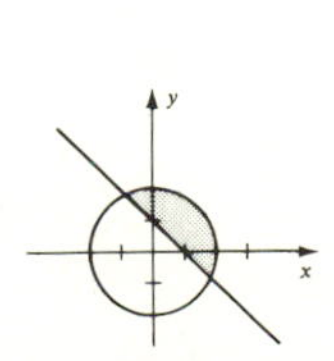

17

19

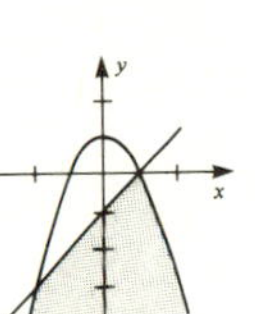

21

23

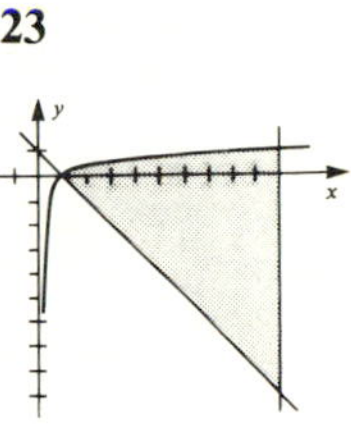

25 Let y denote the number of bats and x the number of balls. Then $x \geq 12$, $y \geq 5$, and $3x + 13y \leq 160$. The graph is the region bounded by the triangle with vertices (12, 5), $(12, \frac{124}{13})$, $(\frac{95}{3}, 5)$.
27 If x and y denote the number of brand A and brand B, respectively, then $x \geq 20$, $y \geq 10$, $x \geq 2y$, and $x + y \leq 100$. The graph is the region bounded by the triangle with vertices (20, 10), (90, 10), $(\frac{200}{3}, \frac{100}{3})$.
29 If x and y denote the amounts in the first and second accounts, respectively, then $x \geq 2000$, $y \geq 2000$, $y \geq 3x$, and $x + y \leq 15000$. The graph is the region bounded by the triangle with vertices (2000, 6000), (2000, 13000), (3750, 11250).

Exercises 7.9, page 371

1 50 Double Fault and 30 Set Point
3 3.5 lb of S and 1 lb of T
5 Send 25 from W_1 to A and 0 from W_1 to B. Send 10 from W_2 to A and 60 from W_2 to B.
7 45 acres of crop A and 50 acres of crop B
9 Minimum cost: 16 oz X, 4 oz Y, 0 oz Z; maximum cost: 0 oz X, 8 oz Y, 12 oz Z

Exercises 7.10, page 373

1 $(\frac{19}{23}, -\frac{18}{23})$ **3** $(-3, 5)$, $(1, -3)$
5 $(2\sqrt{3}, \pm\sqrt{2})$, $(-2\sqrt{3}, \pm\sqrt{2})$ **7** $(\frac{14}{17}, \frac{14}{27})$
9 $(\frac{6}{11}, -\frac{7}{11}, 1)$
11 $(-2c, -3c, c)$ where c is any real number.
13 $(5c - 1, (-19c + 5)/2, c)$ where c is any real number
15 $(-1, \frac{1}{2}, \frac{1}{3})$

17

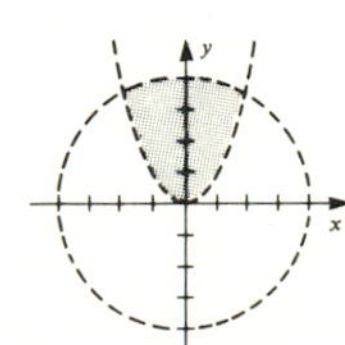

19

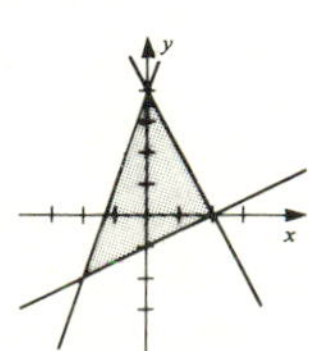

21 -6 **23** 48 **25** -84 **27** 120 **29** 0

31 $a_{11}a_{22}a_{33}\cdots a_{nn}$ **33** $\left(-\frac{1}{2}\right)\begin{bmatrix} 2 & 4 \\ 3 & 5 \end{bmatrix}$

35 $\frac{1}{7}\begin{bmatrix} 2 & 1 & 0 & 0 \\ -1 & 3 & 0 & 0 \\ 0 & 0 & 3 & 2 \\ 0 & 0 & -5 & -1 \end{bmatrix}$ **37** $\begin{bmatrix} 4 & -5 & 6 \\ 4 & -11 & 5 \end{bmatrix}$

39 $\begin{bmatrix} 0 & 4 & -6 \\ 16 & 22 & 1 \\ 12 & 11 & 9 \end{bmatrix}$ **41** $\begin{bmatrix} -12 & 4 & -11 \\ 6 & -11 & 5 \end{bmatrix}$

43 $\begin{bmatrix} a & 3a \\ 2b & 4b \end{bmatrix}$ **45** $\begin{bmatrix} 5 & 9 \\ 13 & 19 \end{bmatrix}$

CHAPTER 8

Exercises 8.2, page 386

1 $a^6 + 6a^5b + 15a^4b^2 + 20a^3b^3 + 15a^2b^4 + 6ab^5 + b^6$
3 $a^8 - 8a^7b + 28a^6b^2 - 56a^5b^3 + 70a^4b^4 - 56a^3b^5 + 28a^2b^6 - 8ab^7 + b^8$
5 $81x^4 - 540x^3y + 1350x^2y^2 - 1500xy^3 + 625y^4$
7 $u^{10} + 20u^8v + 160u^6v^2 + 640u^4v^3 + 1280u^2v^4 + 1024v^5$
9 $r^{-12} - 12r^{-9} + 60r^{-6} - 160r^{-3} + 240 - 192r^3 + 64r^6$
11 $1 + 10x + 45x^2 + 120x^3 + 210x^4 + 252x^5 + 210x^6 + 120x^7 + 45x^8 + 10x^9 + x^{10}$
13 $(3^{25})c^{10} + 25(3^{24})c^{52/5} + 300(3^{23})c^{54/5} + 2300(3^{22})c^{56/5}$
15 $60(3^{14})b^{13} - (3^{15})b^{15}$ **17** $30618a^{10}b^2$
19 $840u^4v^6$ **21** $924x^2y^2$ **23** -61236
25 $24x^2y^6$ **27** $210c^3d^2$
29 $1 + 0.2 + 0.018 + 0.00096 = 1.21896$

Exercises 8.3, page 393

1 $9, 6, 3, 0, -3; -12$ **3** $\frac{1}{2}, \frac{4}{5}, \frac{7}{10}, \frac{10}{17}, \frac{13}{26}; \frac{22}{65}$
5 $9, 9, 9, 9, 9; 9$
7 $1.9, 2.01, 1.999, 2.0001, 1.99999; 2.00000001$
9 $4, -\frac{9}{4}, \frac{5}{3}, -\frac{11}{8}, \frac{6}{5}; -\frac{15}{16}$ **11** $2, 0, 2, 0, 2; 0$
13 $\frac{2}{3}, \frac{2}{3}, \frac{8}{11}, \frac{8}{9}, \frac{32}{27}; \frac{128}{33}$ **15** $1, 2, 3, 4, 5; 8$
17 $2, 1, -2, -11, -38$ **19** $-3, 3^2, 3^4, 3^8, 3^{16}$
21 $5, 5, 10, 30, 120$ **23** $2, 2, 4, 4^3, 4^{12}$
25 $a_n = 2n + \frac{1}{24}(n-1)(n-2)(n-3)(n-4)(a-10)$ (There are many other answers.)
27 -5 **29** 10
31 25 **33** $-\frac{17}{15}$ **35** 61
37 $10,000$ **39** $(2n^3 + 12n^2 + 40/n)/6$
41 $(4n^3 - 12n^2 + 11n)/3$ **43** $\sum_{k=1}^{5} (4k-3)$
45 $\sum_{k=1}^{4} k/(3k-1)$ **47** $1 + \sum_{k=1}^{n} (-1)^k x^{2k}/(2k)$
49 $\sum_{k=1}^{7} (-1)^{k-1}/k$ **51** $\sum_{n=1}^{99} 1/n(n+1)$

Exercises 8.4, page 398

1 $18, 38, 4n - 2$ **3** $1.8, 0.3, 3.3 - 0.3n$
5 $5.4, 20.9, (3.1)n - (10.1)$ **7** $\ln 3^5, \ln 3^{10}, \ln 3^n$
9 -8.5 **11** -9.8 **13** -105 **15** 30
17 530 **19** $\frac{423}{2}$ **21** $60; 12,780$ **23** 24
25 $\frac{10}{3}, \frac{14}{3}, 6, \frac{22}{3}, \frac{26}{3}$ **27** 255
29 $23.25, 22.5, 21.75, 21, 20.25, 19.5, 18.75$

Exercises 8.5, page 404

1 $\frac{1}{2}, \frac{1}{16}; 8(\frac{1}{2})^{n-1} = 2^{4-n}$
3 $0.03, -0.00003; 300(-0.1)^{n-1}$ **5** $3125, 5^8; 5^n$
7 $\frac{81}{4}, -3^7/2^5; 4(-\frac{3}{2})^{n-1}$
9 $x^8, -x^{14}; (-1)^{n-1}x^{2n-2}$
11 $2^{4x+1}, 2^{7x+1}; 2^{nx-x+1}$ **13** $\frac{243}{8}$
15 $a_1 = \frac{1}{81}, S_5 = \frac{211}{1296}$ **17** $-\frac{3}{2}(1 - 3^{10}) = 88,572$
19 $-\frac{1}{3}(1 - 2^{-10})$ **21** $\frac{25}{256}\%$
23 $10,000(\frac{6}{5})^t, 10,000(\frac{6}{5})^{10}$ **25** $\frac{2}{3}$ **27** $\frac{50}{33}$
29 Sum does not exist ($|r| = \sqrt{2} > 1$). **31** $\frac{23}{99}$
33 $\frac{2393}{990}$ **35** $\frac{5141}{999}$ **37** $\frac{16123}{9999}$ **39** 30 meters

Exercises 8.6, page 405

7 $x^{12} - 18x^{10}y + 135x^8y^2 - 540x^6y^3 + 1215x^4y^4 - 1458x^2y^5 + 729y^6$
9 $-\frac{63}{16}b^{12}c^{10}$ **11** $5, -2, -1, -\frac{20}{29}, -\frac{7}{19}$
13 $1, \frac{1}{2}, \frac{5}{4}, \frac{7}{8}, \frac{65}{64}$ **15** $10, \frac{11}{10}, \frac{21}{11}, \frac{32}{21}, \frac{53}{32}$
17 $9, 3, \sqrt{3}, \sqrt[4]{3}, \sqrt[8]{3}$ **19** 75 **21** 1000
23 $\sum_{k=1}^{5} 3k$ **25** $\sum_{k=0}^{4} (-1)^k 5(20-k)$
27 $-5 - 8\sqrt{3}, -5 - 35\sqrt{3}$
29 $-31, 50$ **31** 64
33 $4\sqrt{2}$ **35** 570 **37** 2041 **39** $\frac{5}{7}$

CHAPTER 9

Exercises 9.2, page 414

1 $V(0, 0); F(0, -3); y = 3$ **3** $V(0, 0); F(-\frac{3}{8}, 0); x = \frac{3}{8}$

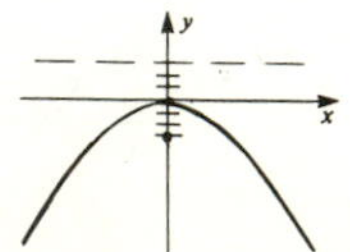

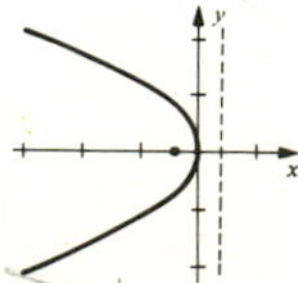

5 $V(0, 0)$; $F(0, \frac{1}{32})$; $y = -\frac{1}{32}$

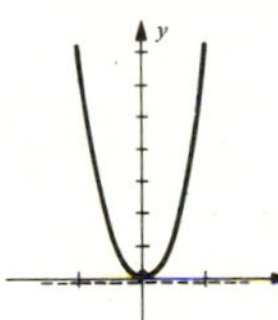

7 $V(-1, 0)$; $F(2, 0)$; $x = -4$

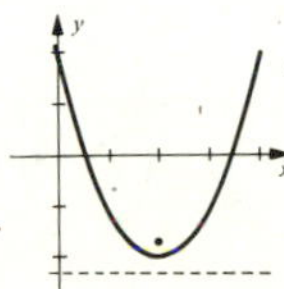

9 $V(2, -2)$; $F(2, -\frac{7}{4})$; $y = -\frac{9}{4}$

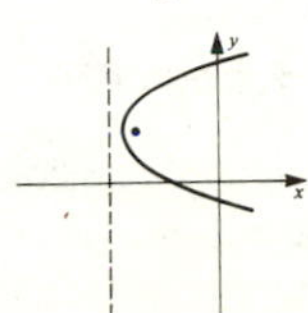

11 $V(-4, 2)$; $F(-\frac{7}{2}, 2)$; $x = -\frac{9}{2}$

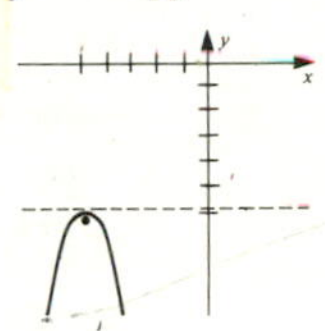

13 $V(-5, -6)$; $F(-5, -\frac{97}{16})$; $y = -\frac{95}{16}$

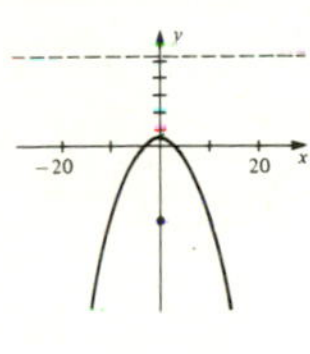

15 $V(0, \frac{1}{2})$; $F(0, -\frac{9}{2})$; $y = \frac{11}{2}$

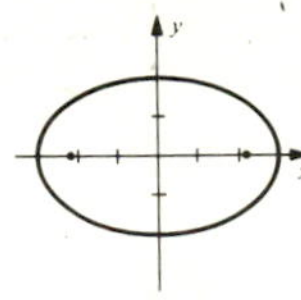

17 $y^2 = 8x$ 19 $(x - 6)^2 = 12(y - 1)$ 21 $3x^2 = -4y$
23 $\frac{9}{16}$ ft from the vertex 25 $y = 2x^2 - 3x + 1$

Exercises 9.3, page 420

1 $V(\pm 3, 0)$; $F(\pm\sqrt{5}, 0)$

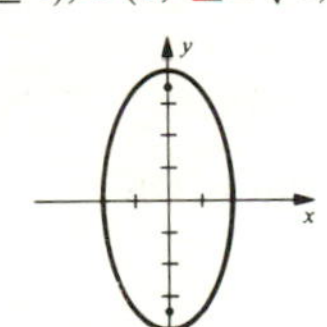

3 $V(0, \pm 4)$; $F(0, \pm 2\sqrt{3})$

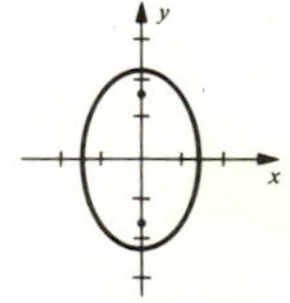

5 $V(0, \pm\sqrt{5})$; $F(0, \pm\sqrt{3})$

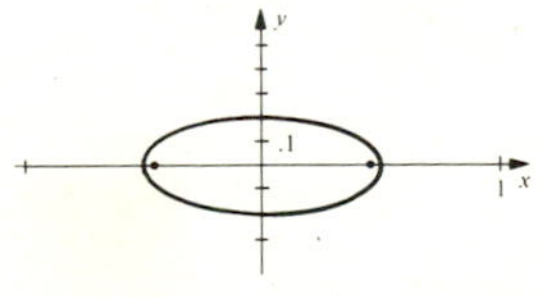

7 $V(\pm\frac{1}{2}, 0)$; $F(\pm\sqrt{21}/10, 0)$

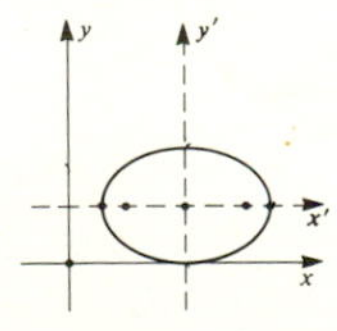

9 Center $(4, 2)$, vertices $(1, 2)$ and $(7, 2)$, $F(4 \pm \sqrt{5}, 2)$; endpoints of minor axis $(4, 4)$ and $(4, 0)$

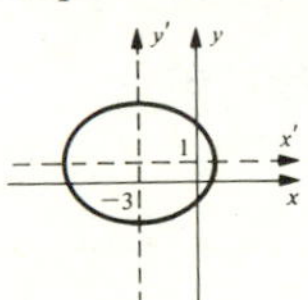

11 Center $(-3, 1)$, vertices $(-7, 1)$ and $(1, 1)$, $F(-3 \pm \sqrt{7}, 1)$; endpoints of minor axis $(-3, 4)$ and $(-3, -2)$

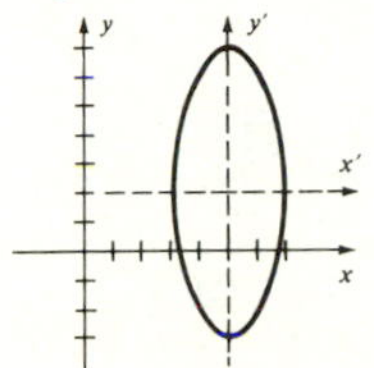

13 Center $(5, 2)$, vertices $(5, 7)$ and $(5, -3)$, $F(5, 2 \pm \sqrt{21})$; endpoints of minor axis $(3, 2)$ and $(7, 2)$

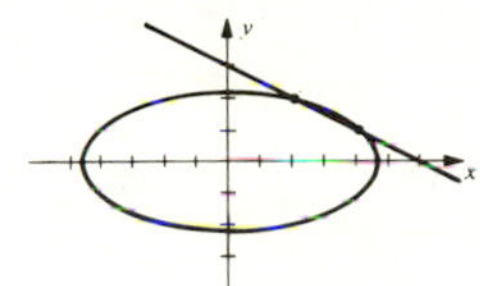

15 $x^2/64 + y^2/39 = 1$ 17 $4x^2/9 + y^2/25 = 1$
19 $8x^2/81 + y^2/36 = 1$
21 $\{(2, 2), (4, 1)\}$ 23 $2\sqrt{21}$ ft

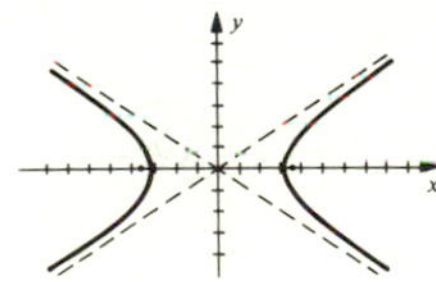

Exercises 9.4, page 426

1 $V(\pm 3, 0)$; $F(\pm\sqrt{13}, 0)$; $y = \pm\frac{2}{3}x$

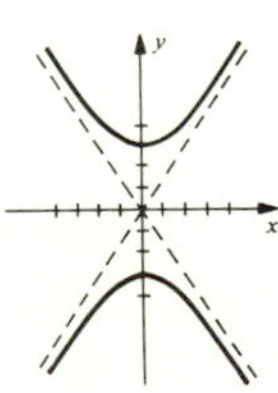

3 $V(0, \pm 3)$; $F(0, \pm\sqrt{13})$; $y = \pm\frac{3}{2}x$

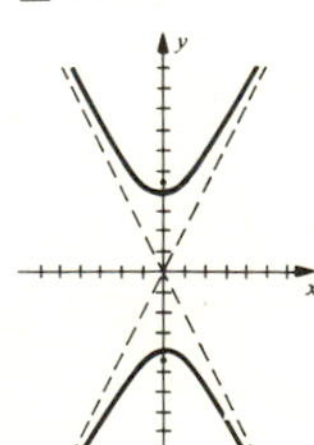

5 $V(0, \pm 4)$; $F(0, \pm 2\sqrt{5})$; $y = \pm 2x$

7 $V(\pm 1, 0)$; $F(\pm\sqrt{2}, 0)$; $y = \pm x$

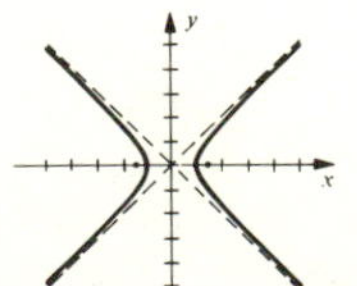

9 $V(\pm 5, 0)$; $F(\pm\sqrt{30}, 0)$; $y = \pm(\sqrt{5}/5)x$

11 $V(0, \pm\sqrt{3})$; $F(0, \pm 2)$; $y = \pm\sqrt{3}x$

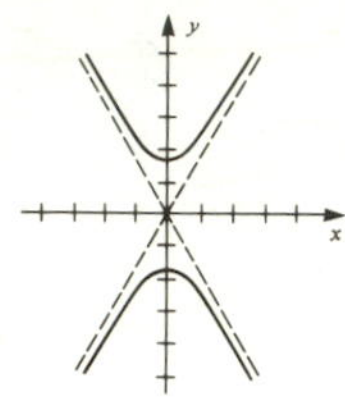

13 Center $(-5, 1)$; vertices $(-5 \pm 2\sqrt{5}, 1)$; $F(-5 \pm \sqrt{205}/2, 1)$; $y - 1 = \pm\frac{5}{4}(x + 5)$

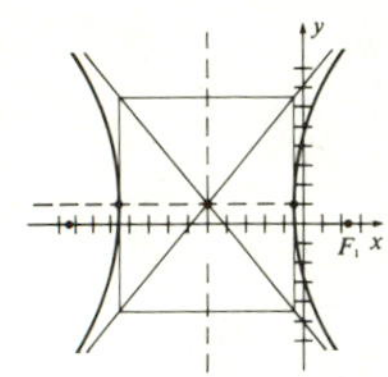

15 Center $(-2, -5)$; vertices $(-2, -2)$ and $(-2, -8)$; $F(-2, -5 \pm 3\sqrt{5})$; $y + 5 = \pm\frac{1}{2}(x + 2)$

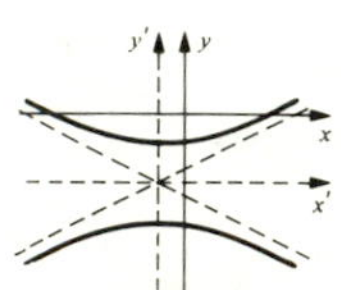

17 Center $(6, 2)$; vertices $(6, 4)$ and $(6, 0)$; $F(6, 2 \pm 2\sqrt{10})$ $y - 2 = \pm\frac{1}{3}(x - 6)$

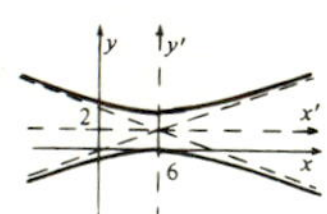

19 $15y^2 - x^2 = 15$ **21** $x^2/9 - y^2/16 = 1$

23 $y^2/21 - x^2/4 = 1$ **25** $x^2/9 - y^2/36 = 1$

27 $\{(0, 4), (\frac{8}{3}, \frac{20}{3})\}$

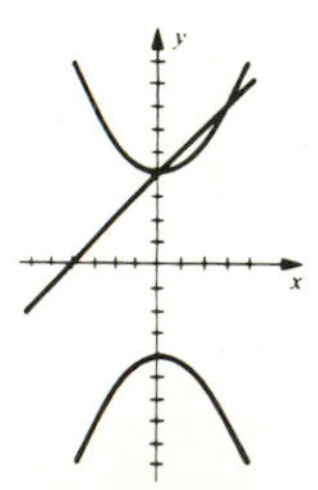

29 Conjugate hyperbolas have the same asymptotes.

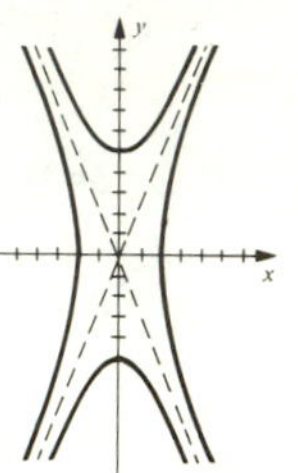

Exercises 9.5, page 430

The following answers contain equations in x' and y' resulting from a rotation of axis.

1 Ellipse, $(x')^2 + 16(y')^2 = 16$

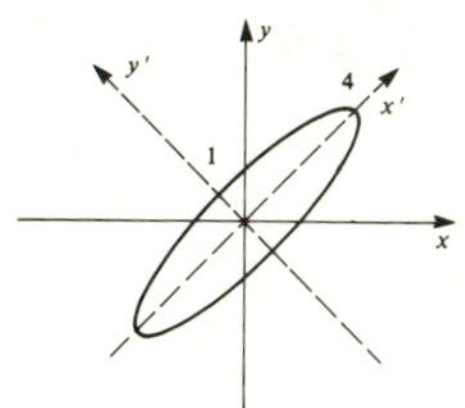

3 Hyperbola, $4(x')^2 - (y')^2 = 1$

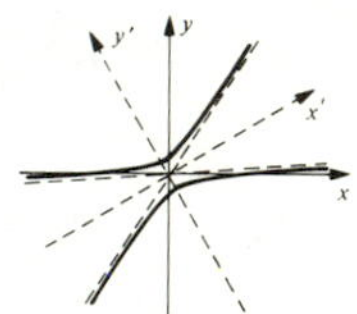

5 Ellipse, $(x')^2 + 9(y')^2 = 9$

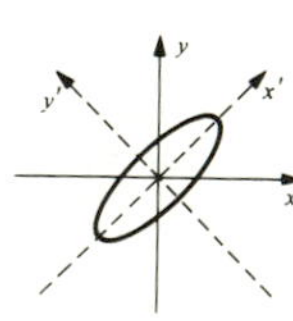

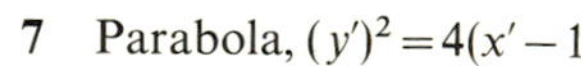

7 Parabola, $(y')^2 = 4(x' - 1)$

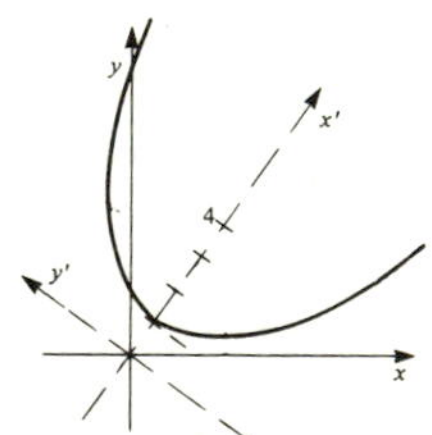

9 Hyperbola, $2(x')^2 - (y')^2 - 4y' - 3 = 0$

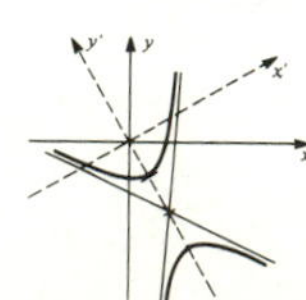

11 Parabola, $(x')^2 - 6x' - 6y' + 9 = 0$

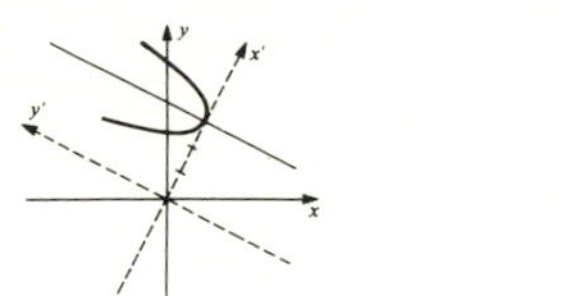

13 Ellipse, $(x')^2 + 4(y')^2 - 4x' = 0$

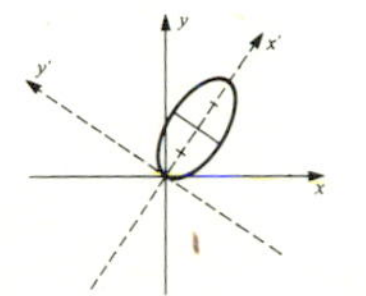

15 Sketch of proof: It can be shown (see page 429) that $B^2 - 4AC = B'^2 - 4A'C'$. For a suitable rotation of axes we obtain $B' = 0$ and the transformed equation has the form $A'x'^2 + C'y'^2 + D'x' + E'y' + F' = 0$. Except for degenerate cases, the graph of the last equation is an ellipse, hyperbola, or parabola if $A'C' > 0$, $A'C' < 0$, or $A'C' = 0$, respectively. However, if $B' = 0$, then $B^2 - 4AC = -4A'C'$ and hence the graph is an ellipse, hyperbola, or parabola if $B^2 - 4AC < 0$, $B^2 - 4AC > 0$, or $B^2 - 4AC = 0$, respectively.

Exercises 9.6, page 436

1 **3** **5**

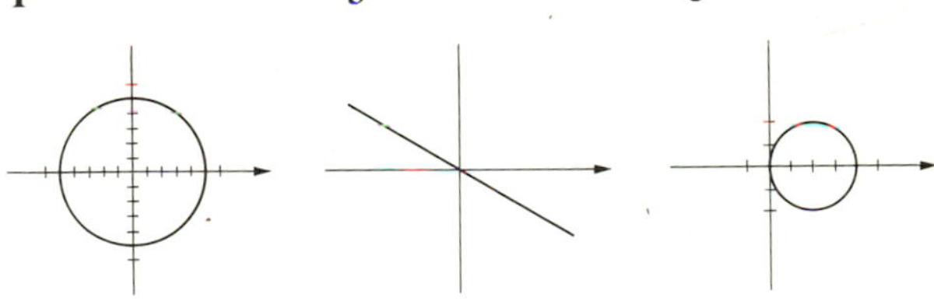

7 **9** **11**

13 **15** **17**

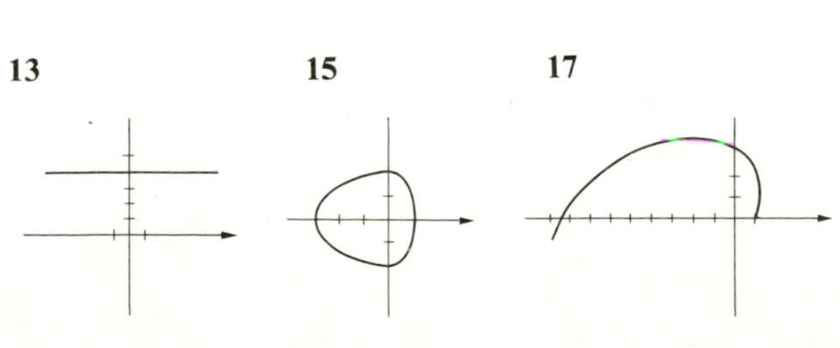

19 **21** **23**

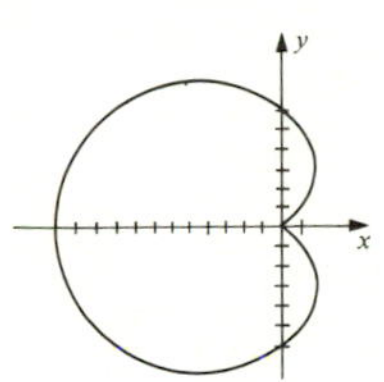

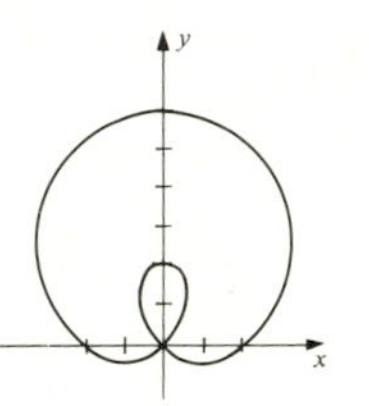

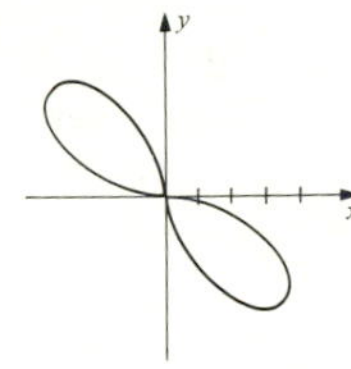

25 $r = -3 \sec \theta$ **27** $r = 4$ **29** $r = 6 \csc \theta$

31 $r^2 = 16 \sec 2\theta$ **33** $x = 5$ **35** $x^2 + y^2 - 6y = 0$

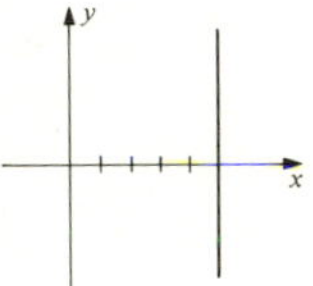

37 $x^2 + y^2 = 4$ **39** $y^2 = x^4/(1 - x^2)$

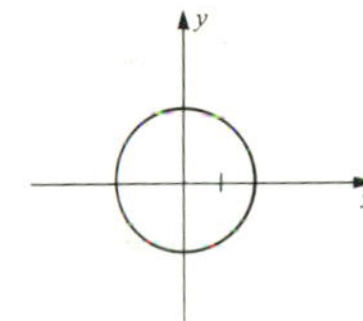

41 $y^2/9 - x^2/4 = 1$ **43** $x^2 - y^2 = 1$

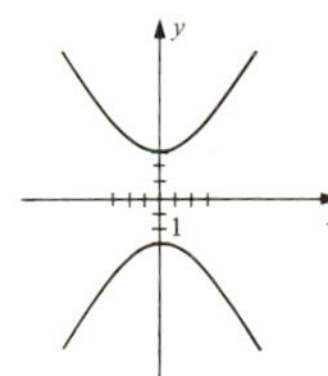

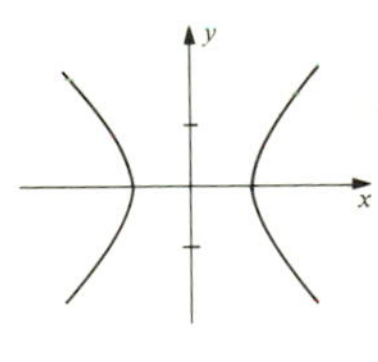

45 $y - 2x = 6$ **47** $y^2 = 1 - 2x$

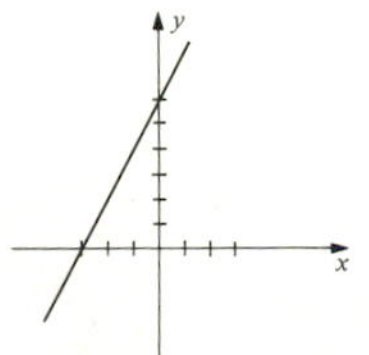

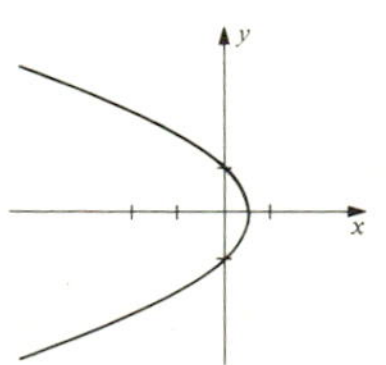

Exercises 9.7, page 441

1 Ellipse: vertices $(\frac{3}{2}, \pi/2)$ and $(3, 3\pi/2)$; foci $(0, 0)$ and $(\frac{3}{2}, 3\pi/2)$

3 Hyperbola; vertices $(-3, 0)$ and $(\frac{3}{2}, \pi)$; foci $(0, 0)$ and $(-\frac{9}{2}, 0)$
5 Parabola; $V(\frac{3}{4}, 0)$; $F(0, 0)$
7 Ellipse; vertices $(-4, 0)$ and $(-\frac{4}{3}, \pi)$; foci $(0, 0)$ and $(-\frac{8}{3}, 0)$
9 Hyperbola (except for the points $(\pm 3, 0)$); vertices $(\frac{6}{5}, \pi/2)$ and $(-6, 3\pi/2)$; foci $(0, 0)$ and $(-\frac{36}{5}, 3\pi/2)$
11 $9x^2 + 8y^2 + 12y - 36 = 0$
13 $y^2 - 8x^2 - 36x - 36 = 0$
15 $4y^2 = 9 - 12x$ **17** $3x^2 + 4y^2 + 8x - 16 = 0$
19 $4x^2 - 5y^2 + 36y - 36 = 0$
21 $r = 2/(1 + \cos\theta)$; $e = 1$; $r = 2 \sec\theta$
23 $r = 4/(1 + 2\sin\theta)$; $e = 2$; $r = 2\csc\theta$
25 $r = 2/(3 + \cos\theta)$; $e = \frac{1}{3}$; $r = 2\sec\theta$
27 $r = 2/(3 + \cos\theta)$ **29** $r = 12/(1 - 4\sin\theta)$
31 $r = 5/(1 + \cos\theta)$ **33** $r = 8/(1 + \sin\theta)$

Exercises 9.8, page 447

1 $y = 2x + 7$ **3** $y = x - 2$ **5** $x = y^2 - 6y + 4$

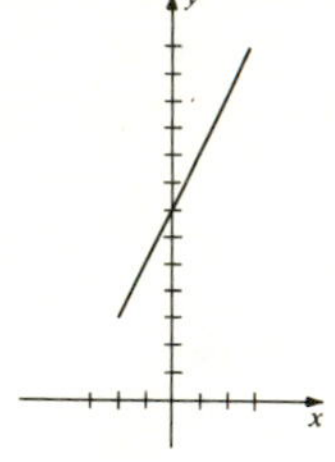
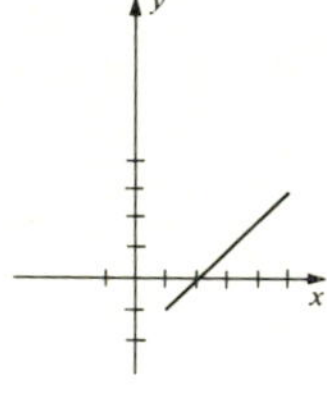
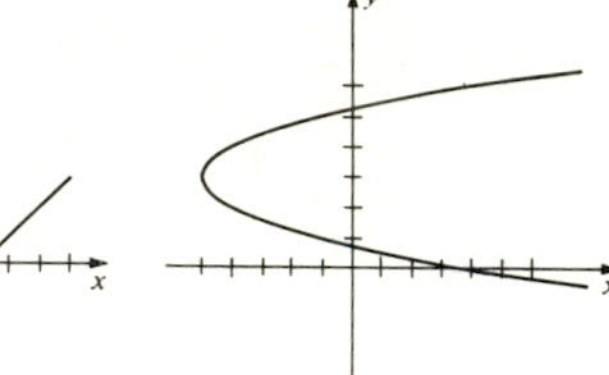

7 $y = 1/x^2$ **9** $x^2/4 + y^2/9 = 1$ **11** $x^2 - y^2 = 1$

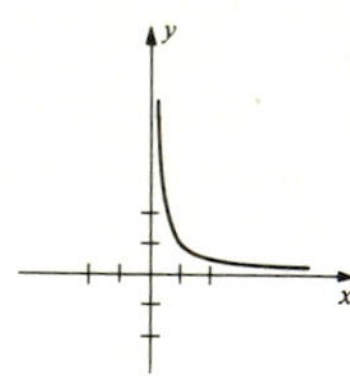
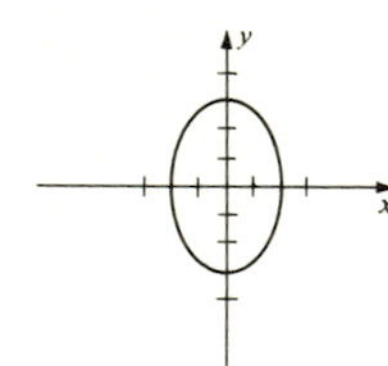
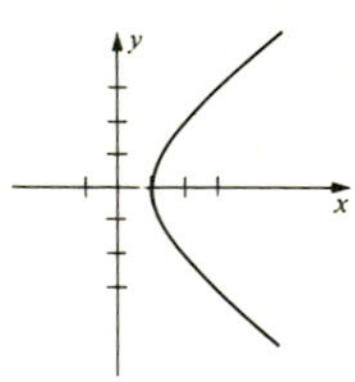

13 $y = \ln x$ **15** $y = 1/x$

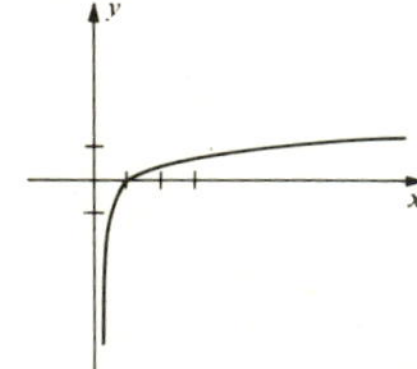
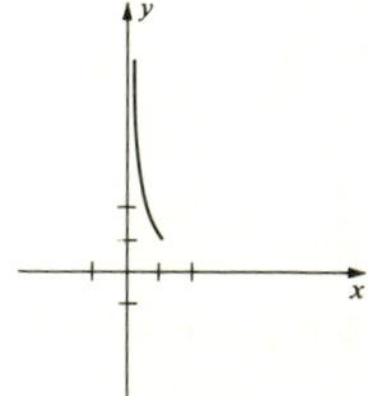

17 $y = \sqrt{x^2 - 1}$ **19** **21**

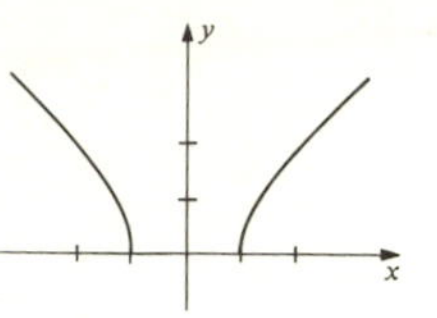
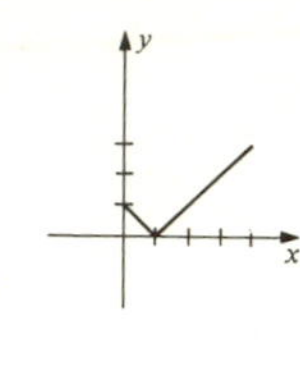
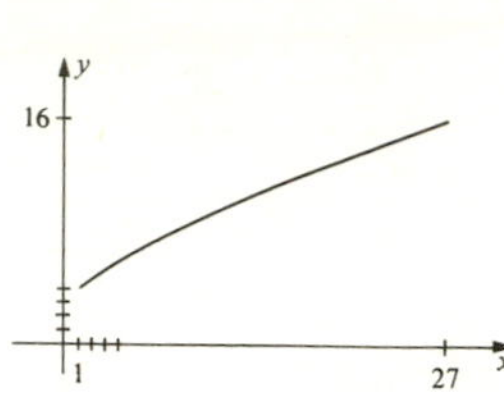

23

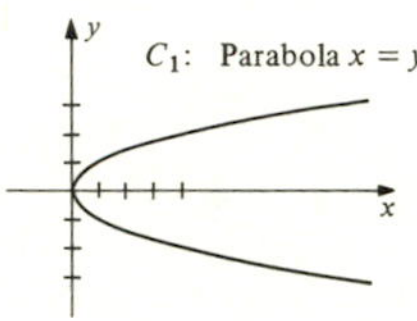

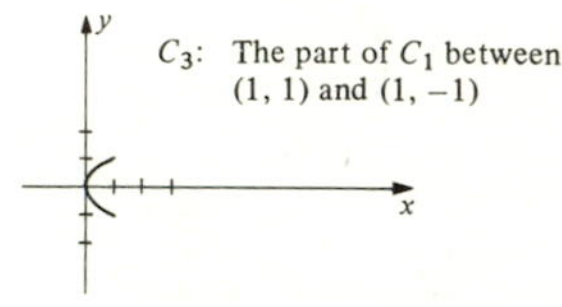

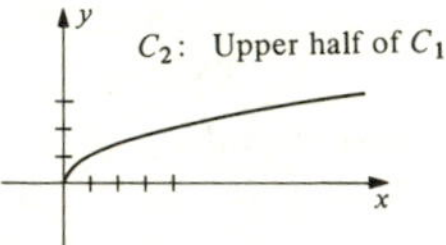

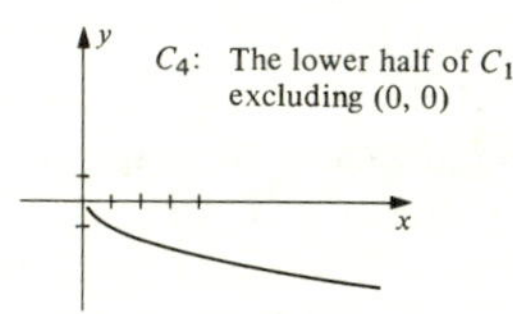

25 $x = (x_2 - x_1)t^n + x_1$, $y = (y_2 - y_1)t^n + y_1$, are parametric equations for l if n is any odd positive integer.

27

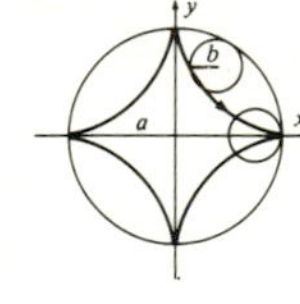

29 $x = 4b\cos t - b\cos 4t$, $y = 4b\sin t - b\sin 4t$

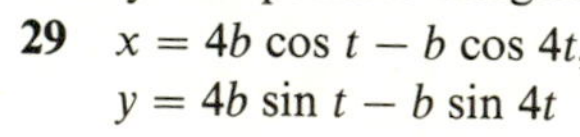
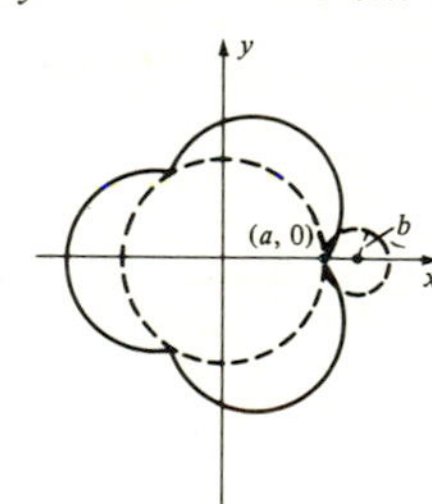

Exercises 9.9, page 449

1 Parabola; $F(16, 0)$; $V(0, 0)$
3 Ellipse; $F(0, \pm\sqrt{7})$; $V(0, \pm 4)$
5 Hyperbola; $F(\pm 2\sqrt{2}, 0)$; $V(\pm 2, 0)$

1 **3** **5**

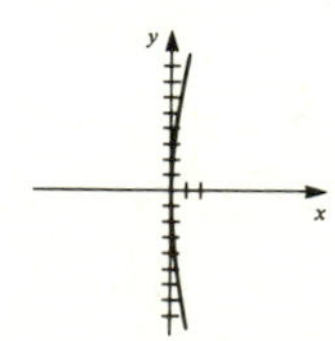
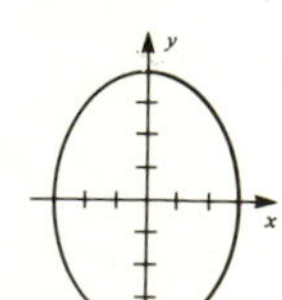
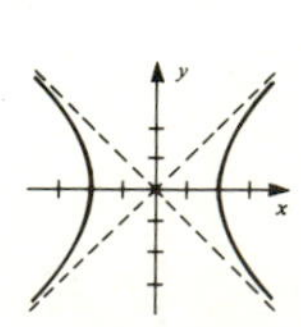

7 Parabola; $F(0, -\frac{9}{4})$; $V(0, 4)$

9 Hyperbola: vertices $(-5, 5)$ and $(-3, 5)$; foci $(-4 \pm \sqrt{10}/3, 5)$

7

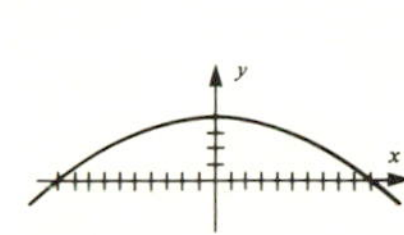

9

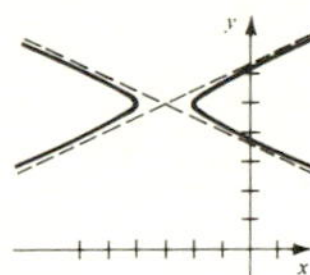

11 $9y^2 - 49x^2 = 441$ **13** $x^2 = -40y$

15 $4x^2 + 3y^2 = 300$ **17** $y^2 - 81x^2 = 36$

19 Ellipse; center $(-3, 2)$, vertices $(-6, 2)$ and $(0, 2)$, endpoints of minor axis $(-3, 0)$ and $(-3, 4)$

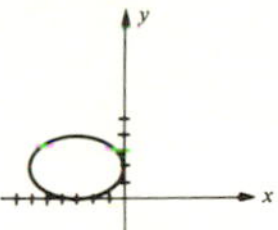

21 Parabola; $V(2, -4)$, $F(4, -4)$

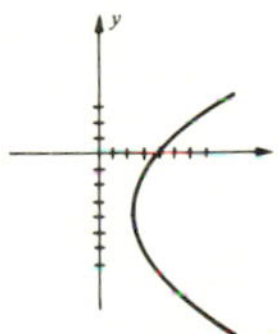

23 Hyperbola; center $(2, -3)$, vertices $(2, -1)$ and $(2, -5)$, endpoints of conjugate axis $(2 \pm \sqrt{2}, -3)$

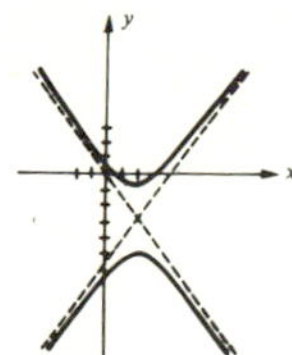

25

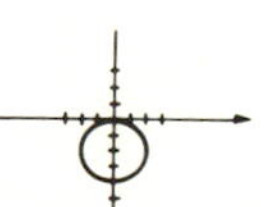

27

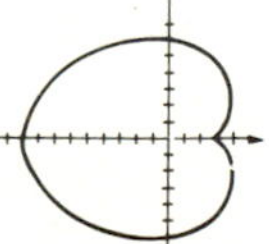

29

31

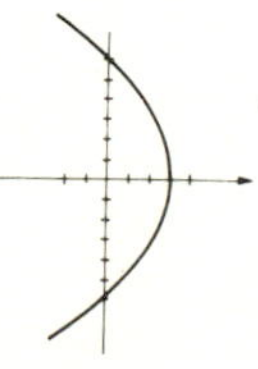

33 $r \sin^2 \theta = 4 \cos \theta$ **35** $r(2 \cos \theta - 3 \sin \theta) = 8$

37 $x^3 + xy^2 = y$ **39** $(x^2 + y^2)^2 = 8xy$

41 $y = 2(x - 1) - 1/(x - 1)$ **43** $y = 2^{-x^2}$

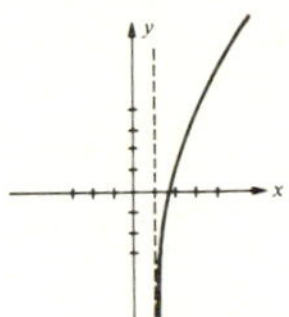

45 Parabola: $(y')^2 - 3x' = 0$ is an equation after a rotation of axes.

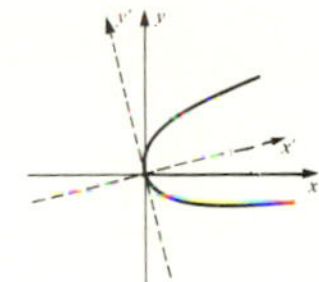

Exercises A.1, page A9

1 2.5403, 7.5403 − 10, 0.5403
3 9.7324 − 10, 2.7324, 5.7324
5 1.7796, 5.7796 − 10, 2.7796
7 3.3044, 0.8261, 6.6956 − 10
9 9.9235 − 10, 0.0765, 9.9915 − 10 **11** 4.0428
13 2.0878 **15** 8.1462 **17** 4,240 **19** 8.85
21 162,000 **23** 0.543 **25** 0.0677 **27** 189
29 0.0237 **31** 1.4062 **33** 3.7294
35 7.1000 − 10 **37** 5.0913 **39** 9.8913 − 10
41 2.5851 **43** 9.9760 − 10 **45** 4.8234
47 8.6356 − 10 **49** 0.4776 **51** 27.78
53 49,330 **55** 0.1467 **57** 0.006536
59 234.7 **61** 1.367 **63** 0.09445 **65** 109,900
67 0.001345 **69** 0.2442 **71** 0.4440
73 −0.1426 **75** 1.032 **77** 0.7911 **79** 0.5217
81 3.436 **83** 0.5315 **85** 1.3521 **87** 0.7703
89 21°33′, 158°27′ **91** 26°45′, 206°45′
93 69°44′, 290°16′ **95** 43°48′, 316°12′

INDEX

Abscissa, 56
Absolute value
  of a complex number, 294
  properties of, 46
  of a real number, 7
Acute angle, 192
Addition formulas, 260, 261
Addition of y-coordinates, 232
Additive identity, 2
Additive inverse, 2
Adjacent side, 198
Algebraic expression, 21
Alternating infinite series, 404
Ambiguous case, 287
Amplitude
  of a complex number, 295
  of a trigonometric function, 229
Angle(s)
  acute, 2
  central, 193
  complementary, 192
  coterminal, 191
  definition of, 190
  degree measure of, 191
  of depression, 238
  of elevation, 237
  initial side of, 191
  negative, 191
  obtuse, 192
  positive, 191
  quadrantal, 191
  radian measure of, 193
  reference, 214
  right, 192
  standard position of, 191
  subtended, 193
  supplementary, 192
  terminal side of, 191
  trigonometric functions of, 198, 203
  vertex of, 191
Angular speed, 241
Arccosine function, 280
Arcsine function, 278
Arctangent function, 280
Argument of a complex number, 295
Arithmetic mean, 397
Arithmetic progression, 395
Arithmetic sequence, 395
Associative properties, 2
Asymptote(s)
  horizontal, 144
  of a hyperbola, 423
  oblique, 149
  vertical, 143
Augmented matrix, 330
Axis (axes)
  conjugate, 423
  coordinate, 56
  of an ellipse, 417
  imaginary, 294
  major, 417
  minor, 417
  of a parabola, 60, 409
  polar, 431
  real, 294
  transverse, 422

Base
of an exponential function, 160
logarithmic, 170, 177
Bearing, 238
Binomial coefficients, 384
Binomial Theorem, 383
Bounds for zeros of polynomials, 133
Branches of a hyperbola, 423

Cardiod, 433
Cartesian coordinate system, 56
Center
of a circle, 62
of an ellipse, 416
of a hyperbola, 422
Central angle, 193
Characteristic of a logarithm, A2
Circle
equation of, 62
unit, 62
Circular functions, 209
Closed curve, 443
Closed interval, 40
Closed operation, 1
Coefficient, 22
Coefficient matrix, 330
Cofactor, 349, 352
Cofunction, 261
Column of a matrix, 329
Column subscript, 330
Common difference, 395
Common logarithm, 182
Common ratio, 399
Commutative properties, 2
Complementary angles, 192
Completing the square, 63, 192
Complex number(s), 121
absolute value of, 294
addition of, 122
amplitude of, 295
argument of, 295
conjugate of, 124
equality of, 122
geometric representation of, 293
imaginary part of, 121
modulus of, 295
polar form of, 295
product of, 122
real part of, 121
trigonometric form of, 295
zero, 123
Complex plane, 293
Components of a vector, 304, 309
Composite function, 97
Compound interest, 163
Conditional equation, 34
Conic section, 407
Conjugate axis, 423
Conjugate of a complex number, 124
Conjugate hyperbolas, 427
Consistent system of equations, 325
Constant, 21
Constant function, 78, 83
Constant polynomial, 22
Constant term, 131
Coordinate axes, 56
Coordinate line, 5
Coordinate plane, 56
Coordinate system, 5, 56
Corespondence, 73
one-to-one, 5, 77
Cosecant function, 198, 203
graph of, 226
Cosine function, 198, 203
graph of, 224
Cotangent function, 198, 203
graph of, 226
Coterminal angles, 191
Cramer's Rule, 360, 361
Cube root, 13
of unity, 127
Curve, 442
Cycloid, 446
curtate, 449
prolate, 449
Damped sine wave, 234
Damping factor, 234
Decimal fraction, A2
Decreasing function, 83
Degenerate conic, 408
Degree
as an angular measurement, 191
relation to radian, 195
Degree of a polynomial, 22
De Moivre's Theorem, 299
Denominator, 3
Dependent system of equations, 325
Descartes' Rule of Signs, 131
Dependent variable, 77
Determinant, 348
Diagonal elements of a matrix, 330
Directed distance, 411
Directed line segment, 302
Direction of a line, 238
Directrix
of a parabola, 409
of a conic, 437
Displacement, 303
Distance
on a coordinate line, 8
directed, 411
Distance Formula, 57
Distributive properties, 2
Divisible polynomial, 112
Division, 3
Algorithm, 114
synthetic, 118
Divisor, 4
Domain
of a function, 74
of a variable, 21
Double-angle formulas, 266
Dot product, 311

e (the number), 166
Eccentricity, 437
Echelon form, 332

Element
of a matrix, 330
of a set, 20
Elementary row transformations of a matrix, 331
Ellipse, 415
center of, 416
eccentricity of, 437
foci of, 415
major axis of, 417
minor axis of, 417
polar equations of, 439
vertices of, 417
Endpoint
of a curve, 443
of an interval, 39
Epicycloid, 448
Equality, 2
Equality
of complex numbers, 122
of functions, 74
of matrices, 339
of polynomials, 22
of sequences, 388
of sets, 21
of vectors, 303
Equation(s), 34
conditional, 34
equivalent, 34, 59
graph of, 59
homogeneous, 335
linear, 69, 321
of lines, 67, 68, 69
quadratic, 93
root of, 34
solution of, 34, 59, 107, 315, 319
system of, 316
trigonometric, 253
Equivalent equations, 34, 59, 315
Equivalent inequalities, 37
Even function, 78
Expansion Theorem for Determinants, 352
Exponential equation, 184, 186
Exponential function, 160
Exponential law of growth, 161
Exponents, 10, 17

Factor, 4, 26
Factor Theorem, 115
Factorial notation, 382
Factoring, 27
Factoring formulas
algebraic, 27
trigonometric, 274
Focus
of an ellipse, 415
of a hyperbola, 421
of a parabola, 409
Four-leafed rose, 434
Force vector, 303
Fraction, 3
Fractional expression, 27
Free vector, 302
Frequency, 242
Function(s)
circular, 209
composite, 97
constant, 78, 83
decreasing, 83
definition of, 74, 79
domain of, 74
equality of, 74
even, 78
exponential, 160
graph of, 82
greatest integer, 89
growth, 161
identity, 78
image under, 74
increasing, 83
inverse, 99
inverse trigonometric, 276
linear, 82
logarithmic, 177
mapping terminology, 74
maximum value of, 84
minimum value of, 84
natural exponential, 166
natural logarithmic, 179
normal distribution, 169
odd, 78
one-to-one, 77
periodic, 210
polynomial, 106
quadratic, 90
range of, 74
rational, 141
trigonometric, 198
value of, 74
zero of, 82
Fundamental Identities, 201, 248
Fundamental Theorem of Algebra, 128
Fundamental Theorem of Arithmetic, 4

Geometric progression, 399
Geometric sequence, 399
Geometric series (infinite), 401
Gompertz growth curve, 169
Graph(s)
of a curve, 82
of an equation, 59
of a function, 82
of an inequality, 362
of a set of real numbers, 39
shifts of, 86
stretching of, 87
of trigonometric functions, 222ff.
Greater than, 6
Greatest integer function, 89
Growth function, 161

Half-angle formulas, 269, 270
Half-open interval, 40
Half-plane, 363
Harmonic motion, 241
Homogeneous system of equations, 335
Horizontal asymptote, 144
Horizontal component, 309
Horizontal line, 66

Horizontal shift, 86
Hyperbola, 421
  asymptotes of, 423
  branches of, 423
  center of, 422
  conjugate, 427
  conjugate axis of, 423
  foci of, 421
  polar equations of, 439
  transverse axis of, 422
  vertices of, 422
Hypocycloid, 448
Hypotenuse, 198

Identity, equation as, 34
Identity
  additive, 2
  multiplicative, 2
Identity function, 78
Identity matrix, 342
Image, 74
Imaginary axis, 294
Imaginary part of a complex number, 121
Inconsistent system of equations, 325
Increasing function, 83
Independent variable, 77
Index of a radical, 14
Index of summation, 390
Induction hypothesis, 377
Induction, mathematical, 376
Inequalities, 6, 37, 362
  equivalent, 37, 362
  graphs of, 362
  linear, 363
  properties of, 37, 38
  solutions of, 37, 362
  systems of, 364
Infinite interval, 41
Infinite sequence, 387
Infinite series, 402
Infinity (∞), 41
Initial point of a vector, 302
Initial side of an angle, 191
Integers, 4
Intercepts of a graph, 59
Interest, compound, 163
Intermediate Value Theorem, 108
Interpolation, linear, A5
Intervals, 39
Inverse
  additive, 2
  of a matrix, 342
  multiplicative, 2
Inverse cosine function, 279
Inverse function, 99
Inverse sine function, 278
Inverse tangent function, 280
Involute of a circle, 449
Irrational number, 4
Irreducible polynomial, 26

Law of Cosines, 290
Law of Sines, 285
Laws of Exponents, 11
Laws of Logarithms, 171
Laws of Radicals, 14
Laws of signs, 7
Leading coefficient, 22
Learning curves, 169
Least common denominator, 29
Length of a line segment, 8
Less than, 6
Limaçon, 433
Line(s)
  equation of, 67, 68, 69
  horizontal, 66
  parallel, 70
  perpendicular, 71
  slope of, 65
  vertical, 66
Linear combination, 309
Linear equation, 69, 321
Linear function, 82
Linear inequalities, 363
Linear interpolation, A5
Linear programming, 367
Logarithmic equation, 185
Logarithmic function, 177
Logarithm(s)
  base of, 170
  change of base, 174
  characteristic of, A2
  common, 182
  definition of, 170
  Laws of, 179
  mantissa of, A2
  natural, 179
  standard form of, A2
  using tables of, A1
Lower bound for solutions, 133

Magnitude of a vector, 303, 304
Major axis of an ellipse, 417
Mantissa, A2
Mathematical induction, 376
Matrix (matrices)
  addition of, 339
  additive inverse of, 340
  augmented, 330
  coefficient, 330
  columns of, 329
  definition of, 330
  echelon form of, 332
  element of, 330
  elementary row transformations of, 331
  equality of, 339
  identity, 342
  inverse of, 342
  main diagonal elements of, 330
  multiplication of, 341
  of order n, 330
  rectangular, 330
  rows of, 329
  square, 330
  subtraction of, 340
  of a system of equations, 330
  zero, 339
Maximum value of a function, 84
Method of elimination, 323
Method of substitution, 317
Midpoint Formula, 58
Minimum value of a function, 84

Minor, 349, 352
Minor axis of an ellipse, 417
Minute, 192
Modulus of a complex number, 295
Multiple-angle formulas, 266
Multiplicative identity, 2
Multiplicative inverse, 2
Multiplicity of a zero, 130

*n*th power, 11
*n*th roots, 13, 298
*n*th roots of unity, 127, 301
Natural exponential function, 166
Natural logarithms, 179
Negative, 2
Negative angle, 191
Negative direction, 6
Negative integers, 4
Negative real numbers, 6
Nontrivial factor, 26
Normal distribution function, 169
Number(s)
  complex, 121
  irrational, 4
  negative, 6
  positive, 6
  prime, 4
  rational, 4
  real, 1
Numerator, 3

Oblique asymptote, 149
Oblique triangle, 284
Obtuse angle, 192
Odd function, 78
One-to-one correspondence, 77
One-to-one function, 77
Open interval, 39
Opposite side, 198
Order of a square matrix, 330
Ordered pair, 55
Ordered triple, 319
Ordinate, 56
Origin, 56

Parabola, 60, 92, 409
  axis of, 60, 409
  directrix of, 409
  focus of, 409
  polar equations of, 439
  vertex of, 60, 409
Parameter, 443
Parametric equations, 443
Partial fractions, 152
Partial sum, 391
Pascal's triangle, 386
Period, 210, 241
Periodic function, 210
Phase difference, 244
Phase shift, 231
Plane curve, 442
Plotting points, 56
Point-slope form, 68
Polar axis, 431
Polar coordinates, 431
Polar equation(s), 432
  of conics, 439
Polar form of a complex number, 295
Pole, 431
Polynomial(s), 22
  bounds for zeros of, 133
  constant, 22
  constant term of, 131
  degree of, 22
  division of, 114
  equality of, 22
  factor of, 26
  function, 106
  irreducible, 26
  leading coefficient of, 22
  number of zeros of, 130
  prime, 26
  rational zeros of, 138
  term of, 22
  zero, 22
  zero of, 128
Positive angle, 191
Positive direction, 5
Positive integers, 4
Positive real numbers, 6

Power, 11
Prime number, 4
Prime polynomial, 26
Principal, 163
Principal root, 4, 13, 125
Product
  of complex numbers, 121
  of real numbers, 2
Product formulas
  algebraic, 25
  trigonometric, 273
Projection of a point, 241

Quadrantal angle, 191
Quadrant, 56
Quadratic equation, 93
Quadratic formula, 94
Quadratic function, 90
Quotient(s)
  in division process, 113, 114
  properties of, 4
  of real numbers, 3

Radian, 193
  relation to degree, 195
Radical, 14
Radicand, 14
Range of a function, 74
Rational exponent, 17
Rational expression, 27
Rational function, 141
Rational number, 4
Rational zeros of polynomials, 138
Rationalizing denominators, 15
Real axis, 294
Real line, 5
Real numbers, 1
Real part of a complex number, 121
Reciprocal, 2
Rectangular coordinate system, 56
Rectangular matrix, 330

Recursive definition, 389
Reduction formulas, 275
Reference angle, 214
Reference number, 214
Remainder, 113, 114
Remainder Theorem, 115
Resultant force, 304
Right angle, 192
Right triangle, 197
Root(s)
  of an equation, 34, 107
  multiplicity of, 130
  square, 4, 13
  of unity, 301
Rotation of axes, 427
Row of a matrix, 329
Row subscript, 330
Row transformations of a matrix, 331

Scalar, 304
Scalar multiple of a vector, 304, 306
Scientific form for numbers, 17
Secant function, 198, 203
  graph of, 226
Second, 192
Sequence(s), infinite, 387
  arithmetic, 395
  equality of, 388
  geometric, 399
Sequence of partial sums, 391
Series, infinite, 402
  alternating, 404
Sets, 20
  equality of, 21
Sign of a real number, 7
Significant figures, 19, 236, A7
Simple closed curve, 443
Simple harmonic motion, 241
Sine function, 198, 203
  graph of, 223
Sine wave, 223
Sketch of a graph, 39
Slope, 65
Slope-intercept form, 69
Solution
  of an equation, 34, 59, 107, 315, 319
  of an inequality, 37, 362
  of a system of equations, 316
Square matrix, 330
Square root, 4, 13, 125
Standard position of an angle, 191
Stretching a graph, 87
Subset, 21
Subscript, 330
Subtended angle, 193
Subtraction, 3
Subtraction formulas, 259, 261
Sum
  of complex numbers, 121
  of matrices, 339
  of real numbers, 1
  of vectors, 306
Summation notation, 390
Summation variable, 390
Supplementary angles, 192
Symmetry
  to an axis, 60, 61
  to the origin, 61
  tests for, 61, 436
Synthetic division, 118
System of equations, 316
System of inequalities, 364

Tangent function, 198, 203
  graph of, 225
Term of a polynomial, 22
Term of a sequence, 388
Terminal point of a vector, 302
Terminal side of an angle, 191
Test value, 48
Theory of Equations, 128
Translation of axes, 411
Transverse axis, 422
Triangle
  oblique, 284
  right, 197
Trigonometric equation, 253
Trigonometric expression, 247
Trigonometric function(s), 198
  addition formulas, 260, 261
  of angles, 198, 203
  double-angle formulas, 266
  factoring formulas, 274
  graphs of, 222
  half-angle formulas, 269, 270
  inverse, 276
  multiple-angle formulas, 266
  period of, 210
  as ratios, 198
  of real numbers, 207
  in terms of a right triangle, 198
  signs of, 206
  in terms of a unit circle, 208
  values of, 213
Trigonometric identities, 247
Trivial factor, 26
Trivial solution, 335

Unit circle, 62
Unit vector, 309
Unity, roots of, 301
Upper bound for solutions, 133

Value of an expression, 21
Value of a function, 74
Variable, 21
  dependent, 77
  domain of, 21
  independent, 77
  summation, 390
Vector(s), 302
  addition of, 306
  components of, 304
  dot product of, 311
  equality of, 302
  force, 303
  free, 302
  horizontal component of, 309
  initial point of, 302
  magnitude of, 303, 304

scalar multiple of, 304, 306
subtraction of, 307
terminal point of, 302
unit, 309
velocity, 303
vertical component of, 309
zero, 307
Velocity vector, 303
Vertex
of an angle, 191
of an ellipse, 417
of a hyperbola, 422
of a parabola, 60, 409
Vertical asymptote, 143
Vertical component, 309
Vertical line, 66
Vertical shift, 85

x-axis, 56
x-coordinate, 56
x-intercept, 59
xy-plane, 56

y-axis, 56
y-coordinate, 56
y-intercept, 59

Zero, 2
Zero of a function, 82
Zero matrix, 339
Zero of multiplicity *m*, 130
Zero polynomial, 22
Zero(s) of a polynomial function, 129ff.
multiplicity of, 130
rational, 138
Zero vector, 307

THE TRIGONOMETRIC FUNCTIONS

I. Of Acute Angles

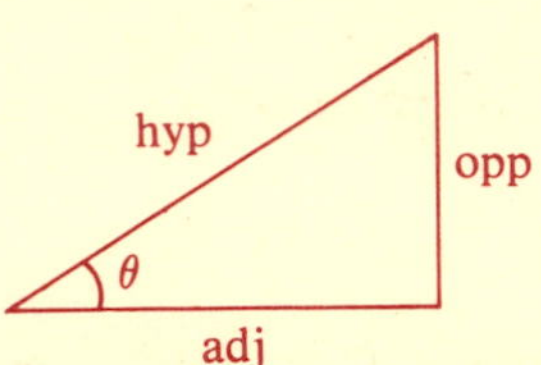

$$\sin\theta = \frac{\text{opp}}{\text{hyp}} \qquad \csc\theta = \frac{\text{hyp}}{\text{opp}}$$

$$\cos\theta = \frac{\text{adj}}{\text{hyp}} \qquad \sec\theta = \frac{\text{hyp}}{\text{adj}}$$

$$\tan\theta = \frac{\text{opp}}{\text{adj}} \qquad \cot\theta = \frac{\text{adj}}{\text{opp}}$$

II. Of Arbitrary Angles

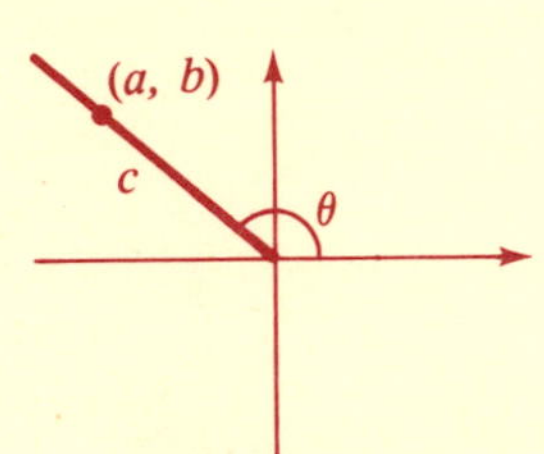

$$\sin\theta = \frac{b}{c} \qquad \csc\theta = \frac{c}{b}$$

$$\cos\theta = \frac{a}{c} \qquad \sec\theta = \frac{c}{a}$$

$$\tan\theta = \frac{b}{a} \qquad \cot\theta = \frac{a}{b}$$

III. Of Real Numbers

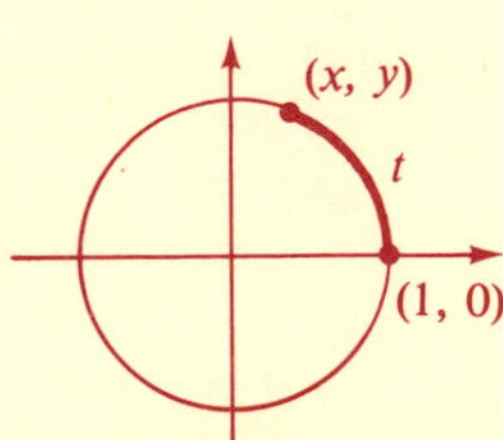

$$\sin t = y \qquad \csc t = \frac{1}{y}$$

$$\cos t = x \qquad \sec t = \frac{1}{x}$$

$$\tan t = \frac{y}{x} \qquad \cot t = \frac{x}{y}$$